Setting Yourself up for...
SUCCESS

MAKE THE MOST OF CLASS

1. Attend class. Regular attendance is a key to success.
2. Make sure you have a book and a notebook for homework and notes.
3. Take notes in class. At a minimum, write down what your instructor writes on the board.
4. If you have tried to solve a problem, but need help, ask a question in class when it is appropriate. Many times other students have the same question.
5. Get to know someone in class. Learning is enhanced when you can talk to someone about math.
6. Did you miss something in class? The **Video Lectures on CD or DVD*** offer a short lecture for each section of your text and are taught by math instructors. These videos will help you review material you may have missed in class or found confusing.

HELP IS HERE WHENEVER YOU WANT IT!

1. Find out your instructor's office number and office hours.
2. Ask if there is a tutor room on campus and find out the hours when tutors are available. The tutors at the Math Tutor Center are available by phone, fax, email, or interactive web, and offer instant, live, one-on-one help with your course. For more information, visit www.mathtutorcenter.com.
3. Before going to office hours or the tutor room, organize your questions. Be sure you try a problem before you ask your instructor or tutor to help you.
4. Don't know why you got the wrong answer? The **Student's Solutions Manual*** contains complete, worked-out solutions (not just answers) to every odd-numbered exercise problem and to ALL the review and test problems in the book.
5. Need some more practice? The **Worksheets for Classroom or Lab Practice*** include extra practice problems for every section of the book.

MASTERING THE EXAM

1. Try the **Pass the Test CD,** found in the front of your book. It contains video clips of an instructor working through the complete solution for all exercises from the chapter tests in your book. **Pass the Test** also includes vocabulary flashcards, tips for time management, and a Spanish glossary.
2. Find a consistent time and place to study every day rather than "pulling an all
3. Keep all your homework and notes in a notebook. Reading your notes and go assignments will make reviewing for the exam easier.
4. Find out from your instructor which topics will be covered on the exam.
5. **Tab Your Way to Success** by using the color-coded Post-It® tabs provided in the back important pages of the text that you may want to return to for review, instructor help, or test preparation.
6. When you think you are ready for the exam, pick a few problems from your assignments *at random*. If you can work them, without any help, you are well on your way to being ready.
7. Carefully show your work whenever possible on your exam.
8. Relax and visualize doing well.

You can SUCCEED in Math!

* These items and others can be purchased at www.mypearsonstore.com

D1413621

Beginning Algebra
with Applications and Visualization
Second Edition

GARY K. ROCKSWOLD

Minnesota State University, Mankato

TERRY A. KRIEGER

Rochester Community and Technical College

PEARSON

Addison
Wesley

Boston San Francisco New York
London Toronto Sydney Tokyo Singapore Madrid
Mexico City Munich Paris Cape Town Hong Kong Montreal

C
ma
elec
miss
ing pe
Educat
MA 02
permissi
1 2 3 4 5 6

Editorial Director	Christine Hoag
Acquisitions Editor	Randy Welch
Executive Project Manager	Kari Heen
Senior Project Editor	Lauren Morse
Assistant Editor	Antonio Arvelo
Senior Managing Editor	Karen Wernholm
Senior Production Supervisor	Kathleen A. Manley
Cover Design	Nathaniel Koven and Barbara T. Atkinson
Photo Researcher	Beth Anderson
Supplements Production	Marianne Groth and Kayla Smith-Tarbox
Media Producers	Ceci Fleming and MiMi Yeh
Software Development	Mary Durnwald, TestGen; Geoff M. Erickson, MathXL
Marketing Managers	Michelle Renda and Marlana Voerster
Marketing Assistant	Nathaniel Koven
Senior Author Support/Technology Specialist	Joe Vetere
Senior Prepress Supervisor	Caroline Fell
Rights and Permissions Advisor	Shannon Barbe
Manufacturing Manager	Evelyn Beaton
Production Coordination	Kathy Diamond
Text Design and Composition	Nesbitt Graphics, Inc.
Illustrations	Techsetters, Inc. and Nesbitt Graphics, Inc.
Cover Photograph	Shutterstock

Photo credits appear on page B-2.

Library of Congress Cataloging-in-Publication Data
Rockswold, Gary K.
  Beginning algebra with applications & visualization / Gary K. Rockswold,
    Terry A. Krieger. –2nd ed.
    p. cm.
  Rev. ed. of: Beginning algebra with applications and visualization. 2005.
  ISBN 0-321-50004-0 (student)
  1. Algebra-- Textbooks. I. Krieger, Terry A. II. Rockswold, Gary K. Beginning algebra with applications and visualization. III. Title. IV. Title: Beginning algebra with applications and visualization.
QA152.3.R59 2008
512–dc22   2007060828

ISBN-13: 978-0-321-50004-5   ISBN-10: 0-321-50004-0

7 8 9 10—QWT—11 10 09 08 07

To

Daniel M. Krieger

CONTENTS

1   INTRODUCTION TO ALGEBRA   1

4 SYSTEMS OF LINEAR EQUATIONS IN TWO VARIABLES   251

7 RATIONAL EXPRESSIONS   433

8 RADICAL EXPRESSIONS   517

9 QUADRATIC EQUATIONS   573

PREFACE

Beginning Algebra with Applications and Visualization, Second Edition, more than any other conventional textbook, connects the real world to mathematics in ways that are both meaningful and motivational to students. Students using this textbook will have no shortage of realistic and convincing answers to the perennial question, "When will I ever use this?" The early introduction of graphs allows instructors to use graphs and other visualizations to marshal a host of applications of mathematical concepts to galvanize students' attention and comprehension. Real data, graphs, and tables play an important role in the course, giving meaning to the numbers and equations that students encounter. This approach increases students' interest and motivation and, consequently, the likelihood of success.

This textbook is one of three textbooks in a series that includes *Intermediate Algebra with Applications and Visualization,* Third Edition, and *Beginning & Intermediate Algebra with Applications and Visualization,* Second Edition.

APPROACH

We introduce mathematical concepts by moving from the concrete to the abstract, using relevant applications to support students' comprehension of abstract mathematical ideas. We have included a diverse collection of unique, up-to-date applications to buttress students' understanding of difficult concepts. Additionally, we present mathematical concepts using multiple representations, giving students a variety of perspectives for learning the same material—something especially helpful because students have various learning styles.

The primary purpose of the text, of course, is to teach mathematical concepts and skills. Building skills is an objective of every section. Problem-solving skills, which will help students succeed in more advanced mathematics courses, are given a strong emphasis. Throughout the text, problem solving is frequently combined with real-world applications to capture and maintain students' interest while reinforcing problem-solving habits. Standard mathematical definitions, theorems, symbolism, and rigor will prepare students for higher-level courses should they choose to pursue them.

CHANGES INCORPORATED INTO THIS EDITION

This edition contains important changes resulting from numerous comments and suggestions from instructors, students, and reviewers. We have made the text more visual by using more color and adding many photos. The following exciting features have been added to enhance students' learning of mathematical skills and concepts. To emphasize how mathematics can be used in everyday life, **A Look into Math** appears at the start of every section and **Real-World Connections** are highlighted throughout the text. **Now Try Exercises** appear after every example, and **Thinking Generally** concept exercises appear in most exercise sets. There are also new **Cumulative Review** exercise sets for chapters that previously did not have them, and **Chapter Tests** have been updated and are now more comprehensive.

The examples and the exercises have also been updated and expanded throughout the text:

- There are **more than 650 new exercises**, including additional graphing calculator exercises.
- **Approximately 50 new examples** have been added, many of them multi-part, with an emphasis on solving word problems, factoring, and radicals.
- **Data in applications have been updated** throughout, and many new topics have been added, including applications about iPods, global warming, and video games, giving instructors familiar with the text a fresh, rich set of practice material.

CHAPTER 1: Starting with this chapter, hundreds of applications involving data have been updated throughout the text to make it even more current for students. More examples and exercises have been added that increase student skills and understanding about fractions, absolute values, and powers of positive and negative of integers.

CHAPTER 2: Throughout this text there is increased emphasis on having students distinguish between expressions and equations. The number and variety of application problems have been expanded in Section 2.3. An optional introduction to interval notation now appears.

CHAPTER 3: Interpretation of graphical and numerical data is expanded throughout this chapter. In Section 3.4, new questions and examples that ask students to interpret slope and intercepts of a graph have been added.

CHAPTER 4: Additional examples and exercises appear throughout this chapter. More emphasis has been given to identifying both the solution to a system of linear equations and the solution to a related application.

CHAPTER 5: New exercises that give students skills to simplify both exponential expressions and products of polynomials now appear.

CHAPTER 6: A new section summarizes factoring techniques and gives students practice factoring a variety of polynomials in the same section. New examples and exercises that require factoring out the greatest common factor first are also included. Additional examples of factoring by grouping and identifying prime trinomials now appear.

CHAPTER 7: A new subsection helps students distinguish between rational expressions and rational equations. Additional explanations and exercises make it easier for students to simplify complex fractions and solve rational equations. More explanation and exercises covering both direct and inverse variation appear.

CHAPTER 8: A new optional subsection allows students to graph equations containing basic radical expressions. More examples and applications have been included.

CHAPTER 9: A new section introduces the important concept of a function in a simple, clear way. More examples, exercises, and applications involving quadratic equations also appear.

FEATURES

CHAPTER OPENERS
Each chapter opens with an application that motivates students by giving them insight into the relevance of that chapter's mathematical concepts.

CHAPTER **3**

Graphing Equations

Global warming has been in the news for years. Scientists are documenting that both the arctic and antarctic ice packs are melting. This phenomenon is an indication that global temperatures may be rising. The amount of light from the sun that Earth reflects into space is an important factor. If more light is reflected, then Earth stays cooler. Astronomer Phil Goode from the New Jersey Institute of Technology has developed a simple way to determine how reflective Earth is by measuring the amount of earthshine illuminating the moon. The area of the moon illuminated by earthshine is the darker portion of the moon that is visible but not as visible as the portion illuminated by the sun. Goode found that the percent change in earthshine (from the average) varied with the month. An excellent way to display these data is with the *line graph* shown in the accompanying figure. We discuss line graphs and other types of graphs in this chapter.

3.1 Introduction to Graphing
3.2 Linear Equations in Two Variables
3.3 More Graphing of Lines
3.4 Slope and Rates of Change
3.5 Slope–Intercept Form
3.6 Point–Slope Form
3.7 Introduction to Modeling

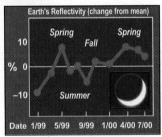

Originality is the essence of true scholarship.
—NNAMDI AZIKIWE

Source: Hannah Hoag, *Discover*, November 2002, p. 11. Graphic by Matt Zang. (Reprinted with permission.)

161

XV

MATH IN THE REAL WORLD

NEW!
A LOOK INTO MATH
Nearly every section begins in the context of the practical world that motivates the math topic which follows.

A LOOK INTO MATH ▷

In this section we discuss factoring polynomials having higher degree. Polynomials of degree 2 or higher are often used in applications. For example, the polynomial

$$0.0013x^3 - 0.085x^2 + 1.6x + 12$$

models natural gas consumption in the United States (in trillions of cubic feet) where $x = 0$ corresponds to 1960, $x = 10$ to 1970, and so on until $x = 40$ corresponds to 2000, as shown in Figure 6.8. In Exercises 75–78 we discuss further the consumption of natural gas. (**Source:** Department of Energy.)

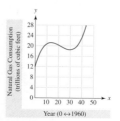

Figure 6.8 Natural Gas Consumption

▶ REAL-WORLD CONNECTION   Suppose that someone leaves a rest stop on an Interstate highway and drives home at a constant speed of 50 miles per hour. The graph in Figure 3.17 reflects the distance of the driver from home at various times. The graph intersects the *y*-axis at 200 miles, which is called the *y-intercept*. In this situation the *y*-intercept represents the initial distance (when $x = 0$) between the driver and home. The graph also intersects the *x*-axis at 4 hours, which is called the *x-intercept*. This intercept represents the elapsed time when the distance of the driver from home is 0 miles.

NEW!
REAL-WORLD CONNECTION
Where appropriate, the authors expand on specific math topics and their connections to the everyday world.

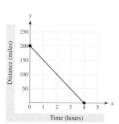

Figure 3.17

GROUP ACTIVITY
WORKING WITH REAL DATA

Directions: Form a group of 2 to 4 people. Select someone to record the group's responses for this activity. All members of the group should work cooperatively to answer the questions. If your instructor asks for the results, each member of the group should be prepared to respond.

Exercises 1–5: In this set of exercises you are to use your knowledge of equations of lines to model the average annual cost of tuition and fees.

1. *Cost of Tuition*   In 2000, the average cost of tuition and fees at *private* four-year colleges was $16,200, and in 2005 it was $20,100. Sketch a line that passes through the points (2000, 16200) and (2005, 20100). (*Source:* The College Board.)

2. *Rate of Change in Tuition*   Calculate the slope of the line in your graph. Interpret this slope as a rate of change.

3. *Modeling Tuition*   Find the slope–intercept form of the line in your sketch. What is the *y*-intercept and does it have meaning in this situation?

4. *Predicting Tuition*   Use your equation to estimate tuition and fees in 2001 and compare it to the known value of $17,300. Estimate tuition and fees in 2010.

5. *Public Tuition*   In 2000, the average cost of tuition and fees at *public* four-year colleges was $3500, and in 2005 it was $5100. Repeat Exercises 1–4 for these data. Note that the known value for 2001 is $3700. (*Source:* The College Board.)

GROUP ACTIVITIES:
WORKING WITH REAL DATA
This feature occurs after select sections (1 or 2 per chapter) and provides an opportunity for students to work collaboratively on a problem that involves real-world data. Most activities can be completed with limited use of class time.

PRACTICE

NEW!
**EXAMPLES WITH
NOW TRY EXERCISES**
Every example directs students to practice exercises patterned after that example for immediate reinforcement of that skill.

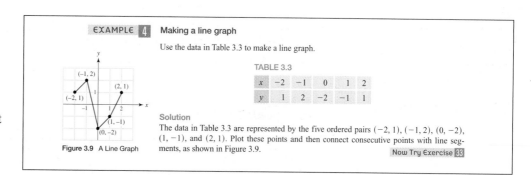

EXAMPLE **4**   Making a line graph

Use the data in Table 3.3 to make a line graph.

TABLE 3.3

x	−2	−1	0	1	2
y	1	2	−2	−1	1

Solution
The data in Table 3.3 are represented by the five ordered pairs $(−2, 1)$, $(−1, 2)$, $(0, −2)$, $(1, −1)$, and $(2, 1)$. Plot these points and then connect consecutive points with line segments, as shown in Figure 3.9.

Figure 3.9   A Line Graph

Now Try Exercise **33**

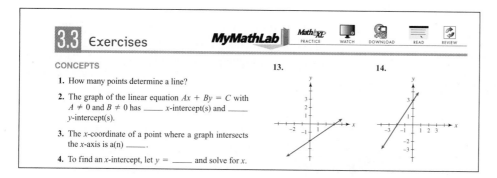

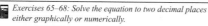

3.3 Exercises

MyMathLab   MathXL PRACTICE   WATCH   DOWNLOAD   READ   REVIEW

CONCEPTS

1. How many points determine a line?

2. The graph of the linear equation $Ax + By = C$ with $A \neq 0$ and $B \neq 0$ has _____ x-intercept(s) and _____ y-intercept(s).

3. The x-coordinate of a point where a graph intersects the x-axis is a(n) _____.

4. To find an x-intercept, let $y =$ _____ and solve for x.

13.

14.

EXERCISES
The exercise sets lead from basic concepts to skill-building, writing, applications, and conceptual mastery.

**GRAPHING CALCULATOR
EXERCISES**
The icon 📷 is used to denote an exercise that requires students to use a graphing calculator.

Exercises 65–68: Solve the equation to two decimal places either graphically or numerically.

65. $\sqrt{x − 1.1} + \pi x = 8$

66. $\sqrt{x − 1} + 0.4x = \sqrt{2x + \pi}$

67. $\sqrt{x + 3.2} + 0.1x = −x^2 + 4$

68. $\sqrt{x − 2.2} = x + 2$

87. **Thinking Generally**   Students sometimes mistakenly apply the "rule" $a^m \cdot b^n \stackrel{?}{=} (ab)^{m+n}$. In general, this equation is *not true*. Find values for a, b, m, and n with $a \neq b$ and $m \neq n$ that will make this equation true.

88. **Thinking Generally**   Students sometimes mistakenly apply the "rule" $(a + b)^n \stackrel{?}{=} a^n + b^n$. In general, this equation is *not true*. Find values for a, b, and n with $a \neq b$ that will make this equation true.

NEW!
THINKING GENERALLY
Most exercise sets include open-ended conceptual questions encouraging students to synthesize what they've learned.

PRACTICE

CONTINUOUS REVIEW EXERCISES

After every two sections, *Checking Basic Concepts* presents exercises that students can use to review and tie together recently studied concepts.

CHECKING BASIC CONCEPTS
SECTIONS 4.1 AND 4.2

1. Determine graphically the x-value in each equation when $y = 2$.
   (a) $y = 1 - \frac{1}{2}x$ (b) $2x - 3y = 6$

2. Determine whether $(-1, 0)$ or $(4, 2)$ is a solution to the system
   $$2x - 5y = -2$$
   $$3x + 2y = 16.$$

3. Solve the system of equations graphically. Check your answer.
   $$x - y = 1$$
   $$2x + y = 5$$

4. Use the method of substitution to solve each system of equations. How many solutions does each have?

   (a) $x + y = -1$
       $y = 2 - x$
   (b) $4x - y = 5$
       $-x + y = -2$
   (c) $x + 2y = 3$
       $-x - 2y = -3$

5. *Room Prices* A hotel rents single and double rooms for $150 and $200, respectively. The hotel receives $55,000 for renting 300 rooms.
   (a) Let x be the number of single rooms rented and let y be the number of double rooms rented. Write a system of linear equations whose solution gives the values of x and y.
   (b) Use the method of substitution to solve the system. Check your answer.

TECHNOLOGY NOTES

Occurring throughout the book, optional Technology Notes offer students guidance, suggestions, and cautions on the use of graphing calculator.

TECHNOLOGY NOTE: Intersection of Graphs and Table of Values

A graphing calculator can be used to find the intersection of the two graphs shown in Figure 4.3. It can also be used to create Table 4.1. The accompanying figures illustrate how a calculator can be used to determine that $y_1 = 10x$ equals $y_2 = 40$ when $x = 4$.

NEW!
TAB YOUR WAY TO SUCCESS

Forty reusable color-coded Post-It® tabs encourage students to mark important pages of the text for review, extra help, test preparation, etc.

Using the Tabs

Customize Your Textbook and Make It Work for You!

These removable and reusable tabs offer you five ways to be successful in your math course by letting you bookmark pages with helpful reminders.

Important! Use these tabs to flag anything your instructor indicates is important.

Review This Mark important definitions, procedures, or key terms to review later.

Ask for Help Not sure of something? Need more instruction? Place these tabs in your textbook to address any questions with your instructor during your next class meeting or with your tutor during your next tutoring session.

On the Test If your instructor alerts you that something will be covered on a test, use these tabs to bookmark it.

Write your own notes or create more of the preceding tabs to help you succeed in your math course.

Tab Your Way to Success

Use these tabs to mark important pages of your textbook for quick reference and review.

CONCEPTS

MATH FROM MULTIPLE PERSPECTIVES
Throughout the text, concepts are presented by means of **verbal**, **graphical**, **numerical**, and **symbolic** representations to facilitate student learning.

CRITICAL THINKING
One or more Critical Thinking exercises are included in most sections. They pose questions that can be used for classroom discussion or homework assignments.

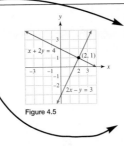

Figure 4.5

Solution
Graphically   Begin by writing each equation in slope–intercept form.

$x + 2y = 4$	First equation	$2x - y = 3$	Second equation
$2y = -x + 4$	Subtract x.	$-y = -2x + 3$	Subtract $2x$.
$y = -\dfrac{1}{2}x + 2$	Divide by 2.	$y = 2x - 3$	Multiply by -1.

The graphs of $y = -\frac{1}{2}x + 2$ and $y = 2x - 3$ are shown in Figure 4.5. Their graphs intersect at the point $(2, 1)$.

Numerically   Table 4.2 shows the equations $y = -\frac{1}{2}x + 2$ and $y = 2x - 3$ evaluated for various values of x. Note that when $x = 2$, both equations have a y-value of 1. Thus $(2, 1)$ is the solution.

CRITICAL THINKING
If the graph of a system of linear equations consists of two parallel lines, how many solutions are there?

TABLE 4.2   **A Numerical Solution**

x	-1	0	1	2	3
$y = -\frac{1}{2}x + 2$	2.5	2	1.5	1	0.5
$y = 2x - 3$	-5	-3	-1	1	3

Now Try Exercise 37

MAKING CONNECTIONS

The Signs in the Binomial Factors

If a trinomial of the form $x^2 + bx + c$ can be factored, the signs of the coefficients in the trinomial can be used to determine the signs in the binomial factors. If c is positive, the binomial factors must have the same signs. If c is negative, the binomial factors must have opposite signs. If b and c represent positive numbers, this can be summarized as follows.

Form of the Trinomial	*Signs in the Binomial Factors*
$x^2 + bx + c$	$(\ +\ )(\ +\ )$
$x^2 - bx + c$	$(\ -\ )(\ -\ )$
$x^2 + bx - c$	$(\ -\ )(\ +\ )$
$x^2 - bx - c$	$(\ -\ )(\ +\ )$

MAKING CONNECTIONS
This feature occurs throughout the text and helps students relate previously learned concepts to new ones, pointing out the interrelatedness of mathematical topics.

PUTTING IT ALL TOGETHER
This helpful feature appears at the end of each section to summarize techniques and reinforce the mathematical concepts presented in the section.

5.2 PUTTING IT ALL TOGETHER

In this section we discussed monomials and polynomials, including how to add, subtract, and evaluate them. The following table summarizes several important concepts related to these topics.

Concept	Explanation	Examples	
Monomial	A number, variable, or product of numbers and variables raised to natural number powers	$4x^2y$	Degree: 3, coefficient: 4
		$-6x^2$	Degree: 2, coefficient: -6
		$-a^4$	Degree: 4, coefficient: -1
	Degree is the sum of the exponents.	x	Degree: 1, coefficient: 1
	Coefficient is the number in a monomial.	-8	Degree: 0, coefficient: -8
Polynomial	A sum of one or more monomials	$4x^2 + 8xy^2 + 3y^2$	Trinomial
		$-9x^4 + 100$	Binomial
		$-3x^2y^3$	Monomial

CHAPTER MASTERY

CHAPTER SUMMARY
For a quick and thorough review, each chapter is condensed into key terms, topics, and procedures accompanied by illuminating examples to assist in test preparation.

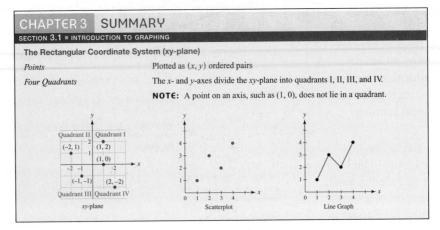

CHAPTER 3   SUMMARY

SECTION 3.1 ■ INTRODUCTION TO GRAPHING

The Rectangular Coordinate System (xy-plane)

Points Plotted as (x, y) ordered pairs

Four Quadrants The x- and y-axes divide the xy-plane into quadrants I, II, III, and IV.

NOTE: A point on an axis, such as $(1, 0)$, does not lie in a quadrant.

CHAPTER 3   REVIEW EXERCISES

SECTION 3.1

1. Identify the coordinates of each point in the graph. Identify the quadrant, if any, in which each point lies.

2. Make a scatterplot by plotting the following four points: $(-2, 3)$, $(-1, -1)$, $(0, 3)$, and $(2, -1)$.

Exercises 3 and 4: Use the table of xy-values to make a line graph.

3.
x	-2	-1	0	1	2
y	-3	2	-1	-2	3

4.
x	-10	-5	0	5	10
y	5	-10	10	-5	0

SECTION 3.2

Exercises 5–8: Determine whether the ordered pair is a solution for the given equation.

5. $y = x - 3$ $(6, 3)$

6. $y = 5 - 2x$ $(-2, 1)$

7. $3x - y = 3$ $(-1, 6)$

8. $\frac{1}{2}x + 2y = -8$ $(-4, -3)$

CHAPTER REVIEW EXERCISES
For extra practice of any topic, students can use these exercises to gain confidence that they've mastered the material.

EXTENDED AND DISCOVERY EXERCISES
These capstone projects at the end of every chapter challenge students to synthesize what they've learned and apply it in other college courses.

CHAPTER 3   EXTENDED AND DISCOVERY EXERCISES

Exercises 1 and 2: The table of real data can be modeled by a line. However, the line may not be an exact model, so answers may vary.

1. *Women in Politics* The table lists percentages P of women in state legislatures during year x.

x	1993	1995	1997	1999
P	20.5	20.6	21.6	22.4

x	2001	2003	2005	2007
P	22.4	22.4	22.7	23.5

x	1960	1970	1980	1990	2000
P	179	203	227	249	281

Source: U.S. Census Bureau.

(a) Make a scatterplot of the data.
(b) Find a point–slope form of a line that models these data.
(c) Use your equation to estimate the U.S. population in 2005.

CREATING GEOMETRIC SHAPES

CHAPTER TEST

Students can reduce math anxiety by using these tests as a rehearsal for the real thing.

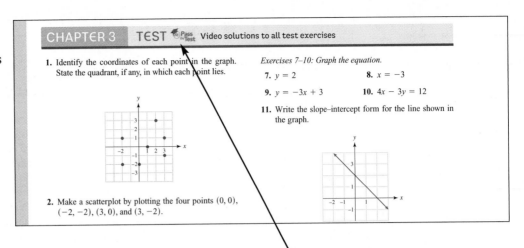

CHAPTER 3 **TEST** Video solutions to all test exercises

1. Identify the coordinates of each point in the graph. State the quadrant, if any, in which each point lies.

2. Make a scatterplot by plotting the four points $(0, 0)$, $(-2, -2)$, $(3, 0)$, and $(3, -2)$.

Exercises 7–10: Graph the equation.

7. $y = 2$

8. $x = -3$

9. $y = -3x + 3$

10. $4x - 3y = 12$

11. Write the slope–intercept form for the line shown in the graph.

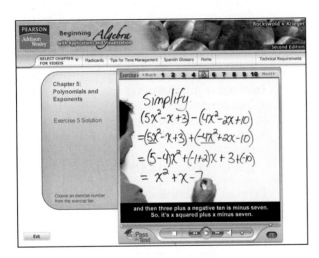

NEW!
PASS THE TEST CD

Included with every new copy of the book, the Pass The Test CD includes the following resources: video clips of an instructor working through the complete solutions for all Chapter Test exercises for each chapter; vocabulary flashcards; tips for time management; and a Spanish glossary.

CUMULATIVE REVIEWS

Starting with Chapter 2 and appearing in all subsequent chapters in this edition, these reviews help students see the big picture of math by staying fresh with topics and skills they've already learned.

CHAPTERS 1–3 **CUMULATIVE REVIEW EXERCISES**

Exercises 1 and 2: Classify the number as prime or composite. If a number is composite, write it as a product of prime numbers.

1. 40

2. 61

Exercises 3 and 4: Translate the phrase into an algebraic expression using the variable n.

3. Ten more than a number

4. A number squared decreased by 2

Exercises 5–8: Evaluate by hand and then simplify to lowest terms.

5. $\frac{4}{3} \cdot \frac{21}{24}$

6. $\frac{3}{4} \div \frac{9}{8}$

7. $\frac{2}{3} + \frac{4}{3}$

8. $\frac{7}{10} - \frac{2}{15}$

Exercises 17 and 18: Evaluate the expression by hand.

17. $-12 \div \left(-\frac{2}{3}\right)$

18. $-\frac{2x}{5y} \div \left(\frac{x}{10y}\right)$

Exercises 19 and 20: Simplify the expression.

19. $3 + 4x - 2 + 3x$

20. $2(x - 1) - (x + 2)$

Exercises 21–24: Solve the equation. Check your solution.

21. $x + 5 = 2$

22. $\frac{1}{3}z = 7$

23. $3t - 5 = 1$

24. $2(x - 3) = -6 - x$

25. Complete the table. Then use the table to solve the equation $6 - 2x = 4$.

x	-2	-1	0	1	2
$6 - 2x$					

STUDENT SUPPLEMENTS

INSTRUCTOR SUPPLEMENTS

Student's Solutions Manual

- Contains solutions for the odd-numbered section-level exercises (excluding Writing About Mathematics and Group Activity exercises), and solutions to all Concepts exercises, Checking Basic Concepts exercises, Chapter Review Exercises, Chapter Test exercises, and Cumulative Review Exercises.

  ISBNs: 0-321-52336-9, 978-0-321-52336-5

Video Lectures on CD or DVD

- Complete set of digitized videos on CD-ROM or DVD for student use at home or on campus.
- Presents a series of lectures correlated directly to the content of each section of the text.
- Features an engaging team of instructors who present material in a format that stresses student interaction, often using examples and exercises from the text.
- Ideal for distance learning and supplemental instruction.
- Video lectures include optional English captioning.

  CD ISBNs: 0-321-52340-7, 978-0-321-52340-2
  DVD ISBNs: 0-321-53462-X, 978-0-321-53462-0

Math XL Tutorials on CD

- Provides algorithmically generated practice exercises that correlate at the objective level to the content of the text.
- Includes an example and a guided solution to accompany every exercise, and video clips for selected exercises.
- Recognizes student errors and provides appropriate feedback; generates printed summaries of students' progress.

  ISBNs: 0-321-52342-3, 978-0-321-52342-6

Pass the Test: Chapter Test Solutions on CD

- Contains video clips of an instructor working through the complete solution for all Chapter Test exercises.
- Additional resources include tips for time management, vocabulary flashcards, and a Spanish glossary.

  Automatically included in every new copy of the textbook.

Math Tutor Center

The Math Tutor Center is staffed by qualified mathematics instructors who provide students with tutoring on examples and odd-numbered exercises from the textbook. Tutoring is available via toll-free telephone, toll-free fax, e-mail, and the Internet. White Board technology allows tutors and students to actually see problems worked while they message in real time over the Internet during tutoring sessions.

Annotated Instructor's Edition

- Contains Teaching Tips and provides answers to every exercise in the textbook excluding the Writing About Mathematics exercises.
- Answers that do not fit on the same page as the exercises themselves are supplied in the Instructor Answer Appendix at the back of the textbook.

  ISBNs: 0-321-51329-0, 978-0-321-51329-8

Instructor's Solutions Manual

- Provides solutions to all section-level exercises (excluding Writing About Mathematics exercises), and solutions to all Checking Basic Concepts exercises, Chapter Review Exercises, Chapter Test exercises, and Cumulative Review Exercises.

  ISBNs: 0-321-52335-0, 978-0-321-52335-8

Instructor and Adjunct Support Manual

- Includes resources designed to help both new and adjunct faculty with course preparation and classroom management.
- Offers helpful teaching tips and additional exercises for selected content.

  ISBNs: 0-321-52341-5, 978-0-321-52341-9

Printed Test Bank and Instructor's Resource Guide

- The Test Bank contains three free-response test forms per chapter, one multiple-choice test form per chapter, and one free-response and multiple-choice final exam.
- The Resource Guide contains three sets of Cumulative Review Exercises that cover Chapters 1–3, 1–6, and 1–9, and notes for presenting graphing calculator topics, as well as supplemental activities.

  ISBNs: 0-321-52334-2, 978-0-321-52334-1

TestGen

- Enables instructors to build, edit, print, and administer tests.
- Features a computerized bank of questions developed to cover all text objectives.
- Creates multiple but equivalent versions of the same question or test with the click of a button.
- Instructors can modify questions or add new questions.
- Tests can be printed or administered online.

The software and testbank are available for download from Pearson Education's online catalog.

Worksheets for Classroom or Lab Practice

- Extra practice exercises for every section of the text with ample space for students to show their work.
- These lab- and classroom-friendly workbooks also list the key topics and vocabulary terms for every text section, along with vocabulary practice problems.

ISBNs: 0-321-54583-4, 978-0-321-54583-1

Pearson Math Adjunct Support Center

The Pearson Math Adjunct Support Center is staffed by qualified mathematics instructors with more than 50 years of combined experience at both the community college and university level. Assistance is provided for faculty in the following areas:

- Suggested syllabus consultation
- Tips on using materials packed with your book
- Book-specific content assistance
- Teaching suggestions including advice on classroom strategies

  For more information, visit
  www.aw-bc.com/tutorcenter/math-adjunct.html

MathXL® www.mathxl.com

MathXL is a powerful online homework, tutorial, and assessment system that accompanies Pearson Education textbooks in mathematics and statistics. With MathXL, instructors can create, edit, and assign online homework and tests using algorithmically generated exercises correlated at the objective level to the textbook. They can also create and assign their own online exercises and import Test-Gen tests for added flexibility. All student work is tracked in MathXL's online gradebook. Students can take chapter tests in MathXL and receive personalized study plans based on their test results. The study plan diagnoses weaknesses and links students directly to tutorial exercises for the objectives they need to study and retest. Students can also access supplemental animations and video clips directly from selected exercises. MathXL is available to qualified adopters. For more information, visit our Web site at www.mathxl.com, or contact your sales representative.

MyMathLab® www.mymathlab.com

MyMathLab is a series of text-specific, easily customizable online courses for Pearson Education textbooks in mathematics and statistics. Powered by CourseCompass™ (our online teaching and learning environment) and MathXL® (our online homework, tutorial, and assessment system), MyMathLab gives instructors the tools they need to deliver all or a portion of their course online, whether students are in a lab or working at home or elsewhere. MyMathLab provides a rich and flexible set of course materials, featuring free-response exercises that are algorithmically generated for unlimited practice and mastery. Students can also use online tools, such as video lectures, animations, and a multimedia textbook, to independently improve their understanding and performance. Instructors can use MyMathLab's homework and test managers to select and assign online exercises correlated directly to the textbook, and they can create and assign their own online exercises and import TestGen tests for added flexibility. MyMathLab's online gradebook—designed specifically for mathematics and statistics—automatically tracks students' homework and test results and gives the instructor control over how to calculate final grades. Instructors can also add offline (paper-and-pencil) grades to the gradebook. MyMathLab is available to qualified adopters. For more information, visit our Web site at www.mymathlab.com, or contact your sales representative.

InterAct Math® Tutorial Web site: www.interactmath.com

Get practice and tutorial help online! This interactive tutorial Web site provides algorithmically generated practice exercises that correlate directly to the exercises in the textbook. Students can do an exercise as many times as they like with new values each time for unlimited practice and mastery. Every exercise is accompanied by an interactive guided solution that provides helpful feedback for incorrect answers, and students can also view a worked-out sample problem that steps them through an exercise similar to the one they're working on.

ACKNOWLEDGMENTS

Many individuals contributed to the development of this textbook. We thank the following reviewers, whose comments and suggestions were invaluable in preparing *Beginning Algebra with Applications and Visualization*, Second Edition.

Susan Bradley, *Angelina College*
John Buoni, *Youngstown State University*
Annette Burden, *Youngstown State University*
Beth Fraser, *Middlesex Community College*
Christian Glardon, *Valencia Community College—East Campus*
Mary Beth Headlee, *Manatee Community College*
Jolene Rhodes, *Valencia Community College—East Campus*
Jennifer Lawhon, *Valencia Community College—East Campus*
Julie Mays, *Angelina College*
Martha Muenks, *Maysville Community and Technical College*
Linda Murphy, *Northern Essex Community College*
Brenda Norman, *Tidewater Community College—Virginia Beach Campus*
Sherry Schulz, *College of the Canyons*
Denise Turcotte, *Manatee Community College*
Jewell Valrie, *Wake Technical Community College*
Richard Watkins, *Tidewater Community College—Virginia Beach Campus*
Myong Wooten, *Ohio State University—Newark*

Elina Niemelä, Bob Bohland, David Atwood, Paul Lorczak, Jessica Rockswold, and Namyong Lee deserve special credit for their help with accuracy checking. Without the excellent cooperation from the professional staff at Addison-Wesley Publishing Company, this project would have been impossible. Thanks go to Greg Tobin and Maureen O'Connor for giving their support. Particular recognition is due Randy Welch and Lauren Morse, who gave essential advice and assistance. The outstanding contributions of Kathy Manley, Joe Vetere, Michelle Renda, Marlana Voerster, Nathaniel Koven, Antonio Arvelo, and Ceci Fleming are greatly appreciated. Special thanks go to Kathy Diamond, who was instrumental in the success of this project.

Thanks go to Wendy Rockswold and Carrie Krieger, whose unwavering encouragement and support made this project possible. We also thank the many students and instructors who used the previous edition of this textbook. Their suggestions were insightful and helpful.

Please feel free to send us your comments and questions at either of the following e-mail addresses: gary.rockswold@mnsu.edu or terry.krieger@roch.edu. Your opinions are important to us.

Gary K. Rockswold
Terry A. Krieger

Introduction to Algebra

J ust over a century ago relatively few people were educated past the elementary years. With the industrial revolution came the need for a more educated workforce that possessed the skills necessary to work in a variety of areas, from new factory assembly lines to the rapidly expanding transportation systems. As the skills needed in the workforce increased, the number of people seeking a high school education rose dramatically.

In 1900 only about one in ten workers was in a professional, technical, or managerial occupation. By 1970, two of every ten workers had a job of this type, and today this proportion is nearly one in three.

The study of mathematics is essential for anyone who wants to keep up with the technological changes that are occurring in nearly every occupation. A person who is mathematically competent is more prepared to handle these technological changes as they occur in the workplace with increasing frequency. Mathematics is the *language of technology.*

Students who avoid mathematics classes throughout their education may find that doing so dramatically affects their vocation, income, and lifestyle. As the gap between the wages of the college educated and those with a high school diploma continues to widen, employers continue to seek employees with higher levels of mathematical skill.

It is just a matter of time before the *majority* of the workforce will need the analytic and problem-solving skills that are taught in mathematics classes every day. No matter what career path students choose, a solid background in mathematics will provide opportunities that are not available to those who lack these important skills.

> *Unless you try to do something beyond what you have already mastered, you will never grow.*
>
> —RONALD E. OSBORN

Source: A. Greenspan, "The Economic Importance of Improving Math-Science Education."

1.1   NUMBERS, VARIABLES, AND EXPRESSIONS

Natural Numbers and Whole Numbers ▪ Prime Numbers and Composite Numbers ▪ Variables, Algebraic Expressions, and Equations ▪ Translating Words to Expressions

A LOOK INTO MATH ▷

Numbers are an important concept in every society. Without numbers it would be impossible to quantify (count) a group of objects. One number system used in southern Africa consisted of only the numbers from 1 to 20. Numbers larger than 20 were named by counting groups of *twenties*. For example, the number 67 was called *three twenties and seven*. This base-20 number system would not work well in today's technologically advanced world. In this section, we will begin by introducing two sets of numbers that are used extensively in the modern world: natural numbers and whole numbers.

Natural Numbers and Whole Numbers

One important set of numbers found in most societies is the set of **natural numbers**. These numbers comprise the *counting numbers* and may be expressed as follows.

$$1, 2, 3, 4, 5, 6, \ldots$$

Because there are infinitely many natural numbers, three dots are used to show that the list continues in the same pattern without end. A second set of numbers is called the **whole numbers**, and may be expressed as follows.

$$0, 1, 2, 3, 4, 5, \ldots$$

Whole numbers include the natural numbers and the number 0.

▶ **REAL-WORLD CONNECTION**   Natural numbers and whole numbers can be used when data are not broken into fractional parts. For example, Table 1.1 lists the recent and projected U.S. population of persons aged 85 and older during selected years. Note that both natural numbers and whole numbers are appropriate to describe these data because a fraction of a person is not possible.

TABLE 1.1   **U.S. Population Aged 85 and Older**

Year	1990	2000	2010	2020
Population	3,080,000	4,240,000	6,123,000	7,269,000

Source: U.S. Census Bureau.

CRITICAL THINKING

Give an example from everyday life of natural number or whole number use.

Prime Numbers and Composite Numbers

When two natural numbers are multiplied, the result is another natural number. For example, if the natural numbers 3 and 4 are multiplied, the result is 12, a natural number. We say that the **product** of 3 and 4 is 12 and express this product as $3 \times 4 = 12$. The numbers 3 and 4 are **factors** of 12. Note that other natural numbers also are factors of 12, such as 2 and 6 or 1 and 12.

A natural number greater than 1 that has *only* itself and 1 as natural number factors is a **prime number**. The number 7 is prime because the only natural number factors of 7 are 1 and 7. The following is a partial list of prime numbers. There are infinitely many prime numbers.

$$2, \quad 3, \quad 5, \quad 7, \quad 11, \quad 13, \quad 17, \quad 19, \quad 23, \quad 29$$

A natural number greater than 1 that is not prime is a **composite number**. The natural number 15 is a composite number because $3 \times 5 = 15$, so 15 has factors other than itself and 1. Any composite number can be written as a product of prime numbers. For example, the natural number 120 is composite because 10 and 12 are natural number factors of 120. We can use the tree diagram shown in Figure 1.1(a) to factor 120 into prime numbers. We start by showing the factors 10 and 12 below 120 and then continue downward, factoring **10** and **12**. The bottom of the tree shows

$$120 = 2 \times 5 \times 3 \times 2 \times 2.$$

The tree is not unique because 120 could be factored as 3×40 instead of 10×12, as illustrated in Figure 1.1(b). However, the **prime factorization** is *always* unique.

CRITICAL THINKING

Suppose that you draw a tree diagram for the prime factorization of a prime number. Describe what the tree will look like.

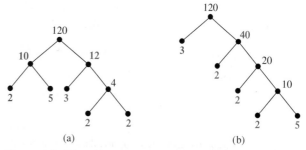

(a)   (b)

Figure 1.1 Prime Factorization of 120

Because the prime factor tree for a given composite number may not be unique, it is customary to write the prime factors in ascending order. For example,

$$120 = 2 \times 2 \times 2 \times 3 \times 5.$$

MAKING CONNECTIONS

Prime Numbers and the Internet

Have you ever ordered something on the Internet by credit card? To ensure the privacy of your credit card number, security codes are used. Large prime numbers often play an essential role in the development of these codes. One such security code is based on the fact that it is relatively easy to multiply two large prime numbers to generate a very large composite number, but it is enormously difficult to find the original two prime numbers (factors) given the large composite number.

EXAMPLE 1   Classifying numbers as prime or composite

Classify each natural number as prime or composite, if possible. If a number is composite, write it as a product of prime numbers.
(a) 31   **(b)** 1   **(c)** 35   **(d)** 200

Solution
(a) The only factors of 31 are itself and 1, so the natural number 31 is prime.
(b) The number 1 is neither prime nor composite because prime and composite numbers must be greater than 1.
(c) The number 35 is composite because 5 and 7 are factors. It can be written as a product of prime numbers as $35 = 5 \times 7$.

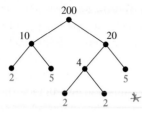

Figure 1.2 Prime Factorization of 200

(d) The number 200 is composite because 10 and 20 are factors. A tree diagram can be used to write 200 as a product of prime numbers, as shown in Figure 1.2. The tree diagram reveals that 200 can be factored as $200 = 2 \times 5 \times 2 \times 2 \times 5$, or it can be factored as $200 = 2 \times 2 \times 2 \times 5 \times 5$.

Now Try Exercises 17, 23, 25

✱ Variables, Algebraic Expressions, and Equations

There are 12 inches in 1 foot, so 5 feet equal $5 \times 12 = 60$ inches. Similarly, in 3 feet there are $3 \times 12 = 36$ inches. If we need to convert feet to inches frequently, Table 1.2 might help.

TABLE 1.2   Converting Feet to Inches

Feet	1	2	3	4	5	6	7
Inches	12	24	36	48	60	72	84

One difficulty with this table is that a needed value might not occur in it. For example, the table is not helpful in converting 11 feet to inches. We could expand Table 1.2 into Table 1.3 to include $11 \times 12 = 132$. However, expanding the table to accommodate every possible value for feet would be impossible.

TABLE 1.3   Converting Feet to Inches

Feet	1	2	3	4	5	6	7	11
Inches	12	24	36	48	60	72	84	132

Variables are often used in mathematics when tables of numbers are inadequate. A **variable** is a symbol, typically an italic letter such as x, y, z, or F used to represent an unknown quantity. In the preceding example, the number of feet to be converted to inches could be represented by the variable F. The corresponding number of inches could be represented by the variable I. The number of inches in F feet is given by the *algebraic expression* $12 \times F$. That is, to calculate the number of inches in F feet, multiply F by **12**. The relationship between feet F and inches I is shown by using the *equation* or *formula*

$$I = 12 \times F.$$

Sometimes a dot ($\cdot$) is used to indicate multiplication because a multiplication sign ($\times$) can be confused with the variable x. Other times the multiplication sign is omitted altogether. Thus all four formulas

$$I = 12 \times F, \quad I = 12 \cdot F, \quad I = 12F, \quad \text{and} \quad I = 12(F)$$

represent the same relationship between feet and inches. For example, if we wish to find the number of inches in 10 feet, we can replace F in one of these formulas with the number 10. If we let $F = 10$ in the second formula,

$$I = 12 \cdot 10 = 120.$$

That is, there are 120 inches in 10 feet.

More formally, an **algebraic expression** consists of numbers, variables, operation symbols, such as $+, -, \times$, and $\div$, and grouping symbols, such as parentheses. An **equation** is a mathematical statement that two algebraic expressions are equal. Equations *always* contain an equals sign. A **formula** is a special type of equation that expresses a relationship between two or more quantities. The formula $I = 12F$ states that to calculate the number of inches in F feet, multiply F by 12.

EXAMPLE 2   Evaluating algebraic expressions with one variable

Evaluate each algebraic expression for $x = 4$.

(a) $x + 5$ **(b)** $5x$ **(c)** $15 - x$ **(d)** $\dfrac{x}{(x - 2)}$

Solution

(a) Replace x with 4 in the expression $x + 5$ to obtain $4 + 5 = 9$.
(b) The expression $5x$ indicates multiplication of 5 and x. Thus $5x = 5 \cdot 4 = 20$.
(c) $15 - x = 15 - 4 = 11$
(d) Perform all arithmetic operations inside parentheses first.

$$\frac{x}{(x - 2)} = \frac{4}{(4 - 2)} = \frac{4}{2} = 2$$

Now Try Exercises 47, 49, 55

▶ **REAL-WORLD CONNECTION**   Some algebraic expressions contain more than one variable. For example, if a car travels 120 miles on 6 gallons of gasoline, then the car's *mileage* is $\frac{120}{6} = 20$ miles per gallon. In general, if a car travels M miles on G gallons of gasoline, then its mileage is given by the expression $\frac{M}{G}$. Note that $\frac{M}{G}$ contains two variables, M and G, whereas $12F$ contains only one variable, F.

EXAMPLE 3   Evaluating algebraic expressions with two variables

Evaluate each algebraic expression for $y = 2$ and $z = 8$.

(a) $3yz$ **(b)** $z - y$ **(c)** $\dfrac{z}{y}$

Solution

(a) Replace y with 2 and z with 8 to obtain $3yz = 3 \cdot 2 \cdot 8 = 48$.
(b) $z - y = 8 - 2 = 6$

(c) $\dfrac{z}{y} = \dfrac{8}{2} = 4$

Now Try Exercises 57, 59

EXAMPLE 4   Evaluating formulas

Find the value of y for $x = 15$ and $z = 10$.

(a) $y = x - 3$ **(b)** $y = \dfrac{x}{5}$ **(c)** $y = 8xz$

Solution

(a) Substitute 15 for x and then evaluate the right side of the formula to find y.

$$y = x - 3 \qquad \text{Given formula}$$
$$= 15 - 3 \qquad \text{Replace } x \text{ with 15.}$$
$$= 12 \qquad \text{Subtract.}$$

(b) $y = \dfrac{x}{5} = \dfrac{15}{5} = 3$

(c) $y = 8xz = 8 \cdot 15 \cdot 10 = 1200$

Now Try Exercises 63, 73

Translating Words to Expressions

Many times in mathematics, algebraic expressions are not given; rather, we must write our own expressions. To accomplish this task, we often translate words to symbols. The symbols $+$, $-$, $\times$, and $\div$ have special mathematical words associated with them. When two numbers are added, the result is called the *sum*. When one number is subtracted from another number, the result is called the *difference*. Similarly, multiplying two numbers results in a *product* and dividing two numbers results in a *quotient*. Table 1.4 illustrates many of the words commonly associated with these operations.

TABLE 1.4   **Words Associated with Arithmetic Symbols**

Symbol	Associated Words
$+$	add, plus, more, sum, total
$-$	subtract, minus, less, difference, fewer
$\times$	multiply, times, twice, double, triple, product
$\div$	divide, divided by, quotient

EXAMPLE 5   **Translating words to expressions**

Translate each phrase to an algebraic expression. Specify what each variable represents.
(a) Twice the cost of a CD
(b) Four more than a number
(c) Ten less than the president's age
(d) A number plus 10, all divided by a different number
(e) The product of 6 and a number
(f) The sum of the heights of two children

Solution
(a) If the cost were \$15, then twice this cost would be $2 \cdot \mathbf{15} = \$30$. In general, if we let C be the cost of a CD, then twice the cost of a CD would be $2 \cdot C$, or $2C$.
(b) If the number were 20, then four more than the number would be $\mathbf{20} + 4 = 24$. If we let n represent the number, then four more than the number would be $n + 4$.
(c) If the president's age were 55, then ten less than the president's age would be $\mathbf{55} - 10 = 45$. If we let A represent the president's age, then ten less would be $A - 10$.
(d) Let x be the first number and y be the other number. Then the expression can be written as $(x + 10) \div y$. Note that parentheses are used around $x + 10$ because it is "all" divided by y. This expression could also be written as $\frac{x + 10}{y}$.
(e) If n represents the number, then the product of 6 and n is $6 \cdot n$ or $6n$.
(f) Let x be the height of one child and y be the height of the other child. The sum of their heights is $x + y$.

Now Try Exercises 75, 79, 83, 89

▶ **REAL-WORLD CONNECTION**   Cities are made up of large amounts of concrete and asphalt that heat up in the daytime from sunlight but do not cool off completely at night. As a result, urban areas tend to be warmer than surrounding rural areas. This effect is called the *urban heat island* and has been documented in cities throughout the world. The next example discusses the impact of this effect in Phoenix, Arizona.

EXAMPLE 6　Translating words to a formula

For each year after 1950, the average nighttime temperature in Phoenix has increased by about 0.1°C.
(a) What was the increase in the nighttime temperature after 20 years, or in 1970?
(b) Write a formula (or equation) that gives the increase T in average nighttime temperature x years after 1950.
(c) Use your formula to estimate the increase in nighttime temperature in 1980.

Solution
(a) The average nighttime temperature has increased by 0.1°C per year, so after **20** years the temperature increase would be $0.1 \times 20 = 2.0$°C.
(b) To calculate the nighttime increase in temperature multiply the number of years past 1950 by 0.1. Thus $T = 0.1x$, where x represents the number of years after 1950.
(c) Because $1980 - 1950 = 30$, let $x = 30$ in the formula $T = 0.1x$ to get

$$T = 0.1(30) = 3.$$

The average nighttime temperature increased by 3°C.　Now Try Exercise 95

Another use of translating words to formulas is in finding the areas of various shapes.

EXAMPLE 7　Finding the area of a rectangle

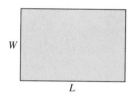

The area A of a rectangle equals its length L times its width W, as illustrated in the accompanying figure.
(a) Write a formula that shows the relationship between these three quantities.
(b) Find the area of a standard sheet of paper that is 8.5 inches wide and 11 inches long.

Solution
(a) The word *times* indicates that the length and width should be multiplied. The formula is given by $A = LW$.
(b) $A = 11 \cdot 8.5 = 93.5$ square inches　Now Try Exercise 101

1.1 PUTTING IT ALL TOGETHER

The following table summarizes some of the topics discussed in this section.

Concept	Comments	Examples
Natural Numbers	Sometimes referred to as the *counting numbers*	1, 2, 3, 4, 5, . . .
Whole Numbers	Includes the natural numbers and 0	0, 1, 2, 3, 4, . . .
Prime Number	A natural number greater than 1 whose only factors are itself and 1; there are infinitely many prime numbers.	2, 3, 5, 7, 11, 13, 17, and 19 are the primes less than 20

continued on next page

continued from previous page

Concept	Comments	Examples
Composite Number	A natural number greater than 1 that is *not* a prime number; there are infinitely many composite numbers.	4, 9, 25, 39, 62, 76, 87, 91, 100
Variable	Represents an unknown quantity	x, y, z, A, F, and T
Algebraic Expression	May consist of variables, numbers, operation symbols, such as $+, -, \times$, and $\div$, and grouping symbols, such as parentheses	$x + 3, \frac{x}{y}, 2z + 5, 12F,$ $x + y + z,$ $x(y + 5)$
Equation	An equation is a statement that two algebraic expressions are equal. An equation always includes an equals sign.	$2 + 3 = 5, x + 5 = 7,$ $I = 12F,$ $y = 0.1x$
Formula	A formula is a special type of equation that expresses a relationship between two or more quantities.	$I = 12F, y = 0.1x,$ $A = LW,$ $F = 3Y$

1.1 Exercises

MyMathLab

 Math XL PRACTICE   WATCH   DOWNLOAD   READ   REVIEW

CONCEPTS

1. The natural numbers are also called the _____ numbers.

2. A whole number is either a natural number or the number _____.

3. The factors of a prime number are itself and _____.

4. If a natural number is greater than 1 and not a prime number, then it is called a(n) _____ number.

5. The number 11 is a (prime/composite) number.

6. The number 9 is a (prime/composite) number.

7. Because $3 \times 6 = 18$, the numbers 3 and 6 are _____ of 18.

8. A _____ is a special type of equation that expresses a relationship between two or more quantities.

9. A symbol or letter used to represent an unknown quantity is called a(n) _____.

10. Equations always contain a(n) _____.

11. When one number is added to another number, the result is called the _____.

12. When one number is multiplied by another number, the result is called the _____.

13. The result of dividing one number by another is called the _____.

14. The result of subtracting one number from another is called the _____.

PRIME NUMBERS AND COMPOSITE NUMBERS

Exercises 15–26: Classify the number as prime, composite, or neither. If the number is composite, write it as a product of prime numbers.

15. 4   16. 36

17. 1   18. 0

19. 29   20. 13

21. 92   22. 69

23. 225

24. 900

25. 149

26. 101

Exercises 27–38: Write the composite number as a product of prime numbers.

27. 6

28. 8

29. 12

30. 20

31. 32

32. 100

33. 39

34. 51

35. 294

36. 175

37. 300

38. 455

Exercises 39–46: State whether the given quantity could accurately be described by the whole numbers.

39. The population of a country

40. The cost of a candy bar in dollars

41. A student's grade point average

42. The Fahrenheit temperature in Antarctica

43. The number of bytes stored on a computer's hard drive

44. The exact winning time in a marathon, measured in hours

45. The number of students in a class

46. The number of bald eagles in the United States

ALGEBRAIC EXPRESSIONS, FORMULAS, AND EQUATIONS

Exercises 47–56: Evaluate the expression for the given value of x.

47. $2x$ $\quad x = 5$

48. $x + 8$ $\quad x = 7$

49. $8 - x$ $\quad x = 1$

50. $7x$ $\quad x = 0$

51. $\dfrac{x}{8}$ $\quad x = 32$

52. $\dfrac{5}{x - 3}$ $\quad x = 8$

53. $3(x + 1)$ $\quad x = 5$

54. $7(6 - x)$ $\quad x = 3$

55. $\left(\dfrac{x}{2}\right) + 1$ $\quad x = 6$

56. $3 - \left(\dfrac{6}{x}\right)$ $\quad x = 2$

49. $8 - (1) = 7$ 50. $7(x) = 7(0) = 0$

55 $\left(\dfrac{6}{2}\right) + 1 = 3 + 1 = 4$ 58 $5(2)(3) = 5 \times 2 \times 3 = 30$

Exercises 57–62: Evaluate the expression for the given values of x and y.

57. $x + y$ $\qquad x = 5, \quad y = 4$

58. $5xy$ $\qquad x = 2, \quad y = 3$

59. $4 \cdot \dfrac{x}{y}$ $\qquad x = 8, \quad y = 4$

60. $y - x$ $\qquad x = 8, \quad y = 11$

61. $y(x - 3)$ $\qquad x = 5, \quad y = 7$

62. $(x + y) - 5$ $\quad x = 6, \quad y = 3$

Exercises 63–66: Find the value of y for the given value of x.

63. $y = x + 1$ $\qquad x = 0$

64. $y = x \cdot x$ $\qquad x = 5$

65. $y = 4x$ $\qquad x = 7$

66. $y = 2(x - 3)$ $\quad x = 3$

Exercises 67–70: Find the value of F for the given value of z.

67. $F = z - 5$ $\qquad z = 12$

68. $F = \dfrac{z}{4}$ $\qquad z = 40$

69. $F = \dfrac{30}{z}$ $\qquad z = 6$

70. $F = z \cdot z \cdot z$ $\qquad z = 5$

Exercises 71–74: Find the value of y for the given values of x and z.

71. $y = x + z$ $\qquad x = 3, \quad z = 15$

72. $y = 3xz$ $\qquad x = 2, \quad z = 0$

73. $y = \dfrac{x}{z}$ $\qquad x = 9, \quad z = 3$

74. $y = x - z$ $\qquad x = 9, \quad z = 1$

TRANSLATING WORDS TO SYMBOLS

Exercises 75–90: Translate the phrase to an algebraic expression. State what each variable represents.

75. Three times the cost of a soda

76. Twice the cost of a gallon of gasoline

77. Five more than a number

78. Five less than a number

79. The sum of a number and 5

80. The quotient of two numbers

81. Triple a number

82. A number plus two

83. Two hundred less than the population of a town

84. The total number of dogs and cats in a city

85. A number divided by six

86. A number divided by another number

87. The product of the speed of a car and the time it has been traveling

88. The difference between 220 and a person's heart rate

89. A number plus seven, all divided by a different number

90. One-fourth of a number plus one-tenth of a different number

APPLICATIONS

91. *Yards to Feet*   Make a table of values that converts y yards to F feet. Let $y = 1, 2, 3, \ldots, 7$. Write a formula that converts y yards to F feet.

92. *Gallons to Quarts*   Make a table of values that converts g gallons to Q quarts. Let $g = 1, 2, 3, \ldots, 6$. Write a formula that converts g gallons to Q quarts.

93. *Dollars to Pennies*   Write a formula that converts D dollars to P pennies.

94. *Quarters to Nickels*   Write a formula that converts Q quarters to N nickels.

95. *NASCAR Speeds*   On the fastest speedways, some NASCAR drivers reach average speeds of 3 miles per minute. Write a formula that gives the number of miles M that such a driver would travel in x minutes. How far would this driver travel in 36 minutes?

96. *NASCAR Speeds*   On slower speedways, some NASCAR drivers reach average speeds of 2 miles per minute. Write a formula that gives the number of miles M that such a driver would travel in x minutes. How far would this driver travel in 42 minutes?

97. Thinking Generally   If there are 6 blims in every drog, is the formula that relates B blims and D drogs $D = 6B$ or $B = 6D$?

98. *Heart Beat*   The resting heart beat of a person is 70 beats per minute. Write a formula that gives the number of beats B that occur in x minutes. How many beats are there in an hour?

99. *Cost of a CD*   The table lists the cost C of buying x compact discs. Write an equation that relates C and x.

Number of CDs (x)	1	2	3	4
Cost (C)	$12	$24	$36	$48

100. *Gallons of Water*   The table lists the gallons G of water coming from a garden hose after m minutes. Write an equation that relates G and m.

Minutes (m)	1	2	3	4
Gallons (G)	4	8	12	16

101. *Area of a Rectangle*   The area of a rectangle equals its length times its width. Find the area of the rectangle shown in the figure.

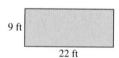

9 ft

22 ft

102. *Area of a Square*   A square is a rectangle whose length and width have equal measures. Find the area of a square with length 14 inches.

WRITING ABOUT MATHEMATICS

103. Give an example where the whole numbers are not sufficient to describe a quantity in real life. Explain your reasoning.

104. Explain what a prime number is. How can you determine whether a number is prime?

105. Explain what a composite number is. How can you determine whether a number is composite?

106. When are variables used? Give an example.

1.2 FRACTIONS

Basic Concepts ▪ Simplifying Fractions to Lowest Terms ▪ Multiplication and Division of Fractions ▪ Addition and Subtraction of Fractions ▪ Applications

A LOOK INTO MATH ▷

Historically, natural and whole numbers have not been sufficient for most societies. Early on, the concept of splitting a quantity into parts was common, and as a result, fractions were developed. For instance, pouring a quart of milk into four cups gives rise to the concept of a fourth of a quart. In this section we discuss fractions and how to add, subtract, multiply, and divide them.

Basic Concepts

If we divide a circular pie into 6 equal slices, as shown in Figure 1.3, then each piece represents one-sixth of the pie and can be represented by the fraction $\frac{1}{6}$. Five slices of the pie would represent five-sixths of the pie and can be represented by the fraction $\frac{5}{6}$.

The parts of a fraction are named as follows.

Numerator $\longrightarrow$ $\dfrac{5}{6}$ $\longleftarrow$ Fraction bar

Denominator

In the fraction $\frac{13}{17}$, the numerator is 13 and the denominator is 17. Sometimes we can represent a general fraction by using variables. The fraction $\frac{a}{b}$ can represent any fraction with numerator a and denominator b. However, the value of b cannot equal 0, which is denoted $b \neq 0$. (The symbol $\neq$ means "not equal to.")

NOTE: The fraction bar represents division. For example, the fraction $\frac{1}{2}$ represents the result when 1 is divided by 2, which is 0.5. We discuss this concept further in Section 1.4.

Figure 1.3

(handwritten annotation) $\frac{?}{0}$ (when there is a ZERO in the numerator what does that mean)

EXAMPLE 1 Identifying numerators and denominators

Give the numerator and denominator of each fraction.

(a) $\dfrac{6}{13}$ **(b)** $\dfrac{ac}{b}$ **(c)** $\dfrac{x-5}{y+z}$

Solution
(a) The numerator is 6, and the denominator is 13.
(b) The numerator is ac, and the denominator is b.
(c) The numerator is $x - 5$, and the denominator is $y + z$.   Now Try Exercise 2

Simplifying Fractions to Lowest Terms

Consider the amount of pizza shown in each of the three pies in Figure 1.4 on the next page. The first pie was cut into sixths and there are three pieces remaining, the second pie was cut into fourths and there are two pieces remaining, and the third pie was cut into only two pieces with one piece remaining. In all three cases half a pizza remains.

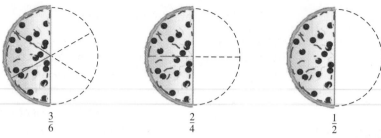

$$\frac{3}{6} \qquad\qquad \frac{2}{4} \qquad\qquad \frac{1}{2}$$

Figure 1.4

Figure 1.4 illustrates that the fractions $\frac{3}{6}$, $\frac{2}{4}$, and $\frac{1}{2}$ are equal. The fraction $\frac{1}{2}$ is in **lowest terms** because its numerator and denominator have no factors in common, whereas the fractions $\frac{3}{6}$ and $\frac{2}{4}$ are not in lowest terms. In the fraction $\frac{3}{6}$, the numerator and denominator have a common factor of 3, and in the fraction $\frac{2}{4}$, the numerator and denominator have a common factor of 2. The fraction $\frac{3}{6}$ can be simplified as follows.

$$\frac{3}{6} = \frac{1 \cdot 3}{2 \cdot 3} \qquad \text{Factor out 3.}$$

$$= \frac{1}{2} \qquad \frac{a \cdot c}{b \cdot c} = \frac{a}{b}$$

To simplify $\frac{3}{6}$, we used the *basic principle of fractions*: The value of a fraction is unchanged if the numerator and denominator of the fraction are multiplied (or divided) by the same nonzero number. We can also simplify the fraction $\frac{2}{4}$ to $\frac{1}{2}$ by using the basic principle of fractions.

$$\frac{2}{4} = \frac{1 \cdot 2}{2 \cdot 2} = \frac{1}{2}$$

When simplifying fractions, we usually factor out the *greatest common factor* (GCF) for the numerator and the denominator. For example, to simplify $\frac{27}{36}$ we first find the greatest common factor for 27 and 36. The **greatest common factor** is the largest factor that is common to both 27 and 36. Because $27 = 9 \times 3$ and $36 = 9 \times 4$, their greatest common factor is 9. Note that 3 is a factor of both 27 and 36, but it is not the *greatest* common factor. Greatest common factors can also be found by determining the largest number that divides evenly into each number. Because 9 is the largest number that divides evenly into 27 and 36, it is the greatest common factor. Thus, simplified to lowest terms, $\frac{27}{36} = \frac{3 \cdot 9}{4 \cdot 9} = \frac{3}{4}$.

SIMPLIFYING FRACTIONS

To simplify a fraction to lowest terms, begin by factoring out the greatest common factor c in the numerator and in the denominator. Then apply the **basic principle of fractions:**

$$\frac{a \cdot c}{b \cdot c} = \frac{a}{b}.$$

handwritten annotation: value of a fraction is unchanged if num + denom of fraction are multiplied by the same ≠0 number

NOTE: This principle is true because multiplying a fraction by $\frac{c}{c}$ or 1 does not change the value of the fraction.

handwritten margin notes:
$\frac{27}{36}$
$(1, 27) (3, 9)$
$(1, 27)$ $(3, 9)$
$(4, 9)(2, 18)(3, 12)(1, 36)$
36
$GCF = 9$

EXAMPLE 2   Finding the greatest common factor

Find the greatest common factor (GCF) for each pair of numbers.
(a) 8, 12   **(b)** 24, 60   **(c)** 495, 735

Solution

(a) Because $8 = 4 \cdot 2$ and $12 = 4 \cdot 3$, the number 4 is the largest factor that is common to both 8 and 12. Thus the GCF of 8 and 12 is 4. (Note that 2 is a *common* factor of 8 and 12 because $8 = 2 \cdot 4$ and $12 = 2 \cdot 6$, but 2 is not the *greatest* common factor.)

(b) When we are working with larger numbers, one way to determine the greatest common factor is to find the prime factorization of each number, as in

$$24 = 4 \cdot 6 = 2 \cdot 2 \cdot 2 \cdot 3 \quad \text{and}$$

$$60 = 6 \cdot 10 = 2 \cdot 3 \cdot 2 \cdot 5 = 2 \cdot 2 \cdot 3 \cdot 5.$$

The prime factorizations have two 2s and one 3 in common. Thus the GCF of 24 and 60 is $2 \cdot 2 \cdot 3 = 12$.

(c) Prime factorization trees for 495 and 735 are shown in Figure 1.5. Thus

$$495 = 3 \cdot 3 \cdot 5 \cdot 11 \quad \text{and}$$

$$735 = 3 \cdot 5 \cdot 7 \cdot 7.$$

The prime factorizations have one 3 and one 5 in common. Thus the GCF of 495 and 735 is $3 \cdot 5 = 15$.   **Now Try Exercises 21, 23**

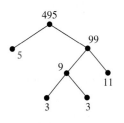
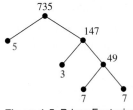

Figure 1.5 Prime Factorizations of 495 and 735

EXAMPLE 3   Simplifying fractions to lowest terms

Simplify each fraction to lowest terms.
(a) $\frac{8}{12}$ **(b)** $\frac{24}{60}$ **(c)** $\frac{42}{105}$

Solution

(a) From Example 2(a), the GCF of 8 and 12 is 4. Thus

$$\frac{8}{12} = \frac{2 \cdot 4}{3 \cdot 4} = \frac{2}{3}.$$

(b) From Example 2(b), the GCF of 24 and 60 is 12. Thus

$$\frac{24}{60} = \frac{2 \cdot 12}{5 \cdot 12} = \frac{2}{5}.$$

(c) The prime factorizations of 42 and 105 are

$$42 = 6 \cdot 7 = 2 \cdot 3 \cdot 7 \quad \text{and} \quad 105 = 5 \cdot 21 = 5 \cdot 3 \cdot 7.$$

The GCF of 42 and 105 is $3 \cdot 7 = 21$. Thus

$$\frac{42}{105} = \frac{2 \cdot 21}{5 \cdot 21} = \frac{2}{5}.$$

Now Try Exercises 31, 39

Find the GCF

23) 50, 75 = 25

6) 36, 48, 72 = 12

Simplifying Fractions in Steps

Sometimes a fraction can be simplified to lowest terms in multiple steps. By using *any* common factor that is not the GCF a new fraction in *lower* terms will result. This new fraction may then be simplified using a common factor of its numerator and denominator. If this process is continued, the result will be the given fraction simplified to lowest terms. The fraction in Example 3(c) could be simplified to lowest terms in two steps.

$$\frac{42}{105} = \frac{14 \cdot 3}{35 \cdot 3} = \frac{14}{35} = \frac{2 \cdot 7}{5 \cdot 7} = \frac{2}{5}$$

Multiplication and Division of Fractions

Suppose we cut *half* an apple into *thirds*, as illustrated in Figure 1.6. Then each piece represents one-sixth of the original apple. One-third of one-half is described by the product

$$\frac{1}{3} \cdot \frac{1}{2} = \frac{1}{6}.$$

Figure 1.6

NOTE: The word "of" in mathematics often indicates multiplication. For example, the phrases "one-fifth of the cookies," "twenty percent of the price," and "half of the money" all suggest that multiplication may be required.

This example demonstrates that the numerator of the product of two fractions is found by multiplying the numerators of the two fractions. Similarly, the denominator of the product of two fractions is found by multiplying the denominators of the two fractions. For example, the product of $\frac{2}{3}$ and $\frac{5}{7}$ is

$$\frac{2}{3} \cdot \frac{5}{7} = \frac{2 \cdot 5}{3 \cdot 7} = \frac{10}{21}.$$

Multiply numerators.

Multiply denominators.

MULTIPLICATION OF FRACTIONS

The product of $\frac{a}{b}$ and $\frac{c}{d}$ is given by

$$\frac{a}{b} \cdot \frac{c}{d} = \frac{ac}{bd},$$

where b and d are not 0.

EXAMPLE 4   **Multiplying fractions**

Multiply each expression and simplify the result when appropriate.

(a) $\frac{4}{5} \cdot \frac{6}{7}$ **(b)** $\frac{8}{9} \cdot \frac{3}{4}$ **(c)** $3 \cdot \frac{5}{9}$ **(d)** $\frac{x}{y} \cdot \frac{z}{3}$

Solution

(a) $\frac{4}{5} \cdot \frac{6}{7} = \frac{4 \cdot 6}{5 \cdot 7} = \frac{24}{35}$

(b) $\frac{8}{9} \cdot \frac{3}{4} = \frac{8 \cdot 3}{9 \cdot 4} = \frac{24}{36}$; the GCF of 24 and 36 is 12, so

$$\frac{24}{36} = \frac{2 \cdot 12}{3 \cdot 12} = \frac{2}{3}.$$

(c) Start by writing 3 as $\frac{3}{1}$.

$$3 \cdot \frac{5}{9} = \frac{3}{1} \cdot \frac{5}{9} = \frac{3 \cdot 5}{1 \cdot 9} = \frac{15}{9}$$

The GCF of 15 and 9 is 3, so

$$\frac{15}{9} = \frac{5 \cdot 3}{3 \cdot 3} = \frac{5}{3}.$$

(d) $\frac{x}{y} \cdot \frac{z}{3} = \frac{x \cdot z}{y \cdot 3} = \frac{xz}{3y}$

When we write the product of a variable and a number, such as $y \cdot 3$, we typically write the number first, followed by the variable. That is, $y \cdot 3 = 3y$.

Now Try Exercises 43, 51, 55

MAKING CONNECTIONS

Multiplying and Simplifying Fractions

When multiplying fractions, sometimes it is possible to change the order of the factors to rewrite the product so that it is easier to simplify. In Example 4(b) the product could be written as

$$\frac{8}{9} \cdot \frac{3}{4} = \frac{3 \cdot 8}{9 \cdot 4} = \frac{3}{9} \cdot \frac{8}{4} = \frac{1}{3} \cdot \frac{2}{1} = \frac{2}{3}.$$

Instead of simplifying $\frac{24}{36}$, which contains larger numbers, the fractions of $\frac{3}{9}$ and $\frac{8}{4}$ were simplified first.

EXAMPLE 5 **Finding fractional parts**

Find each fractional part.
(a) One-fifth of two-thirds **(b)** Four-fifths of three-sevenths **(c)** Three-fifths of ten

Solution

(a) The phrase "one-fifth of" indicates multiplication by one-fifth. The fractional part is

$$\frac{1}{5} \cdot \frac{2}{3} = \frac{1 \cdot 2}{5 \cdot 3} = \frac{2}{15}.$$

(b) $\frac{4}{5} \cdot \frac{3}{7} = \frac{4 \cdot 3}{5 \cdot 7} = \frac{12}{35}$

(c) $\frac{3}{5} \cdot 10 = \frac{3}{5} \cdot \frac{10}{1} = \frac{30}{5} = 6$

Now Try Exercise 59

▶ **REAL-WORLD CONNECTION**   In the next application we use fractions.

EXAMPLE 6   Estimating college completion rates

About four-fifths of the U.S. population over the age of 25 has a high school diploma. About one-third of those people have gone on to complete 4 or more years of college. What fraction of the U.S. population over the age of 25 has completed 4 or more years of college?

Solution
One-third of four-fifths is

$$\frac{1}{3} \cdot \frac{4}{5} = \frac{1 \cdot 4}{3 \cdot 5} = \frac{4}{15}.$$

About four-fifteenths of the U.S. population over the age of 25 has completed 4 or more years of college.

Now Try Exercise 139

TABLE 1.5 **Numbers and Their Reciprocals**

Number	Reciprocal
3	$\frac{1}{3}$
$\frac{1}{4}$	4
$\frac{3}{2}$	$\frac{2}{3}$
$\frac{21}{37}$	$\frac{37}{21}$

The **multiplicative inverse or reciprocal** of a nonzero number a is $\frac{1}{a}$. Table 1.5 lists several numbers and their reciprocals. Note that the product of a number and its reciprocal is always 1. For example, the reciprocal of 2 is $\frac{1}{2}$, and their product is $2 \cdot \frac{1}{2} = 1$.

Suppose that a group of children wants to buy gum from a gum ball machine that costs a half dollar for each gum ball. If the caregiver for the children has 4 dollars, then the number of gum balls that can be bought equals the number of half dollars that there are in 4 dollars. Thus 8 gum balls can be bought.

This calculation is given by $4 \div \frac{1}{2}$. To divide a number by a fraction, multiply the number by the reciprocal of the fraction. That is, "invert and multiply."

$$4 \div \frac{1}{2} = 4 \cdot \frac{2}{1} \qquad \text{Multiply by the reciprocal of } \frac{1}{2}.$$

$$= \frac{4}{1} \cdot \frac{2}{1} \qquad \text{Write 4 as } \frac{4}{1}.$$

$$= \frac{8}{1} \qquad \text{Multiply the fractions.}$$

$$= 8 \qquad \frac{a}{1} = a \text{ for all values of } a.$$

Justification for multiplying by the reciprocal when dividing two fractions is

$$\frac{a}{b} \div \frac{c}{d} = \frac{\dfrac{a}{b}}{\dfrac{c}{d}} = \frac{\dfrac{a}{b} \cdot \dfrac{d}{c}}{\dfrac{c}{d} \cdot \dfrac{d}{c}} = \frac{\dfrac{a}{b} \cdot \dfrac{d}{c}}{1} = \frac{a}{b} \cdot \frac{d}{c}.$$

These results are summarized as follows.

DIVISION OF FRACTIONS

For real numbers a, b, c, and d, with b, c, and d not equal to 0,

$$\frac{a}{b} \div \frac{c}{d} = \frac{a}{b} \cdot \frac{d}{c}.$$

EXAMPLE 7   Dividing fractions

Divide each expression.

(a) $\dfrac{1}{3} \div \dfrac{3}{5}$ (b) $\dfrac{4}{5} \div \dfrac{4}{5}$ (c) $5 \div \dfrac{10}{3}$ (d) $\dfrac{x}{2} \div \dfrac{y}{z}$

Solution

(a) To divide $\frac{1}{3}$ by $\frac{3}{5}$, multiply $\frac{1}{3}$ by $\frac{5}{3}$, which is the reciprocal of $\frac{3}{5}$.

$$\frac{1}{3} \div \frac{3}{5} = \frac{1}{3} \cdot \frac{5}{3} = \frac{1 \cdot 5}{3 \cdot 3} = \frac{5}{9}$$

(b) $\frac{4}{5} \div \frac{4}{5} = \frac{4}{5} \cdot \frac{5}{4} = \frac{4 \cdot 5}{5 \cdot 4} = \frac{20}{20} = 1$. Note that when we divide any nonzero number by itself, the result is 1.

(c) $5 \div \frac{10}{3} = \frac{5}{1} \cdot \frac{3}{10} = \frac{15}{10} = \frac{3}{2}$

(d) $\dfrac{x}{2} \div \dfrac{y}{z} = \dfrac{x}{2} \cdot \dfrac{z}{y} = \dfrac{xz}{2y}$

Now Try Exercises 69, 79, 85

NOTE: In Example 7(c) the answer was left as $\frac{3}{2}$. In arithmetic, $\frac{3}{2}$ is sometimes called an **improper fraction** and is often written as the mixed number $1\frac{1}{2}$. However, in algebra there is nothing "improper" about $\frac{3}{2}$, and fractions are often left as improper fractions.

TECHNOLOGY NOTE

When entering a mixed number into a calculator, it is usually easiest first to convert the mixed number to an improper fraction. For example, enter $2\frac{2}{3}$ as $\frac{8}{3}$. Otherwise, enter $2\frac{2}{3}$ as $2 + \frac{2}{3}$. See the accompanying figure.

```
8/3
        2.666666667
2+2/3
        2.666666667
```

EXAMPLE 8   Writing a problem

Describe a problem for which the solution could be found by dividing 5 by $\frac{1}{6}$.

Solution

If five pies are each cut into sixths, how many pieces of pie are there? (Note that one way to answer this question is to determine how many one-sixths there are in 5, which is the equivalent of multiplying 5 by 6.) See Figure 1.7.

Figure 1.7 Five Pies Cut into Sixths

Now Try Exercise 141

Addition and Subtraction of Fractions

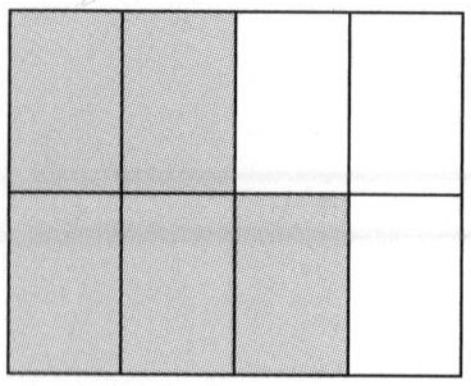

FRACTIONS WITH LIKE DENOMINATORS   Suppose that a person cuts a sheet of paper into eighths. If that person picks up two pieces and another person picks up three pieces, then together they have

$$\frac{2}{8} + \frac{3}{8} = \frac{5}{8}$$

Figure 1.8

of a sheet of paper, as illustrated in Figure 1.8. When the denominator of one fraction is the same as the denominator of a second fraction, the sum of the two fractions can be found by adding their numerators and keeping the common denominator.

Similarly, if someone picks up 5 pieces of paper and gives 2 away, then that person has

$$\frac{5}{8} - \frac{2}{8} = \frac{3}{8}$$

of a sheet of paper. To subtract two fractions with common denominators, subtract their numerators and keep the common denominator.

ADDITION AND SUBTRACTION OF FRACTIONS

To add or subtract fractions with a common denominator d, use the equations

$$\frac{a}{d} + \frac{b}{d} = \frac{a+b}{d} \quad \text{and} \quad \frac{a}{d} - \frac{b}{d} = \frac{a-b}{d},$$

where d is not 0.

EXAMPLE 9   Adding and subtracting fractions with common denominators

Add or subtract as indicated. Simplify your answer to lowest terms when appropriate.
(a) $\frac{5}{13} + \frac{12}{13}$ **(b)** $\frac{11}{8} - \frac{5}{8}$

Solution
(a) Because the fractions have a common denominator, add the numerators and keep the common denominator.

$$\frac{5}{13} + \frac{12}{13} = \frac{5+12}{13} = \frac{17}{13}$$

(b) Because the fractions have a common denominator, subtract the numerators and keep the common denominator.

$$\frac{11}{8} - \frac{5}{8} = \frac{11-5}{8} = \frac{6}{8}$$

The fraction $\frac{6}{8}$ can be simplified to $\frac{3}{4}$.   Now Try Exercise **87**

▶ **REAL-WORLD CONNECTION**   Suppose that one person mows half a large lawn while another person mows a fourth of the lawn. To determine how much they mowed together we need to find the sum $\frac{1}{2} + \frac{1}{4}$. See Figure 1.9(a). Before we can add fractions with unlike denominators, we must write each fraction with a common denominator. The least

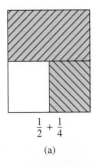

$\frac{1}{2} + \frac{1}{4}$

(a)

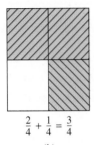

$\frac{2}{4} + \frac{1}{4} = \frac{3}{4}$

(b)

Figure 1.9

common denominator of 2 and 4 is 4. Thus we need to write $\frac{1}{2}$ as $\frac{?}{4}$ by multiplying the numerator and denominator by the *same nonzero number*.

$$\frac{1}{2} = \frac{1}{2} \cdot \frac{2}{2} \qquad \text{Multiply by 1.}$$

$$= \frac{2}{4} \qquad \text{Multiply fractions.}$$

Now we can find the needed sum.

$$\frac{1}{2} + \frac{1}{4} = \frac{2}{4} + \frac{1}{4} = \frac{3}{4}$$

Together the two people mow three-fourths of the lawn, as illustrated in Figure 1.9(b).

FRACTIONS WITH UNLIKE DENOMINATORS   To add or subtract fractions with unlike denominators, we first determine a common denominator. Even though *any* common denominator may be used, it is often best to find the *least* common denominator (LCD). Finding the LCD is described in Example 10. Second, each fraction must be rewritten so that it has the LCD as its denominator. This is accomplished by multiplying each fraction by 1 written in an appropriate form. Example 11 shows how to rewrite each fraction so that its denominator changes to the LCD. Finally, once the rewritten fractions have a common denominator, they may be added or subtracted, as described earlier. This entire process is illustrated in Example 12.

FINDING THE LEAST COMMON DENOMINATOR (LCD)

STEP 1: Find the prime factorization for each denominator.

STEP 2: List each factor that appears in one or more of the factorizations. If a factor is repeated in any of the factorizations, list this factor the maximum number of times that it is repeated.

STEP 3: The product of this list of factors is the LCD.

EXAMPLE 10   Finding the least common denominator

Find the LCD for each set of fractions.
(a) $\frac{5}{6}, \frac{3}{4}$ **(b)** $\frac{5}{12}, \frac{2}{27}$ **(c)** $\frac{1}{4}, \frac{2}{5}, \frac{7}{10}$

Solution
(a) STEP 1: For the fractions $\frac{5}{6}$ and $\frac{3}{4}$ the prime factorizations of the denominators are

$$6 = 2 \cdot 3 \text{ and } 4 = 2 \cdot 2.$$

STEP 2: List the factors: **2, 2, 3**. Note that, because the factor 2 appears a maximum of two times, it is listed twice.

STEP 3: The LCD is the product of this list, or $2 \cdot 2 \cdot 3 = 12$.

NOTE: Finding an LCD is equivalent to finding the smallest number that each denominator divides into evenly. Both 6 and 4 divide into 12 evenly, and 12 is the smallest such number. Thus 12 is the LCD for $\frac{5}{6}$ and $\frac{3}{4}$.

(b) **STEP 1:** For $\frac{5}{12}$ and $\frac{2}{27}$ the prime factorizations of the denominators are

$$12 = 2 \cdot 2 \cdot 3 \text{ and } 27 = 3 \cdot 3 \cdot 3.$$

STEP 2: List the factors: $2, 2, 3, 3, 3$.
STEP 3: The LCD is $2 \cdot 2 \cdot 3 \cdot 3 \cdot 3 = 108$.

(c) **STEP 1:** For $\frac{1}{4}, \frac{2}{5}$, and $\frac{7}{10}$ the prime factorizations of the denominators are

$$4 = 2 \cdot 2, 5 = 5, \text{ and } 10 = 2 \cdot 5.$$

STEP 2: List the factors: $2, 2, 5$.
STEP 3: The LCD is $2 \cdot 2 \cdot 5 = 20$.

Now Try Exercises 93, 95, 101

EXAMPLE 11 Rewriting fractions with the LCD

Rewrite each set of fractions using the LCD.
(a) $\frac{5}{6}, \frac{3}{4}$ **(b)** $\frac{5}{12}, \frac{2}{27}$ **(c)** $\frac{1}{4}, \frac{2}{5}, \frac{7}{10}$

Solution

(a) From Example 10(a) the LCD is 12. To write $\frac{5}{6}$ with a denominator of 12, we multiply the fraction by 1 in the form $\frac{2}{2}$.

$$\frac{5}{6} \cdot \frac{2}{2} = \frac{5 \cdot 2}{6 \cdot 2} = \frac{10}{12}$$

To write $\frac{3}{4}$ with a denominator of 12, we multiply the fraction by 1 in the form $\frac{3}{3}$.

$$\frac{3}{4} \cdot \frac{3}{3} = \frac{3 \cdot 3}{4 \cdot 3} = \frac{9}{12}$$

Thus $\frac{5}{6}$ can be rewritten as $\frac{10}{12}$ and $\frac{3}{4}$ can be rewritten as $\frac{9}{12}$.

(b) From Example 10(b) the LCD is 108. To write $\frac{5}{12}$ with a denominator of 108, multiply the fraction by $\frac{9}{9}$. To write $\frac{2}{27}$ with a denominator of 108, multiply the fraction by $\frac{4}{4}$.

$$\frac{5}{12} \cdot \frac{9}{9} = \frac{5 \cdot 9}{12 \cdot 9} = \frac{45}{108} \quad \text{and} \quad \frac{2}{27} \cdot \frac{4}{4} = \frac{2 \cdot 4}{27 \cdot 4} = \frac{8}{108}.$$

Thus $\frac{5}{12}$ can be rewritten as $\frac{45}{108}$ and $\frac{2}{27}$ can be rewritten as $\frac{8}{108}$.

(c) From Example 10(c) the LCD is 20.

$$\frac{1}{4} \cdot \frac{5}{5} = \frac{5}{20}, \quad \frac{2}{5} \cdot \frac{4}{4} = \frac{8}{20}, \quad \text{and} \quad \frac{7}{10} \cdot \frac{2}{2} = \frac{14}{20}.$$

With an LCD of 20 the fractions are written $\frac{5}{20}, \frac{8}{20},$ and $\frac{14}{20},$ respectively.

Now Try Exercises 105, 111

EXAMPLE 12 Adding and subtracting fractions with unlike denominators

Add or subtract as indicated. Simplify your answer to lowest terms when appropriate.
(a) $\frac{5}{6} + \frac{3}{4}$ **(b)** $\frac{5}{12} - \frac{2}{27}$ **(c)** $\frac{1}{4} + \frac{2}{5} + \frac{7}{10}$

Solution

(a) From Example 10(a) the LCD is 12. Begin by writing each fraction with a denominator of 12 as demonstrated in Example 11(a).

$$\frac{5}{6} + \frac{3}{4} = \frac{5}{6} \cdot \frac{2}{2} + \frac{3}{4} \cdot \frac{3}{3}$$ Change to LCD of 12.

$$= \frac{10}{12} + \frac{9}{12}$$ Multiply the fractions.

$$= \frac{10 + 9}{12}$$ Add the numerators.

$$= \frac{19}{12}$$ Simplify.

(b) Using Example 10(b) and Example 11(b), we perform the following steps.

$$\frac{5}{12} - \frac{2}{27} = \frac{5}{12} \cdot \frac{9}{9} - \frac{2}{27} \cdot \frac{4}{4}$$ Change to LCD of 108.

$$= \frac{45}{108} - \frac{8}{108}$$ Multiply the fractions.

$$= \frac{45 - 8}{108}$$ Subtract the numerators.

$$= \frac{37}{108}$$ Simplify.

(c) Using Example 10(c) and Example 11(c), we perform the following steps.

$$\frac{1}{4} + \frac{2}{5} + \frac{7}{10} = \frac{1}{4} \cdot \frac{5}{5} + \frac{2}{5} \cdot \frac{4}{4} + \frac{7}{10} \cdot \frac{2}{2}$$ Change to LCD of 20.

$$= \frac{5}{20} + \frac{8}{20} + \frac{14}{20}$$ Multiply the fractions.

$$= \frac{5 + 8 + 14}{20}$$ Add the numerators.

$$= \frac{27}{20}$$ Simplify.

Now Try Exercises 113, 123

Applications

▶ **REAL-WORLD CONNECTION**   Fractions occur in a variety of applications.

EXAMPLE 13   **Applying fractions to carpentry**

A board measures $35\frac{3}{4}$ inches and needs to be cut into four equal parts, as depicted in Figure 1.10. Find the length of each piece.

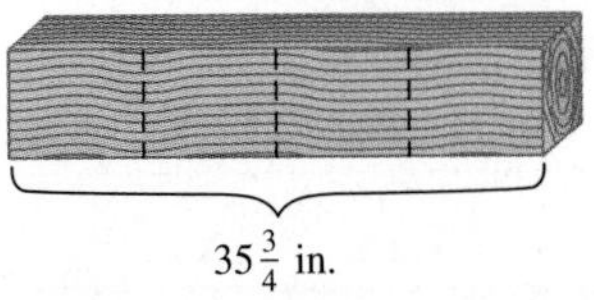

Figure 1.10

Solution

Begin by writing $35\frac{3}{4}$ as the improper fraction $\frac{143}{4}$ ($4 \cdot 35 + 3 = 143$). As the board is to be cut into four equal parts, the length of each piece should be

$$\frac{143}{4} \div 4 = \frac{143}{4} \cdot \frac{1}{4} = \frac{143}{16}, \quad \text{or} \quad 8\frac{15}{16} \text{ inches.}$$

Now Try Exercise 133

EXAMPLE 14   Applying fractions to population

According to a recent study, about $\frac{1}{20}$ of the population in the United States will serve time in prison. Within the part of the population that does serve prison time, $\frac{1}{10}$ will be female. What fraction of the population in the United States is female and will serve time in prison?

(*Source:* Bureau of Justice Statistics.)

CRITICAL THINKING

Think of a situation in every-day life in which fractions would be needed.

Solution

Because $\frac{1}{10}$ of those persons who will serve prison time will be female, we multiply the fraction of the population that will serve prison time, $\frac{1}{20}$, by $\frac{1}{10}$.

$$\frac{1}{20} \cdot \frac{1}{10} = \frac{1}{200}$$

Thus $\frac{1}{200}$ of the population in the United States is female and will serve time in prison.

Now Try Exercise 140

1.2 PUTTING IT ALL TOGETHER

The following table summarizes some important concepts related to fractions.

Topic	Comments	Examples
Fraction	The fraction $\frac{a}{b}$ has numerator a and denominator b.	The fraction $\frac{xy}{2}$ has numerator xy and denominator 2.
Greatest Common Factor (GCF)	The GCF of two numbers equals the largest number that divides into both evenly.	The GCF of 12 and 18 is 6 because 6 is the largest number that divides into 12 and 18 evenly.
Simplifying Fractions	Use the principle $$\frac{a \cdot c}{b \cdot c} = \frac{a}{b}$$ to simplify fractions, where c is the GCF of the numerator and denominator.	The GCF of 24 and 32 is 8, so $$\frac{24}{32} = \frac{3 \cdot 8}{4 \cdot 8} = \frac{3}{4}.$$ The GCF of 20 and 8 is 4, so $$\frac{20}{8} = \frac{5 \cdot 4}{2 \cdot 4} = \frac{5}{2}.$$
Multiplicative Inverse or Reciprocal	The reciprocal of $\frac{a}{b}$ is $\frac{b}{a}$, where a and b are not zero.	The reciprocals of 5 and $\frac{3}{4}$ are $\frac{1}{5}$ and $\frac{4}{3}$, respectively. The product of a number and its reciprocal is 1.
Multiplication and Division of Fractions	$$\frac{a}{b} \cdot \frac{c}{d} = \frac{ac}{bd}$$ $$\frac{a}{b} \div \frac{c}{d} = \frac{a}{b} \cdot \frac{d}{c}$$	$$\frac{3}{5} \cdot \frac{4}{9} = \frac{12}{45} = \frac{4}{15}$$ $$\frac{3}{2} \div \frac{6}{5} = \frac{3}{2} \cdot \frac{5}{6} = \frac{15}{12} = \frac{5}{4}$$

Topic	Comments	Examples
Addition and Subtraction of Fractions with Like Denominators	$\dfrac{a}{d} + \dfrac{c}{d} = \dfrac{a+c}{d}$ and $\dfrac{a}{d} - \dfrac{c}{d} = \dfrac{a-c}{d}$	$\dfrac{3}{5} + \dfrac{4}{5} = \dfrac{3+4}{5} = \dfrac{7}{5}$ and $\dfrac{17}{12} - \dfrac{11}{12} = \dfrac{17-11}{12} = \dfrac{6}{12} = \dfrac{1}{2}$
Least Common Denominator (LCD)	The LCD of two fractions equals the smallest number that both denominators divide into evenly.	The LCD of $\frac{5}{12}$ and $\frac{7}{18}$ is 36 because 36 is the smallest number that both 12 and 18 divide into evenly.
Addition and Subtraction of Fractions with Unlike Denominators	First write each fraction with the least common denominator. Then add or subtract the numerators.	The LCD of $\frac{3}{4}$ and $\frac{7}{10}$ is 20. $$\frac{3}{4} + \frac{7}{10} = \frac{3}{4} \cdot \frac{5}{5} + \frac{7}{10} \cdot \frac{2}{2}$$ $$= \frac{15}{20} + \frac{14}{20}$$ $$= \frac{29}{20}$$

1.2 Exercises

MyMathLab | Math XL PRACTICE | WATCH | DOWNLOAD | READ | REVIEW

CONCEPTS

1. A small pie is cut into 4 pieces. If someone eats 3 of the pieces, what fraction of the pie does the person eat? What fraction of the pie remains?

2. In the fraction $\frac{11}{21}$ the numerator is _____ and the denominator is _____.

3. In the fraction $\frac{a}{b}$, the variable b cannot equal _____.

4. The fraction $\frac{a}{a}$ with $a \neq 0$ equals _____.

5. The numerator of the product of two fractions is found by multiplying the _____ of the two fractions.

6. The denominator of the product of two fractions is found by multiplying the _____ of the two fractions.

7. The fraction $\frac{2}{8}$ simplifies to _____.

8. $\dfrac{ac}{bc} = $ _____

9. In the phrase "two-fifths of one-third," the word *of* indicates that we should _____ the fractions $\frac{2}{5}$ and $\frac{1}{3}$.

10. What fractional part is half of a half?

11. What is the reciprocal of a, provided $a \neq 0$?

12. To divide $\frac{3}{4}$ by 5, multiply $\frac{3}{4}$ by _____.

13. $\dfrac{a}{b} \cdot \dfrac{c}{d} = $ _____

14. $\dfrac{a}{b} \div \dfrac{c}{d} = $ _____

15. $\dfrac{a}{b} + \dfrac{c}{b} = $ _____

16. $\dfrac{a}{b} - \dfrac{c}{b} = $ _____

17. The greatest common factor of 4 and 6 is _____.

18. The least common denominator for the fractions $\frac{1}{4}$ and $\frac{5}{6}$ is _____.

19. What is the least common denominator of $\frac{2}{9}$ and 7?

20. To rewrite the fraction $\frac{3}{4}$ with denominator 24, multiply by 1 in the form_____.

LOWEST TERMS

Exercises 21–26: Find the greatest common factor.

21. 4, 12

22. 3, 27

23. 50, 75

24. 45, 105

25. 100, 60, 70

26. 36, 48, 72

Exercises 27–30: Use the basic principle of fractions to simplify the expression.

27. $\frac{3 \cdot 4}{5 \cdot 4}$

28. $\frac{2 \cdot 7}{9 \cdot 7}$

29. $\frac{3 \cdot 8}{8 \cdot 5}$

30. $\frac{7 \cdot 16}{16 \cdot 3}$

Exercises 31–42: Simplify the fraction to lowest terms.

31. $\frac{4}{8}$

32. $\frac{4}{12}$

33. $\frac{5}{15}$

34. $\frac{3}{27}$

35. $\frac{10}{25}$

36. $\frac{5}{20}$

37. $\frac{12}{36}$

38. $\frac{16}{24}$

39. $\frac{12}{30}$

40. $\frac{60}{105}$

41. $\frac{19}{76}$

42. $\frac{17}{51}$

MULTIPLICATION AND DIVISION OF FRACTIONS

Exercises 43–58: Multiply and simplify to lowest terms when appropriate.

43. $\frac{1}{2} \cdot \frac{1}{3}$

44. $\frac{2}{3} \cdot \frac{7}{5}$

45. $\frac{3}{4} \cdot \frac{1}{5}$

46. $\frac{3}{2} \cdot \frac{5}{8}$

47. $\frac{5}{3} \cdot \frac{3}{5}$

48. $\frac{21}{32} \cdot \frac{32}{21}$

49. $\frac{5}{6} \cdot \frac{18}{25}$

50. $\frac{7}{9} \cdot \frac{3}{14}$

51. $4 \cdot \frac{3}{5}$

52. $5 \cdot \frac{7}{10}$

53. $2 \cdot \frac{3}{8}$

54. $10 \cdot \frac{1}{100}$

55. $\frac{x}{y} \cdot \frac{y}{x}$

56. $\frac{x}{y} \cdot \frac{y}{z}$

57. $\frac{a}{b} \cdot \frac{3}{2}$

58. $\frac{5}{8} \cdot \frac{4x}{5y}$

Exercises 59–64: Find the fractional part.

59. One-fourth of three-fourths

60. Three-sevenths of nine-sixteenths

61. Two-thirds of six

62. Three-fourths of seven

63. One-half of two-thirds

64. Five-elevenths of nine-eighths

Exercises 65–68: Give the reciprocal of each number.

65. (a) 5   (b) 7   (c) $\frac{4}{7}$ (d) $\frac{9}{8}$

66. (a) 3   (b) 2   (c) $\frac{6}{5}$ (d) $\frac{3}{8}$

67. (a) $\frac{1}{2}$ (b) $\frac{1}{9}$ (c) $\frac{12}{101}$ (d) $\frac{31}{17}$

68. (a) $\frac{1}{5}$ (b) $\frac{7}{3}$ (c) $\frac{23}{64}$ (d) $\frac{63}{29}$

Exercises 69–86: Divide and simplify to lowest terms when appropriate.

69. $\frac{1}{2} \div \frac{1}{3}$

70. $\frac{3}{4} \div \frac{1}{5}$

71. $\frac{3}{4} \div \frac{1}{2}$

72. $\frac{6}{7} \div \frac{3}{14}$

73. $\frac{4}{3} \div \frac{1}{6}$

74. $\frac{12}{21} \div \frac{4}{7}$

75. $\frac{32}{27} \div \frac{8}{9}$

76. $\frac{8}{15} \div \frac{2}{25}$

77. $\frac{9}{1} \div \frac{8}{7}$

78. $2 \div \frac{3}{4}$

79. $10 \div \frac{5}{6}$

80. $8 \div \frac{4}{3}$

81. $\frac{9}{10} \div 3$

82. $\frac{32}{27} \div 16$

83. $\frac{a}{b} \div \frac{2}{b}$

84. $\frac{3a}{b} \div \frac{3}{c}$

85. $\frac{x}{y} \div \frac{x}{y}$

86. $\frac{x}{3y} \div \frac{x}{3}$

ADDITION AND SUBTRACTION OF FRACTIONS

Exercises 87–92: Add or subtract. Write each answer in lowest terms.

87. (a) $\frac{2}{3} + \frac{1}{3}$ (b) $\frac{2}{3} - \frac{1}{3}$

88. (a) $\frac{5}{12} + \frac{1}{12}$ (b) $\frac{5}{12} - \frac{1}{12}$

89. (a) $\frac{3}{2} + \frac{1}{2}$ (b) $\frac{3}{2} - \frac{1}{2}$

90. (a) $\frac{18}{29} + \frac{7}{29}$ (b) $\frac{18}{29} - \frac{7}{29}$

91. (a) $\frac{5}{33} + \frac{2}{33}$ (b) $\frac{5}{33} - \frac{2}{33}$

92. (a) $\frac{91}{104} + \frac{17}{104}$ (b) $\frac{91}{104} - \frac{17}{104}$

Exercises 93–104: Find the least common denominator.

93. $\frac{1}{5}, \frac{3}{10}$

94. $\frac{5}{12}, \frac{1}{18}$

95. $\frac{4}{9}, \frac{2}{15}$

96. $\frac{1}{11}, \frac{1}{2}$

97. $\frac{2}{5}, \frac{3}{15}$

98. $\frac{8}{21}, \frac{3}{7}$

99. $\frac{1}{6}, \frac{5}{8}$

100. $\frac{1}{9}, \frac{5}{12}$

101. $\frac{1}{2}, \frac{1}{3}, \frac{1}{4}$

102. $\frac{2}{5}, \frac{2}{3}, \frac{1}{6}$

103. $\frac{1}{4}, \frac{3}{8}, \frac{1}{12}$

104. $\frac{2}{15}, \frac{7}{20}, \frac{1}{30}$

Exercises 105–112: Rewrite each set of fractions with the least common denominator.

105. $\frac{1}{2}, \frac{2}{3}$

106. $\frac{3}{4}, \frac{1}{5}$

107. $\frac{7}{9}, \frac{5}{12}$

108. $\frac{5}{13}, \frac{1}{2}$

109. $\frac{1}{16}, \frac{7}{12}$

110. $\frac{5}{18}, \frac{1}{24}$

111. $\frac{1}{3}, \frac{3}{4}, \frac{5}{6}$

112. $\frac{4}{15}, \frac{2}{9}, \frac{3}{5}$

Exercises 113–128: Add or subtract. Write your answer in lowest terms.

113. $\frac{1}{2} + \frac{1}{3}$

114. $\frac{2}{3} + \frac{1}{4}$

115. $\frac{5}{8} + \frac{3}{16}$

116. $\frac{1}{9} + \frac{2}{15}$

117. $\frac{1}{2} - \frac{1}{4}$

118. $\frac{2}{3} - \frac{2}{9}$

119. $\frac{25}{24} - \frac{7}{8}$

120. $\frac{4}{5} - \frac{1}{4}$

121. $\frac{11}{14} + \frac{2}{35}$

122. $\frac{7}{8} + \frac{4}{15}$

123. $\frac{5}{12} - \frac{1}{18}$

124. $\frac{9}{20} - \frac{7}{30}$

125. $\frac{3}{100} + \frac{1}{300} - \frac{1}{200}$

126. $\frac{43}{36} + \frac{4}{9} + \frac{1}{4}$

127. $\frac{7}{8} - \frac{1}{6} + \frac{5}{12}$

128. $\frac{9}{40} - \frac{3}{50} - \frac{1}{100}$

APPLICATIONS

129. *American Flag* According to Executive Order 10834, the length of an official American flag should be $1\frac{9}{10}$ times the width. If an official flag has a width of $2\frac{1}{2}$ feet, find its length.

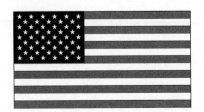

130. *American Flag* The blue rectangle containing the stars on an American flag is called the *union*. On an official American flag, the width of the union should be $\frac{7}{13}$ of the width of the flag. If an official flag has a width of $32\frac{1}{2}$ inches, what is the width of the union?

131. *Accidental Deaths* For the age group 15 to 24, motor vehicle accidents account for $\frac{31}{42}$ of all accidental deaths, and firearms account for $\frac{31}{1260}$ of all accidental deaths. What fraction of all accidental deaths do vehicle accidents *and* firearms account for? (*Source:* National Safety Council.)

132. *Illicit Drug Use* For the age group 18 to 25, the fraction of people who used illicit drugs during their lifetime was $\frac{3}{5}$, whereas the fraction who used illicit drugs during the past year was $\frac{7}{20}$. What fraction of this population has used illicit drugs but not during the past year? (*Source:* Department of Health and Human Services.)

133. *Carpentry* A board measures $64\frac{5}{8}$ inches and needs to be cut in half. Find the length of each half.

134. *Carpentry* A rope measures $15\frac{1}{2}$ feet and needs to be cut in four equal parts. Find the length of each piece.

135. *Geometry* Find the area of the triangle shown with base $1\frac{2}{3}$ yards and height $\frac{3}{4}$ yard. (*Hint:* The area of a triangle equals half the product of its base and height.)

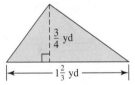

136. *Geometry* Find the area of the rectangle shown.

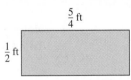

137. *Distance* Use the map to find the distance between Smalltown and Bigtown by traveling through Middletown.

138. *Distance* An athlete jogs $1\frac{3}{8}$ miles, $5\frac{3}{4}$ miles, and $3\frac{5}{8}$ miles. In all, how far does the athlete jog?

139. *Adult Smokers* About $\frac{1}{5}$ of the adult population in the United States smoked in 2005. Within the part of the population that smoked, $\frac{6}{25}$ of them were aged 25 to 44. What fraction of the adult population in the United States was aged 25 to 44 and smoked? (***Source:*** Department of Health and Human Services.)

140. *Smokeless Tobacco Use* About $\frac{3}{50}$ of the teenage population in the United States used smokeless tobacco in 2004. Within the part of the teenage population that used smokeless tobacco, $\frac{54}{61}$ were males. What fraction of the teenage population was male and used smokeless tobacco? (***Source:*** Department of Health and Human Services.)

WRITING ABOUT MATHEMATICS

141. Describe a problem in real life for which the solution could be found by multiplying 30 by $\frac{1}{4}$.

142. Describe a problem in real life for which the solution could be found by dividing $2\frac{1}{2}$ by $\frac{1}{3}$.

CHECKING BASIC CONCEPTS
SECTIONS 1.1 AND 1.2

1. Classify each number as prime, composite, or neither. If a number is composite, write it as a product of primes.
   (a) 19   **(b)** 28   **(c)** 1   **(d)** 180

2. Evaluate $\frac{10}{(x + 2)}$ for $x = 3$.

3. Find y for $x = 5$ if $y = 6x$.

4. Translate the phrase "a number x plus five" into an algebraic expression.

5. Write a formula that converts F feet to I inches.

6. Find the greatest common factor.
   (a) 3, 18   **(b)** 40, 72

7. Simplify each fraction to lowest terms.
   (a) $\frac{25}{35}$ **(b)** $\frac{26}{39}$

8. Give the reciprocal of $\frac{4}{3}$.

9. Evaluate each expression. Write each answer in lowest terms.
   (a) $\frac{2}{3} \cdot \frac{3}{4}$ **(b)** $\frac{5}{6} \div \frac{10}{3}$
   (c) $\frac{3}{10} + \frac{1}{10}$ **(d)** $\frac{3}{4} - \frac{1}{6}$

10. A recipe calls for $1\frac{2}{3}$ cups of flour. How much flour should be used if the recipe is doubled?

1.3 EXPONENTS AND ORDER OF OPERATIONS

Natural Number Exponents ▪ Order of Operations ▪
Translating Words to Expressions

A LOOK INTO MATH ▷ In elementary school you learned addition. Later, you learned that multiplication is a fast way to add. For example, rather than adding

$$3 + 3 + 3 + 3, \quad \longleftarrow 4 \text{ terms}$$

we can multiply $3 \cdot 4$. Similarly, exponents represent a fast way to multiply. Rather than multiplying

$$3 \cdot 3 \cdot 3 \cdot 3, \quad \longleftarrow 4 \text{ factors}$$

we can evaluate the *exponential expression* 3^4. We begin this section by discussing natural numbers as exponents.

Natural Number Exponents

The area of a square equals the length of one of its sides times itself. If the square is 5 inches on a side, then its area is

$$5 \cdot 5 = \overset{\text{Exponent}}{5^2} = 25 \text{ square inches.}$$

The expression 5^2 is an **exponential expression** with *base* 5 and *exponent* 2. Exponential expressions occur in a variety of applications. For example, suppose an investment doubles 3 times. Then, the calculation

$$\underset{\text{Factors}}{\underbrace{2 \cdot 2 \cdot 2}} = 2^{\overset{\text{Exponent 3}}{3}} = 8$$

shows that the final value is 8 times as large as the original investment. For example, if $10 doubles 3 times, it becomes $20, $40, and finally $80, which is 8 times as large as $10. Table 1.6 contains examples of exponential expressions.

TABLE 1.6

Expression	Base	Exponent
$2 \cdot 2 \cdot 2 \cdot 2 = 2^4$	2	4
$4 \cdot 4 \cdot 4 = 4^3$	4	3
$9 \cdot 9 = 9^2$	9	2
$\left(\frac{1}{2}\right)^1 = \frac{1}{2}$	$\frac{1}{2}$	1
$b \cdot b \cdot b \cdot b = b^4$	b	4

Figure 1.11
3 Squared

Read 9^2 as "9 squared," 4^3 as "4 cubed," and 2^4 as "2 to the fourth power." The terms *squared* and *cubed* come from geometry. If the length of a side of a square is 3, then its area is

$$3 \cdot 3 = 3^2 = 9$$

square units, as illustrated in Figure 1.11. Similarly, if the length of an edge of a cube is 3, then its volume is

$$3 \cdot 3 \cdot 3 = 3^3 = 27$$

cubic units, as shown in Figure 1.12.

Figure 1.12
3 Cubed

NOTE: The expressions 3^2 and 3^3 can also be read as "3 to the second power" and "3 to the third power," respectively. In general, the expression x^n is read as "x to the nth power" or "the nth power of x."

EXPONENTIAL NOTATION

The expression b^n, where n is a natural number, means

$$b^n = \underset{n \text{ factors}}{\underbrace{b \cdot b \cdot b \cdot \cdots \cdot b}}.$$

The **base** is b and the **exponent** is n.

EXAMPLE **1**   Writing products in exponential notation

Write each product as an exponential expression.
(a) $7 \cdot 7 \cdot 7 \cdot 7$ **(b)** $\frac{1}{4} \cdot \frac{1}{4} \cdot \frac{1}{4}$ **(c)** $x \cdot x \cdot x \cdot x \cdot x$

Solution
(a) Because there are four factors of 7, the exponent is 4 and the base is 7.
Thus $7 \cdot 7 \cdot 7 \cdot 7 = 7^4$.
(b) Because there are three factors of $\frac{1}{4}$, the exponent is 3 and the base is $\frac{1}{4}$.
Thus $\frac{1}{4} \cdot \frac{1}{4} \cdot \frac{1}{4} = \left(\frac{1}{4}\right)^3$.
(c) Because there are five factors of x, the exponent is 5 and the base is x.
Thus $x \cdot x \cdot x \cdot x \cdot x = x^5$.   Now Try Exercise **17**

EXAMPLE **2**   Evaluating exponential notation

Evaluate each expression.
(a) 3^4 **(b)** 10^3 **(c)** 23^1 **(d)** $\left(\frac{3}{4}\right)^2$

Solution
(a) The exponential expression 3^4 indicates that 3 is to be multiplied times itself 4 times.

$$3^4 = \underbrace{3 \cdot 3 \cdot 3 \cdot 3}_{4 \text{ factors}} = 9 \cdot 9 = 81$$

(b) $10^3 = 10 \cdot 10 \cdot 10 = 1000$
(c) $23^1 = 23$
(d) $\left(\frac{3}{4}\right)^2 = \frac{3}{4} \cdot \frac{3}{4} = \frac{9}{16}$ Now Try Exercise **27**

EXAMPLE **3**   Writing numbers in exponential notation

Use the given base to write each number as an exponential expression. Check your results with a calculator, if one is available.
(a) 100 (base 10)   **(b)** 16 (base 2)   **(c)** 27 (base 3)

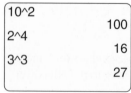

Figure 1.13

Solution
(a) $100 = 10 \cdot 10 = 10^2$
(b) $16 = 4 \cdot 4 = 2 \cdot 2 \cdot 2 \cdot 2 = 2^4$
(c) $27 = 3 \cdot 9 = 3 \cdot 3 \cdot 3 = 3^3$

These values are supported in Figure 1.13, where exponential expressions are evaluated with a calculator by using the "^" key. (Note that some calculators may have a different key for evaluating exponential expressions.)   Now Try Exercises **37, 43**

▶ **REAL-WORLD CONNECTION**   Computer memory is often measured in bytes, with each *byte* capable of storing one letter of the alphabet. For example, it takes four bytes to store the word "math" in a computer. Bytes of computer memory are often manufactured in amounts equal to powers of 2, as illustrated in the next example.

EXAMPLE 4   Analyzing computer memory

In computer technology 1K (kilobyte) of memory equals 2^{10} bytes, and 1MB (megabyte) of memory equals 2^{20} bytes. Determine whether 1K of memory equals one thousand bytes and whether 1MB equals one million bytes.   (**Source:** D. Horn, *Basic Electronics Theory.*)

Solution

Figure 1.14 shows that $2^{10} = 1024$ and $2^{20} = 1,048,576$. Thus 1K represents slightly more than one thousand bytes, and 1MB represents more than one million bytes.

Now Try Exercise 83

Figure 1.14

CRITICAL THINKING

One gigabyte is often referred to as 1 billion bytes. Use Example 4 to write an expression that gives the number of bytes in a gigabyte. If you have a calculator available, determine whether 1 gigabyte is exactly 1 billion bytes.

Order of Operations

```
10-2*3
           4
```

When the expression $10 - 2 \cdot 3$ is evaluated, is the result

$$8 \cdot 3 = 24 \quad \text{or} \quad 10 - 6 = 4?$$

Figure 1.15 shows that a calculator gives a result of 4. The reason is that multiplication is performed before subtraction.

Because arithmetic expressions may contain parentheses, exponents, absolute values, and several operations, it is important to evaluate these expressions consistently. (Absolute value will be discussed in Section 1.4.) To ensure that we all obtain the same result when evaluating an arithmetic expression, the following rules are used.

Figure 1.15

ORDER OF OPERATIONS

Use the following order of operations. First perform all calculations within parentheses and absolute values, or above and below the fraction bar.
1. Evaluate all exponential expressions.
2. Do all multiplication and division from *left to right*.
3. Do all addition and subtraction from *left to right*.

EXAMPLE 5   Evaluating arithmetic expressions

Evaluate each expression by hand.
(a) $10 - 4 - 3$ **(b)** $10 - (4 - 3)$ **(c)** $5 + \frac{12}{3}$ **(d)** $\frac{4}{2 + 6}$

Solution

(a) There are no parentheses, so we evaluate subtraction from *left to right*.

$$10 - 4 - 3 = 6 - 3 = 3$$

(b) Note the similarity between this part and part (a). The difference is the parentheses, so subtraction inside the parentheses must be performed first.

$$10 - (4 - 3) = 10 - 1 = 9$$

(c) In this part we perform division before addition.

$$5 + \frac{12}{3} = 5 + 4 = 9$$

(d) The implication is that both the numerator and the denominator of a fraction have parentheses around them.

$$\frac{4}{2 + 6} = \frac{4}{(2 + 6)} = \frac{4}{8} = \frac{1}{2}$$

Note that the given expression does *not* equal $\frac{4}{2} + 6 = 8$.

Now Try Exercises 51, 53, 65

EXAMPLE 6 **Evaluating arithmetic expressions**

Evaluate each expression.

(a) $25 - 4 \cdot 6$ **(b)** $6 + 7 \cdot 2 - (4 - 1)$ **(c)** $\dfrac{3 + 3^2}{14 - 2}$ **(d)** $5 \cdot 2^3 - (3 + 2)$

Solution

(a) Multiplication is performed before subtraction, so evaluate the expression as follows.

$$25 - 4 \cdot 6 = 25 - 24 \qquad \text{Multiply.}$$
$$= 1 \qquad \text{Subtract.}$$

(b) Start by performing the subtraction within the parentheses first and then perform the multiplication. Finally perform the addition and subtraction from left to right.

$$6 + 7 \cdot 2 - (4 - 1) = 6 + 7 \cdot 2 - 3 \qquad \text{Subtract within parentheses.}$$
$$= 6 + 14 - 3 \qquad \text{Multiply.}$$
$$= 20 - 3 \qquad \text{Add.}$$
$$= 17 \qquad \text{Subtract.}$$

(c) First note that parentheses are implied around the numerator and denominator.

$$\frac{3 + 3^2}{14 - 2} = \frac{(3 + 3^2)}{(14 - 2)} \qquad \text{Insert parentheses.}$$
$$= \frac{(3 + 9)}{(14 - 2)} \qquad \text{Do the exponent first.}$$
$$= \frac{12}{12} \qquad \text{Add and subtract.}$$
$$= 1 \qquad \text{Simplify.}$$

(d) Begin by evaluating the expression inside parentheses.

$$5 \cdot 2^3 - (3 + 2) = 5 \cdot 2^3 - 5 \qquad \text{Add within parentheses.}$$
$$= 5 \cdot 8 - 5 \qquad \text{Evaluate the exponent.}$$
$$= 40 - 5 \qquad \text{Multiply.}$$
$$= 35 \qquad \text{Subtract.}$$

Now Try Exercises 67, 69, 71

Translating Words to Expressions

Sometimes before we can solve a problem we must translate words into mathematical expressions. For example, if the gasoline tank in your car holds 20 gallons and there are 11 gallons in it, then "twenty minus eleven," or $20 - 11 = 9$, is the number of gallons of gasoline needed to fill the tank.

EXAMPLE 7   **Writing and evaluating expressions**

Write each expression and then evaluate it.
(a) Two to the fourth power plus ten
(b) Twenty less five times three
(c) Ten cubed divided by five squared
(d) Sixty divided by the quantity ten minus six

Solution
(a) $2^4 + 10 = 2 \cdot 2 \cdot 2 \cdot 2 + 10 = 16 + 10 = 26$
(b) $20 - 5 \cdot 3 = 20 - 15 = 5$
(c) $\frac{10^3}{5^2} = \frac{1000}{25} = 40$

(d) Here, the word "quantity" indicates that parentheses should be used.

$$60 \div (10 - 6) = \frac{60}{10 - 6} = \frac{60}{4} = 15$$ Now Try Exercise 73

1.3 PUTTING IT ALL TOGETHER

The following table summarizes some topics related to exponential expressions. Remember that exponential expressions are a fast way to multiply.

Topic	Comments	Examples
Exponential Expression	If n is a natural number, then b^n equals $$\underbrace{b \cdot b \cdot b \cdot \;\cdots\; \cdot b}_{n \text{ factors}}$$ and is read "b to the nth power."	$5^1 = 5, 7^2 = 7 \cdot 7 = 49,$ $4^3 = 4 \cdot 4 \cdot 4 = 64,$ and $k^4 = k \cdot k \cdot k \cdot k$
Base and Exponent	The base in b^n is b and the exponent is n.	7^4 has base 7 and exponent 4 and x^3 has base x and exponent 3.

continued on next page

continued from previous page

Without rules for the order of operations, the value of many arithmetic expressions would be ambiguous. For example, without rules of precedence, one person might evaluate $10 - 4 \cdot 2$ to be $6 \cdot 2 = 12$, whereas another person might evaluate it to be $10 - 8 = 2$. However, because multiplication is performed before subtraction, 2 is the correct answer.

To evaluate expressions, use the following order of operations. First perform all calculations within parentheses and absolute values, or above and below the fraction bar.

1. Evaluate all exponential expressions.
2. Do all multiplication and division from *left to right*.
3. Do all addition and subtraction from *left to right*.

1.3 Exercises

MyMathLab

PRACTICE   WATCH   DOWNLOAD   READ   REVIEW

CONCEPTS

1. Multiplication is a fast way to _____.

2. Exponents represent a fast way to _____.

3. In the expression 3^6, how many factors of 3 are being multiplied?

4. In the expression 2^5, there are five factors of _____ being multiplied.

5. The expression $a \cdot a \cdot a \cdot a \cdot a \cdot a$ equals _____.

6. In the expression 5^3, the number 5 is called the _____ and the number 3 is called the _____.

7. Use symbols to write "6 squared."

8. Use symbols to write "8 cubed."

9. Write "four to the fifth power" as an exponential expression.

10. Write "the third power of x" as an exponential expression.

11. When evaluating the expression $5 + 6 \cdot 2$, the result is _____ because _____ is performed before _____.

12. When evaluating the expression $10 - 2^3$, the result is _____ because _____ are evaluated before _____ is performed.

13. The expression $10 - 4 - 2$ equals _____ because subtraction is performed from _____ to _____.

14. The expression $16 \div 4 \div 2$ equals _____ because division is performed from _____ to _____.

15. Are the expressions 2^3 and 3^2 equal? Explain.

16. Is 5^2 equal to $5 \cdot 5$ or $5 \cdot 2$?

NATURAL NUMBER EXPONENTS

Exercises 17–26: Write the product as an exponential expression.

17. $2 \cdot 2 \cdot 2 \cdot 2 \cdot 2$ 18. $4 \cdot 4 \cdot 4$

19. $3 \cdot 3 \cdot 3 \cdot 3$ 20. $10 \cdot 10$

21. $\frac{1}{2} \cdot \frac{1}{2} \cdot \frac{1}{2} \cdot \frac{1}{2}$ 22. $\frac{5}{7} \cdot \frac{5}{7} \cdot \frac{5}{7} \cdot \frac{5}{7} \cdot \frac{5}{7}$

23. $a \cdot a \cdot a \cdot a \cdot a$ 24. $b \cdot b \cdot b \cdot b$

25. $(x + 3) \cdot (x + 3)$

26. $(x - 4) \cdot (x - 4) \cdot (x - 4)$

Exercises 27–36: Use multiplication to rewrite the expression, and then evaluate the result.

27. (a) 2^4 (b) 4^2 28. (a) 3^2 (b) 5^3

29. (a) 6^1 (b) 1^6 30. (a) 17^1 (b) 1^{17}

31. (a) 2^5 (b) 10^3 32. (a) 10^5 (b) 3^4

33. (a) $\left(\frac{2}{3}\right)^2$ (b) $\left(\frac{1}{2}\right)^5$ 34. (a) $\left(\frac{1}{10}\right)^3$ (b) $\left(\frac{4}{3}\right)^1$

35. (a) $\left(\frac{2}{5}\right)^3$ (b) $\left(\frac{9}{7}\right)^2$ 36. (a) $\left(\frac{3}{10}\right)^4$ (b) $\left(\frac{3}{4}\right)^4$

Exercises 37–48: (Refer to Example 3.) Use the given base to write the number as an exponential expression. Check your result if you have a calculator available.

37. 8   (base 2)   **38.** 9   (base 3)

39. 25   (base 5)   **40.** 32   (base 2)

41. 49   (base 7)   **42.** 81   (base 3)

43. 1000   (base 10)   **44.** 256   (base 4)

45. $\frac{1}{16}$ $\left(\text{base } \frac{1}{2}\right)$ **46.** $\frac{9}{25}$ $\left(\text{base } \frac{3}{5}\right)$

47. $\frac{32}{243}$ $\left(\text{base } \frac{2}{3}\right)$ **48.** $\frac{216}{343}$ $\left(\text{base } \frac{6}{7}\right)$

ORDER OF OPERATIONS

Exercises 49–72: Evaluate the expression by hand.

49. $5 + 4 \cdot 6$ **50.** $6 \cdot 7 - 8$

51. $6 \div 3 + 2$ **52.** $20 - 10 \div 5$

53. $100 - \frac{50}{5}$ **54.** $\frac{200}{100} + 6$

55. $10 - 6 - 1$ **56.** $30 - 9 - 5$

57. $20 \div 5 \div 2$ **58.** $500 \div 100 \div 5$

59. $3 + 2^4$ **60.** $10 - 3^2 + 1$

61. $4 \cdot 2^3$ **62.** $100 - 2 \cdot 3^3$

63. $(3 + 2)^3$ **64.** $5 \cdot (3 - 2)^8 - 5$

65. $\frac{4 + 8}{1 + 3}$ **66.** $5 - \frac{3 + 1}{3 - 1}$

67. $\frac{2^3}{4 - 2}$ **68.** $\frac{10 - 3^2}{2 \cdot 4^2}$

69. $10^2 - (30 - 2 \cdot 5)$ **70.** $5^2 + 3 \cdot 5 \div 3 - 1$

71. $\left(\frac{1}{2}\right)^4 + \frac{5 + 4}{3}$ **72.** $\left(\frac{7}{9}\right)^2 - \frac{6 - 5}{3}$

TRANSLATING WORDS TO SYMBOLS

Exercises 73–82: Use symbols to write the expression and then evaluate it.

73. Two cubed minus eight

74. Five squared plus nine

75. Thirty less four times three

76. One hundred plus five times six

77. Four squared divided by two cubed

78. Three cubed times two squared

79. Forty divided by ten, plus two

80. Thirty times ten, minus three

81. One hundred times the quantity two plus three

82. Fifty divided by the quantity eight plus two

APPLICATIONS

83. *iPod Memory* (Refer to Example 4.) Determine the number of bytes on a 512 MB iPod Shuffle.

84. *iPod Memory* Determine the number of bytes on a 60 GB video iPod. (*Hint:* One gigabyte equals 2^{30} bytes.)

85. *Population by Gender* One way to measure the gender balance in a given population is to find the number of males for every 100 females in the population. In 1900, the western region of the United States was significantly out of gender balance. In this region, there were 128 males for every 100 females. (**Source:** U.S. Census Bureau.)
   (a) Find an exponent k so that $2^k = 128$.
   (b) During this time, how many males were there for every 25 females?

86. *Solar Eclipse* In early December 2048 there will be a total solar eclipse visible in parts of Botswana. Find an exponent k so that $2^k = 2048$. (**Source:** NASA.)

87. *Rule of 72* Investors sometimes use the *rule of 72* to determine the time required to double an investment. If 72 is divided by the annual interest rate earned on an investment, the result approximates the number of years needed to double the investment. For example, an investment earning 6% annual interest will double in value approximately every $72 \div 6 = 12$ years.
   (a) Approximate the number of years required to double an investment earning 9% annual interest.
   (b) If an investment of $10,000 earns 12% annual interest, approximate the value of the investment after 18 years.

88. *Doubling Effect* Suppose that a savings account containing $1000 doubles its value every 7 years. How much money will be in the account after 28 years?

WRITING ABOUT MATHEMATICS

89. Explain how exponential expressions are related to multiplication. Give an example.

90. Explain why agreement on the order of operations is necessary.

GROUP ACTIVITY
WORKING WITH REAL DATA

Directions: Form a group of 2 to 4 people. Select someone to record the group's responses for this activity. All members of the group should work cooperatively to answer the questions. If your instructor asks for your results, each member of the group should be prepared to respond.

Converting Temperatures   To convert Celsius degrees C to Fahrenheit degrees F use the formula $F = 32 + \frac{9}{5}C$. This exercise illustrates the importance of understanding the order of operations.

(a) Complete the following table by evaluating the formula in the two ways shown.

(b) At what Celsius temperature does water freeze? At what Fahrenheit temperature does water freeze?

(c) Which column gives the correct Fahrenheit temperatures? Why?

(d) Explain why having an agreed order for operations in mathematics is necessary.

Celsius	$F = \left(32 + \frac{9}{5}\right)C$	$F = 32 + \left(\frac{9}{5}C\right)$
$-40°C$		
$0°C$		
$5°C$	$169°F$	$41°F$
$20°C$		
$30°C$		
$100°C$		

1.4 REAL NUMBERS AND THE NUMBER LINE

ON DAY 2 (week 2) of class we began here!

Signed Numbers ■ Integers and Rational Numbers ■ Square Roots ■ Real and Irrational Numbers ■ The Number Line ■ Absolute Value ■ Inequality

A LOOK INTO MATH ▷

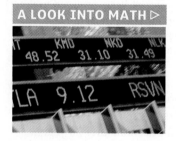

Fractions were invented before decimals and negative numbers. The concept of splitting something into parts was readily understood by most people. However, performing arithmetic with fractions was quite cumbersome, particularly if the fractions had unlike denominators. For example, to find the sum $\frac{1}{2} + \frac{1}{4}$, a common denominator needed to be found. With the invention of decimals, this addition problem became much simpler and could be expressed as $0.50 + 0.25 = 0.75$. One important reason for the invention of decimal numbers is that arithmetic with decimals is much easier to perform than arithmetic with fractions. Even the stock market switched from fractions to decimals in late 2000. In this section we discuss decimal numbers.

Signed Numbers

The idea that numbers could be negative was a difficult concept for many mathematicians. As late as the eighteenth century negative numbers were not readily accepted by everyone. After all, how could a person have -5 oranges?

However, negative numbers make more sense when someone is working with money. If you owe someone 100 dollars, this amount can be thought of as -100, whereas if you have a balance of 100 dollars in your checking account, this amount can be thought of as $+100$. (The positive sign is usually omitted.) The number 0 is neither positive nor negative.

The **opposite,** or **additive inverse,** of a number a is $-a$. For example, the opposite of 25 is -25, the opposite of -5 is $-(-5)$, or 5, and the opposite of 0 is 0. The following double negative rule is helpful in simplifying expressions containing negative signs.

[handwritten: Number a ⇒ Opposite of additive inverse of a = -a]

DOUBLE NEGATIVE RULE

[handwritten: -2 = -(-2) = 2]

Let a be any number. Then $-(-a) = a$.

[handwritten: Double negative Rule = -(-a) = a]

Thus $-(-8) = 8$ and $-\left(-(-10)\right) = -(10) = -10$.

EXAMPLE 1 Finding opposites (or additive inverses)

Find the opposite of each expression.
(a) 13   **(b)** $-\frac{4}{7}$ **(c)** $5 - \frac{4}{2}$ **(d)** $-(-7)$

Solution
(a) The opposite of 13 is -13.
(b) The opposite of $-\frac{4}{7}$ is $\frac{4}{7}$.
(c) $5 - \frac{4}{2} = 5 - 2 = 3$, so the opposite of $5 - \frac{4}{2}$ is -3.
(d) $-(-7) = 7$, so the opposite of $-(-7)$ is -7.   Now Try Exercises **27, 31**

NOTE: To find the opposite of an exponential expression evaluate the exponent first. For example, the opposite of 2^4 is

$$-2^4 = -(2 \cdot 2 \cdot 2 \cdot 2) = -16.$$

EXAMPLE 2 Finding an additive inverse (or opposite)

Find the additive inverse of $-t$, if $t = -\frac{2}{3}$.

Solution
The additive inverse of $-t$ is $t = -\frac{2}{3}$ because $-(-t) = t$ by the double negative rule.
Now Try Exercise **37**

✳ Integers and Rational Numbers

In the opening section to this chapter we discussed natural numbers and whole numbers. Because these sets of numbers do not include negative numbers, fractions, or decimals, other sets of numbers are needed. The **integers** include the natural numbers, zero, and the opposites of the natural numbers. The integers are given by the following.

$$\ldots, \ -3, \ -2, \ -1, \ 0, \ 1, \ 2, \ 3, \ \ldots$$

A **rational number** is any number that can be expressed as the ratio of two integers, $\frac{p}{q}$, where $q \neq 0$. Rational numbers can be written as fractions, and they include all integers. Rational numbers may be positive, negative, or zero. Some examples of rational numbers are

(meaning undefined)

$$\frac{2}{3}, \quad -\frac{3}{5}, \quad \frac{-7}{2}, \quad 1.2, \quad \text{and} \quad 3.$$

The numbers 1.2 and 3 are both rational numbers because they can be written as $\frac{12}{10}$ and $\frac{3}{1}$.

NOTE: $\frac{-7}{2} = \frac{7}{-2} = -\frac{7}{2}$; the position of the negative sign does not affect the value of the fraction.

The fraction bar can be thought of as a division symbol. As a result, rational numbers have decimal equivalents. For example, $\frac{1}{2}$ is equivalent to $1 \div 2$. The division

$$\begin{array}{r} 0.5 \\ 2\overline{)1.0} \end{array}$$

shows that $\frac{1}{2} = 0.5$. In general, a rational number may be expressed in a decimal form that either *repeats* or *terminates*. The fraction $\frac{1}{3}$ may be expressed as $0.\overline{3}$, a repeating decimal, and the fraction $\frac{1}{4}$ may be expressed as 0.25, a terminating decimal. The overbar indicates that $0.\overline{3} = 0.3333333\ldots$

▶ **REAL-WORLD CONNECTION**   Integers and rational numbers are used to describe quantities such as change in population from one decade to the next. Table 1.7 lists the actual change in population along with the percent change from 1990 to 2000 for selected U.S. cities. Note that both positive and negative numbers are used to describe these population changes.

TABLE 1.7   **Change in Population from 1990 to 2000 for Selected Cities**

City	Actual Change	Percent Change
Baltimore	−84,860	−11.5
Cleveland	−27,213	−5.4
Houston	323,078	19.8
Los Angeles	209,263	6.0
New York	685,714	9.4

Source: U.S. Census Bureau.

EXAMPLE 3   Classifying numbers

Classify each number as one or more of the following: natural number, whole number, integer, or rational number.
(a) $\frac{12}{4}$ **(b)** -3 **(c)** 0   **(d)** $-\frac{9}{5}$

Solution
(a) Because $\frac{12}{4} = 3$, the number $\frac{12}{4}$ is a natural number, whole number, integer, and rational number.
(b) The number -3 is an integer and rational number but not a natural number or a whole number.

(c) The number 0 is a whole number, integer, and rational number but not a natural number.

(d) The fraction $-\frac{9}{5}$ is a rational number because it is the ratio of two integers. However, it is not a natural number, a whole number, or an integer.   Now Try Exercises 51, 53, 55

Square Roots

or $2 \cdot 2 = 4$

Square roots are frequently used in algebra. The number b is a **square root** of a number a if $b \cdot b = a$. Every *positive* number has one positive square root and one negative square root. For example, the positive square root of 9 is 3 because $3 \cdot 3 = 9$. The negative square root of 9 is -3. (We will show that $-3 \cdot (-3) = 9$ in Section 1.6.) If a is a nonnegative number (a number that is not negative), then the **principal square root** of a, denoted $\sqrt{a}$, is the nonnegative square root of a. For example, $\sqrt{25} = 5$ because $5 \cdot 5 = 25$ and the number 5 is nonnegative. Note that $\sqrt{0} = 0$.

$9 \Rightarrow \sqrt{9} = \boxed{+3}, -3$

principle Square Root

EXAMPLE 4   Calculating principal square roots

Evaluate each square root. Approximate your answer to three decimal places when appropriate.

(a) $\sqrt{36}$ **(b)** $\sqrt{100}$ **(c)** $\sqrt{5}$

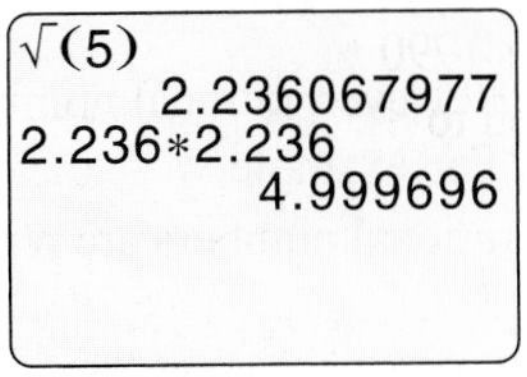

```
√(5)
         2.236067977
2.236*2.236
          4.999696
```

Figure 1.16

Solution

(a) $\sqrt{36} = 6$ because $6 \cdot 6 = 36$ and 6 is nonnegative.

(b) $\sqrt{100} = 10$ because $10 \cdot 10 = 100$ and 10 is nonnegative.

(c) $\sqrt{5}$ is a number between 2 and 3 because $2 \cdot 2 = 4$ and $3 \cdot 3 = 9$. We can estimate the value of $\sqrt{5}$ with a calculator. Figure 1.16 reveals that $\sqrt{5}$ *approximately* equals 2.236. However, 2.236 does not exactly equal $\sqrt{5}$ because $2.236 \times 2.236 = 4.999696$, which does not equal 5.   Now Try Exercise 7

Real and Irrational Numbers

Real numbers can be represented by decimal numbers. That is, if a number can be represented by a decimal number, then it is a real number. Every fraction has a decimal form, so real numbers include rational numbers. However, some real numbers cannot be expressed by fractions. They are called **irrational numbers**. The numbers $\sqrt{2}$, $\sqrt{15}$, and π are examples of irrational numbers. They can be represented by decimals, but not by decimals that either repeat or terminate. Note that for any positive integer a, if $\sqrt{a}$ is not an integer then $\sqrt{a}$ is an irrational number. Examples of real numbers include

$$-17, \quad \frac{4}{5}, \quad -\sqrt{3}, \quad 21\frac{1}{2}, \quad 57.63, \quad \text{and} \quad \sqrt{7}.$$

Any real number may be approximated by a terminating decimal. The symbol $\approx$ represents **approximately equal**. Each of the following real numbers has been approximated to two *decimal places*.

$$\frac{1}{7} \approx 0.14, \quad 2\pi \approx 6.28, \quad \text{and} \quad \sqrt{60} \approx 7.75$$

If you have a calculator, verify these results.

Figure 1.17 shows the relationships of the different sets of numbers. Note that each real number is either a rational number or an irrational number but not both. The natural numbers, whole numbers, and integers are rational numbers.

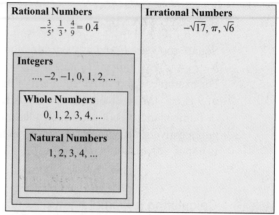

Figure 1.17 The Real Numbers

MAKING CONNECTIONS

Rational and Irrational Numbers

Both rational and irrational numbers can be written as decimals. However, rational numbers can be represented by either terminating or repeating decimals. For example, $\frac{1}{2} = 0.5$ is a terminating decimal and $\frac{1}{3} = 0.333\ldots$ is a repeating decimal. Irrational numbers are represented by decimals that neither terminate nor repeat.

EXAMPLE 5   Classifying numbers

Identify the natural numbers, whole numbers, integers, rational numbers, and irrational numbers in the following list.

$$-\sqrt{5}, \quad 9, \quad -3.8, \quad \sqrt{49}, \quad \frac{11}{4}, \quad \text{and} \quad -41$$

Solution

Natural numbers: 9 and $\sqrt{49} = 7$

Whole numbers: 9 and $\sqrt{49} = 7$

Integers: 9, $\sqrt{49} = 7$, and -41

Rational numbers: 9, -3.8, $\sqrt{49} = 7$, $\frac{11}{4}$, and -41

Irrational number: $-\sqrt{5}$

Now Try Exercises **57, 61, 65**

Even though a data set may contain only integers, we often need decimals to describe it. One common way to describe data is to find their **average**. To find the average of a set of numbers we add the numbers and then divide by how many numbers there are in the set. For example, the average of 6, 13, and 14 equals 11 because

$$\frac{6 + 13 + 14}{3} = \frac{33}{3} = 11.$$

EXAMPLE 6   Analyzing data

Table 1.8 lists the number of higher education institutions, such as colleges and universities, for various years. Find the average number of institutions during this 4-year period. Is the result a natural number, a rational number, or an irrational number?

TABLE 1.8   Higher Education Institutions

Year	2000	2001	2002	2003
Institutions	4182	4197	4168	4236

Source: National Center for Education Statistics.

CRITICAL THINKING

Think of an example in which the sum of two irrational numbers is a rational number.

Solution

The average number of institutions was

$$\frac{4182 + 4197 + 4168 + 4236}{4} = \frac{16,783}{4} = 4195.75.$$

The average of these four natural numbers is an integer divided by an integer, which is a rational number. However, it is neither a natural number nor an irrational number.

Now Try Exercise 107

The Number Line

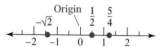

Figure 1.18   The Number Line

The real numbers can be represented visually by using a number line, as shown in Figure 1.18. Each real number corresponds to a unique point on the number line. The point associated with the real number 0 is called the **origin**. The positive integers are equally spaced to the right of the origin, and the negative integers are equally spaced to the left of the origin. The number line extends indefinitely both left and right. Other real numbers can also be located on the number line. For example, the number $\frac{1}{2}$ can be identified by placing a dot halfway between the integers 0 and 1. The numbers $-\sqrt{2} \approx -1.41$ and $\frac{5}{4} = 1.25$ can also be placed (approximately) on this number line.

EXAMPLE 7   Plotting numbers on a number line

Plot each real number on a number line.
(a) $-\frac{3}{2}$ (b) $\sqrt{3}$ (c) π

Figure 1.19   Plotting Real Numbers

Solution

(a) $-\frac{3}{2} = -1.5$. Place a dot halfway between -2 and -1, as shown in Figure 1.19.
(b) A calculator gives $\sqrt{3} \approx 1.73$. Place a dot between 1 and 2 so that it is about three-fourths of the way toward 2, as shown in Figure 1.19.
(c) $\pi \approx 3.14$. Place a dot just past the integer 3, as shown in Figure 1.19.

Now Try Exercises 69, 77

Absolute Value

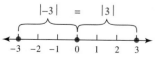

Figure 1.20

The **absolute value** of a real number equals its distance on the number line from the origin. Because distance is never negative, the absolute value of a real number is *never negative*. The absolute value of a real number a is denoted $|a|$ and is read "the absolute value of a." Figure 1.20 shows that the absolute values of -3 and 3 equal 3 because both have distance 3 from the origin. That is, $|-3| = 3$ and $|3| = 3$.

EXAMPLE 8   **Finding the absolute value of a real number**

Write the expression without the absolute value sign.

(a) $|-7|$ (b) $|0|$
(c) $|-a|$, if a is a positive number   (d) $|a|$, if a is a negative number

Solution
(a) $|-7| = 7$ because the distance between the origin and -7 is 7.
(b) $|0| = 0$ because the distance is 0 between the origin and 0.
(c) If a is positive, then $-a$ is negative. Thus $|-a| = a$.
(d) If a is negative, then $-a$ is positive. Thus $|a| = -a$. For example, if $a = -5$, then
     $|-5| = -(-5) = 5$.   Now Try Exercises 79, 81, 87

Our results about absolute value can be summarized as follows.

$$|a| = a, \qquad \text{if } a \text{ is positive or 0.}$$
$$|a| = -a, \qquad \text{if } a \text{ is negative.}$$

Inequality

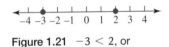

Figure 1.21 $-3 < 2$, or $2 > -3$

If a real number a is located to the left of a real number b on the number line, we say that a is **less than** b and write $a < b$. Similarly, if a real number b is located to the right of a real number a, we say that b is **greater than** a and write $b > a$. Thus $-3 < 2$ because -3 is located to the left of 2, and $2 > -3$ because 2 is located to the right of -3. See Figure 1.21. In general, any negative number will always be less than any positive number, and any positive number will always be greater than any negative number.

We say that a is **less than or equal to** b, denoted $a \leq b$, if either $a < b$ or $a = b$ is true. Similarly, a is **greater than or equal to** b, denoted $a \geq b$, if either $a > b$ or $a = b$ is true.

Inequalities are often used to compare the relative sizes of two quantities. This can be illustrated visually as follows.

EXAMPLE 9   **Ordering real numbers**

List the following numbers from least to greatest. Then plot these numbers on a number line.

$$-2, \quad -\pi, \quad \sqrt{2}, \quad 0, \quad \text{and} \quad 2.5$$

Solution
First note that $-\pi \approx -3.14 < -2$. The two negative numbers are less than 0, and the two positive numbers are greater than 0. Also, $\sqrt{2} \approx 1.41$, so $\sqrt{2} < 2.5$. Listing the numbers from least to greatest results in

$$-\pi, \quad -2, \quad 0, \quad \sqrt{2}, \quad \text{and} \quad 2.5.$$

Figure 1.22

These numbers are plotted on the number line shown in Figure 1.22. Note that these numbers increase from left to right on the number line.   Now Try Exercise 103

1.4 PUTTING IT ALL TOGETHER

Because of the variety of information produced and used by our society, different sets of numbers had to be developed. Without numbers, information can only be described qualitatively, not quantitatively. For example, we might say that it rained a lot last night, but we would not be able to give an actual measurement of the rainfall. The following table summarizes several sets of numbers.

Set of Numbers	Comments	Examples
Integers	Include the natural numbers, their opposites, and 0	$\ldots, -2, -1, 0, 1, 2, \ldots$
Rational Numbers	Include integers, all fractions $\frac{p}{q}$, where p and $q \neq 0$ are integers, and all repeating and terminating decimals	$\frac{1}{2}, -3, \frac{128}{6}, -0.335, 0, 0.25 = \frac{1}{4},$ and $0.\overline{3} = \frac{1}{3}$
Irrational Numbers	Any decimal number that neither terminates nor repeats; a real number that is not rational	$\pi, \sqrt{3},$ and $\sqrt{15}$
Real Numbers	Any number that can be expressed in decimal form; include the rational and irrational numbers	$\pi, \sqrt{3}, -\frac{4}{7}, 0, -10, 0.\overline{6} = \frac{2}{3}, 1000,$ and $\sqrt{15}$

A number line can be used to visualize the real number system, as illustrated in the figure. The point associated with the number 0 is called the origin.

The absolute value of a number a equals its distance on the number line from the origin. If a number a is located to the left of a number b, then a is less than b (written $a < b$). If a is located to the right of b, then a is greater than b (written $a > b$).

Topic	Notation	Examples
Absolute Value	If $a \geq 0$, then $$\lvert a \rvert = a.$$ If $a < 0$, then $$\lvert a \rvert = -a.$$ $\lvert a \rvert$ is *never* negative.	$\lvert 17 \rvert = 17$ $\lvert -12 \rvert = 12$ $\lvert 0 \rvert = 0$
Inequality	Symbols of inequality include: $<, >, \leq, \geq,$ and $\neq$.	$-3 < -2$ less than $6 > 4$ greater than $-5 \leq -5$ less than or equal $18 \geq 0$ greater than or equal $7 \neq 8$ not equal

1.4 Exercises

MyMathLab | Math XL PRACTICE | WATCH | DOWNLOAD | READ | REVIEW

CONCEPTS

1. The opposite of the number b is _____.

2. The additive inverse of -7 is _____.

3. $-(-b) =$ _____.

4. The value of -3^2 is _____.

5. The integers include the natural numbers, zero, and the additive inverses of the _____ numbers.

6. A number that can be written as $\frac{p}{q}$, where p and q are integers with $q \neq 0$, is a(n) _____ number.

7. The equation $4 \cdot 4 = 16$ indicates that $\sqrt{16} =$ _____.

8. The nonnegative square root of a number is called the _____ square root.

9. If a number can be written in decimal form, then it is a(n) _____ number.

10. If a real number is not a rational number, then it is a(n) _____ number.

11. A real number that can be written as a repeating or terminating decimal is a(n) _____ number.

12. A real number that can not be written as a repeating or terminating decimal is a(n) _____ number.

13. Write $0.272727\ldots$ using an overbar.

14. The decimal equivalent for $\frac{1}{4}$ can be found by dividing _____ by _____.

15. Give an example of an irrational number.

16. The symbol $\neq$ is used to indicate that two numbers are _____.

17. The symbol $\approx$ is used to indicate that two numbers are _____.

18. To calculate the average of three numbers, add the three numbers and divide the sum by _____.

19. The origin on the number line corresponds to the number _____.

20. The negative numbers are to the (left/right) of the origin on the number line.

21. The absolute value of a number a gives its distance on the number line from the _____.

22. If $b > 0$, then $|-b| =$ _____.

Exercises 23–26: Insert the symbol $<$, $=$, or $>$ to make the statement true.

23. If a number a is located to the right of a number b on the number line, then a __ b.

24. If $b > 0$ and $a < 0$, then b __ a.

25. If $a \geq b$, then either $a > b$ or a __ b.

26. If $a \geq b$ and $b \geq a$, then a __ b.

SIGNED NUMBERS

Exercises 27–34: Find the opposite of each expression.

27. (a) 9   (b) -9

28. (a) -6 (b) 6

29. (a) $\frac{2}{3}$ (b) $-\frac{2}{3}$

30. (a) $-\left(\frac{-4}{5}\right)$ (b) $-\left(\frac{-4}{-5}\right)$

31. (a) $-(-8)$ (b) $-\left(-(-8)\right)$

32. (a) $-\left(-(-2)\right)$ (b) $-(-2)$

33. (a) a (b) $-a$

34. (a) $-b$ (b) $-(-b)$

35. Find the additive inverse of t, if $-t = 6$.

36. Find the additive inverse of $-t$, if $t = -\frac{4}{5}$.

37. Find the additive inverse of $-b$, if $b = \frac{1}{2}$.

38. Find the additive inverse of b, if $-b = \frac{5}{-6}$.

NUMBERS AND THE NUMBER LINE

Exercises 39–50: Find the decimal equivalent for the rational number.

39. $\frac{1}{4}$ 40. $\frac{3}{5}$

41. $\frac{7}{8}$ 42. $\frac{3}{10}$

43. $\frac{3}{2}$ 44. $\frac{3}{50}$

45. $\frac{1}{20}$ **46.** $\frac{3}{16}$

47. $\frac{2}{3}$ **48.** $\frac{2}{9}$

49. $\frac{7}{9}$ **50.** $\frac{5}{11}$

Exercises 51–56: Classify the number as one or more of the following: natural number, whole number, integer, or rational number.

51. 8   **52.** -8

53. $\frac{16}{4}$ **54.** $\frac{5}{7}$

55. 0   **56.** $-\frac{15}{31}$

Exercises 57–68: Classify the number as one or more of the following: natural number, integer, rational number, or irrational number.

57. -4.5 **58.** π

59. $\frac{3}{7}$ **60.** $\sqrt{25}$

61. $\sqrt{11}$ **62.** $-\sqrt{3}$

63. $\frac{8}{4}$ **64.** -5

65. $\sqrt{49}$ **66.** $3.\overline{3}$

67. $1.\overline{8}$ **68.** $\frac{9}{3}$

Exercises 69–78: Plot each number on a number line.

69. (a) 0   **(b)** -2 **(c)** 3

70. (a) -1 **(b)** -3 **(c)** 4

71. (a) $\frac{1}{2}$ **(b)** $-\frac{1}{2}$ **(c)** 2

72. (a) $-\frac{3}{2}$ **(b)** $\frac{3}{2}$ **(c)** 0

73. (a) 1.3   **(b)** -2.5 **(c)** 0.7

74. (a) 3.2   **(b)** -2.8 **(c)** 0.5

75. (a) -10 **(b)** -20 **(c)** 30

76. (a) 5   **(b)** 10   **(c)** -10

77. (a) π **(b)** $\sqrt{2}$ **(c)** $-\sqrt{3}$

78. (a) $\sqrt{11}$ **(b)** $-\sqrt{5}$ **(c)** $\sqrt{4}$

ABSOLUTE VALUE

Exercises 79–88: Evaluate the expression.

79. $|5.23|$ **80.** $|\pi|$

81. $|-7|$ **82.** $|-\sqrt{2}|$

83. $|2 - 6|$ **84.** $|\frac{2}{3} - \frac{1}{3}|$

85. $|\pi - 3|$ **86.** $|3 - \pi|$

87. $|b|$, if b is negative   **88.** $|-b|$, if b is positive

INEQUALITY

Exercises 89–100: Insert the symbol $>$ or $<$ to make the statement true.

89. 5 ___ 7   **90.** -5 ___ 7

91. -5 ___ -7 **92.** $\frac{3}{5}$ ___ $\frac{2}{5}$

93. $-\frac{1}{3}$ ___ $-\frac{2}{3}$ **94.** $-\frac{1}{10}$ ___ 0

95. -1.9 ___ -1.3 **96.** 5.1 ___ -6.2

97. $|-8|$ ___ 3   **98.** 4 ___ $|-1|$

99. $|-2|$ ___ $|-7|$ **100.** $|-15|$ ___ $|32|$

Exercises 101–106: List the given numbers from least to greatest.

101. $-3, 0, 1, -9, -2^3$

102. $4, -2^3, \frac{1}{2}, -\frac{3}{2}, \frac{3}{2}$

103. $-2, \pi, \frac{1}{3}, -\frac{3}{2}, \sqrt{5}$

104. $9, 14, -\frac{1}{12}, -\frac{3}{16}, \sqrt{7}$

105. $-\frac{4}{7}, -\frac{17}{28}, -4^2, \sqrt{7}, \sqrt{2}$

106. $-3.1, 2^3, -3^3, 6, 9.4$

APPLICATIONS

107. *Household Size*   The table lists the percentage of U.S. households that were 3-person households during selected years.

Year	1970	1980	1990	2000
Percent	17.2	17.4	17.4	16.5

Source: U.S. Census Bureau.

(a) What was this percentage in 2000?

(b) Mentally estimate the average percentage for these selected years.

(c) Calculate the average percentage. Is your mental estimate in reasonable agreement with your calculated result?

108. *Music Sales*   The table lists the percentage of recorded music purchases attributed to children aged 10 to 14 during selected years.

Year	2001	2002	2003	2004
Percent	8.5	8.9	8.6	9.4

Source: Recording Industry Association of America.

(a) What was this percentage in 2001?

(b) Mentally estimate the average percentage for this 4-year period.

(c) Calculate the average percentage. Is your mental estimate in reasonable agreement with your calculated result?

WRITING ABOUT MATHEMATICS

109 What is a rational number? Is every integer a rational number? Why or why not?

110. Explain why $\frac{3}{7} > \frac{1}{3}$. Now explain in general how to determine whether $\frac{a}{b} > \frac{c}{d}$. Assume that a, b, c, and d are natural numbers.

CHECKING BASIC CONCEPTS
SECTIONS 1.3 AND 1.4

1. Write each product as an exponential expression.
   (a) $5 \cdot 5 \cdot 5 \cdot 5$ **(b)** $7 \cdot 7 \cdot 7 \cdot 7 \cdot 7$

2. Evaluate each expression.
   (a) 2^3 **(b)** 10^4 **(c)** $\left(\frac{2}{3}\right)^3$ **(d)** -3^4

3. Use the given base to write the number as an exponential expression.
   (a) 64   (base 4)   **(b)** 64   (base 2)

4. Evaluate each expression without a calculator.
   (a) $6 + 5 \cdot 4$ **(b)** $6 + 6 \div 2$
   (c) $5 - 2 - 1$ **(d)** $\frac{6 - 3}{2 + 4}$
   (e) $12 \div (6 \div 2)$ **(f)** $2^3 - 2\left(2 + \frac{4}{2}\right)$

5. Translate the phrase "five cubed divided by three" to an algebraic expression.

6. Find the opposite of each expression.
   (a) -17 **(b)** a

7. Find the decimal equivalent for the rational number.
   (a) $\frac{3}{20}$ **(b)** $\frac{5}{8}$

8. Classify each number as one or more of the following: natural number, integer, rational number, or irrational number.
   (a) $\frac{10}{2}$ **(b)** -5 **(c)** $\sqrt{5}$ **(d)** $-\frac{5}{6}$

9. Plot each number on the same number line.
   (a) 0   **(b)** -3 **(c)** 2
   (d) $\frac{3}{4}$ **(e)** $-\sqrt{2}$

10. Evaluate each expression.
   (a) $|-12|$ **(b)** $|-a|$, if $a > 0$

11. Insert the symbol $>$ or $<$ to make the statement true.
   (a) $4 \_\_ 9$ **(b)** $-1.3 \_\_ -0.5$
   (c) $|-3| \_\_ |-5|$

12. List the following numbers from least to greatest.
$$\sqrt{3}, \quad -7, \quad 0, \quad \frac{1}{3}, \quad -1.6, \quad 3^2$$

1.5   ADDITION AND SUBTRACTION OF REAL NUMBERS

Addition of Real Numbers ■ Subtraction of Real Numbers ■ Applications

A LOOK INTO MATH ▷   There are four arithmetic operations: addition, subtraction, multiplication, and division. You probably learned to use arithmetic operations on whole numbers in elementary school. In this section we extend those concepts of addition and subtraction of real numbers to include addition and subtraction of positive *and* negative numbers. Both addition and subtraction require two numbers to calculate an answer. As a result, these operations are called **binary operations**. Multiplication and division are also binary operations. Some operations, such as finding the opposite (negation), are called **unary operations** because they require only one number.

Addition of Real Numbers

In an addition problem the two numbers added are called **addends**, and the answer is called the **sum**. In the addition problem $3 + 5 = 8$, the numbers 3 and 5 are the addends and the number 8 is the sum.

The *opposite* (or *additive inverse*) of a real number a is $-a$. When we add opposites, the result is 0. That is, $a + (-a) = 0$ for every real number a.

EXAMPLE 1   Adding opposites

Find the opposite of each number and calculate the sum of the number and its opposite.

(a) 45   **(b)** $\sqrt{2}$ **(c)** $-\frac{1}{2}$

Solution

(a) The opposite of 45 is -45. Their sum is $45 + (-45) = 0$.
(b) The opposite of $\sqrt{2}$ is $-\sqrt{2}$. Their sum is $\sqrt{2} + (-\sqrt{2}) = 0$.
(c) The opposite of $-\frac{1}{2}$ is $\frac{1}{2}$. Their sum is $-\frac{1}{2} + \frac{1}{2} = 0$.   Now Try Exercise 13

When adding real numbers, it may be helpful to think of money. A positive number represents income, and a negative number indicates a debt. The sum $9 + (-5) = 4$ would represent being paid \$9 and owing \$5. In this case \$4 would be left over. Similarly, the sum $-3 + (-6) = -9$ would represent debts of \$3 and \$6, resulting in a total debt of \$9. To add two real numbers we can use the following rules.

ADDITION OF REAL NUMBERS

To add two numbers that are either *both positive* or *both negative*, add their absolute values. Their sum has the same sign as the two numbers.

To add two numbers with *opposite signs*, find the absolute value of each number. Subtract the smaller absolute value from the larger. The sum has the same sign as the sign of the number with the larger absolute value. If the two numbers are opposites, their sum is 0.

EXAMPLE 2   Adding real numbers

Evaluate each expression.
(a) $-2 + (-4)$ **(b)** $-\frac{2}{5} + \frac{7}{10}$ **(c)** $6.2 + (-8.5)$

Solution

(a) The numbers are both negative, so we add the absolute values $|-2|$ and $|-4|$ to obtain 6. The signs of the addends are both negative, so the answer is -6. That is, $-2 + (-4) = -6$. If we owe \$2 and then owe \$4, the total amount owed is \$6.

(b) The numbers have opposite signs, so we subtract their absolute values to obtain

$$\frac{7}{10} - \frac{2}{5} = \frac{7}{10} - \frac{4}{10} = \frac{3}{10}.$$

The sum is positive because $\left|\frac{7}{10}\right|$ is greater than $\left|-\frac{2}{5}\right|$. That is, $-\frac{2}{5} + \frac{7}{10} = \frac{3}{10}$. If we spend \$0.40 $\left(-\frac{2}{5} = -0.4\right)$ and receive \$0.70 $\left(\frac{7}{10} = 0.7\right)$, we have \$0.30 $\left(\frac{3}{10} = 0.3\right)$ left.

(c) $6.2 + (-8.5) = -2.3$ because $|-8.5|$ is 2.3 more than $|6.2|$. If we have \$6.20 and we owe \$8.50, we are short \$2.30.   Now Try Exercises 35, 41, 45

ADDING INTEGERS VISUALLY   One way to add integers visually is to use the symbol ⌒ to represent a positive unit and to use the symbol ∪ to represent a negative unit. Now adding opposites visually results in "zero," as shown.

For example, to add $-3 + 5$, we draw three negative units and five positive units.

Because the "zeros" add no value, the sum is two positive units, or 2.

EXAMPLE **3**   **Adding integers visually**

Perform the addition visually using the symbols ⌒ and ∪.
(a) $3 + 2$ **(b)** $-6 + 4$ **(c)** $2 + (-3)$ **(d)** $(-5) + (-2)$

Solution
(a) Draw three positive units and then draw two more positive units.

Because no zeros were formed, the sum is five positive units, or 5.
(b) Draw six negative units and then draw four positive units.

Ignoring the zeros that were formed, the sum is two negative units, or -2.
(c) Draw two positive and three negative units.

The sum is -1.
(d)

The sum is -7.   Now Try Exercises **21, 25**

Figure 1.23 $4 + (-3) = 1$

Figure 1.24 $-2 + (-3) = -5$

Another way to add integers visually is to use a number line. To add $4 + (-3)$ start at 0 (the origin) and draw an arrow to the right 4 units long from 0 to 4. The number -3 is a negative number, so draw an arrow 3 units long to the left, starting at the tip of the first arrow. See Figure 1.23. The tip of the second arrow is at 1, which equals the sum of 4 and -3.

To find the sum $-2 + (-3)$, draw an arrow 2 units long to the left, starting at the origin. Then draw an arrow 3 units long to the left, starting at the tip of the first arrow, which is located at -2. See Figure 1.24. Because the tip of the second arrow coincides with -5 on the number line, the sum of -2 and -3 is -5.

Subtraction of Real Numbers

▶ **REAL-WORLD CONNECTION**   The answer to a subtraction problem is the **difference**. Addition and subtraction occur every day at grocery stores, where the costs of various items are added to the total and discounts from coupons are subtracted. When subtracting two real numbers, changing the problem to an addition problem may be helpful.

SUBTRACTION OF REAL NUMBERS

For any real numbers a and b,

$$a - b = a + (-b).$$

To subtract b from a, add a and the opposite of b.

EXAMPLE 4   Subtracting real numbers

Find each difference by hand.
(a) $10 - 20$ **(b)** $-5 - 2$ **(c)** $-2.1 - (-3.2)$ **(d)** $\frac{1}{2} - \left(-\frac{3}{4}\right)$

Solution
(a) $10 - 20 = 10 + (-20) = -10$
(b) $-5 - 2 = -5 + (-2) = -7$
(c) $-2.1 - (-3.2) = -2.1 + 3.2 = 1.1$
(d) $\frac{1}{2} - \left(-\frac{3}{4}\right) = \frac{1}{2} + \frac{3}{4} = \frac{2}{4} + \frac{3}{4} = \frac{5}{4}$ Now Try Exercise **53**

In the next example, we show how to add and subtract groups of numbers.

EXAMPLE 5   Adding and subtracting real numbers

Evaluate each expression.
(a) $5 - 4 - (-6) + 1$ **(b)** $\frac{1}{2} - \frac{3}{4} + \frac{1}{3}$ **(c)** $-6.1 + 5.6 - 10.1$

Solution
(a) Rewrite the expression in terms of addition only, and then find the sum.

$$5 - 4 - (-6) + 1 = 5 + (-4) + 6 + 1$$
$$= 1 + 6 + 1$$
$$= 8$$

(b) Begin by rewriting the fractions with the LCD of 12.

$$\frac{1}{2} - \frac{3}{4} + \frac{1}{3} = \frac{6}{12} - \frac{9}{12} + \frac{4}{12}$$
$$= \frac{6}{12} + \left(-\frac{9}{12}\right) + \frac{4}{12}$$
$$= -\frac{3}{12} + \frac{4}{12}$$
$$= \frac{1}{12}$$

(c) The expression can be evaluated by changing subtraction to addition.

$$-6.1 + 5.6 - \mathbf{10.1} = -6.1 + 5.6 + (-\mathbf{10.1})$$
$$= -0.5 + (-10.1)$$
$$= -10.6$$

Now Try Exercises 71, 79

Applications

▶ **REAL-WORLD CONNECTION** In word problems, certain phrases often indicate that we should add two numbers. For example, if a person's age is 10 more than 15, then we add 10 and 15 to obtain 25. The phrase *more than* often indicates addition. Other words that indicate addition are *sum, greater, greater than, plus,* and *increased by.*

Similarly, if the outside temperature is 10°F less than 32°F, then the outside temperature is $32 - 10 = 22°$F. The phrase *less than* often indicates subtraction. Other words that indicate subtraction are *difference, less, minus,* and *decreased by.*

EXAMPLE 6 Calculating temperature differences

The hottest outdoor temperature ever recorded in the shade was 136°F in the Sahara desert, and the coldest outside temperature ever recorded was $-129°$F in Antarctica. Find the temperature difference between these two temperatures.   (***Source:*** *Guinness Book of Records.*)

Solution
The word *difference* indicates subtraction. We must subtract the two temperatures.

$$136 - (-129) = 136 + 129 = 265°\text{F}.$$ Now Try Exercise 97

▶ **REAL-WORLD CONNECTION** Addition of positive and negative numbers occurs at banks if positive numbers represent deposits and negative numbers represent withdrawals.

EXAMPLE 7 Balancing a checking account

The initial balance in a checking account is $285. Find the final balance if the following represents a list of withdrawals and deposits: $-\$15, -\$20, \$500,$ and $-\$100.$

Solution
Find the sum of the five numbers.

$$285 + (-15) + (-20) + 500 + (-100) = 270 + (-20) + 500 + (-100)$$
$$= 250 + 500 + (-100)$$
$$= 750 + (-100)$$
$$= 650$$

```
285+(-15)+(-20)+
500+(-100)
              650
285-15-20+500-10
0
              650
```

Figure 1.25

The final balance is $650. This result may be supported by evaluating the expression with a calculator, as illustrated in Figure 1.25. Note that the expression has been evaluated two different ways.

Now Try Exercise 93

TECHNOLOGY NOTE: Subtraction and Negation

Calculators typically have *different* keys to represent subtraction and negation. Be sure to use the correct key. Many graphing calculators have the following keys for subtraction and negation.

| − | (−) |
| *Subtraction* | *Negation* |

1.5 PUTTING IT ALL TOGETHER

The following table outlines how to add and subtract real numbers.

Operation	Comments	Examples
Addition	To add real numbers, use a number line or follow the rules found in the box on page 45.	$-2 + 8 = 6$ $0.8 + (-0.3) = 0.5$ $-\frac{1}{7} + \left(-\frac{3}{7}\right) = \frac{-1 + (-3)}{7} = -\frac{4}{7}$ $-4 + 4 = 0$ $-3 + 4 + (-2) = -1$
Subtraction	To subtract real numbers, transform the problem to an addition problem by adding the opposite. $$a - b = a + (-b)$$	$6 - 8 = 6 + (-8) = -2$ $-3 - 4 = -3 + (-4) = -7$ $-\frac{1}{2} - \left(-\frac{3}{2}\right) = -\frac{1}{2} + \frac{3}{2} = \frac{2}{2} = 1$ $-5 - (-5) = -5 + 5 = 0$ $9.4 - (-1.2) = 9.4 + 1.2 = 10.6$

1.5 Exercises

CONCEPTS

1. Addition and subtraction are called binary operations because each requires _____ numbers to calculate an answer.

2. In an addition problem, the two numbers added are called _____.

3. The solution to an addition problem is the _____.

4. When you add opposites, the sum is always _____.

5. If two positive numbers are added, the sum is always a _____ number.

6. If two negative numbers are added, the sum is always a _____ number.

7. If two numbers with opposite signs are added, the sum has the same sign as the number with the larger _____.

8. The solution to a subtraction problem is the _____.

9. When subtracting two real numbers, it may be helpful to change the problem to a(n) _____ problem.

10. To subtract b from a, add the _____ of b to a. That is, $a - b = a +$ _____.

11. The words *sum*, *more*, and *plus* indicate that _____ should be performed.

12. The words *difference*, *less than*, and *minus* indicate that _____ should be performed.

ADDITION AND SUBTRACTION OF REAL NUMBERS

Exercises 13–18: Find the opposite of the number and then calculate the sum of the number and its opposite.

13. 25

14. $-\frac{1}{2}$

15. $-\sqrt{21}$

16. $-\pi$

17. 5.63

18. -5^2

Exercises 19–26: Refer to Example 3. Find the sum visually.

19. $1 + 3$

20. $3 + 1$

21. $4 + (-2)$

22. $-4 + 6$

23. $-1 + (-2)$

24. $-2 + (-2)$

25. $-3 + 7$

26. $5 + (-6)$

Exercises 27–34: Use a number line to find the sum.

27. $-1 + 3$

28. $3 + (-1)$

29. $4 + (-5)$

30. $2 + 6$

31. $-10 + 20$

32. $15 + (-5)$

33. $-50 + (-100)$

34. $-100 + 100$

Exercises 35–52: Find the sum.

35. $5 + (-4)$

36. $-9 + 7$

37. $-1 + (-6)$

38. $-10 + (-23)$

39. $\frac{3}{4} + \left(-\frac{1}{2}\right)$

40. $-\frac{5}{12} + \left(-\frac{1}{6}\right)$

41. $-\frac{6}{7} + \frac{3}{14}$

42. $-\frac{1}{5} + \frac{2}{5}$

43. $-\frac{1}{2} + \left(-\frac{3}{4}\right)$

44. $-\frac{2}{9} + \left(-\frac{1}{12}\right)$

45. $0.6 + (-1.7)$

46. $4.3 + (-2.4)$

47. $-52 + 86$

48. $-103 + (-134)$

49. $8 + (-7) + (-2)$

50. $-10 + 20 + (-10)$

51. $\frac{1}{2} + \frac{3}{4} + \left(-\frac{1}{2}\right) + \left(-\frac{3}{4}\right)$

52. $0.1 + (-0.4) + (-1.5) + 0.9$

Exercises 53–68: Find the difference.

53. $5 - 8$

54. $3 - 5$

55. $-2 - (-9)$

56. $-10 - (-19)$

57. $\frac{1}{3} - \left(-\frac{2}{3}\right)$

58. $\frac{3}{4} - \frac{5}{4}$

59. $\frac{6}{7} - \frac{13}{14}$

60. $-\frac{5}{6} - \frac{1}{6}$

61. $-\frac{1}{10} - \left(-\frac{3}{5}\right)$

62. $-\frac{2}{11} - \left(-\frac{5}{11}\right)$

63. $0.8 - (-2.1)$

64. $-9.6 - (-5.7)$

65. $-73 - 91$

66. $201 - 502$

67. $-7 - (-6) - 10$

68. $-20 - 30 - (-40)$

Exercises 69–82: Evaluate the expression.

69. $10 - 19$

70. $5 + (-9)$

71. $19 - (-22) + 1$

72. $53 + (-43) - 10$

73. $-3 + 4 - 6$

74. $-11 + 8 - 10$

75. $100 - 200 + 100 - (-50)$

76. $-50 - (-40) + (-60) + 80$

77. $1.5 - 2.3 + 9.6$

78. $10.5 - (-5.5) + (-1.5)$

79. $-\frac{1}{2} + \frac{1}{4} - \left(-\frac{3}{4}\right)$

80. $\frac{1}{4} - \left(-\frac{2}{5}\right) + \left(-\frac{3}{20}\right)$

81. $|4 - 9| - |1 - 7|$

82. $|-5 - (-3)| - |-6 + 8|$

Exercises 83–92: Write an arithmetic expression for the given phrase and then simplify it.

83. The sum of two and negative five

84. Subtract ten from negative six

85. Negative five increased by seven

86. Negative twenty decreased by eight

87. The additive inverse of the quantity two cubed

88. Five minus the quantity two cubed

89. The difference between negative six and seven (*Hint:* Write the numbers for the subtraction problem in the order given.)

90. The difference between one-half and three-fourths

91. Six plus negative ten minus five

92. Ten minus seven plus negative twenty

APPLICATIONS

93. *Checking Account*   The initial balance in a checking account is $358. Find the final balance resulting from the following withdrawals and deposits: $-\$45$, $\$37$, $\$120$, and $-\$240$.

94. *Savings Account*   A savings account has $1245 in it. Find the final balance resulting from the following withdrawals and deposits: $-\$189$, $\$975$, $-\$226$, and $-\$876$.

95. *Football Stats*   A running back carries the ball five times. Find his total yardage if the carries were 9, -2, -1, 14, and 5 yards.

96. *Football Stats*   A football team starts on the 50 yard line and makes gains of 12, 6, and 22 yards. However, during these three plays the team is assessed penalties of 10 and 15 yards. On what yard line is the team after these plays?

97. *Deepest and Highest* The deepest point in the ocean is the Mariana Trench, which is 35,839 feet below sea level. The highest point on Earth is Mount Everest, which is 29,029 feet above sea level. What is the difference in height between Mount Everest and the Mariana Trench? (*Source:* The Guinness Book of Records.)

98. *Greatest Temperature Ranges* The greatest temperature range on Earth occurs in Siberia, where the temperature can vary between 98°F in the summer and −90°F

in the winter. Find the difference between these two temperatures. (*Source:* The Guinness Book of Records.)

WRITING ABOUT MATHEMATICS

99. Explain how to add two negative numbers. Give an example.

100. Explain how to subtract a negative number from a positive number. Give an example.

1.6 MULTIPLICATION AND DIVISION OF REAL NUMBERS

Multiplication of Real Numbers ▪ Division of Real Numbers ▪ Applications

A LOOK INTO MATH ▷ In some businesses, such as auto-parts stores, sometimes it is necessary to convert fractions to decimal numbers and decimals to fractions. For example, it might be necessary to convert the size of a $2\frac{3}{8}$-inch bolt to a decimal number before it can be ordered. This conversion makes use of division of real numbers. In this section we discuss both multiplication and division of real numbers. (*Source:* NAPA.)

Multiplication of Real Numbers

In a multiplication problem, the numbers multiplied are called the *factors*, and the answer is called the *product*. In the problem $3 \cdot 5 = 15$, the numbers 3 and 5 are factors and 15 is the product. Multiplication is a fast way to perform addition. For example, $5 \cdot 2 = 10$ is equivalent to finding the sum of five 2s, or

$$2 + 2 + 2 + 2 + 2 = 10.$$

Similarly, the product $5 \cdot (-2)$ is equivalent to finding the sum of five −2s, or

$$(-2) + (-2) + (-2) + (-2) + (-2) = -10.$$

Thus $5 \cdot (-2) = -10$. In general, the product of a positive number and a negative number is a negative number.

What sign should the product of two negative numbers have? To answer this question, consider the following patterns.

$$
\begin{array}{ll}
3 \cdot 2 = 6 & 3 \cdot (-2) = -6 \\
2 \cdot 2 = 4 & 2 \cdot (-2) = -4 \\
1 \cdot 2 = 2 & 1 \cdot (-2) = -2 \\
0 \cdot 2 = 0 & 0 \cdot (-2) = 0 \\
-1 \cdot 2 = -2 & -1 \cdot (-2) = ? \\
-2 \cdot 2 = -4 & -2 \cdot (-2) = ?
\end{array}
$$

In the first column, each time the first factor is reduced by 1, the product is found by *subtracting* 2 from the product in the previous row. Note that because multiplication is a fast way to perform addition, the number of 2s being added is one less as we move down the rows in the first column.

In the second column, each time the first factor is reduced by 1, the product is found by *subtracting* -2 from the previous product, which is equivalent to *adding* 2. If the pattern in the second column is to continue, then

$$-1 \cdot (-2) = 2$$
$$-2 \cdot (-2) = 4$$
$$-3 \cdot (-2) = 6,$$

and so on. Note that, in general, the product of two negative numbers is a positive number.

SIGNS OF PRODUCTS

The product of two numbers with *like* signs is positive. The product of two numbers with *unlike* signs is negative.

As a result of these two rules, $4 \cdot 5 = 20, 4 \cdot (-5) = -20, -4 \cdot 5 = -20$, and $-4 \cdot (-5) = 20$. Furthermore, multiplying a number by -1 results in the additive inverse (opposite) of the number. For example, $-1 \cdot 4 = -4$ and -4 is the additive inverse of 4. Similarly, $-1 \cdot (-3) = 3$ and 3 is the additive inverse of -3. In general, $-1 \cdot a = -a$.

EXAMPLE 1   Multiplying real numbers

Find each product by hand.
(a) $-9 \cdot 7$ **(b)** $\frac{2}{3} \cdot \frac{5}{9}$ **(c)** $-2.1(-40)$ **(d)** $(-2.5)(4)(-9)(-2)$

Solution
(a) The resulting product is negative because the factors -9 and 7 have unlike signs. Thus $-9 \cdot 7 = -63$.
(b) The product is positive because both factors are positive.

$$\frac{2}{3} \cdot \frac{5}{9} = \frac{2 \cdot 5}{3 \cdot 9} = \frac{10}{27}$$

(c) As both factors are negative, the product is positive. Thus $-2.1(-40) = 84$.
(d) $(-2.5)(4)(-9)(-2) = (-10)(-9)(-2) = (90)(-2) = -180$

Now Try Exercises 19, 29, 35

MAKING CONNECTIONS

Multiplying More Than Two Negative Factors

Because the product of two negative numbers is positive, it is possible to determine the sign of a product by counting the number of negative factors. For example, the product $-3 \cdot 4 \cdot (-5) \cdot 7 \cdot (-6)$ is negative because there are an odd number of negative factors and the product $2 \cdot (-1) \cdot (-3) \cdot (-5) \cdot (-4)$ is positive because there are an even number of negative factors.

The rules for the signs of products allow us to evaluate powers of signed numbers. However, it is important to note that $(-3)^2$ means that -3 should be multiplied with itself two times, but -3^2 means we should negate the product $3 \cdot 3$. The next example illustrates the difference between powers of signed numbers and the opposite of an exponential expression.

EXAMPLE 2 Evaluating real numbers with exponents

Evaluate each expression by hand.
(a) $(-4)^2$ **(b)** -4^2 **(c)** $(-2)^3$ **(d)** -2^3

Solution
(a) Because the exponent is outside of parentheses, the base of the exponential expression is -4. The expression is evaluated as
$$(-4)^2 = (-4)(-4) = 16.$$

(b) This is the negation of an exponential expression with base 4. Evaluating the exponent before negating results in $-4^2 = -(4)(4) = -16$.

(c) $(-2)^3 = (-2)(-2)(-2) = -8$

(d) $-2^3 = -(2)(2)(2) = -8$

Now Try Exercises 43, 45

MAKING CONNECTIONS

Negative Square Roots

Because the product of two negative numbers is positive, $(-3) \cdot (-3) = 9$. That is, -3 is a square root of 9. As discussed in Section 1.4, every positive number has one positive square root and one negative square root. For a positive number a, the positive square root is called the principal square root and is denoted $\sqrt{a}$. The negative square root is denoted $-\sqrt{a}$. For example, $\sqrt{4} = 2$ and $-\sqrt{4} = -2$.

✱Division of Real Numbers *must also work on long division*

In the division problem $20 \div 4 = 5$, the number 20 is the **dividend**, 4 is the **divisor**, and 5 is the **quotient**. This division problem can also be written in fraction form as $\frac{20}{4} = 5$. Division of real numbers can be defined in terms of multiplication and reciprocals. The reciprocal, or multiplicative inverse, of a real number a is $\frac{1}{a}$. The number 0 has *no reciprocal*.

$$5 \div \tfrac{1}{4} = 5 \cdot \tfrac{4}{1} \text{ or } 5 \cdot 4 = 20$$

TECHNOLOGY NOTE

Try dividing 5 by 0 with a calculator. On most calculators dividing a number by 0 results in an error message. Does it on your calculator?

DIVIDING REAL NUMBERS

For real numbers a and b with $b \neq 0$,
$$\frac{a}{b} = a \cdot \frac{1}{b}.$$
That is, to divide a by b, multiply a by the reciprocal of b.

NOTE: Division by 0 is undefined because 0 has no reciprocal.

EXAMPLE 3 Dividing real numbers

Evaluate each expression by hand.

(a) $-16 \div \tfrac{1}{4}$ **(b)** $\dfrac{\tfrac{3}{5}}{-8}$ **(c)** $\dfrac{-8}{-36}$ **(d)** $9 \div 0$

Solution

(a) $-16 \div \frac{1}{4} = \frac{-16}{1} \cdot \frac{4}{1} = \frac{-64}{1} = -64$

(b) $\frac{\frac{3}{5}}{-8} = \frac{3}{5} \div (-8) = \frac{3}{5} \cdot \left(-\frac{1}{8}\right) = -\frac{3}{40}$

(c) $\frac{-8}{-36} = -8 \cdot \left(-\frac{1}{36}\right) = \frac{8}{36} = \frac{2}{9}$

(d) $9 \div 0$ is undefined. The number 0 has no reciprocal.   Now Try Exercises 51, 59, 71

When determining the sign of a quotient, the following rules may be helpful.

SIGNS OF QUOTIENTS

The quotient of two numbers with *like* signs is positive. The quotient of two numbers with *unlike* signs is negative.

As a result of these two rules, $\frac{20}{4} = 5$, $\frac{-20}{4} = -5$, $\frac{20}{-4} = -5$, and $\frac{-20}{-4} = 5$. You may wonder why a negative number divided by a negative number is a positive number. Remember that division is a fast way to perform subtraction. For example, because

$$20 - 4 - 4 - 4 - 4 - 4 = 0,$$

there are five 4s in 20 and so $\frac{20}{4} = 5$. To check this answer, multiply 5 and 4 to obtain $5 \cdot 4 = 20$. Similarly,

$$-20 - (-4) - (-4) - (-4) - (-4) - (-4) = 0.$$

Thus there are five -4s in -20 and so $\frac{-20}{-4} = 5$. To check this answer, multiply 5 and -4 to obtain $5 \cdot (-4) = -20$.

▶ **REAL-WORLD CONNECTION**   In business, employees may need to convert fractions and mixed numbers to decimal numbers. The next example illustrates this process.

EXAMPLE 4   Converting fractions to decimals

Convert each measurement to a decimal number.
(a) $2\frac{3}{8}$-inch bolt   **(b)** $\frac{15}{16}$-inch diameter   **(c)** $1\frac{1}{3}$-cup flour

Solution

(a) Because $\frac{3}{8}$ indicates the division problem $3 \div 8$, begin by dividing 3 by 8.

$$\begin{array}{r} 0.375 \\ 8\overline{)3.000} \\ \underline{24} \\ 60 \\ \underline{56} \\ 40 \\ \underline{40} \\ 0 \end{array}$$

Thus the mixed number $2\frac{3}{8}$ is equivalent to the decimal number 2.375.

(b) Start by dividing 15 by 16.

$$
\begin{array}{r}
0.9375 \\
16\overline{)15.0000} \\
\underline{144} \\
60 \\
\underline{48} \\
120 \\
\underline{112} \\
80 \\
\underline{80} \\
0
\end{array}
$$

Thus the fraction $\frac{15}{16}$ is equivalent to the decimal number 0.9375.

(c) Start by dividing 1 by 3.

$$
\begin{array}{r}
0.333\ldots \\
3\overline{)1.000} \\
\underline{9} \\
10 \\
\underline{9} \\
10 \\
\underline{9} \\
1
\end{array}
$$

Thus the mixed number $1\frac{1}{3}$ is equivalent to the decimal number $1.333\ldots$, or $1.\overline{3}$.

Now Try Exercises 77, 81, 83

In the next example, numbers expressed as terminating decimals are converted to fractions.

EXAMPLE 5   Converting decimals to fractions

Convert each decimal number to a fraction in lowest terms.
(a) 0.06   **(b)** 0.375   **(c)** 0.0025

Solution

(a) The decimal 0.06 equals six hundredths, or $\frac{6}{100}$. Simplifying this fraction gives

$$
\frac{6}{100} = \frac{3 \cdot 2}{50 \cdot 2} = \frac{3}{50}.
$$

(b) The decimal 0.375 equals three hundred seventy-five thousandths, or $\frac{375}{1000}$. Simplifying this fraction gives

$$
\frac{375}{1000} = \frac{3 \cdot 125}{8 \cdot 125} = \frac{3}{8}.
$$

(c) The decimal 0.0025 equals twenty-five ten thousandths, or $\frac{25}{10,000}$. Simplifying this fraction gives

$$
\frac{25}{10,000} = \frac{1 \cdot 25}{400 \cdot 25} = \frac{1}{400}.
$$

Now Try Exercises 89, 93

FRACTIONS AND CALCULATORS (OPTIONAL) Many calculators have the capability to perform arithmetic on fractions and express the answer as either a decimal or a fraction. The next example illustrates this capability.

EXAMPLE 6 Performing arithmetic operations with technology

Use a calculator to evaluate each expression. Express your answer as a decimal and as a fraction.

(a) $\frac{1}{3} + \frac{2}{5} - \frac{4}{9}$ **(b)** $\left(\frac{4}{9} \cdot \frac{3}{8}\right) \div \frac{2}{3}$

Solution

(a) Figure 1.26(a) shows that

$$\frac{1}{3} + \frac{2}{5} - \frac{4}{9} = 0.2\overline{8}, \text{ or } \frac{13}{45}.$$

NOTE: Generally it is a good idea to put parentheses around fractions when you are using a calculator.

(b) Figure 1.26(b) shows that $\left(\frac{4}{9} \cdot \frac{3}{8}\right) \div \frac{2}{3} = 0.25$, or $\frac{1}{4}$.

```
(1/3)+(2/5)-(4/9
)
         .2888888889
(1/3)+(2/5)-(4/9
)▶Frac
              13/45
```

```
((4/9)*(3/8))/(2
/3)
               .25
((4/9)*(3/8))/(2
/3)▶Frac
               1/4
```

(a)   (b)

Figure 1.26

Now Try Exercise 97

MAKING CONNECTIONS

The Four Arithmetic Operations

If you know how to add real numbers, then you also know how to subtract real numbers because subtraction is defined in terms of addition. That is,

$$a - b = a + (-b).$$

If you know how to multiply real numbers, then you also know how to divide real numbers because division is defined in terms of multiplication. That is,

$$\frac{a}{b} = a \cdot \frac{1}{b}.$$

Applications

There are many instances when we may need to multiply or divide real numbers. Two examples are provided here.

EXAMPLE 7   Comparing top-grossing movies

Even though *Titanic* (1997) was an extremely successful film that set box office records, it ranks only sixth in top-grossing movies of all time when calculated by using estimated total admissions. The total admissions for *Titanic* are $\frac{10}{11}$ of the total admissions for *The Sound of Music* (1965), which ranks third.   (**Source:** *Exhibitor Relations Co., Inc.*)

(a) If *The Sound of Music* had estimated total admissions of 142 million people, find the estimated total admissions for *Titanic*.

(b) The top-grossing movie of all time is *Gone With the Wind* (1939), which had estimated total admissions of 202 million. How many more people saw *Gone With the Wind* than saw *Titanic*?

Solution

(a) To find the total admissions for *Titanic* we multiply the real numbers 142 and $\frac{10}{11}$ to obtain $142 \cdot \frac{10}{11} = \frac{1420}{11} \approx 129.09$.

Total admissions for *Titanic* were about 129 million people.

(b) The difference is $202 - 129 = 73$ million people.   Now Try Exercise 105

EXAMPLE 8   Analyzing the federal budget

In 2005, the federal government used $\frac{9}{125}$ of its budget to pay interest on loans. Write this fraction as a decimal.   (**Source:** *U.S. Office of Management and Budget.*)

Solution

One method for writing the fraction as a decimal is to divide 9 by 125 using long division. An alternative method is to multiply the fraction by $\frac{8}{8}$ so that the denominator becomes 1000. Then write the numerator in the thousandths place in the decimal.

$$\frac{9}{125} \cdot \frac{8}{8} = \frac{72}{1000} = 0.072$$

Now Try Exercise 107

1.6 PUTTING IT ALL TOGETHER

The following table summarizes some of the information presented in this section.

Topic	Comments	Examples	
Multiplication	The product of two numbers with like signs is positive, and the product of two numbers with unlike signs is negative.	$6 \cdot 7 = 42$	Like signs
		$6 \cdot (-7) = -42$	Unlike signs
		$-6 \cdot 7 = -42$	Unlike signs
		$-6 \cdot (-7) = 42$	Like signs

continued on next page

continued from previous page

Topic	Comments	Examples
Division	For real numbers a and b, with $b \neq 0$, $$\frac{a}{b} = a \cdot \frac{1}{b}.$$ The quotient of two numbers with like signs is positive, and the quotient of two numbers with unlike signs is negative.	$\dfrac{42}{6} = 7$ Like signs $\dfrac{-42}{6} = -7$ Unlike signs $\dfrac{42}{-6} = -7$ Unlike signs $\dfrac{-42}{-6} = 7$ Like signs $\dfrac{\frac{3}{2}}{-6} = \dfrac{3}{2} \cdot \left(-\dfrac{1}{6}\right)$ $\qquad = -\dfrac{3}{12}$ Unlike signs $\qquad = -\dfrac{1}{4}$
Converting Fractions to Decimals	The fraction $\frac{a}{b}$ is equivalent to $a \div b$.	The fraction $\frac{2}{9}$ is equivalent to $0.222\ldots$, or $0.\overline{2}$, because the division problem $$\begin{array}{r} 0.222\ldots \\ 9\overline{)2.00000} \\ \underline{18} \\ 20 \\ \underline{18} \\ 20 \\ \underline{18} \\ 2 \end{array}$$ results in a repeating decimal.
Converting Terminating Decimals to Fractions	Write the decimal as a fraction with a denominator equal to a power of 10 and then simplify this fraction.	$0.55 = \dfrac{55}{100}$ $\qquad = \dfrac{11 \cdot 5}{20 \cdot 5}$ $\qquad = \dfrac{11}{20}$

1.6 Exercises

CONCEPTS

1. In a multiplication problem, the numbers multiplied are called _____.

2. The solution to a multiplication problem is the _____.

3. The product of a positive number and a negative number is a _____ number.

4. The product of two negative numbers is _____.

5. The solution to a division problem is the _____.

6. In the problem $15 \div 5$, the number 15 is the _____ and the number 5 is the _____.

7. The reciprocal of a nonzero number a is _____.

8. Division by zero is undefined because zero has no _____.

9. The reciprocal of $-\frac{3}{4}$ is _____.

10. To divide a by b, multiply a by the _____ of b.

11. In general, $-1 \cdot a =$ _____.

12. $(-5)^2 =$ _____

13. $-5^2 =$ _____

14. $\frac{a}{b} = a \cdot$ _____

15. A negative number divided by a negative number is a _____ number.

16. A negative number divided by a positive number is a _____ number.

17. Division is a fast way to perform _____.

18. To convert $\frac{5}{8}$ to a decimal, divide _____ by _____.

MULTIPLICATION AND DIVISION OF REAL NUMBERS

Exercises 19–40: Multiply.

19. $-3 \cdot 4$

20. $-5 \cdot 7$

21. $6 \cdot (-3)$

22. $2 \cdot (-1)$

23. $0 \cdot (-2.13)$

24. $-2 \cdot (-7)$

25. $-6 \cdot (-10)$

26. $-3 \cdot (-1.7) \cdot 0$

27. $-\frac{1}{2} \cdot \left(-\frac{2}{4}\right)$

28. $-\frac{3}{4} \cdot \left(-\frac{5}{12}\right)$

29. $-\frac{3}{7} \cdot \frac{7}{3}$

30. $\frac{5}{8} \cdot \left(-\frac{4}{15}\right)$

31. $-10 \cdot (-20)$

32. $1000 \cdot (-70)$

33. $-50 \cdot 100$

34. $-0.5 \cdot (-0.3)$

35. $-2 \cdot 3 \cdot (-4) \cdot 5$

36. $-3 \cdot (-5) \cdot (-2) \cdot 10$

37. $-6 \cdot \frac{1}{6} \cdot \frac{7}{9} \cdot \left(-\frac{9}{7}\right) \cdot \left(-\frac{3}{2}\right)$

38. $-\frac{8}{5} \cdot \frac{1}{8} \cdot \left(-\frac{5}{7}\right) \cdot -7$

39. $(-1) \cdot (-1) \cdot (-1) \cdot (-1)$

40. $5 \cdot (-2) \cdot (-2) \cdot (-2)$

41. **Thinking Generally** Is the product given by the expression $a \cdot (-a) \cdot (-a) \cdot a \cdot (-a)$ positive or negative if $a > 0$?

42. **Thinking Generally** Is the product given by the expression $a \cdot (-a) \cdot (-a) \cdot a \cdot (-a)$ positive or negative if $a < 0$?

Exercises 43–50: Evaluate the expression.

43. $(-1)^3$

44. $(-6)^2$

45. -2^4

46. $-(-4)^2$

47. $-(-2)^3$

48. $3 \cdot (-3)^2$

49. $5 \cdot (-2)^3$

50. -1^4

Exercises 51–76: Divide.

51. $-10 \div 5$

52. $-8 \div 4$

53. $-20 \div (-2)$

54. $-15 \div (-3)$

55. $\frac{-12}{3}$

56. $\frac{-25}{-5}$

57. $\frac{39}{-13}$

58. $-\frac{100}{-20}$

59. $-16 \div \frac{1}{2}$

60. $10 \div \left(-\frac{1}{3}\right)$

61. $0 \div 3$

62. $\frac{0}{-5}$

63. $\frac{0}{-2}$

64. $\frac{-1}{0}$

65. $\frac{1}{2} \div (-11)$

66. $-\frac{3}{4} \div (-6)$

67. $\frac{-\frac{4}{5}}{-3}$

68. $\frac{\frac{7}{8}}{-7}$

69. $\frac{5}{6} \div \left(-\frac{8}{9}\right)$

70. $-\frac{11}{12} \div \left(-\frac{11}{4}\right)$

71. $-\frac{1}{2} \div 0$

72. $-9 \div 0$

73. $-0.5 \div \frac{1}{2}$

74. $-0.25 \div \left(-\frac{3}{4}\right)$

75. $-\frac{2}{3} \div 0.5$

76. $\frac{1}{6} \div 1.5$

CONVERTING BETWEEN FRACTIONS AND DECIMALS

Exercises 77–88: Write the number as a decimal.

77. $\frac{1}{2}$

78. $\frac{3}{4}$

79. $\frac{3}{16}$

80. $\frac{1}{9}$

81. $3\frac{1}{2}$

82. $2\frac{1}{4}$

83. $5\frac{2}{3}$

84. $6\frac{7}{9}$

85. $1\frac{7}{16}$

86. $\frac{11}{16}$

87. $\frac{7}{8}$

88. $6\frac{1}{12}$

Exercises 89–96: Write the decimal number as a fraction in lowest terms.

89. 0.25

90. 0.8

91. 0.16

92. 0.35

93. 0.625

94. 0.0125

95. 0.6875

96. 0.21875

 Exercises 97–104: Use a calculator to evaluate each expression. Express your answer as a decimal and as a fraction.

97. $\left(\frac{1}{3} + \frac{5}{6}\right) \div \frac{1}{2}$

98. $\frac{4}{9} - \frac{1}{6} + \frac{2}{3}$

99. $\frac{4}{5} \div \frac{2}{3} \cdot \frac{7}{4}$

100. $4 - \frac{7}{4} \cdot 2$

101. $\frac{15}{2} - 4 \cdot \frac{7}{3}$

102. $\frac{1}{6} - \frac{3}{5} + \frac{7}{8}$

103. $\frac{17}{40} + 3 \div 8$

104. $\frac{3}{4} \cdot \left(6 + \frac{1}{2}\right)$

APPLICATIONS

105. *Top-Grossing Movies* (Refer to Example 7.) *The Ten Commandments* (1956) is the fifth top-grossing movie of all time. Find the total admissions for *The Ten Commandments* if they were $\frac{13}{20}$ of the total admissions for the top-grossing movie of all time, *Gone With the Wind* (1939), which had total admissions of 202 million. (*Source: Exhibitor Relations Co., Inc.*)

106. *Planet Climate* Saturn has an average surface temperature of $-220°F$. Neptune has an average surface temperature that is $\frac{3}{2}$ times that of Saturn. Find the average surface temperature on Neptune. (*Source: NASA.*)

107. *Uninsured Americans* In 2005, the fraction of Americans who did not have health insurance coverage was $\frac{39}{250}$. Write this fraction as a decimal. (*Source: U.S. Census Bureau.*)

108. *Uninsured Minnesotans* In 2005, the fraction of Minnesotans who did not have health insurance coverage was $\frac{21}{250}$, the lowest in the country. Write this fraction as a decimal. (*Source: U.S. Census Bureau.*)

WRITING ABOUT MATHEMATICS

109. Division is a fast way to subtract. Consider the division problem $\frac{-6}{-2}$, whose quotient represents the number of -2s in -6. Using this idea, explain why the answer is a positive number.

110. Explain how to determine whether the product of three signed numbers is positive or negative.

CHECKING BASIC CONCEPTS
SECTIONS 1.5 AND 1.6

1. Find each sum.
   (a) $-4 + 4$ **(b)** $-10 + (-12) + 3$

2. Evaluate each expression.
   (a) $\frac{2}{3} - \left(-\frac{2}{9}\right)$ **(b)** $-1.2 - 5.1 + 3.1$

3. Write an arithmetic expression for the given phrase and then simplify it.
   (a) The sum of negative one and five
   (b) The difference between four and negative three

4. The hottest temperature ever recorded at International Falls, Minnesota, was 98°F, and the coldest temperature ever recorded was −46°F. What is the difference between these two temperatures?

5. Find each product.
(a) $-5 \cdot (-7)$ (b) $-\frac{1}{2} \cdot \frac{2}{3} \cdot \left(-\frac{4}{5}\right)$

6. Evaluate each expression.
(a) -3^2 (b) $4 \cdot (-2)^3$ (c) $(-5)^2$

7. Evaluate each expression.
(a) $-5 \div \frac{2}{3}$ (b) $-\frac{5}{8} \div \left(-\frac{4}{3}\right)$

8. What is the reciprocal of $-\frac{7}{6}$?

9. Simplify each expression.
(a) $\frac{-10}{2}$ (b) $\frac{10}{-2}$ (c) $-\frac{10}{2}$ (d) $\frac{-10}{-2}$

10. Convert each fraction or mixed number to a decimal number.
(a) $\frac{3}{5}$ (b) $3\frac{7}{8}$

1.7 PROPERTIES OF REAL NUMBERS

Commutative Properties ▪ Associative Properties ▪ Distributive Properties ▪ Identity and Inverse Properties ▪ Mental Calculations

A LOOK INTO MATH ▷ The order in which you perform actions is often important. For example, putting on your socks and then your shoes is not the same as putting on your shoes and then your socks. In mathematical terms, these two actions are not *commutative*. However, the actions of tying your shoes and putting on a sweatshirt probably are *commutative*. In mathematics some operations are commutative and others are not. In this section we discuss several properties of real numbers.

Commutative Properties

The **commutative property for addition** states that two numbers, a and b, can be added in any order and the result will be the same. That is, $a + b = b + a$. For example, if a person buys 4 CDs and then buys 9 CDs or first buys 9 CDs and then buys 4 CDs, the result is the same. Either way the person buys a total of

$$4 + 9 = 9 + 4 = 13 \text{ CDs.}$$

There is also a **commutative property for multiplication**. It states that two numbers, a and b, can be multiplied in any order and the result will be the same. That is, $a \cdot b = b \cdot a$. For example, if one person has 6 boxes each containing 12 greeting cards and another person has 12 boxes each containing 6 greeting cards, then each person has

$$6 \cdot 12 = 12 \cdot 6 = 72 \text{ greeting cards.}$$

We can summarize these results as follows.

COMMUTATIVE PROPERTIES

For any real numbers a and b,

$$a + b = b + a \qquad \text{Addition}$$

and

$$a \cdot b = b \cdot a. \qquad \text{Multiplication}$$

EXAMPLE **1**   Applying the commutative properties

Use the commutative properties to rewrite each expression.
(a) $15 + 100$ **(b)** $a \cdot 8$ **(c)** $x(y + z)$

Solution
(a) By the commutative property for addition $15 + 100$ can be written as $100 + 15$.
(b) By the commutative property for multiplication $a \cdot 8$ can be written as $8 \cdot a$ or $8a$.
(c) We can apply both commutative properties.

$$x(y + z) = (y + z)x \quad \text{Commutative property for multiplication}$$
$$= (z + y)x \quad \text{Commutative property for addition}$$

Now Try Exercises 15, 21

While there are commutative properties for addition and multiplication, the operations of subtraction and division are *not* commutative. Changing the order in which numbers are subtracted can affect the result. For example, $5 - 3 = 2$ but $3 - 5 = -2$. Similarly, changing the order in which numbers are divided can affect the result. For example, $4 \div 2 = 2$ but $2 \div 4 = \frac{1}{2}$.

Associative Properties

The commutative properties allow us to interchange the order of two numbers when we add or multiply. The *associative properties* allow us to change how numbers are grouped. For example, if a person earns $5, $4, and $6, then the total amount earned can be calculated either as

$$(5 + 4) + 6 = 9 + 6 = 15 \quad \text{or as}$$
$$5 + (4 + 6) = 5 + 10 = 15.$$

In either case we obtain the same answer, $15, which is the result of the **associative property for addition**. We did not change the order of the numbers; we only changed how the numbers were grouped. There is also an **associative property for multiplication**, which is illustrated by

$$(5 \cdot 4) \cdot 6 = 20 \cdot 6 = 120 \quad \text{and}$$
$$5 \cdot (4 \cdot 6) = 5 \cdot 24 = 120.$$

We can summarize these results as follows.

ASSOCIATIVE PROPERTIES

For any real numbers a, b, and c,
$$(a + b) + c = a + (b + c) \quad \text{Addition}$$
and
$$(a \cdot b) \cdot c = a \cdot (b \cdot c). \quad \text{Multiplication}$$

NOTE: Sometimes we omit the multiplication dot. Thus $a \cdot b = ab$ and $5 \cdot x \cdot y = 5xy$.

EXAMPLE 2   Applying the associative properties

Use the associative property to rewrite each expression.
(a) $(5 + 6) + 7$ **(b)** $x(yz)$

Solution
(a) The given expression is equivalent to $5 + (6 + 7)$.
(b) The given expression is equivalent to $(xy)z$.

Now Try Exercises 23, 25

EXAMPLE 3   Identifying properties of real numbers

State the property that each equation illustrates.
(a) $5 \cdot (8y) = (5 \cdot 8)y$ **(b)** $3 \cdot 7 = 7 \cdot 3$ **(c)** $x + yz = yz + x$

Solution
(a) This equation illustrates the associative property for multiplication because the grouping of the numbers has been changed.
(b) This equation illustrates the commutative property for multiplication because the order of the numbers 3 and 7 has been changed.
(c) This equation illustrates the commutative property for addition because the order of the terms x and yz has been changed.

Now Try Exercises 55, 57

While there are associative properties for addition and multiplication, the operations of subtraction and division are *not* associative. Changing the grouping in a subtraction problem can affect the result. For example, $(5 - 3) - 2 = 2 - 2 = 0$ but $5 - (3 - 2) = 5 - 1 = 4$. Similarly, changing the grouping of the numbers in a division problem can affect the result. For example, $(24 \div 4) \div 2 = 6 \div 2 = 3$ but $24 \div (4 \div 2) = 24 \div 2 = 12$.

MAKING CONNECTIONS

Commutative and Associative Properties

Both the commutative and associative properties work for addition and multiplication. However, neither property works for subtraction or division.

Distributive Properties

The distributive properties are used frequently in algebra to simplify expressions. An example is

$$4(2 + 3) = 4 \cdot 2 + 4 \cdot 3.$$

The 4 must be multiplied by *both* the 2 and the 3—not just the 2.

The distributive property remains valid when addition is replaced with subtraction. For example,

$$4(2 - 3) = 4 \cdot 2 - 4 \cdot 3$$

is a true statement.

We illustrate a distributive property geometrically in Figure 1.27. Note that the area of one rectangle that is 4 squares by 5 squares is the same as the area of two rectangles: one that is 4 squares by 2 squares and another that is 4 squares by 3 squares. In either case the total area is 20 square units.

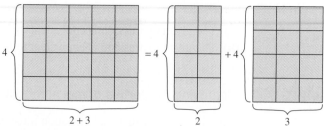

Figure 1.27 $4(2 + 3) = 4 \cdot 2 + 4 \cdot 3$

DISTRIBUTIVE PROPERTIES

For any real numbers a, b, and c,

$$a(b + c) = ab + ac \quad \text{and} \quad a(b - c) = ab - ac.$$

NOTE: Because multiplication is commutative, the distributive properties can also be written as

$$(b + c)a = ba + ca \quad \text{and} \quad (b - c)a = ba - ca.$$

EXAMPLE 4 Applying the distributive properties

Apply a distributive property to each expression.
(a) $3(x + 2)$ **(b)** $-6(a - 2)$ **(c)** $-(x + 7)$ **(d)** $15 - (b + 4)$

Solution
(a) Both the x and the 2 must be multiplied by 3.

$$3(x + 2) = 3 \cdot x + 3 \cdot 2 = 3x + 6$$

(b) $-6(a - 2) = -6 \cdot a - (-6) \cdot 2 = -6a + 12$

(c) $-(x + 7) = (-1)(x + 7)$ In general, $-a = -1 \cdot a$.

$\quad\quad\quad\quad = (-1) \cdot x + (-1) \cdot 7$ Distributive property

$\quad\quad\quad\quad = -x - 7$ Multiply.

(d) $15 - (b + 4) = 15 + (-1)(b + 4)$ Change subtraction to addition.

$\quad\quad\quad\quad\quad = 15 + (-1) \cdot b + (-1) \cdot 4$ Distributive property

$\quad\quad\quad\quad\quad = 15 - b - 4$ Multiply.

$\quad\quad\quad\quad\quad = 11 - b$ Subtract.

NOTE: To simplify the expression $15 - (b + 4)$, we subtract *both* the b and the 4. Thus we can quickly simplify the given expression to

$$15 - (b + 4) = 15 - b - 4.$$

Now Try Exercises 33, 39, 43

EXAMPLE 5 Inserting parentheses using the distributive property

Use the distributive property to insert parentheses in the expression and then simplify the result.
(a) $5a + 2a$ **(b)** $3x - 7x$ **(c)** $-4y + 5y$

Solution
(a) Because a is a factor in both $5a$ and $2a$, we use the distributive property to write the a outside of parentheses.

$$5a + 2a = (5 + 2)a \qquad \text{Distributive property}$$
$$= 7a \qquad \text{Simplify.}$$

(b) $3x - 7x = (3 - 7)x = -4x$
(c) $-4y + 5y = (-4 + 5)y = 1y = y$

Now Try Exercises 47, 49

EXAMPLE 6 Identifying properties of real numbers

State the property or properties illustrated by each equation.
(a) $4(5 - x) = 20 - 4x$ **(b)** $(4 + x) + 5 = x + 9$
(c) $5z + 7z = 12z$ **(d)** $x(y + z) = zx + yx$

Solution
(a) This equation illustrates the distributive property with subtraction.

$$4(5 - x) = 4 \cdot 5 - 4 \cdot x = 20 - 4x$$

(b) This equation illustrates the commutative and associative properties for addition.

$$(4 + x) + 5 = (x + 4) + 5 \qquad \text{Commutative property for addition}$$
$$= x + (4 + 5) \qquad \text{Associative property for addition}$$
$$= x + 9 \qquad \text{Simplify.}$$

(c) This equation illustrates the distributive property with addition.

$$5z + 7z = (5 + 7)z = 12z$$

(d) This equation illustrates a distributive property and commutative properties.

$$x(y + z) = xy + xz \qquad \text{Distributive property}$$
$$= xz + xy \qquad \text{Commutative property for addition}$$
$$= zx + yx \qquad \text{Commutative property for multiplication}$$

Now Try Exercises 59, 61

Identity and Inverse Properties

The **identity property of 0** states that if 0 is added to any real number a, the result is a. The number 0 is called the **additive identity**. Examples include

$$-4 + 0 = -4 \quad \text{and} \quad 0 + 11 = 11.$$

The **identity property of 1** states that if any number a is multiplied by 1, the result is a. The number 1 is called the **multiplicative identity**. Examples include

$$-3 \cdot 1 = -3 \quad \text{and} \quad 1 \cdot 8 = 8.$$

We can summarize these results as follows.

IDENTITY PROPERTIES

For any real number a,

$$a + 0 = 0 + a = a \qquad \text{Additive identity}$$

and

$$a \cdot 1 = 1 \cdot a = a. \qquad \text{Multiplicative identity}$$

The *additive inverse*, or *opposite*, of a number a is $-a$. The number 0 is its own opposite. The opposite of -5 is $-(-5) = 5$ and the opposite of x is $-x$. The sum of a number a and its additive inverse equals the additive identity 0. Thus $-5 + 5 = 0$ and $x + (-x) = 0$.

The *multiplicative inverse*, or *reciprocal*, of a nonzero number a is $\frac{1}{a}$. The number 0 has no multiplicative inverse. The multiplicative inverse of $-\frac{5}{4}$ is $-\frac{4}{5}$. The product of a number and its multiplicative inverse equals the multiplicative identity 1. Thus $-\frac{5}{4} \cdot \left(-\frac{4}{5}\right) = 1$.

INVERSE PROPERTIES

For any real number a,

$$a + (-a) = 0 \quad \text{and} \quad -a + a = 0. \qquad \text{Additive inverse}$$

For any *nonzero* real number a,

$$a \cdot \frac{1}{a} = 1 \quad \text{and} \quad \frac{1}{a} \cdot a = 1. \qquad \text{Multiplicative inverse}$$

EXAMPLE 7   Identifying identity and inverse properties

State the property or properties illustrated by each equation.

(a) $0 + xy = xy$ **(b)** $\frac{36}{30} = \frac{6}{5} \cdot \frac{6}{6} = \frac{6}{5}$

(c) $x + (-x) + 5 = 0 + 5 = 5$ **(d)** $\frac{1}{9} \cdot 9y = 1 \cdot y = y$

Solution
(a) This equation illustrates use of the identity property for 0.
(b) Because $\frac{6}{6} = 1$, these equations illustrate how a fraction can be simplified by using the identity property for 1.
(c) These equations illustrate use of the additive inverse property and the identity property for 0.
(d) These equations illustrate use of the multiplicative inverse property and the identity property of 1.

Now Try Exercises 69, 75

Mental Calculations

Properties of numbers can be used to simplify calculations. For example, to find the sum

$$4 + 7 + 6 + 3$$

we might apply the commutative and associative properties for addition to obtain

$$(4 + 6) + (7 + 3) = 10 + 10 = 20.$$

Similarly, we might calculate

$$14 + 7 + 16 + 33 \quad \text{as}$$

$$(14 + 16) + (7 + 33) = 30 + 40 = 70.$$

Suppose that we are to add $128 + 19$ mentally. One way is to add 20 to 128 and then subtract 1.

$$
\begin{aligned}
128 + 19 &= 128 + (20 - 1) && 19 = 20 - 1 \\
&= (128 + 20) - 1 && \text{Associative property} \\
&= 148 - 1 && \text{Add.} \\
&= 147. && \text{Subtract.}
\end{aligned}
$$

Similarly, we could find the sum $128 + 98$ by first adding 100 to 128 to obtain 228 and then subtracting 2, resulting in 226.

The distributive property can be helpful when we are multiplying mentally. For example, to estimate the number of people in a marching band with 7 columns and 23 rows we need to find the product $7 \cdot 23$. To evaluate the product mentally, think of 23 as $20 + 3$.

$$
\begin{aligned}
7(20 + 3) &= 7 \cdot 20 + 7 \cdot 3 && \text{Distributive property} \\
&= 140 + 21 && \text{Multiply.} \\
&= 161 && \text{Add.}
\end{aligned}
$$

EXAMPLE 8 **Performing calculations mentally**

Use properties of real numbers to calculate each expression mentally.
(a) $21 + 15 + 9 + 5$ **(b)** $\frac{1}{2} \cdot \frac{2}{3} \cdot 2 \cdot \frac{3}{2}$
(c) $523 + 199$ **(d)** $6 \cdot 55$

Solution
(a) Use the commutative and associative properties to group numbers that sum to a multiple of 10.

$$21 + 15 + 9 + 5 = (21 + 9) + (15 + 5) = 30 + 20 = 50$$

(b) Use the commutative and associative properties to group numbers with their reciprocals.

$$\frac{1}{2} \cdot \frac{2}{3} \cdot 2 \cdot \frac{3}{2} = \left(\frac{1}{2} \cdot 2\right) \cdot \left(\frac{2}{3} \cdot \frac{3}{2}\right) = 1 \cdot 1 = 1$$

(c) Instead of adding 199, add 200 and then subtract 1.

$$523 + 200 - 1 = 723 - 1 = 722$$

(d) Think of 55 as $50 + 5$ and then apply the distributive property.

$$6 \cdot (50 + 5) = 300 + 30 = 330$$

Now Try Exercises 79, 91, 95

CRITICAL THINKING
How could you quickly calculate $5283 - 198$ without a calculator?

The next example illustrates how the commutative and associative properties for multiplication can be used together to simplify a product.

EXAMPLE 9   Finding the volume of a swimming pool

An Olympic swimming pool is 50 meters long, 25 meters wide, and 2 meters deep. The volume V of the pool is found by multiplying the numbers 50, 25, and 2. Use the commutative and associative properties for multiplication to calculate the volume of the pool mentally.

Solution
Because $50 \times 2 = 100$ and multiplication by 100 is relatively easy, it may be convenient to order and group the multiplication as $(50 \times 2) \times 25 = 100 \times 25 = 2500$. Thus the pool contains 2500 cubic meters of water.   Now Try Exercise 107

1.7 PUTTING IT ALL TOGETHER

The following table summarizes some properties of real numbers.

Property	Definition	Examples
Commutative	For any real numbers a and b, $$a + b = b + a \quad \text{and}$$ $$a \cdot b = b \cdot a.$$	$4 + 6 = 6 + 4$ $4 \cdot 6 = 6 \cdot 4$
Associative	For any real numbers a, b, and c, $$(a + b) + c = a + (b + c) \quad \text{and}$$ $$(a \cdot b) \cdot c = a \cdot (b \cdot c).$$	$(3 + 4) + 5 = 3 + (4 + 5)$ $(3 \cdot 4) \cdot 5 = 3 \cdot (4 \cdot 5)$
Distributive	For any real numbers a, b, and c, $$a(b + c) = ab + ac \quad \text{and}$$ $$a(b - c) = ab - ac.$$	$5(x + 2) = 5x + 10$ $5(x - 2) = 5x - 10$
Identity (0 and 1)	The identity for addition is 0, and the identity for multiplication is 1. For any real number a, $a + 0 = a$ and $a \cdot 1 = a$.	$5 + 0 = 5 \quad \text{and} \quad 5 \cdot 1 = 5$
Inverse	The additive inverse of a is $-a$, and $a + (-a) = 0$. The multiplicative inverse of a nonzero number a is $\frac{1}{a}$, and $a \cdot \frac{1}{a} = 1$.	$8 + (-8) = 0 \quad \text{and} \quad \frac{2}{3} \cdot \frac{3}{2} = 1$

1.7 Exercises

CONCEPTS

1. $a + b = b + a$ illustrates the _____ property for _____.

2. $a \cdot b = b \cdot a$ illustrates the _____ property for _____.

3. $(a + b) + c = a + (b + c)$ illustrates the _____ property for _____.

4. $(a \cdot b) \cdot c = a \cdot (b \cdot c)$ illustrates the _____ property for _____.

5. Addition and multiplication are commutative but _____ and _____ are not.

6. Addition and multiplication are associative but _____ and _____ are not.

7. $a(b + c) = ab + ac$ illustrates the _____ property.

8. $a(b - c) = ab - ac$ illustrates the _____ property.

9. $3b + 2b = (3 + 2) \cdot$ _____.

10. $4a - 3a = (4 - 3) \cdot$ _____.

11. $a + 0 = 0 + a = a$ illustrates the _____ property for _____.

12. $a \cdot 1 = 1 \cdot a = a$ illustrates the _____ property for _____.

13. The additive inverse of a is _____.

14. The multiplicative inverse, or reciprocal, of a nonzero number a is _____.

PROPERTIES OF REAL NUMBERS

Exercises 15–22: Use a commutative property to rewrite the expression. Do not simplify.

15. $-6 + 10$

16. $23 + 7$

17. $-5 \cdot 6$

18. $25 \cdot (-46)$

19. $a + 10$

20. $b + c$

21. $b \cdot 7$

22. $a \cdot 23$

Exercises 23–30: Use an associative property to rewrite the expression by changing the parentheses. Do not simplify.

23. $(1 + 2) + 3$

24. $-7 + (5 + 15)$

25. $2 \cdot (3 \cdot 4)$

26. $(9 \cdot (-4)) \cdot 5$

27. $(a + 5) + c$

28. $(10 + b) + a$

29. $(x \cdot 3) \cdot 4$

30. $5 \cdot (x \cdot y)$

31. Thinking Generally Use the commutative and associative properties to show that $a + b + c = c + b + a$.

32. Thinking Generally Use the commutative and associative properties to show that $a \cdot b \cdot c = c \cdot b \cdot a$.

Exercises 33–44: Use a distributive property to rewrite the expression. Then simplify the expression.

33. $4(3 + 2)$

34. $5(6 - 9)$

35. $a(b - 8)$

36. $3(x + y)$

37. $-1(t + z)$

38. $-1(a + 6)$

39. $-(5 - a)$

40. $12 - (4u - b)$

41. $(a + 5)3$

42. $(x + y)7$

43. $(6 - z)(-3)$

44. $4x - 2(3y - 5)$

45. Thinking Generally Use properties of real numbers to show that the distributive property can be extended as follows.
$$a \cdot (b + c + d) = ab + ac + ad.$$

46. Thinking Generally Use properties of real numbers to show that the distributive property can be extended as follows.
$$a \cdot (b - c - d) = ab - ac - ad.$$

Exercises 47–54: Use the distributive property to insert parentheses in the expression and then simplify the result.

47. $6x + 5x$

48. $4y - y$

49. $-4b + 3b$

50. $2b + 8b$

51. $3a - a$

52. $-2x + 5x$

53. $13w - 27w$

54. $25a - 21a$

Exercises 55–68: State the property or properties that the equation illustrates.

55. $x \cdot 5 = 5x$

56. $7 + a = a + 7$

57. $(a + 5) + 7 = a + 12$

58. $(9 + a) + 8 = a + 17$

59. $4(5 + x) = 20 + 4x$

60. $3(5 + x) = 3x + 15$

61. $x(3 - y) = 3x - xy$

62. $-(u - v) = -u + v$

63. $6x + 9x = 15x$

64. $9x - 11x = -2x$

65. $3 \cdot (4 \cdot a) = 12a$

66. $(x \cdot 3) \cdot 5 = 15x$

67. $-(t - 7) = -t + 7$

68. $z \cdot 5 - y \cdot 6 = 5z - 6y$

IDENTITY AND INVERSE PROPERTIES

Exercises 69–78: State the property or properties that are illustrated.

69. $0 + x = x$

70. $5x + 0 = 5x$

71. $1 \cdot a = a$

72. $\frac{1}{7} \cdot (7 \cdot a) = a$

73. $\frac{25}{15} = \frac{5}{3} \cdot \frac{5}{5} = \frac{5}{3}$

74. $\frac{50}{40} = \frac{5}{4} \cdot \frac{10}{10} = \frac{5}{4}$

75. $\frac{1}{xy} \cdot xy = 1$

76. $\frac{1}{a + b} \cdot (a + b) = 1$

77. $-xyz + xyz = 0$

78. $\frac{1}{y} + \left(-\frac{1}{y}\right) = 0$

MENTAL CALCULATIONS

Exercises 79–96: Use properties of real numbers to calculate the expression mentally.

79. $4 + 2 + 9 + 8 + 1 + 6$

80. $21 + 32 + 19 + 8$

81. $45 + 43 + 5 + 7$

82. $5 + 7 + 12 + 13 + 8$

83. $129 + 49$

84. $87 + 99$

85. $379 + 98$

86. $4570 + 998$

87. $178 - 99$

88. $500 - 101$

89. $6 \cdot 15$

90. $4 \cdot 56$

91. $8 \cdot 102$

92. $5 \cdot 999$

93. $\frac{1}{2} \cdot \frac{1}{2} \cdot \frac{1}{2} \cdot 2 \cdot 2 \cdot 2$

94. $\frac{1}{2} \cdot \frac{4}{5} \cdot \frac{7}{3} \cdot 2 \cdot \frac{5}{4}$

95. $\frac{7}{6} \cdot \frac{1}{2} \cdot \frac{1}{2} \cdot \frac{1}{2} \cdot \frac{8}{7}$

96. $\frac{4}{11} \cdot \frac{11}{6} \cdot \frac{6}{7} \cdot \frac{7}{4}$

MULTIPLYING AND DIVIDING BY POWERS OF 10 MENTALLY

97. *Multiplying by 10* Multiplying by 10 is easy in the decimal system. To multiply an integer by 10, append one 0 to the number. For example, $10 \times 23 = 230$. Simplify each expression mentally.
(a) 10×41 **(b)** 10×997
(c) -630×10 **(d)** $-14,000 \times 10$

98. *Multiplying by 10* To multiply a decimal number by 10, move the decimal point one place to the right. For example, $10 \times 23.45 = 234.5$. Simplify each expression mentally.
(a) 10×101.68 **(b)** $10 \times (-1.235)$
(c) -113.4×10 **(d)** 0.567×10
(e) 10×0.0045 **(f)** -0.05×10

99. *Multiplying by Powers of 10* To multiply an integer by a power of 10 in the form

$$10^k = \underbrace{100 \ldots 0}_{k \text{ zeros}},$$

append k zeros to the number. Some examples of this are $100 \times 45 = 4500$, $1000 \times 235 = 235,000$, and $10,000 \times 12 = 120,000$. Simplify each expression mentally.
(a) 1000×19 **(b)** $100 \times (-451)$
(c) $10,000 \times 6$ **(d)** $-79 \times 100,000$

100. *Multiplying by Powers of 10* To multiply a decimal number by a power of 10 in the form

$$10^k = \underbrace{100 \ldots 0}_{k \text{ zeros}},$$

move the decimal point k places to the right. For example, $100 \times 1.234 = 123.4$. Simplify each given expression mentally.
(a) 1000×1.2345 **(b)** $100 \times (-5.1)$
(c) 45.67×1000 **(d)** $0.567 \times 10,000$
(e) 100×0.0005 **(f)** $-0.05 \times 100,000$

101. *Dividing by 10* To divide a number by 10, move the decimal point one place to the left. For example, $78.9 \div 10 = 7.89$. Simplify each expression mentally.

(a) $12.56 \div 10$ (b) $9.6 \div 10$
(c) $0.987 \div 10$ (d) $-0.056 \div 10$
(e) $1200 \div 10$ (f) $4578 \div 10$

102. *Dividing by Powers of 10* To divide a decimal number by a power of 10 in the form

$$10^k = 1\underbrace{00 \ldots 0},$$
$$\quad\; k \text{ zeros}$$

move the decimal point k places to the left. For example, $123.4 \div 100 = 1.234$. Simplify each expression mentally.
(a) $78.89 \div 100$ (b) $0.05 \div 1000$
(c) $5678 \div 10{,}000$ (d) $-9.8 \div 1000$
(e) $-101 \div 100{,}000$ (f) $7.8 \div 100$

APPLICATIONS

103. *Earnings* Earning $100 one day and $75 the next day is equivalent to earning $75 the first day and $100 the second day. What property of real numbers does this example illustrate?

104. *Leasing a Car* An advertisement for a lease on a new car states that it costs $2480 down and $201 per month for 20 months. Mentally calculate the cost of the lease. Explain your reasoning.

105. *Gasoline Mileage* A car travels 198 miles on 10 gallons of gasoline. Mentally calculate the number of miles that the car travels on 1 gallon of gasoline.

106. *Gallons of Water* A wading pool is 50 feet long, 20 feet wide, and 1 foot deep.
(a) Mentally determine the number of cubic feet in the pool. (*Hint:* Volume equals length times width times height.)
(b) One cubic foot equals about 7.5 gallons. Mentally calculate the number of gallons of water in the pool.

107. *Dimensions of a Pool* A small pool of water is 13 feet long, 5 feet wide, and 2 feet deep. The volume V of the pool in cubic feet is found by multiplying 13 by 5 by 2.
(a) If you did this calculation mentally would you multiply $(13 \times 5) \times 2$ or $13 \times (5 \times 2)$? Why?
(b) What property allows you to do either calculation and still obtain the correct answer?

108. *Digital Images of Io* Satellites take digital pictures and transmit them back to Earth. The accompanying picture of Jupiter's moon Io is a digital picture, created by using a rectangular pattern of small pixels. This image is 500 pixels wide and 400 pixels high, so the total number of pixels in it is $500 \times 400 = 200{,}000$ pixels. (*Source:* NASA.)

(a) Find the total number of pixels in an image 400 pixels wide and 500 pixels high.
(b) Suppose that a picture is x pixels wide and y pixels high. What property states that it has the same number of pixels as a picture y pixels wide and x pixels high?

WRITING ABOUT MATHEMATICS

109. To determine the cost of tuition, a student tries to compute $16 \times \$96$ with a calculator and gets $15{,}360$. What would you tell this person?

110. A student performs the following computation by hand.

$$20 - 6 - 2 + 8 \div 4 \div 2 \stackrel{?}{=} 20 - 4 + 8 \div 2$$
$$\stackrel{?}{=} 20 - 4 + 4$$
$$\stackrel{?}{=} 20$$

Find any incorrect steps and explain what is wrong. What is the correct answer?

111. The computation $3 + 10^{20} - 10^{20}$ performed on a calculator gives a result of 0. (Try it.) What is the correct answer? Why is it important to know properties of real numbers even though you have a calculator?

112. Does the "distributive property"

$$a + (b \cdot c) \stackrel{?}{=} (a + b) \cdot (a + c)$$

hold for all numbers a, b, and c? Explain your reasoning.

Directions: Form a group of 2 to 4 people. Select someone to record the group's responses for this activity. All members of the group should work cooperatively to answer the questions. If your instructor asks for your results, each member of the group should be prepared to respond.

Winning the Lottery In the multistate lottery game *Powerball,* there are 120,526,770 possible number combinations, only one of which is the grand prize winner. The cost of a single ticket (one number combination) is $1. (**Source:** Powerball.com.)

Suppose that a *very* wealthy person decides to buy tickets for every possible number combination to be assured of winning a $150 million grand prize.

(a) If this individual could purchase one ticket every second, how many hours would it take to buy all of the tickets? How many years is this?

(b) If there were a way for this individual to buy all possible number combinations quickly, discuss reasons why this strategy would probably lose money.

1.8 SIMPLIFYING AND WRITING ALGEBRAIC EXPRESSIONS

Terms ▪ Combining Like Terms ▪ Simplifying Expressions ▪ Writing Expressions

A LOOK INTO MATH ▷ In arithmetic it is common to simplify expressions such as

$$\frac{1}{2} + \frac{1}{4} + \frac{1}{4}.$$

This expression equals 1, and it is easier to perform calculations with the number 1 than with these fractions. However, simplifying expressions requires mathematical skills. In this section we introduce some of these basic skills.

Terms

One way to simplify expressions is to combine like terms. A **term** is a number, a variable, or a *product* of numbers and variables raised to powers. Examples of terms include

$$4, \quad z, \quad 5x, \quad \frac{2}{5}z, \quad -4xy, \quad -x^2, \quad \text{and} \quad 6x^3y^4.$$

Terms do not contain addition or subtraction signs, but they can contain negative signs. The **coefficient** of a term is the number that appears in the term. If no number appears, the coefficient is either -1 or 1. The coefficient of $\frac{1}{2}xy$ is $\frac{1}{2}$, the coefficient of -15 is -15, and the coefficient of $-y^2$ is -1.

EXAMPLE 1 Identifying terms

Determine whether each expression is a term. If it is a term, identify its coefficient.
(a) 51   **(b)** $5a$ **(c)** $2x + 3y$ **(d)** $-3x^2$

Solution
(a) A number is a term. Its coefficient is 51.
(b) The product of a number and a variable is a term. Its coefficient is 5.

(c) The sum (or difference) of two terms is not a term.

(d) The product of a number and a variable with an exponent is a term. Its coefficient is -3.

Now Try Exercises **11, 15, 21**

MAKING CONNECTIONS

Factors and Terms

When variables and numbers are multiplied, they are called *factors*. For example, the expression $4xy$ has factors of 4, x, and y. When variables and numbers are added or subtracted, they are called *terms*. For example, the expression $x - 5xy + 1$ has terms of x, $-5xy$, and 1.

Combining Like Terms

Suppose that we have two boards with lengths $2x$ and $3x$, where the value of x could be any length such as 2 feet. See Figure 1.28. Because $2x$ and $3x$ are *like terms* we can find the total length of the two boards by applying the distributive property and *adding like terms*.

$$2x + 3x = (2 + 3)x = 5x$$

The combined length of the two boards is $5x$ units.

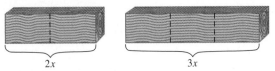

Figure 1.28 $2x + 3x = 5x$

We can also determine how much longer the second board is than the first board by subtracting terms.

$$3x - 2x = (3 - 2)x = 1x = x$$

Thus the second board is x units longer than the first board.

If two terms contain the same variables raised to the same powers, we call them **like terms**. We can add or subtract like terms but not *unlike* terms. If one board has length $2x$ and the other board has length $3y$, then we cannot determine the total length other than to say that it is $2x + 3y$. See Figure 1.29. The reason is that the lengths of x and y might not be equal. The terms $2x$ and $3y$ are unlike terms and *cannot* be combined.

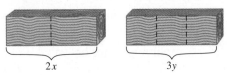

Figure 1.29 $2x + 3y$

EXAMPLE 2   Identifying like terms

Determine whether the terms are like or unlike.

(a) $-4m, 7m$ **(b)** $8x^2, 8y^2$ **(c)** $-x^2, 3x^2$ **(d)** $\frac{1}{2}z, -3z^2$ **(e)** $5, -4n$

Solution

(a) The variable in both terms is m (with power 1), so they are like terms.

(b) The variables are different, so they are unlike terms.

(c) The terms $-x^2$ and $3x^2$ have the same variable raised to the same power (x^2), so they are like terms.

(d) The term $-3z^2$ contains $z^2 = z \cdot z$, whereas the term $\frac{1}{2}z$ contains only z. Thus they are unlike terms.

(e) The term 5 has no variable, whereas the term $-4n$ contains the variable n. They are unlike terms.

Now Try Exercises 23, 29, 33

EXAMPLE   3   **Combining like terms**

Combine terms in each expression, if possible.
(a) $-3x + 5x$ **(b)** $8y - y$ **(c)** $-x^2 + 5x^2$ **(d)** $\frac{1}{2}y - 3y^3$

Solution

(a) Combine terms by applying a distributive property.
$$-3x + 5x = (-3 + 5)x = 2x$$

(b) Note that y can be written as $1y$. They are like terms and can be combined.
$$8y - y = (8 - 1)y = 7y$$

(c) Note that $-x^2$ can be written as $-1x^2$. They are like terms and can be combined.
$$-x^2 + 5x^2 = (-1 + 5)x^2 = 4x^2$$

(d) They are unlike terms, so they cannot be combined.   Now Try Exercises 39, 47, 51

✳ Simplifying Expressions

CRITICAL THINKING

Use rectangles to explain how to add $xy + 2xy$.

The area of a rectangle equals length times width. In Figure 1.30 the area of the first rectangle is $3x$, the area of the second rectangle is $2x$, and the area of the third rectangle is x. The area of the last rectangle equals the total area of the three smaller rectangles. That is,
$$3x + 2x + x = (3 + 2 + 1)x = 6x.$$

The expression $3x + 2x + x$ can be *simplified* to $6x$.

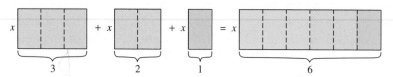

Figure 1.30 $3x + 2x + x = 6x$

EXAMPLE   4   **Simplifying expressions**

Simplify each expression.
(a) $2 + x - 6 + 5x$ **(b)** $2y - (y + 3)$ **(c)** $\dfrac{-1.5x}{-1.5}$ **(d)** $\dfrac{5x}{3x}$

Solution

(a) Combine like terms by applying the properties of real numbers.
$$
\begin{aligned}
2 + x - 6 + 5x &= 2 + (-6) + x + 5x & \text{Commutative property} \\
&= 2 + (-6) + (1 + 5)x & \text{Distributive property} \\
&= -4 + 6x & \text{Add.}
\end{aligned}
$$

(b) $\quad 2y - 1(\overset{\frown}{y + 3}) = 2y + (-1)y + (-1) \cdot 3 \qquad$ Distributive property

$$= 2y - 1y - 3 \qquad \text{Definition of subtraction}$$

$$= (2 - 1)y - 3 \qquad \text{Distributive property}$$

$$= y - 3 \qquad \text{Subtract.}$$

(c) $\qquad \dfrac{-1.5x}{-1.5} = \dfrac{-1.5}{-1.5} \cdot \dfrac{x}{1} \qquad$ Multiplication of fractions

$$= 1 \cdot x \qquad \text{Simplify the fractions.}$$

$$= x \qquad \text{Multiplicative identity}$$

(d) $\qquad \dfrac{5x}{3x} = \dfrac{5}{3} \cdot \dfrac{x}{x} \qquad$ Property of fractions

$$= \dfrac{5}{3} \cdot 1 \qquad \text{Simplify.}$$

$$= \dfrac{5}{3} \qquad \text{Multiplicative identity}$$

Now Try Exercises 57, 69, 83

NOTE: The expressions in Example 4(c) and 4(d) can be simplified directly by using the basic principle of fractions: $\frac{ac}{bc} = \frac{a}{b}$.

EXAMPLE 5   **Simplifying expressions**

Simplify each expression.

(a) $7 - 2(5x - 3)$ **(b)** $3x^3 + 2x^3 - x^3$

(c) $3y^2 - z + 4y^2 - 2z$ **(d)** $\dfrac{12x - 8}{4}$

Solution

(a) $\quad 7 - 2(5x - 3) = 7 + (-2)\big(5x + (-3)\big) \qquad$ Change subtraction to addition.

$$= 7 + (-2)(5x) + (-2)(-3) \qquad \text{Distributive property}$$

$$= 7 - 10x + 6 \qquad \text{Multiply.}$$

$$= 13 - 10x \qquad \text{Combine like terms.}$$

(b) $\quad 3x^3 + 2x^3 - x^3 = (3 + 2 - 1)x^3 \qquad$ Distributive property

$$= 4x^3 \qquad \text{Add and subtract.}$$

(c) $3y^2 - z + 4y^2 - 2z = 3y^2 + 4y^2 + (-1z) + (-2z) \qquad$ Commutative property

$$= (3 + 4)y^2 + \big(-1 + (-2)\big)z \qquad \text{Distributive property}$$

$$= 7y^2 - 3z \qquad \text{Add.}$$

(d) $\qquad \dfrac{12x - 8}{4} = \dfrac{12x}{4} - \dfrac{8}{4} \qquad$ Subtraction of fractions

$$= 3x - 2 \qquad \text{Simplify fractions.}$$

Now Try Exercises 69, 71, 81, 89

Writing Expressions

▶ REAL-WORLD CONNECTION   In real-life situations, we often have to translate words to symbols. For example, to calculate federal and state income tax we might have to multiply taxable income by 0.15 and by 0.05 and then find the sum. If we let x represent taxable income, then the total federal and state income tax is $0.15x + 0.05x$. This expression can be simplified with a distributive property.

$$0.15x + 0.05x = (0.15 + 0.05)x$$
$$= 0.20x$$

Thus the total income tax on a taxable income of $x = \$20,000$ would be

$$0.20(20,000) = \$4000.$$

EXAMPLE 6   **Writing and simplifying an expression**

A sidewalk has a constant width w and comprises several short sections with lengths 12, 6, and 5 feet, as illustrated in Figure 1.31.
(a) Write and simplify an expression that gives the number of square feet of sidewalk.
(b) Find the area of the sidewalk if its width is 3 feet.

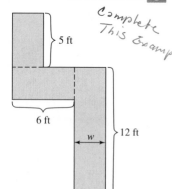

Solution
(a) The area of each sidewalk section equals its length times its width w. The total area of the sidewalk is

$$12w + 6w + 5w = (12 + 6 + 5)w = 23w.$$

(b) When $w = 3$, the area is $23w = 23 \cdot 3 = 69$ square feet.   Now Try Exercise **99**

5 ft

6 ft

w 12 ft

Complete This Example

Figure 1.31

1.8   PUTTING IT ALL TOGETHER

The following table summarizes some of the concepts presented in this section.

Concept	Comments	Examples
Term	A term is a number, variable, or product of numbers and variables raised to powers.	$12, -10, x, -y$ $-3x, 5z, xy, 6x^2$ $10y^3, \frac{3}{4}x, 25xyz^2$
Coefficient of a Term	The coefficient of a term is the number that appears in the term. If no number appears, then the coefficient is either 1 or -1.	12 Coefficient is 12 x Coefficient is 1 $-xy$ Coefficient is -1 $-4x$ Coefficient is -4
Like Terms	Like terms have the same variables raised to the same powers.	$5m$ and $-6m$ x^2 and $-74x^2$ $2xy$ and $-xy$
Combining Like Terms	Like terms can be combined by using a distributive property.	$5x + 2x = (5 + 2)x = 7x$ $y - 3y = (1 - 3)y = -2y$ $8x^2 + x^2 = (8 + 1)x^2 = 9x^2$

1.8 Exercises

MyMathLab
PRACTICE   WATCH   DOWNLOAD   READ   REVIEW

CONCEPTS

1. A _____ is a number, a variable, or a product of numbers and variables raised to powers.

2. The expression $a + b$ (is/is not) a term, whereas the expression ab (is/is not) a term.

3. The number 7 in the term $7x^2y$ is called the _____ of the term.

4. The coefficient of the term $-xy^2$ is _____.

5. The coefficient of the term 6 is _____.

6. When variables and numbers are multiplied, they are called _____. When they are added or subtracted, they are called _____.

7. If two terms contain the same variables raised to the same powers, we call them (like/unlike) terms.

8. The terms $2x$ and $5x$ are (like/unlike) terms, whereas $9x$ and $9z$ are (like/unlike) terms.

9. We can add or subtract (like/unlike) terms.

10. We can combine like terms in an expression by applying a _____ property.

LIKE TERMS

Exercises 11–22: Determine whether the expression is a term. If the expression is a term, identify its coefficient.

11. 91

12. -12

13. $-6b$

14. $9z$

15. $x + 10$

16. $20 - 2y$

17. x^2

18. $4x^3$

19. $4x - 5$

20. $5z + 6x$

21. $-9xyz$

22. $-a^2b^2$

Exercises 23–36: Determine whether the terms are like or unlike.

23. $6, -8$

24. $2x, 19$

25. $5x, -22x$

26. $19y, -y$

27. $14, 14a$

28. $-33b, -3b$

29. $18x, 18y$

30. $-6a, -6b$

31. $x^2, -15x^2, 6x^2$

32. $xyz, 19xyz, -xyz$

33. $3x^2y, 5xy^2$

34. $12xy^2, 4xy^2, -xy^2$

35. $xy, xz, 2xy$

36. $-8x^2y, 2x^2z, x^2y$

37. **Thinking Generally** Are the terms $4ab$ and $-3ba$ like or unlike?

38. **Thinking Generally** Are the terms $-xyz^2$ and $3yz^2x$ like or unlike?

Exercises 39–56: Combine terms, if possible.

39. $3x + 5x$

40. $6x - 8x$

41. $19y - 5y$

42. $22z + z$

43. $28a + 13a$

44. $41b - 17b$

45. $11z - 11z$

46. $4y + 4y$

47. $5x - 7y$

48. $3y + 3z$

49. $5 + 5y$

50. $x + x^2$

51. $5x^2 - 2x^2$

52. $25z^3 - 10z^3$

53. $8xy - 10xy + xy$

54. $4x^2 + x^2 - 5x^2$

55. $12ab - 7ab - 5ab$

56. $100xy^2 + 25xy^2 - 5xy^2$

SIMPLIFYING AND WRITING EXPRESSIONS

Exercises 57–92: Simplify the expression.

57. $5 + x - 3 + 2x$

58. $x - 5 - 5x + 7$

59. $-\frac{3}{4} + z - 3z + \frac{5}{4}$

60. $\frac{4}{3}z - 100 + 200 - \frac{1}{3}z$

61. $4y - y + 8y$

62. $14z - 15z - z$

63. $-3 + 6z + 2 - 2z$

64. $19a - 12a + 5 - 6$

65. $-2(3z - 6y) - z$

66. $6\left(\frac{1}{2}a - \frac{1}{6}b\right) - 3b$

67. $2 - \frac{3}{4}(4x + 8)$

68. $-5 - (5x - 6)$

69. $-x - (5x + 1)$

70. $2x - 4(x + 2)$

71. $1 - \frac{1}{3}(x + 1)$ **72.** $-3 - 3(4 - x)$

73. $\frac{3}{5}(x + y) - \frac{1}{5}(x - 1)$

74. $-5(a + b) - (a + b)$

75. $0.2x^2 + 0.3x^2 - 0.1x^2$ **76.** $32z^3 - 52z^3 + 20z^3$

77. $2x^2 - 3x + 5x^2 - 4x$ **78.** $\frac{5}{6}y^2 - 4 + \frac{1}{12}y^2 + 3$

79. $a + 3b - a - b$ **80.** $2z^2 - z - z^2 + 3z$

81. $8x^3 + 7 - x^3 - 5$ **82.** $4y - 6z + 2y - 3z$

83. $\dfrac{8x}{8}$ **84.** $\dfrac{-0.1y}{-0.1}$

85. $\dfrac{-3y}{-y}$ **86.** $\dfrac{2x}{7x}$

87. $\dfrac{-108z}{-108}$ **88.** $\dfrac{3xy}{-6xy}$

89. $\dfrac{9x - 6}{3}$ **90.** $\dfrac{18y + 9}{9}$

91. $\dfrac{14z + 21}{7}$ **92.** $\dfrac{15x - 20}{5}$

Exercises 93–98: Translate the phrase into a mathematical expression and then simplify. Use the variable x.

93. The sum of five times a number and six times the same number

94. The sum of a number and three times the same number

95. The sum of a number squared and twice the same number squared

96. One-half of a number minus three-fourths of the same number

97. Six times a number minus four times the same number

98. Two cubed times a number minus three squared times the same number

APPLICATIONS

99. *Street Dimensions*   (Refer to Example 6.) A street has a constant width w and comprises several straight sections having lengths 600, 400, 350, and 220 feet.

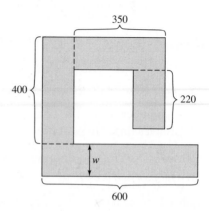

(a) Write a simplified expression that gives the square footage of the street.

(b) Find the area of the street if its width is 42 feet.

100. *Sidewalk Dimensions*   A sidewalk has a constant width w and comprises several short sections having lengths 12, 14, and 10 feet.

(a) Write a simplified expression that gives the number of square feet of sidewalk.

(b) Find the area of the sidewalk if its width is 5 feet.

101. *Snowblowers*   Two snowblowers are being used to clear a driveway. The first blower can remove 20 cubic feet per minute, and the second blower can remove 30 cubic feet per minute.

(a) Write a simplified expression that gives the number of cubic feet of snow removed in x minutes.

(b) Find the total number of cubic feet of snow removed in 48 minutes.

(c) How many minutes would it take to remove the snow from a driveway that measures 30 feet by 20 feet, if the snow is 2 feet deep?

102. *Winding Rope*   Two motors are winding up rope. The first motor can wind up rope at 2 feet per second, and the second motor can wind up rope at 5 feet per second.

(a) Write a simplified expression that gives the length of rope wound by both motors in x seconds.

(b) Find the total length of rope wound up in 3 minutes.

(c) How many minutes would it take to wind up 2100 feet of rope by using both motors?

WRITING ABOUT MATHEMATICS

103. The following expression was simplified *incorrectly*.

$$3(x - 5) + 5x \overset{?}{=} 3x - 5 + 5x$$
$$\overset{?}{=} 3x + 5x - 5$$
$$\overset{?}{=} (3 + 5)x - 5$$
$$\overset{?}{=} 8x - 5$$

Find the error and explain what went wrong. What should the final answer be?

104. Explain how to add like terms. What property of real numbers is used?

CHECKING BASIC CONCEPTS
SECTIONS 1.7 AND 1.8

1. Use the commutative property to rewrite each expression.
  (a) $y \cdot 18$ **(b)** $10 + x$

2. Use the associative and commutative properties to simplify $5 \cdot (y \cdot 4)$.

3. Simplify each expression.
  (a) $10 - (5 + x)$ **(b)** $5(x - 7)$

4. State the property that the equation $5x + 3x = 8x$ illustrates.

5. Simplify $-4xy + 4xy$.

6. Mentally evaluate each expression.
  (a) $32 + 17 + 8 + 3$ **(b)** $\frac{5}{6} \cdot \frac{7}{8} \cdot \frac{6}{5} \cdot 8$
  (c) $567 - 199$

7. Determine whether the terms are like or unlike.
  (a) $-3xy, -3yz$ **(b)** $4x^2, -2x^2$

8. Combine like terms in each expression.
  (a) $5z + 9z$ **(b)** $5y - 4 - 8y + 7$

9. Simplify each expression.
  (a) $2y - (5y + 3)$
  (b) $-4(x + 3y) + 2(2x - y)$
  (c) $\dfrac{20x}{20}$ **(d)** $\dfrac{35x^2}{x^2}$

10. Write "the sum of three times a number and five times the same number" as a mathematical expression with the variable x and then simplify the expression.

CHAPTER 1   SUMMARY
SECTION 1.1 ■ NUMBERS, VARIABLES, AND EXPRESSIONS

Sets of Numbers

Natural Numbers	$1, 2, 3, 4, \ldots$
Whole Numbers	$0, 1, 2, 3, \ldots$

Prime and Composite Numbers

Prime Number   Only natural number factors are itself and 1; must be a natural number greater than 1

   Examples: 2, 3, 5, 7, 11, 13, and 17

Composite Number   A natural number greater than 1 that is not prime; a composite number can be written as a product of two or more prime numbers.

   Examples: $24 = 2 \times 2 \times 2 \times 3$ and $18 = 2 \times 3 \times 3$

Important Terms

Variable

A symbol or letter that represents an unknown quantity

Examples: $a, b, F, x,$ and y

Algebraic Expression

Consists of numbers, variables, arithmetic symbols, and grouping symbols

Examples: $3x + 1$ and $5(x + 2) - y$

Equation

A statement that two algebraic expressions are equal; an equation always contains an equals sign.

Examples: $1 + 2 = 3$ and $z - 7 = 8$

SECTION 1.2 ■ FRACTIONS

Parts of a Fraction

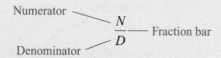

Numerator $\longrightarrow \dfrac{N}{D} \longrightarrow$ Fraction bar

Denominator

Simplifying Fractions $\dfrac{a \cdot c}{b \cdot c} = \dfrac{a}{b}$

Example: $\dfrac{20}{35} = \dfrac{4 \cdot 5}{7 \cdot 5} = \dfrac{4}{7}$

Multiplicative Inverse, or Reciprocal   The reciprocal of a nonzero number a is $\dfrac{1}{a}$.

Examples: The reciprocal of -5 is $-\dfrac{1}{5}$, and the reciprocal of $\dfrac{2x}{y}$ is $\dfrac{y}{2x}$, provided $x \neq 0$ and $y \neq 0$.

Multiplication and Division

$$\frac{a}{b} \cdot \frac{c}{d} = \frac{ac}{bd} \qquad b \text{ and } d \text{ are nonzero.}$$

$$\frac{a}{b} \div \frac{c}{d} = \frac{a}{b} \cdot \frac{d}{c} \qquad b, c, \text{ and } d \text{ are nonzero.}$$

Examples: $\dfrac{3}{4} \cdot \dfrac{7}{5} = \dfrac{21}{20}$ and $\dfrac{3}{4} \div \dfrac{7}{5} = \dfrac{3}{4} \cdot \dfrac{5}{7} = \dfrac{3 \cdot 5}{4 \cdot 7} = \dfrac{15}{28}$

Addition and Subtraction with a Common Denominator

$$\frac{a}{b} + \frac{c}{b} = \frac{a + c}{b} \qquad \text{and} \qquad \frac{a}{b} - \frac{c}{b} = \frac{a - c}{b}$$

Examples: $\dfrac{3}{5} + \dfrac{1}{5} = \dfrac{3 + 1}{5} = \dfrac{4}{5}$ and $\dfrac{3}{5} - \dfrac{1}{5} = \dfrac{3 - 1}{5} = \dfrac{2}{5}$

Addition and Subtraction with Unlike Denominators   Write the expressions with an LCD. Then add or subtract the numerators, keeping the denominator unchanged.

Examples: $\dfrac{2}{9} + \dfrac{1}{6} = \dfrac{2}{9} \cdot \dfrac{2}{2} + \dfrac{1}{6} \cdot \dfrac{3}{3} = \dfrac{4}{18} + \dfrac{3}{18} = \dfrac{7}{18}$ LCD is 18.

$\dfrac{2}{9} - \dfrac{1}{6} = \dfrac{2}{9} \cdot \dfrac{2}{2} - \dfrac{1}{6} \cdot \dfrac{3}{3} = \dfrac{4}{18} - \dfrac{3}{18} = \dfrac{1}{18}$

SECTION 1.3 ■ EXPONENTS AND ORDER OF OPERATIONS

Exponential Expression

$$\text{Base} \longrightarrow 5^3 \longleftarrow \text{Exponent}$$

Example: $3^4 = 3 \cdot 3 \cdot 3 \cdot 3 = 81$

Order of Operations

Use the following order of operations. First perform all calculations within parentheses and absolute values, or above and below the fraction bar.

1. Evaluate all exponential expressions.
2. Do all multiplication and division from *left to right*.
3. Do all addition and subtraction from *left to right*.

NOTE: Negative signs are evaluated after exponents so $-2^4 = -16$.

Examples:
$$
\begin{aligned}
100 - 5^2 \cdot 2 &= 100 - 25 \cdot 2 \\
&= 100 - 50 \\
&= 50
\end{aligned}
\qquad \text{and} \qquad
\begin{aligned}
\frac{4+2}{5-3} \cdot 4 &= \frac{6}{2} \cdot 4 \\
&= 3 \cdot 4 \\
&= 12
\end{aligned}
$$

SECTION 1.4 ■ REAL NUMBERS AND THE NUMBER LINE

Opposite or Additive Inverse   The opposite of the number a is $-a$.

Examples: The opposite of 5 is -5, and the opposite of -8 is $-(-8) = 8$.

Sets of Numbers

Integers $\dots, -3, -2, -1, 0, 1, 2, 3, \dots$

Rational Numbers $\frac{p}{q}$, where p and $q \neq 0$ are integers; rational numbers can be written as decimal numbers that either repeat or terminate.

   Examples: $-\frac{3}{4}, 5, -7.6, \frac{6}{3},$ and $\frac{1}{3}$

Real Numbers   Numbers that can be written as decimal numbers

   Examples: $-\frac{3}{4}, 5, -7.6, \frac{6}{3}, \pi, \sqrt{3},$ and $-\sqrt{7}$

Irrational Numbers   Real numbers that are not rational

   Examples: $\pi, \sqrt{3},$ and $-\sqrt{7}$

Average   To calculate the average of a set of numbers, find the sum of the numbers and then divide the sum by how many numbers there are in the set.

Example: The average of 4, 5, 20, and 11 is

$$\frac{4 + 5 + 20 + 11}{4} = \frac{40}{4} = 10.$$

We divide by 4 because we are finding the average of 4 numbers.

The Number Line

The origin corresponds to the number 0.

Absolute Value   If a is positive or 0, then $|a| = a$, and if a is negative, then $|a| = -a$. The absolute value of a number is *never* negative.

Examples: $|5| = 5$, $|-5| = 5$,   and $|0| = 0$

Inequality   If a is located to the left of b on the number line, then $a < b$. If a is located to the right of b on the number line, then $a > b$.

Examples: $-3 < 6$ and $-1 > -2$

SECTION 1.5 ■ ADDITION AND SUBTRACTION OF REAL NUMBERS

Addition of Opposites

$$a + (-a) = 0$$

Examples: $3 + (-3) = 0$ and $-\frac{5}{7} + \frac{5}{7} = 0$

Addition of Real Numbers

To add two numbers that are either *both positive* or *both negative*, add their absolute values. Their sum has the same sign as the two numbers.

To add two numbers with *opposite signs*, find the absolute value of each number. Subtract the smaller absolute value from the larger. The sum has the same sign as the sign of the number with the larger absolute value. If the two numbers are opposites, their sum is 0.

Examples: $-4 + 5 = 1$, $3 + (-7) = -4$, $-4 + (-2) = -6$,   and $8 + 2 = 10$

Subtraction of Real Numbers   For any real numbers a and b, $a - b = a + (-b)$.

Examples: $5 - 9 = 5 + (-9) = -4$ and $-4 - (-3) = -4 + 3 = -1$

SECTION 1.6 ■ MULTIPLICATION AND DIVISION OF REAL NUMBERS

Important Terms

Factors   Numbers multiplied in a multiplication problem

Example: 5 and 4 are factors of 20 because $5 \cdot 4 = 20$.

Product   The answer to a multiplication problem

Example: The product of 5 and 4 is 20.

Dividend, Divisor, and Quotient   If $\frac{a}{b} = c$, then a is the dividend, b is the divisor, and c is the quotient.

Example: In the division problem $30 \div 5 = \frac{30}{5} = 6$, 30 is the dividend, 5 is the divisor, and 6 is the quotient.

Dividing Real Numbers   For real numbers a and b with $b \neq 0$,

$$\frac{a}{b} = a \cdot \frac{1}{b}.$$

Examples: $\dfrac{8}{\frac{1}{2}} = \dfrac{8}{1} \cdot \dfrac{2}{1} = 16$, $\quad 14 \div \dfrac{2}{3} = \dfrac{14}{1} \cdot \dfrac{3}{2} = \dfrac{42}{2} = 21$, $\quad$ and $\quad 5 \div 0$ is undefined.

Signs of Products or Quotients The product or quotient of two numbers with like signs is positive. The product or quotient of two numbers with unlike signs is negative.

Examples: $-4 \cdot 6 = -24$, $\quad -2 \cdot (-5) = 10$, $\quad \dfrac{-18}{6} = -3$, $\quad$ and $\quad -4 \div -2 = 2$

Writing Fractions as Decimals To write the fraction $\dfrac{a}{b}$ as a decimal, divide b into a.

Example: $\dfrac{4}{9} = 0.\overline{4}$ because division of 4 by 9 gives the repeating decimal $0.4444\ldots$.

$$
\begin{array}{r}
0.444\ldots \\
9\overline{)4.000} \\
\underline{36} \\
40 \\
\underline{36} \\
40 \\
\underline{36} \\
4
\end{array}
$$

SECTION 1.7 ■ PROPERTIES OF REAL NUMBERS

Important Properties

Commutative

$a + b = b + a$ and $a \cdot b = b \cdot a$

Examples: $3 + 4 = 4 + 3$ and $-6 \cdot 3 = 3 \cdot (-6)$

Associative

$(a + b) + c = a + (b + c)$ and $(a \cdot b) \cdot c = a \cdot (b \cdot c)$

Examples: $(2 + 3) + 4 = 2 + (3 + 4)$
$(2 \cdot 3) \cdot 4 = 2 \cdot (3 \cdot 4)$

Distributive

$a(b + c) = ab + ac$ and $a(b - c) = ab - ac$

Examples: $3(x + 5) = 3x + 15$
$4(5 - 2) = 4 \cdot 5 - 4 \cdot 2$

Identity

$a + 0 = 0 + a = a$ $\qquad$ Additive identity is 0.
$a \cdot 1 = 1 \cdot a = a$ $\qquad$ Multiplicative identity is 1.

Examples: $5 + 0 = 0 + 5 = 5$ and $1 \cdot (-4) = -4 \cdot 1 = -4$

Inverse

$a + (-a) = 0$ and $-a + a = 0$
$a \cdot \dfrac{1}{a} = 1$ and $\dfrac{1}{a} \cdot a = 1$

Examples: $5 + (-5) = 0$ and $\dfrac{1}{2} \cdot 2 = 1$

NOTE: The commutative and associative properties apply to addition and multiplication but *not* to subtraction and division.

SECTION 1.8 ■ SIMPLIFYING AND WRITING ALGEBRAIC EXPRESSIONS

Important Terms

Term	A term is a number, a variable, or a product of numbers and variables raised to powers.
	Examples: 5, $-10x$, $3xy$, and x^2
Coefficient	The number portion of a term
	Examples: The coefficients for the terms $3xy$, $-x^2$, and -7 are $3, -1$, and -7, respectively.
Like Terms	Terms containing the same variables raised to the same powers; their coefficients may be different.
	Examples: The following pairs are like terms: $5x$ and $-x$; $6x^2$ and $-2x^2$; $3xy$ and $-\dfrac{1}{2}xy$.

Combining Like Terms   To add or subtract like terms, apply the distributive property.

Examples: $4x + 5x = (4 + 5)x = 9x$ and $5xy - 7xy = (5 - 7)xy = -2xy$

CHAPTER 1   REVIEW EXERCISES

SECTION 1.1

Exercises 1–6: Classify the number as prime, composite, or neither. If the number is composite, write it as a product of prime numbers.

1. 29

2. 27

3. 108

4. 91

5. 0

6. 1

Exercises 7–10: Evaluate the expression for the given values of x and y.

7. $2x - 5$ $x = 4$

8. $7 - \dfrac{10}{x}$ $x = 5$

9. $9x - 2y$ $x = 2$, $y = 3$

10. $\dfrac{2x}{x - y}$ $x = 6$, $y = 4$

Exercises 11–14: Find the value of y for the given values of x and z.

11. $y = x - 5$ $x = 12$

12. $y = xz + 1$ $x = 2$, $z = 3$

13. $y = 4(x - z)$ $x = 7$, $z = 5$

14. $y = \dfrac{x + z}{4}$ $x = 14$, $z = 10$

Exercises 15–20: Translate the phrase into an algebraic expression. State precisely what each variable represents when appropriate.

15. Five times the cost of a CD

16. Five less than a number

17. Three squared increased by five

18. Two cubed divided by the quantity three plus one

19. The product of three and a number

20. The difference between a number and four

SECTION 1.2

21. Use the basic principle of fractions to simplify each expression.
    (a) $\dfrac{5 \cdot 7}{8 \cdot 7}$ (b) $\dfrac{3a}{4a}$

22. Simplify each fraction to lowest terms.
    (a) $\dfrac{9}{12}$ (b) $\dfrac{36}{60}$

Exercises 23–28: Multiply and then simplify the result to lowest terms when appropriate.

23. $\frac{3}{4} \cdot \frac{5}{6}$

24. $\frac{1}{2} \cdot \frac{4}{9}$

25. $\frac{2}{3} \cdot \frac{5}{11} \cdot \frac{9}{10}$

26. $\frac{12}{11} \cdot \frac{22}{23} \cdot \frac{1}{2}$

27. $\frac{x}{3} \cdot \frac{6}{x}$

28. $\frac{2}{3} \cdot \frac{9x}{4y}$

29. Find the fractional part: one-fifth of three-sevenths.

30. Find the reciprocal of each number.
   (a) 8   **(b)** 1   **(c)** $\frac{5}{19}$ **(d)** $\frac{3}{2}$

Exercises 31–34: Divide and then simplify to lowest terms when appropriate.

31. $\frac{3}{2} \div \frac{1}{6}$

32. $\frac{9}{10} \div \frac{7}{5}$

33. $8 \div \frac{2}{3}$

34. $\frac{3}{4} \div 6$

Exercises 35 and 36: Find the least common denominator for the fractions.

35. $\frac{1}{8}, \frac{5}{12}$

36. $\frac{3}{14}, \frac{1}{21}$

Exercises 37–42: Add or subtract and then simplify to lowest terms when appropriate.

37. $\frac{2}{15} + \frac{3}{15}$

38. $\frac{5}{4} - \frac{3}{4}$

39. $\frac{11}{12} - \frac{1}{8}$

40. $\frac{6}{11} - \frac{3}{22}$

41. $\frac{2}{3} - \frac{1}{2} + \frac{1}{4}$

42. $\frac{1}{6} + \frac{2}{3} - \frac{1}{9}$

SECTION 1.3

Exercises 43–48: Write the expression as an exponential expression.

43. $5 \cdot 5 \cdot 5 \cdot 5 \cdot 5 \cdot 5$

44. $\frac{7}{6} \cdot \frac{7}{6} \cdot \frac{7}{6}$

45. $3 \cdot 3 \cdot 3 \cdot 3$

46. $x \cdot x \cdot x \cdot x \cdot x$

47. $(x + 1) \cdot (x + 1)$

48. $(a - 5) \cdot (a - 5) \cdot (a - 5)$

49. Use multiplication to rewrite each expression, and then evaluate the result.
   (a) 4^3 **(b)** 7^2 **(c)** 8^1

50. Find a natural number n such that $2^n = 32$.

Exercises 51–62: Evaluate the expression by hand.

51. $7 + 3 \cdot 6$

52. $15 - 5 - 3$

53. $24 \div 4 \div 2$

54. $30 - 15 \div 3$

55. $18 \div 6 - 2$

56. $\frac{18}{4 + 5}$

57. $9 - 3^2$

58. $2^3 - 8$

59. $2^4 - 8 + \frac{4}{2}$

60. $3^2 - 4(5 - 3)$

61. $7 - \frac{4 + 6}{2 + 3}$

62. $3^3 - 2^3$

SECTION 1.4

Exercises 63 and 64: Find the opposite of each expression.

63. **(a)** -8 **(b)** $-(-(-3))$

64. **(a)** $-\left(\frac{-3}{7}\right)$ **(b)** $\frac{-2}{-5}$

Exercises 65 and 66: Find the decimal equivalent for the rational number.

65. **(a)** $\frac{4}{5}$ **(b)** $\frac{3}{20}$

66. **(a)** $\frac{5}{9}$ **(b)** $\frac{7}{11}$

Exercises 67–72: Classify the number as one or more of the following: natural number, whole number, integer, rational number, or irrational number.

67. 0   **68.** $-\frac{5}{6}$

69. -7 **70.** $\sqrt{17}$

71. π **72.** 3.4

73. Plot each number on the same number line.
   (a) 0   **(b)** -2 **(c)** $\frac{5}{4}$

74. Evaluate each expression.
   (a) $|-5|$ **(b)** $|-\pi|$ **(c)** $|\sqrt{2} - 1|$

75. Insert the symbol $>$ or $<$ to make each statement true.
   (a) $-5 \underline{} 4$ **(b)** $-\frac{1}{2} \underline{} -\frac{5}{2}$
   (c) $-3 \underline{} |-9|$ **(d)** $|-8| \underline{} |-1|$

76. List the numbers $\sqrt{3}, -3, 3, -\frac{2}{3},$ and $\pi - 1$ from least to greatest.

SECTIONS 1.5 AND 1.6

Exercises 77 and 78: (Refer to Example 3 in Section 1.5.) Find the sum visually.

77. $-5 + 9$ **78.** $4 + (-7)$

Exercises 79 and 80: Use a number line to find the sum.

79. $-1 + 2$ **80.** $-2 + (-3)$

Exercises 81–92: Evaluate the expression.

81. $5 + (-4)$

82. $-9 - (-7)$

83. $11 \cdot (-4)$

84. $-8 \cdot (-5)$

85. $11 \div (-4)$

86. $-4 \div \frac{4}{7}$

87. $-\frac{5}{9} - \left(-\frac{1}{3}\right)$

88. $-\frac{1}{2} + \left(-\frac{3}{4}\right)$

89. $-\frac{1}{3} \cdot \left(-\frac{6}{7}\right)$

90. $\dfrac{\frac{4}{5}}{-7}$

91. $-\frac{3}{2} \div \left(-\frac{3}{8}\right)$

92. $\frac{3}{8} \div (-0.5)$

Exercises 93 and 94: Write an arithmetic expression for the given phrase and then simplify.

93. Three plus negative five

94. Subtract negative four from two

Exercises 95 and 96: Write the fraction or mixed number as a decimal.

95. $\frac{7}{9}$

96. $2\frac{1}{5}$

Exercises 97 and 98: Write the decimal number as a fraction in lowest terms.

97. 0.6

98. 0.375

SECTION 1.7

Exercises 99–108: State the property that the equation illustrates.

99. $z \cdot 3 = 3z$

100. $6 + (7 + 5x) = (6 + 7) + 5x$

101. $2(5x - 2) = 10x - 4$

102. $5 + x + 3 = 5 + 3 + x$

103. $1 \cdot a = a$

104. $3 \cdot (5x) = (3 \cdot 5)x$

105. $12 - (x + 7) = 12 - x - 7$

106. $a + 0 = a$

107. $-5x + 5x = 0$

108. $-5 \cdot \left(-\frac{1}{5}\right) = 1$

Exercises 109–114: Use properties of real numbers to evaluate the expression mentally.

109. $7 + 9 + 12 + 8 + 1 + 3$

110. $500 - 199$

111. $25 \cdot 99$

112. $4581 + 1999$

113. 54.98×10

114. $4356 \div 100$

SECTION 1.8

Exercises 115–118: Determine whether the expression is a term. If the expression is a term, identify its coefficient.

115. $55x$

116. $-xy$

117. $9xy + 2z$

118. $x - 7$

Exercises 119–130: Simplify the expression.

119. $-10x + 4x$

120. $19z - 4z$

121. $3x^2 + x^2$

122. $7 + 2x - 6 + x$

123. $-\frac{1}{2} + \frac{3}{2}z - z + \frac{5}{2}$

124. $5(x - 3) - (4x + 3)$

125. $4x^2 - 3 + 5x^2 - 3$

126. $3x^2 + 4x^2 - 7x^2$

127. $\dfrac{35a}{7a}$

128. $\dfrac{0.5c}{0.5}$

129. $\dfrac{15y + 10}{5}$

130. $\dfrac{24x - 60}{12}$

APPLICATIONS

131. *Painting a Wall* Two people are painting a large wall. The first person paints 3 square feet per minute while the second person paints 4 square feet per minute.

   (a) Write a simplified expression that gives the total number of square feet the two people can paint in x minutes.

   (b) Find the number of square feet painted in 1 hour.

   (c) How many minutes would it take for them to paint a wall 8 feet tall and 21 feet wide?

132. *Area of a Triangle* Find the area of the triangle shown.

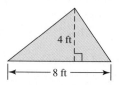

133. *Gallons to Pints* There are 8 pints in 1 gallon. Make a table of values that converts G gallons to P pints. Let $G = 1, 2, 3, \ldots, 6$. Write a formula that converts G gallons to P pints.

134. *Blank CDs*   The table lists the cost C of buying x blank compact discs. Write an equation that relates C and x.

Blank CDs (x)	1	2	3	4
Cost (C)	$0.25	$0.50	$0.75	$1.00

135. *Aging in the United States*   In 2050, about $\frac{1}{5}$ of the population will be aged 65 or over and about $\frac{1}{20}$ of the population will be aged 85 or over. Estimate the fraction of the population that will be between the ages of 65 and 85 in 2050.   (*Source:* U.S. Census Bureau.)

136. *Rule of 72*   (Refer to Exercise 87 in Section 1.3.) If an investment of $25,000 earns 9% annual interest, approximate the value of the investment after 24 years.

137. *Carpentry*   A board measures $5\frac{3}{4}$ feet and needs to be cut in five equal pieces. Find the length of each piece.

138. *Distance*   Over four days an athlete jogs $3\frac{1}{8}$ miles, $4\frac{3}{8}$ miles, $6\frac{1}{4}$ miles, and $1\frac{5}{8}$ miles. How far does the athlete jog in all?

139. *Checking Account*   The initial balance in a checking account is $1652. Find the final balance resulting from the following sequence of withdrawals and deposits: $-$78$, $-$91$, 256, and $-$638$.

140. *Temperature Range*   The highest temperature ever recorded in Amarillo, Texas, was $108°F$ and the coldest was $-16°F$. Find the difference between these two temperatures.   (*Source:* The Weather Almanac.)

141. *Top-Grossing Movies*   *Titanic* (1997) is the sixth top-grossing movie of all time. Find the total admissions for *Titanic* if they were $\frac{16}{25}$ of the total admissions for the top-grossing movie of all time, *Gone With the Wind* (1939), which had total admissions of 202 million.   (*Source:* Exhibitor Relations Co., Inc.)

CHAPTER 1   TEST   *Pass the Test*   Video solutions to all test exercises

1. Classify the number as prime or composite. If the number is composite, write it as a product of prime numbers.
   (a) 29   **(b)** 56

2. Evaluate the expression $\dfrac{5x}{2x-1}$ for $x = -3$.

3. Translate the phrase "four squared decreased by three" to an algebraic expression. Then find the value of the expression.

4. Simplify $\frac{24}{32}$ to lowest terms.

5. Evaluate each expression. Write your answer in lowest terms.
   (a) $\frac{5}{8} + \frac{1}{8}$ **(b)** $\frac{5}{9} - \frac{3}{15}$ **(c)** $\frac{3}{5} \cdot \frac{10}{21}$
   (d) $6 \div \frac{8}{5}$ **(e)** $\frac{5}{12} + \frac{4}{9}$ **(f)** $\frac{10}{13} \div 5$

6. Write $y \cdot y \cdot y \cdot y$ as an exponential expression.

7. Evaluate each expression.
   (a) $6 + 10 \div 5$ **(b)** $4^3 - (3 - 5 \cdot 2)$
   (c) $-6^2 - 6 + \frac{4}{2}$ **(d)** $11 - \frac{1+3}{6-4}$

8. Classify the number as one or more of the following: natural number, whole number, integer, rational number, or irrational number.
   (a) -1 **(b)** $\sqrt{5}$

9. Plot each number on the same number line.
   (a) -2 **(b)** $\frac{1}{3}$ **(c)** $\sqrt{7}$

10. Insert the symbol $>$ or $<$ to make each statement true.
   (a) $2 \underline{\quad} |-5|$ **(b)** $|-1| \underline{\quad} |0|$

11. Evaluate the expression.
   (a) $-5 \div \frac{5}{6}$ **(b)** $-7 \cdot (-3)$

12. State the property or properties that each equation illustrates.
   (a) $6x - 2x = 4x$ **(b)** $12 \cdot (3x) = 36x$
   (c) $4 + x + 8 = 12 + x$

13. Use properties of real numbers to evaluate $17 \cdot 102$ mentally.

14. Simplify each expression.
   (a) $5 - 5z + 7 + z$ **(b)** $12x - (6 - 3x)$
   (c) $5 - 4(x + 6) + \dfrac{15x}{3}$

15. *Mowing a Lawn*   Two people are mowing a lawn. The first person has a riding mower and can mow $\frac{4}{3}$ acres per hour; the second person has a push mower and can mow $\frac{1}{4}$ acre per hour.
   (a) Write a simplified expression that gives the total number of acres that the two people mow in x hours.
   (b) Find the total acreage that they can mow in an 8-hour work day.

16. A wire $7\frac{4}{5}$ feet long is to be cut in 3 equal parts. How long should each part be?

17. *Cost Equation*   The table lists the cost C of buying x tickets to a hockey game.

Tickets (x)	3	4	5	6
Cost (C)	$39	$52	$65	$78

   (a) Find an equation that relates C and x.
   (b) What is the cost of 17 tickets?

18. The initial balance for a savings account is $892. Find the final balance resulting from withdrawals and deposits of $-$57, $150, and $-$345.

CHAPTER 1   EXTENDED AND DISCOVERY EXERCISES

1. *Arithmetic Operations*   Insert one of the symbols $+, -, \times,$ or $\div$ in each blank to obtain the given answer. Do not use any parentheses.

$$2 \_ 2 \_ 2 \_ 2 = 0$$
$$3 \_ 3 \_ 3 \_ 3 = 10$$
$$4 \_ 4 \_ 4 \_ 4 = 1$$
$$6 \_ 6 \_ 6 \_ 6 = 36$$
$$7 \_ 7 \_ 7 \_ 7 = 63$$

2. *Magic Squares*   The following square is called a "magic square" because the numbers in each row, column, and diagonal sum to 15.

8	3	4
1	5	9
6	7	2

Complete the following magic square having 4 rows and 4 columns by arranging the numbers 1 through 16 so that each row, column, and diagonal sums to 34. The four corners will also sum to 34.

	2		13
5			
		6	
4			1

CHAPTER 2

Linear Equations and Inequalities

Mathematics is a unique subject because it does not depend on people making experiments, but it is essential for describing, or modeling, events in the real world. For example, sunbathing is a popular pastime, but ultraviolet light from the sun is responsible for both tanning and burning exposed skin. Mathematics lets us use numbers to describe the intensity of ultraviolet light. The table shows the maximum ultraviolet intensity measured in milliwatts per square meter for various latitudes and dates.

Latitude	Mar. 21	June 21	Sept. 21	Dec. 21
0°	325	254	325	272
10°	311	275	280	220
20°	249	292	256	143
30°	179	248	182	80
40°	99	199	127	34
50°	57	143	75	13

If a student from Chicago, located at a latitude of 42°, spends spring break in Hawaii with a latitude of 20°, the sun's ultraviolet rays in Hawaii will be approximately $\frac{249}{99} \approx 2.5$ times as intense as they are in Chicago. Suppose that you travel to the equator for spring break. How many times as intense will the sun be at the equator compared to where you presently live?

> *Education is not the filling of a pail, but the lighting of a fire.*
> —WILLIAM BUTLER YEATS

Source: J. Williams, *The USA Today Weather Almanac.*

2.1 INTRODUCTION TO EQUATIONS

Basic Concepts ▪ The Addition Property of Equality ▪
The Multiplication Property of Equality

A LOOK INTO MATH ▷

A new and exciting technology is the Global Positioning System (GPS), which consists of 24 satellites that travel around Earth in nearly circular orbits. The GPS can be used to determine locations and velocities of cars, airplanes, and hikers with an amazing degree of accuracy. New cars often come equipped with the GPS, and their drivers can determine their cars' locations to within a few feet. (**Source:** J. Van Sickle, *GPS for Land Surveyors.*)

To create the GPS thousands of equations were solved, and mathematics was essential in finding their solutions. In this section we discuss many of the basic concepts needed to solve equations.

Basic Concepts

▶ **REAL-WORLD CONNECTION**   Suppose that during a storm it rains 2 inches before noon and 1 inch per hour thereafter until 5 P.M. Table 2.1 lists the total rainfall R after various elapsed times x, where $x = 0$ corresponds to noon.

TABLE 2.1   Rainfall *x* Hours Past Noon

Elapsed Time: x (hours)	0	1	2	3	4	5
Total Rainfall: R (inches)	2	3	4	5	6	7

The data suggest that the total rainfall R in inches is **2** more than the elapsed time x. A formula that *models*, or *describes*, the rainfall x hours past noon is given by

$$R = x + 2.$$

For example, **3** hours past noon, or at 3 P.M.,

$$R = 3 + 2 = 5$$

inches of rain have fallen. Even though $x = 4.5$ does not appear in the table, we can calculate the amount of rainfall at 4:30 P.M. with the formula as

$$R = 4.5 + 2 = 6.5 \text{ inches.}$$

The advantage that a formula has over a table of values is that a formula can be used to calculate the rainfall at any time x, not just at the times listed in the table.

At what time have 6 inches of rain fallen? From Table 2.1 the *solution* is 4, or 4 P.M. To find this solution without the table, we can *solve* the equation

$$x + 2 = 6.$$

An equation can be either true or false. For example, the equation $1 + 2 = 3$ is true, whereas the equation $1 + 2 = 4$ is false. When an equation contains a variable, the equation may be true for some values of the variable and false for other values of the variable. Each value of the variable that makes the equation true is called a **solution** to the equation, and the *set of all solutions* is called the **solution set**. *Solving an equation* means finding all of its solutions. Because $4 + 2 = 6$, the solution to the equation

$$x + 2 = 6$$

is 4, and the solution set is $\{4\}$. Note that **braces** $\{\,\}$ are used to denote a set. Sometimes an equation can have more than one solution. For example, the equation $x^2 = 4$ has two solutions, -2 and 2, because

$$(-2)^2 = -2 \cdot (-2) = 4 \quad \text{and} \quad 2^2 = 2 \cdot 2 = 4.$$

For the equation $x^2 = 4$, the solution set is $\{-2, 2\}$.

Many times we cannot solve an equation simply by looking at it. In these situations we must use a step-by-step procedure. During each step an equation is transformed into a different but equivalent equation. **Equivalent equations** are equations that have the same solution set. For example, the equations

$$x + 2 = 5 \quad \text{and} \quad x = 3$$

are equivalent equations because the solution set for both equations is $\{3\}$.

MAKING CONNECTIONS

Equations and Expressions

Although the words "equation" and "expression" occur frequently in mathematics, they are *not* interchangeable. An equation *always* contains an equals sign but an expression *never* contains an equals sign. We often want to *solve* an equation, whereas an expression can sometimes be *simplified*. Furthermore, the equals sign in an equation separates two expressions. For example, $3x - 5 = x + 1$ is an equation where $3x - 5$ and $x + 1$ are each expressions.

When solving equations, it is often helpful to transform a more complicated equation into an equivalent equation that has an obvious solution, such as $x = 3$. Two important principles used to solve equations are the *addition property of equality* and the *multiplication property of equality*. These properties can be used to transform an equation into an equivalent equation that is easier to solve.

✳The Addition Property of Equality

When solving an equation, we have to apply the same operation to each side of the equation. For example, one way to solve the equation

$$x + 2 = 5$$

is to add -2 to each side. This step results in isolating the x on one side of the equation.

$x + 2 = 5$	Given equation
$x + 2 + (-2) = 5 + (-2)$	Add -2 to each side.
$x + 0 = 3$	Addition of real numbers
$x = 3$	Additive identity

These four equations are equivalent, but the solution is most apparent in the last equation. To transform one equation to the next, a step-by-step procedure based on mathematical properties is used. The reasons that justify each step are written at the right in blue. When -2 is added to each side of the equation, the addition property of equality is used.

ADDITION PROPERTY OF EQUALITY

If a, b, and c are real numbers, then

$$a = b \quad \text{is equivalent to} \quad a + c = b + c.$$

That is, adding the same number to each side of an equation results in an equivalent equation.

Generally, the addition property of equality is based on the concept that "if equals are added to equals, the results are equal."

NOTE: Because any subtraction problem can be changed to an addition problem, *the addition property of equality also works for subtraction. That is, if the same number is subtracted from each side of an equation, the result is an equivalent equation.*

EXAMPLE 1   **Using the addition property of equality**

Solve each equation.

(a) $x + 10 = 7$ **(b)** $t - 4 = 3$ **(c)** $\frac{1}{2} = -\frac{3}{4} + y$

Solution

(a) When solving an equation, we try to isolate the variable on one side of the equation. If we add -10 to (or subtract 10 from) each side of the equation, the value of x becomes apparent.

$x + 10 = 7$	Given equation
$x + 10 + (-10) = 7 + (-10)$	Add -10 to each side.
$x + 0 = -3$	Addition of real numbers
$x = -3$	Additive identity

The solution is -3.

(b) To isolate the variable t, add **4** to each side.

$t - 4 = 3$	Given equation
$t - 4 + 4 = 3 + 4$	Add 4 to each side.
$t + 0 = 7$	Addition of real numbers
$t = 7$	Additive identity

The solution is 7.

(c) To isolate the variable y, add $\frac{3}{4}$ to each side.

$\dfrac{1}{2} = -\dfrac{3}{4} + y$	Given equation
$\dfrac{1}{2} + \dfrac{3}{4} = -\dfrac{3}{4} + \dfrac{3}{4} + y$	Add $\frac{3}{4}$ to each side.
$\dfrac{5}{4} = 0 + y$	Addition of real numbers
$\dfrac{5}{4} = y$	Additive identity

The solution is $\frac{5}{4}$.

Now Try Exercises 17, 27

CHECKING A SOLUTION To check a solution, substitute it in the given equation to find out if a true statement results. To check the solution for Example 1(c), substitute $\frac{5}{4}$ for y in the given equation. Note that a question mark is placed over the equals sign when a solution is being checked.

$$\frac{1}{2} = -\frac{3}{4} + y \qquad \text{Given equation}$$

$$\frac{1}{2} \stackrel{?}{=} -\frac{3}{4} + \frac{5}{4} \qquad \text{Replace } y \text{ with } \frac{5}{4}.$$

$$\frac{1}{2} \stackrel{?}{=} \frac{2}{4} \qquad \text{Add fractions.}$$

$$\frac{1}{2} = \frac{1}{2} \qquad \text{The answer checks.}$$

The answer of $\frac{5}{4}$ checks because the left side of the equation equals the right side of the equation.

CRITICAL THINKING

When you are checking a solution, why do you substitute your answer in the *given* equation?

EXAMPLE 2 Solving and checking a solution

Solve the equation $-5 + y = 3$ and then check the solution.

Solution
Isolate y by adding 5 to each side.

$$-5 + y = 3 \qquad \text{Given equation}$$

$$5 + (-5) + y = 5 + 3 \qquad \text{Add 5 to each side.}$$

$$0 + y = 8 \qquad \text{Addition of real numbers}$$

$$y = 8 \qquad \text{Additive identity}$$

The solution is 8. To check this answer substitute **8** for y in the given equation.

$$-5 + y = 3 \qquad \text{Given equation}$$

$$-5 + 8 \stackrel{?}{=} 3 \qquad \text{Replace } y \text{ with 8.}$$

$$3 = 3 \qquad \text{The answer checks.}$$

Now Try Exercise **31**

MAKING CONNECTIONS

Equations and Scales

Think of an equation as an old-fashioned scale, where two pans must balance, as illustrated in the accompanying figure. If two weights initially balance the pans, then adding an equal amount of weight to each pan (or subtracting an equal amount from each pan) results in the pans remaining balanced.

The Multiplication Property of Equality

The multiplication property of equality is another important property used to solve equations. We can illustrate this property by considering a formula that converts yards to feet. Because there are 3 feet in 1 yard, the formula $F = 3Y$ computes F, the number of feet in Y yards. For example, if $Y = 5$ yards, then $F = 3 \cdot 5 = 15$ feet.

Now consider the reverse, converting 27 feet to yards. The answer to this conversion corresponds to the solution to

$$27 = 3Y.$$

To find the solution, multiply each side of the equation by the reciprocal of 3, or $\frac{1}{3}$.

$$27 = 3Y \qquad \text{Given equation}$$
$$\frac{1}{3} \cdot 27 = \frac{1}{3} \cdot 3 \cdot Y \qquad \text{Multiply each side by } \frac{1}{3}.$$
$$9 = 1 \cdot Y \qquad \text{Multiplication of fractions}$$
$$9 = Y \qquad \text{Multiplicative identity}$$

Thus 27 feet are equivalent to 9 yards.

MULTIPLICATION PROPERTY OF EQUALITY

If a, b, and c are real numbers with $c \neq 0$, then

$$a = b \quad \text{is equivalent to} \quad ac = bc.$$

That is, multiplying each side of an equation by the same nonzero number results in an equivalent equation.

Generally, the multiplication property of equality is based on the concept that "if equals are multiplied by equals, the results are equal."

NOTE: Because any division problem can be changed to a multiplication problem, *the multiplication property of equality also works for division.* That is, if each side of an equation is divided by the same nonzero number, the result is an equivalent equation.

EXAMPLE 3   Using the multiplication property of equality

Solve each equation.
(a) $\frac{1}{3}x = 4$ **(b)** $-4y = 8$ **(c)** $5 = \frac{3}{4}z$

Solution
(a) We start by multiplying each side of the equation by **3**, the reciprocal of $\frac{1}{3}$.

$$\frac{1}{3}x = 4 \qquad \text{Given equation}$$
$$3 \cdot \frac{1}{3}x = 3 \cdot 4 \qquad \text{Multiply each side by 3.}$$
$$1 \cdot x = 12 \qquad \text{Multiplication of real numbers}$$
$$x = 12 \qquad \text{Multiplicative identity}$$

The solution is 12.

(b) The coefficient of the y-term is -4, so we can either multiply each side of the equation by $-\frac{1}{4}$ or divide each side by -4. This step will make the coefficient of y equal to 1.

$$-4y = 8 \qquad \text{Given equation}$$

$$\frac{-4y}{-4} = \frac{8}{-4} \qquad \text{Divide each side by } -4.$$

$$y = -2 \qquad \text{Simplify fractions.}$$

The solution is -2.

(c) To change the coefficient of z from $\frac{3}{4}$ to 1, multiply each side of the equation by $\frac{4}{3}$, the reciprocal of $\frac{3}{4}$.

$$5 = \frac{3}{4}z \qquad \text{Given equation}$$

$$\frac{4}{3} \cdot 5 = \frac{4}{3} \cdot \frac{3}{4}z \qquad \text{Multiply each side by } \frac{4}{3}.$$

$$\frac{20}{3} = 1 \cdot z \qquad \text{Multiplication of real numbers}$$

$$\frac{20}{3} = z \qquad \text{Multiplicative identity}$$

The solution is $\frac{20}{3}$.

Now Try Exercises 41, 49

EXAMPLE 4   Solving and checking a solution

Solve the equation $\frac{3}{4} = -\frac{3}{7}t$ and then check the solution.

Solution
Multiply each side of the equation by $-\frac{7}{3}$, the reciprocal of $-\frac{3}{7}$.

$$\frac{3}{4} = -\frac{3}{7}t \qquad \text{Given equation}$$

$$-\frac{7}{3} \cdot \frac{3}{4} = -\frac{7}{3} \cdot \left(-\frac{3}{7}\right)t \qquad \text{Multiply each side by } -\frac{7}{3}.$$

$$-\frac{7}{4} = 1 \cdot t \qquad \text{Multiplication of real numbers}$$

$$-\frac{7}{4} = t \qquad \text{Multiplicative identity}$$

The solution is $-\frac{7}{4}$. To check this answer, substitute $-\frac{7}{4}$ for t in the given equation.

$$\frac{3}{4} = -\frac{3}{7}t \qquad \text{Given equation}$$

$$\frac{3}{4} \stackrel{?}{=} -\frac{3}{7} \cdot \left(-\frac{7}{4}\right) \qquad \text{Replace } t \text{ with } -\frac{7}{4}.$$

$$\frac{3}{4} = \frac{3}{4} \qquad \text{The answer checks.}$$

Now Try Exercise 57

▶ **REAL-WORLD CONNECTION**   Alaska's major glaciers are melting at an alarming rate. From the mid 1950s to the mid 1990s, these glaciers melted at a rate of 13 cubic miles per year. Recently, Alaska's glaciers have been melting at a rate of 24 cubic miles per year, almost double the earlier rate. Ecologists are concerned that this phenomenon is a sign of global warming.   (**Source:** USA Today.)

EXAMPLE **5**   Estimating glacial melting in Alaska

The glaciers in Alaska are currently melting at a rate of 24 cubic miles per year.
(a) Write a formula that gives the cubic miles of ice I that will melt in x years.
(b) At this rate, determine how long it will take for 300 cubic miles of ice to melt.

Solution
(a) In 1 year $24 \cdot 1 = 24$ cubic miles will melt, in 2 years $24 \cdot 2 = 48$ cubic miles will melt, and in x years $24 \cdot x = 24x$ cubic miles will melt. Thus $I = \mathbf{24x}$, where x is in years and I is in cubic miles.
(b) To determine how long it will take for 300 cubic miles to melt let $I = \mathbf{300}$ in the formula.

$$I = 24x \qquad \text{Formula from part (a)}$$
$$\mathbf{300} = 24x \qquad \text{Let } I = 300.$$
$$\frac{300}{24} = \frac{24x}{24} \qquad \text{Divide each side by 24.}$$
$$12.5 = x \qquad \text{Simplify.}$$

At current rates, it will take 12.5 years for 300 cubic miles of ice to melt.

Now Try Exercise **65**

2.1   PUTTING IT ALL TOGETHER

The following table summarizes some of the topics discussed in this section.

Concept	Comments	Examples
Equation	An equation is a mathematical statement that two expressions are equal. An equation can be either true or false.	The equation $2 + 3 = 5$ is true. The equation $1 + 3 = 7$ is false.
Solution	A value for a variable that makes an equation a true statement	The solution to $x + 5 = 20$ is 15, and the solutions to $x^2 = 9$ are -3 and 3.
Solution Set	The set of all solutions to an equation	The solution set to $x + 5 = 20$ is $\{15\}$, and the solution set to $x^2 = 9$ is $\{-3, 3\}$.
Equivalent Equations	Two equations are equivalent if they have the same solution set.	The equations $$2x = 14 \quad \text{and} \quad x = 7$$ are equivalent because the solution set to both equations is $\{7\}$.

Concept	Comments	Examples
Addition Property of Equality	The equations $a = b$ and $a + c = b + c$ are equivalent. This property is used to solve equations.	To solve $x - 3 = 8$ add 3 to each side of the equation. $x - 3 + 3 = 8 + 3$ $x = 11$ The solution is 11.
Multiplication Property of Equality	The equations $a = b$ and $a \cdot c = b \cdot c$ with $c \neq 0$ are equivalent. This property is used to solve equations.	To solve $\frac{1}{5}x = 10$ multiply each side of the equation by 5. $5 \cdot \frac{1}{5}x = 5 \cdot 10$ $x = 50$ The solution is 50.
Checking a Solution	Substitute the solution in the given equation and then simplify each side to check it.	To show that 8 is a solution to $x + 12 = 20$ substitute 8 for x. $8 + 12 \overset{?}{=} 20$ $20 = 20$ True

2.1 Exercises

CONCEPTS

1. The equation $1 + 3 = 4$ is (true/false).

2. The equation $2 + 3 = 6$ is (true/false).

3. Each value of a variable that makes an equation true is called a(n) _____.

4. The _____ is the set of all solutions to an equation.

5. To solve an equation, find all _____.

6. _____ equations have the same solution sets.

7. If $a = b$, then $a + c =$ _____.

8. Because any subtraction problem can be changed to an addition problem, the addition property of equality also works for _____.

9. If $a = b$ and $c \neq 0$, then $ac =$ _____.

10. Because any division problem can be changed to a multiplication problem, the multiplication property of equality also works for _____.

11. To solve an equation, transform the equation into a(n) _____ equation that is easier to solve.

12. To check an answer, substitute it in the _____ equation.

THE ADDITION PROPERTY OF EQUALITY

13. To solve $x - 22 = 4$, add _____ to each side.

14. To solve $\frac{5}{6} = \frac{1}{6} + x$, add _____ to each side.

15. To solve $x + 3 = 13$, subtract _____ from each side.

16. To solve $\frac{3}{4} = \frac{1}{4} + x$, subtract _____ from each side.

Exercises 17–34: Solve the equation. Check your answer.

17. $x + 5 = 0$

18. $x + 3 = 7$

19. $x - 7 = 1$

20. $x - 23 = 0$

21. $a + 41 = 7$

22. $a + 30 = 4$

23. $a - 12 = -3$

24. $a - 19 = -11$

25. $9 = y - 8$

26. $97 = -23 + y$

27. $\frac{1}{2} = z - \frac{3}{2}$

28. $\frac{3}{4} + z = -\frac{1}{2}$

29. $t - 0.8 = 4.2$

30. $4 = -9 + t$

31. $25 + x = 10$

32. $85 = x - 20$

33. $1989 = 26 + y$

34. $y - 1.23 = -0.02$

35. Thinking Generally   To solve $x - a = b$ for x, add _____ to each side.

36. Thinking Generally   To solve $x + a = b$ for x, subtract _____ from each side.

THE MULTIPLICATION PROPERTY OF EQUALITY

37. To solve $5x = 4$, multiply each side by _____.

38. To solve $\frac{4}{3}y = 8$, multiply each side by _____.

39. To solve $6x = 11$, divide each side by _____.

40. To solve $0.2x = 4$, divide each side by _____.

Exercises 41–58: Solve the equation. Check your answer.

41. $5x = 15$

42. $-2x = 8$

43. $-7x = 0$

44. $25x = 0$

45. $-5a = -35$

46. $-4a = -32$

47. $3a = -18$

48. $10a = -70$

49. $\frac{1}{2}x = \frac{3}{2}$

50. $\frac{3}{4}x = \frac{5}{8}$

51. $\frac{1}{2} = \frac{2}{5}z$

52. $-\frac{3}{4} = -\frac{1}{8}z$

53. $25 = 5z$

54. $-10 = -4z$

55. $0.5t = 3.5$

56. $2.2t = -9.9$

57. $\frac{3}{8} = \frac{1}{4}y$

58. $1.2 = 0.3y$

59. Thinking Generally   To solve $\frac{1}{a} \cdot x = b$ for x, multiply each side by _____.

60. Thinking Generally   To solve $ax = b$, where $a \neq 0$, for x, divide each side by _____.

APPLICATIONS

61. *Rainfall*   On a stormy day it rains 3 inches before noon and $\frac{1}{2}$ inch per hour thereafter until 6 P.M.
   (a) Make a table that shows the total rainfall R in inches, x hours past noon, ending at 6 P.M.
   (b) Write a formula that calculates R.
   (c) Use your formula to calculate the total rainfall at 3 P.M. Does the answer agree with the value in your table from part (a)?
   (d) How much rain has fallen by 2:15 P.M.?

62. *Cold Weather*   A furnace is turned on at midnight when the temperature inside a cabin is $0°$F. The cabin warms at a rate of $10°$F per hour until 7 A.M.
   (a) Make a table that shows the cabin temperature T in degrees Fahrenheit, x hours past midnight, ending at 7 A.M.
   (b) Write a formula that calculates T.
   (c) Use your formula to calculate the temperature at 5 A.M. Does the answer agree with the value in your table from part (a)?
   (d) Find the cabin temperature at 2:45 A.M.

63. *Football Field*   A football field is 300 feet long.
   (a) Write a formula that gives the length L of x football fields in feet.
   (b) Use your formula to write an equation whose solution gives the number of football fields in 870 feet.
   (c) Solve your equation from part (b).

64. *Acreage*   An acre equals 43,560 square feet.
   (a) Write a formula that converts A acres to S square feet.
   (b) Use your formula to write an equation whose solution gives the number of acres in 871,200 square feet.
   (c) Solve your equation from part (b).

65. *Glacial Melting in Alaska*   (Refer to Example 5.) The Alaskan glaciers are melting at a rate of 24 cubic miles per year. At this rate, how many years will it take for 420 cubic miles of the glacier to melt? (*Source:* USA Today.)

66. *Arctic Ice Cap*   The Arctic ice cap contains 680,000 cubic miles of ice. If this ice cap melted at a rate of 50 cubic miles per year, how many years would it take for the entire ice cap to melt?   (*Source:* Department of the Interior, Geological Survey.)

67. *Cost of a Car*   When the cost of a car is multiplied by 0.07 the result is $1750. Find the cost of the car.

68. *Raise in Salary*   If an employee's salary is multiplied by 1.06, which corresponds to a 6% raise, the result is $58,300. Find the employee's current salary.

WRITING ABOUT MATHEMATICS

69. A student solves an equation as follows.

$$x + 30 = 64$$
$$x \stackrel{?}{=} 64 + 30$$
$$x \stackrel{?}{=} 94$$

Identify the student's mistake. What is the solution?

70. What is a good first step for solving the equation $\frac{a}{b}x = 1$, where a and b are natural numbers? What is the solution? Explain your answers.

2.2 LINEAR EQUATIONS

Basic Concepts ▪ Solving Linear Equations ▪ Applying the Distributive Property ▪ Clearing Fractions and Decimals ▪ Equations with No Solutions or Infinitely Many Solutions

A LOOK INTO MATH ▷

Solving equations is an important concern in mathematics. Billions of dollars are spent each year to solve equations that lead to the creation of better products. If our society could not solve equations, we would not have TV sets, DVD players, satellites, fiber optics, CAT scans, computers, or accurate weather forecasts. In this section we discuss linear equations and some of their applications. One of the simplest types of equations, linear equations, can always be solved by hand.

Basic Concepts

Suppose that a bicyclist is 5 miles from home, riding *away* from home at 10 miles per hour, as shown in Figure 2.1. The distance between the bicyclist and home for various elapsed times is shown in Table 2.2.

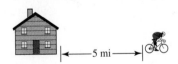

Figure 2.1 Distance from Home

TABLE 2.2   **Distance from Home**

Elapsed Time (hours)	0	1	2	3
Distance (miles)	5	15	25	35

10 miles   10 miles   10 miles

The bicyclist is moving at a constant speed, so the distance increases by 10 miles every hour. The distance D from home after x hours could be calculated by the formula

$$D = 10x + 5.$$

For example, after **2** hours the distance is

$$D = 10(2) + 5 = 25 \text{ miles.}$$

Table 2.2 verifies that the bicyclist is 25 miles from home after 2 hours. However, the table is less helpful if we want to find the elapsed time when the bicyclist is 18 miles from home. To answer this question, we could begin by substituting **18** for D in the formula to obtain the equation

$$18 = 10x + 5.$$

This equation can be written in a different form by applying the addition property of equality. Subtracting 18 from each side allows us to write an equivalent equation.

$$18 - 18 = 10x + 5 - 18 \qquad \text{Subtract 18 from each side.}$$
$$0 = 10x - 13 \qquad \text{Simplify.}$$
$$10x - 13 = 0 \qquad \text{Rewrite the equation.}$$

The equation $10x - 13 = 0$ is an example of a *linear equation* and is solved in Example 3(a). Linear equations often model applications in which things either move or change at a constant rate.

LINEAR EQUATION IN ONE VARIABLE

A **linear equation** in one variable is an equation that can be written in the form

$$ax + b = 0,$$

where a and b are constants with $a \neq 0$.

If an equation is linear, writing it in the form $ax + b = 0$ should not require any properties or processes other than the following:

- using the distributive property to clear any parentheses,
- combining like terms,
- applying the addition property of equality.

For example, the equation $18 = 10x + 5$ is linear because applying the addition property of equality results in $10x - 13 = 0$.

Table 2.3 gives examples of linear equations and values for a and b.

TABLE 2.3   Linear Equations

$ax + b = 0$	a	b
$x - 1 = 0$	1	-1
$-5x + 1 = 0$	-5	1
$2.5x = 0$	2.5	0

To determine whether an equation in one variable is linear we must determine whether the equation can be written in the form $ax + b = 0$. An equation *cannot* be written in this form if after clearing parentheses and combining like terms, any of the following statements are true.

1. The variable has an exponent other than 1.
2. The variable appears in a denominator of a fraction.
3. The variable appears under the symbol $\sqrt{}$ or within an absolute value.

EXAMPLE 1   **Determining whether an equation is linear**

Determine whether the equation is linear. If the equation is linear, give values for a and b.
(a) $4x + 5 = 0$ **(b)** $5 = -\frac{3}{4}x$ **(c)** $4x^2 + 6 = 0$ **(d)** $\frac{3}{x} + 5 = 0$

Solution
(a) The equation is linear because it is in the form $ax + b = 0$ with $a = 4$ and $b = 5$.
(b) The equation can be rewritten as follows.

$$5 = -\frac{3}{4}x \qquad \text{Given equation}$$

$$\frac{3}{4}x + 5 = \frac{3}{4}x + \left(-\frac{3}{4}x\right) \qquad \text{Add } \tfrac{3}{4}x \text{ to each side.}$$

$$\frac{3}{4}x + 5 = 0 \qquad \text{Additive inverse.}$$

This equation is linear because it is in the form $ax + b = 0$ with $a = \frac{3}{4}$ and $b = 5$.

NOTE: If 5 had been subtracted from each side, the result would be $0 = -\frac{3}{4}x - 5$, which is an equivalent linear equation with $a = -\frac{3}{4}$ and $b = -5$.

(c) The equation is *not* linear because it cannot be written in the form $ax + b = 0$. The variable has an exponent other than 1.
(d) The equation is *not* linear because it cannot be written in the form $ax + b = 0$. The variable appears in the denominator of a fraction.   Now Try Exercises 13, 19, 21

Solving Linear Equations

Because a linear equation can be written in the form $ax + b = 0$ with $a \neq 0$, it has *exactly one* solution. Showing that this is true is left as an exercise. Solving a linear equation means finding the value of the variable that makes the equation true.

One way to solve a linear equation is to make a table. Making a table is an organized way of checking possible values of the variable to see if any value makes the equation true. For example, if we want to solve the equation $2x - 5 = -7$, we substitute various values for x in the left side of the equation. If one of these values results in -7, the right side of the equation, the value makes the equation true and is the solution. In the next example a table is used to solve this equation.

EXAMPLE 2   **Using a table to solve an equation**

Complete Table 2.4 for the given values of x. Then solve the equation $2x - 5 = -7$.

TABLE 2.4

x	-3	-2	-1	0	1	2	3
$2x - 5$	-11						

Solution
To complete the table, substitute $x = -2, -1, 0, 1, 2,$ and 3 into the expression $2x - 5$. For example, if $x = -2$, then $2x - 5 = 2(-2) - 5 = -9$. The other values shown in Table 2.5 on the next page can be found similarly.

Now Try Exercise 27

TECHNOLOGY NOTE:
**Graphing Calculators
and Tables**

Many graphing calculators
have the capability to make
tables. Table 2.5 is shown in
the accompanying figure.

X	Y1
-3	-11
-2	-9
-1	-7
0	-5
1	-3
2	-1
3	1

Y1■2X−5

TABLE 2.5

x	−3	−2	−1	0	1	2	3
$2x - 5$	−11	−9	−7	−5	−3	−1	1

From the table, $2x - 5$ equals -7 when $x = -1$. Thus the solution to $2x - 5 = -7$ is -1.

Although tables can be used to solve some linear equations, the process of creating a table that contains the solution can take a significant amount of time. For some equations, choosing appropriate values to be substituted for the variable can be very difficult. For example, the solution to the equation $9x - 4 = 0$ is $\frac{4}{9}$. However, creating a table that reveals this solution would be quite challenging.

A method for solving equations that is more efficient than making a table is to use the following strategy, which involves the addition and multiplication properties of equality.

STEPS FOR SOLVING A LINEAR EQUATION

STEP 1: Use the distributive property to clear any parentheses on each side of the equation. Combine any like terms on each side.

STEP 2: Use the addition property of equality to get all of the terms containing the variable on one side of the equation and all other terms on the other side of the equation. Combine any like terms on each side.

STEP 3: Use the multiplication property of equality to isolate the variable by multiplying each side of the equation by the reciprocal of the number in front of the variable (or divide each side by that number).

STEP 4: Check the solution by substituting it in the given equation.

When a linear equation does not contain parentheses, we can start with the second step in the strategy shown above. This is the case for the equations in the next example.

EXAMPLE 3 Solving linear equations

Solve each linear equation. Check the answer for part (b).
(a) $10x - 13 = 0$ **(b)** $\frac{1}{2}x + 3 = 6$ **(c)** $5x + 7 = 2x + 3$

Solution
(a) First, isolate the x-term on the left side of the equation by adding 13 to each side.

$$10x - 13 = 0 \qquad \text{Given equation}$$
$$10x - 13 + 13 = 0 + 13 \qquad \text{Add 13 to each side. (Step 2)}$$
$$10x = 13 \qquad \text{Add the real numbers.}$$

To obtain a coefficient of 1 on the x-term, divide each side by 10.

$$\frac{10x}{10} = \frac{13}{10} \qquad \text{Divide each side by 10. (Step 3)}$$
$$x = \frac{13}{10} \qquad \text{Simplify.}$$

The solution is $\frac{13}{10}$.

NOTE: In the second equation of the solution the addition property is used, and in the fourth equation of the solution the multiplication property is used.

(b) Start by subtracting 3 from each side.

$$\frac{1}{2}x + 3 = 6 \qquad \text{Given equation}$$

$$\frac{1}{2}x + 3 - 3 = 6 - 3 \qquad \text{Subtract 3 from each side. (Step 2)}$$

$$\frac{1}{2}x = 3 \qquad \text{Subtract the real numbers.}$$

$$2 \cdot \frac{1}{2}x = 2 \cdot 3 \qquad \text{Multiply each side by 2. (Step 3)}$$

$$x = 6 \qquad \text{Multiply the real numbers.}$$

The solution is 6. To check it substitute 6 for x in the given equation.

$$\frac{1}{2} \cdot 6 + 3 \stackrel{?}{=} 6 \qquad \text{Replace } x \text{ with 6. (Step 4)}$$

$$3 + 3 \stackrel{?}{=} 6 \qquad \text{Multiply.}$$

$$6 = 6 \qquad \text{The answer checks.}$$

(c) In this equation there are two x-terms. Generally it is a good idea to move all x-terms to one side of the equation and all real numbers to the other side. To accomplish this task, begin by subtracting $2x$ from each side.

$$5x + 7 = 2x + 3 \qquad \text{Given equation}$$
$$5x - 2x + 7 = 2x - 2x + 3 \qquad \text{Subtract } 2x \text{ from each side. (Step 2)}$$
$$3x + 7 = 3 \qquad \text{Combine like terms.}$$
$$3x + 7 - 7 = 3 - 7 \qquad \text{Subtract 7 from each side. (Step 2)}$$
$$3x = -4 \qquad \text{Simplify.}$$
$$\frac{3x}{3} = \frac{-4}{3} \qquad \text{Divide each side by 3. (Step 3)}$$
$$x = -\frac{4}{3} \qquad \text{Simplify the fractions.}$$

The solution is $-\frac{4}{3}$.

Now Try Exercises 33, 39

▶ **REAL-WORLD CONNECTION** The next example involves a linear equation.

EXAMPLE 4 Estimating numbers of worldwide Internet users

The number of Internet users I in millions during year x can be approximated by the formula

$$I = 116x - 231{,}627,$$

where $x \geq 2000$. Estimate the year when there were 950 million Internet users. (*Source:* Internet World Stats.)

Solution

Let $I = 950$ in the formula $I = 116x - 231{,}627$ and solve for x.

$$950 = 116x - 231{,}627 \qquad \text{Equation to be solved}$$

$$232{,}577 = 116x \qquad \text{Add 231,627 to each side.}$$

$$\frac{232{,}577}{116} = \frac{116x}{116} \qquad \text{Divide each side by 116.}$$

$$\frac{232{,}577}{116} = x \qquad \text{Simplify.}$$

$$x \approx 2005 \qquad \text{Approximate with a calculator.}$$

During 2005 the number of Internet users reached 950 million.   Now Try Exercise 75

Applying the Distributive Property

Sometimes the distributive property is helpful in solving linear equations. The next example demonstrates how to apply the distributive property in such situations. Use of the distributive property was indicated in Step 1 of the strategy for solving linear equations discussed earlier.

EXAMPLE 5   **Applying the distributive property**

Solve each linear equation. Check the answer for part (a).
(a) $4(x - 3) + x = 0$
(b) $2(3z - 4) + 1 = 3(z + 1)$

Solution
(a) Begin by applying the distributive property.

$$4(x - 3) + x = 0 \qquad \text{Given equation}$$

$$4x - 12 + x = 0 \qquad \text{Distributive property (Step 1)}$$

$$5x - 12 = 0 \qquad \text{Combine like terms.}$$

$$5x - 12 + 12 = 0 + 12 \qquad \text{Add 12 to each side. (Step 2)}$$

$$5x = 12 \qquad \text{Add the real numbers.}$$

$$\frac{5x}{5} = \frac{12}{5} \qquad \text{Divide each side by 5. (Step 3)}$$

$$x = \frac{12}{5} \qquad \text{Simplify.}$$

To check whether $\frac{12}{5}$ is the solution, substitute $\frac{12}{5}$ for x in the given equation.

$$4\left(\frac{12}{5} - 3\right) + \frac{12}{5} \overset{?}{=} 0 \qquad \text{Replace } x \text{ with } \tfrac{12}{5}. \text{ (Step 4)}$$

$$4\left(\frac{12}{5} - \frac{15}{5}\right) + \frac{12}{5} \overset{?}{=} 0 \qquad \text{Common denominator}$$

$$4\left(-\frac{3}{5}\right) + \frac{12}{5} \overset{?}{=} 0 \qquad \text{Subtract within parentheses.}$$

$$-\frac{12}{5} + \frac{12}{5} \overset{?}{=} 0 \qquad \text{Multiply.}$$

$$0 = 0 \qquad \text{The answer checks.}$$

(b) Begin by applying the distributive property to each side of the equation. Then move z-terms to the left side and terms containing only real numbers to the right side.

$$2(3z - 4) + 1 = 3(z + 1) \qquad \text{Given equation}$$
$$6z - 8 + 1 = 3z + 3 \qquad \text{Distributive property (Step 1)}$$
$$6z - 7 = 3z + 3 \qquad \text{Add the real numbers.}$$
$$3z - 7 = 3 \qquad \text{Subtract } 3z \text{ from each side. (Step 2)}$$
$$3z = 10 \qquad \text{Add 7 to each side. (Step 2)}$$
$$\frac{3z}{3} = \frac{10}{3} \qquad \text{Divide each side by 3. (Step 3)}$$
$$z = \frac{10}{3} \qquad \text{Simplify.}$$

Now Try Exercises **43, 49**

Clearing Fractions and Decimals

Because most people find it easier to do hand calculations without fractions or decimals, clearing an equation of fractions or decimals before solving for the variable is often helpful. To clear fractions or decimals we often multiply each side of the equation by the least common denominator (LCD).

EXAMPLE 6   Clearing fractions from linear equations

Solve each linear equation.

(a) $\frac{1}{7}x - \frac{5}{7} = \frac{3}{7}$ (b) $\frac{2}{3}x - \frac{1}{6} = x$

Solution

(a) Multiply each side of the equation by the LCD 7 to clear (remove) fractions from the equation.

$$\frac{1}{7}x - \frac{5}{7} = \frac{3}{7} \qquad \text{Given equation}$$
$$7\left(\frac{1}{7}x - \frac{5}{7}\right) = 7 \cdot \frac{3}{7} \qquad \text{Multiply each side by 7.}$$
$$x - 5 = 3 \qquad \text{Distributive property}$$
$$x = 8 \qquad \text{Add 5 to each side.}$$

The solution is 8.

(b) The LCD for 3 and 6 is 6. Multiply each side of the equation by 6.

$$\frac{2}{3}x - \frac{1}{6} = x \qquad \text{Given equation}$$
$$6\left(\frac{2}{3}x - \frac{1}{6}\right) = 6 \cdot x \qquad \text{Multiply each side by 6.}$$
$$4x - 1 = 6x \qquad \text{Distributive property}$$
$$-1 = 2x \qquad \text{Subtract } 4x \text{ from each side.}$$
$$-\frac{1}{2} = x \qquad \text{Divide each side by 2.}$$

The solution is $-\frac{1}{2}$.

Now Try Exercise **55**

EXAMPLE **7**   Clearing decimals from a linear equation

Solve each linear equation.
(a) $0.2x - 0.7 = 0.4$ **(b)** $0.01x - 0.42 = -0.2x$

Solution

(a) The least common denominator for 0.2, 0.7, and 0.4 $\left(\text{or } \frac{2}{10}, \frac{7}{10}, \text{ and } \frac{4}{10}\right)$ is 10. Multiply each side by 10. When multiplying by 10, move the decimal point 1 place to the right.

$$0.2x - 0.7 = 0.4 \qquad \text{Given equation}$$
$$10(0.2x - 0.7) = 10(0.4) \qquad \text{Multiply each side by 10.}$$
$$2x - 7 = 4 \qquad \text{Distributive property}$$
$$2x = 11 \qquad \text{Add 7 to each side.}$$
$$x = \frac{11}{2} \qquad \text{Divide each side by 2.}$$

The solution is $\frac{11}{2}$, or 5.5.

(b) The least common denominator for 0.01, 0.2, and 0.42 $\left(\text{or } \frac{1}{100}, \frac{2}{10}, \text{ and } \frac{42}{100}\right)$ is 100. Multiply each side by 100. To do this move the decimal point 2 places to the right.

$$0.01x - 0.42 = -0.2x \qquad \text{Given equation}$$
$$100(0.01x - 0.42) = 100(-0.2x) \qquad \text{Multiply each side by 100.}$$
$$x - 42 = -20x \qquad \text{Distributive property}$$
$$x - 42 + 20x + 42 = -20x + 20x + 42 \qquad \text{Add 20x and 42.}$$
$$21x = 42 \qquad \text{Combine like terms.}$$
$$x = 2 \qquad \text{Divide each side by 21.}$$

The solution is 2.   Now Try Exercise **53**

Equations with No Solutions or Infinitely Many Solutions

Some equations that appear to be linear are not because when they are written in the form $ax + b = 0$ the value of a is 0 and no x-term appears. This type of equation can have no solutions or infinitely many solutions. An example of an equation that has no solutions is

$$x = x + 1$$

because a number x cannot equal itself plus 1. If we attempt to solve this equation by subtracting x from each side, we obtain the equation $1 = 0$, which is *always false*, indicating there are no solutions. This equation has no x-term.

An example of an equation with infinitely many solutions is

$$5x = 2x + 3x,$$

because the equation simplifies to

$$5x = 5x,$$

which is true for any real number x. If $5x$ is subtracted from each side the result is $0 = 0$, which has no x-term and is *always true*.

NOTE: An equation that is always true is called an **identity** and an equation that is always false is called a **contradiction**.

EXAMPLE 8   Determining numbers of solutions

Determine whether the equation has no solutions, one solution, or infinitely many solutions.

(a) $3x = 2(x + 1) + x$ **(b)** $2x - (x + 1) = x - 1$ **(c)** $5x = 2(x - 4)$

Solution

(a) Start by applying the distributive property.

$3x = 2(x + 1) + x$	Given equation
$3x = 2x + 2 + x$	Distributive property
$3x = 3x + 2$	Combine like terms.
$0 = 2$	Subtract $3x$ from each side.

Because the equation $0 = 2$ is always false, it is a contradiction and there are no solutions.

(b) Start by applying the distributive property.

$2x - (x + 1) = x - 1$	Given equation
$2x - x - 1 = x - 1$	Distributive property
$x - 1 = x - 1$	Combine like terms.
$x = x$	Add 1 to each side.
$0 = 0$	Subtract x from each side.

Because the equation $0 = 0$ is always true, it is an identity and there are infinitely many solutions. Note that the solution set contains all real numbers.

(c) Start by applying the distributive property.

$5x = 2(x - 4)$	Given equation
$5x = 2x - 8$	Distributive property
$3x = -8$	Subtract $2x$ from each side.
$x = -\dfrac{8}{3}$	Divide each side by 3.

Thus there is one solution.

Now Try Exercises 63, 67, 69

CRITICAL THINKING

What must be true about b and d for the equation

$$bx - 2 = dx + 7$$

to have no solutions? What must be true about b and d for this equation to have exactly one solution?

MAKING CONNECTIONS

Number of Solutions

When solving the general form $ax + b = 0$, the resulting equivalent equation will indicate whether the given equation has no solutions, one solution, or infinitely many solutions.

No Solutions: The result is an equation such as $4 = 0$ or $3 = 2$, which is *always false* for any value of the variable.

One Solution: The result is an equation such as $x = 1$ or $x = -12$, which is true for *only one* value of the variable.

Infinitely Many Solutions: The result is an equation such as $0 = 0$ or $-3 = -3$, which is *always true* for any value of the variable.

2.2 PUTTING IT ALL TOGETHER

In this section a four-step approach to solving linear equations was presented on page 102. Some of the other topics discussed in this section are summarized in the following table.

Concept	Comments	Examples
Linear Equation	Can be written as $$ax + b = 0,$$ where $a \neq 0$; has one solution	The equation $5x - 8 = 0$ is linear, with $a = 5$ and $b = -8$. The equation $2x^2 + 4 = 0$ is not linear.
Solving Linear Equations	Use the addition and multiplication properties of equality to isolate the variable.	$5x - 8 = 0$ Given equation $5x = 8$ Add 8 to each side. $x = \dfrac{8}{5}$ Divide each side by 5.
Equations with No Solutions	Some equations that appear to be linear have no solutions. Solving will result in an equivalent equation that is always false.	The equation $$x = x + 5$$ has no solutions because a number cannot equal itself plus 5.
Equations with Infinitely Many Solutions	Some equations that appear to be linear have infinitely many solutions. Solving will result in an equivalent equation that is always true.	The equation $$2x = x + x$$ has infinitely many solutions because the equation is true for all values of x.

2.2 Exercises

CONCEPTS

1. A linear equation can be written in the form _____ with $a \neq 0$.

2. In the linear equation $3x + 2 = 0$, $a =$ _____ and $b =$ _____.

3. How many solutions does a linear equation in one variable have?

4. The equation $x^2 - 1 = 0$ (is/is not) a linear equation.

5. What two properties of equality are frequently used to solve linear equations?

6. What property justifies that $4(x - 3) = 4x - 12$ is true for any value of x?

7. To clear fractions from an equation, multiply each side by the _____.

8. To clear decimals from $0.3x + 1.2 = 0.01$, multiply each side by _____.

9. If solving an equation results in $0 = 4$, how many solutions does it have?

10. If solving an equation results in $0 = 0$, how many solutions does it have?

11. How many solutions does $3x = 2x + x$ have?

12. How many solutions does $x = x + 10$ have?

IDENTIFYING LINEAR EQUATIONS

Exercises 13–26: (Refer to Example 1.) Determine whether the equation is linear. If it is linear, give values for a and b so that the given equation can be written in the form $ax + b = 0$.

13. $3x - 7 = 0$ **14.** $-2x + 1 = 4$

15. $\frac{1}{2}x = 0$ **16.** $-\frac{3}{4}x = 0$

17. $4x^2 - 6 = 11$ **18.** $-2x^2 + x = 4$

19. $\frac{6}{x} - 4 = 2$ **20.** $2\sqrt{x} - 1 = 0$

21. $1.1x = 0.9$ **22.** $-5.7x = -3.4$

23. $2(x - 3) = 0$ **24.** $\frac{1}{2}(x + 4) = 0$

25. $|3x| + 2 = 1$ **26.** $3x = 4x^3$

SOLVING LINEAR EQUATIONS

Exercises 27–30: Evaluate the expression for each value of x in the table. Then use the table to solve the given equation.

27. $-3x + 7 = 1$

x	0	1	2	3	4
$-3x + 7$	7				

28. $5x - 2 = 3$

x	-1	0	1	2	3
$5x - 2$	-7				

29. $4 - 2x = 6$

x	-2	-1	0	1	2
$4 - 2x$	8				

30. $9 - (x + 3) = 4$

x	-2	-1	0	1	2
$9 - (x + 3)$	8				

Exercises 31–60: Solve the equation and check the solution.

31. $11x = 3$ **32.** $-5x = 15$

33. $x - 18 = 5$ **34.** $8 = 5 + 3x$

35. $2x - 1 = 13$ **36.** $4x + 3 = 39$

37. $5x + 5 = -6$ **38.** $-7x - 4 = 31$

39. $3z + 2 = z - 5$ **40.** $z - 5 = 5z - 3$

41. $12y - 6 = 33 - y$ **42.** $-13y + 2 = 22 - 3y$

43. $4(x - 1) = 5$ **44.** $-2(2x + 7) = 1$

45. $1 - (3x + 1) = 5 - x$

46. $6 + 2(x - 7) = 10 - 3(x - 3)$

47. $(5t - 6) + 2(t + 1) = 0$

48. $-2(t - 7) - (t + 5) = 5$

49. $3(4z - 1) - 2(z + 2) = 2(z + 1)$

50. $-(z + 4) + (3z + 1) = -2(z + 1)$

51. $7.3x - 1.7 = 5.6$ **52.** $5.5x + 3x = 51$

53. $-9.5x - 0.05 = 10.5x + 1.05$

54. $0.04x + 0.03 = 0.02x - 0.1$

55. $\frac{1}{2}x - \frac{3}{2} = \frac{5}{2}$ **56.** $-\frac{1}{4}x + \frac{5}{4} = \frac{3}{4}$

57. $-\frac{3}{8}x + \frac{1}{4} = \frac{1}{2}x + \frac{1}{8}$ **58.** $\frac{1}{3}x + \frac{1}{4} = \frac{1}{6} - x$

59. $4y - 2(y + 1) = 0$

60. $(15y + 20) - 5y = 5 - 10y$

61. **Thinking Generally** A linear equation has exactly one solution. Find the solution to the equation $ax + b = 0$, where $a \neq 0$, by solving for x.

62. **Thinking Generally** Solve the linear equation $\frac{1}{a}x - b = 0$ for x.

Exercises 63–72: Determine whether the equation has no solutions, one solution, or infinitely many solutions.

63. $5x = 5x + 1$ **64.** $2(x - 3) = 2x - 6$

65. $8x = 0$ **66.** $9x = x + 1$

67. $5(2x + 7) - (10x + 5) = 30$

68. $4(x + 2) - 2(2x + 3) = 10$

69. $4x = 5(x + 3) - x$

70. $x - (3x + 2) = 15 - 2x$

71. $2x - (x + 5) = x - 5$

72. $5x = 15 - 2(x + 7)$

APPLICATIONS

73. *Distance Traveled*   A bicyclist is 4 miles from home, riding away from home at 8 miles per hour.
 (a) Make a table that shows the bicyclist's distance D from home after 0, 1, 2, 3, and 4 hours.
 (b) Write a formula that calculates D after x hours.
 (c) Use your formula to determine D when $x = 3$ hours. Does your answer agree with the value found in your table?
 (d) Find x when $D = 22$ miles. Interpret the result.

74. *Distance Traveled*   An athlete is 16 miles from home, running *toward* home at 6 miles per hour.
 (a) Write a formula that calculates the distance D that the athlete is from home after x hours.
 (b) Determine D when $x = 1.5$ hours.
 (c) Find x when $D = 5.5$ miles. Interpret the result.

75. *Internet Users*   (Refer to Example 4.) The number of Internet users I in millions during year x, where $x \geq 2000$, can be approximated by the formula

$$I = 116x - 231{,}627.$$

Approximate the year in which there were 490 million Internet users for the first time.

76. *HIV Infections*   The cumulative number of HIV infections N in thousands for the United States in year x can be approximated by the formula

$$N = 42x - 83{,}197,$$

where $x \geq 2000$. Approximate the year when this number reached 970 thousand.   (*Source:* Centers for Disease Control and Prevention.)

77. *State and Federal Inmates*   The number N of state and federal inmates in millions during year x, where $x \geq 2002$, can be approximated by the formula $N = 0.03x - 58.62$. Determine the year in which there were 1.5 million inmates.   (*Source:* Bureau of Justice.)

78. *Government Costs*   From 1960 to 2000 the cost C (in billions of 1992 dollars) to regulate social and economic programs could be approximated by the formula $C = 0.35x - 684$ during year x. Estimate the year in which the cost reached \$6.6 billion.   (*Source:* Center for the Study of American Business.)

WRITING ABOUT MATHEMATICS

79. A student says that the equation $4x - 1 = 1 - x$ is not a linear equation because it is not in the form $ax + b = 0$. Is the student correct? Explain.

80. A student solves a linear equation as follows.

$$4(x + 3) = 5 - (x + 3)$$
$$4x + 3 \stackrel{?}{=} 5 - x + 3$$
$$4x + 3 \stackrel{?}{=} 8 - x$$
$$5x \stackrel{?}{=} 5$$
$$x \stackrel{?}{=} 1$$

Identify and explain the errors that the student made. What is the correct answer?

CHECKING BASIC CONCEPTS
SECTIONS 2.1 AND 2.2

1. Determine whether the equation is linear.
 (a) $4x^3 - 2 = 0$ (b) $2(x + 1) = 4$

2. Evaluate $4x - 3$ for each value of x in the table. Then use the table to solve $4x - 3 = 13$.

x	3	3.5	4	4.5	5
$4x - 3$	9				17

3. Solve each equation and check your answer.
 (a) $x - 12 = 6$
 (b) $\frac{3}{4}z = \frac{1}{8}$
 (c) $0.6t + 0.4 = 2$
 (d) $5 - 2(x - 2) = 3(4 - x)$

4. Determine whether each equation has no solutions, one solution, or infinitely many solutions.
 (a) $x - 5 = 6x$
 (b) $-2(x - 5) = 10 - 2x$
 (c) $-(x - 1) = -x - 1$

5. *Distance Traveled* A driver is 300 miles from home and is traveling toward home on a freeway at a constant speed of 75 miles per hour.
 (a) Write a formula to calculate the distance D that the driver is from home after x hours.
 (b) Write an equation whose solution gives the hours needed for the driver to reach home.
 (c) Solve the equation from part (b).

2.3 INTRODUCTION TO PROBLEM SOLVING

Steps for Solving a Problem ▪ Percent Problems ▪ Distance Problems ▪ Other Types of Problems

A LOOK INTO MATH ▷

One important characteristic of human beings is their ability to solve problems. For example, throughout history people have wanted to predict the weather but have found it difficult to do. After all, how could anyone predict the future? Today, with the help of mathematics, science, computers, and years of data collection, meteorologists are providing increasingly accurate weather forecasts. Solving problems requires problem-solving skills, which we introduce in this section.

Steps for Solving a Problem

Word problems are challenging because formulas and equations are not usually given. To solve such problems we need a strategy. The following steps are based on George Polya's (1888–1985) four-step process for problem solving.

> ### STEPS FOR SOLVING A PROBLEM
>
> **STEP 1:** Read the problem carefully and be sure that you understand it. (You may need to read the problem more than once.). Assign a variable to what you are being asked to find. If necessary, write other quantities in terms of this variable.
>
> **STEP 2:** Write an equation that relates the quantities described in the problem. You may need to sketch a diagram or refer to known formulas.
>
> **STEP 3:** Solve the equation. Use the solution to determine the solution(s) to the original problem. Include any necessary units.
>
> **STEP 4:** Look back and check your solution in the given problem. Does it seem reasonable?

Even if we understand the problem that we are trying to solve, we may not be able to find a solution if we cannot write an appropriate equation. In the next example, we practice the second step in the four-step process by translating sentences into equations.

EXAMPLE 1   Translating sentences into equations

Translate the sentence into an equation using the variable x. Then solve the resulting equation.
(a) Three times a number minus 6 is equal to 18.
(b) The sum of half a number and 5 is zero.
(c) Sixteen is 4 less than twice a number.

Solution

(a) The phrase "Three times a number" indicates that we multiply x by 3 to get $3x$. The word "minus" indicates that we then subtract 6 from $3x$ to get $3x - 6$. This expression "equals" 18, so the equation is $3x - 6 = 18$. The solution is 6 as shown here.

$$3x - 6 = 18 \qquad \text{Equation to be solved}$$
$$3x = 24 \qquad \text{Add 6 to each side.}$$
$$\frac{3x}{3} = \frac{18}{3} \qquad \text{Divide each side by 3.}$$
$$x = 6 \qquad \text{Simplify the fractions.}$$

(b) The word "sum" indicates that we add "half a number" and 5 to get $\frac{1}{2}x + 5$. The word "is" implies equality, so the equation is $\frac{1}{2}x + 5 = 0$. The solution is -10 as shown here.

$$\frac{1}{2}x + 5 = 0 \qquad \text{Equation to be solved}$$
$$\frac{1}{2}x = -5 \qquad \text{Subtract 5 from each side.}$$
$$x = -10 \qquad \text{Multiply each side by 2.}$$

(c) To translate "4 less than twice a number" into a mathematical expression, we write $2x - 4$. If this seems backwards, consider how you would calculate "4 less than your age." The equation is $16 = 2x - 4$. The solution is 10 as shown here.

$$16 = 2x - 4 \qquad \text{Equation to be solved}$$
$$20 = 2x \qquad \text{Add 4 to each side.}$$
$$\frac{20}{2} = \frac{2x}{2} \qquad \text{Divide each side by 2.}$$
$$x = 10 \qquad \text{Simplify and rewrite.}$$

Now Try Exercises 11, 13

In the next example we apply the four-step process to a word problem that involves three unknown numbers.

EXAMPLE 2   **Solving a number problem**

The sum of three consecutive natural numbers is 81. Find the three numbers.

Solution

STEP 1: Start by assigning a variable n to an unknown quantity.

n: smallest of the three natural numbers

Next, write the other two natural numbers in terms of n.

$n + 1$: next consecutive natural number

$n + 2$: largest of the three consecutive natural numbers

STEP 2: Write an equation that relates these unknown quantities. The sum of the three consecutive natural numbers is 81, so the needed equation is

$$n + (n + 1) + (n + 2) = 81.$$

STEP 3: Solve the equation in STEP 2.

$$n + (n + 1) + (n + 2) = 81 \quad \text{Equation to be solved}$$
$$(n + n + n) + (1 + 2) = 81 \quad \text{Commutative and associative properties}$$
$$3n + 3 = 81 \quad \text{Combine like terms.}$$
$$3n = 78 \quad \text{Subtract 3 from each side.}$$
$$n = 26 \quad \text{Divide each side by 3.}$$

The smallest of the three numbers is 26, so the three numbers are 26, 27, and 28.

STEP 4: To check this solution we can add the three numbers to find out if their sum is 81.

$$26 + 27 + 28 = 81$$

The solution checks.   Now Try Exercise 19

▶ **REAL-WORLD CONNECTION**   One way to measure the quality of health care available to babies in various countries is to calculate the infant mortality rate. This rate measures the number of deaths of infants under one year of age per 1000 live births in the same year. A high infant mortality rate may indicate a lack of good quality health care for infants. In the next example, a number problem is solved to find the infant mortality rate in Iceland, which has the lowest rate in the world.

EXAMPLE 3   Solving a number problem

Sierra Leone, in Western Africa, has the highest infant mortality rate in the world at 165. This rate is 15 more than 50 times the rate in Iceland. Find the infant mortality rate in Iceland. (*Source:* World Population Prospectus.)

Solution

If we let x represent the infant mortality rate in Iceland, then "15 more than 50 times the rate in Iceland" can be written $50x + 15$. Because this quantity is equal to 165, the equation that must be solved is $50x + 15 = 165$.

$$50x + 15 = 165 \quad \text{Equation to be solved}$$
$$50x = 150 \quad \text{Subtract 15 from each side.}$$
$$x = 3 \quad \text{Divide each side by 50.}$$

The infant mortality rate in Iceland is 3 deaths per 1000 live births.   Now Try Exercise 31

Percent Problems

▶ **REAL-WORLD CONNECTION**   Problems involving percentages occur in everyday life. Wage increases, sales tax, and government data all make use of percentages. For example, it is estimated that 10% of the world's population will be older than 65 in 2025. As 1% (one percent) represents one part in one hundred, 10 of every 100 people in 2025 will be older than 65. Percent notation can be changed to either fraction or decimal notation. (*Source:* World Health Organization.)

PERCENT NOTATION

The expression $x\%$ represents the fraction $\frac{x}{100}$ or the decimal number $x \times 0.01$.

NOTE: To write $x\%$ as a decimal number, move the decimal point in the number x two places to the *left* and then remove the % symbol.

EXAMPLE 4   Converting percent notation

Convert each percentage to fraction and decimal notation.
(a) 23%   **(b)** 5.2%   **(c)** 0.3%

Solution
(a) *Fraction Notation*: To convert to fraction notation divide 23 by 100. Thus $23\% = \frac{23}{100}$.
Decimal Notation: Move the decimal point two places to the left and then remove the % symbol: $23\% = 0.23$.
(b) *Fraction Notation*: $5.2\% = \frac{5.2}{100} = \frac{52}{1000} = \frac{13 \cdot 4}{250 \cdot 4} = \frac{13}{250}$
Decimal Notation: Move the decimal point two places to the left and then remove the % symbol: $5.2\% = 0.052$.
(c) *Fraction Notation*: $0.3\% = \frac{0.3}{100} = \frac{3}{1000}$
Decimal Notation: Move the decimal point two places to the left and then remove the % symbol: $0.3\% = 0.003$.   Now Try Exercise **35**

EXAMPLE 5   Converting to percent notation

Convert each real number to a percentage.
(a) 0.234   **(b)** $\frac{1}{4}$ **(c)** 2.7

Solution
(a) Move the decimal point two places to the *right* and then insert the % symbol to obtain $0.234 = 23.4\%$.
(b) $\frac{1}{4} = 0.25$, so $\frac{1}{4} = 25\%$.
(c) Move the decimal point two places to the right and then insert the % symbol to obtain $2.7 = 270\%$. Note that percentages can be greater than 100%.   Now Try Exercise **43**

PERCENT CHANGE   When prices increase (or decrease), the actual amount of the increase (or decrease) is often not as significant as the *percent change* in the price. For example, if the price of a new home increases from $250,000 to $251,000, this price increase may not seem as significant as an increase in college tuition from $4500 to $5500. Even though both prices increased by $1000, the percent increase in the price of the home is four-tenths of 1% and the percent increase in the cost of tuition is more than 22%. The increase in tuition is much more dramatic than the increase in the price of the home.

If an amount changes from an **old value** to a **new value**, the **percent change** is given by

$$\frac{\text{new value} - \text{old value}}{\text{old value}} \times 100.$$

The reason for multiplying by 100 is to change the decimal representation to a percentage.

NOTE: A positive percent change corresponds to an increase and a negative percent change corresponds to a decrease.

EXAMPLE 6   Calculating percent increase

From 1995 to 2005 consumer credit increased from $1.1 trillion to $2.5 trillion. Calculate the percent increase in consumer credit from 1995 to 2005.   (*Source:* Federal Reserve Bulletin.)

Solution

The **old value** is **1.1** and the **new value** is **2.5**. The percent increase in consumer credit is

$$\frac{\textbf{new value} - \textbf{old value}}{\textbf{old value}} \times 100 = \frac{\textbf{2.5} - \textbf{1.1}}{\textbf{1.1}} \times 100 \approx 127\%.$$

Now Try Exercise 55

As we previously indicated, percentages frequently occur in applications.

EXAMPLE 7   Solving a percent problem

During the 2003–2004 academic year, tuition and fees at public colleges and universities were $4694, on average, and increased by 9.3% during the next year. Find the average cost of tuition and fees during the 2004–2005 academic year. Round your answer to the nearest dollar.

Solution

STEP 1: Let T represent the tuition and fees during 2004–2005.

STEP 2: T equals $4694 plus 9.3% of $4694 and 9.3% = 0.093, so

$$T = 4694 + (0.093)4694.$$

NOTE: The word "of" in percent problems often indicates multiplication.

STEP 3: To find T, evaluate the expression on the right side to get

$$T = 4694 + (0.093)4694 \approx 5131.$$

Tuition and fees were $5131.

STEP 4: To check the answer, calculate the percent increase. Tuition and fees went up by

$$\frac{5131 - 4694}{4694} \times 100 \approx 9.3\%,$$

so the answer checks.

Now Try Exercise 57

CRITICAL THINKING

If your salary increased 200%, by what factor did it increase?

EXAMPLE 8   Solving a percent problem

In 2025, 10% of the world's population, or 800 million people, will be older than 65. Find the estimated population of the world in 2025.

Solution

STEP 1: Let P represent the world's population in millions in 2025.

STEP 2: 10% of P equals 800 million. As 10% = 0.10, this information is described by

$$0.10P = 800.$$

STEP 3: To solve the equation in STEP 2, multiply each side by 10.

$$0.10P = 800 \qquad \text{Equation to be solved}$$
$$P = 8000 \qquad \text{Multiply by 10.}$$

In 2025 the estimated world population is 8000 million, or 8 billion.

STEP 4: To check the answer, determine whether 10% of 8 billion is 800 million.

$$(0.10)8{,}000{,}000{,}000 = 800{,}000{,}000$$

The answer checks.

Now Try Exercise **61**

Distance Problems

If a person drives on an interstate highway at 70 miles per hour for 3 hours, then the total distance traveled is $70 \cdot 3 = 210$ miles. In general, $d = rt$, where d is the distance traveled, r is the rate (or speed), and t is time. In this example, the distance is in miles, the time is in hours, and the rate is expressed in miles per hour. In general, the rate in a distance problem is expressed in units of distance per unit of time.

EXAMPLE **9**   Solving a distance problem

A person drives for 2 hours and 30 minutes at a constant speed and travels 180 miles. See Figure 2.2. Find the speed of the car in miles per hour.

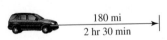
180 mi
2 hr 30 min

Figure 2.2

Solution

STEP 1: Let r represent the car's rate, or speed, in miles per hour.

STEP 2: The rate is to be given in miles per hour, so change 2 hours and 30 minutes to 2.5 or $\frac{5}{2}$ hours. Because $d = 180$ and $t = \frac{5}{2}$, the equation $d = rt$ becomes

$$180 = \frac{5}{2} \cdot r.$$

STEP 3: Solve the equation in STEP 2 for r by multiplying each side of the equation by $\frac{2}{5}$, which is the reciprocal of $\frac{5}{2}$.

$$180 = r \cdot \frac{5}{2} \qquad \text{Equation to be solved}$$

$$\frac{2}{5} \cdot 180 = r \cdot \frac{2}{5} \cdot \frac{5}{2} \qquad \text{Multiply each side by } \tfrac{2}{5}.$$

$$72 = r \qquad \text{Simplify.}$$

The speed of the car is 72 miles per hour.

STEP 4: Because 2 hours and 30 minutes is equivalent to $\frac{5}{2}$ hours, traveling for 2 hours and 30 minutes at a constant rate of 72 miles per hour results in a distance of

$$d = rt = 72 \cdot \frac{5}{2} = 180 \text{ miles.}$$

The answer checks.

Now Try Exercise **71**

EXAMPLE **10**   Solving a distance problem

An athlete jogs at two speeds, covering a distance of 7 miles in $\frac{3}{4}$ hour. If the athlete runs $\frac{1}{4}$ hour at 8 miles per hour, find the second speed.

Solution

STEP 1: Let r represent the second speed of the jogger in miles per hour.

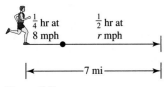

$\frac{1}{4}$ hr at 8 mph $\frac{1}{2}$ hr at r mph

|←——— 7 mi ———→|

Figure 2.3

STEP 2: The total time spent jogging is $\frac{3}{4}$ hour, so the time spent jogging at the second speed must be $\frac{3}{4} - \frac{1}{4} = \frac{1}{2}$ hour. The total distance of 7 miles is the result of jogging at 8 miles per hour for $\frac{1}{4}$ hour and at r miles per hour for $\frac{1}{2}$ hour. See Figure 2.3. The distance traveled at 8 miles per hour for $\frac{1}{4}$ hour is given by $8 \cdot \frac{1}{4}$ and the distance traveled at r miles per hour for $\frac{1}{2}$ hour is given by $r \cdot \frac{1}{2}$. The sum of these distances must equal 7 miles. Thus

$$8 \cdot \frac{1}{4} + r \cdot \frac{1}{2} = 7.$$

STEP 3: Solve the equation in STEP 2 for r.

$$8 \cdot \frac{1}{4} + r \cdot \frac{1}{2} = 7 \qquad \text{Equation to be solved}$$

$$2 + \frac{r}{2} = 7 \qquad \text{Simplify.}$$

$$\frac{r}{2} = 5 \qquad \text{Subtract 2 from each side.}$$

$$r = 10 \qquad \text{Multiply each side by 2.}$$

The athlete's second speed is 10 miles per hour.

STEP 4: Jogging at a rate of 8 miles per hour for $\frac{1}{4}$ hour results in a distance of $8 \cdot \frac{1}{4} = 2$ miles. Jogging at a rate of 10 miles per hour for $\frac{1}{2}$ hour results in a distance of $10 \cdot \frac{1}{2} = 5$ miles. The total distance is $2 + 5 = 7$ miles and the total time is $\frac{1}{4} + \frac{1}{2} = \frac{3}{4}$ hour, so the answer checks.   Now Try Exercise 75

Other Types of Problems

Many applied problems involve linear equations. For example, right after people have their wisdom teeth pulled, they may need to rinse their mouth with salt water. In the next example, we use linear equations to determine how much water must be added to dilute a concentrated saline solution.

EXAMPLE 11   Diluting a saline solution

A solution contains 4% salt. How much pure water should be added to 30 ounces of the solution to dilute it to a 1.5% solution?

Solution
STEP 1: Assign a variable x as follows.

x: ounces of pure water (0% salt solution)

30: ounces of 4% salt solution

$x + 30$: ounces of 1.5% salt solution

In Figure 2.4 on the next page, three beakers illustrate this situation.

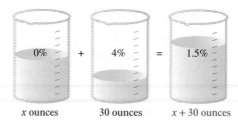

x ounces   30 ounces   x + 30 ounces

Figure 2.4 Mixing a Saline Solution

STEP 2: Note that the amount of salt in the first two beakers must equal the amount of salt in the third beaker. We use Table 2.6 to organize our calculations. The amount of salt in a solution equals the concentration times the solution amount, as shown in the last column of the table.

TABLE 2.6   **Mixing a Saline Solution**

Solution Type	Concentration (as a decimal)	Solution Amount (ounces)	Salt (ounces)
Pure Water	0% = 0.00	x	$0.00x$
Initial Solution	4% = 0.04	30	0.04(30)
Final Solution	1.5% = 0.015	$x + 30$	$0.015(x + 30)$

The amount of salt in the first two beakers is

$$0.00x + 0.04(30) = 0 + 1.2 = 1.2 \text{ ounces.}$$

The amount of salt in the final beaker is

$$0.015(x + 30) \text{ ounces.}$$

Because the amounts of salt in the solutions before and after mixing must be equal, the following equation must hold.

$$0.015(x + 30) = 1.2$$

STEP 3: Solve the equation in STEP 2.

$$0.015(x + 30) = 1.2 \qquad \text{Equation to be solved}$$
$$0.015x + 0.45 = 1.2 \qquad \text{Distributive property}$$
$$0.015x = 0.75 \qquad \text{Subtract 0.45 from each side.}$$
$$\frac{0.015x}{0.015} = \frac{0.75}{0.015} \qquad \text{Divide each side by 0.015.}$$
$$x = 50 \qquad \text{Simplify fractions.}$$

Fifty ounces of water should be added.

STEP 4: Adding 50 ounces of water will yield $50 + 30 = 80$ ounces of water containing $0.04(30) = 1.2$ ounces of salt. The concentration is $\frac{1.2}{80} = 0.015$ or 1.5%, so the answer checks.   Now Try Exercise 79

▶ **REAL-WORLD CONNECTION**   Many times interest rates for student loans vary. In the next example, we present a situation in which a student has to borrow money at two different interest rates.

EXAMPLE 12   Calculating interest on college loans

A student takes out a loan for a limited amount of money at 5% interest and then must pay 7% for any additional money. If the student borrows $2000 more at 7% than at 5%, then the total interest for one year is $440. How much does the student borrow at each rate?

Solution

STEP 1: Assign a variable x as follows.

$$x: \text{loan amount at 5\% interest}$$
$$x + 2000: \text{loan amount at 7\% interest}$$

STEP 2: The amount of interest paid for the 5% loan is 5% of x, or $0.05x$. The amount of interest paid for the 7% loan is 7% of $x + 2000$, or $0.07(x + 2000)$. The total interest equals $440, so we solve the equation

$$0.05x + 0.07(x + 2000) = 440.$$

STEP 3: Solve the equation in STEP 2 for x.

$0.05x + 0.07(x + 2000) = 440$	Equation to be solved
$5x + 7(x + 2000) = 44{,}000$	Multiply by 100 to clear decimals.
$5x + 7x + 14{,}000 = 44{,}000$	Distributive property
$12x + 14{,}000 = 44{,}000$	Combine like terms.
$12x = 30{,}000$	Subtract 14,000 from each side.
$x = 2500$	Divide each side by 12.

The student borrows $2500 at 5% and $2500 + 2000 = \$4500$ at 7%.

STEP 4: The amount of interest on $4500 at 5% is $0.05(2500) = \$125$ and the amount of interest on $4500 at 7% is $0.07(4500) = \$315$. Thus the total interest is given by $125 + 315 = \$440$. Furthermore, the amount borrowed at 7% is $2000 more than the amount borrowed at 5%. The answer checks.   Now Try Exercise 81

2.3 PUTTING IT ALL TOGETHER

In this section on page 111 we presented a four-step approach to problem solving. However, because no approach works in every situation, solving mathematical problems takes time, effort, and creativity.

Concept	Comments	Examples
Percent Notation	$x\%$ represents either the fraction $\frac{x}{100}$ or the decimal $x \times 0.01$.	$17\% = \frac{17}{100} = 0.17$ $1.5\% = \frac{1.5}{100} = \frac{15}{1000} = 0.015$ $234\% = \frac{234}{100} = 2.34$
Converting Fractions to Percent Notation	Divide the numerator by the denominator, multiply by 100, and insert the % symbol.	$\frac{1}{2} = 0.5 = 50\%$ $\frac{2}{3} = 0.\overline{6} \approx 66.7\%$ $\frac{8}{5} = 1.6 = 160\%$

continued on next page

continued from previous page

Concept	Comments	Examples
Percent Change	If a quantity changes from an old value to a new value, then the percent change is $$\frac{\text{(new value)} - \text{(old value)}}{\text{(old value)}} \times 100.$$	If a price changes from \$1.50 to \$1.35, the percent change is $$\frac{1.35 - 1.50}{1.50} \times 100 = -10\%.$$ The price decreases by 10%.
Distance Problems	Distance d equals rate r times time t, or $$d = r \cdot t.$$	If a car travels 60 mph for 3 hours, then the distance d is $$d = r \cdot t = 60 \cdot 3 = 180 \text{ miles.}$$

2.3 Exercises

 MyMathLab  Math XL PRACTICE WATCH DOWNLOAD READ REVIEW

CONCEPTS

1. When you are solving a word problem, what is the last step?

2. Given an integer n, what are the next two consecutive integers?

3. The expression $x\%$ equals the fraction _____ .

4. The expression $x\%$ equals the decimal $x \times$ _____.

5. To write 63.2% as a decimal, move the decimal point 2 places (left/right) and then remove the % symbol.

6. To write 0.349 as a percentage, move the decimal point 2 places (left/right) and then insert the % symbol.

7. If a price changes from P_1 to P_2, then the percent change equals _____ .

8. A positive percent change corresponds to a(n) _____ and a negative percent change corresponds to a(n) _____.

9. If a car travels at speed r for time t, then distance is given by $d = $ _____.

10. In general, the rate in a distance problem is expressed in units of _____ per unit of _____.

NUMBER PROBLEMS

Exercises 11–18: Using the variable x, translate the sentence into an equation. Solve the resulting equation.

11. The sum of 2 and a number is 12.

12. Twice a number plus 7 equals 9.

13. A number divided by 5 equals the number decreased by 24.

14. 25 times a number is 125.

15. If a number is increased by 5 and then divided by 2, the result is 7.

16. A number subtracted from 8 is 5.

17. The quotient of a number and 2 is 17.

18. The product of 5 and a number equals 95.

Exercises 19–28: Find the number or numbers.

19. The sum of three consecutive natural numbers is 96.

20. The sum of three consecutive integers is -123.

21. Three times a number equals 102.

22. A number plus 18 equals twice the number.

23. Five times a number is 24 more than twice the number.

24. Three times a number is 18 less than the number.

25. Six times a number divided by 7 equals 18.

26. Two less than twice a number, divided by 5 equals 4.

27. Four times the sum of a number and 5 equals 64.

28. The opposite of the sum of a number and −5 equals 24.

29. *Finding Age*   In 10 years, a child will be 3 years older than twice her current age. What is the current age of the child?

30. *Weight Loss*   After losing 30 pounds, an individual weighs 110 pounds more than one-third his previous weight. Find the previous weight of the individual.

31. *Hazardous Waste*   In 2005, the number of federal hazardous waste sites in California was 2 less than twice the number of sites in Washington. How many hazardous waste sites were there in Washington if there were 24 such sites in California? (**Source:** Environmental Protection Agency.)

32. *Endangered Species*   There were 77 birds on the endangered species list in 2005. This is 7 more than 5 times the number of reptiles on the list. How many reptiles were on the endangered species list in 2005? (**Source:** U.S. Fish and Wildlife Service.)

33. *Energy Production*   In 1995, U.S. hydroelectric power production hit an all-time high of 311 billion kilowatthours. This is 23 billion kilowatt-hours more than three times the 1950 production level. Find the amount of hydroelectric power produced in 1950. (**Source:** U.S. Department of Energy.)

34. *Open Heart Surgery*   In 2002, open heart surgery was performed 368 thousand times in the United States on persons over the age of 65. This is 22 thousand fewer than 13 times the number of this type of surgery performed on persons under the age of 15. How many times was open heart surgery performed on persons under the age of 15 in 2002? (**Source:** U.S. Department of Health and Human Services.)

PERCENT PROBLEMS

Exercises 35–42: Convert the percentage to fraction and decimal notation.

35. 37%

36. 52%

37. 148%

38. 252%

39. 6.9%

40. 8.1%

41. 0.05%

42. 0.12%

Exercises 43–54: Convert the number to a percentage.

43. 0.45

44. 0.08

45. 1.8

46. 2.97

47. 0.006

48. 0.0001

49. $\frac{2}{5}$

50. $\frac{1}{3}$

51. $\frac{3}{4}$

52. $\frac{7}{20}$

53. $\frac{5}{6}$

54. $\frac{53}{50}$

55. *Voter Turnout*   In the 1980 election for president there were 86.5 million voters, whereas in 2004 there were 122.3 million voters. Find the percent change in the number of voters.

56. *College Degrees*   In 2000, about 457,000 people received a master's degree, and by 2005 this number had increased to 506,000. Find the percent change in the number of master's degrees received over this time period.   (**Source:** The College Board.)

57. *Wages*   A part-time instructor is receiving $950 per credit taught. If the instructor receives a 4% increase, how much will the new per credit compensation be?

58. *Tuition Increase*   Tuition is currently $125 per credit. There are plans to raise tuition by 8% for next year. What will the new tuition be per credit?

59. *Income Tax Returns*   In 2004, the Internal Revenue Service processed 131 million individual income tax returns, up 0.46% from the preceding year. How many individual income tax returns were processed in 2003? (**Source:** Internal Revenue Service.)

60. *Medicare Enrollment*   In 2004, 41.2 million people were enrolled in Medicare, an increase of 5.1% from 2000. How many people were enrolled in Medicare in 2000?   (**Source:** Health Care Financing Administration.)

61. *AIDS Deaths*   There were 18,017 AIDS deaths in 2003, or 34.5% of the 1995 AIDS deaths. Determine the number of AIDS deaths in 1995.   (*Source:* Centers for Disease Control and Prevention.)

62. *Rural Forestland*   There are about 13.6 million acres of rural forestland in Wisconsin. This is about 38% of the state's total area. Approximate the total area of Wisconsin in millions of acres.   (*Source:* U.S. Department of Agriculture.)

63. Thinking Generally   Calculate the percent change if the price of an item increases from $1.20 to $1.50. Now calculate the percent change if the price of the item decreases from $1.50 to $1.20.

64. Thinking Generally   Refer to the previous exercise. Suppose an amount increases from A_1 to A_2 and the percent change is calculated to be 30%. If that amount now decreases from A_2 back to A_1, will the percent change be -30%?

DISTANCE PROBLEMS

Exercises 65–70: Use the formula $d = rt$ to find the value of the missing variable.

65. $r = 4$ mph, $t = 2$ hours

66. $r = 70$ mph, $t = 2.5$ hours

67. $d = 1000$ feet, $t = 50$ seconds

68. $d = 1250$ miles, $t = 5$ days

69. $d = 200$ miles, $r = 40$ mph

70. $d = 1700$ feet, $r = 10$ feet per second

71. *Driving a Car*   A person drives a car at a constant speed for 4 hours and 15 minutes, traveling 255 miles. Find the speed of the car in miles per hour.

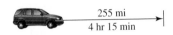
255 mi
4 hr 15 min

72. *Flying an Airplane*   A pilot flies a plane at a constant speed for 5 hours and 30 minutes, traveling 715 miles. Find the speed of the plane in miles per hour.

73. *Jogging Speeds*   One runner passes another runner traveling in the same direction on a hiking trail. The faster runner is jogging 2 miles per hour faster than the slower runner. Determine how long it will be before the faster runner is $\frac{3}{4}$ mile ahead of the slower runner.

74. *Distance Running*   An athlete runs 8 miles, first at a slower speed and then at a faster speed. The total time spent running is 1 hour. If the athlete runs $\frac{1}{3}$ hour at 6 miles per hour, find the second speed.

75. *Jogging Speeds*   At first an athlete jogs at 5 miles per hour and then jogs at 8 miles per hour, traveling 7 miles in 1.1 hours. How long does the athlete jog at each speed? (*Hint:* Let t represent the amount of time the athlete jogs at 5 mph. Then $1.1 - t$ represents the amount of time the athlete jogs at 8 mph.)

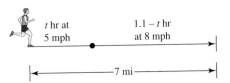
t hr at
5 mph

$1.1 - t$ hr
at 8 mph

7 mi

76. *Distance and Time*   A bus is 160 miles east of the North Dakota–Montana border and is traveling west at 70 miles per hour. How long will it take for the bus to be 295 miles west of the border?

77. *Distance and Time*   A plane is 300 miles west of Chicago, Illinois, and is flying west at 500 miles per hour. How long will it take for the plane to be 2175 miles west of Chicago?

500 mph   300 mi
Chicago
2175 mi

78. *Finding Speeds*   Two cars pass on a straight highway while traveling in opposite directions. One car is traveling 6 miles per hour faster than the other car. After 1.5 hours the two cars are 171 miles apart. Find the speed of each car.

OTHER TYPES OF PROBLEMS

79. *Saline Solution*   (Refer to Example 11.) A solution contains 3% salt. How much water should be added to 20 ounces of this solution to make a 1.2% solution?

80. *Acid Solution*   A solution contains 15% hydrochloric acid. How much water should be added to 50 milliliters of this solution to dilute it to a 2% solution?

81. *College Loans*   (Refer to Example 12.) A student takes out two loans, one at 5% interest and the other at 6% interest. The 5% loan is $1000 more than the 6% loan, and the total interest for 1 year is $215. How much is each loan?

82. *Bank Loans* Two bank loans, one for $5000 and the other for $3000, cost a total of $550 in interest for one year. The $5000 loan has an interest rate 3% lower than the interest rate for the $3000 loan. Find the interest rate for each loan.

83. *Mixing Antifreeze* How many gallons of 70% antifreeze should be mixed with 10 gallons of 30% antifreeze to obtain a 45% antifreeze mixture?

84. *Mixing Antifreeze* How many gallons of 65% antifreeze and how many gallons of 20% antifreeze should be mixed to obtain 50 gallons of a 56% mixture of antifreeze? (*Hint:* Let x represent the number of gallons of 65% antifreeze. Then $50 - x$ represents the amount of 20% antifreeze.)

85. *Hydrocortisone Cream* A pharmacist needs to make a 1% hydrocortisone cream. How many grams of 2.5% hydrocortisone cream should be added to 15 grams of cream base (0% hydrocortisone) to make the 1% cream?

86. *Credit Card Debt* A person carries a balance on two credit cards, one with a monthly interest rate of 1.5% and the other with a monthly rate of 1.75%. The balance on the 1.5% card is $600 less than the balance on the 1.75% card. If the total interest for the month is $49.50, what is the balance on each card?

WRITING ABOUT MATHEMATICS

87. State the four steps for solving a word problem.

88. The cost of living has increased about 600% during the past 50 years. Does this percent change correspond to a cost of living increase of 6 times? Explain.

GROUP ACTIVITY
WORKING WITH REAL DATA

Directions: Form a group of 2 to 4 people. Select someone to record the group's responses for this activity. All members of the group should work cooperatively to answer the questions. If your instructor asks for your results, each member of the group should be prepared to respond.

Exercises 1–5: In this set of exercises you are to use your mathematical problem-solving skills to find the thickness of a piece of aluminum foil without measuring it directly.

1. *Area of a Rectangle* The area of a rectangle equals length times width. Find the area of a rectangle with length 12 centimeters and width 11 centimeters.

2. *Volume of a Box* The volume of a box equals length times width times height. Find the volume of the box shown, which is 12 centimeters long, 11 centimeters wide, and 5 centimeters high.

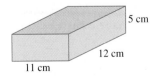

3. *Height of a Box* Suppose that a box has a volume of 100 cubic centimeters and that the area of the bottom of the box is 50 square centimeters. Find the height of the box.

4. *Volume of Aluminum Foil* One cubic centimeter of aluminum weighs 2.7 grams. If a piece of aluminum foil weighs 5.4 grams, find the volume of the aluminum foil.

5. *Thickness of Aluminum Foil* A rectangular sheet of aluminum foil is 50 centimeters long and 20 centimeters wide, and weighs 5.4 grams. Find the thickness of the aluminum foil in centimeters.

2.4 FORMULAS

Basic Concepts ▪ Formulas from Geometry ▪ Solving for a Variable ▪ Other Formulas

A LOOK INTO MATH ▷

Have you ever wondered how the registrar's office calculates your grade point average (GPA)? A formula is used that involves the number of credits earned at each possible grade. Once a formula has been derived, it can be used over and over by any number of people. This fact makes formulas a convenient and easy way to solve certain types of recurring problems. In this section we discuss several types of formulas.

Basic Concepts

A *formula* is an equation that can be used to calculate an unknown quantity by using known values of other quantities. Formulas also establish a relationship between different quantities. For example, to calculate the area A of a rectangular room with length L and width W, the formula

$$A = LW$$

can be used. If a dormitory room is 16 feet by 10 feet, then its area is

$$A = 16 \cdot 10 = 160 \text{ square feet.}$$

The formula $A = LW$ calculates area by using the quantities of length and width.

Another useful formula is $M = \frac{D}{G}$, which can be used to calculate a car's gas mileage M after it has traveled D miles on G gallons of gasoline.

EXAMPLE 1   Calculating mileage of a trip

A tourist starts a trip with a full tank of gas and an odometer that reads 45,682 miles. At the end of the trip, it takes 9.7 gallons of gas to fill the tank, and the odometer reads 45,903 miles. Find the gas mileage for the car.

Solution

The distance traveled is $D = 45,903 - 45,682 = \mathbf{221}$ miles and the number of gallons used is $G = \mathbf{9.7}$. Thus

$$M = \frac{D}{G} = \frac{\mathbf{221}}{\mathbf{9.7}} \approx 22.8 \text{ miles per gallon.} \qquad \boxed{\text{Now Try Exercise } 81}$$

Formulas from Geometry

Formulas from geometry are frequently used in various fields, including surveying and construction. In this subsection we discuss several important formulas from geometry.

TRIANGLES   If a triangle has base b and height h, where b and h have the *same* units, then its area is $A = \frac{1}{2}bh$, as shown in Figure 2.5(a). If a triangle has a base of 4 feet and a height of 18 inches, then its area is

$$A = \frac{1}{2}bh = \frac{1}{2}(4)(\mathbf{1.5}) = 3 \text{ square feet,}$$

as illustrated in Figure 2.5(b). Note that 18 inches needs to be converted to **1.5** feet so that all the units of length are the same. If the base had been converted to 48 inches, then the area would have been

$$A = \frac{1}{2}bh = \frac{1}{2}(48)(18) = 432 \text{ square } inches.$$

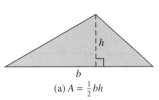

(a) $A = \frac{1}{2}bh$ (b)

Figure 2.5 Triangles

EXAMPLE 2   Calculating area of a region

A residential lot is shown in Figure 2.6. It comprises a rectangular region and an adjacent triangular region.
(a) Find the area of this lot.
(b) An acre contains 43,560 square feet. How many acres are there in this lot?

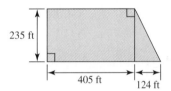

Figure 2.6

Solution
(a) The rectangular portion of the lot has length **405** feet and width **235** feet. Its area A_R is

$$A_R = LW = 405 \cdot 235 = 95{,}175 \text{ square feet.}$$

The triangular region has base **124** feet and height **235** feet. Its area A_T is

$$A_T = \frac{1}{2}bh = \frac{1}{2} \cdot 124 \cdot 235 = 14{,}570 \text{ square feet.}$$

The total area A of the lot equals the sum of A_R and A_T.

$$A = 95{,}175 + 14{,}570 = 109{,}745 \text{ square feet}$$

(b) Each acre equals 43,560 square feet, so divide 109,745 by 43,560 to calculate the number of acres.

$$\frac{109{,}745}{43{,}560} \approx 2.5 \text{ acres} \qquad \text{Now Try Exercise } 29$$

ANGLES   Angles are often measured in degrees. A **degree** (°) is $\frac{1}{360}$ of a revolution, so there are 360° in one complete revolution. In any triangle, the sum of the measures of the angles equals 180°. In Figure 2.7, triangle ABC has angles with measures x, y, and z. Therefore

$$x + y + z = 180°.$$

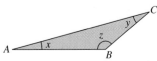

Figure 2.7

EXAMPLE 3   Finding angles in a triangle

In a triangle the two smaller angles are equal in measure and are half the measure of the largest angle. Find the measure of each angle.

Solution

Let x represent the measure of each of the two smaller angles, as illustrated in Figure 2.8. Then the measure of the largest angle is $2x$, and the sum of the measures of the three angles is given by

$$x + x + 2x = 180°.$$

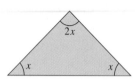

Figure 2.8

This equation can be solved as follows.

$$x + x + 2x = 180° \qquad \text{Equation to be solved}$$
$$4x = 180° \qquad \text{Combine like terms.}$$
$$\frac{4x}{4} = \frac{180°}{4} \qquad \text{Divide each side by 4.}$$
$$x = 45° \qquad \text{Divide the real numbers.}$$

The measure of the largest angle is $2x = 2 \cdot 45° = 90°$. Thus the measures of the three angles are 45°, 45°, and 90°.   Now Try Exercise 35

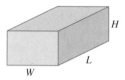

Figure 2.9
Rectangular Box

BOXES   The box in Figure 2.9 has length L, width W, and height H. Its volume V is given by

$$V = LWH.$$

The surface of the box comprises six rectangular regions: top and bottom, front and back, and left and right sides. The total surface area S of the box is given by

$$S = LW + LW + WH + WH + LH + LH.$$
$$\text{(top + bottom + front + back + left side + right side)}$$

When we combine like terms, this expression simplifies to

$$S = 2LW + 2WH + 2LH.$$

EXAMPLE 4   Finding the volume and surface area of a box

Find the volume and surface area of the box shown in Figure 2.10.

5 in.

10 in.

8 in.

Figure 2.10

Solution

Figure 2.10 shows that the box has length $L = 10$ inches, width $W = 8$ inches, and height $H = 5$ inches. The volume of the box is

$$V = LWH = 10 \cdot 8 \cdot 5 = 400 \text{ cubic inches.}$$

The surface area of the box is

$$S = 2LW + 2WH + 2LH$$
$$= 2(10)(8) + 2(8)(5) + 2(10)(5)$$
$$= 160 + 80 + 100$$
$$= 340 \text{ square inches.}$$ Now Try Exercise 43

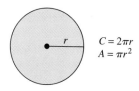

Figure 2.11 Circle

$C = 2\pi r$
$A = \pi r^2$

CIRCLES   The *radius* of a circle is the distance from its center to the perimeter of the circle. The perimeter of a circle is called its **circumference** C and is given by $C = 2\pi r$, where r is the radius of the circle. The area A of a circle is given by $A = \pi r^2$. See Figure 2.11. The distance across a circle, through its center, is called the *diameter*. Note that a circle's radius is half of its diameter. (Recall that $\pi \approx 3.14$.)

EXAMPLE 5   **Finding the circumference and area of a circle**

A circle has a diameter of 25 inches. Find its circumference and area.

Solution
The radius is half the diameter, or **12.5** inches.

$$\textit{Circumference: } C = 2\pi r = 2\pi(\mathbf{12.5}) = 25\pi \approx 78.5 \text{ inches.}$$
$$\textit{Area: } A = \pi r^2 = \pi(\mathbf{12.5})^2 = 156.25\pi \approx 491 \text{ square inches.}$$

Now Try Exercise 39

CRITICAL THINKING

Write a formula for the circumference and area of a circle having diameter d.

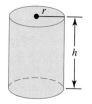

Figure 2.12
Cylinder

CYLINDERS   A soup can is usually made in the shape of a cylinder. The volume of a cylinder having radius r and height h is $V = \pi r^2 h$. See Figure 2.12.

EXAMPLE 6   **Calculating the volume of a soda can**

A cylindrical soda can has a radius of $1\frac{1}{4}$ inches and a height of $4\frac{3}{8}$ inches.
(a) Find the volume of the can.
(b) If 1 cubic inch equals 0.554 fluid ounces, find the number of fluid ounces in the can.

Solution
(a) Changing mixed numbers to improper fractions gives the radius as $r = \frac{5}{4}$ inches and the height as $h = \frac{35}{8}$ inches.

$$V = \pi r^2 h \qquad\qquad \text{Volume of the soda can}$$
$$= \pi \left(\frac{5}{4}\right)^2\left(\frac{35}{8}\right) \qquad \text{Substitute.}$$
$$= \pi\left(\frac{875}{128}\right) \qquad\qquad \text{Multiply the fractions.}$$
$$\approx 21.48 \text{ cubic inches} \qquad \text{Approximate.}$$

(b) To calculate the number of fluid ounces in 21.48 cubic inches, multiply by 0.554.

$$21.48(0.554) \approx 11.9 \text{ fluid ounces}$$

Note that a typical aluminum soda can holds 12 fluid ounces.   Now Try Exercise 51

Solving for a Variable

A formula establishes a relationship between two or more variables (or quantities). Sometimes a formula is not solved for the needed variable. For example, if the area A and the width W

of a rectangular region are given, then its length L can be found by solving the formula $A = LW$ for L.

$$A = LW \qquad \text{Area formula}$$

$$\frac{A}{W} = \frac{LW}{W} \qquad \text{Divide each side by } W.$$

$$\frac{A}{W} = L \qquad \text{Simplify the fraction.}$$

$$L = \frac{A}{W} \qquad \text{Rewrite the equation.}$$

If the area A of a rectangle is 400 square inches and its width W is 16 inches, then the rectangle's length L is

$$L = \frac{A}{W} = \frac{400}{16} = 25 \text{ inches.}$$

Once the area formula has been solved for L, the resulting formula can be used to find the length of *any* rectangle whose area and width are known.

▶ **REAL-WORLD CONNECTION**   The use of formulas is common in almost any field. For example, there is a useful formula for calculating how far away a bolt of lightning is. It is based on the fact that sometimes we see a flash of lighting before we hear the thunder. This phenomenon happens because light travels much faster than sound. The farther away the lightning, the greater is the delay between our seeing the flash and hearing the thunder. In general, if the delay is x seconds, then the lightning is $D = \frac{x}{5}$ miles away. For instance, if there is a 10-second delay, then the lightning is $D = \frac{10}{5} = 2$ miles away.

EXAMPLE 7   **Calculating the delay between lightning and thunder**

Doppler radar shows an electrical storm 3.5 miles away. If you see lightning from this storm, how long will it be before you hear the thunder?

Solution
Solve the formula $D = \frac{x}{5}$ for x.

$$D = \frac{x}{5} \qquad \text{Given formula}$$

$$5D = x \qquad \text{Multiply each side by 5.}$$

If $D = 3.5$ miles, then the delay is

$$x = 5D = 5(3.5) = 17.5 \text{ seconds.}$$ Now Try Exercise 83

In the next example we apply the area formula for a trapezoid.

EXAMPLE 8   **Finding the base of a trapezoid**

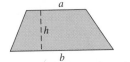

Figure 2.13 Trapezoid

The area of the trapezoid shown in Figure 2.13 is given by

$$A = \frac{1}{2}(a + b)h,$$

where a and b are the bases of the trapezoid and h is the height.

(a) Solve the formula for b.

(b) A trapezoid has area $A = 36$ square inches, height $h = 4$ inches, and base $a = 8$ inches. Find b.

Solution

(a) To clear the equation of the fraction, multiply each side by 2.

$$A = \frac{1}{2}(a + b)h \qquad \text{Area formula}$$

$$2A = (a + b)h \qquad \text{Multiply each side by 2.}$$

$$\frac{2A}{h} = a + b \qquad \text{Divide each side by } h.$$

$$\frac{2A}{h} - a = b \qquad \text{Subtract } a \text{ from each side.}$$

$$b = \frac{2A}{h} - a \qquad \text{Rewrite the formula.}$$

(b) Let $A = 36$, $h = 4$, and $a = 8$ in $b = \frac{2A}{h} - a$. Then

$$b = \frac{2(36)}{4} - 8 = 18 - 8 = 10 \text{ inches.} \qquad \boxed{\text{Now Try Exercise } 57}$$

EXAMPLE 9 Solving for a variable

Solve each equation for the indicated variable.

(a) $c = \frac{a + b}{2}$ for b **(b)** $ab - bc = ac$ for c

Solution

(a) To clear the equation of the fraction, multiply each side by 2.

$$c = \frac{a + b}{2} \qquad \text{Given formula}$$

$$2c = a + b \qquad \text{Multiply each side by 2.}$$

$$2c - a = b \qquad \text{Subtract } a.$$

The formula solved for b is $b = 2c - a$.

(b) In this equation c appears in two terms. We will combine the terms containing c by using the distributive property. Begin by moving the term on the left side containing c to the right side of the equation.

$$ab - bc = ac \qquad \text{Given formula}$$

$$ab - bc + bc = ac + bc \qquad \text{Add } bc \text{ to each side.}$$

$$ab = (a + b)c \qquad \text{Combine terms; distributive property.}$$

$$\frac{ab}{a + b} = \frac{(a + b)c}{(a + b)} \qquad \text{Divide each side by } (a + b).$$

$$\frac{ab}{a + b} = c \qquad \text{Simplify the fraction.}$$

CRITICAL THINKING

Are the formulas $c = \frac{1}{a - b}$ and $c = \frac{-1}{b - a}$ equivalent? Why?

The formula solved for c is $c = \frac{ab}{a + b}$. $\qquad \boxed{\text{Now Try Exercise } 61}$

Other Formulas

To calculate a student's GPA, the number of credits earned with a grade of A, B, C, D, and F must be known. If a, b, c, d, and f represent these credit counts respectively, then

$$\text{GPA} = \frac{4a + 3b + 2c + d}{a + b + c + d + f}.$$

This formula is based on the assumption that a 4.0 GPA is an A, a 3.0 GPA is a B, and so on.

EXAMPLE 10   Calculating a student's GPA

A student has earned 16 credits of A, 32 credits of B, 12 credits of C, 2 credits of D, and 5 credits of F. Calculate the student's GPA to the nearest hundredth.

Solution
Let $a = 16$, $b = 32$, $c = 12$, $d = 2$, and $f = 5$. Then

$$\text{GPA} = \frac{4 \cdot 16 + 3 \cdot 32 + 2 \cdot 12 + 2}{16 + 32 + 12 + 2 + 5} = \frac{186}{67} \approx 2.78.$$

The student's GPA is 2.78.

Now Try Exercise 69

EXAMPLE 11   Converting temperature scales

In the United States, temperature is measured with either the Fahrenheit or the Celsius temperature scales. To convert Fahrenheit degrees F to Celsius degrees C, the formula $C = \frac{5}{9}(F - 32)$ can be used.
(a) Solve the formula for F to find a formula that converts Celsius degrees to Fahrenheit degrees.
(b) If the outside temperature is $20°C$, find the equivalent Fahrenheit temperature.

Solution
(a) The reciprocal of $\frac{5}{9}$ is $\frac{9}{5}$, so multiply each side by $\frac{9}{5}$.

$$C = \frac{5}{9}(F - 32) \qquad \text{Given equation}$$

$$\frac{9}{5}C = \frac{9}{5} \cdot \frac{5}{9}(F - 32) \qquad \text{Multiply each side by } \tfrac{9}{5}.$$

$$\frac{9}{5}C = F - 32 \qquad \text{Multiplicative inverses}$$

$$\frac{9}{5}C + 32 = F \qquad \text{Add 32 to each side.}$$

The required formula is $F = \frac{9}{5}C + 32$.

(b) If $C = 20°C$, then $F = \frac{9}{5}(20) + 32 = 36 + 32 = 68°F$.

Now Try Exercise 73

2.4 PUTTING IT ALL TOGETHER

In this section we discussed formulas and how to solve for a variable. The following table summarizes some of these formulas.

Concept	Formula	Examples
Area of a Rectangle	$A = LW$, where L is the length and W is the width.	If $L = 10$ feet and $W = 5$ feet, then the area is $$A = 10 \cdot 5 = 50 \text{ square feet.}$$
Gas Mileage	$M = \frac{D}{G}$, where D is the distance and G is the gasoline used.	If a car travels 100 miles on 5 gallons of gasoline, then its mileage is $$M = \frac{100}{5} = 20 \text{ miles per gallon.}$$
Area of a Triangle	$A = \frac{1}{2}bh$, where b is the base and h is the height.	If $b = 5$ inches and $h = 6$ inches, then the area is $$A = \frac{1}{2}(5)(6) = 15 \text{ square inches.}$$
Angle Measure in a Triangle	$x + y + z = 180°$, where x, y, and z are the angle measures.	If $x = 40°$ and $y = 60°$, then $z = 80°$ because $$40° + 60° + 80° = 180°.$$
Volume and Surface Area of a Box	If a box has length L, width W, and height H, then its volume is $$V = LWH$$ and its surface area is $$S = 2LW + 2WH + 2LH.$$	If a box has dimensions $L = 4$ feet, $W = 3$ feet, and $H = 2$ feet, then $$V = 4(3)(2) = 24 \text{ cubic feet}$$ and $$S = 2(4)(3) + 2(3)(2) + 2(4)(2)$$ $$= 52 \text{ square feet.}$$

continued on next page

continued from previous page

Concept	Formula	Examples
Circumference and Area of a Circle	If a circle has radius r, then its circumference is $$C = 2\pi r$$ and its area is $$A = \pi r^2.$$	If $r = 6$ inches, $C = 2\pi(6) = 12\pi \approx 37.7$ inches and $A = \pi(6)^2 = 36\pi \approx 113.1$ square inches.
Volume of a Cylinder	$V = \pi r^2 h$, where r is the radius and h is the height.	If $r = 5$ inches and $h = 20$ inches, then the volume is $V = \pi(5^2)(20) = 500\pi$ cubic inches.
Distance from Lightning	$D = \frac{x}{5}$, where x is the delay in seconds between seeing the flash and hearing the thunder. D is in miles.	If $x = 15$ seconds, then the lightning is $D = \frac{15}{5} = 3$ miles away.
Area of a Trapezoid	$A = \frac{1}{2}(a + b)h$, where a and b are the bases and h is the height.	If $a = 4$, $b = 6$, and $h = 3$, then the area is $A = \frac{1}{2}(4 + 6)(3) = 15$ square units.
Calculating Grade Point Average (GPA)	GPA is calculated by $$\frac{4a + 3b + 2c + d}{a + b + c + d + f},$$ where a, b, c, d, and f represent the credits earned with grades of A, B, C, D, and F, respectively.	10 credits of A, 8 credits of B, 6 credits of C, 12 credits of D, and 8 credits of F results in a GPA of $$\frac{4(10) + 3(8) + 2(6) + 12}{10 + 8 + 6 + 12 + 8} = 2.0.$$
Converting Between Fahrenheit and Celsius Degrees	$$C = \frac{5}{9}(F - 32)$$ $$F = \frac{9}{5}C + 32$$	212°F is equivalent to $$C = \frac{5}{9}(212 - 32) = 100°C.$$ 100°C is equivalent to $$F = \frac{9}{5}(100) + 32 = 212°F.$$

2.4 Exercises

CONCEPTS

1. A(n) _____ can be used to calculate one quantity by using known values of other quantities.

2. The area A of a rectangle with length L and width W is $A =$ _____.

3. If a car that travels D miles uses G gallons of gasoline, the mileage is $M =$ _____.

4. The area A of a triangle with base b and height h is $A =$ _____.

5. One degree equals _____ of a revolution.

6. There are _____ degrees in one revolution.

7. The sum of the measures of the angles in a triangle equals _____ degrees.

8. The volume V of a box with length L, width W, and height H is $V =$ _____.

9. The surface area S of a box with length L, width W, and height H is $S =$ _____.

10. The circumference C of a circle with radius r is $C =$ _____.

11. The area A of a circle with radius r is $A =$ _____.

12. The volume V of a cylinder with radius r and height h is $V =$ _____.

13. If thunder is heard x seconds after lightning is seen, the distance in miles to the storm is $D =$ _____.

14. The area A of a trapezoid with height h and bases a and b is $A =$ _____.

FORMULAS FROM GEOMETRY

Exercises 15–22: Find the area of the region shown.

15.

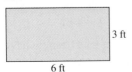

6 ft
3 ft

16.

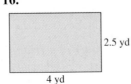

4 yd
2.5 yd

17.

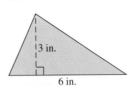

3 in.
6 in.

18.

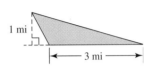

1 mi
3 mi

19.

4 ft

20.

6 in.

21.

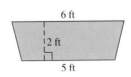

6 ft
2 ft
5 ft

22.

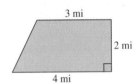

3 mi
2 mi
4 mi

23. Find the area of a rectangle having a 7-inch width and a 13-inch length.

24. Find the area of a rectangle having a 5-foot width and a 7-yard length.

25. Find the area of a triangle having a 12-inch base and a 6-inch height.

26. Find the area of a triangle having a 9-foot base and a 72-inch height.

27. Find the circumference of a circle having an 8-inch diameter.

28. Find the area of a circle having a 9-foot radius.

29. *Area of a Lot* Find the area of the lot shown, which consists of a square and a triangle.

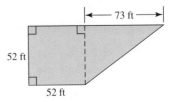

73 ft
52 ft
52 ft

30. *Area of a Lot*   Find the area of the lot shown, which consists of a rectangle and two triangles.

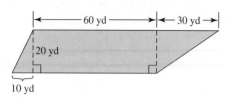

Exercises 31 and 32: Angle Measure   *Find the measure of the third angle in the triangle.*

31.

32.

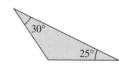

33. A triangle contains two angles having measures of 23° and 76°. Find the measure of the third angle.

34. The measures of the angles in an *equilateral triangle* are equal. Find their measure.

35. The measures of the angles in a triangle are x, $2x$, and $3x$. Find the value of x.

36. The measures of the angles in a triangle are $3x$, $4x$, and $11x$. Find the value of x.

37. In a triangle the two smaller angles are equal in measure and are each one third of the measure of the larger angle. Find the measure of each angle.

38. In a triangle the two larger angles differ by 10°. The smaller angle is 50° less than the largest angle. Find the measure of each angle.

39. The diameter of a circle is 12 inches. Find its circumference and area.

40. The radius of a circle is $\frac{5}{4}$ feet. Find its circumference and area.

41. The circumference of a circle is 2π inches. Find its radius and area.

42. The circumference of a circle is 13π feet. Find its radius and area.

Exercises 43–46: A box with a top has length L, width W, and height H. Find the volume and surface area of the box.

43. $L = 22$ inches, $W = 12$ inches, $H = 10$ inches

44. $L = 5$ feet, $W = 3$ feet, $H = 6$ feet

45. $L = \frac{2}{3}$ yard, $W = \frac{2}{3}$ foot, $H = \frac{3}{2}$ feet

46. $L = 1.2$ meters, $W = 0.8$ meter, $H = 0.6$ meter

Exercises 47–50: Use the formula $V = \pi r^2 h$ to find the volume of a cylindrical container for the given r and h. Leave your answer in terms of π.

47. $r = 2$ inches, $h = 5$ inches

48. $r = \frac{1}{2}$ inch, $h = \frac{3}{2}$ inches

49. $r = 5$ inches, $h = 2$ feet

50. $r = 2.5$ feet, $h = 1.5$ yards

51. *Volume of a Barrel*   (Refer to Example 6.) A cylindrical barrel has a diameter of $1\frac{3}{4}$ feet and a height of 3 feet. Find the volume of the barrel.

52. *Volume of a Can*   (Refer to Example 6.)
 (a) Find the volume of a can with a radius of $\frac{3}{4}$ inch and a height of $2\frac{1}{2}$ inches.
 (b) Find the number of fluid ounces in the can if one cubic inch equals 0.554 fluid ounces.

SOLVING FOR A VARIABLE

Exercises 53–66: Solve the formula for the given variable.

53. $A = LW$ for W **54.** $A = \frac{1}{2}bh$ for b

55. $V = \pi r^2 h$ for h **56.** $V = \frac{1}{3}\pi r^2 h$ for h

57. $A = \frac{1}{2}(a + b)h$ for a **58.** $C = 2\pi r$ for r

59. $V = LWH$ for W **60.** $P = 2x + 2y$ for y

61. $s = \dfrac{a + b + c}{2}$ for b **62.** $t = \dfrac{x - y}{3}$ for x

63. $\dfrac{a}{b} - \dfrac{c}{b} = 1$ for b **64.** $\dfrac{x}{y} + \dfrac{z}{y} = 5$ for z

65. $ab = cd + ad$ for a

66. $S = 2LW + 2LH + 2WH$ for W

67. *Perimeter of a Rectangle*   The perimeter of a rectangle equals the sum of the lengths of its four sides. If the width of a rectangle is 5 inches and its perimeter is 40 inches, find the length of the rectangle.

68. *Perimeter of a Triangle*   Two sides of a triangle have lengths of 5 feet and 7 feet. If the triangle's perimeter is 21 feet, what is the length of the third side?

OTHER FORMULAS AND APPLICATIONS

Exercises 69–72: (Refer to Example 10.) Let a represent the number of credits with a grade of A, b the number of credits with a grade of B, and so on. Calculate the corresponding grade point average (GPA). Round your answer to the nearest hundredth.

69. $a = 30, b = 45, c = 12, d = 4, f = 4$

70. $a = 70, b = 35, c = 5, d = 0, f = 0$

71. $a = 0, b = 60, c = 80, d = 10, f = 6$

72. $a = 3, b = 5, c = 8, d = 0, f = 22$

Exercises 73–76: (Refer to Example 11.) Convert the Celsius temperature to an equivalent Fahrenheit temperature.

73. $25°C$ **74.** $100°C$

75. $-40°C$ **76.** $0°C$

Exercises 77–80: (Refer to Example 11.) Convert the Fahrenheit temperature to an equivalent Celsius temperature.

77. $23°F$ **78.** $98.6°F$

79. $-4°F$ **80.** $-31°F$

81. *Gas Mileage* A truck driver leaves a gas station with a full tank of gas and the odometer showing 87,625 miles. At the next gas stop, it takes 38 gallons to fill the tank and the odometer reads 88,043 miles. Find the gas mileage for the truck.

82. *Gas Mileage* A car that gets 34 miles per gallon is driven 578 miles. How many gallons of gasoline are used on this trip?

83. *Lightning* (Refer to Example 7.) The time delay between a flash of lightning and the sound of thunder is 12 seconds. How far away is the lightning?

84. *Lightning* (Refer to Example 7.) Doppler radar shows an electrical storm 2.5 miles away. If you see the lightning from this storm, how long will it be before you hear the thunder?

WRITING ABOUT MATHEMATICS

85. A student solves the formula $A = \frac{1}{2}bh$ for h and obtains the formula $h = \frac{1}{2}bA$. Explain the error that the student is making. What is the correct answer?

86. Give an example of a formula that you have used and explain how you used it.

CHECKING BASIC CONCEPTS
SECTIONS 2.3 AND 2.4

1. Translate the sentence into an equation containing the variable x. Then solve the resulting equation.
  (a) The product of 3 and a number is 36.
  (b) A number subtracted from 35 is 43.

2. The sum of three consecutive integers is -93. Find the three integers.

3. Convert 9.5% to a decimal.

4. Convert $\frac{5}{4}$ to a percentage.

5. *Serious Crime* In 2003, New York City experienced 2652 serious crimes per 100,000 people. This figure represented a 25% decrease from the number in 2000. Find the rate for serious crimes per 100,000 people in 2000.

6. *Driving a Car* How many hours does it take the driver of a car to travel 390 miles at 60 miles per hour?

7. *College Loans* A student takes out two loans, one at 6% and the other at 7%. The 6% loan is $2000 more than the 7% loan, and the total interest for one year is $510. Find the amount of each loan.

8. *Gas Mileage* A car that gets 28 miles per gallon is driven 504 miles. How many gallons of gasoline are used on this trip?

9. *Height of a Triangle* The area of a triangle having a base of 6 inches is 36 square inches. Find the height of the triangle.

continued on next page

continued from previous page

10. *Area and Circumference*   Find the area and cir-cumference of the circle shown.

3 ft

11. *Angles*   Find the value of x in the triangle shown.

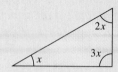

$2x$

$3x$

x

12. *Solving a Formula*   Solve $A = \pi r^2 + \pi r l$ for l.

2.5 LINEAR INEQUALITIES

Solutions and Number Line Graphs ▪ The Addition Property of Inequalities ▪ The Multiplication Property of Inequalities ▪ Applications

A LOOK INTO MATH ▷

At an amusement park, a particular ride might be restricted to people at least 48 inches tall. A child who is x inches tall may go on the ride if $x \geq 48$ but may not go on the ride if $x < 48$. A height of 48 inches represents the boundary between being allowed on the ride and being denied access to the ride. A posted height restriction, or *boundary*, allows parents to easily determine if their child may go on the ride.

Solving linear inequalities is closely related to solving linear equations because equality is the boundary between *greater than* and *less than*. In this section we discuss techniques used to solve linear inequalities.

Solutions and Number Line Graphs

A **linear inequality** results whenever the equals sign in a linear equation is replaced with any one of the symbols $<, \leq, >,$ or $\geq$. Examples of linear equations include

$$x = 5, \quad 2x + 1 = 0, \quad 1 - x = 6, \quad \text{and} \quad 5x + 1 = 3 - 2x.$$

Therefore examples of linear inequalities include

$$x > 5, \quad 2x + 1 < 0, \quad 1 - x \geq 6, \quad \text{and} \quad 5x + 1 \leq 3 - 2x.$$

A **solution** to an inequality is a value of the variable that makes the statement true. The set of all solutions is called the **solution set.** Two inequalities are *equivalent* if they have the same solution set. Inequalities frequently have infinitely many solutions. For example, the solution set for the inequality $x > 5$ includes all real numbers greater than 5.

A number line can be used to graph the solution set for an inequality. The graph of all real numbers satisfying $x < 2$ is shown in Figure 2.14(a), and the graph of all real numbers satisfying $x \leq 2$ is shown in Figure 2.14(b). (The symbol $\leq$ is read "less than or equal to." Similarly, the symbol $\geq$ is read "greater than or equal to.") A parenthesis ")" is used to show that 2 is not included in Figure 2.14(a), and a bracket "]" is used to show that 2 is included in Figure 2.14(b).

(a) $x < 2$ (b) $x \leq 2$

Figure 2.14

EXAMPLE 1 Graphing inequalities on a number line

Use a number line to graph the solution set to each inequality.
(a) $x > 0$ **(b)** $x \geq 0$ **(c)** $x \leq -1$ **(d)** $x < 3$

Solution
(a) First locate $x = 0$ (or the origin) on a number line. Numbers greater than 0 are located to the right of the origin, so shade the number line to the right of the origin. Because $x > 0$, the number 0 is not included, so place a parenthesis "(" at 0, as shown in Figure 2.15(a).
(b) Figure 2.15(b) is similar to the graph in part (a) except that a bracket "[" is placed at the origin because 0 is included in the solution set.
(c) First locate $x = -1$ on the number line. Numbers less than -1 are located to the left of -1. Because -1 is included, a bracket "]" is placed at -1, as shown in Figure 2.15(c).
(d) Real numbers less than 3 are graphed in Figure 2.15(d).

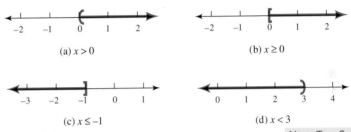

(a) $x > 0$ (b) $x \geq 0$

(c) $x \leq -1$ (d) $x < 3$

Figure 2.15

Now Try Exercises 13, 17

INTERVAL NOTATION (OPTIONAL) Each number line graph in Figure 2.15 represents an interval of real numbers that corresponds to the solution set to an inequality. These solution sets can also be represented in a convenient notation called **interval notation**. Rather than draw the entire number line, we can use brackets or parentheses to indicate the interval of values that represent the solution set. For example, the number line shown in Figure 2.15(a) can be represented by the interval $(0, \infty)$, and the number line shown in Figure 2.15(b) can be represented by the interval $[0, \infty)$. The symbol ∞ refers to infinity and is used to indicate that the values increase without bound. Similarly, $-\infty$ can be used when the values decrease without bound. The number lines shown in Figures 2.15(c) and (d) can be represented by $(-\infty, -1]$ and $(-\infty, 3)$, respectively.

EXAMPLE 2 Writing solution sets in interval notation

Write the solution set to each inequality in interval notation.
(a) $x > 4$ **(b)** $y \leq -3$ **(c)** $z \geq -1$

Solution
(a) Real numbers greater than 4 are represented by the interval $(4, \infty)$.
(b) Real numbers less than or equal to -3 are represented by the interval $(-\infty, -3]$.
(c) The solution set is represented by the interval $[-1, \infty)$.   Now Try Exercise 27

CHECKING SOLUTIONS We can check possible solutions to an inequality in the same way that we checked possible solutions to an equation. For example, to check whether 5 is a solution to $2x + 3 = 13$, we substitute $x = 5$ in the equation.

$$2(5) + 3 \stackrel{?}{=} 13 \qquad \text{Replace } x \text{ with 5.}$$
$$13 = 13 \qquad \text{A true statement}$$

Thus 5 is a solution to this equation. Similarly, to check whether 7 is a solution to $2x + 3 > 13$, we substitute $x = 7$ in the inequality.

$$2(7) + 3 \overset{?}{>} 13 \qquad \text{Replace } x \text{ with 7.}$$
$$17 > 13 \qquad \text{A true statement}$$

Thus 7 is a solution to the inequality.

.EXAMPLE 3   Checking possible solutions

Determine whether the given value of x is a solution to the inequality.
(a) $3x - 4 < 10, \quad x = 6$ **(b)** $4 - 2x \leq 8, \quad x = -2$

Solution

(a) Substitute 6 for x and simplify.

$$3(6) - 4 \overset{?}{<} 10 \qquad \text{Replace } x \text{ with 6.}$$
$$14 \overset{?}{<} 10 \qquad \text{A false statement}$$

Thus 6 is *not* a solution to the inequality.

(b) Substitute -2 for x and simplify.

$$4 - 2(-2) \overset{?}{\leq} 8 \qquad \text{Replace } x \text{ with } -2.$$
$$8 \leq 8 \qquad \text{A true statement}$$

Thus -2 is a solution to the inequality.

Now Try Exercise 35

Just as solving a linear equation means finding the value of the variable that makes the equation true, solving a linear inequality means finding the *values* of the variable that make the inequality true.

Making a table is an organized way of checking possible values of the variable to see if any value makes an inequality true. For example, if we want to find some of the solutions to the inequality $2x - 6 < 0$, we substitute various values for x in the left side of the inequality. If a value gives a result that is less than 0, the value makes the inequality true and is a solution. In the next example, we use a table to find solutions to an equation and related inequalities.

EXAMPLE 4   Finding solutions to equations and inequalities

In Table 2.7 the expression $2x - 6$ has been evaluated for several values of x. Use the table to determine any solutions to each equation or inequality.
(a) $2x - 6 = 0$ **(b)** $2x - 6 > 0$ **(c)** $2x - 6 \geq 0$ **(d)** $2x - 6 < 0$

TABLE 2.7

x	0	1	2	3	4	5	6
$2x - 6$	-6	-4	-2	0	2	4	6

Solution

(a) From Table 2.7, $2x - 6$ equals 0 when $x = 3$.
(b) The values of x in the table that make the expression $2x - 6$ greater than 0 are 4, 5, and 6. These values are all greater than 3, which is the solution found in part (a). It follows that $2x - 6 > 0$ when $x > 3$.
(c) The values of x in the table that make the expression $2x - 6$ greater than or equal to 0 are 3, 4, 5, and 6. It follows that $2x - 6 \geq 0$ when $x \geq 3$.
(d) The expression $2x - 6$ is less than 0 when $x < 3$.

Now Try Exercise 43

The Addition Property of Inequalities

▶ **REAL-WORLD CONNECTION** Suppose that the speed limit on a country road is 55 miles per hour, which is 25 miles per hour faster than the speed limit in town. If x represents lawful speeds in town, then x satisfies the inequality

$$x + 25 \leq 55.$$

To solve this inequality we **add** -25 to each side of the inequality.

$$x + 25 + (-25) \leq 55 + (-25) \qquad \text{Add } -25 \text{ to each side.}$$
$$x \leq 30 \qquad \text{Add the real numbers.}$$

Thus drivers are obeying the speed limit in town when they travel at 30 miles per hour or less. To solve this inequality the addition property of inequalities was used.

ADDITION PROPERTY OF INEQUALITIES

Let a, b, and c be expressions that represent real numbers. The inequalities

$$a < b \quad \text{and} \quad a + c < b + c$$

are equivalent. That is, the same number may be added to (or subtracted from) each side of an inequality. Similar properties exist for $>$, $\leq$, and $\geq$.

To solve some inequalities, we apply the addition property of inequalities to obtain a simpler, equivalent inequality.

EXAMPLE 5 Applying the addition property of inequalities

Solve each inequality. Then graph the solution set.
(a) $x - 1 > 4$ **(b)** $3 + 2x \leq 5 + x$

Solution
(a) Begin by adding 1 to each side of the inequality.

$$x - 1 > 4 \qquad \text{Given inequality}$$
$$x - 1 + 1 > 4 + 1 \qquad \text{Add 1 to each side.}$$
$$x > 5 \qquad \text{Add the real numbers.}$$

The solution set is given by $x > 5$ and is graphed as follows.

```
◄──┼───┼───┼───┼──(┼───┼──►
  -2   0   2   4   6   8
```

(b) Begin by subtracting x from (or adding $-x$ to) each side of the inequality.

$$3 + 2x \leq 5 + x \qquad \text{Given inequality}$$
$$3 + 2x - x \leq 5 + x - x \qquad \text{Subtract } x \text{ from each side.}$$
$$3 + x \leq 5 \qquad \text{Combine like terms.}$$
$$3 + x - 3 \leq 5 - 3 \qquad \text{Subtract 3 from each side.}$$
$$x \leq 2 \qquad \text{Subtract the real numbers.}$$

The solution set is given by $x \leq 2$ and is graphed as follows.

```
◄──┼───┼───┼───┼───┼───►
  -2  -1   0   1   2   3
```

MAKING CONNECTIONS

The addition property of inequalities can be illustrated by an old-fashioned pan balance. In the figure on the left, the weight on the right pan is heavier than the weight on the left because the right pan rests lower than the left pan. If we add the same amount of weight to (or subtract the same amount of weight from) both pans, the right pan still weighs more than the left pan by the same amount, as illustrated in the figure on the right.

In the next example we apply the addition property of inequalities and graph the solution set.

EXAMPLE 6   **Applying the addition property of inequalities**

Solve $5 + \frac{1}{2}x \le 3 - \frac{1}{2}x$. Then graph the solution set.

Solution
Begin by subtracting 5 from each side of the inequality.

$$5 + \frac{1}{2}x \le 3 - \frac{1}{2}x \qquad \text{Given inequality}$$

$$\frac{1}{2}x \le -\frac{1}{2}x - 2 \qquad \text{Subtract 5 from each side.}$$

$$\frac{1}{2}x + \frac{1}{2}x \le -\frac{1}{2}x + \frac{1}{2}x - 2 \qquad \text{Add } \tfrac{1}{2}x \text{ to each side.}$$

$$x \le -2 \qquad \text{Combine like terms.}$$

The solution set is given by $x \le -2$ and is graphed as follows.

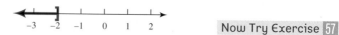

Now Try Exercise 57

The Multiplication Property of Inequalities

The multiplication property of inequalities differs from the multiplication property of *equality*. When we multiply each side of an inequality by the same nonzero number, we may need to reverse the inequality symbol to make sure that the resulting inequality remains true. Table 2.8 shows various results that occur when each side of a true inequality is multiplied by the same nonzero number.

TABLE 2.8 Determining When the Inequality Symbol Should Be Reversed

True Statement	Multiply Each Side By	Resulting Inequality	Is the Result True or False?	Reverse the Inequality Symbol
$-3 < 5$	4	$-12 \overset{?}{<} 20$	True	Not needed
$-4 \leq 2$	-5	$20 \overset{?}{\leq} -10$	False	$20 \geq -10$
$7 > -1$	-2	$-14 \overset{?}{>} 2$	False	$-14 < 2$
$0 \geq -6$	8	$0 \overset{?}{\geq} -48$	True	Not needed
$-2 > -5$	3	$-6 \overset{?}{>} -15$	True	Not needed
$4 < 9$	-11	$-44 \overset{?}{<} -99$	False	$-44 > -99$

Table 2.8 indicates that the inequality symbol must be reversed when each side of the given inequality is multiplied by a negative number. This result is summarized in the following.

MULTIPLICATION PROPERTY OF INEQUALITIES

Let a, b, and c be expressions that represent real numbers with $c \neq 0$.

1. If $c > 0$, then the inequalities $a < b$ and $ac < bc$ are equivalent. That is, each side of an inequality may be multiplied (or divided) by the same positive number.
2. If $c < 0$, then the inequalities $a < b$ and $ac > bc$ are equivalent. That is, each side of an inequality may be multiplied (or divided) by the same negative number, provided the inequality symbol is reversed.

Note that similar properties exist for $\leq$ and $\geq$.

NOTE: It is important to reverse the inequality symbol when either multiplying or *dividing* by a negative number.

EXAMPLE **7**   **Applying the multiplication property of inequalities**

Solve each inequality. Then graph the solution set.
(a) $3x < 18$ **(b)** $-7 \leq -\frac{1}{2}x$

Solution
(a) To solve for x, divide each side by 3.

$$3x < 18 \qquad \text{Given inequality}$$

$$\frac{3x}{3} < \frac{18}{3} \qquad \text{Divide each side by 3.}$$

$$x < 6 \qquad \text{Simplify fractions.}$$

The solution set is given by $x < 6$ and is graphed as follows.

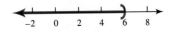

(b) To isolate x in $-7 \le -\frac{1}{2}x$, multiply each side by -2 and reverse the inequality symbol.

$$-7 \le -\frac{1}{2}x \qquad \text{Given inequality}$$

$$-2(-7) \ge -2\left(-\frac{1}{2}\right)x \qquad \text{Multiply by } -2; \text{ reverse the inequality.}$$

$$14 \ge 1 \cdot x \qquad \text{Multiply the real numbers.}$$

$$x \le 14 \qquad \text{Rewrite the inequality.}$$

The solution set is given by $x \le 14$ and is graphed as follows.

Now Try Exercise 61

SET-BUILDER NOTATION (OPTIONAL)   Because $x \le 14$ is an inequality with infinitely many solutions and is not itself a set of solutions, a mathematical notation called **set-builder notation** has been devised for writing the solutions to an inequality as a set. For example, the solution set consisting of "all real numbers x such that x is less than or equal to 14" can be written $\{x \mid x \le 14\}$. The vertical line segment "$\mid$" is read "such that."

In the next example, the solution sets are expressed in set-builder notation. However, this notation is not widely used throughout this text.

EXAMPLE 8   Applying both properties of inequalities

Solve each inequality. Write the solution set in set-builder notation.
(a) $4x - 7 \ge -6$ **(b)** $-8 + 4x \le 5x + 3$ **(c)** $0.4(2x - 5) < 1.1x + 2$

Solution
(a) Start by adding 7 to each side.

$$4x - 7 \ge -6 \qquad \text{Given inequality}$$

$$4x \ge 1 \qquad \text{Add 7 to each side.}$$

$$x \ge \frac{1}{4} \qquad \text{Divide each side by 4.}$$

In set-builder notation, the solution set is $\left\{x \mid x \ge \frac{1}{4}\right\}$.
(b) Begin by adding 8 to each side.

$$-8 + 4x \le 5x + 3 \qquad \text{Given inequality}$$

$$4x \le 5x + 11 \qquad \text{Add 8 to each side.}$$

$$4x - 5x \le 5x + 11 - 5x \qquad \text{Subtract } 5x \text{ from each side.}$$

$$-x \le 11 \qquad \text{Combine like terms.}$$

$$-1 \cdot (-x) \ge -1 \cdot 11 \qquad \text{Multiply by } -1; \text{ reverse the inequality.}$$

$$x \ge -11 \qquad \text{Simplify.}$$

The solution set is $\{x \mid x \ge -11\}$.

(c) Begin by multiplying each side by 10 to eliminate decimals.

$$0.4(2x - 5) < 1.1x + 2 \qquad \text{Given inequality}$$
$$4(2x - 5) < 11x + 20 \qquad \text{Multiply each side by 10.}$$
$$8x - 20 < 11x + 20 \qquad \text{Distributive property}$$
$$8x < 11x + 40 \qquad \text{Add 20 to each side.}$$
$$-3x < 40 \qquad \text{Subtract } 11x \text{ from each side.}$$
$$x > -\frac{40}{3} \qquad \text{Divide by } -3; \text{ reverse the inequality.}$$

The solution set is $\left\{x \mid x > -\frac{40}{3}\right\}$.

Now Try Exercises 71, 81, 91

CRITICAL THINKING

Solve $-5 - 3x > -2x + 7$ without having to reverse the inequality symbol.

Applications

To solve applications involving inequalities, we often have to translate words to mathematical statements. For example, the phrase "at least" can be translated to "greater than or equal to" and the phrase "at most" can be translated to "less than or equal to."

EXAMPLE 9 Translating words to inequalities

Translate each phrase to an inequality. Let the variable be x.
(a) A number that is more than 30
(b) An age that is at least 18
(c) A grade point average that is at most 3.25

Solution
(a) The inequality $x > 30$ represents a number x that is more than 30.
(b) The inequality $x \geq 18$ represents an age x that is at least 18.
(c) The inequality $x \leq 3.25$ represents a grade point average x that is at most 3.25.

Now Try Exercises 101, 107

▶ **REAL-WORLD CONNECTION** In the lower atmosphere the air temperature generally becomes colder as the altitude increases. One mile above Earth's surface the temperature is about 19°F colder than the ground-level temperature. As the air cools, there is an increased chance of clouds forming. In the next example we estimate the altitudes where clouds may form. (**Source:** A. Miller and R. Anthes, *Meteorology.*)

EXAMPLE 10 Finding the altitude of clouds

If the ground temperature is 79°F, then the temperature T above Earth's surface is given by the formula $T = 79 - 19x$, where x is the altitude in miles. Suppose that clouds form only where the temperature is 3°F or colder. Determine the heights at which clouds may form.

Solution
Clouds may form at altitudes at which the temperature T is less than or equal to 3°F. Thus we must solve the inequality $79 - 19x \leq 3$.

$$79 - 19x \leq 3 \qquad \text{Inequality to be solved}$$
$$-19x \leq -76 \qquad \text{Subtract 79 from each side.}$$
$$\frac{-19x}{-19} \geq \frac{-76}{-19} \qquad \text{Divide by } -19; \text{ reverse the inequality.}$$
$$x \geq 4 \qquad \text{Simplify the fraction.}$$

Clouds may form at 4 miles or higher.

Now Try Exercise 123

EXAMPLE 11   Modeling AIDS Research

AIDS research funding F in billions of dollars increased from 1995 to 2005 and can be modeled by the formula $F = 0.153x - 303.85$, where x is the year. Estimate when the funding F was greater than or equal to $2.15 billion.   (**Source:** National Institute of Health.)

Solution
We must solve the inequality $F \geq 2.15$.

$$0.153x - 303.85 \geq 2.15 \qquad \text{Inequality to be solved}$$

$$0.153x \geq 306 \qquad \text{Add 303.85 to each side.}$$

$$x \geq \frac{306}{0.153} \qquad \text{Divide each side by 0.153.}$$

Because $\frac{306}{0.153} = 2000$, funding for AIDS was greater than or equal to $2.15 billion during 2000 and after.

Now Try Exercise 125

EXAMPLE 12   Calculating revenue, cost, and profit

For a computer company, the cost to produce one laptop computer is $1320 plus a one-time fixed cost of $200,000 for research and development. The revenue received from selling one laptop computer is $1850.
(a) Write a formula that gives the cost C of producing x laptop computers.
(b) Write a formula that gives the revenue R from selling x laptop computers.
(c) Profit equals revenue minus cost. Write a formula that calculates the profit P from selling x laptop computers.
(d) How many computers need to be sold to yield a positive profit?

Solution
(a) The cost of producing the first laptop is

$$1320 \times 1 + 200{,}000 = \$201{,}320.$$

The cost of producing two laptops is

$$1320 \times 2 + 200{,}000 = \$202{,}640.$$

And, in general, the cost of producing x laptops is

$$1320 \times x + 200{,}000 = 1320x + 200{,}000.$$

Thus $C = 1320x + 200{,}000$.
(b) Because the company receives $1850 for each laptop, the revenue for x laptops is given by $R = 1850x$.
(c) Profit equals revenue minus cost, so

$$P = R - C$$

$$= 1850x - (1320x + 200{,}000)$$

$$= 530x - 200{,}000.$$

Thus $P = 530x - 200{,}000$.

(d) To determine how many laptops need to be sold to yield a positive profit, we must solve the inequality $P > 0$.

$$530x - 200{,}000 > 0 \qquad \text{Inequality to be solved}$$
$$530x > 200{,}000 \qquad \text{Add 200,000 to each side.}$$
$$x > \frac{200{,}000}{530} \qquad \text{Divide each side by 530.}$$

Because $\frac{200{,}000}{530} \approx 377.4$, the company must sell at least 378 laptops. Note that the company cannot sell a fraction of a laptop. **Now Try Exercise 119**

2.5 PUTTING IT ALL TOGETHER

In this section we discussed linear inequalities and how to solve them. A linear inequality has infinitely many solutions and can be solved by using the addition and multiplication properties of inequalities. When multiplying or dividing an inequality by a negative number, we must reverse the inequality symbol. The following table summarizes some of the concepts presented in this section.

Concept	Comments	Examples
Linear Inequality	If the equals sign in a linear equation is replaced with $<$, $>$, $\le$, or $\ge$, a linear inequality results.	*Linear Equation* *Linear Inequality* $4x - 1 = 0$ $4x - 1 > 0$ $2 - x = 3x$ $2 - x \le 3x$ $4(x+3) = 1 - x$ $4(x+3) < 1 - x$ $-6x + 3 = 5$ $-6x + 3 \ge 5$
Solution to an Inequality	A value for a variable that makes the inequality a true statement	5 is a solution to $2x > 5$ because $2(5) > 5$ is a true statement.
Set-Builder Notation	A notation that can be used to identify the solution set to an inequality	The solution set for $x - 2 < 5$ can be written as $\{x \mid x < 7\}$ and is read "the set of real numbers x such that x is less than 7."
Solution Set to an Inequality	The set of all solutions to an inequality	The solution set to $x + 1 > 5$ is given by $x > 4$ and can be written in set-builder notation as $\{x \mid x > 4\}$.
Number Line Graphs	The solutions to an inequality can be graphed on a number line.	$x < 2$ is graphed as follows. $x \ge -1$ is graphed as follows.

continued on next page

continued from previous page

Concept	Comments	Examples	
Addition Property of Inequalities	$a < b$ is equivalent to $a + c < b + c,$ where a, b, and c represent real number expressions.	$x - 5 \geq 6$ $x \geq 11$	Given inequality Add 5.
		$3x > 5 + 2x$ $x > 5$	Given inequality Subtract $2x$.
Multiplication Property of Inequalities	$a < b$ is equivalent to $ac < bc$ when $c > 0$, and is equivalent to $ac > bc$ when $c < 0$.	$\dfrac{1}{2}x \geq 6$ $x \geq 12$	Given inequality Multiply by 2.
		$-3x > 5$ $x < -\dfrac{5}{3}$	Given inequality Divide by -3; reverse the inequality symbol.

2.5 Exercises

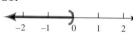 

CONCEPTS

1. A linear inequality results whenever the _____ in a linear equation is replaced by any one of the symbols _____, _____, _____, or _____.

2. Equality is the boundary between _____ and _____.

3. A(n) _____ is a value of the variable that makes an inequality statement true.

4. Two linear inequalities are _____ if they have the same solution set.

5. When a linear equation is solved, the solution set contains $\underline{\text{(one/infinitely many)}}$ solution(s).

6. When a linear inequality is solved, the solution set contains $\underline{\text{(one/infinitely many)}}$ solution(s).

7. The solution set to a linear inequality can be graphed by using a _____.

8. The value of 5 $\underline{\text{(is/is not)}}$ a solution to the inequality $3x < 10$.

9. The addition property of inequalities states that if $a > b$, then $a + c$ _____ $b + c$.

10. The multiplication property of inequalities states that if $a < b$ and $c > 0$, then ac _____ bc.

11. The multiplication property of inequalities states that if $a < b$ and $c < 0$, then ac _____ bc.

12. Are $-4x < 8$ and $x < -2$ equivalent inequalities? Explain.

SOLUTIONS AND NUMBER LINE GRAPHS

Exercises 13–20: Use a number line to graph the solution set to the inequality.

13. $x < 0$

14. $x > -2$

15. $x > 1$

16. $x < -\frac{5}{2}$

17. $x \leq 1.5$

18. $x \geq -3$

19. $z \geq -2$

20. $z \leq -\pi$

Exercises 21–26: Express the set of real numbers graphed on the number line with an inequality.

21.

22.

23.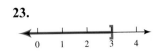

24.

25.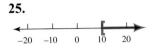

26.

Exercises 27–32: Write the solution set to the inequality in interval notation.

27. $x \geq 6$

28. $x < 3$

29. $y > -2$

30. $y \geq 1$

31. $z \leq 7$

32. $z < -5$

Exercises 33–42: Determine whether the given value of the variable is a solution to the inequality.

33. $x + 5 > 5 \qquad x = 4$

34. $x - 7 < 0 \qquad x = 6$

35. $5x \geq 25 \qquad x = 5$

36. $-3x \leq -8 \qquad x = -2$

37. $4y - 3 \leq 5 \qquad y = -3$

38. $3y + 5 \geq -8 \qquad y = -3$

39. $5(z + 1) < 3z - 7 \qquad z = -7$

40. $-(z + 7) > 3(6 - z) \qquad z = 2$

41. $\frac{3}{2}t - \frac{1}{2} \geq 1 - t \qquad t = -2$

42. $2t - 3 > 5t - (2t + 1) \qquad t = 5$

TABLES AND LINEAR INEQUALITIES

Exercises 43–46: Use the table to solve the inequality.

43. $3x + 6 > 0$

x	−4	−3	−2	−1	0
$3x + 6$	−6	−3	0	3	6

44. $6 - 3x \leq 0$

x	1	2	3	4	5
$6 - 3x$	3	0	−3	−6	−9

45. $-2x + 7 > 5$

x	−1	0	1	2	3
$-2x + 7$	9	7	5	3	1

46. $5(x - 3) \leq 4$

x	3.2	3.4	3.6	3.8	4
$5(x - 3)$	1	2	3	4	5

Exercises 47–50: Complete the table. Then use the table to solve the inequality.

47. $-2x + 6 \leq 0$

x	1	2	3	4	5
$-2x + 6$	4				−4

48. $3x - 1 < 8$

x	0	1	2	3	4
$3x - 1$	−1				

49. $5 - x > x + 7$

x	−3	−2	−1	0	1
$5 - x$	8				4
$x + 7$	4				8

50. $2(3 - x) \geq -3(x - 2)$

x	−2	−1	0	1	2
$2(3 - x)$					
$-3(x - 2)$					

SOLVING LINEAR INEQUALITIES

Exercises 51–58: Use the addition property of inequalities to solve the inequality. Then graph the solution set.

51. $x - 3 > 0$

52. $x + 6 < 3$

53. $3 - y \leq 5$

54. $8 - y \geq 10$

55. $12 < 4 + z$

56. $2z \leq z + 17$

57. $5 - 2t \geq 10 - t$

58. $-2t > -3t + 1$

Exercises 59–66: Use the multiplication property of inequalities to solve the inequality. Then graph the solution set.

59. $2x < 10$

60. $3x > 9$

61. $-\frac{1}{2}t \geq 1$

62. $-5t \leq -6$

63. $\frac{3}{4} > -5y$

64. $10 \geq -\frac{1}{7}y$

65. $-\frac{2}{3} \leq \frac{1}{7}z$

66. $-\frac{3}{10}z < 11$

Exercises 67–72: Solve the linear inequality and write the solution in set-builder notation.

67. $x + 6 > 7$

68. $x + 4 < 1$

69. $-3x \leq 21$

70. $4x \geq -20$

71. $2x - 3 < 9$

72. $-5x + 4 < 44$

Exercises 73–100: Solve the linear inequality.

73. $3x + 1 < 22$

74. $4 + 5x \leq 9$

75. $5 - \frac{3}{4}x \geq 6$

76. $10 - \frac{2}{5}x > 0$

77. $45 > 6 - 2x$

78. $69 \geq 3 - 11x$

79. $5x - 2 \leq 3x + 1$

80. $12x + 1 < 25 - 3x$

81. $-x + 24 < x + 23$

82. $6 - 4x \leq x + 1$

83. $-(x + 1) \geq 3(x - 2)$

84. $5(x + 2) > -2(x - 3)$

85. $3(2x + 1) > -(5 - 3x)$

86. $4x \geq -3(7 - 2x) + 1$

87. $-(7x + 5) + 1 \geq 3x - 1$

88. $3(2 - x) - 5 > -4(5 - x)$

89. $1.6x + 0.4 \leq 0.4x$

90. $-5.1x + 1.1 < 0.1 - 0.1x$

91. $0.8x - 0.5 < x + 1 - 0.5x$

92. $0.1(x + 1) - 0.1 \leq 0.2x - 0.5$

93. $-\frac{1}{2}\left(\frac{2}{3}x + 4\right) \geq x$

94. $-5x > \frac{4}{5}\left(\frac{10}{3}x + 10\right)$

95. $\frac{3}{7}x + \frac{2}{7} > -\frac{1}{7}x - \frac{5}{14}$

96. $\frac{5}{6} - \frac{1}{3}x \geq -\frac{1}{3}\left(\frac{5}{6}x - 1\right)$

97. $\frac{x}{3} + \frac{5x}{6} \leq \frac{2}{3}$

98. $\frac{3x}{4} - \frac{x}{2} < 1$

99. $\frac{6x}{7} < \frac{1}{3}x + 1$

100. $\frac{5x}{8} - \frac{3x}{4} \leq 8$

TRANSLATING PHRASES TO INEQUALITIES

Exercises 101–108: Translate each phrase to an inequality. Let x be the variable.

101. A speed that is greater than 60 miles per hour

102. A speed that is at most 60 miles per hour

103. An age that is at least 21 years old

104. An age that is less than 21 years old

105. A salary that is more than $40,000

106. A salary that is less than or equal to $40,000

107. A speed that does not exceed 70 miles per hour

108. A speed that is not less than 70 miles per hour

APPLICATIONS

109. *Geometry*   Find values for x so that the perimeter of the rectangle is less than 50 feet.

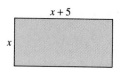

110. *Geometry*   A rectangle is twice as long as it is wide. If the rectangle is to have a perimeter of at least 36 inches, what values for the width are possible?

111. *Geometry*   A triangle with height 12 inches is to have area less than 120 square inches. What must be true about the base of the triangle?

112. *Geometry*   A trapezoid with height 6 inches is to have area not more than 120 square inches. What must be true about the sum of the bases of the trapezoid? (*Hint:* $A = \frac{1}{2}h(a + b)$, where a and b are the bases of the trapezoid.)

113. *Grade Average*   A student scores 74 out of 100 on a test. If the maximum score on the next test is also 100 points, what score does the student need to maintain at least an average of 80?

114. *Grade Average*   A student scores 65 and 82 on two different 100-point tests. If the maximum score on the next test is also 100 points, what score does the student need to maintain at least an average of 70?

115. *Parking Rates*   Parking in a student lot costs $2 for the first half hour and $1.25 for each hour thereafter. A partial hour is charged the same as a full hour. What is the longest time that a student can park in this lot for $8?

116. *Parking Rates* Parking in a student lot costs $2.50 for the first hour and $1 for each hour thereafter. A nearby lot costs $1.25 for each hour. In both lots a partial hour is charged as a full hour. In which lot can a student park the longest for $5? For $11?

117. *Car Rental* A rental car costs $25 per day plus $0.20 per mile. If someone has $200 to spend and needs to drive the car 90 miles each day, for how many days can that person rent the car? Assume that the car cannot be rented for part of a day.

118. *Car Rental* One car rental agency charges $20 per day plus $0.25 per mile. A different agency charges $37 per day with unlimited mileage. For what mileages is the second rental agency a better deal?

119. *Revenue and Cost* (Refer to Example 12.) The cost to produce one compact disc is $1.50 plus a one-time fixed cost of $2000. The revenue received from selling one compact disc is $12.
   (a) Write a formula that gives the cost C of producing x compact discs. Be sure to include the fixed cost.
   (b) Write a formula that gives the revenue R from selling x compact discs.
   (c) Profit equals revenue minus cost. Write a formula that calculates the profit P from selling x compact discs.
   (d) What numbers of compact discs need to be sold to yield a positive profit?

120. *Revenue and Cost* The cost to produce one laptop computer is $890 plus a one-time fixed cost of $100,000 for research and development. The revenue received from selling one laptop computer is $1520.
   (a) Write a formula that gives the cost C of producing x laptop computers.
   (b) Write a formula that gives the revenue R from selling x laptop computers.
   (c) Profit equals revenue minus cost. Write a formula that calculates the profit P from selling x laptop computers.
   (d) How many computers need to be sold to yield a positive profit?

121. *Distance and Time* Two athletes are jogging in the same direction along an exercise path. After x minutes the first athlete's distance in miles from a parking lot is given by $\frac{1}{6}x$ and the second athlete's distance is given by $\frac{1}{8}x + 2$.
   (a) When are the athletes the same distance from the parking lot?
   (b) When is the first athlete farther from the parking lot than the second?

122. *Sales of CDs and Cassettes* From 1987 to 2004, the percentage of U.S. music sales in CD format can be modeled by

$$C = 4.64(x - 1987) + 11.5,$$

and the percentage of sales in cassette tape format can be modeled by

$$T = -3.58(x - 1987) + 62.5.$$

In both expressions x represents the year. (*Source:* Recording Industry Association of America.)
   (a) Estimate the years when cassette tape sales were greater than or equal to CD sales.
   (b) What happened to sales of CDs and cassette tapes after 1995?

123. *Altitude and Temperature* (Refer to Example 10.) If the temperature on the ground is 90°F, then the air temperature x miles high is given by $T = 90 - 19x$. Determine the altitudes at which the air temperature is less than 4.5°F. (*Source:* A. Miller.)

124. *Altitude and Dew Point* If the dew point on the ground is 65°F, then the dew point x miles high is given by $D = 65 - 5.8x$. Determine the altitudes at which the dew point is greater than 36°F. (*Source:* A. Miller.)

125. *Life Expectancy* Because of medical advances and improved health care, people in the U.S. can expect to live longer. From 1980 to 2005, the number of years Y that a 65-year-old man could expect to live beyond age 65 (years remaining), can be approximated by

$$Y = 0.11(x - 1980) + 14.1,$$

where x represents the year. Determine the years when a 65-year-old man could expect to live an additional 16.3 years or more. (*Source:* Department of Health and Human Services.)

126. *Size and Weight of a Fish*   If the length of a bass is between 20 and 25 inches, its weight W in pounds can be estimated by the formula $W = 0.96x - 14.4$, where x is the length of the fish.   (*Source:* Minnesota Department of Natural Resources.)

   (a) What length of bass is likely to weigh 7.2 pounds?

   (b) What lengths of bass are likely to weigh less than 7.2 pounds?

WRITING ABOUT MATHEMATICS

127. Explain each of the terms and give an example.
   (a) Linear equation
   (b) Linear inequality

128. Suppose that a student says that a linear equation and a linear inequality can be solved the same way. How would you respond?

CHECKING BASIC CONCEPTS
SECTION 2.5

1. Use a number line to graph the solution set to the inequality $x + 1 \geq -1$.

2. Express the set of real numbers graphed on the number line by using an inequality.

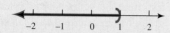

3. Determine whether -3 is a solution to the inequality $4x - 5 \leq -15$.

4. Complete the table. Then use the table to solve the inequality $5 - 2x \leq 7$.

x	-2	-1	0	1	2
$5 - 2x$					1

5. Solve each inequality.
   (a) $x + 5 > 8$
   (b) $-\frac{5}{7}x \leq 25$
   (c) $3x \geq -2(1 - 2x) + 3$

6. Translate the phrase "a price that is not more than $12" to an inequality using the variable x.

7. *Geometry*   The length of a rectangle is 5 inches longer than twice its width. If the perimeter of the rectangle is more than 88 inches, find possible widths for the rectangle.

CHAPTER 2   SUMMARY
SECTION 2.1 ■ INTRODUCTION TO EQUATIONS

Equations   Every equation contains an equals sign. An equation can either be true or false.

Important Terms

Solution	A value for a variable that makes the equation true
Solution Set	The set of all solutions
Equivalent Equations	Equations that have the same solution set
Checking a Solution	Substituting the solution in the given equation to verify that the equation is true

   Example: 3 is the solution to $4x - 2 = 10$ because $4(3) - 2 = 10$ is a true statement.

Properties of Equality

Addition Property $a = b$ is equivalent to $a + c = b + c$.

Example: $x - 3 = 0$ and $x - 3 + 3 = 0 + 3$ are equivalent equations.

Multiplication Property $a = b$ is equivalent to $ac = bc$, provided $c \neq 0$.

Example: $2x = 5$ and $\frac{1}{2} \cdot 2x = \frac{1}{2} \cdot 5$ are equivalent equations.

SECTION 2.2 ■ LINEAR EQUATIONS

Linear Equation   Can be written in the form $ax + b = 0$, where $a \neq 0$

Examples: $3x - 5 = 0$ is linear, whereas $5x^2 + 2x = 0$ is *not* linear.

Solving Linear Equations   The following steps can be used as a guide for solving linear equations.

STEP 1: Use the distributive property to clear any parentheses on each side of the equation. Combine any like terms on each side.

STEP 2: Use the addition property of equality to get all of the terms containing the variable on one side of the equation and all other terms on the other side of the equation. Combine any like terms on each side.

STEP 3: Use the multiplication property of equality to isolate the variable by multiplying each side of the equation by the reciprocal of the number in front of the variable (or divide each side by that number).

STEP 4: Check the solution by substituting it in the given equation.

Distributive Properties $a(b + c) = ab + ac$ or $a(b - c) = ab - ac$

Examples: $5(2x + 3) = 10x + 15$ and $5(2x - 3) = 10x - 15$

Clearing Fractions and Decimals   When fractions or decimals appear in an equation, multiplying each side by the least common denominator can be helpful.

Examples: Multiply each side of $\frac{1}{3}x - \frac{1}{6} = \frac{2}{3}$ by 6 to obtain $2x - 1 = 4$.

Multiply each side of $0.04x + 0.1 = 0.07$ by 100 to obtain $4x + 10 = 7$.

Number of Solutions   Equations that can be written in the form $ax + b = 0$, where a and b are *any* real number, can have no solutions, one solution, or infinitely many solutions.

Examples: $x + 3 = x$ is equivalent to $3 = 0$.   (No solutions)

$2y + 1 = 9$ is equivalent to $y = 4$.   (One solution)

$z + z = 2z$ is equivalent to $0 = 0$.   (Infinitely many solutions)

SECTION 2.3 ■ INTRODUCTION TO PROBLEM SOLVING

Steps for Solving a Problem   The following steps can be used as a guide for solving word problems.

STEP 1: Read the problem carefully and be sure that you understand it. (You may need to read the problem more than once.) Assign a variable to what you are being asked to find. If necessary, write other quantities in terms of this variable.

STEP 2: Write an equation that relates the quantities described in the problem. You may need to sketch a diagram or refer to known formulas.

STEP 3: Solve the equation. Use the solution to determine the solution(s) to the given problem. Include any necessary units.

STEP 4: Look back and check your solution in the given problem. Does it seem reasonable?

Percent Problems

The expression x%

Represents the fraction $\frac{x}{100}$ or the decimal number given by $x \times 0.01$.

Examples: $45\% = \dfrac{45}{100} = 0.45$

$7.1\% = \dfrac{7.1}{100} = \dfrac{71}{1000} = 0.071$

Percent Change

If a quantity changes from an old value to a new value, then the

percent change equals $\dfrac{(\text{new value}) - (\text{old value})}{(\text{old value})} \times 100$.

Example: If a price increases from \$2 to \$3, then the percent change
equals $\dfrac{3 - 2}{2} \times 100 = 50\%$.

Distance Problems   If an object travels at speed (rate) r for time t, then the distance d traveled is calculated by $d = rt$.

Example: A car moving at 65 mph for 2 hours travels

$$d = rt = 65 \cdot 2 = 130 \text{ miles.}$$

SECTION 2.4 ■ FORMULAS

Formula   A formula is an equation that can be used to calculate a quantity by using known values of other quantities.

Example: The formula $M = \dfrac{D}{G}$ can be used to calculate the gas mileage M obtained by a car traveling D miles on G gallons of gasoline.

Formulas from Geometry

Area of a Rectangle $A = LW$, where L is the length and W is the width.

Area of a Triangle $A = \frac{1}{2}bh$, where b is the base and h is the height.

Degree Measure   There are $360°$ in one complete revolution.

Angle Measure   The sum of the angles in a triangle equals $180°$.

Volume of a Box	$V = LWH$, where L is the length, W is the width, and H is the height.

Surface Area of a Box $S = 2LW + 2WH + 2LH$, where L is the length, W is the width, and H is the height.

Circumference $C = 2\pi r$, where r is the radius.

Area of a Circle $A = \pi r^2$, where r is the radius.

Volume of a Cylinder $V = \pi r^2 h$, where r is the radius and h is the height.

Area of a Trapezoid $A = \frac{1}{2}(a + b)h$, where h is the height and a and b are the bases of the trapezoid.

Other Formulas

Distance from Lightning $D = \frac{x}{5}$, where x is the number of seconds between seeing the flash and hearing the thunder.

GPA $GPA = \frac{4a + 3b + 2c + d}{a + b + c + d + f}$, where a represents the number of A credits earned, b the number of B credits earned, and so on.

Temperature Scales $F = \frac{9}{5}C + 32$ and $C = \frac{5}{9}(F - 32)$, where F is the Fahrenheit temperature and C is the Celsius temperature.

See Putting It All Together in Section 2.4 for examples.

SECTION 2.5 ■ LINEAR INEQUALITIES

Linear Inequality When the equals sign in a linear equation is replaced with any one of the symbols $<, \leq, >,$ or $\geq$, a linear inequality results.

Examples: $x > 0$, $6 - \frac{2}{3}x \leq 7$, and $4(x - 1) < 3x - 1$

Number Line Graphs A number line can be used to graph the solution set to a linear inequality.

Example: The graph of $x \leq 1$ is shown in the figure.

Properties of Inequality

Addition Property $a < b$ is equivalent to $a + c < b + c$.

Example: $x - 3 < 0$ and $x - 3 + 3 < 0 + 3$ are equivalent inequalities.

Multiplication Property When $c > 0$, $a < b$ is equivalent to $ac < bc$.
When $c < 0$, $a < b$ is equivalent to $ac > bc$.

Examples: $2x < 6$ is equivalent to $2x(\frac{1}{2}) < 6(\frac{1}{2})$ or $x < 3$.
$-2x < 6$ is equivalent to $-2x(-\frac{1}{2}) > 6(-\frac{1}{2})$ or $x > -3$.

CHAPTER 2   REVIEW EXERCISES

SECTION 2.1

Exercises 1–8: Solve the equation. Check your solution.

1. $x + 9 = 3$

2. $x - 4 = -2$

3. $x - \frac{3}{4} = \frac{3}{2}$

4. $x + 0.5 = 0$

5. $4x = 12$

6. $3x = -7$

7. $-0.5x = 1.25$

8. $-\frac{1}{3}x = \frac{7}{6}$

SECTION 2.2

Exercises 9–12: Decide whether the equation is linear. If the equation is linear, give values for a and b so that it can be written in the form ax + b = 0.

9. $5x - 3 = 0$

10. $-4x + 3 = 2$

11. $\frac{1}{x} + 3 = 0$

12. $\frac{3}{8}x^2 - x = \frac{1}{4}$

Exercises 13–22: Solve the equation. Check the solution.

13. $4x - 5 = 3$

14. $7 - \frac{1}{2}x = -4$

15. $5(x - 3) = 12$

16. $3 + x = 2x - 4$

17. $2(x - 1) = 4(x + 3)$

18. $1 - (x - 3) = 6 + 2x$

19. $3.4x - 4 = 5 - 0.6x$

20. $-\frac{1}{3}(3 - 6x) = -(x + 2) + 1$

21. $\frac{2}{3}x - \frac{1}{6} = \frac{5}{12}$

22. $2y - 3(2 - y) = 5 + y$

Exercises 23–26: Determine whether the equation has no solutions, one solution, or infinitely many solutions.

23. $4(3x - 2) = 2(6x + 5)$

24. $5(3x - 1) = 15x - 5$

25. $8x = 5x + 3x$

26. $9x - 2 = 8x - 2$

Exercises 27 and 28: Complete the table. Then use the table to solve the given equation.

27. $-2x + 3 = 0$

x	0.5	1.0	1.5	2.0	2.5
$-2x + 3$	2				

28. $-(x + 1) + 3 = 2$

x	-2	-1	0	1	2
$-(x + 1) + 3$					0

SECTION 2.3

Exercises 29–32: Using the variable x, translate the sentence into an equation. Solve the resulting equation.

29. The product of a number and 6 is 72.

30. The sum of a number and 18 is −23.

31. Twice a number minus 5 equals the number plus 4.

32. The sum of a number and 4 equals the product of the number and 3.

Exercises 33 and 34: Find the number or numbers.

33. The sum of four consecutive natural numbers is 70.

34. The sum of three consecutive integers is −153.

Exercises 35–38: Convert the percentage to fraction and decimal notation.

35. 85%

36. 5.6%

37. 0.03%

38. 342%

Exercises 39–42: Convert the number to a percentage.

39. 0.89

40. 0.005

41. 2.3

42. 1

Exercises 43–46: Use the formula d = rt to find the value of the missing variable.

43. $r = 8$ miles per hour, $t = 3$ hours

44. $r = 70$ feet per second, $t = 55$ seconds

45. $d = 500$ yards, $t = 20$ seconds

46. $d = 125$ miles, $r = 15$ miles per hour

SECTION 2.4

Exercises 47 and 48: Find the area of the region shown.

47.

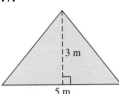

3 m

5 m

48.

6 ft

49. Find the area of a rectangle having a 24-inch width and a 3-foot length.

50. Find the area of a triangle having a 13-inch base and a 7-inch height.

51. Find the circumference of a circle having an 18-foot diameter.

52. Find the area of a circle having a 5-inch radius.

53. Find the measure of the third angle in the triangle.

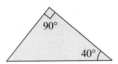

90°

40°

54. The angles in a triangle have measures x, $3x$, and $4x$. Find the value of x.

55. If a cylinder has radius 5 inches and height 25 inches, find its volume. (*Hint:* $V = \pi r^2 h$.)

56. Find the area of a trapezoid with height 5 feet and bases 3 feet and 18 inches. (*Hint:* $A = \frac{1}{2}(a + b)h$.)

Exercises 57 and 58: Find the area of the figure shown.

57.

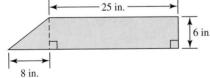

25 in.

6 in.

8 in.

58.

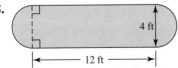

4 ft

12 ft

Exercises 59–64: Solve the formula for the specified variable.

59. $a = x + y$ for x

60. $P = 2x + 2y$ for x

61. $z = 2xy$ for y

62. $S = \dfrac{a + b + c}{3}$ for b

63. $T = \dfrac{a}{3} + \dfrac{b}{4}$ for b

64. $cd = ab + bc$ for c

Exercises 65 and 66: Let a represent the number of credits with a grade of A, b the number of credits with a grade of B, and so on. Calculate the grade point average (GPA). Round your answer to the nearest hundredth.

65. $a = 20, b = 25, c = 12, d = 4, f = 4$

66. $a = 64, b = 32, c = 20, d = 10, f = 3$

67. Convert 15°C to an equivalent Fahrenheit temperature.

68. Convert 113°F to an equivalent Celsius temperature.

SECTION 2.5

Exercises 69–72: Use a number line to graph the solution set to the inequality.

69. $x < 2$

70. $x > -1$

71. $y \geq -\frac{3}{2}$

72. $y \leq 2.5$

Exercises 73 and 74: Express the set of real numbers graphed on the number line with an inequality.

73.

0 1 2 3 4

74.

-2 -1 0 1 2

Exercises 75–78: Determine whether the given value of x is a solution to the inequality.

75. $2x + 1 \leq 5$ $x = -3$

76. $5 - \frac{1}{2}x > -1$ $x = 4$

77. $1 - (x + 3) \geq x$ $x = -2$

78. $4(x + 1) < -(5 - x)$ $x = -1$

Exercises 79 and 80: Complete the table and then use the table to solve the inequality.

79. $5 - x > 3$

x	0	1	2	3	4
$5 - x$	5				

80. $2x - 5 \leq 0$

x	1	1.5	2	2.5	3
$2x - 5$	-3				

Exercises 81–86: Solve the inequality.

81. $x - 3 > 0$ **82.** $-2x \leq 10$

83. $5 - 2x \geq 7$ **84.** $3(x - 1) < 20$

85. $5x \leq 3 - (4x + 2)$

86. $3x - 2(4 - x) \geq x + 1$

Exercises 87–90: Translate the phrase to an inequality. Let x be the variable.

87. A speed that is less than 50 miles per hour

88. A salary that is at most $45,000

89. An age that is at least 16 years old

90. A year before 1995

APPLICATIONS

91. *Rainfall*   On a stormy day 2 inches of rain fall before noon and $\frac{3}{4}$ inch per hour fall thereafter until 5 P.M.
   (a) Make a table that shows the total rainfall at each hour starting at noon and ending at 5 P.M.
   (b) Write a formula that calculates the rainfall R in inches, x hours past noon.
   (c) Use your formula to calculate the total rainfall at 5 P.M. Does your answer agree with the value in your table from part (a)?
   (d) How much rain had fallen at 3:45 P.M.?

92. *Cost of a Laptop*   A 5% sales tax on a laptop computer amounted to $106.25. Find the cost of the laptop.

93. *Distance Traveled*   At noon a bicyclist is 50 miles from home, riding toward home at 10 miles per hour.
   (a) Make a table that shows the bicyclist's distance D from home after 1, 2, 3, 4, and 5 hours.
   (b) Write a formula that calculates the distance D from home after x hours.
   (c) Use your formula to determine D when $x = 3$ hours. Does your answer agree with the value shown in your table?
   (d) For what times was the bicyclist at least 20 miles from home? Assume that $0 \leq x \leq 5$.

94. *High School Graduates*   The number of high school graduates N in millions during year x, $x \geq 1995$, can be approximated by the formula

$$N = \frac{1}{15}x - 130.4.$$

Estimate the year during which this number reached 2.8 million.   (**Source:** Department of Education.)

95. *Hazardous Waste*   In 2005 the number of federal hazardous waste sites in Wisconsin was 12 less than three times the number of sites in Texas. How many hazardous waste sites were there in Texas if there were no federal hazardous waste sites in Wisconsin? (**Source:** Environmental Protection Agency.)

96. *Master's Degree*   In 1971, about 230,500 people received a master's degree, and in 1997, about 419,400 did. Find the percent change in the number of master's degrees received between 1971 and 1997.

97. *Car Speeds*   One car passes another car on a freeway. The faster car is traveling 12 miles per hour faster than the slower car. Determine how long it will be before the faster car is 2 miles ahead of the slower car.

98. *Gas Mileage*   An SUV that gets 18 miles per gallon is driven 504 miles. How many gallons of gasoline are used on this trip?

99. *Saline Solution*   A saline solution contains 3% salt. How much water should be added to 100 milliliters of this solution to dilute it to a 2% solution?

100. *Investment Money*   A student invests two sums of money, $500 and $800, at different interest rates, receiving a total of $55 in interest after one year. The $500 investment receives an interest rate 2% lower than the interest rate for the $800 investment. Find the interest rate for each investment.

101. *Dimensions of a Rectangle*   The width of a rectangle is 10 inches less than its length. If the perimeter is 112 inches, find the dimensions of the rectangle.

102. *Lightning*   The time delay between a flash of lightning and the sound of thunder is 9 seconds. How far away is the lightning?

103. *Geometry*   A triangle with height 8 inches is to have an area that is not more than 100 square inches. What lengths are possible for the base of the triangle?

104. *Grade Average* A student scores 75 and 91 on two different tests of 100 points. If the maximum score on the next test is also 100 points, what score does the student need to maintain an average of at least 80?

105. *Parking Rates* Parking in a lot costs $2.25 for the first hour and $1.25 for each hour thereafter. A partial hour is charged the same as a full hour. What is the longest time that someone can park for $9?

106. *Profit* The cost to produce one DVD player is $85 plus a one-time fixed cost of $150,000. The revenue received from selling one DVD player is $225.

(a) Write a formula that gives the cost C of producing x DVD players.

(b) Write a formula that gives the revenue R from selling x DVD players.

(c) Profit equals revenue minus cost. Write a formula that calculates the profit P from selling x DVD players.

(d) What numbers of DVD players sold will result in a loss? (*Hint:* A loss corresponds to a negative profit.)

CHAPTER 2 TEST Pass the Test Video solutions to all test exercises

Exercises 1–4: Solve the equation. Check your solution.

1. $9 = 3 - x$

2. $4x - 3 = 7$

3. $4x - (2 - x) = -3(2x + 6)$

4. $\frac{1}{12}x - \frac{2}{3} = \frac{1}{2}\left(\frac{3}{4} - \frac{1}{3}x\right)$

5. Determine the number of solutions to the equation

$$6(2x - 1) = -4(3 - 3x).$$

6. Complete the table. Then use the table to solve the equation $6 - 2x = 0$.

x	0	1	2	3	4
$6 - 2x$	6				

Exercises 7 and 8: Translate the sentence into an equation, using the variable x. Then solve the resulting equation.

7. The sum of a number and -7 is 6.

8. Twice a number plus 6 equals the number minus 7.

9. The sum of three consecutive natural numbers is 336. Find the three numbers.

10. Convert 3.2% to fraction and decimal notation.

11. Convert 0.345 to a percentage.

12. Find 7.5% of $500.

13. *Speed of Sound* Sound can travel one mile (5280 feet) in 5 seconds. Find the speed of sound in feet per second.

14. *Area* Find the area of the triangle shown.

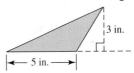

15. Find the circumference and area of a circle with a 30-inch diameter.

16. The measures of the angles in a triangle are x, $2x$, and $3x$. Find the value of x.

Exercises 17 and 18: Solve the formula for x.

17. $z = y - 3xy$

18. $R = \frac{x}{4} + \frac{y}{5}$

Exercises 19 and 20: Solve the inequality.

19. $-3x + 9 \geq x - 15$

20. $3(6 - 5x) < 20 - x$

21. *Snowfall* Suppose 5 inches of snow fall before noon and 2 inches per hour fall thereafter until 10 P.M.

(a) Write a formula that calculates the snowfall S in inches, x hours past noon.

(b) Use your formula to calculate the total snowfall at 8 P.M.

(c) How much snow had fallen at 6:15 P.M.?

22. *Mixing Acid*   A solution is 45% hydrochloric acid. How much water should be added to 1000 milliliters of this solution to dilute it to a 15% solution?

23. *Medical Malpractice*   Malpractice insurance premiums at Thomas Jefferson University Hospital in Philadelphia increased from $8 million in 1998 to $32 million in 2003. Calculate the percent change in malpractice insurance premiums over this 5-year period.   (*Source: The New York Times.*)

CHAPTER 2   EXTENDED AND DISCOVERY EXERCISES

Exercises 1–4: Average Speed   If someone travels a distance d in time t, then the person's average speed is $\frac{d}{t}$. Use this fact to solve the problem.

1. A driver travels at 50 mph for the first hour and then travels at 70 mph for the second hour. What is the average speed of the car?

2. A bicyclist rides 1 mile uphill at 5 mph and then rides 1 mile downhill at 10 mph. Find the average speed of the bicyclist. Does your answer agree with what you expected?

3. At a 3-mile cross-country race an athlete runs 2 miles at 8 mph and 1 mile at 10 mph. What is the athlete's average speed?

4. A pilot flies an airplane between two cities and travels half the distance at 200 mph and the other half at 100 mph. Find the average speed of the airplane.

5. *A Puzzle About Coins*   Suppose that seven coins look exactly alike but that one coin weighs less than any of the other six coins. If you have only a balance with two pans, devise a plan to find the lighter coin. What is the minimum number of weighings necessary? Explain your answer.

6. *Global Warming*   If the global climate were to warm significantly as a result of the greenhouse effect or other climatic change, the Arctic ice cap would start to melt. This ice cap contains the equivalent of some 680,000 cubic miles of water. More than 200 million people live on land that is less than 3 feet above sea level. In the United States several large cities have low average elevations. Three examples are Boston (14 feet), New Orleans (4 feet), and San Diego (13 feet). In this exercise you are to estimate the rise in sea level if the Arctic ice cap were to melt and to determine whether this event would have a significant impact on people living in coastal areas.

(a) The surface area of a sphere is given by the formula $4\pi r^2$, where r is its radius. Although the shape of Earth is not exactly spherical, it has an average radius of 3960 miles. Estimate the surface area of Earth.

(b) Oceans cover approximately 71% of the total surface area of Earth. How many square miles of Earth's surface are covered by oceans?

(c) Approximate the potential rise in sea level by dividing the total volume of the water from the ice cap by the surface area of the oceans. Convert your answer from miles to feet.

(d) Discuss the implications of your calculation. How would cities such as Boston, New Orleans, and San Diego be affected?

(e) The Antarctic ice cap contains some 6,300,000 cubic miles of water. Estimate how much the sea level would rise if this ice cap melted.   (*Source: Department of the Interior, Geological Survey.*)

CHAPTERS 1–2   CUMULATIVE REVIEW EXERCISES

Exercises 1 and 2: Classify the number as prime or composite. If the number is composite, write it as a product of prime numbers.

1. 45

2. 37

Exercises 3 and 4: Multiply or divide and then simplify to lowest terms when appropriate.

3. $\frac{4}{3} \cdot \frac{3}{8}$

4. $\frac{2}{3} \div 6$

Exercises 5 and 6: Add or subtract and then simplify to lowest terms when appropriate.

5. $\frac{11}{12} - \frac{3}{8}$

6. $\frac{2}{3} + \frac{1}{5}$

Exercises 7 and 8: Classify the number as one or more of the following: natural number, whole number, integer, rational number, or irrational number.

7. -1

8. $\sqrt{3}$

Exercises 9–14: Evaluate the expression by hand.

9. $15 - 4 \cdot 3$

10. $30 \div 6 \cdot 2$

11. $23 - 4^2 \div 2$

12. $11 - \frac{3+1}{6-4}$

13. $-14 - (-7)$

14. $-\frac{2}{3} \cdot \left(-\frac{9}{14}\right)$

Exercises 15 and 16: Simplify the expression.

15. $5x^3 - x^3$

16. $4 + 2x - 1 + 3x$

Exercises 17–20: Solve the equation.

17. $x - 3 = 11$

18. $4x - 6 = -22$

19. $5(6y + 2) = 25$

20. $11 - (y + 2) = 3y + 5$

Exercises 21 and 22: Determine whether the equation has no solutions, one solution, or infinitely many solutions.

21. $6x + 2 = 2(3x + 1)$

22. $2(3x - 4) = 6(x - 1)$

23. Find three consecutive integers whose sum is 90.

24. Convert 4.7% to decimal notation.

25. Convert 0.17 to a percentage.

26. Find the speed of a car that travels 325 miles in 5 hours.

27. The angles in a triangle have measures $2x$, $3x$, and $4x$. Find the value of x.

28. Find the area of a circle having a 10-inch diameter.

Exercises 29 and 30: Solve the formula for x.

29. $a = 3xy - 4$

30. $A = \dfrac{x + y + z}{3}$

Exercises 31 and 32: Solve the inequality and graph the solution set.

31. $7 - 3x > 4$

32. $6x \le 5 - (x - 9)$

33. *Yards to Inches*   There are 36 inches in 1 yard. Write a formula that converts Y yards to I inches.

34. *Checking Account*   The initial balance in a checking account is $468. Find the final balance resulting from the following sequence of withdrawals and deposits: $-\$14$, $\$200$, $-\$73$, $-\$21$, and $\$58$.

35. *Acid Solution*   How much of a 4% acid solution should be added to 150 milliliters of a 10% acid solution to dilute it to a 6% acid solution?

36. *Bank Loans*   An individual has two low-interest loans, one at 4% interest and the other at 6% interest. The amount borrowed at 6% is $250 more than the amount borrowed at 4%. If the total interest for one year is $165, how much money is borrowed at each rate?

Graphing Equations

Global warming has been in the news for years. Scientists are documenting that both the arctic and antarctic ice packs are melting. This phenomenon is an indication that global temperatures may be rising. The amount of light from the sun that Earth reflects into space is an important factor. If more light is reflected, then Earth stays cooler. Astronomer Phil Goode from the New Jersey Institute of Technology has developed a simple way to determine how reflective Earth is by measuring the amount of earthshine illuminating the moon. The area of the moon illuminated by earthshine is the darker portion of the moon that is visible but not as visible as the portion illuminated by the sun. Goode found that the percent change in earthshine (from the average) varied with the month. An excellent way to display these data is with the *line graph* shown in the accompanying figure. We discuss line graphs and other types of graphs in this chapter.

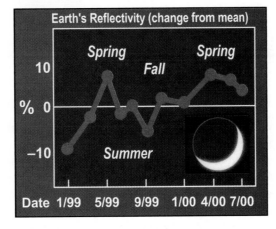

Originality is the essence of true scholarship.

—NNAMDI AZIKIWE

Source: Hannah Hoag, *Discover*, November 2002, p. 11. Graphic by Matt Zang. (Reprinted with permission.)

3.1 INTRODUCTION TO GRAPHING

The Rectangular Coordinate System ▪ Scatterplots and Line Graphs

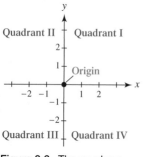

Table 3.1 lists the per capita (per person) personal income for the United States for selected years. These income amounts have *not* been adjusted for inflation. When a table contains only four data values, as in Table 3.1, we can easily see that income increased significantly from 1990 to 2005. If a table listed income for every year from 1990 to 2005, seeing trends in the data would be more difficult. And if a table contained 1000 data values, determining trends in the data would be extremely difficult. In mathematical problems, there are frequently infinitely many data points!

Rather than always using tables to display data, presenting data on a graph is often more useful. For example, the data in Table 3.1 are graphed in Figure 3.1. This line graph is more visual than the table and shows the trend at a glance.

TABLE 3.1
U.S. Per Capita Income

Year	Amount
1990	$18,667
1995	$23,562
2000	$29,469
2005	$34,586

Source: Bureau of Economic Analysis.

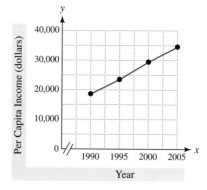

Figure 3.1 Per Capita Income

NOTE: In Figure 3.1 the symbol ─//─ on the x-axis indicates a break in the x-axis data values. The years before 1990 have been skipped.

The Rectangular Coordinate System

Figure 3.2 The *xy*-plane

One common way to graph data is to use the **rectangular coordinate system**, or **xy-plane**. In the *xy*-plane the horizontal axis is the **x-axis**, and the vertical axis is the **y-axis**. The axes can be thought of as intersecting number lines. The point of intersection is called the **origin** and is associated with zero on each axis. Negative values are located left of the origin on the *x*-axis and below the origin on the *y*-axis. Similarly, positive values are located right of the origin on the *x*-axis and above the origin on the *y*-axis. The axes divide the *xy*-plane into four regions called **quadrants**, which are numbered I, II, III, and IV counterclockwise, as shown in Figure 3.2.

Before we can plot data, we must first understand the concept of an **ordered pair** (x, y). In Table 3.1 we can let *x*-values correspond to the year and *y*-values correspond to the per capita income. Then the fact that the per capita income in **1990** was **$18,667** can be summarized by the ordered pair (**1990, 18667**). Similarly, the ordered pair (2000, 29469) indicates that the per capita income was $29,469 in 2000.

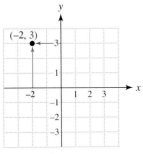

Figure 3.3 Plotting a Point

Order is important in an ordered pair. The ordered pairs given by $(1950, 2025)$ and $(2025, 1950)$ are different. The first ordered pair indicates that the per capita income in 1950 was \$2025, whereas the second ordered pair indicates that the per capita income in 2025 will be \$1950.

To plot the ordered pair $(-2, 3)$ in the xy-plane, begin at the origin and move left to locate $x = -2$ on the x-axis. Then move upward until a height of $y = 3$ is reached. Thus the point $(-2, 3)$ is located 2 units left of the origin and 3 units above the origin. In Figure 3.3, the point $(-2, 3)$ is plotted in quadrant II.

NOTE: A point that lies on an axis is not located in a quadrant.

EXAMPLE 1   Plotting points

Plot the following ordered pairs on the same xy-plane. State the quadrant in which each point is located, if possible.

(a) $(3, 2)$ **(b)** $(-2, -3)$ **(c)** $(-3, 0)$

Solution

(a) The point $(3, 2)$ is located in quadrant I, 3 units to the right of the origin and 2 units above the origin. See Figure 3.4.

(b) The point $(-2, -3)$ is located in quadrant III, 2 units to the left of the origin and 3 units below the origin. See Figure 3.4.

(c) The point $(-3, 0)$ is not in any quadrant because it is located 3 units left of the origin on the x-axis. See Figure 3.4.

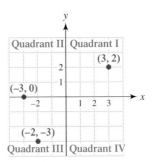

Figure 3.4

Now Try Exercises 11, 15

Choosing appropriate scales for the axes is important when plotting points and making graphs. This can be accomplished by looking at the *coordinates* of the ordered pairs to be plotted. When plotting in the xy-plane, the first value in an ordered pair is called the **x-coordinate** and the second value is called the **y-coordinate**. In Example 1, the x-coordinates of the three points are 3, -2, and -3 and the y-coordinates are 2, -3, and 0. Because there are both positive and negative values in these lists and no coordinate is more than 4 units from the origin, the scale shown in Figure 3.4 is appropriate.

In Figure 3.5 of the next example, the *increment* on the x-axis is 2 because each step from one vertical grid line to the next represents a change of 2 years. Similarly, the increment on the y-axis is 0.5 because each step from one horizontal grid line to the next represents a \$0.5 billion change in sales. This example demonstrates that the scale and increment on one axis are not always the same as those on the other axis.

EXAMPLE 2   Reading a graph

Frozen pizza makers have improved their pizzas to taste more like homemade. As a result, their sales have increased. Figure 3.5 shows U.S. retail sales of frozen pizzas in billions of dollars. Use the graph to estimate frozen pizza sales in 1994 and in 2000.   (*Source:* Business Trend Analyst.)

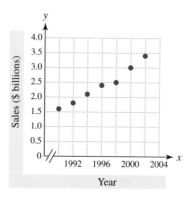

Figure 3.5 Frozen Pizza Sales

Solution

To estimate frozen pizza sales in 1994, first locate 1994 on the x-axis. Then move upward to the data point and approximate its y-coordinate. Figure 3.6(a) shows that there were about $2.1 billion in frozen pizza sales in 1994. Similarly, frozen pizza sales in 2000 were about $3 billion, as shown in Figure 3.6(b).

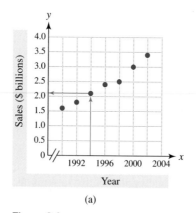

(a)

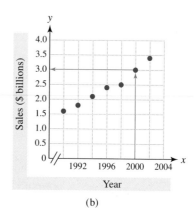

(b)

Figure 3.6

Now Try Exercise 39

Scatterplots and Line Graphs

If distinct points are plotted in the xy-plane, then the resulting graph is called a **scatterplot**. Figure 3.5 is an example of a scatterplot that shows frozen pizza sales. A different scatterplot is shown in Figure 3.7, in which the points $(1, 3)$, $(2, 2)$, $(3, 1)$, $(4, 4)$, and $(5, 1)$ are plotted.

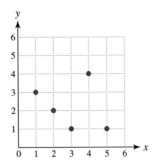

Figure 3.7 A Scatterplot

The next example illustrates how to make a scatterplot by using real data.

EXAMPLE 3 **Making a scatterplot of gasoline prices**

Table 3.2 lists the average price of a gallon of gasoline for selected years. Make a scatterplot of the data. These prices have *not* been adjusted for inflation.

TABLE 3.2 **Average Price of Gasoline**

Year	1950	1960	1970	1980	1990	2000
Cost (per gal)	27¢	31¢	36¢	119¢	115¢	156¢

Source: Department of Energy.

Solution

The data point (**1950**, **27**) can be used to indicate that the average cost of a gallon of gasoline in **1950** was **27**¢. Plot the six data points (1950, 27), (1960, 31), (1970, 36), (1980, 119), (1990, 115), and (2000, 156) in the xy-plane. The x-values vary from 1950 to 2000, so label the x-axis from 1950 to 2000 every 10 years. The y-values vary from 27 to 156, so label the y-axis from 0 to 175 every 25¢. Note that the x- and y-scales must be large enough to accommodate every data point. Figure 3.8 shows the scatterplot.

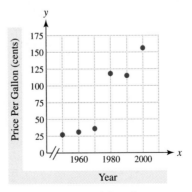

Figure 3.8 Price of Gasoline   Now Try Exercise 23

Sometimes it is helpful to connect consecutive data points in a scatterplot with line segments. This type of graph visually emphasizes changes in the data and is called a **line graph**. When making a line graph, be sure to plot all of the given data points *before* connecting the points with line segments. The points should be connected consecutively, from left to right, on the scatterplot even if the data are given "out of order" in a table.

EXAMPLE 4   Making a line graph

Use the data in Table 3.3 to make a line graph.

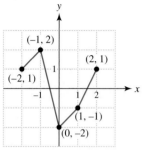

Figure 3.9 A Line Graph

TABLE 3.3

x	-2	-1	0	1	2
y	1	2	-2	-1	1

Solution
The data in Table 3.3 are represented by the five ordered pairs $(-2, 1)$, $(-1, 2)$, $(0, -2)$, $(1, -1)$, and $(2, 1)$. Plot these points and then connect consecutive points with line segments, as shown in Figure 3.9.

Now Try Exercise 33

TECHNOLOGY NOTE: Scatterplots and Line Graphs

Graphing calculators are capable of creating both line graphs and scatterplots. The line graph in Figure 3.9 is shown to the left and the corresponding scatterplot is shown to the right.

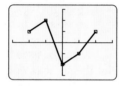   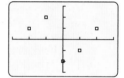

EXAMPLE 5   Analyzing a line graph

The line graph in Figure 3.10 shows the per capita energy consumption in the United States. Units are in millions of Btu, where 1 Btu equals the amount of heat necessary to raise 1 pound of water 1°F.   (**Source:** Department of Energy.)

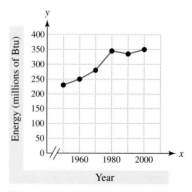

Figure 3.10 U.S. Energy Consumption

(a) Did energy consumption ever decrease during this time period? Explain.
(b) Estimate the energy consumption in 1960 and in 2000.
(c) Estimate the percent change in energy consumption from 1960 to 2000.

Solution
(a) Yes, energy consumption decreased slightly between 1980 and 1990.
(b) From the graph, per capita energy consumption in 1960 was about 250 million Btu. In 2000 it was about 350 million Btu.

NOTE: When different people read a graph, values they obtain may vary slightly.

(c) The percent change from 1960 to 2000 was

$$\frac{350 - 250}{250} \times 100 = 40,$$

so the increase was 40%. (To review the percent change formula, refer to Section 2.3.)

Now Try Exercise 47

CRITICAL THINKING
When analyzing data, do you prefer a table of values, a scatterplot, or a line graph? Explain your answer.

3.1 PUTTING IT ALL TOGETHER

The rectangular coordinate plane, or xy-plane, can be used to plot ordered pairs in the form (x, y). Application data can frequently be represented by ordered pairs. For example, the ordered pair $(12, 45)$ might indicate that the average high temperature in the twelfth month (December) is $45°F$.

Concept	Explanation	Examples
Ordered Pair	Has the form (x, y), where the order of x and y is important	$(1, 2), (-2, 3), (2, 1)$ and $(-4, -2)$ are distinct ordered pairs.
Rectangular Coordinate System, or xy-plane	Consists of a horizontal x-axis and a vertical y-axis Can be used to graph ordered pairs	The points $(1, 2), (0, -2)$, and $(-1, 1)$ are plotted in the graph. 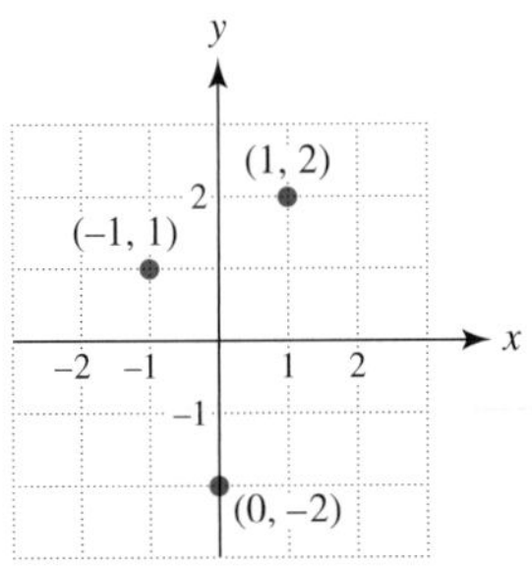

continued on next page

continued from previous page

Concept	Explanation	Examples
Scatterplot	Individual points that are plotted in the xy-plane	
Line Graph	Similar to a scatterplot except that line segments are drawn between consecutive data points	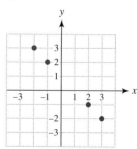

3.1   Exercises

MyMathLab   Math XL PRACTICE   WATCH   DOWNLOAD   READ   REVIEW

CONCEPTS

1. Another name for the rectangular coordinate system is the _____.

2. The intersection point of the x-axis and y-axis is the _____.

3. The origin corresponds to the point _____ in the xy-plane.

4. How many quadrants are there in the xy-plane?

5. The point $(-2, 3)$ is located in quadrant _____.

6. If both x and y are negative, then the point (x, y) is located in quadrant _____.

7. A point that lies on one of the _____ is not located in a quadrant.

8. In the xy-plane, the first value in an ordered pair is called the _____ -coordinate and the second value is called the _____ -coordinate.

9. If distinct points are plotted in the xy-plane, the resulting graph is called a _____.

10. If the consecutive points in a scatterplot are connected with line segments, the resulting graph is called a _____ graph.

CARTESIAN COORDINATE PLANE

Exercises 11–14: Identify the coordinates of each point in the graph.

11.

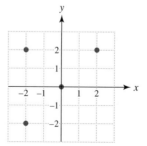

12.

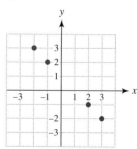

13.

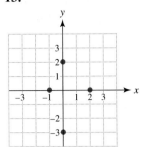

14.

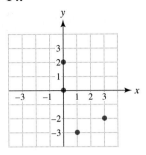

31. $(0, 0.1), (0.2, -0.3), (-0.1, 0.4)$

32. $(1.5, 2.5), (-1, -1.5), (-2.5, 0)$

Exercises 33–38: Use the table to make a line graph.

33.

x	-2	-1	0	1	2
y	2	1	0	-1	-2

34.

x	-4	-2	0	2	4
y	4	-2	3	-1	2

35.

x	-10	-5	0	5	10
y	20	-10	10	0	-20

36.

x	1	2	3	4	5
y	2	3	1	5	4

37.

x	-5	5	-10	10	0
y	10	20	30	20	40

38.

x	3	-2	2	1	-3
y	4	3	3	-2	-3

Exercises 15–20: Identify the quadrant, if any, in which each point is located.

15. (a) $(1, 4)$ **(b)** $(-1, -4)$

16. (a) $(-2, 3)$ **(b)** $(2, -3)$

17. (a) $(7, 0)$ **(b)** $(0.1, 7)$

18. (a) $(100, -3)$ **(b)** $(-100, -3)$

19. (a) $\left(-\frac{1}{2}, \frac{3}{4}\right)$ **(b)** $\left(\frac{3}{4}, -\frac{1}{2}\right)$

20. (a) $(1.2, 0)$ **(b)** $(0, -1.2)$

21. Thinking Generally   Which of the four quadrants contain points whose x- and y-coordinates have the same sign?

22. Thinking Generally   Which of the four quadrants contain points whose x- and y-coordinates have different signs?

Exercises 23–32: Make a scatterplot by plotting the given points. Be sure to label each axis.

23. $(0, 0), (1, 2), (-3, 2), (-1, -2)$

24. $(0, -3), (-2, 1), (2, 2), (-4, -4)$

25. $(-1, 0), (4, -3), (0, -1), (3, 4)$

26. $(1, 1), (-2, 2), (-3, -3), (4, -4)$

27. $(2, 4), (-4, 4), (0, -4), (-6, 2)$

28. $(4, 8), (8, 4), (-8, -4), (-4, 0)$

29. $(5, 0), (5, -5), (-10, -20), (10, -10)$

30. $(10, 30), (-20, 10), (40, 0), (-30, -10)$

Exercises 39 and 40: Identify the coordinates of each point in the graph. Then explain what the coordinates of the first point indicate.

39. Cigarette consumption in the United States   (**Source:** U.S. Department of Health and Human Services.)

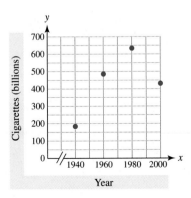

Year

40. Number of unhealthy days per year in Pittsburgh from 2000 to 2004, based on air quality *(Source: Environmental Protection Agency.)*

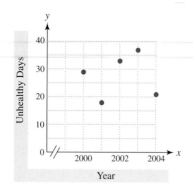

GRAPHING REAL DATA

Exercises 41–46: Graphing Real Data   The table contains real data.

(a) *Make a line graph of the data. Be sure to label the axes.*

(b) *Comment on any trends in the data.*

41. Federal income tax receipts *I* in billions during year *t*

t	1970	1980	1990	2000
I	90	244	467	1003

Source: Office of Management and Budget.

42. Percent *P* of total music sales that were digital downloads during year *t*

t	2001	2002	2003	2004	2005
P	0.2%	0.5%	1.3%	2.9%	5.7%

Source: Recording Industry Association of America.

43. Welfare beneficiaries *B* in millions during year *t*

t	1970	1980	1990	2000
B	7	11	12	6

Source: Administration for Children and Families.

44. U.S. cotton production *C* in millions of bales during year *t*

t	2002	2003	2004	2005
C	17	18	23	24

Source: U.S. Department of Agriculture.

45. Number of farms in Iowa *F* in thousands during year *x*

x	1999	2001	2003	2005
F	95	92	90	89

Source: U.S. Department of Agriculture.

46. U.S. Internet users *y* in millions during year *x*

x	2001	2002	2003	2004	2005
y	143	158	162	189	201

Source: Department of Commerce.

47. The line graph shows the U.S. infant mortality rate for selected years.

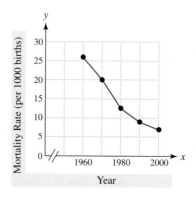

(a) Comment on any trends in the data.

(b) Estimate the infant mortality rate in 1990.

(c) Estimate the percent change in the infant mortality rate from 1960 to 2000.

48. The line graph shows the population in millions of the midwestern states for selected years.

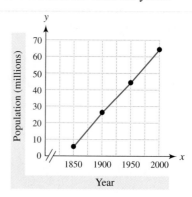

(a) Comment on any trends in the data.

(b) Estimate this population in 1900.

(c) Estimate the percent change in population from 1850 to 2000.

WRITING ABOUT MATHEMATICS

49. Explain how to identify the quadrant that a point lies in if it has coordinates (x, y).

50. Explain the difference between a scatterplot and a line graph. Give an example of each.

3.2 LINEAR EQUATIONS IN TWO VARIABLES

Basic Concepts ▪ Tables of Solutions ▪ Graphing Linear Equations in Two Variables

A LOOK INTO MATH ▷

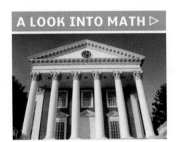

Figure 3.11 shows a scatterplot of average college tuition and fees at public colleges and universities, together with a line that models the data. If we could find an equation for this line, then we could use it to estimate tuition and fees for years without data points. In this section we discuss linear equations, whose graphs are lines. Linear equations and lines are often used to approximate data.

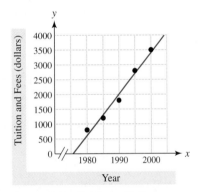

Figure 3.11 Public Institutions

Basic Concepts

Equations can have any number of variables. In Chapter 2 we solved equations having one variable. The following are examples of equations with two variables.

$$y = 2x, \quad 3x + 2y = 4, \quad z = t^2, \quad \text{and} \quad a - b = 1$$

A solution to an equation with one variable is one number that makes the statement true. For example, the solution to $x - 1 = 3$ is 4 because $4 - 1 = 3$ is a true statement. A solution to an equation with two variables consists of two numbers, one for each variable, which can be expressed as an ordered pair. For example, one solution to the equation $y = 5x$ is given by $x = 1$ and $y = 5$ because $5 = 5(1)$ is a true statement. This solution can be expressed as the ordered pair $(1, 5)$. Another solution to $y = 5x$ is $(2, 10)$, because $10 = 5(2)$ is also a true statement. In fact the equation $y = 5x$ has infinitely many solutions because, for every value of x that we choose, there is a corresponding value of y that makes the equation true.

EXAMPLE 1   Testing solutions to equations

Determine whether the given ordered pair is a solution to the given equation.
(a) $y = x + 3$, $(1, 4)$ **(b)** $2x - y = 5$, $\left(\frac{1}{2}, -4\right)$ **(c)** $-4x + 5y = 20$, $(-5, 1)$

Solution
(a) Let $x = 1$ and $y = 4$ in the given equation.

$$y = x + 3 \qquad \text{Given equation}$$
$$4 \stackrel{?}{=} 1 + 3 \qquad \text{Substitute.}$$
$$4 = 4 \qquad \text{A true statement}$$

The ordered pair $(1, 4)$ is a solution.

(b) Let $x = \frac{1}{2}$ and $y = -4$ in the given equation.

$$2x - y = 5 \qquad \text{Given equation}$$
$$2\left(\frac{1}{2}\right) - (-4) \stackrel{?}{=} 5 \qquad \text{Substitute.}$$
$$1 + 4 \stackrel{?}{=} 5 \qquad \text{Simplify the left side.}$$
$$5 = 5 \qquad \text{A true statement}$$

The ordered pair $\left(\frac{1}{2}, -4\right)$ is a solution.

(c) Let $x = -5$ and $y = 1$ in the given equation.

$$-4x + 5y = 20 \qquad \text{Given equation}$$
$$-4(-5) + 5(1) \stackrel{?}{=} 20 \qquad \text{Substitute.}$$
$$20 + 5 \stackrel{?}{=} 20 \qquad \text{Simplify the left side.}$$
$$25 \stackrel{?}{=} 20 \qquad \text{A } \textit{false} \text{ statement}$$

The ordered pair $(-5, 1)$ is *not* a solution.   Now Try Exercises 11, 15

Tables of Solutions

A table can be used to list solutions to an equation. For example, Table 3.4 lists solutions to $x + y = 5$, where the sum of each xy-pair equals 5.

TABLE 3.4   **x + y = 5**

x	-2	-1	0	1	2
y	7	6	5	4	3

Most equations in two variables have infinitely many solutions, so it is impossible to list all solutions in a table. However, when you are graphing an equation, having a table that lists a few solutions to the equation is often helpful. The next two examples demonstrate how to complete a table for a given equation.

EXAMPLE 2   Completing a table of solutions

Complete the table for the equation $y = 2x - 3$.

x	-4	-2	0	2
y				

Solution

Start by determining the corresponding y-value for each x-value in the table. For example, when $x = -2$, the equation $y = 2x - 3$ implies that $y = 2(-2) - 3 = -4 - 3 = -7$. Filling in the y-values results in Table 3.5.

TABLE 3.5 $y = 2x - 3$

x	-4	-2	0	2
y	-11	-7	-3	1

Now Try Exercise **19**

EXAMPLE 3   Making a table of solutions

Use $y = 0, 5, 10$, and 15 to make a table of solutions to $5x + 2y = 10$.

Solution

Begin by listing the required y-values in the table. Next determine the corresponding x-values for each y-value by using the equation $5x + 2y = 10$.

TABLE 3.6 $5x + 2y = 10$

x	2	0	-2	-4
y	0	5	10	15

When $y = 0$,
$$5x + 2(0) = 10$$
$$5x + 0 = 10$$
$$5x = 10$$
$$x = 2$$

When $y = 5$,
$$5x + 2(5) = 10$$
$$5x + 10 = 10$$
$$5x = 0$$
$$x = 0$$

When $y = 10$,
$$5x + 2(10) = 10$$
$$5x + 20 = 10$$
$$5x = -10$$
$$x = -2$$

When $y = 15$,
$$5x + 2(15) = 10$$
$$5x + 30 = 10$$
$$5x = -20$$
$$x = -4$$

Filling in the x-values results in Table 3.6.

Now Try Exercise **23**

▶ **REAL-WORLD CONNECTION**   Formulas can sometimes be difficult for people to understand. As a result, newspapers, magazines, and books often list numbers in a table rather than presenting a formula for the reader to use. The next example illustrates a situation in which a table might be preferable to a formula.

EXAMPLE 4   Calculating appropriate lengths of crutches

People who sustain leg injuries often require crutches. An appropriate crutch length L in inches for an injured person who is t inches tall is estimated by $L = 0.72t + 2$.
(**Source:** Journal of the American Physical Therapy Association.)
(a) Complete the table. Round values to the nearest inch.

t	60	65	70	75	80
L					

(b) Use the table to determine the appropriate crutch length for a person 5 feet 10 inches tall.

Solution

(a) For the formula $L = 0.72t + 2$, if $t = 60$, then $L = 0.72(60) + 2 = 43.2 + 2 = 45.2$, or about 45 inches. If $t = 65$, then $L = 0.72(65) + 2 = 46.8 + 2 = 48.8 \approx 49$. Other values in Table 3.7 are found similarly. (You may want to use a calculator to help complete the table.)

TABLE 3.7   **Crutch Lengths**

t	60	65	70	75	80
L	45	49	52	56	60

(b) A person who is 5 feet 10 inches tall is $5 \cdot 12 + 10 = 70$ inches tall. Table 3.7 reveals that a person 70 inches tall needs crutches that are about 52 inches long.

Now Try Exercise 69

Graphing Linear Equations in Two Variables

In the preceding section we showed that a line graph visually displays data. Many times graphs are used in mathematics to make concepts easier to understand.

EXAMPLE 5   Graphing an equation with two variables

Make a table of values for the equation $y = 2x$, and then use the table to graph this equation.

Solution
Start by selecting a few convenient values for x, such as $x = -1, 0, 1$, and 2. Then complete the table by doubling each x-value to obtain the corresponding y-value.

NOTE: A table of values can be either horizontal or vertical. Table 3.8 is presented in a vertical format.

TABLE 3.8 $y = 2x$

x	y
-1	-2
0	0
1	2
2	4

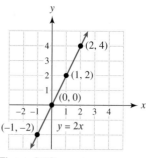

Figure 3.12

To graph the equation $y = 2x$, start by plotting the points $(-1, -2)$, $(0, 0)$, $(1, 2)$, and $(2, 4)$ given in Table 3.8, as shown in Figure 3.12. Note that all four points appear to lie on the same (straight) line. Because there are infinitely many points that satisfy the equation, we can draw a line through these points.

Now Try Exercise 35

An equation whose graph is a single, straight line is called a *linear equation in two variables*. Linear equations in two variables can be written in the following **standard form**.

LINEAR EQUATION IN TWO VARIABLES

A **linear equation in two variables** can be written as

$$Ax + By = C,$$

where A, B, and C are fixed numbers (constants) and A and B are not both equal to 0. The graph of a linear equation in two variables is a line.

NOTE: In mathematics a line is always *straight*.

In Example 5, the equation $y = 2x$ is a linear equation because it can be written in standard form by adding $-2x$ to each side.

$$y = 2x \qquad \text{Given equation}$$
$$-2x + y = -2x + 2x \qquad \text{Add } -2x \text{ to each side.}$$
$$-2x + y = 0 \qquad \text{Simplify.}$$

The equation $-2x + y = 0$ is a linear equation in two variables because it is in the form $Ax + By = C$ with $A = -2$, $B = 1$, and $C = 0$. Hence, the graph of $y = 2x$ is a line.

EXAMPLE 6 Graphing linear equations

Graph each linear equation.
(a) $y = \frac{1}{2}x - 1$ **(b)** $x + y = 4$

Solution

(a) Because $y = \frac{1}{2}x - 1$ can be written in standard form as $-\frac{1}{2}x + y = -1$, it is a linear equation in two variables and its graph is a line. Two points determine a line. However, it is a good idea to plot three points to be sure that the line is graphed correctly. Start by choosing three values for x and then calculate the corresponding y-values, as shown in Table 3.9. In Figure 3.13, the points $(-2, -2)$, $(0, -1)$, and $(2, 0)$ are plotted and the line passing through these points is drawn.

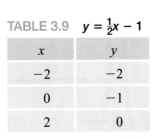

TABLE 3.9 $y = \frac{1}{2}x - 1$

x	y
-2	-2
0	-1
2	0

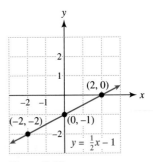

Figure 3.13

(b) The equation $x + y = 4$ is a linear equation in two variables, so its graph is a line. If an ordered pair (x, y) is a solution to the given equation, then the sum of x and y is 4. Table 3.10 shows three examples. In Figure 3.14, the points $(0, 4)$, $(2, 2)$, and $(4, 0)$ are plotted with the line passing through each one.

TABLE 3.10   **x + y = 4**

x	y
0	4
2	2
4	0

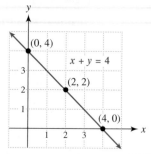

Figure 3.14

Now Try Exercises 39, 49

TECHNOLOGY NOTE: Graphing Equations

Graphing calculators can be used to graph equations. Before graphing the equation $x + y = 4$, solve the equation for y to obtain $y = 4 - x$. A calculator graph of Figure 3.14 is shown, except that the three points have not been plotted.

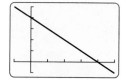

When a linear equation in two variables is given in standard form, it is sometimes difficult to create a table of solutions. Solving such equations for y often makes it easier to select x-values for the table that will make the y-values simpler to calculate. This is demonstrated in the next example.

EXAMPLE 7   Solving for *y* and then graphing

Graph each linear equation by solving for y first.
(a) $4x - 3y = 12$ **(b)** $-2x + 4y = 8$

Solution
(a) First solve the given equation for y.

$$4x - 3y = 12 \qquad \text{Given equation}$$

$$-3y = -4x + 12 \qquad \text{Subtract } 4x \text{ from each side.}$$

$$\frac{-3y}{-3} = \frac{-4x + 12}{-3} \qquad \text{Divide each side by } -3.$$

$$\frac{-3y}{-3} = \frac{-4x}{-3} + \frac{12}{-3} \qquad \text{Property of fractions, } \frac{a + b}{c} = \frac{a}{c} + \frac{b}{c}$$

$$y = \frac{4}{3}x - 4 \qquad \text{Simplify fractions.}$$

Note that dividing each side of an equation by -3 is equivalent to dividing each term in the equation by -3.

Select multiples of 3 (the denominator of $\frac{4}{3}$) as x-values for the table of solutions. For example, if $x = 6$ is chosen, $y = \frac{4}{3}(6) - 4 = \frac{24}{3} - 4 = 8 - 4 = 4$. Table 3.11 lists the solutions $(0, -4)$, $(3, 0)$, and $(6, 4)$, which are plotted in Figure 3.15 with the line passing through each one.

TABLE 3.11 $y = \frac{4}{3}x - 4$

x	y
0	-4
3	0
6	4

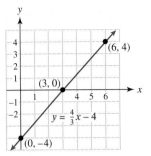

Figure 3.15

(b) First solve the given equation for y.

$$-2x + 4y = 8 \qquad \text{Given equation}$$
$$4y = 2x + 8 \qquad \text{Add } 2x \text{ to each side.}$$
$$\frac{4y}{4} = \frac{2x}{4} + \frac{8}{4} \qquad \text{Divide each side by 4.}$$
$$y = \frac{1}{2}x + 2 \qquad \text{Simplify fractions.}$$

Table 3.12 lists the solutions $(-2, 1)$, $(0, 2)$, and $(2, 3)$, which are plotted in Figure 3.16 with the line passing through each one. Note that the selected x-values are multiples of 2.

TABLE 3.12 $y = \frac{1}{2}x + 2$

x	y
-2	1
0	2
2	3

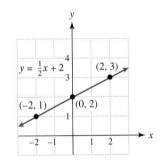

Figure 3.16   Now Try Exercise 57

MAKING CONNECTIONS

Graphs and Solution Sets

A graph visually depicts the set of solutions to an equation. Each point on the graph represents one solution to the equation.

NOTE: In a linear equation, each x-value determines a unique y-value. Because there are infinitely many x-values, there are infinitely many points located on the graph of a linear equation. Thus the graph of a linear equation is a *continuous* line with no breaks.

3.2 PUTTING IT ALL TOGETHER

The following table summarizes several important topics covered in this section.

Concept	Explanation	Examples
Equation in Two Variables	An equation that has two variables	$y = 4x + 5, 4x - 5y = 20$, and $u - v = 100$
Solution to an Equation in Two Variables	An ordered pair (x, y) whose x- and y-values satisfy the equation Equations in two variables often have infinitely many solutions.	$(1, 3)$ is a solution to $3x + y = 6$ because $3(1) + 3 = 6$ is a true statement. The equation $y = 2x$ has infinitely many solutions, such as $(1, 2)$, $(2, 4)$, $(3, 6)$, and so on.
Linear Equation in Two Variables	Can be written as $$Ax + By = C,$$ where A, B, and C are fixed numbers and A and B are not both equal to 0	$3x + 4y = 5$ $y = 4 - 3x$ (or $3x + y = 4$) $x = 2y + 1$ (or $x - 2y = 1$) The graph of each equation is a line.

3.2 Exercises

CONCEPTS

1. $4x + 6y = 24$ is an equation in _____ variables.

2. A solution to an equation with two variables consists of two numbers expressed as a(n) _____.

3. The ordered pair $(1, 3)$ is a(n) _____ to the equation $y = 3x$.

4. A(n) _____ equation in two variables can be written in the form $Ax + By = C$.

5. The equation $4x - 3y = 10$ (is/is not) a linear equation in two variables.

6. A linear equation in two variables has infinitely many _____.

7. An equation's _____ visually depicts its solution set.

8. The graph of a linear equation in two variables is a(n) _____.

SOLUTIONS TO EQUATIONS

Exercises 9–18: Determine whether the ordered pair is a solution to the given equation.

9. $y = x + 1, (5, 6)$ 10. $y = 4 - x, (6, 2)$

11. $y = 4x + 7, (2, 13)$ 12. $y = -3x + 2, (-2, 8)$

13. $4x - y = -13, (-2, 3)$

14. $3y + 2x = 0, (-2, 3)$

15. $y - 6x = -1, \left(\frac{1}{2}, 2\right)$ 16. $\frac{1}{2}x + \frac{3}{2}y = 0, \left(-\frac{3}{2}, \frac{1}{2}\right)$

17. $0.31x - 0.42y = -9, (100, 100)$

18. $0.5x - 0.6y = 4, (20, 10)$

Exercises 19–24: Complete the table for the given equation.

19. $y = 4x$

x	-2	-1	0	1	2
y	-8				

20. $y = \frac{1}{2}x - 1$

x	0	1	2	3	4
y	−1				

21. $y = x + 4$

x	−8				
y	−4	0	4	8	12

22. $2x - y = 1$

x	−1				
y	−3	−1	0	1	3

23. $3y + 2x = 6$

x					
y	−2	0	2	4	8

24. $3x - 5y = 30$

x	−5	0	5	10	15
y					

Exercises 25–32: Use the given values of the variable to make a table of solutions for the equation.

25. $y = 3x$ $\qquad x = -3, 0, 3, 6$

26. $y = 1 - 2x$ $\qquad x = 0, 1, 2, 3$

27. $y = \dfrac{x + 4}{2}$ $\qquad x = -8, -4, 0, 4$

28. $y = \dfrac{x}{3} - 1$ $\qquad x = 0, 2, 4, 6$

29. $x + y = 6$ $\qquad y = -2, 0, 2, 4$

30. $2x - 3y = 9$ $\qquad y = -3, 0, 1, 2$

31. $y - 4x = 0$ $\qquad y = -2, -1, 0, 1$

32. $-4x = 6y - 4$ $\qquad y = -1, 0, 1, 2$

33. Thinking Generally If a student wishes to avoid fractional y-values when making a table of solutions for $y = \frac{3}{5}x - 7$, what must be true about any selected integer x-values?

34. Thinking Generally If a student wishes to avoid fractional y-values when making a table of solutions for $y = \frac{a}{b}x$, where a and b are natural numbers, what must be true about any selected integer x-values?

Exercises 35–40: Make a table of solutions for the equation, and then use the table to graph the equation.

35. $y = -2x$ $\qquad\qquad$ **36.** $y = 2x - 1$

37. $x = 3 - y$ $\qquad\qquad$ **38.** $x = y + 1$

39. $x + 2y = 4$ $\qquad\quad$ **40.** $2x - y = 1$

GRAPHING EQUATIONS

Exercises 41–56: Graph the equation.

41. $y = x$ $\qquad\qquad$ **42.** $y = \frac{1}{2}x$

43. $y = \frac{1}{3}x$ $\qquad\qquad$ **44.** $y = -2x$

45. $y = x + 3$ $\qquad\quad$ **46.** $y = x - 2$

47. $y = x - 4$ $\qquad\quad$ **48.** $y = x + 2$

49. $y = 2x + 1$ $\qquad\quad$ **50.** $y = \frac{1}{2}x - 1$

51. $y = 4 - 2x$ $\qquad\quad$ **52.** $y = 2 - 3x$

53. $y = 7 + x$ $\qquad\quad$ **54.** $y = 2 + 2x$

55. $y = -\frac{1}{2}x + \frac{1}{2}$ $\qquad$ **56.** $y = -\frac{3}{4}x + 2$

Exercises 57–68: Graph the linear equation by solving for y first.

57. $2x + 3y = 6$ $\qquad\quad$ **58.** $3x + 2y = 6$

59. $x + 4y = 4$ $\qquad\quad$ **60.** $4x + y = -4$

61. $-x + 2y = 8$ $\qquad\quad$ **62.** $-2x + 6y = 12$

63. $y - 2x = 7$ $\qquad\quad$ **64.** $3y - x = 2$

65. $5x - 4y = 20$ $\qquad\quad$ **66.** $4x - 5y = -20$

67. $3x + 5y = -9$ $\qquad\quad$ **68.** $5x - 3y = 10$

APPLICATIONS

69. *U.S. Population* For the years 2010 to 2050, the projected percentage P of the U.S. population that will be over the age of 65 during year t is estimated by $P = 0.178t - 344.6$. (*Source: U.S. Census Bureau.*)

(a) Complete the table. Round values to the nearest tenth.

t	2010	2020	2030	2040	2050
P					

(b) Use the table to find the year when the percentage of the population over the age of 65 is expected to reach 16.7%.

70. *U.S. Population* For the years 2010 to 2050, the projected percentage P of the U.S. population that will be 18 to 24 years old during year t is estimated by $P = -0.025t + 60.35$. (*Source:* U.S. Census Bureau.)
  (a) Complete the table. Round values to the nearest tenth.

t	2010	2020	2030	2040	2050
P					

  (b) Use the table to find the year when the percentage of the population that is 18 to 24 years old is expected to be 9.4%.

71. *Solid Waste in the Past* In 1960 the amount A of garbage in pounds produced after t days by the average American is given by $A = 2.7t$. (*Source:* Environmental Protection Agency.)
  (a) Graph the equation for $t \geq 0$.
  (b) How many days did it take for the average American in 1960 to produce 100 pounds of garbage?

72. *Solid Waste Today* Today the amount A of garbage in pounds produced after t days by the average American is given by $A = 4.5t$. (*Source:* Environmental Protection Agency.)
  (a) Graph the equation for $t \geq 0$.
  (b) How many days does it take for the average American to produce 100 pounds of garbage today?

73. *Compact Disc Sales* From 1985 to 2000 the percent P of total music sales with a compact disc format is modeled by $P = 5.9t - 11,709$, where t is the year. (*Source:* Recording Industry Association of America.)
  (a) Evaluate P for $t = 1985$ and for $t = 2000$.
  (b) Use your results from part (a) to graph the equation from 1985 to 2000.
  (c) In what year was $P = 79.2\%$?

74. *HIV Infections in the United States* The cumulative number of HIV infections I in thousands during year t is modeled by the equation $I = 42t - 83,197$ where $t \geq 2000$. (*Source:* Department of Health and Human Services.)
  (a) Evaluate I for $t = 2000$ and for $t = 2005$.
  (b) Use your results from part (a) to graph the equation from 2000 to 2005.
  (c) In what year was $I = 971$?

WRITING ABOUT MATHEMATICS

75. The number of welfare beneficiaries B in millions during year t is shown in the table. Discuss whether a linear equation might work to approximate these data from 1970 to 2000.

t	1970	1980	1990	2000
B	7	11	12	6

Source: Administration for Children and Families.

76. The Asian-American population P in millions during year t is shown in the table. Discuss whether a linear equation might model these data from 2000 to 2004.

t	2000	2002	2004
P	11.2	12.0	12.8

Source: U.S. Census Bureau.

CHECKING BASIC CONCEPTS
SECTIONS 3.1 AND 3.2

1. Identify the coordinates of the four points in the graph. State the quadrant, if any, in which each point lies.

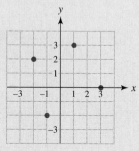

2. Make a scatterplot of the five points $(-2, -2)$, $(-1, -3)$, $(0, 0)$, $(1, 2)$, and $(2, 3)$.

3. *U.S. Population* The table gives the percentage P of the U.S. population that was over the age of 85 during year t. Make a line graph of the data in the table, and then comment on any trends. (*Source:* U.S. Census Bureau.)

t	1970	1980	1990	2000
P	0.7	1.0	1.2	1.5

4. Determine whether $(-2, -3)$ is a solution to the equation $-2x - y = 7$.

5. Complete the table for the equation $y = -2x + 1$.

x	-2	-1	0	1	2
y	5				

6. Graph the equation.
   (a) $y = \frac{1}{2}x$ (b) $4x + 6y = 12$

7. *Total Federal Receipts* The total amount of money A in $ trillions collected by the federal government in year t from 1995 to 2001 can be approximated by $A = 0.115t - 228$. (**Source:** *Internal Revenue Service.*)
   (a) Find A when $t = 1995$ and $t = 2001$. Interpret each result.
   (b) Use your results from part (a) to graph the equation from 1995 to 2001. Be sure to label the axes.
   (c) In what year were the total receipts equal to $1.54 trillion?

3.3 MORE GRAPHING OF LINES

Finding Intercepts ▪ Horizontal Lines ▪ Vertical Lines

A LOOK INTO MATH ▷ The graph of a linear equation is a line, which can be used to model data. Two points, such as those associated with the x- and y-intercepts, can be used to sketch such a line. We begin this section by discussing intercepts and their significance in applications and then consider horizontal and vertical lines.

Finding Intercepts

▶ **REAL-WORLD CONNECTION** Suppose that someone leaves a rest stop on an Interstate highway and drives home at a constant speed of 50 miles per hour. The graph in Figure 3.17 reflects the distance of the driver from home at various times. The graph intersects the y-axis at 200 miles, which is called the *y-intercept*. In this situation the y-intercept represents the initial distance (when $x = 0$) between the driver and home. The graph also intersects the x-axis at 4 hours, which is called the *x-intercept*. This intercept represents the elapsed time when the distance of the driver from home is 0 miles.

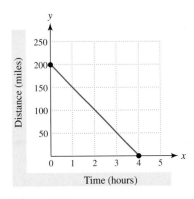

Figure 3.17

FINDING *x*- AND *y*-INTERCEPTS

The *x*-coordinate of a point where a graph intersects the *x*-axis is an **x-intercept**.
To find an *x*-intercept, let $y = 0$ in the equation and solve for *x*.

The *y*-coordinate of a point where a graph intersects the *y*-axis is a **y-intercept**.
To find a *y*-intercept, let $x = 0$ in the equation and solve for *y*.

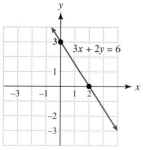

Figure 3.18

The graph of the linear equation $3x + 2y = 6$ is shown in Figure 3.18. The *x*-intercept is 2 and the *y*-intercept is 3. These two intercepts can be found without a graph. To find the *x*-intercept let $y = 0$ in the equation and solve for *x*.

$$3x + 2(0) = 6 \qquad \text{Let } y = 0.$$
$$3x = 6 \qquad \text{Simplify.}$$
$$x = 2 \qquad \text{Divide each side by 3.}$$

To find the *y*-intercept let $x = 0$ in the equation and solve for *y*.

$$3(0) + 2y = 6 \qquad \text{Let } x = 0.$$
$$2y = 6 \qquad \text{Simplify.}$$
$$y = 3 \qquad \text{Divide each side by 2.}$$

Note that the *x*-intercept **2** corresponds to the point $(2, 0)$ on the graph and the *y*-intercept **3** corresponds to the point $(0, 3)$ on the graph.

CRITICAL THINKING

If a line has no *x*-intercept, what can you say about the line?
If a line has no *y*-intercept, what can you say about the line?

EXAMPLE 1   Using intercepts to graph a line

Use intercepts to graph $2x - 6y = 12$.

Solution
The *x*-intercept is found by letting $y = 0$.

$$2x - 6(0) = 12 \qquad \text{Let } y = 0.$$
$$x = 6 \qquad \text{Solve for } x.$$

The *y*-intercept is found by letting $x = 0$.

$$2(0) - 6y = 12 \qquad \text{Let } x = 0.$$
$$y = -2 \qquad \text{Solve for } y.$$

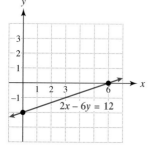

Figure 3.19

Therefore the graph passes through the points $(6, 0)$ and $(0, -2)$, as shown in Figure 3.19.

Now Try Exercise 25

In the next example a table of solutions is used to determine the x- and y-intercepts.

EXAMPLE 2 Using a table to find intercepts

Complete the table. Then determine the x-intercept and y-intercept for the graph of the equation $x - y = -1$.

x	-2	-1	0	1	2
y					

Solution

Substitute -2 for x in $x - y = -1$ to find the corresponding y-value.

$$-2 - y = -1 \qquad \text{Let } x = -2.$$
$$-y = 1 \qquad \text{Add 2 to each side.}$$
$$y = -1 \qquad \text{Multiply each side by } -1.$$

The other y-values can be found similarly. See Table 3.13.

The x-intercept corresponds to a point on the graph whose y-coordinate is 0. Table 3.13 reveals that the y-coordinate is 0 when $x = -1$. Thus the x-intercept is -1. Similarly, the y-intercept corresponds to a point on the graph whose x-coordinate is 0. Table 3.13 reveals that the x-coordinate is 0 when $y = 1$. Thus the y-intercept is 1.

TABLE 3.13 $x - y = -1$

x	-2	-1	0	1	2
y	-1	0	1	2	3

Now Try Exercise 21

EXAMPLE 3 Modeling the velocity of a toy rocket

A toy rocket is shot vertically into the air. Its velocity v in feet per second after t seconds is given by $v = 160 - 32t$. Assume that $t \geq 0$ and $t \leq 5$.
(a) Graph the equation by finding the intercepts. Let t correspond to the horizontal axis (x-axis) and v correspond to the vertical axis (y-axis).
(b) Interpret each intercept.

Solution
(a) To find the t-intercept let $v = 0$.

$$0 = 160 - 32t \qquad \text{Let } v = 0.$$
$$32t = 160 \qquad \text{Add } 32t \text{ to each side.}$$
$$t = 5 \qquad \text{Divide each side by 32.}$$

To find the v-intercept let $t = 0$.

$$v = 160 - 32(0) \qquad \text{Let } t = 0.$$
$$v = 160 \qquad \text{Simplify.}$$

Therefore the graph passes through $(5, 0)$ and $(0, 160)$, as shown in Figure 3.20.

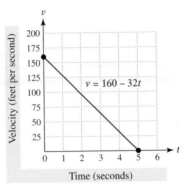

Figure 3.20

(b) The t-intercept indicates that the rocket had a velocity of 0 feet per second after 5 seconds. The v-intercept indicates that the rocket's initial velocity was 160 feet per second.

Now Try Exercise 85

Horizontal Lines

▶ **REAL-WORLD CONNECTION**   Suppose that someone drives a car on a freeway at a constant speed of 70 miles per hour. Table 3.14 shows the speed y after x hours.

TABLE 3.14   Speed of a Car

x	1	2	3	4	5
y	70	70	70	70	70

We can make a scatterplot of the data by plotting the five points $(1, 70)$, $(2, 70)$, $(3, 70)$, $(4, 70)$, and $(5, 70)$, as shown in Figure 3.21(a). The speed is always 70 miles per hour and the graph of the car's speed is a horizontal line, as shown in Figure 3.21(b). The equation of this line is $y = 70$ with y-intercept 70. There are no x-intercepts.

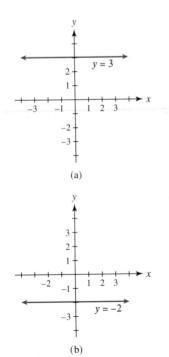

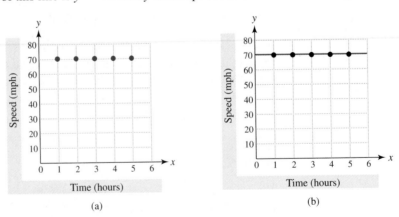

Figure 3.21

In general, the equation of a horizontal line is $y = b$, where b is a constant that corresponds to the y-intercept. Examples of horizontal lines are shown in Figure 3.22. Note that every point on the graph of $y = 3$ in Figure 3.22(a) has a y-coordinate of 3, and that every point on the graph of $y = -2$ in Figure 3.22(b) has a y-coordinate of -2.

Figure 3.22

HORIZONTAL LINE

The equation of a horizontal line with y-intercept b is $y = b$.

The equation $y = b$ is an example of a linear equation in the form $Ax + By = C$ with $A = 0, B = 1$, and $C = b$. (Note that in general B and b do not represent the same number.)

EXAMPLE 4   Graphing a horizontal line

Graph the equation $y = -1$ and identify its y-intercept.

Solution
The graph of $y = -1$ is a horizontal line passing through the point $(0, -1)$, as shown in Figure 3.23. Its y-intercept is -1.

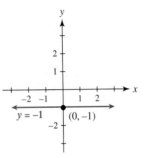

Figure 3.23

Now Try Exercise 47(a)

Vertical Lines

TABLE 3.15
Falling Object

x (seconds)	y (feet)
1	16
1	15
1	18
1	17

▶ REAL-WORLD CONNECTION   A scientist is experimentally determining the distance y that an object falls in 1 second. Because it is an experiment, the distances vary slightly on each trial. Table 3.15 shows the results.

We can make a scatterplot of the data by plotting the points $(1, 16)$, $(1, 15)$, $(1, 18)$, and $(1, 17)$, as shown in Figure 3.24(a). In each case the time is always 1 second and each point lies on the graph of a vertical line, as shown in Figure 3.24(b). This vertical line has the equation $x = 1$ because each point on the line has an x-coordinate of 1 and there are no restrictions on the y-coordinate. This line has x-intercept 1 but no y-intercept.

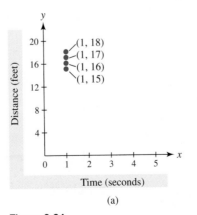

(a)

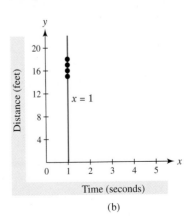

(b)

Figure 3.24

In general, the graph of a vertical line is $x = k$, where k is a constant that corresponds to the x-intercept. Examples of vertical lines are shown in Figure 3.25. Note that every point on the graph of $x = 3$ in Figure 3.25(a) has an x-coordinate of 3 and that every point on the graph of $x = -2$ shown in Figure 3.25(b) has an x-coordinate of -2.

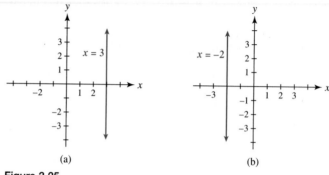

(a)   (b)

Figure 3.25

VERTICAL LINE

The equation of a vertical line with x-intercept k is $x = k$.

The equation $x = k$ is an example of a linear equation in the form $Ax + By = C$ with $A = 1, B = 0$, and $C = k$.

EXAMPLE 5   Graphing a vertical line

Graph the equation $x = -3$, and identify its x-intercept.

Solution
The graph of $x = -3$ is a vertical line passing through the point $(-3, 0)$, as shown in Figure 3.26. Its x-intercept is -3.

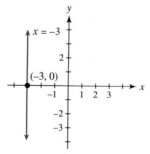

Figure 3.26   Now Try Exercise 47(b)

EXAMPLE 6 **Writing equations of horizontal and vertical lines**

Write the equation of the line shown in each graph.

(a)

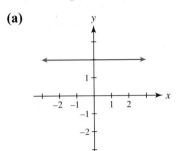

(b)

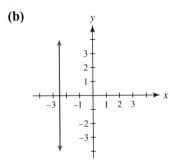

Solution

(a) The graph is a horizontal line with y-intercept 2. Its equation is $y = 2$.

(b) The graph is a vertical line with x-intercept -2.5. Its equation is $x = -2.5$.

Now Try Exercises 55, 57

EXAMPLE 7 **Writing equations of horizontal and vertical lines**

Find an equation for a line satisfying the given conditions.

(a) Vertical, passing through $(2, -3)$

(b) Horizontal, passing through $(3, 1)$

(c) Perpendicular to $x = 3$, passing through $(-1, 2)$

Solution

(a) A vertical line passing through $(2, -3)$ has x-intercept 2, as shown in Figure 3.27(a). The equation of a vertical line with x-intercept 2 is $x = 2$.

(b) A horizontal line passing through $(3, 1)$ has y-intercept 1, as shown in Figure 3.27(b). The equation of a horizontal line with y-intercept 1 is $y = 1$.

(c) Because the line $x = 3$ is vertical, a line that is perpendicular to this line is horizontal, as shown in Figure 3.27(c). The equation of a horizontal line with y-intercept 2 is $y = 2$.

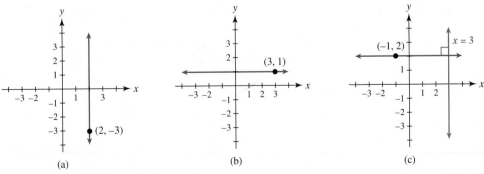

Figure 3.27 Horizontal and Vertical Lines

Now Try Exercises 73, 75, 77

MAKING CONNECTIONS

Lines and Linear Equations

The equation of any line can be written in the standard form $Ax + By = C$.

1. If $A = 0$ and $B \neq 0$, the line is horizontal.
2. If $A \neq 0$ and $B = 0$, the line is vertical.
3. If $A \neq 0$ and $B \neq 0$, the line is neither horizontal nor vertical.

3.3   PUTTING IT ALL TOGETHER

The following table summarizes several important topics presented in this section.

Concept	Explanation	Examples
x- and y-Intercepts	The x-coordinate of a point at which a graph intersects the x-axis is called an x-intercept. The y-coordinate of a point at which a graph intersects the y-axis is called a y-intercept.	 x-intercept, -3 y-intercept, 2
Finding Intercepts	To find x-intercepts let $y = 0$ in the equation and solve for x. To find y-intercepts let $x = 0$ in the equation and solve for y.	Let $4x - 5y = 20$. x-intercept: $4x - 5(0) = 20$ $x = 5$ y-intercept: $4(0) - 5y = 20$ $y = -4$ The x-intercept is 5, and the y-intercept is -4.
Horizontal Line	Has equation $y = b$, where b is a constant Has y-intercept b and no x-intercept if $b \neq 0$	 y-intercept, b

Concept	Explanation	Examples
Vertical Line	Has equation $x = k$, where k is a constant Has x-intercept k and no y-intercept if $k \neq 0$	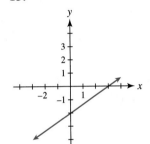 x-intercept, k

3.3 Exercises

MyMathLab

CONCEPTS

1. How many points determine a line?

2. The graph of the linear equation $Ax + By = C$ with $A \neq 0$ and $B \neq 0$ has _____ x-intercept(s) and _____ y-intercept(s).

3. The x-coordinate of a point where a graph intersects the x-axis is a(n) _____.

4. To find an x-intercept, let $y =$ _____ and solve for x.

5. The y-coordinate of a point at which a graph intersects the y-axis is a(n) _____.

6. To find a y-intercept, let $x =$ _____ and solve for y.

7. The graph of $y = 3$ has y-intercept _____.

8. The graph of the linear equation $Ax + By = C$ with $A = 0$ and $B = 1$ is a(n) _____ line.

9. A horizontal line with y-intercept b has equation _____.

10. The graph of $x = 3$ has x-intercept _____.

11. The graph of the linear equation $Ax + By = C$ with $A = 1$ and $B = 0$ is a(n) _____ line.

12. A vertical line with x-intercept k has equation _____.

FINDING INTERCEPTS

Exercises 13–20: Identify any x-intercepts and y-intercepts in the graph.

13.

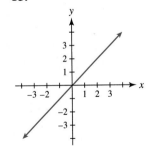

14.

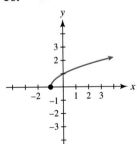

15.

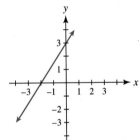

16.

17.

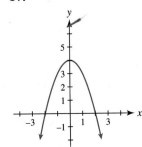

18.

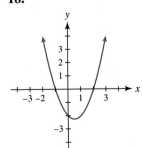

19.

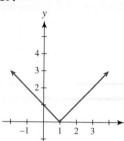

20.

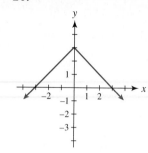

Exercises 21–24: Complete the table. Then determine the x-intercept and the y-intercept for the graph of the equation.

21. $y = x + 2$

x	−2	−1	0	1	2
y					

22. $y = 2x − 4$

x	−2	−1	0	1	2
y					

23. $−x + y = −2$

x	−4	−2	0	2	4
y					

24. $x + y = 1$

x	−2	−1	0	1	2
y					

Exercises 25–44: Find any intercepts. Then graph the linear equation.

25. $−2x + 3y = −6$

26. $4x + 3y = 12$

27. $3x − 5y = 15$

28. $2x − 4y = −4$

29. $x − 3y = 6$

30. $5x + y = −5$

31. $6x − y = −6$

32. $5x + 7y = −35$

33. $3x + 7y = 21$

34. $−3x + 8y = 24$

35. $40y − 30x = −120$

36. $10y − 20x = 40$

37. $\frac{1}{2}x − y = 2$

38. $x − \frac{1}{2}y = 4$

39. $−\frac{x}{4} + \frac{y}{3} = 1$

40. $\frac{x}{3} − \frac{y}{4} = 1$

41. $\frac{x}{3} + \frac{y}{2} = 1$

42. $\frac{x}{5} − \frac{y}{4} = 1$

43. $0.6y − 1.5x = 3$

44. $0.5y − 0.4x = 2$

45. Thinking Generally Find any intercepts for the graph of $Ax + By = C$.

46. Thinking Generally Find any intercepts for the graph of $\frac{x}{A} + \frac{y}{B} = 1$.

HORIZONTAL AND VERTICAL LINES

Exercises 47–54: Graph each equation.

47. (a) $y = 2$ **(b)** $x = 2$

48. (a) $y = −2$ **(b)** $x = −2$

49. (a) $y = −4$ **(b)** $x = −4$

50. (a) $y = 0$ **(b)** $x = 0$

51. (a) $y = −1$ **(b)** $x = −1$

52. (a) $y = −\frac{1}{2}$ **(b)** $x = −\frac{1}{2}$

53. (a) $y = \frac{3}{2}$ **(b)** $x = \frac{3}{2}$

54. (a) $y = −1.5$ **(b)** $x = −1.5$

Exercises 55–62: Write an equation for the line shown in the graph.

55.

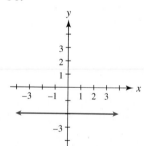

56.

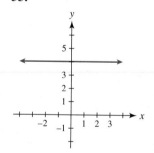

57.

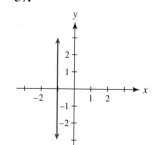

58.

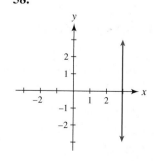

59.

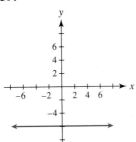

60.

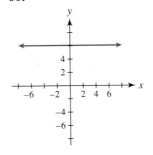

61.

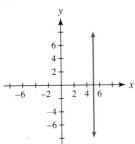

62.

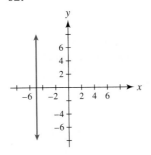

Exercises 63–66: Write an equation for the line that passes through the points shown in the table.

63.

x	−2	−1	0	1	2
y	1	1	1	1	1

64.

x	0	1	2	3	4
y	−10	−10	−10	−10	−10

65.

x	−6	−6	−6	−6	−6
y	5	4	3	2	1

66.

x	20	20	20	20	20
y	−2	−1	0	1	2

Exercises 67–72: Write the equations of a horizontal line and a vertical line that pass through the given point. (Hint: Make a sketch.)

67. $(1, 2)$

68. $(-3, 4)$

69. $(20, -45)$

70. $(-5, 12)$

71. $(0, 5)$

72. $(-3, 0)$

Exercises 73–80: Find an equation for a line satisfying the following conditions.

73. Vertical, passing through $(-1, 6)$

74. Vertical, passing through $(2, -7)$

75. Horizontal, passing through $\left(\frac{3}{4}, -\frac{5}{6}\right)$

76. Horizontal, passing through $(5.1, 6.2)$

77. Perpendicular to $y = \frac{1}{2}$, passing through $(4, -9)$

78. Perpendicular to $x = 2$, passing through $(3, 4)$

79. Parallel to $x = 4$, passing through $\left(-\frac{2}{3}, \frac{1}{2}\right)$

80. Parallel to $y = -2.1$, passing through $(7.6, 3.5)$

81. **Thinking Generally** Write the equation of the x-axis. (*Hint:* The x-axis is a horizontal line.)

82. **Thinking Generally** Write the equation of the y-axis. (*Hint:* The y-axis is a vertical line.)

APPLICATIONS

*Exercises 83 and 84: **Distance**   The distance of a driver from home is illustrated in the graph.*

   (a) Find the intercepts.
   (b) Interpret each intercept.

83.

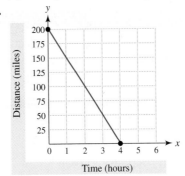

84.

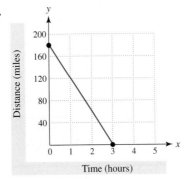

Exercises 85 and 86: Modeling a Toy Rocket   *(Refer to Example 3.) The velocity v of a toy rocket in feet per second after t seconds of flight is given, where t ≥ 0.*

   (a) *Find the intercepts and then graph the equation.*
   (b) *Interpret each intercept.*

85. $v = 128 - 32t$ **86.** $v = 96 - 32t$

Exercises 87 and 88: Water in a Pool   *The amount of water in a swimming pool is depicted in the graph.*

   (a) *Find the intercepts.*
   (b) *Interpret each intercept.*

87.

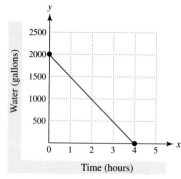

88.

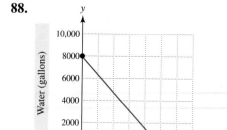

WRITING ABOUT MATHEMATICS

89. Given an equation, explain how to find an x-intercept and a y-intercept.

90. The form $\frac{x}{a} + \frac{y}{b} = 1$ is called the **intercept form** of a linear equation. Explain how you can use this equation to find the intercepts. (*Hint:* Graph $\frac{x}{2} + \frac{y}{3} = 1$ and find its intercepts.)

GROUP ACTIVITY
WORKING WITH REAL DATA

Directions: Form a group of 2 to 4 people. Select someone to record the group's responses for this activity. All members of the group should work cooperatively to answer the questions. If your instructor asks for the results, each member of the group should be prepared to respond.

1. *Radio Stations*   The approximate number of radio stations on the air for selected years from 1950 to 2000 is shown in the table.

x (year)	1950	1960	1970
y (stations)	2800	4100	6800

x (year)	1980	1990	2000
y (stations)	8600	10,800	12,600

Source: M. Street Corporation, 2001.

Make a line graph of the data. Be sure to label both axes.

2. *Estimation*   Discuss ways to estimate the number of radio stations on the air in 1975. Compare your estimates with the actual value of 7700 stations. Repeat this estimate for 1985 and compare it to the actual value of 10,400. Discuss your results.

3. *Modeling Equation*   Substitute each x-value from the table into the equation $y = 196x - 379,400$ and determine the corresponding y-value. Do these y-values give reasonable approximations to the y-values in the table? Explain your answer.

4. *Making Estimates*   Use $y = 196x - 379,400$ to estimate the number of radio stations on the air in 1975 and 1985. Compare the results to your answer in Exercise 2.

3.4 SLOPE AND RATES OF CHANGE

Finding Slopes of Lines ▪ Slope as a Rate of Change

Figure 3.28 shows some graphs of lines, where the horizontal axis represents time.

Which graph might represent the distance traveled by you if you are walking?

Which graph might represent the temperature in your freezer?

Which graph might represent the amount of gasoline in your car's tank while you are driving?

To be able to answer these questions, you probably used the concept of slope. In mathematics, slope is a real number that measures the "tilt" or "angle" of a line. In this section we discuss slope and how it is used in applications.

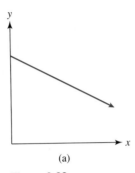
(a)

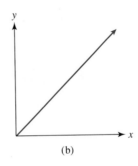
(b)

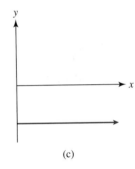
(c)

Figure 3.28

Finding Slopes of Lines

▶ REAL-WORLD CONNECTION   The graph shown in Figure 3.29 illustrates the cost of parking for x hours. The graph tilts upward from left to right, which indicates that the cost increases as the number of hours increases. Note that, for each hour of parking, the cost increases by \$2. The graph *rises* 2 units for every unit of *run*, and the ratio $\frac{\text{rise}}{\text{run}}$ equals the *slope* of the line. The slope m of this line is 2, which indicates that the cost of parking is \$2 per hour.

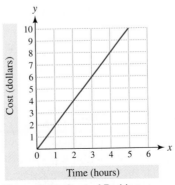

Figure 3.29 Cost of Parking

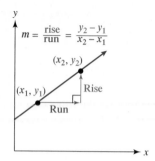

Figure 3.30 Slope m of a Line

A more general case of the slope of a line is shown in Figure 3.30 where a line passes through the points (x_1, y_1) and (x_2, y_2). The **rise**, or *change in y*, is $y_2 - y_1$, and the **run**, or *change in x*, is $x_2 - x_1$. Slope m is given by $m = \frac{y_2 - y_1}{x_2 - x_1}$.

NOTE: The symbol x_1 has a **subscript** of 1 and is read "x sub one" or "x one". Thus x_1 and x_2 are used to denote two different x-values. Similar comments apply to y_1 and y_2.

SLOPE

The **slope** m of the line passing through the points (x_1, y_1) and (x_2, y_2) is

$$m = \frac{\text{rise}}{\text{run}} = \frac{y_2 - y_1}{x_2 - x_1},$$

where $x_1 \neq x_2$. That is, slope equals rise over run.

NOTE: If $x_1 = x_2$, the slope is undefined.

EXAMPLE 1   Calculating the slope of a line

Use the two points labeled in Figure 3.31 to find the slope of the line. What are the rise and run between these two points? Interpret the slope in terms of rise and run.

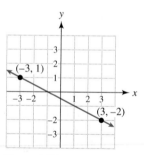

Figure 3.31

Solution
The line passes through the points $(-3, 1)$ and $(3, -2)$, so let $(x_1, y_1) = (-3, 1)$ and $(x_2, y_2) = (3, -2)$. The slope is

$$m = \frac{y_2 - y_1}{x_2 - x_1} = \frac{-2 - 1}{3 - (-3)} = \frac{-3}{6} = -\frac{1}{2}.$$

Starting at the point $(-3, 1)$, count 3 units downward and then 6 units to the right to return to the graph at the point $(3, -2)$. Thus the "rise" is -3 units and the run is 6 units. See Figure 3.32(a). The ratio $\frac{\text{rise}}{\text{run}}$ is $\frac{-3}{6}$, or $-\frac{1}{2}$. Figure 3.32(b) shows an alternate way of finding this slope. Starting at the point $(3, -2)$, count 3 units upward and then 6 units to the left to return to the graph at the point $(-3, 1)$. Here, the rise is 3 units and the run is -6 units so that the ratio $\frac{\text{rise}}{\text{run}}$ is $\frac{3}{-6}$, or $-\frac{1}{2}$. In either case, the slope is $-\frac{1}{2}$, indicating that the graph falls 1 unit for every 2 units of run (to the right).

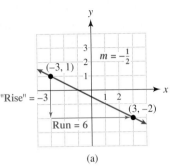 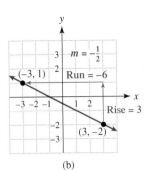

(a)   (b)

Figure 3.32

Now Try Exercise **25**

NOTE: In Example 1 the same slope would result if we let $(x_1, y_1) = (3, -2)$ and $(x_2, y_2) = (-3, 1)$. In this case the calculation would be

$$m = \frac{y_2 - y_1}{x_2 - x_1} = \frac{1 - (-2)}{-3 - 3} = \frac{3}{-6} = -\frac{1}{2}.$$

Two points determine a line and its slope, as demonstrated in the next example.

EXAMPLE 2   Calculating the slope of a line

Calculate the slope of the line passing through each pair of points. Graph the line.
(a) $(-2, -2), (0, 2)$ **(b)** $(-2, 3), (2, 1)$ **(c)** $(-1, 3), (2, 3)$ **(d)** $(-3, 3), (-3, -2)$

Solution

(a) $m = \frac{y_2 - y_1}{x_2 - x_1} = \frac{2 - (-2)}{0 - (-2)} = \frac{4}{2} = 2$. This slope indicates that the line rises 2 units for every 1 unit of horizontal run, as shown in Figure 3.33(a).

(b) $m = \frac{y_2 - y_1}{x_2 - x_1} = \frac{1 - 3}{2 - (-2)} = \frac{-2}{4} = -\frac{1}{2}$. This slope indicates that the line falls 1 unit for every 2 units of horizontal run, as shown in Figure 3.33(b).

(c) $m = \frac{y_2 - y_1}{x_2 - x_1} = \frac{3 - 3}{2 - (-1)} = \frac{0}{3} = 0$. The line is horizontal, as shown in Figure 3.33(c).

(d) Because $x_1 = x_2 = -3$, the slope formula does not apply. If we try to use it, we obtain $m = \frac{y_2 - y_1}{x_2 - x_1} = \frac{-2 - 3}{-3 - (-3)} = \frac{-5}{0}$, which is an undefined expression. The line has undefined slope and is vertical, as shown in Figure 3.33(d).

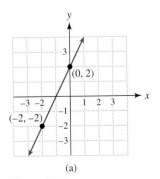 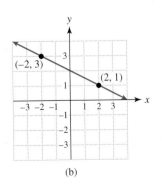 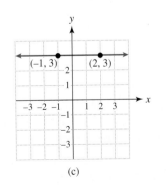 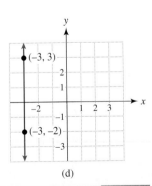

(a)   (b)   (c)   (d)

Figure 3.33

Now Try Exercises **37, 39, 41**

If a line has **positive slope**, it *rises from left to right*, as shown in Figure 3.34(a). If a line has **negative slope**, it *falls from left to right*, as shown in Figure 3.34(b). Slope 0 indicates that a line is horizontal, which is shown in Figure 3.34(c). Any two points on a vertical line have the same x-coordinate so the run always equals 0. Thus the slope is undefined for a vertical line, which is shown in Figure 3.34(d).

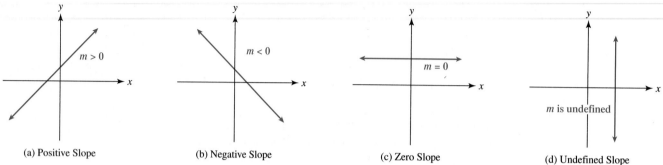

(a) Positive Slope   (b) Negative Slope   (c) Zero Slope   (d) Undefined Slope

Figure 3.34

SLOPE OF A LINE

1. A line that rises *from left to right* has positive slope.
2. A line that falls *from left to right* has negative slope.
3. A horizontal line has slope 0.
4. A vertical line has undefined slope.

EXAMPLE 3   Finding slope from a graph

Find the slope of each line.

(a)

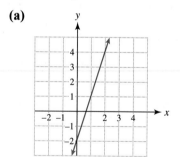

(b)

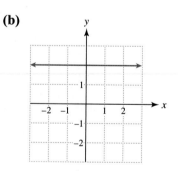

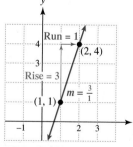

Figure 3.35

Solution

(a) The graph rises 3 units for each unit of run. For example, the graph passes through $(1, 1)$ and $(2, 4)$, so the line rises $4 - 1 = 3$ units with a $2 - 1 = 1$ unit run, as shown in Figure 3.35. Therefore the slope is

$$m = \frac{\text{rise}}{\text{run}} = \frac{3}{1} = 3.$$

(b) The line is horizontal, so the slope is 0.

Now Try Exercises 15, 19

A point and a slope also determine a line, as illustrated in the next example.

EXAMPLE 4

Sketching a line with a given slope

Sketch a line passing through the point $(1, 4)$ and having slope $-\frac{2}{3}$.

Solution

Start by plotting the point $(1, 4)$. A slope of $-\frac{2}{3}$ indicates that the y-values *decrease* 2 units each time the x-values increase by 3 units. That is, the line *falls* 2 units for every 3-unit increase in the run. Because the line passes through $(1, 4)$, a 2-unit decrease in y and a 3-unit increase in x results in the line passing through the point $(1 + 3, 4 - 2)$ or $(4, 2)$. See Figure 3.36.

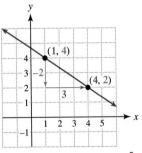

Figure 3.36 Slope $m = -\frac{2}{3}$

Now Try Exercise 55

EXAMPLE 5

Sketching a line with a given y-intercept

Sketch a line with slope -2 and y-intercept 3.

Solution

For the y-intercept of 3, plot the point $(0, 3)$. The slope is -2, so the y-values decrease 2 units for each unit increase in x. Increasing the x-value in the point $(0, 3)$ by 1 and decreasing the y-value by 2 results in the point $(0 + 1, 3 - 2)$ or $(1, 1)$. Plot $(0, 3)$ and $(1, 1)$ and then sketch the line, as shown in Figure 3.37.

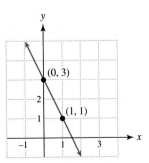

Figure 3.37 Slope: -2;
y-intercept: 3

Now Try Exercise 51

If we know the slope of a line and a point on the line, we can complete a table of points, as demonstrated in the next example.

EXAMPLE 6   Completing a table of values

A line has slope 2 and passes through the first point listed in the table. Complete the table so that each point lies on the line.

x	-2	-1	0	1
y	1			

Solution

Slope 2 indicates that $\frac{\text{rise}}{\text{run}} = 2$. Because consecutive x-values in the table increase by 1 unit, the run from one point in the table to the next is 1 unit. Substituting **1** for the **run** in the slope equation results in $\frac{\text{rise}}{1} = 2$. Thus the **rise** is **2** and consecutive y-values shown in Table 3.16 increase by 2 units.

TABLE 3.16

x	-2	-1	0	1
y	1	3	5	7

Now Try Exercise 63

Slope as a Rate of Change

▶ REAL-WORLD CONNECTION   When lines are used to model physical quantities in applications, their slopes provide important information. Slope measures the **rate of change** in a quantity. We illustrated this concept in the next four examples.

EXAMPLE 7   Interpreting slope

The distance y in miles that an athlete training for a marathon is from home after x hours is shown in Figure 3.38.
(a) Find the y-intercept. What does the y-intercept represent?
(b) The graph passes through the point $(1, 10)$. Discuss the meaning of this point.
(c) Find the slope of this line. Interpret the slope as a rate of change.

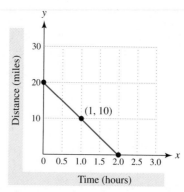

Figure 3.38 Distance from Home

Solution

(a) On the graph the y-intercept is 20, so the athlete is initially 20 miles from home.
(b) The point $(1, 10)$ means that after 1 hour the athlete is 10 miles from home.

(c) The line passes through the points $(0, 20)$ and $(1, 10)$. Its slope is

$$m = \frac{10 - 20}{1 - 0} = -10.$$

Slope -10 means that the athlete is running *toward* home at 10 miles per hour. A *negative* slope indicates that the distance between the runner and home is decreasing.

Now Try Exercise 93

EXAMPLE 8   Interpreting slope

When a company manufactures 2000 MP3 players, its profit is \$10,000, and when it manufactures 4500 MP3 players, its profit is \$35,000.
(a) Find the slope of the line passing through $(2000, 10000)$ and $(4500, 35000)$.
(b) Interpret the slope as a rate of change.

Solution
(a) $m = \dfrac{35,000 - 10,000}{4500 - 2000} = \dfrac{25,000}{2500} = 10$
(b) Profit increases, *on average*, by \$10 for each additional MP3 player made.

Now Try Exercise 99

EXAMPLE 9   Analyzing tetanus cases

Table 3.17 lists numbers of reported cases of tetanus in the United States for selected years.

TABLE 3.17

Year	1950	1960	1970	1980	1990	2000
Cases of Tetanus	486	368	148	95	64	45

Source: Department of Health and Human Services.

(a) Make a line graph of the data.
(b) Find the slope of each line segment.
(c) Interpret each slope as a rate of change.

Solution
(a) A line graph connecting the points $(1950, 486)$, $(1960, 368)$, $(1970, 148)$, $(1980, 95)$, $(1990, 64)$, and $(2000, 45)$ is shown in Figure 3.39.

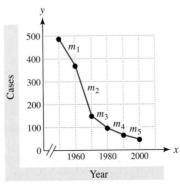

Figure 3.39 Tetanus Cases

(b) The slope of each line segment may be calculated as follows.

$$m_1 = \frac{368 - 486}{1960 - 1950} = -11.8, \qquad m_2 = \frac{148 - 368}{1970 - 1960} = -22.0,$$

$$m_3 = \frac{95 - 148}{1980 - 1970} = -5.3, \qquad m_4 = \frac{64 - 95}{1990 - 1980} = -3.1,$$

$$m_5 = \frac{45 - 64}{2000 - 1990} = -1.9$$

(c) Slope $m_1 = -11.8$ indicates that, *on average*, the number of tetanus cases *decreased* by 11.8 cases per year between 1950 and 1960. The other four slopes can be interpreted similarly.

> **NOTE:** The number of tetanus cases did not decrease by *exactly* 11.8 cases per year between 1950 and 1960. Why? However, the yearly *average* decrease was 11.8 cases.

Now Try Exercise 97

EXAMPLE 10   Sketching a model

During a storm, rain falls at the rate of 2 inches per hour from 1 A.M. to 3 A.M., 1 inch per hour from 3 A.M. to 4 A.M., and $\frac{1}{2}$ inch per hour from 4 A.M. to 6 A.M.
(a) Sketch a graph that shows the total accumulation of rainfall from 1 A.M. to 6 A.M.
(b) What does the slope of each line segment represent?

CRITICAL THINKING

An athlete runs 10 miles per hour for 30 minutes away from home and then jogs back home at 5 miles per hour. Sketch a graph that shows the distance between the athlete and home. What does the slope of each line segment represent?

Solution

(a) At 1 A.M. the accumulated rainfall is 0, so place a point at $(1, 0)$. Rain falls at a constant rate of 2 inches per hour for the next 2 hours, so at 3 A.M. the total rainfall is 4 inches. Place a point at $(3, 4)$. Because the rainfall is constant, sketch a line segment from $(1, 0)$ to $(3, 4)$, as shown in Figure 3.40. Similarly, during the next hour 1 inch of rain falls, so draw a line segment from $(3, 4)$ to $(4, 5)$. Finally, 1 inch of rain falls from 4 A.M. to 6 A.M., so draw a line segment from $(4, 5)$ to $(6, 6)$.

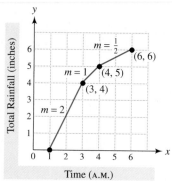

Figure 3.40 Total Rainfall

(b) The slope of each line segment represents the rate at which rain is falling. For example, the first segment has slope 2 because rain falls at a rate of 2 inches per hour during that period of time.

Now Try Exercise 89

3.4 PUTTING IT ALL TOGETHER

The "tilt" of a line is called the slope and equals rise over run. A positive slope indicates that the line *rises* from left to right, whereas a negative slope indicates that the line *falls* from left to right. A horizontal line has slope 0, and a vertical line has undefined slope. When a quantity can be approximated with a line, its slope indicates a rate of change in the quantity. The following table summarizes some basic concepts about slope.

Concept	Comments	Example
Rise, Run, and Slope	Rise is a vertical change in a line, and run is a horizontal change in a line. The ratio $\frac{\text{rise}}{\text{run}}$ is the slope m when run is nonzero.	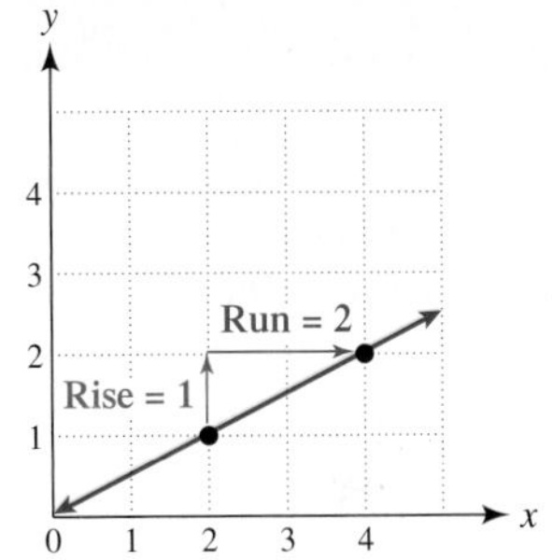 $$m = \frac{\text{rise}}{\text{run}} = \frac{1}{2}$$
Calculating Slope	For any two points (x_1, y_1) and (x_2, y_2) slope m is $$m = \frac{y_2 - y_1}{x_2 - x_1},$$ where $x_1 \neq x_2$.	The slope of the line passing through $(-2, 3)$ and $(1, 5)$ is $$m = \frac{5 - 3}{1 - (-2)} = \frac{2}{3}.$$ The line rises 2 units for every 3 units of run along the x-axis.
Slope as a Rate of Change	Slope indicates how fast the graph of a line is changing.	The graph shows that water is being pumped from a swimming pool. Slope $m = -66\frac{2}{3}$ indicates that water is *leaving* the pool at the rate of $66\frac{2}{3}$ gallons per hour. 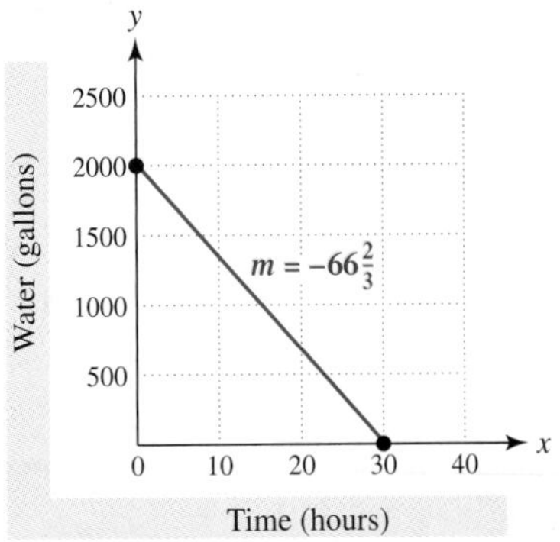

3.4 Exercises

CONCEPTS

1. Run is the change in the (horizontal/vertical) distance along a line.

2. Rise is the change in the (horizontal/vertical) distance along a line.

3. Slope m of a line is _____ over _____.

4. Slope 0 indicates that a line is _____.

5. Undefined slope indicates that a line is _____.

6. If a line passes through (x_1, y_1) and (x_2, y_2), then $m =$ _____.

7. A line that rises from left to right has _____ slope.

8. A line that falls from left to right has _____ slope.

Exercises 9–14: State whether the slope of the line is positive, negative, zero, or undefined.

9. **10.**

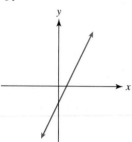

11. **12.**

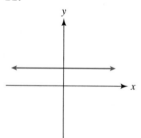 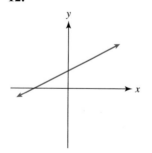

13. **14.**

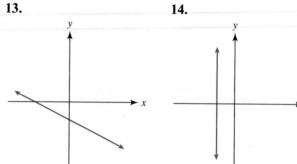

FINDING SLOPES OF LINES

Exercises 15–26: If possible, find the slope of the line. Interpret the slope in terms of rise and run.

15. **16.**

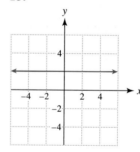 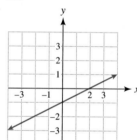

17. **18.**

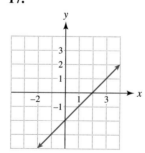 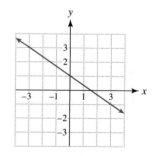

19.

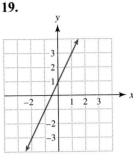

20.

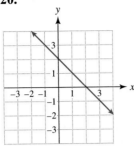

21.

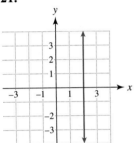

22.

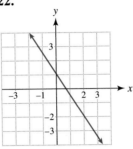

23.

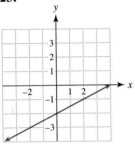

24.

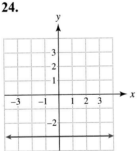

25.

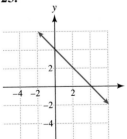

26.

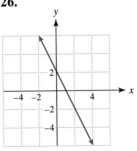

Exercises 27–34: If possible, find the slope of the line passing through the two points. Graph the line.

27. $(1, 2), (2, 4)$

28. $(-4, 7), (7, -4)$

29. $(2, 1), (2, 4)$

30. $(-1, 3), (-1, -1)$

31. $(1, 3), (-2, 5)$

32. $(0, -4), (-4, 6)$

33. $(2, -1), (-2, -1)$

34. $(-2, 3), (1, 3)$

Exercises 35–50: If possible, find the slope of the line passing through the two points.

35. $(4, -2), (-3, -7)$

36. $(15, -3), (20, 9)$

37. $(-3, 4), (4, -2)$

38. $(1, -3), (3, -5)$

39. $(-3, 5), (2, 5)$

40. $(-3, 3), (-5, 3)$

41. $(-1, 6), (-1, -4)$

42. $\left(\frac{1}{2}, -\frac{2}{7}\right), \left(\frac{1}{2}, \frac{13}{17}\right)$

43. $(1980, 5), (2000, 18)$

44. $(1989, 10), (1999, 16)$

45. $(1950, 6.1), (2000, 10.6)$

46. $(1900, 10), (1950, 35)$

47. $\left(\frac{1}{3}, -\frac{2}{7}\right), \left(-\frac{2}{3}, \frac{3}{7}\right)$

48. $(-1.3, 5.6), (-2.6, -2.5)$

49. $(12, -34), (14, 64)$ **50.** $(-25, 105), (60, 55)$

Exercises 51–58: (Refer to Example 4.) Sketch a line passing through the point and having slope m.

51. $(0, 2), m = -1$

52. $(0, -1), m = 2$

53. $(1, 1), m = 3$

54. $(1, -1), m = -2$

55. $(-2, 3), m = -\frac{1}{2}$

56. $(-1, -2), m = \frac{3}{4}$

57. $(-3, 1), m = \frac{1}{2}$

58. $(-2, 2), m = -3$

Exercises 59–62: The table lists points located on a line. Find the slope, x-intercept, and y-intercept of the line.

59.

x	0	1	2	3
y	-2	0	2	4

60.

x	-1	0	1	2
y	0	5	10	15

61.

x	-2	-1	0	1
y	9	0	-9	-18

62.

x	-4	-2	0	2
y	6	3	0	-3

Exercises 63–68: (Refer to Example 6.) A line has the given slope m and passes through the first point listed in the table. Complete the table so that each point in the table lies on the line.

63. $m = 2$

x	0	1	2	3
y	−4			

64. $m = -\frac{1}{2}$

x	0	1	2	3
y	2			

65. $m = -3$

x	1	2	3	4
y	4			

66. $m = -1$

x	−1	0	1	2
y	10			

67. $m = \frac{3}{2}$

x	−4	−2	0	2
y	0			

68. $m = 3$

x	−2	0	2	4
y	−4			

Exercises 69–76: Do the following.
   (a) *Graph the equation.*
   (b) *Find the slope of the line.*

69. $y = 2x - 1$ **70.** $y = x - 2$

71. $3x + y = 2$ **72.** $-\frac{1}{2}x + y = -1$

73. $-x + 3y = 0$ **74.** $2x + y = 0$

75. $y = 2$ **76.** $y = -3$

SLOPE AS A RATE OF CHANGE

*Exercises 77–80: **Modeling**   Choose the graph (a.–d.) in the next column that models the situation best.*

77. Cost of buying x gum balls at a price of 25¢ each

78. Total number of movies in VHS format (videotape) purchased during the past 5 years

79. Average cost of a new car over the past 30 years

80. Height of the Empire State Building after x people have entered it

a.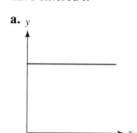

b.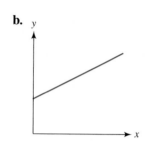

c.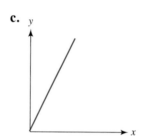

d.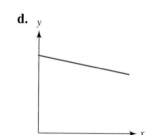

*Exercises 81 and 82: **Modeling**   The line graph represents the gallons of water in a small swimming pool after x hours. Assume that there is a pump that can either add water to or remove water from the pool.*

   (a) *Estimate the slope of each line segment.*
   (b) *Interpret each slope as a rate of change.*
   (c) *Describe what happened to the amount of water in the pool.*

81.

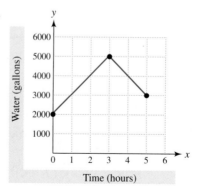

82.

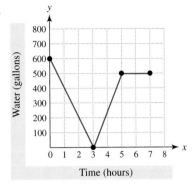

Exercises 83 and 84: Modeling An individual is driving a car along a straight road. The graph shows the distance that the driver is from home after x hours.

(a) Find the slope of each line segment in the graph.
(b) Interpret each slope as a rate of change.
(c) Describe both the motion and location of the car.

83.

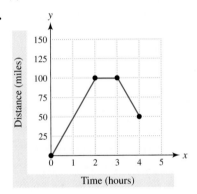

84.

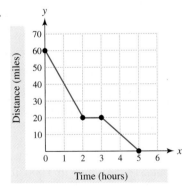

85. Thinking Generally If a line is used to model physical data where the x-axis is labeled "Time (minutes)" and the y-axis is labeled "Volume (cubic meters)," what are the units for the rate of change represented by the slope of the line?

86. Thinking Generally If a line is used to model physical data where the x-axis is labeled "Cookies" and the y-axis is labeled "Chocolate Chips," what are the units for the rate of change represented by the slope of the line?

Exercises 87–90: Sketching a Model Sketch a graph that models the given situation.

87. The distance that a boat is from a harbor if the boat is initially 6 miles from the harbor and arrives at the harbor after sailing at a constant speed for 3 hours

88. The distance that a person is from home if the person starts at home, walks away from home at 4 miles per hour for 90 minutes, and then walks back home at 3 miles per hour

89. The distance that an athlete is from home if the athlete jogs for 1 hour to a park that is 7 miles away, stays for 30 minutes, and then jogs home at the same pace

90. The amount of oil in a 55-gallon drum that is initially full, then is drained at a rate of 5 gallons per minute for 4 minutes, is left for 6 minutes, and then is emptied at a rate of 7 gallons per minute

APPLICATIONS

91. *Older Mothers* The number of children born to mothers 40 years old or older in 2000 was 70 thousand and in 2004 it was 110 thousand.   (**Source:** National Center for Health Statistics.)
(a) Calculate the slope of the line passing through (2000, 70) and (2004, 110).
(b) Interpret the slope as a rate of change.

92. *Profit from Laptops* When a company manufactures 500 laptop computers, its profit is $100,000, and when it manufactures 1500 laptop computers, its profit is $400,000.
(a) Find the slope of the line passing through the points (500, 100000) and (1500, 400000).
(b) Interpret the slope as a rate of change.

93. *Revenue*   The graph shows revenue received from selling *x* screwdrivers.

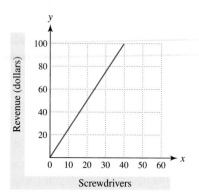

(a) Find the slope of the line shown.
(b) Interpret the slope as a rate of change.

94. *Electricity*   The graph shows how voltage is related to amperage in an electrical circuit. The slope corresponds to the resistance in ohms. Find the resistance in this electrical circuit.

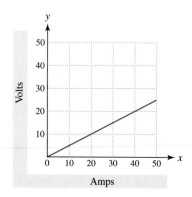

95. *Two-Cycle Engines*   Two-cycle engines used in jet skis, snowmobiles, chain saws, and outboard motors require a mixture of gas and oil to run properly. For certain engines the gallons of oil *O* that should be added to *G* gallons of gasoline is given by $O = \frac{1}{50}G$.
(*Source:* Johnson Outboard Motor Company.)
(a) How many gallons of oil should be added to 50 gallons of gasoline and to 100 gallons of gasoline?
(b) Find the slope of the graph of the equation.
(c) Interpret the slope as a rate of change.

96. *Plant Growth*   The graph shows the seasonal growth of a type of grass for various amounts of rainfall.

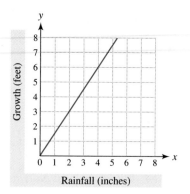

(a) Find the slope of the line shown.
(b) Interpret the slope as a rate of change.

97. *Walking for Charities*   The table lists the amount of money *M* in dollars raised for walking various distances in miles for a charity.

x	0	5	10	15
M	0	100	250	450

(a) Make a line graph of the data.
(b) Calculate the slope of each line segment.
(c) Interpret each slope as a rate of change.

98. *Insect Population*   The table lists the number *N* of black flies in thousands per acre after *x* weeks.

x	0	2	4	6
N	3	4	10	18

(a) Make a line graph of the data.
(b) Calculate the slope of each line segment.
(c) Interpret each slope as a rate of change.

99. *Median Household Income*   In 1980, median family income was about $18,000, and in 2000 it was about $42,000.   (*Source:* Department of the Treasury.)
(a) Find the slope of the line passing through the points (1980, 18000) and (2000, 42000).
(b) Interpret the slope as a rate of change.
(c) If this trend continues, estimate the median family income in 2005.

100. *Minimum Wage* In 1990, the minimum wage was about \$3.50 per hour, and in 2005 it was \$5.15. (*Source:* Department of Labor.)
   (a) Find the slope of the line passing through the points (1990, 3.5) and (2005, 5.15).
   (b) Interpret the slope as a rate of change.
   (c) If this trend continues, estimate the minimum wage in 2015.

101. *Rate of Change* Suppose that $y = -2x + 10$ is graphed in the first quadrant of the xy-plane where the x-axis is labeled "Time (minutes)" and the y-axis is labeled "Distance (feet)." If this graph represents the distance y that an ant is from a stone after x minutes, answer each of the following.
   (a) Is the ant moving toward or away from the stone?
   (b) Initially, how far from the stone is the ant?
   (c) At what rate is the ant moving?
   (d) What is the value of x (time) when the ant reaches the stone?

102. *Rate of Change* Suppose that $y = 15x + 8$ is graphed in the first quadrant of the xy-plane where the x-axis is labeled "Time (minutes)" and the y-axis is labeled "Distance (feet)." If this graph represents the distance y that a frog is from a tree after x minutes, answer each of the following.
   (a) Is the frog moving toward or away from the tree?
   (b) Initially, how far from the tree is the frog?
   (c) At what rate is the frog moving?
   (d) What is the value of x when the frog is 53 feet from the tree?

WRITING ABOUT MATHEMATICS

103. If you are given two points and the slope formula $m = \frac{y_2 - y_1}{x_2 - x_1}$, does it matter which point is (x_1, y_1) and which point is (x_2, y_2)? Explain.

104. Suppose that a line approximates the distance y in miles that a person drives in x hours. What does the slope of the line represent? Give an example.

105. Describe the information that the slope m gives about a line. Be as complete as possible.

106. Could one line have two slopes? Explain.

CHECKING BASIC CONCEPTS
SECTIONS 3.3 AND 3.4

1. Identify the x- and y-intercepts in the graph.

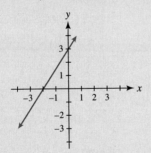

2. Complete the table for the equation $2x - y = 2$. Then determine the x- and y-intercepts.

x	-2	-1	0	1	2
y	-6				

3. Find any intercepts for the graphs of the equations and then graph each linear equation.
   (a) $x - 2y = 6$ **(b)** $y = 2$ **(c)** $x = -1$

4. Write the equations of a horizontal line and a vertical line that pass through the point $(-2, 4)$.

5. If possible, find the slope of the line passing through each pair of points.
   (a) $(-2, 3), (2, 6)$ **(b)** $(-5, 3), (0, 3)$
   (c) $(1, 5), (1, 8)$

6. Find the slope of the line shown.

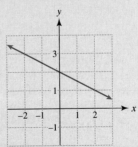

7. Sketch a line passing through the point $(-3, 1)$ and having slope 2.

continued on next page

continued from previous page

8. *Modeling* The line graph shows the depth of water in a small pond before and after a rain storm.
 (a) Estimate the slope of each line segment.
 (b) Interpret each slope as a rate of change.
 (c) Describe what happened to the amount of water in the pond.

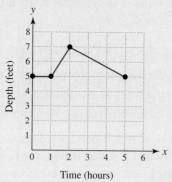

Time (hours)

9. *Minnesota County Population* Dodge County has an area of 450 square miles with a population of 18,100. Meeker County has an area of 600 square miles with a population of 22,300.
(*Source:* U.S. Census Bureau.)
 (a) Find the slope of the line passing through the points (450, 18100) and (600, 22300).
 (b) Interpret the slope as a rate of change.
 (c) Is this rate of change a valid predictor of the population of a county with an area of 700 square miles? Explain your reasoning.

3.5 SLOPE–INTERCEPT FORM

Finding Slope–Intercept Form ▪ Parallel and Perpendicular Lines

A LOOK INTO MATH ▷ For any two points in the *xy*-plane, we can draw a unique line passing through them, as illustrated in Figure 3.41(a). Another way we can determine a unique line is to know the *y*-intercept and the slope. For example, if a line has *y*-intercept 2 and slope $m = 1$, then the resulting line is shown in Figure 3.41(b). In this section we discuss how to find the equation of a nonvertical line given its slope and *y*-intercept.

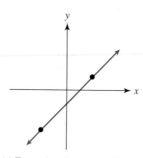

(a) Two points determine a line.

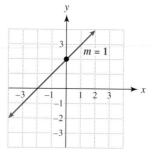

(b) One point and a slope determine a line.

Figure 3.41

Finding Slope–Intercept Form

The graph of $y = 2x + 3$ passes through $(0, 3)$ and $(1, 5)$, as shown in Figure 3.42. The slope of this line is

$$m = \frac{5 - 3}{1 - 0} = 2.$$

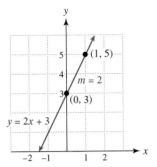

Figure 3.42 Slope 2, y-intercept 3

If $x = 0$ in $y = 2x + 3$, then $y = 2(0) + 3 = 3$. Thus the graph of $y = 2x + 3$ has slope 2 and y-intercept 3. In general, the graph of $y = mx + b$ has slope m and y-intercept b. The form $y = mx + b$ is called the *slope–intercept form*.

SLOPE–INTERCEPT FORM

The line with slope m and y-intercept b is given by

$$y = mx + b,$$

the **slope–intercept form** of a line.

The graph of the equation $y = 4x - 3$ has slope 4 and y-intercept -3, and the graph of the equation $y = -\frac{1}{2}x + 8$ has slope $-\frac{1}{2}$ and y-intercept 8.

EXAMPLE 1   Using a graph to write the slope–intercept form

For each graph write the slope–intercept form of the line.

(a)

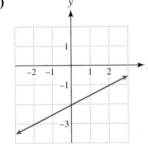

(b)

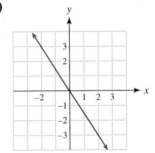

Solution

(a) The graph intersects the y-axis at -2, so the y-intercept is -2. Because the graph rises 1 unit for each 2-unit increase in x, the slope is $\frac{1}{2}$. The slope–intercept form of the line is $y = \frac{1}{2}x - 2$.

(b) The graph intersects the y-axis at 0, so the y-intercept is 0. Because the graph falls 3 units for each 2-unit increase in x, the slope is $-\frac{3}{2}$. The slope–intercept form of the line is $y = -\frac{3}{2}x + 0$, or $y = -\frac{3}{2}x$.

Now Try Exercise 17

EXAMPLE 2   Sketching a line

Sketch a line with slope $-\frac{1}{2}$ and y-intercept 1. Write its slope–intercept form.

Solution

For the y-intercept of 1 plot the point $(0, 1)$. Slope $-\frac{1}{2}$ indicates that the graph falls 1 unit for each 2-unit increase in x. Thus the line passes through the point $(0 + 2, 1 - 1)$, or $(2, 0)$, as shown in Figure 3.43. The slope–intercept form of this line is $y = -\frac{1}{2}x + 1$.

Now Try Exercise 27

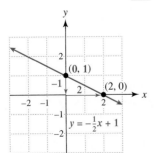

Figure 3.43

When a linear equation is given in slope–intercept form, $y = mx + b$, the coefficient of the x-term gives the slope and the constant term gives the y-intercept for the graph of the line. For example, the graph of $y = -2x + 5$ has slope -2 and y-intercept 5. However, when a linear equation is not given in slope–intercept form, the coefficient of the x-term may not represent the slope and the constant term may not represent the y-intercept. For example, the graph of $2x + 3y = 12$ does *not* have slope 2 and it does *not* have y-intercept 12. To find the correct slope and y-intercept, the equation can first be written in slope–intercept form. This is demonstrated in the next example.

EXAMPLE 3   Writing an equation in slope–intercept form

Write each equation in slope–intercept form. Then give the slope and y-intercept of the line.
(a) $2x + 3y = 12$ **(b)** $x = 2y + 4$

Solution
(a) To write the equation in slope–intercept form, solve for y.

$$2x + 3y = 12 \qquad \text{Given equation}$$
$$3y = -2x + 12 \qquad \text{Subtract } 2x \text{ from each side.}$$
$$y = -\frac{2}{3}x + \mathbf{4} \qquad \text{Divide each side by 3.}$$

The slope of the line is $-\frac{2}{3}$, and the y-intercept is **4**.

(b) This equation is *not* in slope–intercept form because it is solved for x, not y.

$$x = 2y + 4 \qquad \text{Given equation}$$
$$x - 4 = 2y \qquad \text{Subtract 4 from each side.}$$
$$\frac{1}{2}x - 2 = y \qquad \text{Divide each side by 2.}$$
$$y = \frac{1}{2}x - \mathbf{2} \qquad \text{Rewrite the equation.}$$

The slope of the line is $\frac{1}{2}$, and the y-intercept is $\mathbf{-2}$.

Now Try Exercise 43

EXAMPLE 4   Graphing an equation in slope–intercept form

Write the equation $y = 3 - 2x$ in slope–intercept form and then graph it.

Solution
First write the given equation in slope–intercept form.

$$y = 3 - 2x \qquad \text{Given equation}$$
$$y = 3 + (-2x) \qquad \text{Change subtraction to addition.}$$
$$y = -2x + 3 \qquad \text{Commutative property}$$

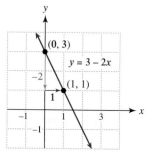

Figure 3.44

The slope–intercept form is $y = -2x + 3$, with slope -2 and y-intercept 3. To graph this equation plot the point $(0, 3)$. The line falls **2** units for each **1** unit increase in x, so plot the point $(0 + 1, 3 - 2) = (1, 1)$. Sketch a line passing through $(0, 3)$ and $(1, 1)$, as shown in Figure 3.44.

Now Try Exercise 59

EXAMPLE 5   **Modeling cell phone costs**

Roaming with a cell phone costs $5 for the initial connection and $0.50 per minute.
(a) If someone talks for 23 minutes, what is the charge?
(b) Write the slope–intercept form that gives the cost of talking for x minutes.
(c) If the charge is $8.50, how long did the person talk?

Solution
(a) The charge for 23 minutes at $0.50 per minute plus $5 would be

$$0.50 \times 23 + 5 = \$16.50.$$

(b) The rate of increase is $0.50 per minute with an initial cost of $5. Let $y = 0.5x + 5$, where the slope or rate of change is **0.5** and the y-intercept is **5**.
(c) To determine how long a person can talk for $8.50, we can solve the following equation.

$$0.5x + 5 = 8.5 \qquad \text{Equation to solve}$$
$$0.5x = 3.5 \qquad \text{Subtract 5 from each side.}$$
$$x = \frac{3.5}{0.5} \qquad \text{Divide each side by 0.5.}$$
$$x = 7 \qquad \text{Simplify.}$$

The person talked for 7 minutes. Note that this solution is based on the assumption that the phone company did not round up a fraction of a minute.   Now Try Exercise 77

Parallel and Perpendicular Lines

Slope is an important concept for determining whether two lines are parallel. If two lines have the same slope, they are parallel. For example, the lines $y = 2x$ and $y = 2x - 1$ are parallel because they both have slope **2**, as shown in Figure 3.45.

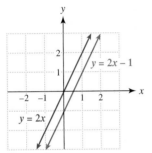

Figure 3.45

PARALLEL LINES

Two lines with the same slope are parallel.

Two nonvertical parallel lines have the same slope.

NOTE: Two vertical lines are parallel and the slope of each is undefined.

EXAMPLE 6   **Finding parallel lines**

Find the slope–intercept form of a line parallel to $y = -2x + 3$ and passing through the point $(-2, 3)$. Sketch a graph of each line.

Solution
Because the line $y = -2x + 3$ has slope -2, any parallel line also has slope -2 with slope–intercept form $y = -2x + b$ for some b. The value of b can be found by substituting the point $(-2, 3)$ in the slope–intercept form.

$$y = -2x + b \qquad \text{Slope–intercept form}$$
$$3 = -2(-2) + b \qquad \text{Let } x = -2 \text{ and } y = 3.$$
$$3 = 4 + b \qquad \text{Multiply.}$$
$$-1 = b \qquad \text{Subtract 4 from each side.}$$

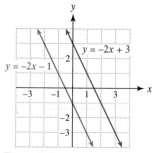

Figure 3.46

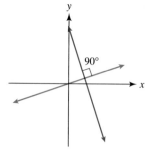

Figure 3.47 Perpendicular Lines

The y-intercept is -1, and so the slope–intercept form is $y = -2x - 1$. The graphs of the equations $y = -2x + 3$ and $y = -2x - 1$ are shown in Figure 3.46. Note that they are parallel lines, both with slope -2 but with different y-intercepts, 3 and -1.

Now Try Exercise 67

The lines shown in Figure 3.47 are perpendicular because they intersect at a 90° angle. Rather than measure the angle between two intersecting lines, we can determine whether two lines are perpendicular from their slopes. The slopes of perpendicular lines satisfy the following properties.

PERPENDICULAR LINES

If two perpendicular lines have nonzero slopes m_1 and m_2, then $m_1 \cdot m_2 = -1$.

If two lines have slopes m_1 and m_2 such that $m_1 \cdot m_2 = -1$, then they are perpendicular lines.

NOTE: A vertical line and a horizontal line are perpendicular.

Table 3.18 shows examples of slopes m_1 and m_2 that result in perpendicular lines. Note that $m_1 \cdot m_2 = -1$ and that $m_2 = -\frac{1}{m_1}$. That is, the product of the two slopes is -1, and the two slopes are negative reciprocals of each other.

TABLE 3.18 **Slopes of Perpendicular Lines**

m_1	1	$-\frac{1}{2}$	-4	$\frac{2}{3}$	$\frac{3}{4}$
m_2	-1	2	$\frac{1}{4}$	$-\frac{3}{2}$	$-\frac{4}{3}$

We can use these concepts to find equations of perpendicular lines, as illustrated in the next two examples.

EXAMPLE 7 Finding perpendicular lines

Find the slope–intercept form of a line passing through the origin that is perpendicular to each line.

(a) $y = 3x$ **(b)** $y = -\frac{2}{5}x + 5$ **(c)** $-3x + 4y = 24$

Solution

(a) If a line passes through the origin, then its y-intercept is 0 with slope–intercept form $y = mx$. The given line $y = 3x$ has slope $m_1 = 3$, so a line perpendicular to it has slope

$$m_2 = -\frac{1}{m_1} = -\frac{1}{3}.$$

The required slope–intercept form is $y = -\frac{1}{3}x$.

(b) The given line $y = -\frac{2}{5}x + 5$ has slope $m_1 = -\frac{2}{5}$, so a line perpendicular to it has slope $m_2 = \frac{5}{2}$. The required slope–intercept form is $y = \frac{5}{2}x$.

(c) To determine the slope of the given line, first write the equation in slope–intercept form.

$$-3x + 4y = 24 \qquad \text{Given equation}$$

$$4y = 3x + 24 \qquad \text{Add } 3x \text{ to each side.}$$

$$y = \frac{3}{4}x + 6 \qquad \text{Divide each side by 4.}$$

The slope of the given line is $m_1 = \frac{3}{4}$, so a line perpendicular to it has slope $m_2 = -\frac{4}{3}$. The required slope–intercept form is $y = -\frac{4}{3}x$.

Now Try Exercise 71

EXAMPLE 8   Finding a perpendicular line

Find the slope–intercept form of the line perpendicular to $y = -\frac{1}{2}x + 1$ and passing through the point $(1, -1)$. Sketch each line in the same xy-plane.

Solution
The line $y = -\frac{1}{2}x + 1$ has slope $m_1 = -\frac{1}{2}$. Any line perpendicular to it has slope $m_2 = 2$ with slope–intercept form $y = 2x + b$ for some b. The value of b can be found by substituting the point $(1, -1)$ in the slope–intercept form.

$$y = 2x + b \qquad \text{Slope–intercept form}$$

$$-1 = 2(1) + b \qquad \text{Let } x = 1 \text{ and } y = -1.$$

$$-1 = 2 + b \qquad \text{Multiply.}$$

$$-3 = b \qquad \text{Subtract 2 from each side.}$$

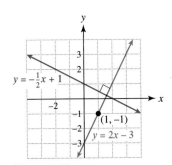

Figure 3.48

The slope–intercept form is $y = 2x - 3$. The graphs of $y = -\frac{1}{2}x + 1$ and $y = 2x - 3$ are shown in Figure 3.48. Note that the point $(1, -1)$ lies on the graph of $y = 2x - 3$.

Now Try Exercise 73

3.5 PUTTING IT ALL TOGETHER

The following table shows important forms of an equation of a line.

Concept	Comments	Example
Slope–Intercept Form $y = mx + b$	A unique equation for a line, determined by the slope m and the y-intercept b	An equation of the line with slope $m = 3$ and y-intercept $b = -5$ is $y = 3x - 5$.
Parallel Lines	$y = m_1x + b_1$ and $y = m_2x + b_2$, where $m_1 = m_2$ Nonvertical parallel lines have the same slope. Two vertical lines are parallel.	The lines $y = 2x - 1$ and $y = 2x + 2$ are parallel because they both have slope 2.

continued on next page

continued from previous page

Concept	Comments	Example
Perpendicular Lines	$y = m_1x + b_1$ and $y = m_2x + b_2$, where $m_1m_2 = -1$ Perpendicular lines which are neither vertical nor horizontal have slopes whose product equals -1. A vertical line and a horizontal line are perpendicular.	The lines $y = 3x - 1$ and $y = -\frac{1}{3}x + 2$ are perpendicular because $m_1m_2 = 3\left(-\frac{1}{3}\right) = -1.$ 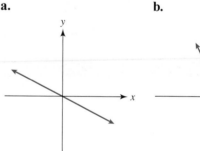

3.5 Exercises

MyMathLab   Math XL PRACTICE   WATCH   DOWNLOAD   READ   REVIEW

CONCEPTS

1. The slope–intercept form of a line is _____.

2. In the slope–intercept form of a line, m represents the _____ of the line.

3. In the slope–intercept form of a line, b represents the _____ of the line.

4. If $m = 0$ in the slope–intercept form of a line, then its graph is a _____ line.

5. If $b = 0$ in the slope–intercept form of a line, then its graph passes through the _____.

6. If a line passes through $(0, 5)$ with slope 2, then its slope–intercept form is _____.

7. Two lines with the same slope are _____.

8. Two nonvertical parallel lines have the same _____.

9. If m_1 and m_2 are the slopes of two lines where $m_1 \cdot m_2 = -1$, the lines are _____.

10. If two perpendicular lines have nonzero slopes, the slopes are negative _____.

Exercises 11–16: Match the description with its graph (a.–f.).

11. A line with positive slope and negative y-intercept

12. A line with positive slope and positive y-intercept

13. A line with negative slope and y-intercept 0

14. A line with negative slope and nonzero y-intercept

15. A line with no x-intercept

16. A line with no y-intercept

a.

b.
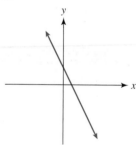

c.
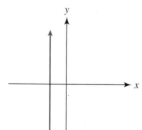

d.
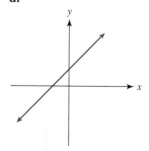

e.

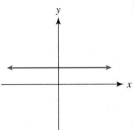

f.

SLOPE–INTERCEPT FORM

Exercises 17–26: Write the slope–intercept form for the line shown.

17.

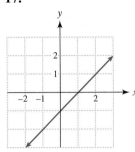

18.

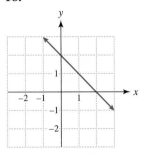

19.

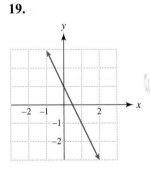

20.

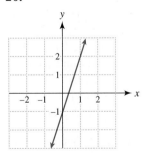

21.

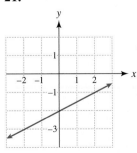

22.

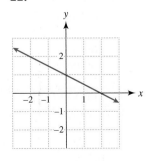

23.

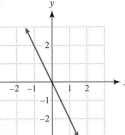

24.

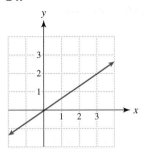

25.

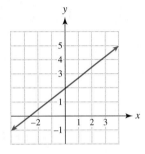

26.

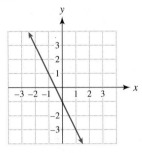

Exercises 27–36: Sketch a line with the given slope m and y-intercept b. Write its slope–intercept form.

27. $m = 1, b = 2$ **28.** $m = -1, b = 3$

29. $m = 2, b = -1$ **30.** $m = -3, b = 2$

31. $m = -\frac{1}{2}, b = -2$ **32.** $m = -\frac{2}{3}, b = 0$

33. $m = \frac{1}{3}, b = 0$ **34.** $m = 3, b = -3$

35. $m = -2, b = 1$ **36.** $m = \frac{1}{2}, b = -1$

Exercises 37–48: Do the following.
  (a) Write the equation in slope–intercept form.
  (b) Give the slope and y-intercept of the line.

37. $x + y = 4$ **38.** $x - y = 6$

39. $2x + y = 4$ **40.** $-4x + y = 8$

41. $x - 2y = -4$ **42.** $x + 3y = -9$

43. $2x - 3y = 6$ **44.** $4x + 5y = 20$

45. $x = 4y - 6$ **46.** $x = -3y + 2$

47. $\frac{1}{2}x + \frac{3}{2}y = 1$ **48.** $-\frac{3}{4}x + \frac{1}{2}y = \frac{1}{2}$

Exercises 49–60: Graph the equation.

49. $y = -3x + 2$ **50.** $y = \frac{1}{2}x - 1$

51. $y = \frac{1}{3}x$

52. $y = -2x$

53. $y = 2$

54. $y = -3$

55. $y = -x + 3$

56. $y = \frac{2}{3}x - 2$

57. $y = -\frac{1}{2}x + 1$

58. $y = -2x + 2$

59. $y = 2 - x$

60. $y = 2 - 3x$

Exercises 61–64: The table shows points that all lie on the same line. Find the slope–intercept form for the line.

61.

x	0	1	2
y	2	4	6

62.

x	−1	0	1
y	4	8	12

63.

x	−2	0	2
y	−4	−2	0

64.

x	0	2	4
y	6	3	0

PARALLEL AND PERPENDICULAR LINES

Exercises 65–74: Find the slope–intercept form of the line satisfying the given conditions.

65. Slope $\frac{4}{7}$, y-intercept 3

66. Slope $-\frac{1}{2}$, y-intercept −7

67. Parallel to $y = 3x + 1$, passing through $(0, 0)$

68. Parallel to $y = -2x$, passing through $(0, 1)$

69. Parallel to $2x + 4y = 5$, passing through $(1, 2)$

70. Parallel to $-x - 3y = 9$, passing through $(-3, 1)$

71. Perpendicular to $y = -\frac{1}{2}x - 3$, passing through $(0, 0)$

72. Perpendicular to $y = \frac{3}{4}x - \frac{1}{2}$, passing through $(3, -2)$

73. Perpendicular to $x = -\frac{1}{3}y$, passing through $(-1, 0)$

74. Perpendicular to $6x - 3y = 18$, passing through $(4, -3)$

APPLICATIONS

75. *Rental Cars*   Driving a rental car x miles costs $y = 0.25x + 25$ dollars.
(a) How much would it cost to rent the car but not drive it?

(b) How much does it cost to drive the car 1 *additional* mile?
(c) What is the y-intercept of $y = 0.25x + 25$? What does it represent?
(d) What is the slope of $y = 0.25x + 25$? What does it represent?

76. *Calculating Rainfall*   The total rainfall y in inches that fell x hours past noon is given by $y = \frac{1}{2}x + 3$.
(a) How much rainfall was there at noon?
(b) At what rate was rain falling in the afternoon?
(c) What is the y-intercept of $y = \frac{1}{2}x + 3$? What does it represent?
(d) What is the slope of $y = \frac{1}{2}x + 3$? What does it represent?

77. *Long-Distance Phone Service*   Long-distance phone service costs $3.95 per month plus $0.07 per minute. (Assume that a partial minute is not rounded up.)
(a) During July, a person talks a total of 50 minutes. What is the charge?
(b) Write an equation in slope–intercept form that gives the monthly cost C of talking long distance for x minutes.
(c) If the long-distance charge for one month is $8.64, how much time did the person spend talking on the phone?

78. *Electrical Rates*   Electrical service costs $8 per month plus $0.10 per kilowatt-hour of electricity used. (Assume that a partial kilowatt-hour is not rounded up.)
(a) If the resident of an apartment uses 650 kilowatt-hours in 1 month, what is the charge?
(b) Write an equation in slope–intercept form that gives the cost C of using x kilowatt-hours in 1 month.
(c) If the monthly electrical bill for the apartment's resident is $43, how many kilowatt-hours were used?

79. *Cost of Driving*   The cost of driving a car includes both fixed costs and mileage costs. Assume that it costs $164.30 per month for insurance and car payments and $0.35 per mile for gasoline, oil, and routine maintenance.
(a) Find values for m and b so that $y = mx + b$ models the monthly cost of driving the car x miles.
(b) What does the value of b represent?

80. *Antarctic Ozone Layer*   The ozone layer occurs in Earth's atmosphere between altitudes of 12 and 18 miles and is an important filter of ultraviolet light from the sun. The thickness of the ozone layer is frequently

measured in Dobson units. An average value is 300 Dobson units. In 1991, the reported minimum in the antarctic *ozone hole* was about 110 Dobson units.
(*Source:* R. Huffman, *Atmospheric Ultraviolet Remote Sensing.*)

(a) The equation $T = 0.01D$ describes the thickness T in millimeters of an ozone layer that is D Dobson units. How many millimeters thick was the ozone layer over the antarctic in 1991?

(b) What is the average thickness of the ozone layer in millimeters?

WRITING ABOUT MATHEMATICS

81. Explain how the values of m and b can be used to graph the equation $y = mx + b$.

82. Explain how to find the value of b in the equation $y = 2x + b$ if the point $(3, 4)$ lies on the line.

GROUP ACTIVITY
WORKING WITH REAL DATA

Directions: Form a group of 2 to 4 people. Select someone to record the group's responses for this activity. All members of the group should work cooperatively to answer the questions. If your instructor asks for the results, each member of the group should be prepared to respond.

Exercises 1–5: In this set of exercises you are to use your knowledge of equations of lines to model the average annual cost of tuition and fees.

1. *Cost of Tuition* In 2000, the average cost of tuition and fees at *private* four-year colleges was $16,200, and in 2005 it was $20,100. Sketch a line that passes through the points (2000, 16200) and (2005, 20100).
(*Source:* The College Board.)

2. *Rate of Change in Tuition* Calculate the slope of the line in your graph. Interpret this slope as a rate of change.

3. *Modeling Tuition* Find the slope–intercept form of the line in your sketch. What is the y-intercept and does it have meaning in this situation?

4. *Predicting Tuition* Use your equation to estimate tuition and fees in 2001 and compare it to the known value of $17,300. Estimate tuition and fees in 2010.

5. *Public Tuition* In 2000, the average cost of tuition and fees at *public* four-year colleges was $3500, and in 2005 it was $5100. Repeat Exercises 1–4 for these data. Note that the known value for 2001 is $3700.
(*Source:* The College Board.)

3.6 POINT–SLOPE FORM

Derivation of Point–Slope Form ■ Finding Point–Slope Form ■ Applications

A LOOK INTO MATH ▷ In 1995, there were 690 female officers in the Marine Corps, and by 2003 this number had increased to about 1090. This growth is illustrated in Figure 3.49 on the next page, where the line passes through the points (1995, 690) and (2003, 1090). Because two points determine a unique line, we can find the equation of this line and use it to *estimate* the number of female officers in other years. In this section we discuss how to find this equation by using the *point–slope form*, rather than the slope–intercept form of the equation of a line.
(*Source:* Department of Defense.)

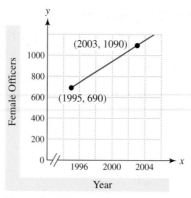

Figure 3.49 Female Officers in the Marine Corps

Derivation of Point–Slope Form

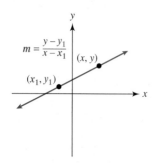

Figure 3.50 Slope of a Line

If we know the slope and y-intercept of a line, we can write its slope–intercept form, $y = mx + b$, which is an example of an **equation of a line**. The point–slope form is a different type of equation of a line.

Suppose that a (nonvertical) line with slope m passes through the point (x_1, y_1). If (x, y) is a different point on this line, then $m = \dfrac{y - y_1}{x - x_1}$. See Figure 3.50. Using this slope formula, we can find the point–slope form.

$$m = \frac{y - y_1}{x - x_1} \qquad \text{Slope formula}$$

$$m \cdot (x - x_1) = \frac{y - y_1}{x - x_1} \cdot (x - x_1) \qquad \text{Multiply each side by } (x - x_1).$$

$$m(x - x_1) = y - y_1 \qquad \text{Simplify.}$$

$$y - y_1 = m(x - x_1) \qquad \text{Rewrite the equation.}$$

$$y = m(x - x_1) + y_1 \qquad \text{Add } y_1 \text{ to each side.}$$

The equation $y - y_1 = m(x - x_1)$ is traditionally called the *point–slope form*. An equivalent form that is helpful when graphing is $y = m(x - x_1) + y_1$. Both equations are in *point–slope form*.

> ### POINT–SLOPE FORM
>
> The line with slope m passing through the point (x_1, y_1) is given by
>
> $$y - y_1 = m(x - x_1), \quad \text{or equivalently,}$$
>
> $$y = m(x - x_1) + y_1,$$
>
> the **point–slope form** of a line.

Finding Point–Slope Form

In the next example we find a point–slope form for a line. Note that *any* point that lies on the line can be used in its point–slope form.

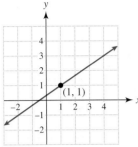

Figure 3.51

EXAMPLE 1 Finding a point–slope form

Use the labeled point in Figure 3.51 to write a point–slope form for the line and then simplify it to the slope–intercept form.

Solution

The graph rises 2 units for each 3 units of horizontal run, so the slope is $\frac{2}{3}$. Let $m = \frac{2}{3}$ and $(x_1, y_1) = (1, 1)$ in the point–slope form.

$$y - y_1 = m(x - x_1) \qquad \text{Point–slope form}$$

$$y - 1 = \frac{2}{3}(x - 1) \qquad \text{Let } m = \frac{2}{3}, x_1 = 1, \text{ and } y_1 = 1.$$

A point–slope form for the line is $y - 1 = \frac{2}{3}(x - 1)$.

To simplify to the slope–intercept form, apply the distributive property.

$$y - 1 = \frac{2}{3}x - \frac{2}{3} \qquad \text{Distributive property}$$

$$y = \frac{2}{3}x + \frac{1}{3} \qquad \text{Add 1, or } \frac{3}{3}, \text{ to each side.}$$

The slope–intercept form is $y = \frac{2}{3}x + \frac{1}{3}$.

Now Try Exercise **15**

EXAMPLE 2 Finding a point–slope form

Find a point–slope form for a line passing through the point $(-2, 3)$ with slope $-\frac{1}{2}$. Does the point $(2, 1)$ lie on this line?

Solution

Let $m = -\frac{1}{2}$ and $(x_1, y_1) = (-2, 3)$ in the point–slope form.

$$y - y_1 = m(x - x_1) \qquad \text{Point–slope form}$$

$$y - 3 = -\frac{1}{2}\left(x - (-2)\right) \qquad x_1 = -2, y_1 = 3, \text{ and } m = -\frac{1}{2}$$

$$y - 3 = -\frac{1}{2}(x + 2) \qquad \text{Simplify.}$$

To determine whether $(2, 1)$ lies on the line, substitute $x = 2$ and $y = 1$ in the equation.

$$1 - 3 \stackrel{?}{=} -\frac{1}{2}(2 + 2) \qquad \text{Let } x = 2 \text{ and } y = 1.$$

$$-2 \stackrel{?}{=} -\frac{1}{2}(4) \qquad \text{Simplify.}$$

$$-2 = -2 \qquad \text{A true statement}$$

The point $(2, 1)$ lies on the line because it satisfies the point–slope form.

Now Try Exercises **9**, **19**

In the next example we use the point–slope form to find the equation of a line passing through two points.

EXAMPLE 3   Finding an equation of a line

Use the point–slope form to find an equation of the line passing through the points $(1, -4)$ and $(-2, 5)$.

Solution

Before we can apply the point–slope form, we must find the slope.

$$m = \frac{y_2 - y_1}{x_2 - x_1} \qquad \text{Slope formula}$$

$$= \frac{5 - (-4)}{-2 - 1} \qquad \begin{array}{l} x_1 = 1, y_1 = -4, x_2 = -2, \\ \text{and } y_2 = 5 \end{array}$$

$$= -3 \qquad \text{Simplify.}$$

We can let either the point $(1, -4)$ or the point $(-2, 5)$ be (x_1, y_1) in the point–slope form. If $(x_1, y_1) = (1, -4)$, then the equation of the line becomes the following.

$$y - y_1 = m(x - x_1) \qquad \text{Point–slope form}$$
$$y - (-4) = -3(x - 1) \qquad \begin{array}{l} x_1 = 1, y_1 = -4, \text{ and} \\ m = -3 \end{array}$$
$$y + 4 = -3(x - 1) \qquad \text{Simplify.}$$

If we let $(x_1, y_1) = (-2, 5)$, the point–slope form becomes

$$y - 5 = -3(x + 2). \qquad \boxed{\text{Now Try Exercise } 25}$$

Although the two point–slope forms in Example 3 might appear to be different, they actually are equivalent because they simplify to the same slope–intercept form.

$y + 4 = -3(x - 1)$	$y - 5 = -3(x + 2)$	Point–slope forms
$y = -3(x - 1) - 4$	$y = -3(x + 2) + 5$	Addition property
$y = -3x + 3 - 4$	$y = -3x - 6 + 5$	Distributive property
$y = -3x - 1$	$y = -3x - 1$	Identical slope–intercept forms

MAKING CONNECTIONS

Slope–Intercept and Point–Slope Forms

The slope–intercept form, $y = mx + b$, is unique because any nonvertical line has one slope m and one y-intercept b. The point–slope form, $y - y_1 = m(x - x_1)$, is *not* unique because (x_1, y_1) can be any point that lies on the line. However, any point–slope form can be simplified to a unique slope–intercept form.

EXAMPLE 4 Finding equations of lines

Find the slope–intercept form for the line that satisfies the conditions.
(a) Slope $\frac{1}{2}$, passing through $(-2, 4)$
(b) x-intercept -3, y-intercept 2
(c) Perpendicular to $y = -\frac{2}{3}x$, passing through $\left(\frac{2}{3}, 3\right)$

Solution
(a) Substitute $m = \frac{1}{2}$, $x_1 = -2$, and $y_1 = 4$ in the point–slope form.

$$y - y_1 = m(x - x_1) \qquad \text{Point–slope form}$$

$$y - 4 = \frac{1}{2}(x + 2) \qquad \text{Substitute and simplify.}$$

$$y - 4 = \frac{1}{2}x + 1 \qquad \text{Distributive property}$$

$$y = \frac{1}{2}x + 5 \qquad \text{Add 4 to each side.}$$

(b) The line passes through the points $(-3, 0)$ and $(0, 2)$. The slope of the line is

$$m = \frac{2 - 0}{0 - (-3)} = \frac{2}{3}.$$

Because the line has slope $\frac{2}{3}$ and y-intercept 2, the slope–intercept form is

$$y = \frac{2}{3}x + 2.$$

(c) The slope of the given line is $m_1 = -\frac{2}{3}$, so the slope of a line perpendicular to it is $\frac{3}{2}$. Let $m = \frac{3}{2}$, $x_1 = \frac{2}{3}$, and $y_1 = 3$ in the point–slope form.

$$y - y_1 = m(x - x_1) \qquad \text{Point–slope form}$$

$$y - 3 = \frac{3}{2}\left(x - \frac{2}{3}\right) \qquad \text{Substitute.}$$

$$y - 3 = \frac{3}{2}x - 1 \qquad \text{Distributive property}$$

$$y = \frac{3}{2}x + 2 \qquad \text{Add 3 to each side.}$$

Now Try Exercises 45, 49, 53

In the next example the point–slope form is used to find the slope–intercept form of a line that passes through several points given in a table.

EXAMPLE 5 Using a table to find slope–intercept form

The points in the table lie on a line. Find the slope–intercept form of the line.

x	2	4	6	8
y	2	1	0	-1

Solution
The y-values in the table decrease one unit for every two-unit increase in the x-values so the line has a "rise" of -1 when the run is 2. The slope is $m = \frac{\text{rise}}{\text{run}} = \frac{-1}{2} = -\frac{1}{2}$. Because the

x	y
2	2
4	1
6	0
8	−1

y-intercept is not given in the table, the value of b in the equation $y = mx + b$ is not immediately available. However, *any* point from the table, which is repeated in vertical form in the margin, can be used to obtain a point–slope form of the line, which can then be simplified to slope–intercept form. Letting $(x_1, y_1) = (2, 2)$ and $m = -\frac{1}{2}$ in the point–slope form yields the following result.

$$y - y_1 = m(x - x_1) \qquad \text{Point–slope form}$$

$$y - 2 = -\frac{1}{2}(x - 2) \qquad \text{Substitute.}$$

$$y - 2 = -\frac{1}{2}x + 1 \qquad \text{Distributive property}$$

$$y = -\frac{1}{2}x + 3 \qquad \text{Add 2 to each side.}$$

This result can be checked by substituting each x-value from the table in the equation. For example, when $x = 4$, the corresponding y-value is $y = -\frac{1}{2}(4) + 3 = -2 + 3 = 1$. This agrees with the table.

Now Try Exercise **55**

NOTE: The equation in Example 5 can be obtained without using the point–slope form. The value of b can be found by letting $(x, y) = (2, 2)$ and $m = -\frac{1}{2}$ in the slope–intercept form.

$$y = mx + b \qquad \text{Slope–intercept form}$$

$$2 = -\frac{1}{2}(2) + b \qquad \text{Substitute.}$$

$$2 = -1 + b \qquad \text{Multiply.}$$

$$3 = b \qquad \text{Add 1 to each side.}$$

Because $m = -\frac{1}{2}$ and $b = 3$, the slope–intercept form is $y = -\frac{1}{2}x + 3$.

Applications

In the next example we find the equation of the line that models the Marine Corps data presented at the beginning of this section.

EXAMPLE **6**   Modeling numbers of female officers

In 1995, there were 690 female officers in the Marine Corps, and by 2003 this number had increased to about 1090. Refer to Figure 3.49 at the beginning of this section.
(a) Use the point (1995, 690) to find a point–slope form of the line shown in Figure 3.49.
(b) Interpret the slope as a rate of change.
(c) Use Figure 3.49 to estimate the number of female officers in 2000. Then use your equation from part (a) to approximate this number. How do your answers compare?

Solution
(a) The slope of the line passing through (1995, 690) and (2003, 1090) is

$$m = \frac{1090 - 690}{2003 - 1995} = 50.$$

If we let $x_1 = 1995$ and $y_1 = 690$, then the point–slope form becomes

$$y - 690 = 50(x - 1995) \quad \text{or} \quad y = 50(x - 1995) + 690.$$

(b) Slope $m = 50$ indicates that the number of female officers increased, *on average,* by about 50 officers per year.

(c) From Figure 3.49, it appears that the number of female officers in 2000 was about 950. To estimate this value let $x = 2000$ in the equation found in part (a).

$$y = 50(2000 - 1995) + 690 = 940$$

Although the graphical estimate and calculated answers are not exactly equal, they are approximately equal. Estimations made from a graph usually are not exact.

Now Try Exercise 67

In the next example we review several concepts of lines.

EXAMPLE 7   Modeling water in a pool

A small swimming pool is being emptied by a pump that removes water at a constant rate. After 1 hour the pool contains 5000 gallons, and after 3 hours it contains 3000 gallons.
(a) How fast is the pump removing water?
(b) Find the slope–intercept form of a line that models the amount of water in the pool. Interpret the slope.
(c) Find the y-intercept and the x-intercept. Interpret each.
(d) Sketch a graph of the amount of water in the pool during the first 6 hours.
(e) The point $(2, 4000)$ lies on the graph. Explain its meaning.

Solution

CRITICAL THINKING

Suppose that a line models the amount of water in a swimming pool. What does a positive slope indicate? What does a negative slope indicate?

(a) The pump removes $5000 - 3000 = 2000$ gallons of water in 2 hours, or 1000 gallons per hour.
(b) The line passes through the points $(1, 5000)$ and $(3, 3000)$, so the slope is

$$m = \frac{3000 - 5000}{3 - 1} = -1000.$$

One way to find the slope–intercept form is to use the point–slope form.

$$y - y_1 = m(x - x_1) \qquad \text{Point–slope form}$$
$$y - 5000 = -1000(x - 1) \qquad m = -1000, x_1 = 1, \text{ and } y_1 = 5000$$
$$y - 5000 = -1000x + 1000 \qquad \text{Distributive property}$$
$$y = -1000x + 6000 \qquad \text{Add 5000 to each side.}$$

Slope -1000 means that the pump is *removing* 1000 gallons per hour.
(c) The y-intercept is 6000 and indicates that the pool initially contained 6000 gallons. To find the x-intercept let $y = 0$ in the slope–intercept form.

$$0 = -1000x + 6000 \qquad \text{Let } y = 0.$$
$$1000x = 6000 \qquad \text{Add 1000x to each side.}$$
$$x = \frac{6000}{1000} \qquad \text{Divide by 1000.}$$
$$x = 6 \qquad \text{Simplify.}$$

An x-intercept of 6 indicates that the pool is empty after 6 hours.
(d) The x-intercept is **6**, and the y-intercept is **6000**. Sketch a line passing through $(6, 0)$ and $(0, 6000)$, as shown in Figure 3.52.
(e) The point $(2, 4000)$ indicates that after 2 hours the pool contains 4000 gallons of water.

Now Try Exercise 63

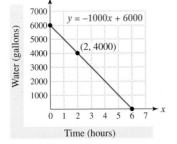

Figure 3.52 Water in a Pool

3.6 PUTTING IT ALL TOGETHER

The following table summarizes the point–slope form of a line. Note that the slope–intercept form is unique for a given line, whereas the point–slope form depends on the point used for (x_1, y_1).

Concept	Comments	Example
Point–Slope Form $$y - y_1 = m(x - x_1) \quad \text{or}$$ $$y = m(x - x_1) + y_1$$	Used to find an equation of a line, given two points or one point and the slope Can always be simplified to slope–intercept form	For two points $(1, 2)$ and $(3, 5)$, first compute $m = \frac{5 - 2}{3 - 1} = \frac{3}{2}$. An equation of this line is $$y - 2 = \frac{3}{2}(x - 1) \quad \text{or}$$ $$y = \frac{3}{2}(x - 1) + 2.$$

3.6 Exercises

CONCEPTS

1. How many lines are determined by two distinct points?

2. How many lines are determined by a point and a slope?

3. Give the slope–intercept form of a line.

4. Give the point–slope form of a line.

5. If the point–slope form is used to write an equation of a line that passes through $(1, 3)$, then $x_1 = $ _____ and $y_1 = $ _____.

6. To write a point–slope equation in slope–intercept form, use the _____ property to clear the parentheses.

7. Is the slope–intercept form of a line unique? Explain.

8. Is the point–slope form of a line unique? Explain.

POINT–SLOPE FORM

Exercises 9–14: Determine whether the given point lies on the line.

9. $(-3, 3)$ $y - 1 = -\frac{2}{3}x$

10. $(4, 0)$ $y + 1 = \frac{1}{4}x$

11. $(1, 4)$ $y - 3 = -(x - 1)$

12. $(3, -11)$ $y + 1 = -2(x + 3)$

13. $(0, 4)$ $y = \frac{1}{2}(x + 4) + 2$

14. $(2, -8)$ $y = 3(x - 5) - 1$

Exercises 15–18: Use the labeled point to write a point–slope form for the line.

15.

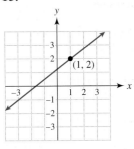

16.

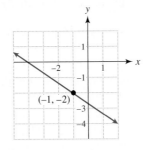

17.

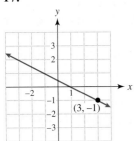

18.

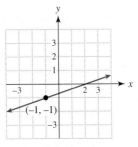

Exercises 19–30: Find a point–slope form for the line that satisfies the stated conditions. When two points are given, use the first point in the point–slope form.

19. Slope 4, passing through $(-3, 1)$

20. Slope -3, passing through $(1, -2)$

21. Slope $\frac{1}{2}$, passing through $(-5, -3)$

22. Slope $-\frac{2}{3}$, passing through $(-3, 6)$

23. Slope 1.5, passing through $(2000, 30)$

24. Slope -10, passing through $(2004, 100)$

25. Passing through $(2, 4)$ and $(-1, -3)$

26. Passing through $(3, -1)$ and $(1, -4)$

27. Passing through $(5, 0)$ and $(0, -3)$

28. Passing through $(-2, 0)$ and $(0, -1)$

29. Passing through $(1990, 15)$ and $(2000, 65)$

30. Passing through $(1999, 5)$ and $(2004, 30)$

Exercises 31–42: Write the point–slope form in slope–intercept form.

31. $y - 4 = 3(x - 2)$ **32.** $y - 3 = -2(x + 1)$

33. $y + 2 = \frac{1}{3}(x + 6)$ **34.** $y - 1 = \frac{2}{5}(x + 10)$

35. $y - \frac{3}{4} = \frac{2}{3}(x - 1)$ **36.** $y + \frac{2}{3} = -\frac{1}{6}(x - 2)$

37. $y = -2(x - 2) + 5$ **38.** $y = 4(x + 3) - 7$

39. $y = \frac{3}{5}(x - 5) + 1$ **40.** $y = -\frac{1}{2}(x + 4) - 6$

41. $y = -16(x + 1.5) + 5$

42. $y = -15(x - 1) + 100$

43. Thinking Generally   Find the y-intercept of the line given by $y - y_1 = m(x - x_1)$.

44. Thinking Generally   Find the x-intercept of the line given by $y - y_1 = m(x - x_1)$.

Exercises 45–54: Find the slope–intercept form for the line satisfying the conditions.

45. Slope -2, passing through $(4, -3)$

46. Slope $\frac{1}{5}$, passing through $(-2, 5)$

47. Passing through $(3, -2)$ and $(2, -1)$

48. Passing through $(8, 3)$ and $(-7, 3)$

49. x-intercept 3, y-intercept $\frac{1}{3}$

50. x-intercept 2, y-intercept -3

51. Parallel to $y = 2x - 1$, passing through $(2, -3)$

52. Parallel to $y = -\frac{3}{2}x$, passing through $(0, 20)$

53. Perpendicular to $y = -\frac{1}{2}x + 3$, passing through the point $(6, -3)$

54. Perpendicular to $y = \frac{3}{5}(x + 1) + 3$, passing through the point $(1, -2)$

Exercises 55–58: The points in the table lie on a line. Find the slope–intercept form of the line.

55.

x	1	2	3	4
y	-3	-5	-7	-9

56.

x	2	3	4	5
y	5	8	11	14

57.

x	-1	1	3	5
y	-3	-2	-1	0

58.

x	-1	5	11	17
y	1	-3	-7	-11

GRAPHICAL INTERPRETATION

59. *Distance and Speed*   A person is driving a car along a straight road. The graph shows the distance y in miles that the driver is from home after x hours.
  (a) Is the person traveling toward or away from home?
  (b) The graph passes through (1, 250) and (4, 100). Discuss the meaning of these points.
  (c) How fast is the driver traveling?
  (d) Find the slope–intercept form of the line. Interpret the slope as a rate of change.

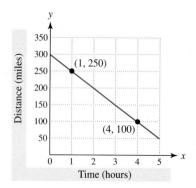
Time (hours)

60. *Distance and Speed*   A person rides a bicycle at 10 miles per hour, first away from home for 1 hour and then toward home for 1 hour. Sketch a graph that shows the distance d between the bicyclist and home after x hours.

61. *Water and Flow*   The graph shows the amount of water y in a 500-gallon tank after x minutes have elapsed.
  (a) Is water entering or leaving the tank? How much water is in the tank after 4 minutes?
  (b) Find the y-intercept. Explain its meaning.
  (c) Find the slope–intercept form of the line. Interpret the slope as a rate of change.
  (d) After how many minutes will the tank be full?

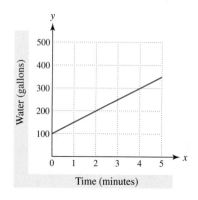
Time (minutes)

62. *Water and Flow*   A hose is used to fill a 100-gallon barrel. If the hose delivers 5 gallons of water per minute, sketch a graph of the amount A of water in the barrel during the first 20 minutes.

63. *Change in Temperature*   The outside temperature was 40°F at 1 A.M. and 15°F at 6 A.M. Assume that the temperature changed at a constant rate.
  (a) At what rate did the temperature change?
  (b) Find the slope–intercept form of a line that models the temperature T at x A.M. Interpret the slope as a rate of change.
  (c) Assuming that your equation is valid for times after 6 A.M., find and interpret the x-intercept.
  (d) Sketch a graph that shows the temperature from 1 A.M. to 9 A.M.
  (e) The point (4, 25) lies on the graph. Explain its meaning.

64. *Cost of Fuel*   The cost of buying 5 gallons of fuel oil is $12 and the cost of buying 15 gallons of fuel oil is $36.
  (a) What is the cost of a gallon of fuel oil?
  (b) Find the slope–intercept form of a line that models the cost of buying x gallons of fuel oil. Interpret the slope as a rate of change.
  (c) Find and interpret the x-intercept.
  (d) Sketch a graph that shows the cost of buying 20 gallons or less of fuel oil.
  (e) The point (11, 26.40) lies on the graph. Explain its meaning.

APPLICATIONS

65. *Inmates*   From 1995 to 2005, the number of state and federal inmates y in millions can be modeled by $y = 0.032(x - 1995) + 1.13$, where x is the year. (***Source:*** Bureau of Justice.)
  (a) Find the number of inmates in 1995 and 2005.
  (b) What is the slope of the graph of y? Interpret the slope as a rate of change.

66. *Municipal Waste*   From 1960 to 2000, municipal solid waste y in millions of tons can be modeled by $y = 3.6(x - 1960) + 87.8$, where x is the year. (***Source:*** Environmental Protection Agency.)
  (a) Find the tons of waste in 1999.
  (b) What is the slope of the graph of y? Interpret the slope as a rate of change.

67. *Tuition and Fees*   The graph models average tuition and fees at public four-year colleges from 2000 to 2006. (**Source:** The College Board.)

   (a) The line passes through the points (2002, 4081) and (2004, 5133). Explain the meaning of each point.

   (b) Find a point–slope form for this line. Interpret the slope as a rate of change.

   (c) Use the graph to estimate tuition and fees in 2005. Then use your equation from part (b) to estimate tuition and fees in 2005.

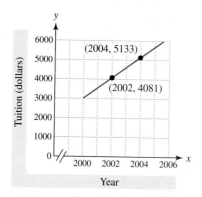

Year

68. *Western Population*   In 1950 the western region of the United States had a population of 20.2 million, and in 1960 this population was 28.4 million. (**Source:** U.S. Census Bureau.)

   (a) Find the slope–intercept form of a line passing through (1950, 20.2) and (1960, 28.4).

   (b) Use your equation from part (a) to estimate the western population in 1967.

69. *HIV Infection Rates*   In 2004, there were an estimated 39 million HIV infections worldwide with an annual infection rate of 4.9 million. (**Source:** U.S. Centers for Disease Control and Prevention.)

   (a) Find the slope–intercept form of a line that approximates the cumulative number of HIV infections in millions during year x, where $x \geq 2004$.

   (b) Estimate the number of HIV infections in 2010.

70. *Farm Pollution*   In 1978, the number of farm pollution incidents reported in England and Wales was about 1500, and in 1988 it was about 4000. (**Source:** C. Mason, *Biology of Freshwater Pollution*.)

   (a) Find a slope–intercept form of a line passing through (1978, 1500) and (1988, 4000).

   (b) Use your equation from part (a) to estimate the number of incidents in 1986.

WRITING ABOUT MATHEMATICS

71. Explain how to find the equation of a line passing through two points with coordinates (x_1, y_1) and (x_2, y_2). Give an example.

72. Explain how slope is related to rate of change. Give an example.

CHECKING BASIC CONCEPTS
SECTIONS 3.5 AND 3.6

1. Write the slope–intercept form for the line shown in the graph.

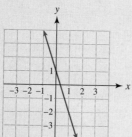

2. Write the equation $4x - 5y = 20$ in slope–intercept form. Give the slope and y-intercept.

3. Graph $y = \frac{1}{2}x - 3$.

4. Write the slope–intercept form of a line that satisfies the following.

   (a) Slope 3, passing through $(0, -2)$

   (b) Perpendicular to $y = \frac{2}{3}x$, passing through the point $(-2, 3)$

   (c) Passing through $(1, -4)$ and $(-2, 3)$

5. Write a point–slope form for a line with slope -2, passing through $(-1, 3)$.

6. Write the equation $y + 3 = -2(x - 2)$ in slope–intercept form.

continued on next page

continued from previous page

7. Find the slope–intercept form of the line passing through the points in the table.

x	−3	−1	1	3
y	−3	1	5	9

8. *Distance and Speed*   A bicyclist is riding at a constant speed and is 36 miles from home at 1 P.M. Two hours later the bicyclist is 12 miles from home.
   (a) Find the slope–intercept form of a line passing through (1, 36) and (3, 12).

(b) How fast is the bicyclist traveling?
(c) When will the bicyclist arrive home?
(d) How far was the bicyclist from home at noon?

9. *Snowfall*   The total amount of snowfall S in inches t hours past noon is given by $S = 2t + 5$.
   (a) How many inches of snow fell by noon?
   (b) At what rate did snow fall in the afternoon?
   (c) What is the S-intercept for the graph of this equation? What does it represent?
   (d) What is the slope for the graph of this equation? What does it represent?

3.7   INTRODUCTION TO MODELING

Basic Concepts ■ Modeling Linear Data

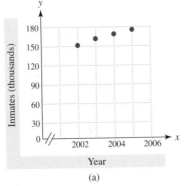

A LOOK INTO MATH ▷

Five-Day Forecast

63°	58°	56°	60°	62°
Mon	Tues	Wed	Thu	Fri

For centuries people have tried to understand the world around them by creating models. For example, a weather forecast is based on a model. Mathematics is used to create these weather models, which often contain thousands of equations.

A model is an *abstraction* of something that people observed. Not only should a good model describe *known* data, but it should also be able to predict *future* data. In this section we discuss linear models, which are used to describe data that have a constant rate of change.

Basic Concepts

▶ **REAL-WORLD CONNECTION**   Figure 3.53(a) shows a scatterplot of the number of inmates in the federal prison system from 2002 to 2005. The four points in the graph appear to be "nearly" collinear. That is, they appear almost to lie on the same line. Using mathematical modeling, we can find an equation for such a line. Once we have found it, we can use it to make estimates about the federal inmate population. An example of such a line is shown in Figure 3.53(b). (See Exercise 51 at the end of this section.)

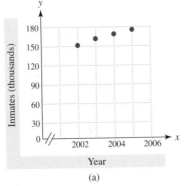
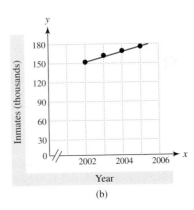

Figure 3.53 Federal Prison Population

Generally, mathematical models are not exact representations of data. Data that might appear to be linear may not be. Although the line in Figure 3.53(b) appears to touch every point, it does not pass through all four points exactly. Figure 3.54(a) shows data modeled *exactly* by a line, whereas Figure 3.54(b) shows data modeled *approximately* by a line. In applications a model is more apt to be approximate than exact.

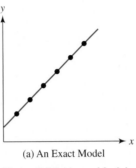

(a) An Exact Model   (b) An Approximate Model

Figure 3.54 Linear Models

EXAMPLE **1**   **Determining whether a model is exact**

A person can vote in the United States at age 18 or over. Table 3.19 shows the voting-age population P in millions for selected years x. Does the equation $P = 3.25x - 6297$ model the data exactly? Explain.

TABLE 3.19

Voting-Age Population

x	2000	2002	2004
P	203	211	216

Source: U.S. Census Bureau.

Solution

To determine whether the equation models the data exactly, let $x = 2000$, 2002, and 2004 in the given equation.

$$x = \mathbf{2000}: \qquad P = 3.25(\mathbf{2000}) - 6297 = 203$$

$$x = \mathbf{2002}: \qquad P = 3.25(\mathbf{2002}) - 6297 = 209.5$$

$$x = \mathbf{2004}: \qquad P = 3.25(\mathbf{2004}) - 6297 = 216$$

The model is *not exact* because it does not predict a voting-age population of 211 million in 2002.

Now Try Exercise **15**

Modeling Linear Data

A line can model linear data. In the next example we use a line to model gas mileage.

EXAMPLE 2   **Determining gas mileage**

TABLE 3.20

x	2	4	6	8
y	30	60	90	120

Table 3.20 shows the number of miles y traveled by an SUV on x gallons of gasoline.
(a) Plot the data in the xy-plane. Be sure to label each axis.
(b) Sketch a line that models the data. (You may want to use a ruler.)
(c) Find the equation of the line and interpret the slope of the line.
(d) How far could this SUV travel on 11 gallons of gasoline?

Solution
(a) Plot the points (2, 30), (4, 60), (6, 90), and (8, 120), as shown Figure 3.55(a).

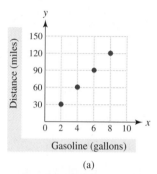 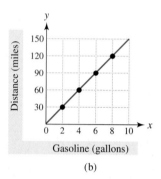

(a)   (b)

Figure 3.55 Gas Mileage of an SUV

(b) Sketch a line similar to the one shown in Figure 3.55(b). This particular line passes through each data point.
(c) First find the slope m of the line by choosing two points that the line passes through, such as (**2, 30**) and (8, 120).

$$m = \frac{120 - 30}{8 - 2} = \frac{90}{6} = 15$$

Now find the equation of the line passing through (2, 30) with slope 15.

$y - y_1 = m(x - x_1)$	Point–slope form
$y - 30 = 15(x - 2)$	$x_1 = 2, y_1 = 30$, and $m = 15$
$y - 30 = 15x - 30$	Distributive property
$y = 15x$	Add 30 to each side.

The data are modeled by the equation $y = 15x$. Slope 15 indicates that the mileage of this SUV is 15 miles per gallon.
(d) On 11 gallons of gasoline the SUV could go $y = 15(11) = 165$ miles.

Now Try Exercise 53

EXAMPLE 3   **Modeling linear data**

Table 3.21 contains ordered pairs that can be modeled approximately by a line.
(a) Plot the data. Could a line pass through all five points?
(b) Sketch a line that models the data and then determine its equation.

TABLE 3.21

x	1	2	3	4	5
y	3	5	6	10	11

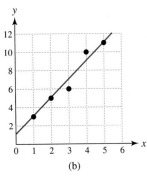

Figure 3.56

Solution

(a) Plot the ordered pairs $(1, 3)$, $(2, 5)$, $(3, 6)$, $(4, 10)$, and $(5, 11)$, as shown in Figure 3.56(a). The points are not collinear, so it is impossible to sketch a line that passes through *all* five points.

(b) One possibility for a line is shown in Figure 3.56(b). This line passes through three of the five points, and is above one point and below another point. To determine the equation of this line, pick two points that the line passes through. For example, the points $(1, 3)$ and $(5, 11)$ lie on the line. The slope of this line is

$$m = \frac{11 - 3}{5 - 1} = \frac{8}{4} = 2.$$

The equation of the line passing through $(1, 3)$ with slope 2 can be found as follows.

$y - y_1 = m(x - x_1)$	Point–slope form
$y - 3 = 2(x - 1)$	$x_1 = 1, y_1 = 3$, and $m = 2$
$y - 3 = 2x - 2$	Distributive property
$y = 2x + 1$	Add 3 to each side.

The equation of this line is $y = 2x + 1$.

Now Try Exercise 33

NOTE: The equation found in Example 3 represents one possible linear model. Other linear models are possible.

▶ **REAL-WORLD CONNECTION** When a quantity increases at a constant rate, it can be modeled with the linear equation $y = mx + b$. This concept is illustrated in the next example.

EXAMPLE 4 Modeling worldwide HIV/AIDS cases in children

At the beginning of 2004, a total of 2.5 million children (under age 15) were living with HIV/AIDS. The rate of new infections was 0.7 million per year.

(a) Write a linear equation $C = mx + b$ that models the total number of children C in millions that were living with HIV/AIDS, x years after January 1, 2004.

(b) Estimate C at the beginning of 2010.

Solution

(a) In the equation $C = mx + b$, the rate of change in HIV infections corresponds to the slope m, and the initial number of cases at the beginning of 2004 corresponds to b. Therefore the equation $C = 0.7x + 2.5$ models the data.

(b) The beginning of 2010 is 6 years after January 1, 2004, so let $x = 6$.

$$C = 0.7(6) + 2.5 = 6.7 \text{ million}$$

NOTE: There were a total of 2.5 million children living with HIV/AIDS at the beginning of 2004 with an infection rate of 0.7 million per year. After 6 years there would be an additional $0.7(6) = 4.2$ million infected children, raising the total number to $2.5 + 4.2 = 6.7$ million.

Now Try Exercise 49

EXAMPLE **5**   **Modeling with linear equations**

Find a linear equation in the form $y = mx + b$ that models the quantity y after x days.
(a) A quantity y is initially 500 and increases at a rate of 6 per day.
(b) A quantity y is initially 1800 and decreases at a rate of 25 per day.
(c) A quantity y is initially 10,000 and remains constant.

Solution
(a) In the equation $y = mx + b$, the y-intercept b represents the initial amount and the slope m represents the rate of change. Therefore $y = 6x + 500$.
(b) The quantity y is decreasing at the rate of 25 per day with an initial amount of 1800, so $y = -25x + 1800$.
(c) The quantity is constant, so $m = 0$. The equation is $y = 10,000$.

Now Try Exercises   27, 29, 31

MODELING WITH A LINEAR EQUATION

To model a quantity y that has a constant rate of change, use the equation

$$y = mx + b,$$

where $m = $ (constant rate of change) and $b = $ (initial amount).

3.7   PUTTING IT ALL TOGETHER

The following table summarizes important concepts related to linear modeling.

Concept	Comments	Example
Linear Model	Used to model a quantity that has a constant rate of change	If a total of 2 inches of rain falls before noon, and if rain falls at the rate of $\frac{1}{2}$ inch per hour, then $y = \frac{1}{2}x + 2$ models the total rainfall x hours past noon.
Modeling Linear Data with a Line	1. Plot the data. 2. Sketch a line that either passes through or nearly through the points. 3. Pick two points on the line and find the equation of the line.	To model $(0, 4)$, $(1, 3)$, and $(2, 2)$, plot the points and sketch a line as shown in the accompanying figure. Many times one line cannot pass through all the points. The equation of the line is $y = -x + 4$.

3.7 Exercises

CONCEPTS

1. A model is a(n) _____ of something that people observe.

2. A good model should be able to _____ future data.

3. Linear data are modeled by a(n) _____ equation.

4. If a line passes through all the data points, it is a(n) _____ model.

5. If a line passes near, but not through each data point, it is a(n) _____ model.

6. Linear models are used to describe data that have a(n) _____ rate of change.

7. If a quantity is modeled by the equation $y = mx + b$, then m represents the _____.

8. If a quantity is modeled by the equation $y = mx + b$, then b represents the _____.

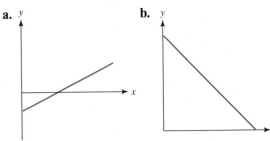

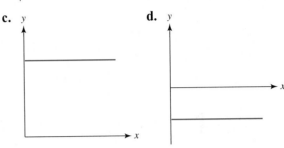

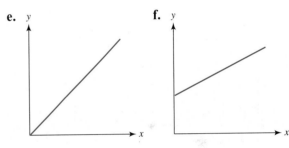

Exercises 9–14: Modeling   Match the situation to the graph (a.–f. in the right-hand column) that models it best.

9. College tuition from 1980 to 2000

10. Yearly average temperature in degrees Celsius at the North Pole

11. Profit from selling boxes of candy if it costs $200 to make the candy

12. Height of Mount Hood in Oregon

13. Total amount of water delivered by a garden hose if it flows at a constant rate

14. Sales of 8-track music tapes from 1970 to 1980 (*Hint:* 8-track tapes are obsolete.)

MODELING LINEAR DATA

Exercises 15–20: State whether the ordered pairs in the table are modeled exactly by the linear equation.

15. $y = 2x + 2$

x	0	1	2
y	2	4	6

16. $y = -2x + 5$

x	0	1	2
y	5	3	0

17. $y = -4x$

x	-1	0	1
y	4	0	-8

18. $y = 5 - x$

x	-2	1	4
y	7	4	1

19. $y = 1.4x - 4$

x	0	5	10
y	-4	3	9

20. $y = -\frac{4}{3}x - \frac{13}{3}$

x	-7	-4	-1
y	5	1	-3

Exercises 21–26: State whether the linear model in the graph is exact or approximate. Then find the equation of the line.

21.

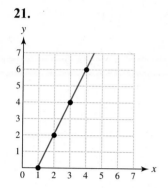

22.

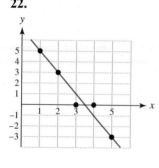

23.

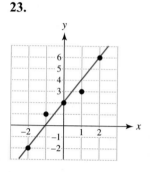

24.

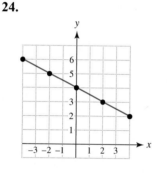

25.

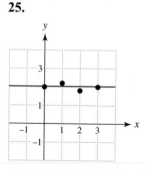

26.

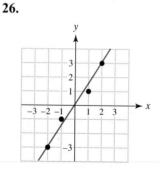

Exercises 27–32: Find an equation $y = mx + b$ that models the quantity y after x units of time.

27. A quantity y is initially 40 and increases at a rate of 5 per minute.

28. A quantity y is initially -60 and increases at a rate of 1.7 per minute.

29. A quantity y is initially -5 and decreases at a rate of 20 per day.

30. A quantity y is initially 5000 and decreases at a rate of 35 per day.

31. A quantity y is initially 8 and remains constant.

32. A quantity y is initially -45 and remains constant.

*Exercises 33–40: **Modeling Data**   For the ordered pairs in the table, do the following.*

   (a) Plot the data.   Could a line pass through all five points?

   (b) Sketch a line that models the data.

   (c) Determine an equation of the line. For data that is not exactly linear, answers may vary.

33.

x	0	1	2	3	4
y	4	2	0	-2	-4

34.

x	0	1	2	2	3
y	7	6	5	4	3

35.

x	-2	-1	0	1	2
y	4	1	0	-1	-4

36.

x	-2	-1	0	1	2
y	7	5	1	-3	-5

37.

x	-6	-4	-2	0	2
y	1	0	-1	-2	-3

38.

x	-4	-2	0	2	4
y	1	2	3	4	5

39.

x	-6	-3	0	3	6
y	-3	-2	-0.5	0	0.5

40.

x	-3	-2	-1	0	1
y	1	2	3	4	5

APPLICATIONS

Exercises 41–48: Write an equation that models the described quantity. Specify what each variable represents.

41. A barrel contains 200 gallons of water and is being filled at a rate of 5 gallons per minute.

42. A barrel contains 40 gallons of gasoline and is being drained at a rate of 3 gallons per minute.

43. An athlete has run 5 miles and is jogging at 6 miles per hour.

44. A new car has 141 miles on its odometer and is traveling at 70 miles per hour.

45. A worker has already earned $200 and is being paid $8 per hour.

46. A gambler has lost $500 and is losing money at a rate of $150 per hour.

47. A carpenter has already shingled 5 garage roofs and is shingling roofs at a rate of 1 per day.

48. A hard drive has been spinning for 2 minutes and is spinning at 7200 revolutions per minute.

49. *Kilimanjaro Glacier*   Mount Kilimanjaro is located in Tanzania, Africa, and has an elevation of 19,340 feet. In 1912, the glacier on this peak covered 5 acres. By 2002 this glacier had melted to only 1 acre. (*Source:* NBC News.)
   (a) Assume that this glacier melted at a constant rate each year. Find this *yearly* rate.
   (b) Use your answer in part (a) to write a linear equation that gives the acreage A of this glacier t years past 1912.

50. *World Population*   In 1987, the world's population reached 5 billion people, and by 2012 the world's population is expected to reach 7 billion people.
   (*Source:* U.S. Census Bureau.)
   (a) Find the average yearly increase in the world's population from 1987 to 2012.
   (b) Write a linear equation that estimates the world's population P in billions x years after 1987.

51. *Prison Population*   The data points in Figure 3.53 are (2002, 152), (2003, 162), (2004, 169), and (2005, 176) where the y-coordinates are in thousands.
   (a) Use the first and last data points to determine a line that models the data. Write the equation in slope–intercept form.
   (b) Use the line to estimate the population in 2009.

52. *Niagara Falls*   The average flow of water over Niagara Falls is 212,000 cubic feet of water per second.
   (a) Write an equation that gives the number of cubic feet of water F that flow over the falls in x seconds.

(b) How many cubic feet of water flow over Niagara Falls in 1 minute?

53. *Gas Mileage*   (Refer to Example 2.) The table shows the number of miles y traveled by a car on x gallons of gasoline.

x (gallons)	3	6	9	12
y (miles)	60	120	180	240

   (a) Plot the data in the xy-plane. Be sure to label each axis.
   (b) Sketch a line that models these data. (You may want to use a ruler.)
   (c) Calculate the slope of the line. Interpret the slope.
   (d) Find an equation of the line.
   (e) How far could this car travel on 7 gallons of gasoline?

54. *Air Temperature*   Generally, the air temperature becomes colder as the altitude above the ground increases. The table lists typical air temperatures x miles high when the ground temperature is $80°F$.

x (miles)	0	1	2	3
y (°F)	80	61	42	23

   (a) Plot the data in the xy-plane. Be sure to label each axis.
   (b) Sketch a line that models these data. (You may want to use a ruler.)
   (c) Calculate the slope of the line. Interpret the slope.
   (d) Find the slope–intercept form of the line.
   (e) Estimate the air temperature 5 miles high.

WRITING ABOUT MATHEMATICS

55. In Example 2 the gas mileage of an SUV is modeled with a linear equation. Explain why it is reasonable to use a linear equation to model this situation. (*Hint*: Compare the amounts of gasoline that the SUV uses traveling 30 miles and 60 miles.)

56. Explain the steps for finding the equation of a line that models a table of data points.

CHECKING BASIC CONCEPTS
SECTION 3.7

1. State whether the ordered pairs shown in the table are modeled exactly by $y = -5x + 10$.

x	-2	-1	0	1
y	20	15	10	5

2. State whether the linear model shown in the graph is exact or approximate. Then find the equation of the line.

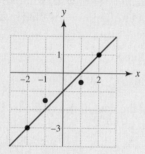

3. Find an equation, $y = mx + b$, that models the quantity y after x units of time.
   (a) A quantity y is initially 50 pounds and increases at a rate of 10 pounds per day.
   (b) A quantity y is initially 200°F and decreases at a rate of 2°F per minute.

4. The table contains ordered pairs.
   (a) Plot the data. Could a line pass through all four points?
   (b) Sketch a line that models the data.
   (c) Determine the equation of the line.

x	-2	0	2	4
y	2	1	0	-1

5. *Global Warming* Since 1945 the average annual recorded temperature on the Antarctic Peninsula has increased by 0.075°F per year.
   (a) Write an equation that models the average temperature *increase* T, x years after 1945.
   (b) Use your equation to calculate the temperature increase between 1945 and 2005.

CHAPTER 3   SUMMARY
SECTION 3.1 ■ INTRODUCTION TO GRAPHING

The Rectangular Coordinate System (xy-plane)

Points   Plotted as (x, y) ordered pairs

Four Quadrants   The x- and y-axes divide the xy-plane into quadrants I, II, III, and IV.

NOTE: A point on an axis, such as $(1, 0)$, does not lie in a quadrant.

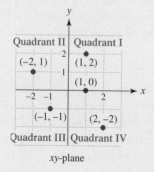

xy-plane

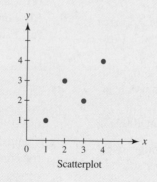

Scatterplot

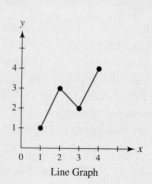

Line Graph

SECTION 3.2 ■ LINEAR EQUATIONS IN TWO VARIABLES

Equations in Two Variables   An equation with two variables and possibly some constants

Examples: $y = 3x + 7$ and $x + y = 100$

Solution to an Equation in Two Variables   The solution to an equation in two variables is an ordered pair that makes the equation a true statement.

Example: $(1, 2)$ is a solution to $2x + y = 4$ because $2(1) + 2 = 4$ is true.

Graphing a Linear Equation in Two Variables

Linear Equation $y = mx + b$ or $Ax + By = C$

Graphing   Plot at least three points and connect them with a line.

Example: $y = 2x - 1$

x	y
0	−1
1	1
2	3

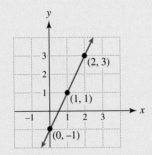

SECTION 3.3 ■ MORE GRAPHING OF LINES

Intercepts

x-Intercept   The *x*-coordinate of a point at which a graph intersects the *x*-axis; to find an *x*-intercept, let $y = 0$ in the equation and solve for *x*.

y-Intercept   The *y*-coordinate of a point at which a graph intersects the *y*-axis; to find a *y*-intercept, let $x = 0$ in the equation and solve for *y*.

Example: $x + 3y = 3$

x-intercept: Solve $x + 3(0) = 3$ to find the *x*-intercept of 3.

y-intercept: Solve $0 + 3y = 3$ to find the *y*-intercept of 1.

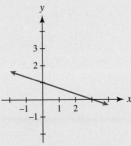

x-intercept: 3, *y*-intercept: 1

Horizontal and Vertical Lines

The equation of a horizontal line with y-intercept b is $y = b$.

The equation of a vertical line with x-intercept k is $x = k$.

Example: The horizontal line $y = -1$ has y-intercept -1 and no x-intercepts. The vertical line
$x = 2$ has x-intercept 2 and no y-intercepts.

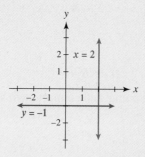

SECTION 3.4 ■ SLOPE AND RATES OF CHANGE

Slope   The ratio $\frac{\text{rise}}{\text{run}}$, or $\frac{\text{change in } y}{\text{change in } x}$, is the slope m of a line when run (change in x) is nonzero. A positive slope indicates that a line rises from left to right, and a negative slope indicates that a line falls from left to right.

Example: Slope $\frac{2}{3}$ indicates that a line rises 2 units for every 3 units of run along the x-axis. The line shown in the graph has slope $\frac{2}{3}$.

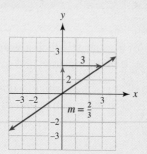

Calculating Slope   A line passing through (x_1, y_1) and (x_2, y_2) has slope

$$m = \frac{y_2 - y_1}{x_2 - x_1}, \text{ where } x_1 \neq x_2.$$

Example: The line through $(-2, 2)$ and $(3, 4)$ has slope

$$m = \frac{4 - 2}{3 - (-2)} = \frac{2}{5}.$$

Horizontal Line   Has slope 0

Vertical Line   Has undefined slope

Slope as a Rate of Change   Slope indicates how fast a line is rising or falling. In applications, slope can indicate how fast a quantity is changing.

Example: The line shown in the graph has slope -2 and depicts an initial outside temperature of $6°F$. Slope -2 indicates that the temperature is *decreasing* at a rate of $2°F$ per hour.

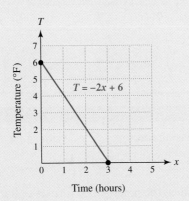

SECTION 3.5 ■ SLOPE–INTERCEPT FORM

Slope–Intercept Form

$$y = mx + b \qquad \text{Slope } m, y\text{-intercept } b$$

Example: $y = -\frac{1}{2}x + 2$ has slope $-\frac{1}{2}$ and y-intercept 2, as shown in the graph.

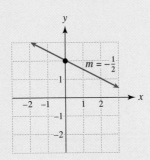

Parallel Lines   Lines with the same slope are parallel; nonvertical parallel lines have the same slope. Two vertical lines are parallel.

Example: The equations $y = -2x + 1$ and $y = -2x$ determine parallel lines because $m_1 = m_2 = -2$. See the graph on the next page.

Perpendicular Lines

If two perpendicular lines have nonzero slopes m_1 and m_2, then $m_1 \cdot m_2 = -1$.

If two lines have nonzero slopes satisfying $m_1 \cdot m_2 = -1$, then they are perpendicular.

A vertical line and a horizontal line are perpendicular.

Example: The equations $y = -\frac{1}{2}x$ and $y = 2x - 2$ determine perpendicular lines because $m_1 \cdot m_2 = -\frac{1}{2} \cdot 2 = -1$. See the graph on the next page.

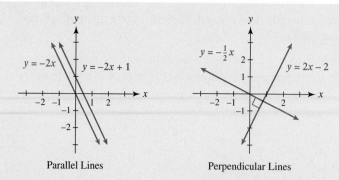

Parallel Lines   Perpendicular Lines

SECTION 3.6 ■ POINT–SLOPE FORM

Point–Slope Form   An equation of the line passing through (x_1, y_1) with slope m is

$$y - y_1 = m(x - x_1) \quad \text{or} \quad y = m(x - x_1) + y_1.$$

Example: If $m = -2$ and $(x_1, y_1) = (-2, 3)$, then the point–slope form is either

$$y - 3 = -2(x + 2) \quad \text{or} \quad y = -2(x + 2) + 3.$$

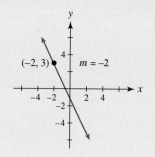

Example: The slope of the line passing through $(-2, 5)$ and $(4, 2)$ is

$$m = \frac{2 - 5}{4 - (-2)} = \frac{-3}{6} = -\frac{1}{2}.$$

Either $(-2, 5)$ or $(4, 2)$ may be used in the point–slope form. The point $(4, 2)$ results in the equation

$$y - 2 = -\frac{1}{2}(x - 4) \quad \text{or} \quad y = -\frac{1}{2}(x - 4) + 2.$$

SECTION 3.7 ■ INTRODUCTION TO MODELING

Mathematical Modeling   Mathematics is used to describe or approximate the behavior of real-life phenomena.

Exact Model   The equation describes the data precisely without error.

Example: $y = 3x$ models the data in the table exactly.

x	0	1	2	3
y	0	3	6	9

Approximate Model   The equation describes the data approximately. An approximate model occurs most often in applications.

Example: The line in the graph models the data approximately.

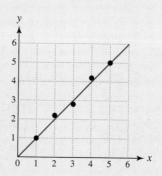

Modeling with a Linear Equation   To model a quantity y that has a constant rate of change, use the equation $y = mx + b$, where

$$m = \text{(constant rate of change)} \quad \text{and} \quad b = \text{(initial amount)}.$$

Example: If the temperature is initially $100°$F and cools at $5°$F per hour, then

$$T = -5x + 100$$

models the temperature T after x hours.

CHAPTER 3   REVIEW EXERCISES

SECTION 3.1

1. Identify the coordinates of each point in the graph. Identify the quadrant, if any, in which each point lies.

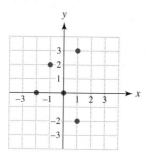

2. Make a scatterplot by plotting the following four points: $(-2, 3)$, $(-1, -1)$, $(0, 3)$, and $(2, -1)$.

Exercises 3 and 4: Use the table of xy-values to make a line graph.

3.

x	-2	-1	0	1	2
y	-3	2	-1	-2	3

4.

x	-10	-5	0	5	10
y	5	-10	10	-5	0

SECTION 3.2

Exercises 5–8: Determine whether the ordered pair is a solution for the given equation.

5. $y = x - 3$ $(6, 3)$

6. $y = 5 - 2x$ $(-2, 1)$

7. $3x - y = 3$ $(-1, 6)$

8. $\frac{1}{2}x + 2y = -8$ $(-4, -3)$

Exercises 9 and 10: Complete the table for the given equation.

9. $y = -3x$

x	-2	-1	0	1	2
y					

10. $2x + y = 5$

x					
y	-3	-1	0	1	3

Exercises 11–14: Use the given values of the variable to find solutions for the equation. Put them in a table.

11. $y = 3x + 2$ $x = -2, 0, 2, 4$

12. $y = 7 - x$ $x = 1, 2, 3, 4$

13. $y - 2x = 0$ $y = -1, 0, 1, 2$

14. $2y + x = 1$ $y = 1, 2, 3, 4$

Exercises 15–22: Graph the equation.

15. $y = 2x$ **16.** $y = x + 1$

17. $y = \frac{1}{2}x - 1$ **18.** $y = -3x + 2$

19. $x + y = 2$ **20.** $3x - 2y = 6$

21. $-4x + y = 8$ **22.** $2x + 3y = 12$

SECTION 3.3

Exercises 23 and 24: Identify the x- and y-intercepts.

23.   **24.**

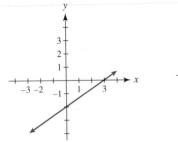   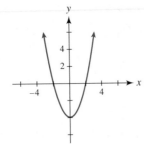

Exercises 25 and 26: Complete the table. Then determine the x- and y-intercepts for the graph of the equation.

25. $y = 2 - x$

x	-2	-1	0	1	2
y					

26. $x - 2y = 4$

x	-4	-2	0	2	4
y					

Exercises 27–30: Find any intercepts for the graph of the equation and then graph the linear equation.

27. $2x - 3y = 6$ **28.** $5x - y = 5$

29. $0.1x - 0.2y = 0.4$ **30.** $\frac{x}{2} + \frac{y}{3} = 1$

31. Graph each equation.
     (a) $y = 1$ **(b)** $x = -3$

32. Write an equation for each line shown in the graph.

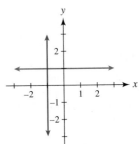

Exercises 33 and 34: Write an equation for the line that passes through the points shown in the table.

33.

x	-2	-1	0	1	2
y	1	1	1	1	1

34.

x	3	3	3	3	3
y	-2	-1	0	1	2

35. Write the equations of a horizontal line and a vertical line that pass through the point $(-2, 3)$.

36. *Distance* The distance a driver is from home is illustrated in the graph.
     (a) Find the intercepts.
     (b) Interpret each intercept.

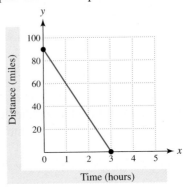

SECTION 3.4

Exercises 37–40: Find the slope, if possible, of the line passing through the two points.

37. $(2, 3), (4, 7)$

38. $(-3, 1), (2, -1)$

39. $(2, 1), (5, 1)$

40. $(-5, 6), (-5, 10)$

Exercises 41 and 42: Find the slope of the line shown in the graph.

41.

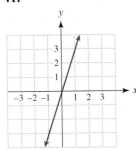

42.

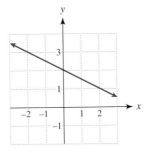

Exercises 43–46: Do the following.

(a) Graph the linear equation.

(b) What are the slope and y-intercept of the line?

43. $y = -2x$

44. $y = x - 1$

45. $x + 2y = 4$

46. $2x - 3y = -6$

Exercises 47–50: Sketch a line passing through the given point and having slope m.

47. $(0, -3), m = 2$

48. $(0, 1), m = -\frac{1}{2}$

49. $(-1, 1), m = -\frac{2}{3}$

50. $(2, 2), m = 1$

51. The table lists points located on a line. Find the slope and intercepts of the line.

x	-1	0	1	2
y	-4	-2	0	2

52. A line with slope $\frac{1}{2}$ passes through the first point shown in the table. Complete the table so that each point in the table lies on the line.

x	0	1	2	3
y	1			

SECTION 3.5

Exercises 53 and 54: Write the slope–intercept form for the line shown in the graph.

53.

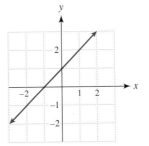

54.

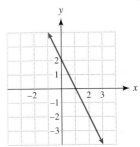

Exercises 55 and 56: Sketch a line with slope m and y-intercept b. Write its slope–intercept form.

55. $m = 2, b = -2$

56. $m = -\frac{3}{4}, b = 3$

Exercises 57–60: Do the following.

(a) Write the equation in slope–intercept form.

(b) Give the slope and y-intercept of the line.

57. $x + y = 3$

58. $-3x + 2y = -6$

59. $20x - 10y = 200$

60. $5x - 6y = 30$

Exercises 61–68: Graph the equation.

61. $y = \frac{1}{2}x + 1$

62. $y = 3x - 2$

63. $y = -\frac{1}{3}x$

64. $y = 3x$

65. $y = 2$

66. $y = -1$

67. $y = 4 - x$

68. $y = 2 - \frac{2}{3}x$

Exercises 69 and 70: All the points shown in the table lie on the same line. Find the slope–intercept form for the line.

69.

x	0	1	2
y	-5	0	5

70.

x	-1	0	1
y	2	0	-2

Exercises 71–74: Find the slope–intercept form for the line satisfying the given conditions.

71. Slope $-\frac{5}{6}$, y-intercept 2

72. Parallel to $y = -2x + 1$, passing through $(1, -5)$

73. Perpendicular to $y = -\frac{3}{2}x$, passing through $(3, 0)$

74. Perpendicular to $y = 5x - 3$, passing through the point $(0, -2)$

SECTION 3.6

Exercises 75 and 76: Determine whether the given point lies on the line.

75. $(-3, 1)$ $y - 1 = 2(x + 3)$

76. $(3, -8)$ $y = -3(x - 1) + 2$

Exercises 77–84: Find the slope–intercept form for the line that satisfies the conditions given.

77. Slope 5, passing through $(1, 2)$

78. Slope 20, passing through $(3, -5)$

79. Passing through $(-2, 1)$ and $(1, -1)$

80. Passing through $(20, -30)$ and $(40, 30)$

81. x-intercept 3, y-intercept -4

82. x-intercept $\frac{1}{2}$, y-intercept -1

83. Parallel to $y = 2x$, passing through $(5, 7)$

84. Perpendicular to $y - 4 = \frac{3}{2}(x + 1)$, passing through $(-1, 0)$

Exercises 85–88: Write the given point–slope form in slope–intercept form.

85. $y - 2 = 3(x + 1)$ **86.** $y - 9 = \frac{1}{3}(x - 6)$

87. $y = 2(x + 3) + 5$ **88.** $y = -\frac{1}{4}(x - 8) + 1$

SECTION 3.7

89. State whether the ordered pairs shown in the table are modeled exactly by $y = -x + 4$.

x	0	1	2	2
y	4	3	2	1

90. State whether the linear model shown in the graph is exact or approximate. Then find the equation of the line.

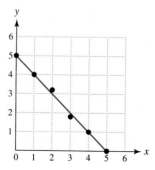

Exercises 91–94: Find an equation $y = mx + b$ that models y after x units of time.

91. y is initially 40 pounds and decreases at a rate of 2 pounds per minute.

92. y is initially 200 gallons and increases at 20 gallons per hour.

93. y is initially 50 and remains constant.

94. y is initially 20 feet *below* sea level and rises at 5 feet per second.

Exercises 95 and 96: For the ordered pairs in the table, do the following.

   (a) Plot the data. Could a line pass through all five points?

   (b) Sketch a line that models the data.

   (c) Determine an equation of the line. For data that is not exactly linear, answers may vary.

95.

x	0	1	2	3	4
y	10	6	2	-2	-6

96.

x	-4	-2	0	2	4
y	1	2.1	3	3.9	5

APPLICATIONS

97. *Graphing Real Data*   The table contains real data on divorces D in millions during year t.

   (a) Make a line graph of the data. Label the axes.

   (b) Comment on any trends in the data.

t	1960	1970	1980	1990	2000
D	0.4	0.7	1.2	1.2	1.2

Source: National Center for Health Statistics.

98. *Water Usage*   The average American uses 100 gallons of water each day.

   (a) Write an equation that gives the gallons G of water that a person uses in t days.

   (b) Graph the equation for $t \geq 0$.

   (c) How many days does it take for the average American to use 5000 gallons of water?

99. *Modeling a Toy Rocket*   The velocity v of a toy rocket in feet per second after t seconds of flight is given by $v = 160 - 32t$, where $t \geq 0$.

   (a) Graph the equation.

   (b) Interpret each intercept.

100. *Modeling* The accompanying line graph represents the insect population on 1 acre of land after x weeks. During this time a farmer sprayed pesticides on the land.

(a) Estimate the slope of each line segment.

(b) Interpret each slope as a rate of change.

(c) Describe what happened to the insect population.

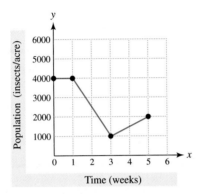

Time (weeks)

101. *Sketching a Graph* An athlete jogs 4 miles away from home at a constant rate of 8 miles per hour and then turns around and jogs back home at the same speed. Sketch a graph that shows the distance that the athlete is from home. Be sure to label each axis.

102. *Nursing Homes* In 1985, there were 19,100 nursing homes, and in 2005, there were 17,000. (*Source:* National Center for Health Statistics.)

(a) Calculate the slope of the line passing through (1985, 19100) and (2005, 17000).

(b) Interpret the slope as a rate of change.

103. *School Enrollment* In 1995, school enrollment at all levels was 65 million students, and in 2005, it was 71 million. (*Source:* National Center for Education Statistics.)

(a) Find the slope of the line passing through the points (1995, 65) and (2005, 71).

(b) Interpret the slope as a rate of change.

(c) If this trend continues, estimate the school enrollment in 2030.

104. *Rental Cars* The cost C in dollars for driving a rental car x miles is $C = 0.2x + 35$.

(a) How much would it cost to rent the car but not drive it?

(b) How much does it cost to drive the car one *additional* mile?

(c) What is the C-intercept of the graph of $C = 0.2x + 35$? What does it represent?

(d) What is the slope of the graph of $C = 0.2x + 35$? What does it represent?

105. *Distance and Speed* A person is driving a car along a straight road. The accompanying graph shows the distance y in miles that the driver is from home after x hours.

(a) Is the person traveling toward or away from home? Why?

(b) The graph passes through (1, 200) and (3, 100). Discuss the meaning of these points.

(c) Find the slope–intercept form of the line. Interpret the slope as a rate of change.

(d) Use the graph to estimate the distance from home after 2 hours. Then check your answer by using your equation from part (c).

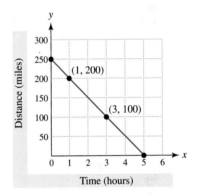

Time (hours)

106. *Arctic Glaciers* Owing to global warming the arctic ice cap has been breaking up. As a result there has been an increase in the number of icebergs floating in the Arctic Ocean. The table gives an iceberg count I for various years t. (*Source:* NBC News.)

t	1970	1980	2000
I	400	600	1000

Source: National Broadcasting Company.

(a) Make a scatterplot of the data.

(b) Find the slope–intercept form of a line that models the number of icebergs I in year t. Interpret the slope of this line.

(c) Is the line you found in part (b) an exact model for the data in the table?

(d) If current trends continue, what will be the iceberg count in 2005?

107. *Deaths from Pneumonia* From 1998 to 2002, deaths D in thousands from pneumonia can be modeled by $D = -6.05(x - 1998) + 90.1$, where x is the year. (*Source: National Center for Health Statistics.*)
  (a) Find the number of deaths from pneumonia in 1998 and 2002.
  (b) What is the slope of the graph of D? Interpret the slope as a rate of change.

108. *Gas Mileage* The table shows the number of miles y traveled by a car on x gallons of gasoline.

x	2	4	8	10
y	40	79	161	200

(a) Plot the data in the xy-plane. Be sure to label each axis.
(b) Sketch a line that models these data. (You may want to use a ruler.)
(c) Calculate the slope of the line. Interpret the slope of this line.
(d) Find the equation of the line. Is your line an *exact* model?
(e) How far could this car travel on 9 gallons of gasoline?

CHAPTER 3   TEST ☞ Pass the Test Video solutions to all test exercises

1. Identify the coordinates of each point in the graph. State the quadrant, if any, in which each point lies.

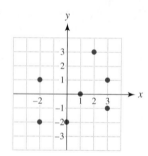

2. Make a scatterplot by plotting the four points $(0, 0)$, $(-2, -2)$, $(3, 0)$, and $(3, -2)$.

3. Complete the table for the equation $y = 2x - 4$. Then determine the x- and y-intercepts for the graph of the equation.

x	−2	−1	0	1	2
y					

4. Determine whether the ordered pair $(1, -3)$ is a solution for the equation $2x - y = 5$.

5. Sketch a line passing through the point $(2, 1)$ and having slope $-\frac{1}{2}$.

6. Find the x- and y-intercepts for the graph of the equation $5x - 3y = 15$.

Exercises 7–10: Graph the equation.

7. $y = 2$ **8.** $x = -3$

9. $y = -3x + 3$ **10.** $4x - 3y = 12$

11. Write the slope–intercept form for the line shown in the graph.

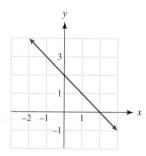

12. Write the equation $-4x + 2y = 1$ in slope–intercept form. Give the slope and the y-intercept.

13. Write the equation of a horizontal line and a vertical line passing through the point $(1, -5)$.

14. Find the slope of a line passing through the points $(-4, 3)$ and $(5, 1)$.

Exercises 15–18: Find the slope–intercept form for the line that satisfies the given conditions.

15. Slope $-\frac{4}{3}$, y-intercept -5

16. Parallel to $y = 3x - 1$, passing through $(2, -5)$

17. Perpendicular to $y = \frac{1}{3}x$, passing through $(1, 2)$

18. Passing through $(-4, 2)$ and $(2, -1)$

19. Write $y - 3 = \frac{1}{2}(x + 4)$ in slope–intercept form.

20. State whether the linear model shown is exact or approximate. Then find the equation of the line.

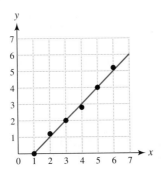

21. *Sketching a Graph* A cyclist rides a bicycle at 10 miles per hour for 2 hours and then at 8 miles per hour for 1 hour. Sketch a graph that shows the total distance d traveled after x hours. Be sure to label each axis.

22. *Modeling* The line graph represents the total fish population P in a small lake after x years. One winter the lake almost froze solid.
  (a) Estimate the slope of each line segment.
  (b) Interpret each slope as a rate of change.
  (c) Describe what happened to the fish population.

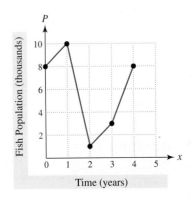

23. *Modeling Insects* Write an equation in slope–intercept form that models the number of insects N after x days if there are initially 2000 insects and they increase at a rate of 100 insects per day.

CHAPTER 3   EXTENDED AND DISCOVERY EXERCISES

Exercises 1 and 2: The table of real data can be modeled by a line. However, the line may not be an exact model, so answers may vary.

1. *Women in Politics* The table lists percentages P of women in state legislatures during year x.

x	1993	1995	1997	1999
P	20.5	20.6	21.6	22.4

x	2001	2003	2005	2007
P	22.4	22.4	22.7	23.5

Source: National Women's Political Caucus.

  (a) Make a scatterplot of the data.
  (b) Find a point–slope form of a line that models these data.
  (c) Use your equation to estimate the percentage of women in state legislatures in 2010.

2. *U.S. Population* The population P of the United States in millions is shown in the table during year x.

x	1960	1970	1980	1990	2000
P	179	203	227	249	281

Source: U.S. Census Bureau.

  (a) Make a scatterplot of the data.
  (b) Find a point–slope form of a line that models these data.
  (c) Use your equation to estimate the U.S. population in 2005.

CREATING GEOMETRIC SHAPES

Exercises 3–6: Many geometric shapes can be created with intersecting lines. For example, a triangle is formed when three lines intersect at three distinct points called vertices. Find equations of lines that satisfy the characteristics described. Sketching a graph may be helpful.

3. *Triangle* A triangle has vertices $(0, 0)$, $(2, 3)$, and $(3, 6)$.
  (a) Find slope–intercept forms for three lines that pass through each pair of points.

(b) Graph the three lines. Is a triangle formed by the line segments connecting the three points?

4. *Parallelogram*   A parallelogram has four sides, with opposite sides parallel. Three sides of a parallelogram are given by the equations $y = 2x + 2$, $y = 2x - 1$, and $y = -x - 2$.

   (a) If the fourth side of the parallelogram passes through the point $(-2, 3)$, find its equation.

   (b) Graph all four lines. Is a parallelogram formed?

5. *Rectangle*   The two vertices $(0, 0)$ and $(4, 2)$ determine one side of a rectangle. The side parallel to this side passes through the point $(0, 3)$.

   (a) Find slope–intercept forms for the four lines that determine the sides of the rectangle.

   (b) Graph all four lines. Is a rectangle formed? (*Hint:* If you use a graphing calculator be sure to set a square window.)

6. *Square*   Three vertices of a square are $(1, 2)$, $(4, 2)$, and $(4, 5)$.

   (a) Find the fourth vertex.

   (b) Find equations of lines that correspond to the four sides of the square.

   (c) Graph all four lines. Is a square formed?

CHAPTERS 1–3   CUMULATIVE REVIEW EXERCISES

Exercises 1 and 2: Classify the number as prime or composite. If a number is composite, write it as a product of prime numbers.

1. 40

2. 61

Exercises 3 and 4: Translate the phrase into an algebraic expression using the variable n.

3. Ten more than a number

4. A number squared decreased by 2

Exercises 5–8: Evaluate by hand and then simplify to lowest terms.

5. $\frac{4}{3} \cdot \frac{21}{24}$

6. $\frac{3}{4} \div \frac{9}{8}$

7. $\frac{2}{3} + \frac{4}{3}$

8. $\frac{7}{10} - \frac{2}{15}$

Exercises 9–12: Evaluate the expression by hand.

9. $20 - 2 \cdot 3$

10. $14 - 5 - 2$

11. $\frac{1 + 4}{1 + 2}$

12. -3^2

Exercises 13 and 14: Classify the number as one or more of the following: natural number, whole number, integer, rational number, or irrational number.

13. $-\frac{4}{5}$

14. $\sqrt{3}$

15. Plot each number on the same number line.
   (a) 0   **(b)** -1 **(c)** $\frac{3}{2}$

16. Evaluate $|3 - 5|$.

Exercises 17 and 18: Evaluate the expression by hand.

17. $-12 \div \left(-\frac{2}{3}\right)$

18. $-\frac{2x}{5y} \div \left(\frac{x}{10y}\right)$

Exercises 19 and 20: Simplify the expression.

19. $3 + 4x - 2 + 3x$

20. $2(x - 1) - (x + 2)$

Exercises 21–24: Solve the equation. Check your solution.

21. $x + 5 = 2$

22. $\frac{1}{3}z = 7$

23. $3t - 5 = 1$

24. $2(x - 3) = -6 - x$

25. Complete the table. Then use the table to solve the equation $6 - 2x = 4$.

x	-2	-1	0	1	2
$6 - 2x$					

26. Translate the sentence "Twice a number increased by 2 equals the number decreased by 5" to an equation, using the variable n. Then solve the equation.

27. The sum of four consecutive integers is -98. Find the integers.

28. If $r = 10$ mph and $d = 80$ miles, use the formula $d = rt$ to find t.

Exercises 29 and 30: Find the area of the figure shown.

29.
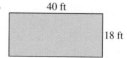
40 ft

18 ft

30.

5 ft

8 ft

Exercises 31 and 32: Solve the formula for the specified variable.

31. $A = \frac{1}{2}bh$ for b

32. $P = 2W + 2L$ for L

33. Use an inequality to express the set of real numbers graphed.

34. Complete the table. Then use the table to solve the linear inequality $2x - 3 \le 1$.

x	0	1	2	3	4
$2x - 3$					

Exercises 35 and 36: Solve the inequality. Write the solution set in set-builder notation.

35. $3 - 6x < 3$

36. $2x \le 1 - (2x - 1)$

37. Identify the coordinates of each point in the graph. State the quadrant, if any, in which each point lies.

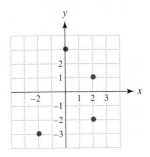

38. Make a scatterplot and a line graph with the points $(-2, 1), (-1, 3), (0, 0), (1, -2),$ and $(2, 3)$.

39. Determine whether $(5, -1)$ is a solution to the equation $x - 3y = 8$.

40. Complete the table for the equation $x + 2y = 4$.

x	-2	-1	0	1	2
y					

Exercises 41–46: Graph the equation.

41. $y = -\frac{1}{2}x + 1$

42. $x + y = 3$

43. $2x + 3y = -6$

44. $5x - 2y = 10$

45. $x = 3$

46. $y = -\frac{3}{2}$

Exercises 47 and 48: Identify the x- and y-intercepts. Then write the slope–intercept form of the line.

47.

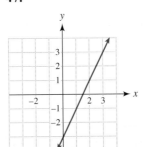

48.

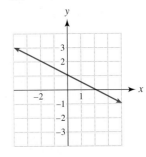

49. Determine the intercepts for the graph of the equation $-4x + 5y = 40$.

50. Write an equation for each line shown in the graph.

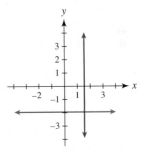

51. Sketch a line with slope $m = -2$, passing through the point $(-1, 2)$.

52. Sketch a line with undefined slope, passing through the point $(2, -1)$.

53. The table lists points located on a line. Write the slope–intercept form of the line.

x	-1	0	1	2
y	-6	-3	0	3

54. Write the equation $3x - 5y = 15$ in slope–intercept form. Graph the equation.

Exercises 55–58: Find the slope–intercept form for the line satisfying the given conditions.

55. Parallel to $y = -\frac{1}{3}x - 5$, passing through $(-3, 8)$

56. Perpendicular to $3x - 2y = 6$, passing through the point $(0, -3)$

57. Passing through $(-1, 3)$ and $(2, -3)$

58. x-intercept -2, y-intercept $\frac{1}{2}$

59. State whether the linear model shown in the graph is exact or approximate. Then find the equation of the line.

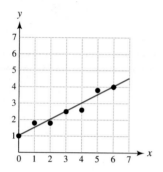

60. Let the initial value of y be 100 pounds, increasing at 5 pounds per hour. Find an equation in slope–intercept form that models y after x hours.

61. An insect population is initially 20,000 and increasing at 5000 per day. Write an equation that gives the number of insects I after x days.

62.

x	0	1	2	3
y	5	3	1	−1

(a) Plot the data in the table. Could a line pass through all four points?
(b) Sketch a line that models the data.
(c) Determine the equation of the line.

APPLICATIONS

63. *Pizzas*   The table lists the cost C of buying x large pepperoni pizzas. Write an equation that relates C and x.

x	2	3	4
C	$16	$24	$32

64. *U.S. Postal Service*   In 2000, about $\frac{11}{20}$ of the mail consisted of first-class mail and periodicals. For every periodical there were 9 pieces of first-class mail.

Estimate the fraction of the mail that was first-class mail.   (*Source:* U.S. Postal Service, *2000 Annual Report.*)

65. *U.S. Per Capita Income*   The per capita income I in dollars from 1970 to 2000 can be modeled by

$$I = 807.4x - 1{,}587{,}300,$$

where x is the year.   (*Source:* U.S. Census Bureau.)
(a) What has been the yearly average increase in per capita income?
(b) Estimate the year when this income reached $19,400.

66. *Income Tax Returns*   In 2000, about 127 million income tax returns were filed. By 2004, this number had increased to about 131 million. Find the percent change in the number of income tax returns filed between 2000 and 2004.   (*Source:* Internal Revenue Service.)

67. *Investment Money*   A student invests two sums of money at 3% and 4% interest, receiving a total of $110 in interest after 1 year. Twice as much money is invested at 4% than at 3%. Find the amount invested at each interest rate.

68. *School Lunch Program*   In 1970, 22.4 million students participated in the national school lunch program. By 2000, this number increased to 27.2 million students.   (*Source:* Department of Agriculture.)
(a) Find the slope of the line passing through the points (1970, 22.4) and (2000, 27.2).
(b) Interpret the slope as a rate of change.
(c) If this trend continues, estimate the number of participants in 2005.

69. *Rental Cars*   The cost C in dollars for driving a rental car x miles is $C = 0.3x + 25$.
(a) How much does it cost to drive the car 200 miles?
(b) How much does it cost to rent the car but not drive it?
(c) How much does it cost to drive the car 1 *additional* mile?

Systems of Linear Equations in Two Variables

T wo factors that have a critical effect on plant growth are temperature and precipitation. If a region has too little precipitation, it will be a desert. Forests often grow in regions where trees can exist at relatively low temperatures and there is sufficient rainfall. At other levels of temperature and precipitation, grasslands may prevail. The accompanying figure illustrates the relationships among forests, grasslands, and deserts suggested by average annual temperature T in degrees Fahrenheit and precipitation P in inches. For example, Casa Grande, Arizona, has an average annual temperature of about 70°F and average annual precipitation of about 8 inches. What does the graph predict about the plant growth near Casa Grande?

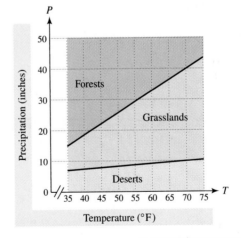

This graph illustrates systems of linear inequalities, which we discuss in this chapter. Mathematics is essential to creating these types of graphs and solving application problems.

We can do anything we want to do if we stick to it long enough.

—HELEN KELLER

Source: A. Miller and J. Thompson, *Elements of Meteorology.*

4.1 SOLVING SYSTEMS OF LINEAR EQUATIONS GRAPHICALLY AND NUMERICALLY

Basic Concepts ▪ Solutions to Systems of Equations

A LOOK INTO MATH ▷ Many equations in everyday life involve more than one variable. To calculate the area of a rectangle we need to know its width and its length. To calculate wind chill we need to know the outside temperature and the wind velocity. In this section we consider linear equations in two variables. The concepts discussed in this section are used in many applications.

Basic Concepts

In Chapter 3 we showed that the graph of $y = mx + b$ is a line with slope m and y-intercept b, as illustrated in Figure 4.1. Each point on this line represents a solution to the equation $y = mx + b$. Because there are infinitely many points on a line, there are infinitely many solutions to this equation. However, many applications require that we find one particular solution to a linear equation. One way to find such a solution is to graph a second line in the same xy-plane and determine the point of intersection (if one exists).

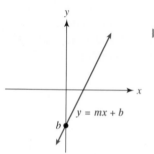

Figure 4.1

▶ **REAL-WORLD CONNECTION** Consider the following application of a line. If renting a moving truck for one day costs \$25 plus \$0.50 per mile driven, then the equation $C = 0.5x + 25$ represents the cost C in dollars of driving the rental truck x miles. The graph of this line is shown in Figure 4.2(a) for $x \geq 0$.

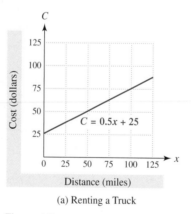

(a) Renting a Truck

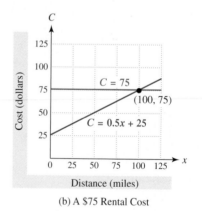

(b) A \$75 Rental Cost

Figure 4.2

Suppose that we want to determine the number of miles that the truck is driven when the rental cost is \$75. One way to solve this problem *graphically* is to graph both $C = 0.5x + 25$ and $C = 75$ in the same coordinate plane, as shown in Figure 4.2(b). The lines intersect at the point (**100, 75**), which is a solution to $C = 0.5x + 25$ and to $C = 75$. That is, if the rental cost is **\$75**, then the mileage must be **100** miles. This graphical technique of solving two equations is sometimes called the **intersection-of-graphs method**. To find a solution with this method, we locate a point where two graphs intersect.

EXAMPLE 1   Solving an equation graphically

The equation $P = 10x$ calculates an employee's pay for working x hours at \$10 per hour. Use the intersection-of-graphs method to find the number of hours that the employee worked if the amount paid is \$40.

Solution
Begin by graphing the equations $P = 10x$ and $P = 40$, as illustrated in Figure 4.3.

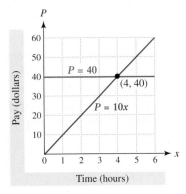

Figure 4.3 Wages

The graphs intersect at the point $(4, 40)$, which indicates that the employee must work 4 hours to earn **\$40**.   Now Try Exercise 59(a), (b)

   Equations can be solved in more than one way. In Example 1 we determined graphically that $P = 10x$ is equal to 40 when $x = 4$. Because we want to know the value of x that makes $P = 40$, we could solve this problem *symbolically* by substituting 40 for P in the equation $P = 10x$. The resulting equation, $40 = 10x$, can be solved using methods discussed in Chapter 2. By dividing each side by 10 and simplifying, we obtain $x = 4$. We could also solve this problem by making a table of values, as illustrated by Table 4.1. Note that, when $x = 4$, $P = \$40$. A table of values provides a *numerical solution*.

TABLE 4.1   Wages Earned at \$10 per hour

x (hours)	0	1	2	3	4	5	6
P (pay)	\$0	\$10	\$20	\$30	\$40	\$50	\$60

NOTE: Although graphical, symbolic, and numerical methods are different, all methods should give the same solution. Exceptions might occur, for example, when reading a graph precisely is difficult, or when a needed value does not appear in a table.

TECHNOLOGY NOTE: Intersection of Graphs and Table of Values
A graphing calculator can be used to find the intersection of the two graphs shown in Figure 4.3. It can also be used to create Table 4.1. The accompanying figures illustrate how a calculator can be used to determine that $y_1 = 10x$ equals $y_2 = 40$ when $x = 4$.

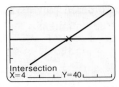

For the remainder of this section we focus on solving two equations by graphing and by making a table of values. Methods for solving two equations symbolically will be discussed later in the chapter. The next example illustrates a graphical approach.

EXAMPLE 2   Solving an equation graphically

Use a graph to find the x-value when $y = 3$.
(a) $y = 2x - 1$ **(b)** $-3x + 2y = 12$

Solution

(a) Begin by graphing the equations $y = 2x - 1$ and $y = 3$. The graph of $y = 2x - 1$ is a line with slope 2 and y-intercept -1. The graph of $y = 3$ is a horizontal line with y-intercept 3. In Figure 4.4(a) their graphs intersect at the point $(2, 3)$. Therefore an x-value of 2 corresponds to a y-value of 3.

(b) One way to graph $-3x + 2y = 12$ is to write this equation in slope–intercept form.

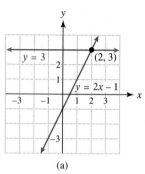

$$-3x + 2y = 12 \qquad \text{Given equation}$$
$$2y = 3x + 12 \qquad \text{Add } 3x \text{ to each side.}$$
$$y = \frac{3}{2}x + 6 \qquad \text{Divide each side by 2.}$$

The line has slope $\frac{3}{2}$ and y-intercept 6. Its graph and the graph of $y = 3$ are shown in Figure 4.4(b). Their graphs intersect at $(-2, 3)$. Therefore an x-value of -2 corresponds to a y-value of 3.   Now Try Exercises 13, 15

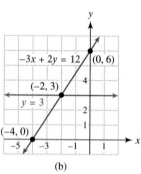

(b)

Figure 4.4

MAKING CONNECTIONS

Different Ways to Graph the Same Line

In Example 2(b) the line $-3x + 2y = 12$ was graphed by finding its slope–intercept form. A second way to graph this line is to find the x- and y-intercepts of the line. If $y = 0$, then $x = -4$ makes this equation true, and if $x = 0$, then $y = 6$ makes this equation true. Thus the x-intercept is -4, and the y-intercept is 6. Note that the (slanted) line in Figure 4.4(b) passes through the points $(-4, 0)$ and $(0, 6)$, which could be used to graph the line.

Solutions to Systems of Equations

In Example 2(b) we determined the x-value when $y = 3$ in the equation $-3x + 2y = 12$. This problem can be thought of as solving the following *system of two equations in two variables.*

$$-3x + 2y = 12$$
$$y = 3$$

The solution is the ordered pair $(-2, 3)$, which indicates that when $x = -2$ and $y = 3$, each equation is a true statement.

$$-3(-2) + 2(3) = 12 \qquad \text{A true statement}$$
$$3 = 3 \qquad \text{A true statement}$$

Suppose that the sum of two numbers is 10 and that their difference is 4. If we let x and y represent the two numbers, then the equations

$$x + y = 10 \qquad \text{Sum is 10.}$$
$$x - y = 4 \qquad \text{Difference is 4.}$$

describe this situation. Each equation is a linear equation in two variables, so we call these equations a **system of linear equations in two variables**. Its graph typically consists of two lines. A **solution to a system** of two equations is an ordered pair (x, y) that makes *both* equations true. This ordered pair gives the coordinates of a point where the two lines intersect.

NOTE: When two lines do not coincide (their graphs are two distinct lines), there can be no more than one intersection point. If such an intersection point exists, the ordered pair corresponding to it represents the only solution to the system of linear equations. In this case, we say the ordered pair is *the* solution to the system of equations.

EXAMPLE 3   **Testing for solutions**

Determine whether $(4, 6)$ or $(7, 3)$ is a solution to

$$x + y = 10$$
$$x - y = 4.$$

Solution   To determine whether $(4, 6)$ is a solution, substitute $x = 4$ and $y = 6$ in each equation. It must make *both* equations true.

$x + y = 10$	$x - y = 4$	Given equations
$4 + 6 \stackrel{?}{=} 10$	$4 - 6 \stackrel{?}{=} 4$	Let $x = 4, y = 6$.
$10 = 10$ (True)	$-2 = 4$ (False)	Second equation is false.

Because $(4, 6)$ does not satisfy *both* equations, it is not a solution for the system of equations. Next let $x = 7$ and $y = 3$ to determine whether $(7, 3)$ is a solution.

$x + y = 10$	$x - y = 4$	Given equations
$7 + 3 \stackrel{?}{=} 10$	$7 - 3 \stackrel{?}{=} 4$	Let $x = 7, y = 3$.
$10 = 10$ (True)	$4 = 4$ (True)	Both are true.

Because $(7, 3)$ makes *both* equations true, it is a solution for the system of equations.

Now Try Exercise 17

In the next example we find the solution to a system of linear equations graphically and numerically.

EXAMPLE 4   **Solving a system graphically and numerically**

Solve the system of linear equations

$$x + 2y = 4$$
$$2x - y = 3$$

with a graph and with a table of values.

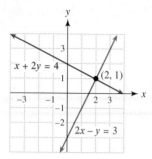

Figure 4.5

Solution

Graphically   Begin by writing each equation in slope–intercept form.

$x + 2y = 4$	First equation	$2x - y = 3$	Second equation
$2y = -x + 4$	Subtract x.	$-y = -2x + 3$	Subtract $2x$.
$y = -\dfrac{1}{2}x + 2$	Divide by 2.	$y = 2x - 3$	Multiply by -1.

The graphs of $y = -\frac{1}{2}x + 2$ and $y = 2x - 3$ are shown in Figure 4.5. Their graphs intersect at the point $(2, 1)$.

Numerically   Table 4.2 shows the equations $y = -\frac{1}{2}x + 2$ and $y = 2x - 3$ evaluated for various values of x. Note that when $x = 2$, both equations have a y-value of **1**. Thus **(2, 1)** is the solution.

TABLE 4.2   A Numerical Solution

x	-1	0	1	2	3
$y = -\frac{1}{2}x + 2$	2.5	2	1.5	1	0.5
$y = 2x - 3$	-5	-3	-1	1	3

Now Try Exercise **37**

CRITICAL THINKING

If the graph of a system of linear equations consists of two parallel lines, how many solutions are there?

EXAMPLE **5**   Solving a system graphically

Solve the system of equations graphically.

$$y = 2x$$
$$2x + y = 4$$

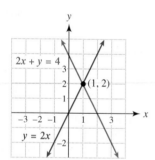

Figure 4.6

Solution

The equation $y = 2x$ can be written in slope–intercept form as $y = 2x + 0$. Its graph is a line passing through the origin with slope 2, as shown in Figure 4.6. The equation $2x + y = 4$ can be graphed by finding the intercepts. If we let $x = 0$, the resulting equation is $y = 4$. If we let $y = 0$, the resulting equation is $2x = 4$ which is equivalent to $x = 2$. Thus the y-intercept is 4 and the x-intercept is 2. This line is also graphed in Figure 4.6. Because the intersection point is $(1, 2)$, the solution to the system of equations is the ordered pair $(1, 2)$.

Now Try Exercise **51**

In the next example, we use a four-step process to solve an application involving a system of linear equations. These steps are based on the four-step process discussed in Section 2.3, with Step 3 split into two parts to emphasize the importance of using the solution to the system of equations to determine the solution to the given problem.

EXAMPLE **6**   Traveling to watch sports

In 2002, about 50 million Americans traveled to watch either football or basketball. About 10 million more people traveled to watch football than basketball. How many Americans traveled to watch each sport?   (*Source: Sports Travel Magazine.*)

Solution

STEP 1: *Identify each variable.*

x: millions of Americans who traveled to watch football
y: millions of Americans who traveled to watch basketball

STEP 2: *Write a system of equations.* The total number of Americans who watched either sport is 50 million, so we know that $x + y = 50$. Because 10 million more people watched football than basketball, we also know that $x - y = 10$. Thus a system of equations representing this situation is

$$x + y = 50$$
$$x - y = 10.$$

STEP 3A: *Solve the system of equations.* To solve this system graphically, write each equation in slope–intercept form.

$$y = -x + 50$$
$$y = \ \ x - 10$$

Their graphs intersect at the point (30, 20), as shown in Figure 4.7.

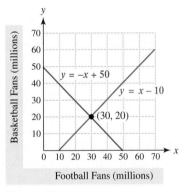

Figure 4.7

STEP 3B: *Determine the solution to the problem.* The point (30, 20) corresponds to $x = 30$ and $y = 20$. Thus about 30 million Americans traveled to watch football and 20 million Americans traveled to watch basketball.

STEP 4: *Check your solution.* Note that $30 + 20 = 50$ million Americans traveled to watch either football or basketball and that $30 - 20 = 10$ million more Americans watched football than basketball.   Now Try Exercise 61

TECHNOLOGY NOTE: Checking Solutions

The solution to the system in Example 6 can be checked with a graphing calculator, as shown in the accompanying figure where the graphs of $y_1 = -x + 50$ and $y_2 = x - 10$ intersect at the point (30, 20). However, a graphing calculator cannot read Example 6 and write down the system of equations. A human mind is needed for these tasks.

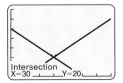

4.1 PUTTING IT ALL TOGETHER

Some types of situations are described by using two equations in two variables. For example, suppose that a person walks a total of 8 miles in 2 days by walking 2 miles farther on the first day than on the second day. If x represents the number of miles walked on the first day and if y represents the number of miles walked on the second day, then the system of linear equations

$$x + y = 8$$
$$x - y = 2$$

describes this situation. The methods discussed in this section reveal that the solution to this system is (5, 3), which indicates that the person walked 5 miles on the first day and 3 miles on the second day. The following table summarizes some important concepts of systems of equations.

Concept	Explanation	Example
System of Linear Equations in Two Variables	Can be written as $Ax + By = C$ $Dx + Ey = F$	$x + y = 8$ $x - y = 2$
Solution to a System of Equations	An ordered pair (x, y) that satisfies *both* equations	The solution to the preceding system is (5, 3) because, when $x = 5$ and $y = 3$ are substituted, both equations are true. $5 + 3 \overset{?}{=} 8$ True $5 - 3 \overset{?}{=} 2$ True
Graphical Solution to a System of Equations	Graph each equation. A point of intersection represents a solution.	The graphs of $y = -x + 8$ and $y = x - 2$ intersect at (5, 3).
Numerical Solution to a System of Equations	Make a table for each equation. A solution occurs when one x-value gives the same y-values in both equations.	Make a table for $y = -x + 8$ and $y = x - 2$. When $x = 5$, $y = 3$ in both equations, so (**5, 3**) is the solution. $\begin{array}{c\|ccc} x & 4 & 5 & 6 \\ \hline y = -x + 8 & 4 & 3 & 2 \\ y = x - 2 & 2 & 3 & 4 \end{array}$

4.1 Exercises

CONCEPTS

1. A solution to a system of two equations in two variables is a(n) _____ pair.

2. A graphical technique for solving a system of two equations in two variables is the _____ method.

3. When two lines are not identical, there can be no more than _____ intersection point(s).

4. The graphs of the equations shown intersect at $(11, 9)$.

$$x + y = 20$$
$$x - y = 2$$

What is the solution to the system?

5. Is $(0, 0)$ a solution to the given system?

$$2x + 3y = 8$$
$$5x - 4y = -3$$

6. To find a numerical solution to a system, start by creating a _____ of values for the equations.

7. If a graphical method and a numerical method (table of values) are used to solve the same system of equations, then the two solutions should be (the same/different).

8. One way to graph a line is to write its equation in slope–intercept form. A second method is to find the x- and y-_____.

SOLVING SYSTEMS OF EQUATIONS

Exercises 9–16: Determine graphically the x-value when y = 2 in the given equation.

9. $y = 2x$

10. $y = \frac{1}{3}x$

11. $y = 4 - x$

12. $y = -2 - x$

13. $y = -\frac{1}{2}x + 1$

14. $y = 3x - 1$

15. $2x + y = 6$

16. $-3x + 4y = 11$

Exercises 17–22: Determine which ordered pair is a solution to the system of equations.

17. $(0, 0), (1, 1)$

$$x + y = 2$$
$$x - y = 0$$

18. $(-1, 2), (1, -2)$

$$2x + y = 0$$
$$x - 2y = -5$$

19. $(-1, -1), (2, -3)$

$$2x + 3y = -5$$
$$4x - 5y = 23$$

20. $(2, -1), (-2, -2)$

$$-x + 4y = -6$$
$$6x - 7y = 19$$

21. $(2, 0), (-1, -3)$

$$-5x + 5y = -10$$
$$4x + 9y = 8$$

22. $\left(\frac{1}{2}, \frac{3}{2}\right), \left(\frac{3}{4}, \frac{5}{4}\right)$

$$x + y = 2$$
$$3x - y = 0$$

Exercises 23–28: The graphs of two equations are shown. Use the intersection-of-graphs method to identify the solution to both equations. Then check your answer.

23.

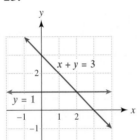

24.

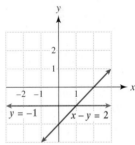

25.

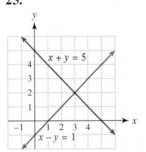

26.

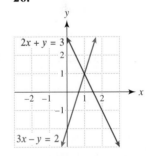

27.

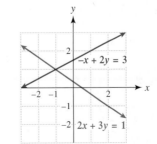

28.

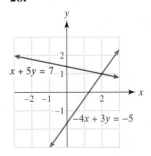

Exercises 29–32: A table for two equations is given. Identify the solution to both equations.

29.

x	1	2	3	4
$y = 2x$	2	4	6	8
$y = 4$	4	4	4	4

30.

x	2	3	4	5
$y = 6 - x$	4	3	2	1
$y = x - 2$	0	1	2	3

31.

x	1	2	3	4
$y = 4 - x$	3	2	1	0
$y = x - 2$	-1	0	1	2

32.

x	1	2	3	4
$y = 6 - 3x$	3	0	-3	-6
$y = 2 - x$	1	0	-1	-2

Exercises 33 and 34: Complete the table for each equation. Then identify the solution to both equations.

33.

x	0	1	2	3
$y = x + 2$				
$y = 4 - x$				

34.

x	-5	-4	-3	-2
$y = 2x + 1$				
$y = x - 1$				

Exercises 35–40: Use the specified method to solve the system of equations.
(a) *Graphically*
(b) *Numerically (table of values)*

35. $y = 2x + 3$
$y = 1$

36. $y = 2 - x$
$y = 0$

37. $y = 4 - x$
$y = x - 2$

38. $y = 2x$
$y = -\frac{1}{2}x$

39. $y = 3x$
$y = x + 2$

40. $y = 2x - 3$
$y = -x + 3$

Exercises 41–52: Solve the system of equations graphically.

41. $x + y = -3$
$x - y = 1$

42. $x - y = 3$
$2x + y = 3$

43. $2x - y = 3$
$3x + y = 2$

44. $x + 2y = 6$
$-x + 3y = 4$

45. $-4x + 2y = 0$
$x - y = -1$

46. $4x - y = 2$
$y = 2x$

47. $2x - y = 4$
$x + 2y = 7$

48. $x - y = 2$
$\frac{1}{2}x + y = 4$

49. $x = -y + 4$
$x = 3y$

50. $x = 2y$
$y = -\frac{1}{2}x$

51. $x + y = 3$
$x = \frac{1}{2}y$

52. $2x - 4y = 8$
$\frac{1}{2}x + y = -4$

*Exercises 53–58: **Number Problems**   For each problem, complete each of the following.*
(a) *Write a system of equations for the problem.*
(b) *Find the unknown numbers by solving the system of equations graphically.*

53. The sum of two numbers is 4 and their difference is 0.

54. The sum of two numbers is −5 and their difference is 1.

55. The sum of twice a number and another number is 7. Their difference is 2.

56. Three times a number subtracted from another number results in 1. Their sum is 5.

57. One number is triple another number. Their difference is 4.

58. Half of a number added to another number equals 5. Their difference is 1.

APPLICATIONS

59. *Renting a Truck*   A rental truck costs $50 plus $0.50 per mile.
   (a) Write an equation that gives the cost of driving the truck x miles.
   (b) Use the intersection-of-graphs method to determine the number of miles that the truck is driven if the rental cost is $80.
   (c) Solve part (b) numerically with a table of values.

60. *Renting a Car* A rental car costs $25 plus $0.25 per mile.
  (a) Write an equation that gives the cost of driving the car x miles.
  (b) Use the intersection-of-graphs method to determine the number of miles that the car is driven if the rental cost is $100.
  (c) Solve part (b) numerically with a table of values.

61. *Recorded Music* In 2004, rock and hip hop music accounted for 36% of all music sales. Rock music sales were 2 times greater than hip hop music sales. (*Source:* Recording Industry Association of America.)
  (a) Let x be the percentage of sales due to rock music and let y be the percentage of music sales due to hip hop music. Write a system of two equations that describes the given information.
  (b) Solve your system graphically.

62. *Sales of Radios* During 2000 and 2001, about 37 thousand radios were sold, excluding car radios. The 2000 sales exceeded the 2001 sales by 3 thousand radios. (*Source:* M. Street Corporation.)
  (a) Let y be the radio sales in thousands during 2000 and x be the radio sales in thousands during 2001.

Write a system of two equations that describes the given information.
  (b) Solve your system graphically.

63. *Dimensions of a Rectangle* A rectangle is 4 inches longer than it is wide. Its perimeter is 28 inches.
  (a) Write a system of two equations in two variables that describes this information. Be sure to specify what each variable means.
  (b) Solve your system graphically. Interpret your results.

64. *Dimensions of a Triangle* An isosceles triangle has a perimeter of 17 inches with its two shorter sides equal in length. The longest side measures 2 inches more than either of the shorter sides.
  (a) Write a system of two equations in two variables that describes this information. Be sure to specify what each variable means.
  (b) Solve your system graphically. Explain what your results mean.

WRITING ABOUT MATHEMATICS

65. Use the intersection-of-graphs method to help explain why you typically expect a linear system in two variables to have one solution.

66. Could a system of two linear equations in two variables have exactly two solutions? Explain your reasoning.

67. Give one disadvantage of using a table to solve a system of equations. Explain your answer.

68. Do the equations $y = 2x + 1$ and $y = 2x - 1$ have a common solution? Explain your answer.

4.2 SOLVING SYSTEMS OF LINEAR EQUATIONS BY SUBSTITUTION

The Method of Substitution ▪ Types of Systems of Linear Equations ▪ Applications

A LOOK INTO MATH ▷ In Section 4.1 we solved systems of linear equations by using graphs and tables. A disadvantage of a graph is that reading the graph precisely can be difficult. A disadvantage of using a table is that locating the solution can be difficult when it is either a fraction or a large number. In this section we introduce the method of substitution, in which we solve systems of equations using only symbols. The advantage of this method is that the *exact* solution can always be found (provided it exists).

The Method of Substitution

If you and a friend earned $120 together, then there is no way to determine how much each one earned. You might have earned $70 while your friend earned $50, or vice versa. If x represents how much your friend earned and y represents how much you earned, then the equation $x + y = 120$ describes this situation. However, if we know that you earned twice as much as your friend, then we can include a second equation, $y = 2x$. The amount that each of you earned can now be determined by *substituting* $2x$ for y in the first equation.

$$x + y = 120 \qquad \text{First equation}$$
$$x + 2x = 120 \qquad \text{Substitute } 2x \text{ for } y.$$
$$3x = 120 \qquad \text{Combine like terms.}$$
$$x = 40 \qquad \text{Divide each side by 3.}$$

Thus your friend earned $40, and you earned twice as much, or $80.

This technique of substituting an expression for a variable and solving the resulting equation is called the **method of substitution**.

EXAMPLE 1   Using the method of substitution

Solve each system of equations.
(a) $2x + y = 10$ **(b)** $-2x + 3y = -8$
        $y = 3x$ $x = 3y + 1$

Solution
(a) From the second equation, substitute $3x$ for y in the first equation.

$$2x + y = 10 \qquad \text{First equation}$$
$$2x + 3x = 10 \qquad \text{Substitute } 3x \text{ for } y.$$
$$5x = 10 \qquad \text{Combine like terms.}$$
$$x = 2 \qquad \text{Divide each side by 5.}$$

The solution to this system is an *ordered pair*, so we must also find y. Because $y = 3x$ and $x = 2$, it follows that $y = 3(2) = 6$. The solution is $(2, 6)$. (Check it.)
(b) The second equation, $x = 3y + 1$, is solved for x. Substitute $(3y + 1)$ for x in the first equation. Be sure to include parentheses around the expression $3y + 1$ since this entire expression is to be multiplied by -2.

$$-2x + 3y = -8 \qquad \text{First equation}$$
$$-2(3y + 1) + 3y = -8 \qquad \text{Substitute } (3y + 1) \text{ for } x.$$
$$-6y - 2 + 3y = -8 \qquad \text{Distributive property}$$
$$-3y - 2 = -8 \qquad \text{Combine like terms.}$$
$$-3y = -6 \qquad \text{Add 2 to each side.}$$
$$y = 2 \qquad \text{Divide each side by } -3.$$

To find x, substitute **2** for y in $x = 3y + 1$ to obtain $x = 3(2) + 1 = 7$. The solution is $(7, 2)$. (Check it.)   Now Try Exercises 11, 19

NOTE: When an expression contains two or more terms, it is usually best to place parentheses around it when substituting it for a single variable in an equation. In Example 1(b), the distributive property would not have been applied correctly without the parentheses.

Sometimes it is necessary to solve for a variable before substitution can be used, as demonstrated in the next example.

EXAMPLE 2   **Using the method of substitution**

Solve each system of equations.

(a) $\begin{aligned} x + y &= 8 \\ 2x - 3y &= 6 \end{aligned}$ (b) $\begin{aligned} 3a - 2b &= 2 \\ a + 4b &= 3 \end{aligned}$

Solution

(a) Neither equation is solved for a variable, but we can easily solve the first equation for y.

$$x + y = 8 \qquad \text{First equation}$$
$$y = 8 - x \qquad \text{Subtract } x \text{ from each side.}$$

Now we can substitute $(8 - x)$ for y in the second equation.

$$2x - 3y = 6 \qquad \text{Second equation}$$
$$2x - 3(8 - x) = 6 \qquad \text{Substitute } (8 - x) \text{ for } y.$$
$$2x - 24 + 3x = 6 \qquad \text{Distributive property}$$
$$5x = 30 \qquad \text{Combine terms; add 24.}$$
$$x = 6 \qquad \text{Divide each side by 5.}$$

Because $y = 8 - x$ and $x = 6$, $y = 8 - 6 = 2$. The solution is $(6, 2)$.

(b) Although we could solve either equation for either variable, solving the second equation for a is easiest because the coefficient of a is 1.

$$a + 4b = 3 \qquad \text{Second equation}$$
$$a = 3 - 4b \qquad \text{Subtract } 4b \text{ from each side.}$$

Now substitute $(3 - 4b)$ for a in the first equation.

$$3a - 2b = 2 \qquad \text{First equation}$$
$$3(3 - 4b) - 2b = 2 \qquad \text{Substitute } (3 - 4b) \text{ for } a.$$
$$9 - 12b - 2b = 2 \qquad \text{Distributive property}$$
$$-14b = -7 \qquad \text{Combine terms; subtract 9.}$$
$$b = \frac{1}{2} \qquad \text{Divide each side by } -14.$$

To find a, substitute $b = \frac{1}{2}$ in $a = 3 - 4b$ to obtain $a = 1$. The solution is $\left(1, \frac{1}{2}\right)$.

Now Try Exercises 23, 29

NOTE: When a system of equations contains variables other than x and y, we will list them alphabetically in an ordered pair.

Types of Systems of Linear Equations

A system of linear equations typically has exactly one solution. However, in the next example we solve two systems of equations that do not have exactly one solution.

EXAMPLE 3   Solving other types of systems

If possible, use substitution to solve the system of equations. Then use graphing to help explain the result.

(a) $3x + y = 4$ (b) $x + y = 2$
    $6x + 2y = 2$ $2x + 2y = 4$

Solution

(a) Solve the first equation for y to obtain $y = 4 - 3x$. Next substitute $(4 - 3x)$ for y in the second equation.

$$6x + 2y = 2 \qquad \text{Second equation}$$
$$6x + 2(4 - 3x) = 2 \qquad \text{Substitute } (4 - 3x) \text{ for } y.$$
$$6x + 8 - 6x = 2 \qquad \text{Distributive property}$$
$$8 = 2 \text{ (False)} \qquad \text{Combine terms.}$$

The equation $8 = 2$ is *always false*, which indicates that there are *no solutions*. One way to graph each equation is to write the equations in slope–intercept form.

$3x + y = 4$ First equation $\mid$ $6x + 2y = 2$ Second equation
$\quad y = -3x + 4$ Subtract 3x. $\mid$ $\quad y = -3x + 1$ Subtract 6x, divide by 2.

The graphs of these equations are parallel lines with slope -3, as shown in Figure 4.8(a). Because the lines *do not intersect* there are *no solutions* to the system of equations.

(b) Solve the first equation for y to obtain $y = 2 - x$. Now substitute $(2 - x)$ for y in the second equation.

$$2x + 2y = 4 \qquad \text{Second equation}$$
$$2x + 2(2 - x) = 4 \qquad \text{Substitute } (2 - x) \text{ for } y.$$
$$2x + 4 - 2x = 4 \qquad \text{Distributive property}$$
$$4 = 4 \text{ (True)} \qquad \text{Combine terms.}$$

The equation $4 = 4$ is *always true*, which means that there are *infinitely many solutions*. One way to graph these equations is to write them in slope–intercept form first.

$x + y = 2$ First equation $\mid$ $2x + 2y = 4$ Second equation
$\quad y = -x + 2$ Subtract x. $\mid$ $\quad y = -x + 2$ Subtract 2x, divide by 2.

Because the equations have the same slope–intercept form, their graphs are identical, resulting in a single line, as shown in Figure 4.8(b). *Every point* on this line *represents a solution* to the system of equations, so there are infinitely many solutions.

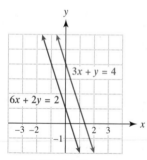

(a) No Solutions

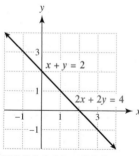

(b) Infinitely Many Solutions

Figure 4.8

Now Try Exercises 53, 55

TYPES OF EQUATIONS AND NUMBER OF SOLUTIONS

A system of linear equations can have no solutions, one solution, or infinitely many solutions. It cannot have any other number of solutions. If a system has

1. no solutions, it is an **inconsistent system**. Graphing the equations results in parallel lines.
2. one solution, it is a **consistent system**, and its equations are **independent equations**. Graphing the equations results in two lines that intersect at one point.
3. infinitely many solutions, it is a **consistent system**, and its equations are **dependent equations**. Graphing the equations results in identical lines.

EXAMPLE 4   Identifying types of equations

Graphs of two equations are shown. State the number of solutions to each system of equations. Then state whether the system is consistent or inconsistent. If it is consistent, state whether the equations are dependent or independent.

(a) 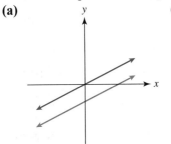   (b) 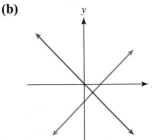   (c)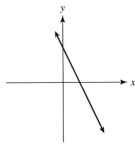

Solution

(a) The lines are parallel, so there are no solutions. The system is inconsistent.
(b) The lines intersect at one point, so there is one solution. The system is consistent, and the equations are independent.
(c) There is only one line, which indicates that the graphs are identical, or coincide, so there are infinitely many solutions. The system is consistent and the equations are dependent.

Now Try Exercises 39, 41, 43

Applications

In the next two examples we use the method of substitution to solve applications.

EXAMPLE 5   Determining pizza sales

In 2002, combined sales of frozen and ready-to-eat pizza reached $28.9 billion. Ready-to-eat pizza sales were 7.5 times frozen pizza sales. Find the amount of sales for each type of pizza.   (**Source:** *Business Trend Analyst.*)

Solution

STEP 1: *Identify each variable.* Clearly identify what each variable represents.

x: sales of ready-to-eat pizza, in billions of dollars
y: sales of frozen pizza, in billions of dollars

STEP 2: *Write a system of equations.*

$$x + y = 28.9 \qquad \text{Sales total \$28.9 billion.}$$

$$x = 7.5y \qquad \begin{array}{l}\text{Ready-to-eat pizza sales } x \text{ are} \\ \text{7.5 times frozen pizza sales } y.\end{array}$$

STEP 3A: *Solve the system of linear equations.* Substitute $7.5y$ for x in the first equation.

$$x + y = 28.9 \qquad \text{First equation}$$

$$7.5y + y = 28.9 \qquad \text{Substitute } 7.5y \text{ for } x.$$

$$8.5y = 28.9 \qquad \text{Combine like terms.}$$

$$y = \frac{28.9}{8.5} \qquad \text{Divide each side by 8.5.}$$

$$y = 3.4 \qquad \text{Simplify.}$$

Because $x = 7.5y$, it follows that $x = 7.5(3.4) = 25.5$.

STEP 3B: *Determine the solution to the problem.* The solution $y = 3.4$ and $x = 25.5$ indicates that frozen pizza sales were \$3.4 billion and ready-to-eat pizza sales were \$25.5 billion in 2002.

STEP 4: *Check the solution.* The sum of these sales was $25.5 + 3.4 = \$28.9$ billion, and ready-to-eat pizza sales were $\frac{25.5}{3.4} = 7.5$ times frozen pizza sales. The answer checks.   Now Try Exercise **87**

EXAMPLE **6**   Determining airplane speed and wind speed

An airplane flies 2400 miles into (or against) the wind in 8 hours. The return trip takes 6 hours. Find the speed of the airplane with no wind and the speed of the wind.

Solution

STEP 1: *Identify each variable.*

x: the speed of the airplane without wind
y: the speed of the wind

STEP 2: *Write a system of equations.* The speed of the airplane against the wind is $\frac{2400}{8} = 300$ miles per hour, because it traveled 2400 miles in 8 hours. The wind slowed the plane, so $x - y = 300$. Similarly, the airplane flew $\frac{2400}{6} = 400$ miles per hour with the wind because it traveled 2400 miles in 6 hours. The wind made the plane fly faster, so $x + y = 400$.

$$x - y = 300 \qquad \text{Speed against the wind}$$

$$x + y = 400 \qquad \text{Speed with the wind}$$

STEP 3A: *Solve the system of linear equations.* Solve the first equation for x to obtain $x = y + 300$. Substitute $(y + 300)$ for x in the second equation.

$$x + y = 400 \qquad \text{Second equation}$$

$$(y + 300) + y = 400 \qquad \text{Substitute } (y + 300) \text{ for } x.$$

$$2y = 100 \qquad \text{Combine like terms; subtract 300.}$$

$$y = 50 \qquad \text{Divide each side by 2.}$$

Because $x = y + 300$, it follows that $x = 50 + 300 = 350$.

CRITICAL THINKING

A boat travels 10 miles per hour upstream and 16 miles per hour downstream. How fast is the current?

STEP 3B: *Determine the solution to the problem.* The solution $x = 350$ and $y = 50$ indicates that the airplane can fly 350 miles per hour with no wind, and the wind speed is 50 miles per hour.

STEP 4: *Check the solution.* The plane flies $350 - 50 = 300$ miles per hour into the wind, taking $\frac{2400}{300} = 8$ hours. The plane flies $350 + 50 = 400$ miles per hour with the wind, taking $\frac{2400}{400} = 6$ hours. The answers check.   Now Try Exercise 83

4.2 PUTTING IT ALL TOGETHER

A system of linear equations can be solved symbolically by using the method of substitution. This method always provides the exact solution (provided one exists). A system of linear equations can have no solutions, one solution, or infinitely many solutions. These concepts are summarized in the following table.

Concept	Explanation	Example
Method of Substitution	Can be used to solve a system of equations **STEP 1:** Solve one equation for one variable. **STEP 2:** Substitute the result in the other equation and solve. **STEP 3:** Use the solution for the first variable to find the other variable. **STEP 4:** Check the solution.	$x + y = 5$ Sum is 5. $x - y = 1$ Difference is 1. **STEP 1:** Solve for x in the second equation. $$x = y + 1$$ **STEP 2:** Substitute $(y + 1)$ in the first equation for x. $$(y + 1) + y = 5$$ $$2y = 4$$ $$y = 2$$ **STEP 3:** $x = 2 + 1 = 3$ **STEP 4:** $3 + 2 = 5$ $3 - 2 = 1$ (3, 2) checks.
Inconsistent System	System of linear equations with no solutions Graphs result in parallel lines.	 No Solutions

continued on next page

continued from previous page

Concept	Explanation	Example
Consistent System with Independent Equations	System of linear equations with 1 solution Graphs result in intersecting lines.	 One Solution
Consistent System with Dependent Equations	System of linear equations with infinitely many solutions Graphs result in identical lines, or lines that coincide.	 Infinitely Many Solutions

4.2 Exercises

MyMathLab   Math XL   PRACTICE   WATCH   DOWNLOAD   READ   REVIEW

CONCEPTS

1. The technique of substituting an expression for a variable in an equation is called the method of _____.

2. A system of linear equations can have _____, _____, or _____ solutions.

3. One advantage of solving a linear system using the method of substitution rather than graphical or numerical methods is that the _____ solution can always be found (provided it exists).

4. When substituting an expression that contains two or more terms for a single variable in an equation, it is usually best to place _____ around it.

5. Suppose that the method of substitution results in the equation $1 = 1$. What does this indicate about the number of solutions to the system of equations?

6. Suppose that the method of substitution results in the equation $0 = 1$. What does this indicate about the number of solutions to the system of equations?

7. If a system of linear equations has at least one solution, then it is a(n) (consistent/inconsistent) system.

8. If a system of linear equations has no solutions, then it is a(n) (consistent/inconsistent) system.

9. If a system of linear equations has exactly one solution, then the equations are (dependent/independent).

10. If a system of linear equations has infinitely many solutions, then the equations are (<u>dependent/independent</u>).

SOLVING SYSTEMS OF EQUATIONS

Exercises 11–38: Use the method of substitution to solve the system of linear equations.

11. $x + y = 9$
$\quad y = 2x$

12. $x + y = -12$
$\quad y = -3x$

13. $x + 2y = 4$
$\quad x = 2y$

14. $-x + 3y = -12$
$\quad x = 5y$

15. $2x + y = -2$
$\quad y = x + 1$

16. $-3x + y = -10$
$\quad y = x - 2$

17. $x + 3y = 3$
$\quad x = y + 3$

18. $x - 2y = -5$
$\quad x = 4 - y$

19. $3x + 2y = \frac{3}{2}$
$\quad y = 2x - 1$

20. $-3x + 5y = 4$
$\quad y = 2 - 3x$

21. $2x - 3y = -12$
$\quad x = 2 - \frac{1}{2}y$

22. $\frac{3}{4}x + \frac{1}{4}y = -\frac{7}{4}$
$\quad x = 1 - 2y$

23. $2x - 3y = -4$
$\quad 3x - y = 1$

24. $\frac{1}{2}x - y = -1$
$\quad 2x - \frac{1}{2}y = \frac{13}{2}$

25. $x - 5y = 26$
$\quad 2x + 6y = -12$

26. $4x - 3y = -4$
$\quad x + 7y = -63$

27. $\frac{1}{2}y - z = 5$
$\quad y - 3z = 13$

28. $3y - 7z = -2$
$\quad 5y - z = 2$

29. $10r - 20t = 20$
$\quad r + 60t = -29$

30. $-r + 10t = 22$
$\quad -10r + 5t = 30$

31. $3x + 2y = 9$
$\quad 2x - 3y = -7$

32. $5x - 2y = -5$
$\quad 2x - 5y = 19$

33. $2a - 3b = 6$
$\quad -5a + 4b = -8$

34. $-5a + 7b = -1$
$\quad 3a + 2b = 13$

35. $-\frac{1}{2}x + 3y = 5$
$\quad 2x - \frac{1}{2}y = 3$

36. $3x - \frac{1}{2}y = 2$
$\quad -\frac{1}{2}x + 5y = \frac{19}{2}$

37. $3a + 5b = 16$
$\quad -8a + 2b = 34$

38. $5a - 10b = 20$
$\quad 10a + 5b = 15$

Exercises 39–44: The graphs of two equations are shown. State the number of solutions to each system of equations. Then state whether the system is consistent or inconsistent. If it is consistent, state whether the equations are dependent or independent.

39.

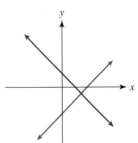

40.

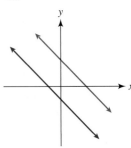

41.

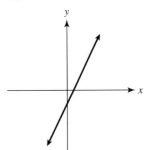

42.

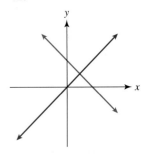

43.

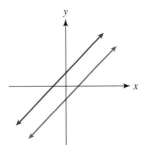

44.

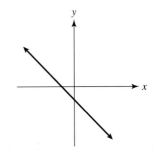

Exercises 45–52: Use the method of substitution and the intersection-of-graphs method to solve the system of linear equations. Then state whether the system is consistent or inconsistent. If it is consistent, state whether the equations are dependent or independent.

45. $x + y = 4$
$\quad x + y = 2$

46. $x + y = 10$
$\quad x - y = 2$

47. $2x - y = 3$
$\quad x - 2y = 0$

48. $-2x + y = 0$
$\quad y = 2x + 1$

49. $x - y = 1$
$2x - 2y = 2$

50. $x + 2y = 4$
$-x - 2y = -4$

51. $x - 2y = 4$
$x = 2y$

52. $-3x + y = 4$
$y = 3x + 4$

Exercises 53–70: Use the method of substitution to solve the system of linear equations. These systems may have no solutions, one solution, or infinitely many solutions.

53. $x + y = 9$
$x + y = 7$

54. $x - y = 8$
$x - y = 4$

55. $x - y = 4$
$2x - 2y = 8$

56. $2x + y = 5$
$4x + 2y = 10$

57. $x + y = 4$
$x - y = 2$

58. $x - y = 3$
$2x - y = 7$

59. $x - y = 7$
$-x + y = -7$

60. $2u - v = 6$
$-4u + 2v = -12$

61. $u - 2v = 5$
$2u - 4v = -2$

62. $3r + 3t = 9$
$2r + 2t = 4$

63. $2r + 3t = 1$
$r - 3t = -5$

64. $5x - y = -1$
$2x - 7y = 7$

65. $y = 5x$
$y = -3x$

66. $a = b + 1$
$a = b - 1$

67. $5a = 4 - b$
$5a = 3 - b$

68. $3y = x$
$3y = 2x$

69. $2x + 4y = 0$
$3x + 6y = 5$

70. $-5x + 10y = 3$
$\frac{1}{2}x - y = 1$

71. Thinking Generally If the method of substitution results in an equation that is always true, what can be said about the graphs of the two equations?

72. Thinking Generally If the method of substitution results in an equation that is always false, what can be said about the graphs of the two equations?

APPLICATIONS

73. *Rectangle* A rectangular garden is 10 feet longer than it is wide. Its perimeter is 72 feet.

(a) Let W be the width of the garden and L be the length. Write a system of linear equations whose solution gives the width and length of the garden.

(b) Use the method of substitution to solve the system. Check your answer.

74. *Isosceles Triangle* The measures of the two smaller angles in a triangle are equal and their sum equals the largest angle.

(a) Let x be the measure of each of the two smaller angles and y be the measure of the largest angle. Write a system of linear equations whose solution gives the measures of these angles.

(b) Use the method of substitution to solve the system. Check your answer.

75. *Complementary Angles* The smaller of two complementary angles is half the measure of the larger angle.

(a) Let x be the measure of the smaller angle and y be the measure of the larger angle. Write a system of linear equations whose solution gives the measures of these angles.

(b) Use the method of substitution to solve your system.

(c) Use graphing to solve the system.

76. *Supplementary Angles* The smaller of two supplementary angles is one-fourth the measure of the larger angle.

(a) Let x be the measure of the smaller angle and y be the measure of the larger angle. Write a system of linear equations whose solution gives the measures of these angles.

(b) Use the method of substitution to solve the system.

77. *Average Room Prices* In 2002, the average room price for a hotel chain was $1.68 less than in 2001. The 2002 room price was 98% of the 2001 room price. (*Source:* Smith Travel Research.)

(a) Let x be the average room price in 2001 and y be this price in 2002. Write a system of linear equations whose solution gives the average room prices for each year.

(b) Use the method of substitution to solve the system.

78. *Ticket Prices*   Two hundred tickets were sold for a baseball game, which amounted to $840. Student tickets cost $3, and adult tickets cost $5.

   (a) Let x be the number of student tickets sold and y be the number of adult tickets sold. Write a system of linear equations whose solution gives the number of each type of ticket sold.

   (b) Use the method of substitution to solve the system.

79. *NBA Basketball Court*   An official NBA basketball court is 44 feet longer than it is wide. If its perimeter is 288 feet, find its dimensions.

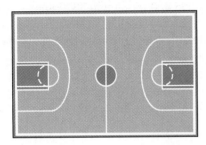

80. *Football Field*   A U.S. football field is 139.5 feet longer than it is wide. If its perimeter is 921 feet, find its dimensions.

81. *Number Problem*   The sum of two numbers is 70. The larger number is two more than three times the smaller number. Find the two numbers.

82. *Number Problem*   The difference of two numbers is 12. The larger number is one less than twice the smaller number. Find the two numbers.

83. *Speed Problem*   A tugboat goes 120 miles upstream in 15 hours. The return trip downstream takes 10 hours. Find the speed of the tugboat without a current and the speed of the current.

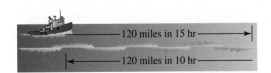

120 miles in 15 hr

120 miles in 10 hr

84. *Speed Problem*   An airplane flies 1200 miles into the wind in 3 hours. The return trip takes 2 hours. Find the speed of the airplane without a wind and the speed of the wind.

85. *Mixture Problem*   A chemist has 20% and 50% solutions of acid available. How many liters of each solution should be mixed to obtain 10 liters of a 40% acid solution?

86. *Mixture Problem*   A mechanic needs a radiator to have a 40% antifreeze solution. The radiator currently is filled with 4 gallons of a 25% antifreeze solution. How much of the antifreeze mixture should be drained from the car if the mechanic replaces it with pure antifreeze?

87. *Great Lakes*   Together, Lake Superior and Lake Michigan cover 54 thousand square miles. Lake Superior is approximately 10 thousand square miles larger than Lake Michigan. Find the size of each lake.
(***Source:*** National Oceanic and Atmospheric Administration.)

88. *Longest Rivers*   The two longest rivers in the world are the Nile and the Amazon. Together, they are 8145 miles with the Amazon being 145 miles shorter than the Nile. Find the length of each river.   (***Source:*** National Oceanic and Atmospheric Administration.)

WRITING ABOUT MATHEMATICS

89. State one advantage that the method of substitution has over the intersection-of-graphs method. Explain your answer.

90. When applying the method of substitution, how do you know that there are no solutions?

91. When applying the method of substitution, how do you know that there are infinitely many solutions?

92. When applying the intersection-of-graphs method, how do you know that there are no solutions?

CHECKING BASIC CONCEPTS
SECTIONS 4.1 AND 4.2

1. Determine graphically the x-value in each equation when $y = 2$.
   (a) $y = 1 - \frac{1}{2}x$ (b) $2x - 3y = 6$

2. Determine whether $(-1, 0)$ or $(4, 2)$ is a solution to the system

$$2x - 5y = -2$$
$$3x + 2y = 16.$$

3. Solve the system of equations graphically. Check your answer.

$$x - y = 1$$
$$2x + y = 5$$

4. Use the method of substitution to solve each system of equations. How many solutions does each have?

   (a) $x + y = -1$
   $y = 2 - x$
   (b) $4x - y = 5$
   $-x + y = -2$
   (c) $x + 2y = 3$
   $-x - 2y = -3$

5. *Room Prices*   A hotel rents single and double rooms for $150 and $200, respectively. The hotel receives $55,000 for renting 300 rooms.
   (a) Let x be the number of single rooms rented and let y be the number of double rooms rented. Write a system of linear equations whose solution gives the values of x and y.
   (b) Use the method of substitution to solve the system. Check your answer.

4.3 SOLVING SYSTEMS OF LINEAR EQUATIONS BY ELIMINATION

The Elimination Method ■ Recognizing Other Types of Systems ■ Applications

A LOOK INTO MATH ▷ Two methods for solving systems of linear equations are the intersection-of-graphs method and the substitution method. The intersection-of-graphs method is a graphical method because it involves the use of graphs to find the solution. The substitution method is a symbolic method because it involves the use of only symbols and algebra to find the solution. In this section we introduce a second symbolic method called the *elimination method*. This method is very efficient for solving some types of systems of linear equations.

The Elimination Method

The elimination method is based on the addition property of equality. Simply put, it is based on the concept that "if equals are added to equals, the results are equal." That is, if

$$a = b \quad \text{and} \quad c = d,$$

then

$$a + c = b + d.$$

For example, if the sum of two numbers is 20 and their difference is 4, then the system of equations

$$x + y = 20$$
$$x - y = 4$$

describes these two numbers. By the addition property of equality, the sum of the left sides of these equations equals the sum of their right sides.

$$(x + y) + (x - y) = 20 + 4 \quad \text{Add left sides of these equations.}$$
$$\text{Add right sides of these equations.}$$
$$2x = 24 \quad \text{Combine terms.}$$

Note that the y-variable is eliminated by adding the left sides. The resulting equation, $2x = 24$, simplifies to $x = 12$. Thus the value of x in the solution is 12. The value of y in the solution can be found by substituting 12 for x in either of the given equations. Substituting 12 for x in the first equation, $x + y = 20$, results in $12 + y = 20$ or $y = 8$. The solution to the system of equations is $x = 12, y = 8$, which can be written as the ordered pair (12, 8).

To organize the elimination method better, we can carry out the addition vertically.

$$x + y = 20$$
$$\underline{x - y = 4}$$
$$2x + 0y = 24 \quad \text{Add left sides and right sides.}$$
$$2x = 24 \quad \text{Simplify.}$$
$$x = 12 \quad \text{Divide each side by 2.}$$

Once the value of one variable is known, in this case $x = 12$, don't forget to find the value of the other variable by substituting this known value in either of the given equations. By substituting 12 for x in the second equation, we obtain $12 - y = 4$ or $y = 8$. The solution is the ordered pair (12, 8).

EXAMPLE 1 Applying the elimination method

Solve each system of equations. Check each solution.
(a) $2x + y = 1$ **(b)** $-2a + b = -3$
$ 3x - y = 9$ $ 2a + 3b = 7$

Solution
(a) Adding these two equations eliminates the y-variable.

$$2x + y = 1 \quad \text{First equation}$$
$$\underline{3x - y = 9} \quad \text{Second equation}$$
$$5x = 10, \quad \text{or} \quad x = 2 \quad \text{Add and solve for } x.$$

To find y, substitute 2 for x in either of the *given* equations.

$$2(2) + y = 1 \quad \text{Let } x = 2 \text{ in first equation.}$$
$$y = -3 \quad \text{Subtract 4 from each side.}$$

The solution is the *ordered pair* $(2, -3)$, which can be checked by substituting 2 for x and -3 for y in the given equations.

$$2x + y = 1 \qquad 3x - y = 9 \qquad \text{Given equations}$$
$$2(2) + (-3) \overset{?}{=} 1 \qquad 3(2) - (-3) \overset{?}{=} 9 \qquad \text{Let } x = 2 \text{ and } y = -3.$$
$$4 - 3 \overset{?}{=} 1 \qquad 6 + 3 \overset{?}{=} 9 \qquad \text{Simplify.}$$
$$1 = 1 \qquad 9 = 9 \qquad \text{The solution checks.}$$

(b) Adding these two equations eliminates the *a*-variable.

$$-2a + b = -3 \qquad \text{First equation}$$
$$\underline{2a + 3b = 7} \qquad \text{Second equation}$$
$$4b = 4, \quad \text{or} \quad b = 1 \qquad \text{Add and solve for } b.$$

To find *a*, substitute 1 for *b* in either of the *given* equations.

$$2a + 3(1) = 7 \qquad \text{Let } b = 1 \text{ in second equation.}$$
$$2a = 4 \qquad \text{Subtract 3 from each side.}$$
$$a = 2 \qquad \text{Divide each side by 2.}$$

The solution is the *ordered pair* $(2, 1)$, which can be checked by substituting 2 for *a* and 1 for *b* in the given equations.

$$-2a + b = -3 \qquad 2a + 3b = 7 \qquad \text{Given equations}$$
$$-2(2) + 1 \overset{?}{=} -3 \qquad 2(2) + 3(1) \overset{?}{=} 7 \qquad \text{Let } a = 2 \text{ and } b = 1.$$
$$-4 + 1 \overset{?}{=} -3 \qquad 4 + 3 \overset{?}{=} 7 \qquad \text{Simplify.}$$
$$-3 = -3 \qquad 7 = 7 \qquad \text{The solution checks.}$$

Now Try Exercises **17, 19**

Adding two equations does not always eliminate a variable. For example, adding the following equations eliminates neither variable.

$$3x - 2y = 11 \qquad \text{First equation}$$
$$\underline{4x + y = 11} \qquad \text{Second equation}$$
$$7x - y = 22 \qquad \text{Add the equations.}$$

However, by the multiplication property of equality, we can multiply the second equation by 2. Then adding the equations eliminates the *y*-variable.

$$3x - 2y = 11 \qquad \text{First equation}$$
$$\underline{8x + 2y = 22} \qquad \text{Multiply the second equation by 2.}$$
$$11x = 33, \quad \text{or} \quad x = 3 \qquad \text{Add and solve for } x.$$

In the next example, we use the multiplication property of equality.

EXAMPLE **2** **Multiplying before applying elimination**

Solve each system of equations.

(a) $5x - y = -11$ **(b)** $3x + 2y = 1$
        $2x + 3y = -1$ $2x - 3y = 5$

Solution

(a) We multiply the first equation by 3 and then add to eliminate the *y*-variable.

$$15x - 3y = -33 \qquad \text{Multiply first equation by 3.}$$
$$\underline{2x + 3y = -1} \qquad \text{Second equation}$$
$$17x = -34, \quad \text{or} \quad x = -2 \qquad \text{Add and solve for } x.$$

We can find y by substituting -2 for x in the second equation.

$$2(-2) + 3y = -1 \qquad \text{Let } x = -2 \text{ in second equation.}$$
$$3y = 3 \qquad \text{Add 4 to each side.}$$
$$y = 1 \qquad \text{Divide each side by 3.}$$

The solution is $(-2, 1)$.

(b) For this system we must apply the multiplication property to both equations. If we multiply the first equation by 3 and the second equation by 2, then the coefficients of the y-variables will be opposites. Adding eliminates the y-variable.

$$9x + 6y = 3 \qquad \text{Multiply the first equation by 3.}$$
$$\underline{4x - 6y = 10} \qquad \text{Multiply the second equation by 2.}$$
$$13x = 13, \quad \text{or} \quad x = 1 \qquad \text{Add and solve for } x.$$

To find y, substitute 1 for x in the first *given* equation.

$$3(1) + 2y = 1 \qquad \text{Let } x = 1 \text{ in first equation.}$$
$$2y = -2 \qquad \text{Subtract 3 from each side.}$$
$$y = -1 \qquad \text{Divide each side by 2.}$$

The solution is $(1, -1)$.   **Now Try Exercises** `29` `33`

In part (b) of the previous example the y-variable was eliminated. However, the system of equations could have been solved by first eliminating the x-variable. In practice, it is possible to eliminate *either* variable from a system of linear equations. It is often best to choose the variable that requires the least amount of computation to complete the elimination. In the next example, we solve a system of equations twice—first by using the multiplication property of equality to eliminate the x-variable and then using it to eliminate the y-variable.

EXAMPLE `3`   **Multiplying before applying elimination**

Solve the system of equations two times, first by eliminating x and then by eliminating y.

$$2y = -6 - 5x$$
$$2x = -5y + 6$$

Solution
It is best to write each equation in the standard form: $Ax + By = C$.

$$5x + 2y = -6 \qquad \text{First equation in standard form}$$
$$2x + 5y = 6 \qquad \text{Second equation in standard form}$$

Eliminate x   If we multiply the first equation in standard form by -2 and the second equation in standard form by 5, then we can eliminate the x-variable by adding.

$$-10x - 4y = 12 \qquad \text{Multiply by } -2.$$
$$\underline{10x + 25y = 30} \qquad \text{Multiply by 5.}$$
$$21y = 42, \quad \text{or} \quad y = 2 \qquad \text{Add and solve for } y.$$

To find x, substitute 2 for y in the first given equation, $2y = -6 - 5x$.

$$2(2) = -6 - 5x \qquad \text{Let } y = 2 \text{ in first equation.}$$
$$10 = -5x \qquad \text{Add 6 to each side.}$$
$$-2 = x \qquad \text{Divide by } -5.$$

The solution is $(-2, 2)$.

Eliminate y   If we multiply the first equation in standard form by -5 and the second equation in standard form by 2, then we can eliminate the y-variable by adding.

$$\begin{array}{ll} -25x - 10y = 30 & \text{Multiply by } -5. \\ \underline{4x + 10y = 12} & \text{Multiply by 2.} \\ -21x = 42, \quad \text{or} \quad x = -2 & \text{Add and solve for } x. \end{array}$$

To find y, substitute -2 for x in the second given equation, $2x = -5y + 6$.

$$2(-2) = -5y + 6 \qquad \text{Let } x = -2 \text{ in second equation.}$$
$$-10 = -5y \qquad \text{Subtract 6 from each side.}$$
$$2 = y \qquad \text{Divide by } -5.$$

The solution is $(-2, 2)$.

Now Try Exercise **31**

In the next example we use three different methods to solve a system of equations.

EXAMPLE 4   Solving a system with different methods

Solve the system of equations symbolically, graphically, and numerically.

$$x + y = 2$$
$$x - 3y = 6$$

Solution

Symbolic Solution   Both the method of substitution and the elimination method are symbolic methods. The elimination method is used here. We can solve the system by multiplying the second equation by -1 and adding to eliminate the x-variable.

$$\begin{array}{ll} x + y = 2 & \text{First equation} \\ \underline{-x + 3y = -6} & \text{Multiply by } -1. \\ 4y = -4, \quad \text{or} \quad y = -1 & \text{Add and solve for } y. \end{array}$$

We can find x by substituting -1 for y in the first equation, $x + y = 2$.

$$x + (-1) = 2 \qquad \text{Let } y = -1 \text{ in first equation.}$$
$$x = 3 \qquad \text{Add 1 to each side.}$$

The solution is $(3, -1)$.

Graphical Solution   For a graphical solution, we solve each equation for y to obtain the slope–intercept form.

$$\begin{array}{ll} x + y = 2 & \text{First equation} \\ y = -x + 2 & \text{Subtract } x. \end{array} \qquad \begin{array}{ll} x - 3y = 6 & \text{Second equation} \\ -3y = -x + 6 & \text{Subtract } x. \\ y = \dfrac{1}{3}x - 2 & \text{Divide by } -3. \end{array}$$

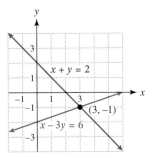

Figure 4.9

The graphs of $y = -x + 2$ and $y = \frac{1}{3}x - 2$ are shown in Figure 4.9. They intersect at $(3, -1)$. (These graphs could also be obtained by finding the x- and y-intercepts for each equation.)

Numerical Solution A numerical solution consists of a table of values, as shown in Table 4.3. Note that when $x = 3$, both y-values equal -1. Therefore the solution is $(3, -1)$.

TABLE 4.3

x	0	1	2	3	4
$y = -x + 2$	2	1	0	-1	-2
$y = \frac{1}{3}x - 2$	-2	$-\frac{5}{3}$	$-\frac{4}{3}$	-1	$-\frac{2}{3}$

Now Try Exercise **43**

Recognizing Other Types of Systems

In Section 4.2 we discussed how a system of linear equations can have no solutions, one solution, or infinitely many solutions. Elimination can also be used on systems that have no solutions or infinitely many solutions.

EXAMPLE **5**   Solving other types of systems

Solve each system of equations by using the elimination method. Then graph the system.
(a) $\quad x - 2y = 4$
$\quad\quad -2x + 4y = -8$
(b) $3x + 3y = 6$
$\quad\ x + \ y = 1$

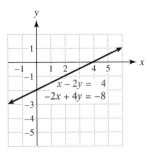

Figure 4.10

Solution
(a) We multiply the first equation by 2 and then add, which eliminates both variables.

$2x - 4y = \ \ \ 8$	Multiply first equation by 2.
$\underline{-2x + 4y = -8}$	Second equation
$\quad\quad 0 = \ \ \ 0 \quad$ (True)	Add.

The equation $0 = 0$ is *always true*, which indicates that the system has *infinitely many solutions*. A graph of the two equations is shown in Figure 4.10. The two lines are identical so there actually is only one line, and *every point on this line represents a solution*. For example, $(0, -2)$ and $(4, 0)$ lie on the line and are both solutions.
(b) We multiply the second equation by -3 and add, eliminating both variables.

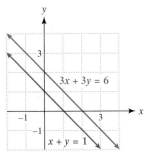

Figure 4.11

$3x + 3y = \ \ \ 6$	First equation
$\underline{-3x - 3y = -3}$	Multiply second equation by -3.
$\quad\quad 0 = \ \ \ 3 \quad$ (False)	Add.

The equation $0 = 3$ is *always false*, which indicates that the system has *no solutions*. A graph of the two equations is shown in Figure 4.11. Note that the two lines are parallel and thus do not intersect.

Now Try Exercises **53, 57**

MAKING CONNECTIONS

Numerical and Graphical Solutions

The left-hand figure shows a numerical solution for Example 5(a), where each equation is solved for y to obtain

$$Y_1 = (4 - X)/(-2) \quad \text{and} \quad Y_2 = (-8 + 2X)/4.$$

Note that $y_1 = y_2$ for each value of x, indicating that the graphs of y_1 and y_2 are the same line. Similarly, the right-hand figure shows a numerical solution for Example 5(b), where

$$Y_1 = (6 - 3X)/3 \quad \text{and} \quad Y_2 = 1 - X.$$

Note that $y_1 \neq y_2$ and the difference $y_1 - y_2$ is 1 for every value of x, indicating that the graphs of y_1 and y_2 are parallel lines that do not intersect.

X	Y1	Y2
0	-2	-2
1	-1.5	-1.5
2	-1	-1
3	-.5	-.5
4	0	0
5	.5	.5
6	1	1

Y1≣(4−X)/(-2)

X	Y1	Y2
-1	3	2
0	2	1
1	1	0
2	0	-1
3	-1	-2
4	-2	-3
5	-3	-4

Y1≣(6−3X)/3

Applications

▶ **REAL-WORLD CONNECTION**   In the next two examples we use elimination to solve applications relating to new cancer cases and burning calories during exercise.

EXAMPLE 6   ### Determining new cancer cases

In 2005, there were 1,373,000 new cancer cases. Men accounted for 47,000 more new cases than women. How many new cases of cancer were there for each gender?   (*Source:* American Cancer Society.)

Solution

STEP 1: *Identify each variable.*

x: new cancer cases for men in 2005
y: new cancer cases for women in 2005

STEP 2: *Write a system of equations.*

$$x + y = 1,373,000$$
$$x - y = 47,000$$

STEP 3A: *Solve the system of linear equations.* Add the two equations to eliminate the y-variable.

$$x + y = 1,373,000$$
$$\underline{x - y = 47,000}$$
$$2x = 1,420,000, \quad \text{or} \quad x = 710,000$$

Substituting 710,000 for x in the first equation results in $710,000 + y = 1,373,000$ or $y = 663,000$. The solution is $x = 710,000$ and $y = 663,000$.

STEP 3B: *Determine the solution to the problem.* There were 710,000 new cases of cancer for men and 663,000 new cases for women in 2005.

STEP 4: *Check the solution.* The total number of cases was

$$710,000 + 663,000 = 1,373,000.$$

The number of new cases for men exceeded the number of new cases for women by

$$710,000 - 663,000 = 47,000.$$

The answer checks.   Now Try Exercise 59

EXAMPLE 7

Burning calories during exercise

During strenuous exercise, an athlete can burn 10 calories per minute on a rowing machine and 11.5 calories per minute on a stair climber. If an athlete burns 433 calories in a 40-minute workout, how many minutes does the athlete spend on each machine? (*Source: Runner's World.*)

Solution

STEP 1: *Identify each variable.*

x: number of minutes on a rowing machine
y: number of minutes on a stair climber

STEP 2: *Write a system of equations.* The total workout takes 40 minutes, so $x + y = 40$. The athlete burns $10x$ calories on the rowing machine and $11.5y$ calories on the stair climber. Because the total number of calories equals 433, it follows that $10x + 11.5y = 433$.

$$
\begin{aligned}
x + \quad y &= 40 \qquad &\text{Workout is 40 minutes.}\\
10x + 11.5y &= 433 \qquad &\text{Total calories is 433.}
\end{aligned}
$$

STEP 3A: *Solve the system of linear equations.* Multiply the first equation by -10 and add the two equations.

$$
\begin{aligned}
-10x - \quad 10y &= -400 \qquad &\text{Multiply by } -10.\\
\underline{10x + 11.5y &= \quad 433} \qquad &\text{Second equation}\\
1.5y &= \quad 33, \quad \text{or} \quad y = \frac{33}{1.5} = 22 \qquad &\text{Add and solve for } y.
\end{aligned}
$$

Because $x + y = 40$ and $y = 22$, it follows that $x = 18$.

STEP 3B: *Determine the solution to the problem.* The athlete spends 18 minutes on the rowing machine and 22 minutes on the stair climber.

STEP 4: *Check your answer.* Because $18 + 22 = 40$, the athlete works out for 40 minutes. Also,

$$10(18) + 11.5(22) = 433,$$

so the athlete burns 433 calories. The answer checks.   Now Try Exercise 61

4.3 PUTTING IT ALL TOGETHER

The method of substitution and the elimination method are two symbolic techniques that can be used to solve systems of linear equations. The elimination method makes use of the addition and multiplication properties of equality and is based on the idea that "if equals are added to equals, the results are equal." The following table highlights the important aspects of elimination.

Concept	Explanation	Example
Elimination Method	If $a = b$ and $c = d$, then $$a + c = b + d.$$ May be used to solve systems of equations	$\begin{array}{l} x + y = 5 \\ \underline{x - y = -1} \\ 2x = 4, \quad \text{or} \quad x = 2 \quad \text{Add.} \end{array}$ Because $x + y = 5$ and $x = 2$, it follows that $y = 3$. The solution is $(2, 3)$.
Other Types of Systems	Elimination can be used to recognize systems having **1.** no solutions **2.** infinitely many solutions.	**1.** $\begin{array}{l} -x - y = -4 \\ \underline{x + y = 2} \\ 0 = -2 \quad \text{Add.} \end{array}$ Because $0 = -2$ is always false, there are no solutions. **2.** $\begin{array}{l} x + y = 4 \\ 2x + 2y = 8 \end{array}$ Multiply the first equation by -2. $\begin{array}{l} -2x - 2y = -8 \\ \underline{2x + 2y = 8} \\ 0 = 0 \quad \text{Add.} \end{array}$ Because $0 = 0$ is always true, there are infinitely many solutions.

4.3 Exercises

PRACTICE   WATCH   DOWNLOAD   READ   REVIEW

CONCEPTS

1. Name two symbolic methods for solving a system of linear equations.

2. The elimination method is based on the _____ property of equality.

3. The addition property of equality states that if $a = b$ and $c = d$, then $a + c$ _____ $b + d$.

4. The multiplication property of equality states that if $a = b$, then ca _____ cb.

5. When you are using elimination to solve

$$2x + y = 6$$
$$x - y = 2,$$

what is a good first step?

6. When you are using elimination to solve

$$x + 2y = 8$$
$$3x - 5y = 2,$$

what is a good first step?

7. Suppose that the elimination method results in the equation $1 = 1$. What does this indicate about the number of solutions to the system of equations?

8. Suppose that the elimination method results in the equation $0 = 1$. What does this indicate about the number of solutions to the system of equations?

USING ELIMINATION

Exercises 9–16: If possible, use the given graph to solve the system of equations. Then use the elimination method to verify your answer.

9. $x - y = 0$
   $x + y = 2$

10. $x + y = 6$
   $2x - y = 3$

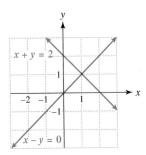

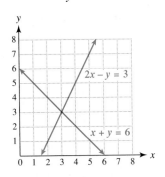

11. $2x + 3y = -1$
   $2x - 3y = -7$

12. $-2x + y = -3$
   $4x - 3y = 7$

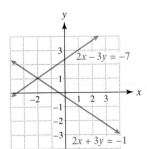

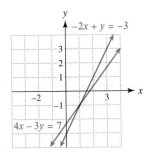

13. $x + y = 3$
   $x + y = -1$

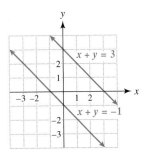

14. $2x - y = 4$
   $-2x + y = -4$

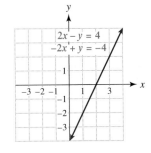

15. $2x + 2y = 6$
   $x + y = 3$

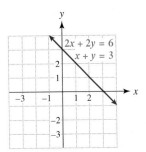

16. $-x + 3y = 4$
   $x - 3y = 3$

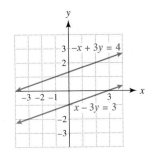

Exercises 17–38: Use the elimination method to solve the system of equations.

17. $x + y = 7$
   $x - y = 5$

18. $x - y = 8$
   $x + y = 4$

19. $-x + y = 5$
   $x + y = 3$

20. $x - y = 10$
   $-x - y = 20$

21. $2x + y = 8$
   $3x - y = 2$

22. $-x + 2y = 3$
   $x + 6y = 5$

23. $-2x + y = -3$
   $2x - 4y = 0$

24. $2x + 6y = -5$
   $7x - 6y = -4$

25. $a + 6b = 2$
   $a + 3b = -1$

26. $5a - 6b = -2$
   $5a + 5b = 9$

27. $3r - t = 7$
   $2r - t = 2$

28. $-r + 2t = 0$
   $3r + 2t = 8$

29. $3u + 2v = -16$
   $2u + v = -9$

30. $5u - v = 0$
   $3u + 3v = -18$

31. $5x - 7y = 5$
$-2x + 2y = -2$

32. $2x + 7y = 6$
$4x - 3y = -22$

33. $5x - 3y = 4$
$3x + 2y = 10$

34. $-3x - 8y = 1$
$2x + 5y = 0$

35. $\frac{1}{2}x - y = 3$
$\frac{3}{2}x + y = 5$

36. $x - \frac{1}{4}y = 4$
$-4x + \frac{1}{4}y = -9$

37. $-5x - 10y = -22$
$10x + 15y = 35$

38. $-15x + 4y = -20$
$5x + 7y = 90$

Exercises 39–42: A table of values is given for two linear equations. Use the table to solve this system.

39.

x	0	1	2	3	4
$y = -x + 5$	5	4	3	2	1
$y = 2x - 4$	-4	-2	0	2	4

40.

x	-3	-2	-1	0	1
$y = x + 1$	-2	-1	0	1	2
$y = -x - 3$	0	-1	-2	-3	-4

41.

x	-2	-1	0	1	2
$y = 3x + 1$	-5	-2	1	4	7
$y = -x + 1$	3	2	1	0	-1

42.

x	-2	-1	0	1	2
$y = 2x$	-4	-2	0	2	4
$y = -x$	2	1	0	-1	-2

USING MORE THAN ONE METHOD

Exercises 43–48: Solve the system of equations
 (a) symbolically,
 (b) graphically, and
 (c) numerically.

43. $2x + y = 5$
$x - y = 1$

44. $-x + y = 2$
$3x + y = -2$

45. $2x + y = 5$
$x + y = 1$

46. $-x + y = 2$
$3x - y = -2$

47. $6x + 3y = 6$
$-2x + 2y = -2$

48. $-x + 2y = 5$
$2x + 2y = 8$

ELIMINATION AND OTHER TYPES OF SYSTEMS

Exercises 49–58: Use elimination to determine whether the system of equations has no solutions, one solution, or infinitely many solutions. Then graph the system.

49. $2x - 2y = 4$
$-x + y = -2$

50. $-2x + y = 4$
$4x - 2y = -8$

51. $x - y = 0$
$x + y = 0$

52. $x - y = 2$
$x + y = 2$

53. $x - y = 4$
$x - y = 1$

54. $-2x + 3y = 5$
$4x - 6y = 10$

55. $x - y = 5$
$2x - y = 4$

56. $6x + 9y = 18$
$4x + 6y = 12$

57. $4x - 8y = 24$
$6x - 12y = 36$

58. $x - 3y = 2$
$-x + 3y = 4$

APPLICATIONS

59. *Skin Cancer*   In 2005, there were 66,000 new cases of skin cancer in the United States. Men represented 10,000 more cases than women. How many new cases of skin cancer were there for men and for women?   (*Source:* American Cancer Society.)

60. *Health Care Expenses* In 2002, out-of-pocket expenses for an elderly person in poor health were $3353 more than for an elderly person in good health. Combined out-of-pocket expenses for one person in poor health and one person in good health were $6213. Find the out-of-pocket expenses for each type of person.   (*Source:* "Trends in Medicare-Choice Benefits and Premiums, 1999–2002." *Commonwealth Fund.*)

61. *Burning Calories*   During strenuous exercise an athlete can burn 9 calories per minute on a stationary bicycle and 11.5 calories per minute on a stair climber. In a 30-minute workout an athlete burns 300 calories. How many minutes does the athlete spend on each type of exercise equipment?   (*Source:* Runner's World.)

62. *Distance Running* An athlete runs at 9 mph and then at 12 mph, covering 10 miles in 1 hour. How long does the athlete run at each speed?

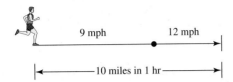

63. *River Current* A riverboat takes 8 hours to travel 64 miles downstream and 16 hours for the return trip. What is the speed of the current and the speed of the riverboat in still water?

64. *Airplane Speed* An airplane travels 3000 miles with the wind in 5 hours and takes 6 hours for the return trip into the wind. What is the speed of the wind and the speed of the airplane without any wind?

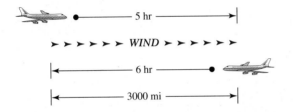

65. *Investments* A total of $5000 is invested at 3% and 5% annual interest. After 1 year the total interest equals $210. How much money is invested at each interest rate?

66. *Mixing Antifreeze* A car radiator holds 2 gallons of fluid and initially is empty. If a mixture of water and antifreeze contains 70% antifreeze and another mixture contains 15% antifreeze, how much of each should be combined to fill the radiator with a 50% antifreeze mixture?

67. *Number Problem* The sum of two integers is -17, and their difference is -69. Find the two integers.

68. *Supplementary Angles* The measures of two supplementary angles differ by $74°$. Find the two angles.

69. *Picture Dimensions* The figure at the top of the next column shows a red graph that gives possible dimensions for a rectangular picture frame with perimeter 120 inches. The blue graph shows possible dimensions for a rectangular frame whose length L is twice its width W.

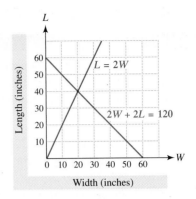

(a) Use the figure to determine the dimensions of a frame with a perimeter of 120 inches and a length that is twice the width.

(b) Solve this problem symbolically.

70. *Sales of CDs and Tapes* A company sells compact discs d and cassette tapes t. The figure shows a red graph of $d + t = 2000$. The blue graph shows a revenue of $15,000 received from selling d compact discs at $12 each and t tapes at $6 each.

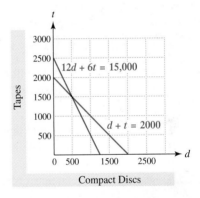

(a) If the total number of discs and tapes sold is 2000, determine how many of each were sold to obtain a revenue of $15,000.

(b) Solve this problem symbolically.

WRITING ABOUT MATHEMATICS

71. Suppose that a system of linear equations is solved symbolically, numerically, and graphically. How do the solutions from each method compare? Explain your answer.

72. When you are solving a system of linear equations by elimination, how can you recognize that the system has no solutions?

GROUP ACTIVITY
WORKING WITH REAL DATA

Directions: Form a group of 2 to 4 people. Select someone to record the group's responses for this activity. All members of the group should work cooperatively to answer the questions. If your instructor asks for your results, each member of the group should be prepared to respond.

*Exercises 1–4: **Per Capita Income**   In 2004, the average of the per capita (per person) incomes for Massachusetts and Maine was $36,500. The per capita income in Massachusetts exceeded the per capita income in Maine by $11,000.*

1. Set up a system of equations whose solution gives the per capita income in each state. Identify what each variable represents.

2. Use substitution to solve this system. Interpret the result.

3. Use elimination to solve this system.

4. Solve this system graphically. Do all your answers agree?

4.4   SYSTEMS OF LINEAR INEQUALITIES

Basic Concepts ▪ Solutions to One Inequality ▪ Solutions to Systems of Inequalities ▪ Applications

A LOOK INTO MATH ▷

Although there is no *ideal* weight for a person, government agencies and insurance companies sometimes recommend a *range* of weights for various heights. *Inequalities* are used with these recommendations. One example is shown in Figure 4.12, where the blue region contains ordered pairs (w, h) that give recommended weight–height combinations. Describing this region mathematically requires an understanding of systems of linear inequalities, which we discuss in this section.   (***Source:*** Department of Agriculture.)

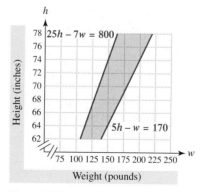

Figure 4.12

Basic Concepts

▶ REAL-WORLD CONNECTION Suppose that a college student works both at the library and at a department store. The library pays $10 per hour, and the department store pays $8 per hour. The equation $A = 10L + 8D$ calculates the amount of money earned from working L hours at the library and D hours at the department store. If the cost of one college credit is $80, then solutions to the equation

$$10L + 8D = 80$$

are ordered pairs (L, D) that result in the student earning enough to pay for one credit. Its graph is the line shown in Figure 4.13(a). The point (4, 5) lies on this line, which indicates that, if the student works 4 hours at the library and 5 hours at the department store, then the pay is $80.

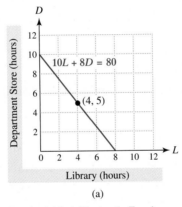
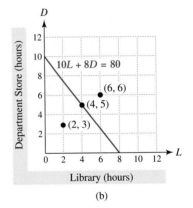

(a) (b)

Figure 4.13 A Student's Earnings

There are many situations in which the student can make more than $80. For example, the **test point** (6, 6) lies above the line in Figure 4.13(b), indicating that, if the student works 6 hours at both the library and the department store, then the pay is more than $80. In fact, any point *above* the line results in pay *greater than* $80. The region **above** the line is described by the inequality

$$10L + 8D > 80.$$

The test point (**6, 6**) represents earnings of $108 and *satisfies* this inequality because

$$10(6) + 8(6) > 80$$

is a true statement. Similarly, any point *below* the line gives an ordered pair (L, D) that results in earnings *less than* $80. The point (**2, 3**) in Figure 4.13(b) lies below the line and represents earnings of $44. That is,

$$10(2) + 8(3) < 80.$$

The region **below** the line is described by the inequality

$$10L + 8D < 80.$$

Solutions to One Inequality

Any linear equation in two variables can be written in standard form as

$$Ax + By = C,$$

where A, B, and C are constants. When the equals sign is replaced with $<$, $>$, $\leq$, or $\geq$, a **linear inequality in two variables** results. Examples of linear *equations* in two variables include

$$2x + 3y = 10 \quad \text{and} \quad y = \frac{1}{2}x - 5,$$

and so examples of linear *inequalities* in two variables include

$$2x + 3y < 10 \quad \text{and} \quad y \geq \frac{1}{2}x - 5.$$

A *solution* to a linear inequality in two variables is an ordered pair (x, y) that makes the inequality a true statement. The *solution set* is the set of all solutions to the inequality. The solution set to an inequality in two variables is typically a region in the xy-plane, which means that there are infinitely many solutions.

EXAMPLE 1   Writing a linear inequality

Write a linear inequality that describes each shaded region.

(a)

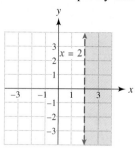

(b)

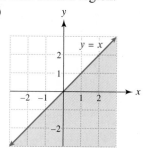

Solution

(a) The shaded region is bounded by the line $x = 2$. The *dashed* line indicates that the line is not included in the solution set. Only points with x-coordinates **greater than** 2 are shaded. Thus every point in the shaded region satisfies $x > 2$.

(b) The solution set includes all points that are on or **below** the line $y = x$. An inequality that describes this region is $y \leq x$, which can also be written as $-x + y \leq 0$.

Now Try Exercises 25, 27

TECHNOLOGY NOTE: Shading an Inequality

Graphing calculators can be used to shade a solution set to an inequality. The left-hand screen shows how to enter the equation from Example 1(b), and the right-hand screen shows the resulting graph.

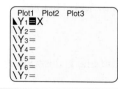

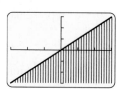

EXAMPLE 2 Graphing a linear inequality

Shade the solution set for each inequality.
(a) $y \le 1$ **(b)** $x + y < 3$ **(c)** $-x + 2y \ge 2$

Solution
(a) First graph the horizontal line $y = 1$, where a solid line indicates that the line is included in the solution set. Next decide whether to shade above or below this line. The inequality $y \le 1$ indicates that a solution (x, y) must have a y-coordinate less than or equal to 1. There are no restrictions on the x-coordinate. The shaded region **below** the line $y = 1$ in Figure 4.14(a) depicts all ordered pairs (x, y) satisfying $y \le 1$. The horizontal line $y = 1$ is solid because it is included in the solution set.

(b) Graph the line $x + y = 3$, as shown in Figure 4.14(b). Because the inequality is $<$ and not $\le$, the line is not included and is dashed rather than solid. To decide whether to shade above or below this dashed line, select a *test point* in either region. For example, the point $(0, 0)$ satisfies the inequality $x + y < 3$, because $0 + 0 < 3$ is a true statement. Therefore shade the region that does contain $(0, 0)$, which is *below* the line.

(c) Graph $-x + 2y = 2$ as a solid line, as shown in Figure 4.14(c). To decide whether to shade above or below the solid line, use the test point $(0, 0)$ again. (Note that other test points can be used.) The point $(0, 0)$ does not satisfy the inequality $-x + 2y \ge 2$ because $-0 + 2(0) \ge 2$ is a false statement. Therefore shade the region that does not contain $(0, 0)$, which is *above* the line.

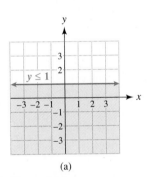
(a)

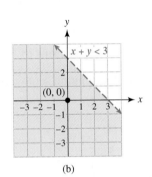
(b)

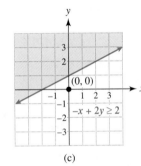
(c)

Figure 4.14

Now Try Exercises 33, 39, 41

NOTE: In both parts (b) and (c) of Example 2, the test point $(0, 0)$ was used to determine which region should be shaded. However, *any* point that is not on the solid or dashed line can be used as a test point. When the line does not pass through the origin, it is often convenient to use $(0, 0)$ as a test point because substituting 0 for both x and y in an inequality results in a very simple computation.

GRAPHING A LINEAR INEQUALITY

1. Replace the inequality symbol with an equals sign and graph the resulting line. If the inequality is $<$ or $>$, use a dashed line, and if it is $\le$ or $\ge$, use a solid line.
2. Pick a test point that does *not* lie on the line. Substitute this point in the given inequality. Determine whether the resulting statement is true or false.
3. If the statement is true, shade the region containing the test point. If the statement is false, shade the region not containing the test point.

Solutions to Systems of Inequalities

Sometimes a solution set must satisfy two inequalities. The point (2, 1) satisfies both of the inequalities in the system of inequalities given by

$$x > 1$$
$$y < 2.$$

Figure 4.15(a) shows the solution set to $x > 1$ in blue, and Figure 4.15(b) shows the solution set to $y < 2$ in red. The two regions are shaded together in Figure 4.15(c) where the blue and red regions intersect to form a purple region. The solution set to the system of inequalities includes all points in the purple region, as shown in Figure 4.15(d). Points in this region satisfy *both* inequalities.

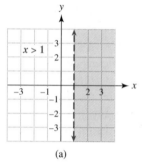

(a)

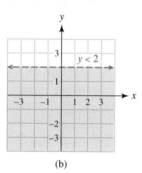

(b)

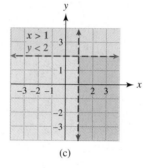

(c)

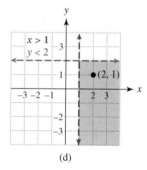

(d)

Figure 4.15

EXAMPLE 3   **Graphing a system of linear inequalities**

Shade the solution set to the system of inequalities.

$$x > -1$$
$$x + y \leq \ \ 1$$

Solution
In this example, we start by graphing the solution set to each inequality. The solution set to $x > -1$ is the blue region to the right of the dashed vertical line $x = -1$, as shown in Figure 4.16(a). The solution set to $x + y \leq 1$ includes the solid line $x + y = 1$ and the red region that lies below it, as shown in Figure 4.16(b). For a point to satisfy the *system* of inequalities it must satisfy *both* inequalities. Therefore the solution set is the *intersection* of the blue and red regions, shown as the purple region in Figure 4.16(c). Note that the test point (0, 0), located in the shaded region, satisfies both inequalities.

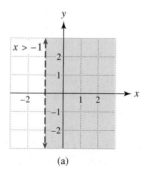

(a)

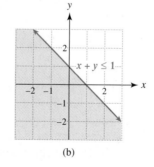

(b)

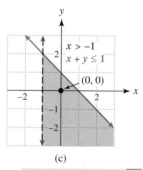

(c)

Figure 4.16

Now Try Exercise 65

EXAMPLE 4   Graphing a system of linear inequalities

Shade the solution set to the system of inequalities.

$$x + 2y < -2$$
$$2x + y \geq 2$$

Solution

In this example we use test points to determine the solution set. We start by graphing the dashed line $x + 2y = -2$ and the solid line $2x + y = 2$, as shown in Figure 4.17(a). Note that these two lines divide the xy-plane into 4 regions, numbered 1, 2, 3, and 4. If we let $(0, 0)$ be a test point, it does not satisfy either inequality. Therefore we do not shade region 2, which contains $(0, 0)$. However, there are still 3 possible regions. If we try the test point $(4, -4)$ in region 4, it satisfies both the given inequalities.

$$4 + 2(-4) < -2 \qquad \text{A true statement}$$
$$2(4) + (-4) \geq 2 \qquad \text{A true statement}$$

Thus we shade region 4, as shown in Figure 4.17(b). Once a test point is found that makes both inequalities true, there is no need to check test points in other regions.

CRITICAL THINKING

Does the solution set in Figure 4.17(b) include the point of intersection, $(2, -2)$? Explain your reasoning.

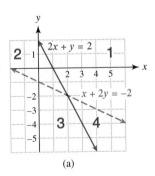

(a)

Figure 4.17

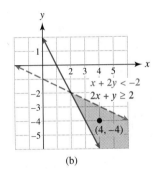

(b)

Now Try Exercise 61

MAKING CONNECTIONS

Solving for y and shading inequalities

Another way to solve the system of inequalities in Example 4 that does not involve test points is to solve each inequality for y to obtain

$$y < -\frac{1}{2}x - 1$$
$$y \geq -2x + 2.$$

The solution set is the region in Figure 4.17(b) that lies *below* the line $y = -\frac{1}{2}x - 1$ and *above and including* the line $y = -2x + 2$.

Applications

▶ REAL-WORLD CONNECTION   The next two examples illustrate applications of inequalities.

EXAMPLE ⑤   Manufacturing radios and CD players

A business manufactures radios and CD players. Because every CD player contains a radio, it must produce at least as many radios as CD players. In addition, the total number of radios and CD players produced each day cannot exceed 50 because of limited resources. Shade the region that shows the numbers of radios R and CD players P that can be produced within these restrictions. Label the horizontal axis R and the vertical axis P.

Solution

Because the company must produce *at least* as many radios R as CD players P, we have $R \geq P$, which can also be written as $P \leq R$. The total number of radios and CD players *cannot exceed* 50 so $R + P \leq 50$. To shade the solution set for

$$P \leq R$$
$$R + P \leq 50,$$

we first graph the lines $P = R$ and $R + P = 50$, as shown in Figure 4.18(a). Because the number of radios and CD players cannot be negative, the graph includes only quadrant I. These lines divide this quadrant into four regions, and we can determine the correct region to shade by selecting one test point from each region. The region containing the test point satisfying both inequalities is the one to be shaded. For example, the test point (20, 10) with $R = 20, P = 10$ satisfies both inequalities.

$$10 \leq 20 \qquad \text{A true statement; } P \leq R$$
$$20 + 10 \leq 50 \qquad \text{A true statement; } R + P \leq 50$$

The solution set is shaded in Figure 4.18(b).

NOTE: An alternative solution is to write the inequalities as $P \leq R$ and $P \leq -R + 50$. Then the solution set lies *below both lines*. This region is shaded in Figure 4.18(b).

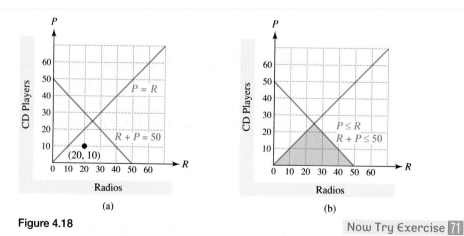

Figure 4.18

Now Try Exercise 71

The next example discusses the application from the beginning of this section.

EXAMPLE 6   Finding weight–height combinations

Figure 4.19 shows a shaded region containing recommended weights w for heights h.

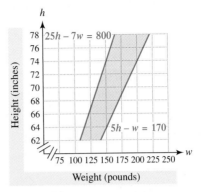

Figure 4.19

(a) What does this graph indicate about someone who is 68 inches tall and weighs 150 pounds?

(b) The shaded region in Figure 4.19 is determined by the following system of inequalities.

$$25h - 7w \leq 800$$
$$5h - w \geq 170$$

Verify that $h = 68$ and $w = 150$ satisfies the system of inequalities.

(c) What do ordered pairs (w, h) to the left of the shaded region indicate?

CRITICAL THINKING

In Example 6 what do ordered pairs (w, h) to the right of the shaded region represent? Explain your answer.

Solution

(a) The point $(150, 68)$ lies in the shaded region. Therefore someone who is 68 inches tall and weighs 150 pounds falls within the recommended guidelines.

(b) Both inequalities are satisfied by $h = 68$, $w = 150$.

$$25(68) - 7(150) = 650 \leq 800$$
$$5(68) - 150 = 190 \geq 170$$

(c) To the left of the shaded region are ordered pairs (w, h) that represent smaller weights and larger heights. This region corresponds to people who weigh less than recommended.

Now Try Exercise 75

4.4 PUTTING IT ALL TOGETHER

The following table summarizes important concepts related to linear inequalities in two variables.

Concept	Explanation	Examples
Linear Inequality in Two Variables	An inequality that can be written as $$Ax + By < C,$$ where $<$ can also be $\leq$, $>$, or $\geq$	$3x + y \geq 10$, $-x + 3y < 5$, $y \leq 5 - x$, and $x > 5$

continued on next page

continued from previous page

Concept	Explanation	Examples
Solution	A solution (x, y) makes the inequality a true statement.	The point $(0, 0)$ satisfies $$2x - y < 2,$$ so it is a solution to the inequality.
Solution Set	The set of all solutions Usually a region in the xy-plane	The solution set to $x + y > 2$ is all points above the line $x + y = 2$.
System of Linear Inequalities in Two Variables	Solutions to systems must satisfy both inequalities. The solution set usually includes infinitely many solutions.	The point $(0, 0)$ is a solution to $$x + y \leq 2$$ $$2x - y > -4,$$ because both inequalities are true when $x = 0$ and $y = 0$.

4.4 Exercises

CONCEPTS

1. Describe the graph of the solution set to $y \leq k$ for some number k.

2. Describe the graph of the solution set to $x > k$ for some number k.

3. Describe the graph of the solution set to $y \geq x$.

4. When graphing the solution set to a linear inequality, one way to determine which region to shade is to use a _____ point.

5. When graphing a linear inequality containing either $<$ or $>$, use a _____ line.

6. When graphing a linear inequality containing either $\leq$ or $\geq$, use a _____ line.

7. When graphing the linear inequality $Ax + By < C$, a first step is to graph the line _____.

8. A solution to a system of two inequalities must make (both inequalities/one inequality) true.

9. If two shaded regions represent the solution sets for two inequalities in a system, then the solution set for the system is where these two shaded regions _____.

10. If a test point is found that satisfies both inequalities in a system, do other test points still need to be checked?

SOLUTIONS TO LINEAR INEQUALITIES

Exercises 11–22: Determine whether the test point is a solution to the linear inequality.

11. $(3, 1), x > 2$

12. $(-3, 4), x \le -3$

13. $(0, 0), y \ge 2$

14. $(0, 0), y < -3$

15. $(5, 4), y \ge x$

16. $(-1, 2), y < x$

17. $(3, 0), y < x - 1$

18. $(0, 5), y > 2x + 4$

19. $(-2, 6), x + y \le 4$

20. $(2, -4), x - y \ge 7$

21. $(-1, -1), 2x + y \ge -1$

22. $(0, 1), -x - 5y \ge -1$

Exercises 23–30: Write a linear inequality that describes the shaded region.

23.

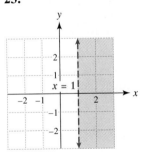

24.

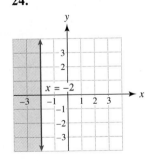

25.

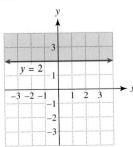

26.

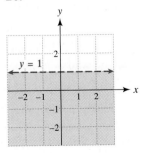

27.

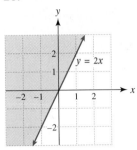

28.

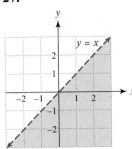

29.

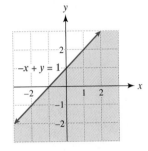

30.

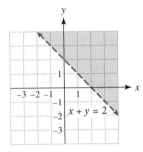

Exercises 31–42: Shade the solution set to the inequality.

31. $x \le -1$

32. $x > 3$

33. $y < -2$

34. $y \ge 0$

35. $y > x$

36. $y \le x$

37. $y \ge 3x$

38. $y < -2x$

39. $x + y \le 1$

40. $x + y \ge -2$

41. $2x - y > 2$

42. $-x - y < 1$

Exercises 43–48: Determine if the test point is a solution to the system of linear inequalities.

43. $(3, 1)$
$x - y < 3$
$x + y > 3$

44. $(0, 0)$
$x - 2y < 1$
$2x - y > -1$

45. $(-2, 3)$
$3x - 2y \ge 1$
$-x + 3y > 3$

46. $(1, 2)$
$2x - 2y < 5$
$x - y > -1$

47. $(4, -2)$
$x - 2y \ge 8$
$-2x - 5y > 0$

48. $(-1, -2)$
$x + y < 0$
$-2x - 3y \le -1$

Exercises 49–52: The graphs of two equations are shown with four test points labeled. Use these points to decide which region should be shaded to solve the given system of inequalities.

49. $x \le 2$
$x + y \ge 2$

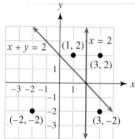

50. $y \ge 1$
$2x - y \ge 3$

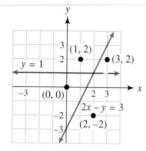

51. $x + y \le 3$
$y \le 2x$

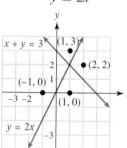

52. $y \le x$
$y \ge -x$

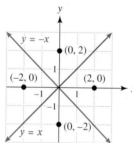

Exercises 53–70: Shade the solution set to the system of inequalities.

53. $x > 2$
$y < 3$

54. $x \le -1$
$y \ge 3$

55. $x \le -2$
$y < 2x$

56. $y > 2$
$y \ge -x$

57. $y \le x$
$y > -x$

58. $y \le \frac{1}{2}x$
$y \ge -2x$

59. $x + y \le 3$
$-x + y \le 1$

60. $x + y > 2$
$x - y < 2$

61. $2x + y > -3$
$x + y \le -1$

62. $-x + y \ge 3$
$2x - y \ge -2$

63. $2x + y \ge -3$
$x + y > -1$

64. $-x + y \ge 2$
$3x - y \ge -2$

65. $y > -2$
$x + 2y \le -4$

66. $x \ge 2$
$3y < x - 3$

67. $x + 2y > -4$
$2x + y \le 3$

68. $x + 3y \ge 3$
$3x - 2y \ge 6$

69. $3x + 4y \le 12$
$5x + 3y \ge 15$

70. $-2x + y \ge 6$
$x - 2y \ge -4$

APPLICATIONS

71. *Radios and CDs*   (Refer to Example 5.) A business manufactures at least two radios for each CD player. The total number of radios and CD players must be less than 90. Shade the region that represents the number of radios R and CD players P that can be produced within these restrictions. Put P on the horizontal axis.

72. *Working on Two Projects*   An employee is required to spend more time on project X than on project Y. The employee can work at most 40 hours on these two projects. Shade the region in the *xy*-plane that represents the number of hours that the employee can spend on each project.

73. *Maximum Heart Rate*   When exercising, people often try to maintain target heart rates that are a percentage of their maximum heart rate. Maximum heart rate R is $R = 220 - A$, where A is the person's age and R is the heart rate in beats per minute.
 (a) Find R for a person 20 years old; 70 years old.
 (b) Sketch a graph of $R \le 220 - A$. Assume that A is between 20 and 70 and put A on the horizontal axis of your graph.
 (c) Interpret your graph.

74. *Target Heart Rate*   (Refer to the preceding exercise.) A target heart rate T that is half a person's maximum heart rate is given by $T = 110 - \frac{1}{2}A$, where A is a person's age.
 (a) What is T for a person 30 years old? 50 years old?
 (b) Sketch a graph of the system of inequalities.

$$T \ge 110 - \frac{1}{2}A$$
$$T \le 220 - A$$

 Assume that A is between 20 and 60.
 (c) Interpret your graph.

75. *Height and Weight*   Use Figure 4.19 in Example 6 to determine the range of recommended weights for a person who is 74 inches tall.

76. *Height and Weight*   Use Figure 4.19 in Example 6 to determine the range of recommended heights for a person who weighs 150 pounds.

WRITING ABOUT MATHEMATICS

77. What is the solution set to the following system of inequalities? Explain your reasoning.

$$y > x$$
$$y < x - 1$$

78. Write down a system of linear inequalities whose solution set is the entire xy-plane. Explain your reasoning.

CHECKING BASIC CONCEPTS
SECTIONS 4.3 AND 4.4

1. Use elimination to solve the system of equations.

$$2x + 3y = 5$$
$$x - 7y = -6$$

2. Use elimination to solve each system of equations. How many solutions are there in each case?

(a) $\quad x + y = -1$
$\quad\quad x - 2y = 2$

(b) $\quad 5x - 6y = 4$
$\quad -5x + 6y = 1$

(c) $\quad x - 3y = 0$
$\quad\; 2x - 6y = 0$

3. Solve the system of equations symbolically, graphically, and numerically.

$$-2x + y = 0$$
$$y = 2x$$

4. Shade the solution set to each inequality.
(a) $y < -1$ (b) $x + y < 1$

5. Shade the solution set to the given system of inequalities.

$$x \le -1$$
$$-2x + y > -3$$

6. *Large Cities in the United States* The combined population of New York and Chicago was 11 million people in 2005. The population of New York exceeded the population of Chicago by 5 million people.
(a) Let x be the population of New York and y be the population of Chicago. Write a system of equations whose solution gives the population of each city in 2005.
(b) Solve the system of equations.

CHAPTER 4 SUMMARY

SECTION 4.1 ■ SOLVING SYSTEMS OF LINEAR EQUATIONS GRAPHICALLY AND NUMERICALLY

System of Linear Equations

Solution — An ordered pair (x, y) that satisfies *both* equations

Solution Set — The set of all solutions

Graphical Solution — Graph each equation. A point of intersection is a solution. (Sometimes determining the exact answer when estimating from a graph may be difficult.)

Numerical Solution — Solve each equation for y and make a table for each equation. A solution occurs when two y-values are equal for a given x-value.

Example: The ordered pair $(3, 1)$ is the solution to the following system.

$$x + y = 4 \qquad 3 + 1 = 4 \text{ is a true statement.}$$
$$x - y = 2 \qquad 3 - 1 = 2 \text{ is a true statement.}$$

A Graphical Solution

The point of intersection, (3, 1), is the solution to the system of equations.

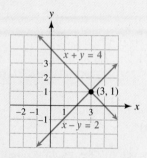

A Numerical Solution

The ordered pair (3, 1) is the solution. When $x = 3$, both y-values equal 1.

	x	1	2	3	4
$y = 4 - x$		3	2	1	0
$y = x - 2$		−1	0	1	2

SECTION 4.2 ■ SOLVING SYSTEMS OF LINEAR EQUATIONS BY SUBSTITUTION

Method of Substitution   This method can be used to solve a system of equations symbolically and always gives the exact solution, provided one exists.

Example: $-2x + y = -3$
$x + y = 3$

STEP 1: Solve one of the equations for a convenient variable.

$$x + y = 3 \quad \text{becomes} \quad y = 3 - x.$$

STEP 2: Substitute this result in the other equation and then solve.

$$-2x + (3 - x) = -3 \qquad \text{Substitute } (3 - x) \text{ for } y.$$
$$-3x = -6 \qquad \text{Combine like terms; subtract 3.}$$
$$x = 2 \qquad \text{Divide each side by } -3.$$

STEP 3: Find the value of the other variable. Because $y = 3 - x$ and $x = 2$, it follows that $y = 3 - 2 = 1$.

STEP 4: Check to determine that (2, 1) is the solution.

$$-2(2) + (1) \overset{?}{=} -3 \qquad \text{A true statement}$$
$$2 + 1 \overset{?}{=} 3 \qquad \text{A true statement}$$

The solution (2, 1) checks.

Types of Systems of Linear Equations   Can have no solutions, one solution, or infinitely many solutions

No solutions	Inconsistent (parallel lines)
One solution	Consistent (independent equations)
Infinitely many solutions	Consistent (dependent equations)

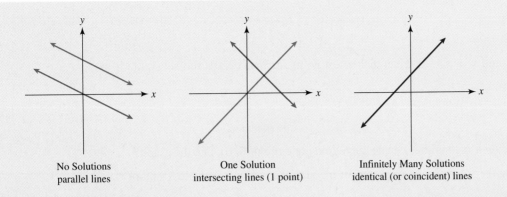

| No Solutions | One Solution | Infinitely Many Solutions |
| parallel lines | intersecting lines (1 point) | identical (or coincident) lines |

SECTION 4.3 ■ SOLVING SYSTEMS OF LINEAR EQUATIONS BY ELIMINATION

Method of Elimination   This method can be used to solve a system of linear equations symbolically and always gives the exact solution, provided one exists.

Example:
$$x + 3y = 1$$
$$\underline{-x + \ y = 3}$$
$$4y = 4, \quad \text{or} \quad y = 1 \qquad \text{Add and solve for } y.$$

Substitute $y = 1$ in either of the given equations: $x + 3(1) = 1$ implies that $x = -2$, so $(-2, 1)$ is the solution.

NOTE: To eliminate a variable, it may be necessary to multiply one or both equations by a constant before adding.

Recognizing Types of Systems

No solutions	Final equation is always false, such as $0 = 1$.
One solution	Final equation has one solution, such as $x = 1$.
Infinitely many solutions	Final equation is always true, such as $0 = 0$.

SECTION 4.4 ■ SYSTEMS OF LINEAR INEQUALITIES

Graphing a Linear Inequality in Two Variables

1. Replace the inequality symbol with an equals sign and graph the resulting line. If the inequality is $<$ or $>$ use a dashed line, and if it is $\leq$ or $\geq$ use a solid line.
2. Pick a *test point* that does *not* lie on the line. Substitute this point in the given inequality. Determine whether the resulting statement is true or false.
3. If the statement is true, shade the region containing the test point. If the statement is false, shade the region on the other side of the line.

Solving a System of Linear Inequalities

1. Perform Step 1 above for each inequality in the system.
2. Pick a test point from one region and substitute it in the given inequalities.
3. If the resulting statements are true, shade the region containing the test point. If not, pick a test point from a different region and substitute it in the given inequalities. Repeat this step until the region to be shaded is found.

Example: $x \le 1$

$x + y \le 2$

Graph the lines $x = 1$ and $x + y = 2$. Then pick a test point, such as $(0, 0)$, and substitute it in each inequality.

$$0 \le 1 \qquad \text{A true statement}$$

$$0 + 0 \le 2 \qquad \text{A true statement}$$

Because $(0, 0)$ satisfies *both* inequalities, shade the region containing $(0, 0)$. See the graph.

NOTE: When shading the solution set to a *system* of inequalities, you may need to try more than one test point.

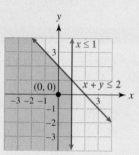

NOTE: An alternative way to determine the solution set is to shade the region to the *left* of the line $x = 1$ (because $x \le 1$) and *below* the line $y = -x + 2$ (because $y \le -x + 2$.)

CHAPTER 4   REVIEW EXERCISES

SECTION 4.1

Exercises 1 and 2: Determine graphically the x-value for the equation when $y = 3$.

1. $y = 2x - 3$

2. $y = \frac{3}{2}x$

Exercises 3–6: Determine which ordered pair is a solution to the system of equations.

3. $(0, 1), (1, 2)$
$x + 2y = 5$
$x - y = -1$

4. $(5, 2), (4, 0)$
$2x - y = 8$
$x + 3y = 11$

5. $(2, 2), (4, 3)$
$\frac{1}{2}x = y - 1$
$2x = 3y - 1$

6. $(2, -4), (-1, 2)$
$5x - 2y = 18$
$y = -2x$

Exercises 7 and 8: The graphs for two equations are shown. Use the intersection-of-graphs method to identify the solution to both equations. Then check your result.

7.

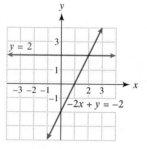

8.

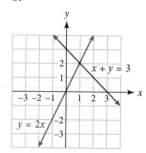

Exercises 9 and 10: A table for two equations is given. Identify the solution to both equations.

9.

x	1	2	3	4
$y = 3x$	3	6	9	12
$y = 6$	6	6	6	6

10.

x	−1	0	1	2
$y = 2x - 1$	−3	−1	1	3
$y = 2 - x$	3	2	1	0

Exercises 11–16: Solve the system of linear equations graphically.

11. $y = -3$
$x + y = 1$

12. $x = 1$
$x - y = -1$

13. $2x + y = 3$
$-x + y = 0$

14. $y = 2x$
$2x + y = 4$

15. $x + 2y = 3$
$2x + y = 3$

16. $-3x - y = 7$
$2x + 3y = -7$

SECTION 4.2

Exercises 17–22: Use the method of substitution to solve the system of linear equations.

17. $x + y = 8$
$y = 3x$

18. $x - 2y = 22$
$y = -5x$

19. $2x + y = 5$
$-3x + y = 0$

20. $3x - y = 5$
$x - y = -5$

21. $x + 3y = 1$
$-2x + 2y = 6$

22. $3x - 2y = -4$
$2x - y = -4$

Exercises 23–26: The graphs of two equations are shown.
(a) *State the number of solutions to the system of equations.*
(b) *Is the system consistent or inconsistent? If the system is consistent, state whether the equations are dependent or independent.*

23.

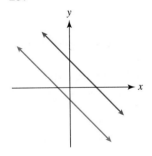

24.

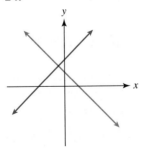

25.

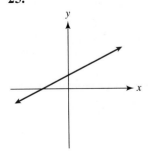

26.

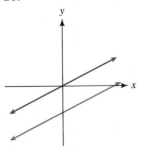

Exercises 27–30: Use the method of substitution to solve the system of linear equations. Then solve the system graphically. Note that these systems may have no solutions, one solution, or infinitely many solutions.

27. $x + y = 2$
$y = -x$

28. $x + y = -2$
$x + y = 3$

29. $-x + 2y = 2$
$x - 2y = -2$

30. $-x - y = -2$
$2x - y = 1$

SECTION 4.3

Exercises 31 and 32: Use the graph to solve the system of equations. Then use the elimination method to verify your answer.

31. $x + y = 3$
$x - y = 1$

32. $2x + 3y = 4$
$x - 2y = -5$

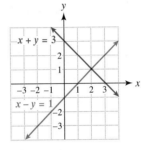

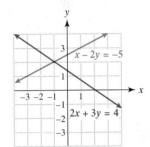

Exercises 33–40: Use the elimination method to solve the system of equations.

33. $x + y = 10$
$x - y = 12$

34. $2x - y = 2$
$3x + y = 3$

35. $-2x + 2y = -1$
$x - 3y = -3$

36. $2x - 5y = 0$
$2x + 4y = 9$

37. $2a + b = 3$
$-3a - 2b = -1$

38. $a - 3b = 2$
$3a + b = 26$

39. $5r + 3t = -1$
$-2r - 5t = -11$

40. $5r + 2t = 5$
$3r - 7t = 3$

Exercises 41 and 42: Solve the system of equations
(a) *symbolically,* **(b)** *graphically, and* **(c)** *numerically.*

41. $3x + y = 6$
$x - y = -2$

42. $2x + y = 3$
$-x + 2y = -4$

Exercises 43–46: Use elimination to determine whether the system of equations has no solutions, one solution, or infinitely many solutions.

43. $x - y = 5$
   $-x + y = -5$

44. $3x - 3y = 0$
   $-x + y = 0$

45. $-2x + y = 3$
   $2x - y = 3$

46. $-2x + y = 2$
   $3x - y = 3$

SECTION 4.4

Exercises 47–50: Determine whether the test point is a solution to the linear inequality.

47. $(5, -3)$ $y \le 2$

48. $(-1, 3)$ $x > -1$

49. $(1, 2)$ $x + y < -2$

50. $(1, -4)$ $2x - 3y \ge 2$

Exercises 51 and 52: Write a linear inequality that describes the shaded region.

51.

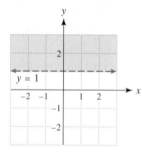

52.

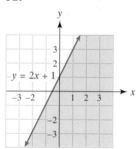

Exercises 53–58: Shade the solution set for the inequality.

53. $x \le 1$

54. $y > 2$

55. $y > 3x$

56. $x \ge 2y$

57. $y < x + 1$

58. $2x + y \ge -2$

Exercises 59 and 60: Determine whether the test point is a solution to the system of linear inequalities.

59. $(1, -2)$
   $x - 2y > 3$
   $2x + y < 3$

60. $(4, -3)$
   $x - y \ge 1$
   $4x + 3y \le 4$

Exercises 61 and 62: The graphs of two equations are shown at the top of the next column with four test points labeled. Use these points to decide which region should be shaded to solve the system of inequalities.

61. $y \le 1$
   $2x + y \ge -1$

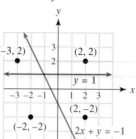

62. $y \ge x$
   $x + y \ge 2$

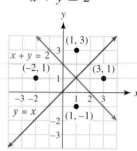

Exercises 63–68: Shade the solution set for the system of inequalities.

63. $x > -1$
   $y < -2$

64. $y \le x$
   $y \ge -2x$

65. $x + y \le 3$
   $y \ge -x$

66. $2x + y < 3$
   $y > x$

67. $\frac{1}{2}x + y \ge 2$
   $x - 2y \le 0$

68. $2x - y < 3$
   $4x + 2y > -6$

APPLICATIONS

69. *Motor Vehicle Fatalities*   The number of motor vehicle deaths increased by 13.75 times from 1912 to 2003. There were 39,525 more deaths in 2003 than in 1912. Find the number of motor vehicle deaths in each of the two years. Note that the number of motor vehicles on the road increased from 1 million to 230 million between 1912 and 2003.   (*Source:* Department of Health and Human Services.)

70. *Lung Cancer*   In 2005, 185,000 new cases of lung cancer were reported. There were 20,000 more new cases for men than for women. How many new cases of lung cancer were there for men and for women? (*Source:* American Cancer Society.)

71. *Renting a Car*   A rental car costs $40 plus $0.20 per mile that it is driven.
   (a) Write an equation that gives the cost C of driving the car x miles.
   (b) Use the intersection-of-graphs method to determine the number of miles that the car is driven if the rental cost is $90.
   (c) Solve part (b) numerically with a table of values.

72. *Supplementary Angles*   The smaller of two supplementary angles is 30° less than the measure of the larger angle. Find each angle.

73. *Triangle*   In an isosceles triangle, the measures of the two smaller angles are equal and their sum is 40° more than the larger angle.
   (a) Let x be the measure of each of the two smaller angles and y be the measure of the larger angle. Write a system of linear equations whose solution gives the measures of these angles.
   (b) Use the method of substitution to solve the system.
   (c) Use the method of elimination to solve the system.

74. *Dimensions of a Garden*   A rectangular garden has 88 feet of fencing around it. The garden is 4 feet longer than it is wide. Find the dimensions of the garden.

75. *Room Prices*   Ten rooms are rented at rates of $80 and $120 per night. The total collected for the 10 rooms is $920.
   (a) Write a system of linear equations whose solution gives the number of each type of room rented. Be sure to state what each variable represents.
   (b) Solve the system of equations.

76. *Mixture Problem*   One type of candy sells for $2 per pound, and another type sells for $3 per pound. An order for 18 pounds of candy costs $47. How much of each type of candy was bought?

77. *Burning Calories*   An athlete burns 9 calories per minute on a stationary bicycle and 11 calories per minute on a stair climber. In a 60-minute workout the athlete burns 590 calories. How many minutes does the athlete spend on each type of exercise equipment?
   (**Source:** *Runner's World.*)

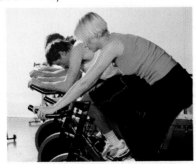

78. *River Current*   A riverboat travels 140 miles downstream in 10 hours, and the return trip takes 14 hours. What is the speed of the current?

79. *Garage Dimensions*   The blue graph shown in the figure gives possible dimensions for a rectangular garage with perimeter 80 feet. The red graph shows possible dimensions for a garage that has width W two-thirds of its length L.

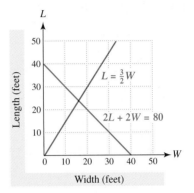

Width (feet)

   (a) Use the graph to estimate the dimensions of a garage with perimeter 80 feet and width two-thirds its length.
   (b) Solve this problem symbolically.

80. *Wheels and Trailers*   A business manufactures at least two wheels for each trailer it makes. The total number of trailers and wheels manufactured cannot exceed 30 per week. Shade the region that represents numbers of wheels W and trailers T that can be produced each week within these restrictions. Label the horizontal axis W and the vertical axis T.

81. *Target Heart Rate*   A target heart rate T that is 70% of a person's maximum heart rate is approximated by $T = 150 - 0.7A$, where A is a person's age.
   (a) What is T for a person 20 years old? 60 years old?
   (b) Sketch a graph of $T \geq 150 - 0.7A$. Assume that A is between 20 and 60.
   (c) Interpret this graph.

CHAPTER 4   TEST Pass the Test Video solutions to all test exercises

1. Determine which ordered pair is a solution to the system of equations.

$$(3, -1), (1, 2)$$
$$3x + 2y = 7$$
$$2x - y = 0$$

2. The graphs for two equations are shown. Use the intersection-of-graphs method to identify the solution. Then check your solution.

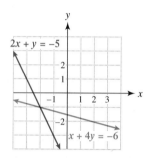

3. A table for two equations is given. Identify the solution to both equations.

x	-2	-1	0	1
$y = 2x$	-4	-2	0	2
$y = 3x + 1$	-5	-2	1	4

4. Solve the system of equations graphically.

$$x + 2y = 4$$
$$x + y = 1$$

5. Use the method of substitution to solve the system of linear equations.

$$3x + 2y = 9$$
$$y = 3x$$

6. Use the method of substitution to solve the system of linear equations. How many solutions are there? Is the system consistent or inconsistent?

(a) $x + 3y = 5$
    $3x - 2y = 4$

(b) $-x + \frac{1}{2}y = 12$
    $2x - y = -4$

Exercises 7 and 8: The graphs of two equations are shown.
(a) *State the number of solutions to the system of equations.*
(b) *Is the system consistent or inconsistent? If the system is consistent, state whether the equations are dependent or independent.*

7.
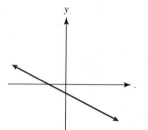

8.
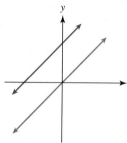

Exercises 9–12: Use the elimination method to solve the system of equations. Note that these systems may have no solutions, one solution, or infinitely many solutions.

9. $x + 2y = 5$
    $3x - 2y = -17$

10. $2x - 2y = 3$
     $-x + y = 5$

11. $x - 2y = 3$
     $-3x + 6y = -9$

12. $4x + 3y = 5$
     $3x - 2y = -9$

13. Write a linear inequality that describes the shaded region.

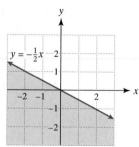

14. Determine whether the test point $(4, -3)$ is a solution to the system of linear inequalities.

$$2x + y > 3$$
$$x - y \geq 7$$

Exercises 15 and 16: Shade the solution set for the given inequality.

15. $x \leq 4$

16. $x + y > 2$

Exercises 17 and 18: Shade the solution set for the given system of inequalities.

17. $x > 2$
$\quad y < 2x$

18. $2x + y \le 3$
$\quad\quad x - y \ge 0$

19. *IRS Collections*   In 2003 and 2004, the IRS collected a total of $3.9 trillion in taxes. The IRS collected $0.1 trillion more in 2004 than in 2003. How much did the IRS collect in each year?   (**Source:** Internal Revenue Service.)

20. *Jogging Speed*   An athlete jogs at 6 miles per hour and at 9 miles per hour for a total time of 1 hour, covering a distance of 7 miles. How long does the athlete jog at each speed?

CHAPTER 4   EXTENDED AND DISCOVERY EXERCISES

Exercises 1–4: Plant Growth   Before doing these exercises, read the introduction to this chapter. Then refer to the figure shown here.

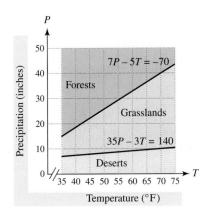

1. The equation of the line that separates grasslands and forests is

$$7P - 5T = -70.$$

Write an inequality that describes temperatures and amounts of precipitation that correspond to forested regions. (Include the line.)

2. The equation of the line that separates grasslands and deserts is

$$35P - 3T = 140.$$

Write an inequality that describes temperatures and amounts of precipitation that correspond to desert regions. (Include the line.)

3. Using the information from Exercises 1 and 2, write a system of inequalities that describes temperatures and amounts of precipitation that correspond to grassland regions.

4. Cheyenne, Wyoming, has an average annual temperature of about $50°F$ and an average annual precipitation of about 14 inches. Use the graph to predict the type of plant growth you might expect near Cheyenne. Then check to determine whether $T = 50$ and $P = 14$ satisfy the proper inequalities.

Exercises 5–8: Numerical Solutions   If a solution to a linear equation does not appear in a table of values, can you still find it? Use only the table to solve the equation. Then explain how you got your answer.

5. $2x + 1 = 0$

x	-2	-1	0	1	2
$y = 2x + 1$	-3	-1	1	3	5

6. $4x + 3 = 5$

x	-2	-1	0	1	2
$y = 4x + 3$	-5	-1	3	7	11

7. $\frac{1}{2}x + 3 = 3.75$

x	-2	-1	0	1	2
$y = \frac{1}{2}x + 3$	2	2.5	3	3.5	4

8. $3x - 1 = 0$

x	-2	-1	0	1	2
$y = 3x - 1$	-7	-4	-1	2	5

CHAPTERS 1–4   CUMULATIVE REVIEW EXERCISES

1. Write 120 as a product of prime numbers.

2. Evaluate the expression by hand.
   (a) $2^3 \div \frac{5+7}{9-3}$ (b) $-\frac{2}{5} \cdot (5-25)$

3. Classify the number as rational or irrational.
   (a) -6.9 (b) $\sqrt{14}$

4. Insert $>$ or $<$ to make each statement true.
   (a) $-5 \underline{\hspace{1cm}} |-5|$ (b) $|7| \underline{\hspace{1cm}} |-1|$

5. State the property of real numbers illustrated by the equation $3 \cdot (2 \cdot 7) = (3 \cdot 2) \cdot 7$.

6. Use properties of real numbers to evaluate $30 \cdot 102$ mentally.

7. Simplify the expression.
   (a) $5x^2 - x^2$ (b) $3 - 2x + 7x - 5$

Exercises 8 and 9: Solve the equation.

8. $5(2x + 1) = 7 + x$ 9. $1 - (x + 1) = x - 1$

10. Determine whether the equation $2(5x + 1) = 10x - 3$ has no solutions, one solution, or infinitely many solutions.

11. Find four consecutive integers whose sum is 50.

12. Find the area of a rectangle having a 36-inch length and a 1-foot width.

13. Solve the formula $W = 3x - 7y$ for x.

14. Solve the inequality $3 - (2x - 7) \le 8x$.

15. Use the table of xy-values to make a line graph.

x	-2	-1	0	1	2
y	-2	2	0	3	-1

16. Graph the equation.
    (a) $y = -2x + 2$ (b) $3y + 2x = 6$

17. Find the x- and y-intercepts for the graph of $4x - y = 8$.

18. Sketch a line with the given slope that passes through the given point.
    (a) $m = -2$, $(1, 1)$ (b) $m = \frac{4}{3}$, $(-3, 2)$

19. Write the slope-intercept form for the line.
    (a)   (b)

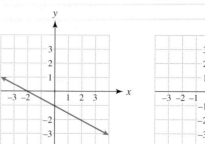 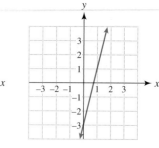

20. Find the slope–intercept form of the line passing through $(4, 2)$ and $(-4, 4)$.

21. Write the slope–intercept form of the line perpendicular to $2x - 6y = 7$, passing through $(1, 1)$.

22. Determine whether $(3, 1)$ or $(4, 4)$ is a solution to the system of equations.

$$3x - y = 8$$
$$2x + y = 12$$

Exercises 23–26: Solve the system of equations. Note that these systems may have no solutions, one solution, or infinitely many solutions.

23. $x - y = 4$ 24. $2x + 3y = 4$
    $-2x - y = 1$ $-4x - 6y = 7$

25. $3x - 4y = 8$ 26. $7x + 2y = -3$
    $-15x + 20y = -40$ $-5x - 3y = -1$

Exercises 27 and 28: Shade the solution set for the system of inequalities.

27. $y < x + 1$ 28. $3x + y \ge 6$
    $x + y \le 3$ $x - 3y \le 3$

29. *Rods to Feet*   There are 16.5 feet in 1 rod. Write a formula that converts R rods to F feet.

30. *Temperature Change*   Find the temperature change for a package of frozen carrots that is removed from a freezer at $-11°C$ and placed in water at $83°C$.

31. *Cost of a Digital Camera*   A 7% sales tax on a digital camera amounts to $17.15. Find the cost of the digital camera.

32. *Tuition Increase* If tuition is currently $145 per credit and it is going to be increased by 9%, what will the new tuition per credit be?

33. *Gasoline Consumption* Write an equation in slope–intercept form that models the number of gallons G of gas in a truck's tank after x hours if the tank initially contains 30 gallons of gas and the truck uses 3 gallons every hour.

34. *Bank Loans* An individual has two low-interest loans totaling $2400. One loan charges 5% interest and the other charges 6% interest. If the interest for one year is $132, how much money is borrowed at each rate?

Polynomials and Exponents

D igital images were first sent between New York and London by cable in the early 1920s. Unfortunately the transmission time was 3 hours and the quality was poor. Digital photography was developed further by NASA in the 1960s because ordinary pictures were subject to interference when transmitted through space. Digital pictures remain crystal clear even if they travel millions of miles. The following digital picture shows the planet Mars.

Digital images comprise tiny units called pixels, which are represented in a computer or camera by numbers. As a result, mathematics plays an important role in digital images. In this chapter we illustrate some of the ways mathematics is used to describe digital pictures. We also use mathematics to model heart rate, computer sales, motion of the planets, and interest on money.

> If you want to do
> something, do it!
>
> —PLAUTUS

Source: NASA. (Photograph reprinted with permission.)

5.1 RULES FOR EXPONENTS

Review of Bases and Exponents ▪ Zero Exponents ▪ The Product Rule ▪ Power Rules

A LOOK INTO MATH ▷

Exponents occur throughout mathematics. Regardless of how many mathematics courses you plan to take, exponents are sure to play an important role in each. Because exponents are so important, this section is essential for your becoming successful in mathematics. It takes practice, so set aside some extra time.

Review of Bases and Exponents

The expression 5^3 is an exponential expression with *base* 5 and *exponent* 3. Its value is

$$5 \cdot 5 \cdot 5 = 125.$$

In general, b^n is an exponential expression with base b and exponent n. If n is a natural number, it indicates the number of times the base b is to be multiplied with itself.

$$\underset{\text{Base} \quad\longrightarrow}{\overset{\text{Exponent} \quad\longrightarrow}{b^n}} = \underbrace{b \cdot b \cdot b \cdots \cdot b}_{n \text{ times}}$$

When evaluating expressions, evaluate exponents *before* performing addition, subtraction, multiplication, division, or negation. The following may be helpful when evaluating expressions.

EVALUATING EXPRESSIONS

When evaluating expressions, use the following order of operations.

1. Evaluate exponents.
2. Perform negation.
3. Do multiplication and division from left to right.
4. Do addition and subtraction from left to right.

EXAMPLE 1 Evaluating exponential expressions

Evaluate each expression.

(a) $1 + \dfrac{2^4}{4}$ **(b)** $3\left(\dfrac{1}{3}\right)^2$ **(c)** -2^4 **(d)** $(-2)^4$

Solution

(a) Evaluate the exponent first.

$$1 + \frac{2^4}{4} = 1 + \frac{\overbrace{2 \cdot 2 \cdot 2 \cdot 2}^{4 \text{ factors}}}{4} = 1 + \frac{16}{4} = 1 + 4 = 5$$

(b) $3\left(\dfrac{1}{3}\right)^2 = 3\left(\overbrace{\dfrac{1}{3} \cdot \dfrac{1}{3}}^{2 \text{ factors}}\right) = 3 \cdot \dfrac{1}{9} = \dfrac{3}{9} = \dfrac{1}{3}$

(c) Because exponents are evaluated before negation is performed,

$$-2^4 = -\overbrace{(2 \cdot 2 \cdot 2 \cdot 2)}^{4 \text{ factors}} = -16.$$

(d) $(-2)^4 = \overbrace{(-2)(-2)(-2)(-2)}^{4 \text{ factors}} = 16.$

Now Try Exercises 11, 13, 15, 25

NOTE: Parts (c) and (d) of Example 1 appear to be very similar. However, the placement of the negation sign inside the parentheses in part (d) means that the base for the exponential expression is -2. In part (c) no parentheses are used, indicating that the base of the exponential expression is 2. In general, operations within parentheses should be evaluated *before* using the order of operations.

TECHNOLOGY NOTE: Evaluating Exponents

Exponents can often be evaluated on calculators by using the ^ key. The four expressions from Example 1 are evaluated with a calculator and the results are shown in the two figures. When evaluating the last two expressions on your calculator, remember to use the negation key rather than the subtraction key.

```
1+2^4/4
                5
3(1/3)²►Frac
              1/3
```

```
-2^4
              -16
(-2)^4
               16
```

Zero Exponents

So far we have discussed the meaning of natural number exponents. What if an exponent is equal to 0? What does 2^0 equal? To answer these questions consider Table 5.1, which shows values for decreasing powers of 2. Note that each time the power of 2 decreases by 1, the resulting value is divided by 2. For this pattern to continue, we need to define 2^0 to be 1 because if we divide 2 by 2, the result is 1.

This discussion suggests that $2^0 = 1$, and is generalized as follows.

TABLE 5.1 **Powers of 2**

Power of 2	Value
2^3	8
2^2	4
2^1	2
2^0	?

ZERO EXPONENT

For any nonzero real number b,

$$b^0 = 1.$$

The expression 0^0 is undefined.

EXAMPLE 2 Evaluating zero exponents

Evaluate each expression. Assume that all variables represent nonzero numbers.

(a) 7^0 **(b)** $3\left(\frac{4}{9}\right)^0$ **(c)** $\left(\frac{x^2 y^5}{3z}\right)^0$

Solution
(a) $7^0 = 1$
(b) $3\left(\frac{4}{9}\right)^0 = 3(1) = 3$. (Note that the exponent 0 does not apply to 3.)
(c) All variables are nonzero, so the expression inside the parentheses is also nonzero. Thus $\left(\frac{x^2 y^5}{3z}\right)^0 = 1$.

Now Try Exercises 17, 35, 79

The Product Rule

We can calculate products of exponential expressions *provided their bases are the same*. For example,

$$4^3 \cdot 4^2 = \underbrace{(4 \cdot 4 \cdot 4) \cdot (4 \cdot 4)}_{5 \text{ factors}} = 4^5.$$

The expression $4^3 \cdot 4^2$ has a total of $3 + 2 = 5$ factors of 4, so the result is $4^{3+2} = 4^5$. To multiply exponential expressions with the *same* base, we add exponents and the base does not change. We generalize this discussion as the following rule.

> **THE PRODUCT RULE**
>
> For any real number a and natural numbers m and n,
>
> $$a^m \cdot a^n = a^{m+n}.$$

NOTE: The product $2^4 \cdot 3^5$ cannot be simplified by using the product rule because the exponential expressions have different bases: **2** and **3**.

EXAMPLE 3 Using the product rule

Multiply and simplify.
(a) $2^3 \cdot 2^2$ (b) $x^4 x^5$ (c) $2x^2 \cdot 5x^6$ (d) $x^3(2x + 3x^2)$

Solution
(a) $2^3 \cdot 2^2 = 2^{3+2} = 2^5 = 32$
(b) $x^4 x^5 = x^{4+5} = x^9$
(c) Begin by applying the commutative property of multiplication to write the product in a more convenient order.

$$2x^2 \cdot 5x^6 = 2 \cdot 5 \cdot x^2 \cdot x^6 = 10x^{2+6} = 10x^8$$

(d) To simplify this expression, first apply a distributive property.

$$x^3(2x + 3x^2) = x^3 \cdot 2x + x^3 \cdot 3x^2 = 2x^4 + 3x^5$$

Exponent is 1.

Now Try Exercises 29, 33, 39, 83

If an exponent does not appear on an expression, it is assumed to be 1. For example, x can be written as x^1 and $(x + y)$ can be written as $(x + y)^1$.

EXAMPLE 4 Applying the product rule

Multiply and simplify.
(a) $x \cdot x^3$ **(b)** $(a + b)(a + b)^4$

Solution
(a) Begin by writing x as x^1. Then $x^1 \cdot x^3 = x^{1+3} = x^4$.
(b) First write $(a + b)$ as $(a + b)^1$. Then

$$(a + b)^1 \cdot (a + b)^4 = (a + b)^{1+4} = (a + b)^5.$$

Now Try Exercises 27, 75

VISUALIZING EXPONENTS (OPTIONAL) We can visualize exponents by stacking wooden blocks having different lengths. For example, the product $2 \cdot 2 \cdot 2$ can be thought of as three blocks with length 2, as illustrated in Figure 5.1. Because they all have the same length (same base), we can stack them and the result represents 2^3.

Figure 5.1 Visualizing $2 \cdot 2 \cdot 2 = 2^3$

In Figure 5.2 the expression $3^2 \cdot 3^3$ can be simplified by stacking blocks of length 3. Note that the stack is 5 blocks high, so the result represents 3^5.

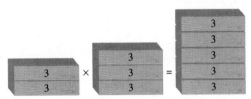

Figure 5.2 Visualizing $3^2 \cdot 3^3 = 3^5$

However, Figure 5.3 reveals that $2^2 \cdot 3^3$ cannot be stacked properly because the blocks have different lengths (bases).

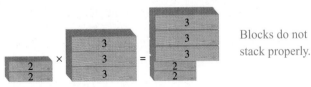

Blocks do not stack properly.

Figure 5.3 Visualizing $2^2 \cdot 3^3 = ?$

Power Rules

How should $(4^3)^2$ be evaluated? To answer this question consider

$$(4^3)^2 = \underbrace{4^3 \cdot 4^3}_{2 \text{ factors}} = 4^{\overset{\overset{3+3=3\cdot2}{\frown}}{3+3}} = 4^6.$$

Similarly,

$$\overbrace{(a^5)^3 = \underbrace{a^5 \cdot a^5 \cdot a^5}_{3 \text{ factors}} = a^{\overbrace{5+5+5}^{5+5+5=5\cdot3}} = a^{15}.}$$

This discussion suggests that to raise a power to a power, we multiply the exponents.

RAISING A POWER TO A POWER

For any real number a and natural numbers m and n,

$$(a^m)^n = a^{mn}.$$

EXAMPLE 5   Raising a power to a power

Simplify the expression.
(a) $(3^2)^4$ **(b)** $(a^3)^2$

Solution
(a) $(3^2)^4 = 3^{2\cdot4} = 3^8$ **(b)** $(a^3)^2 = a^{3\cdot2} = a^6$

Now Try Exercises 43, 45

To decide how to simplify the expression $(2x)^3$, consider

$$(2x)^3 = \underbrace{2x \cdot 2x \cdot 2x}_{3 \text{ factors}} = \underbrace{(2 \cdot 2 \cdot 2)}_{3 \text{ factors}} \cdot \underbrace{(x \cdot x \cdot x)}_{3 \text{ factors}} = 2^3 x^3.$$

To raise a product to a power, we raise each factor to the power, which we generalize as follows.

RAISING A PRODUCT TO A POWER

For any real numbers a and b and natural number n,

$$(ab)^n = a^n b^n.$$

EXAMPLE 6   Raising a product to a power

Simplify the expression.
(a) $(3z)^2$ **(b)** $(-2x^2)^3$ **(c)** $4(x^2 y^3)^5$ **(d)** $(-2^2 a^5)^3$

Solution
(a) $(3z)^2 = 3^2 z^2 = 9z^2$
(b) $(-2x^2)^3 = (-2)^3 (x^2)^3 = -8x^6$
(c) $4(x^2 y^3)^5 = 4(x^2)^5 (y^3)^5 = 4x^{10} y^{15}$
(d) $(-2^2 a^5)^3 = (-4a^5)^3 = (-4)^3 (a^5)^3 = -64a^{15}$ Now Try Exercises 49, 51, 55

The following equation illustrates a third power rule.

$$\left(\frac{2}{3}\right)^4 = \frac{2}{3} \cdot \frac{2}{3} \cdot \frac{2}{3} \cdot \frac{2}{3} = \frac{2 \cdot 2 \cdot 2 \cdot 2}{3 \cdot 3 \cdot 3 \cdot 3} = \frac{2^4}{3^4}$$

That is, to raise a quotient to a power, raise both the numerator and the denominator to the power, which we generalize as follows.

RAISING A QUOTIENT TO A POWER

For any real numbers a and b and natural number n,

$$\left(\frac{a}{b}\right)^n = \frac{a^n}{b^n}. \qquad b \neq 0$$

EXAMPLE 7   Raising a quotient to a power

Simplify the expression.

(a) $\left(\dfrac{2}{3}\right)^3$ **(b)** $\left(\dfrac{a}{b}\right)^9$ **(c)** $\left(\dfrac{a+b}{5}\right)^2$

Solution

(a) $\left(\dfrac{2}{3}\right)^3 = \dfrac{2^3}{3^3} = \dfrac{8}{27}$ **(b)** $\left(\dfrac{a}{b}\right)^9 = \dfrac{a^9}{b^9}$

(c) Because the numerator is an expression with more than one term, we must place parentheses around it before raising it to the power 2.

$$\left(\frac{a+b}{5}\right)^2 = \frac{(a+b)^2}{5^2} = \frac{(a+b)^2}{25}$$

Now Try Exercises 63, 65, 67

MAKING CONNECTIONS

Raising a Sum or Difference to a Power

Although there are power rules for products and quotients, there are not similar rules for sums and differences. In general, $(a+b)^n \neq a^n + b^n$ and $(a-b)^n \neq a^n - b^n$. For example, $(3+4)^2 = 7^2 = 49$ but $3^2 + 4^2 = 9 + 16 = 25$. Similarly, $(4-1)^3 = 3^3 = 27$ but $4^3 - 1^3 = 64 - 1 = 63$.

Simplification of some expressions may require the application of more than one rule of exponents. This is demonstrated in the next example.

EXAMPLE 8   Combining rules for exponents

Simplify the expression.

(a) $(2a)^2(3a)^3$ **(b)** $\left(\dfrac{a^2b^3}{c}\right)^4$ **(c)** $(2x^3y)^2(-4x^2y^3)^3$

Solution
(a) $(2a)^2(3a)^3 = 2^2a^2 \cdot 3^3a^3$ Raising a product to a power

$\qquad\qquad\qquad = 4 \cdot 27 \cdot a^2 \cdot a^3$ Evaluate powers; commutative property

$\qquad\qquad\qquad = 108a^5$ Product rule

(b) $\left(\dfrac{a^2 b^3}{c}\right)^4 = \dfrac{(a^2)^4 (b^3)^4}{c^4}$ Raising a quotient to a power; raising a product to a power

$= \dfrac{a^8 b^{12}}{c^4}$ Raising a power to a power

(c) $(2x^3 y)^2 (-4x^2 y^3)^3 = 2^2 (x^3)^2 y^2 (-4)^3 (x^2)^3 (y^3)^3$ Raising a product to a power

$= 4x^6 y^2 (-64) x^6 y^9$ Raising a power to a power

$= 4(-64) x^6 x^6 y^2 y^9$ Commutative property

$= -256 x^{12} y^{11}$ Product rule

Now Try Exercises 61, 73

EXAMPLE 9   Calculating growth of an investment

If a parcel of property increases in value by about 11% each year for 20 years, then its value will double three times.
(a) Write an exponential expression that represents "doubling three times."
(b) If the property is initially worth $25,000, how much will it be worth if it doubles 3 times?

Solution
(a) Doubling three times is represented by 2^3.
(b) $2^3 (25,000) = 8(25,000) = \$200,000$

Now Try Exercise 95

5.1 PUTTING IT ALL TOGETHER

The following table summarizes properties of exponents.

Concept	Explanation	Examples
Bases and Exponents	In the expression b^n, b is the base and n is the exponent. If n is a natural number, then $$b^n = \underbrace{b \cdot b \cdots\cdots b}_{n \text{ times}}.$$	2^3 has base 2 and exponent 3. $9^1 = 9,$ $3^2 = 3 \cdot 3 = 9,$ $4^3 = 4 \cdot 4 \cdot 4 = 64,$ and $-6^2 = -(6 \cdot 6) = -36$
Zero Exponents	$b^0 = 1$ for any nonzero number b.	$5^0 = 1,$ $x^0 = 1,$ and $(xy^3)^0 = 1$
The Product Rule	$a^m \cdot a^n = a^{m+n}$, where m and n are natural numbers.	$2^4 \cdot 2^3 = 2^{4+3} = 2^7,$ $x \cdot x^2 \cdot x^6 = x^{1+2+6} = x^9,$ and $(x+1) \cdot (x+1)^2 = (x+1)^3$
Raising a Power to a Power	$(a^m)^n = a^{mn}$, where m and n are natural numbers.	$(2^4)^2 = 2^{4 \cdot 2} = 2^8,$ $(x^2)^5 = x^{2 \cdot 5} = x^{10},$ and $(a^4)^3 = a^{4 \cdot 3} = a^{12}$

Concept	Explanation	Examples
Raising a Product to a Power	$(ab)^n = a^n b^n$, where n is a natural number.	$(3x)^3 = 3^3 x^3 = 27x^3$, $(x^2 y)^4 = (x^2)^4 y^4 = x^8 y^4$, and $(-xy)^6 = (-x)^6 y^6 = x^6 y^6$
Raising a Quotient to a Power	$\left(\dfrac{a}{b}\right)^n = \dfrac{a^n}{b^n}$, where $b \neq 0$ and n is a natural number.	$\left(\dfrac{x}{y}\right)^5 = \dfrac{x^5}{y^5}$ and $\left(\dfrac{a^2 b}{d^3}\right)^4 = \dfrac{(a^2)^4 b^4}{(d^3)^4} = \dfrac{a^8 b^4}{d^{12}}$

5.1 Exercises

MyMathLab · Math XL PRACTICE · WATCH · DOWNLOAD · READ · REVIEW

CONCEPTS

1. In the expression b^n, b is the _____ and n is the _____.

2. The expression $b^0 =$ _____ for any nonzero number b.

3. Write $\frac{1}{2} \cdot \frac{1}{2} \cdot \frac{1}{2}$ by using exponents.

4. Write $x \cdot x \cdot x \cdot x$ by using exponents.

5. $-1^2 =$ _____

6. $(-1)^2 =$ _____

7. $a^m \cdot a^n =$ _____

8. $(a^m)^n =$ _____

9. $(ab)^n =$ _____

10. $\left(\dfrac{a}{b}\right)^n =$ _____

PROPERTIES OF EXPONENTS

Exercises 11–26: Evaluate the expression.

11. 8^2

12. 4^3

13. $(-2)^3$

14. $(-3)^4$

15. -2^3

16. -3^4

17. 6^0

18. $(-0.5)^0$

19. $2 \cdot 4^2$

20. $-3 \cdot 2^4$

21. $1 + 5^2$

22. $5^2 - 4^2$

23. $\dfrac{4^2}{2}$

24. $\left(\dfrac{-4}{2}\right)^2$

25. $4 \cdot \dfrac{1}{2^3}$

26. $\dfrac{2}{2^3} - \dfrac{1}{2^3}$

Exercises 27–86: Simplify the expression. Assume that all variables represent nonzero numbers.

27. $3 \cdot 3^2$

28. $x^3 \cdot x$

29. $4^2 \cdot 4^6$

30. $5^3 \cdot 5^3$

31. $2^3 \cdot 2^2$

32. $10^4 \cdot 10^3$

33. $x^3 \cdot x^6$

34. $a^5 \cdot a^2$

35. $z^0 z^4$

36. $a^2 a^3 a^1$

37. $x^2 x^2 x^2$

38. $y^7 y^3 y^0$

39. $4x^2 \cdot 5x^5$

40. $-2y^6 \cdot 5y^2$

41. $3(-xy^3)(x^2 y)$

42. $(a^2 b^3)(-ab^2)$

43. $(2^3)^2$

44. $(10^3)^4$

45. $(n^3)^4$

46. $(z^7)^3$

47. $x(x^3)^2$

48. $(z^3)^2(5z^5)$

49. $(-7b)^2$

50. $(-4z)^3$

51. $(ab)^3$

52. $(xy)^8$

53. $(2x^2)^2$

54. $(3a^2)^4$

55. $(-4b^2)^3$

56. $(-3r^4 t^3)^2$

57. $(x^2 y^3)^7$

58. $(rt^2)^5$

59. $(y^3)^2(x^4 y)^3$

60. $(ab^3)^2(ab)^3$

61. $(a^2b)^2(a^2b^2)^3$

62. $(x^3y)(x^2y^4)^2$

63. $\left(\frac{1}{3}\right)^3$

64. $\left(\frac{5}{2}\right)^2$

65. $\left(\frac{a}{b}\right)^5$

66. $\left(\frac{x}{2}\right)^4$

67. $\left(\frac{x-y}{3}\right)^3$

68. $\left(\frac{4}{x+y}\right)^2$

69. $\left(\frac{5}{a+b}\right)^2$

70. $\left(\frac{a-b}{2}\right)^3$

71. $\left(\frac{2x}{5}\right)^3$

72. $\left(\frac{3y}{2}\right)^4$

73. $\left(\frac{3x^2}{5y^4}\right)^3$

74. $\left(\frac{a^2b^3}{3}\right)^5$

75. $(x+y)(x+y)^3$

76. $(a-b)^2(a-b)$

77. $(a+b)^2(a+b)^3$

78. $(x-y)^5(x-y)^4$

79. $6(x^4y^6)^0$

80. $\left(\frac{xy}{z^2}\right)^0$

81. $a(a^2+2b^2)$

82. $x^3(3x-5y^4)$

83. $3a^3(4a^2+2b)$

84. $2x^2(5-4y^3)$

85. $(r+t)(rt)$

86. $(x-y)(x^2y^3)$

87. Thinking Generally Students sometimes mistakenly apply the "rule" $a^m \cdot b^n \overset{?}{=} (ab)^{m+n}$. In general, this equation is *not true*. Find values for a, b, m, and n with $a \neq b$ and $m \neq n$ that will make this equation true.

88. Thinking Generally Students sometimes mistakenly apply the "rule" $(a+b)^n \overset{?}{=} a^n + b^n$. In general, this equation is *not true*. Find values for a, b, and n with $a \neq b$ that will make this equation true.

APPLICATIONS

89. *Area of a Rectangle*   Find the area of the rectangle.

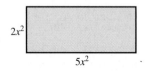

90. *Area of a Square*   Find the area of the square whose sides have length $2ab$.

91. *Volume of a Box*   A box has the dimensions shown. Write a simplified expression for its volume.

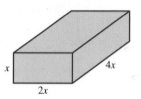

92. *Volume of a Cube*   Each side of a cube is $3a^2$. Write a simplified expression for its volume.

93. *Area of a Circle*   Write a simplified expression for the area of the circle shown.

94. *Volume of a Sphere*   The volume V of a sphere with radius r is given by $V = \frac{4}{3}\pi r^3$. Find a simplified expression for the volume of a sphere with radius $2b$.

95. *Compound Interest*   If P dollars are deposited in an account that pays 5% annual interest, then the amount of money in the account after 3 years is $P(1 + 0.05)^3$. Find the amount when $P = \$1000$.

96. *Compound Interest*   If P dollars are deposited in an account that pays 9% annual interest, then the amount of money in the account after 4 years is $P(1 + 0.09)^4$. Find the amount when $P = \$500$.

Exercises 97–100: (Refer to the discussion about visualizing exponents found in this section.) Make a sketch of blocks that represents the given exponential equation.

97. $4^2 \cdot 4^1 = 4^3$ **98.** $2^3 \cdot 2^3 = 2^6$

99. $2^3 \cdot 3^2 \cdot 3^1 \cdot 2^2 = 2^5 \cdot 3^3$

100. $4^2 \cdot 2^2 \cdot 4^2 = 4^5$ (*Hint:* Write 2^2 with a different base.)

WRITING ABOUT MATHEMATICS

101. Are the expressions $(4x)^2$ and $4x^2$ equal? Explain your answer.

102. Are the expressions $3^3 \cdot 2^3$ and 6^6 equal? Explain your answer.

5.2 ADDITION AND SUBTRACTION OF POLYNOMIALS

Monomials and Polynomials ▪ Addition of Polynomials ▪ Subtraction of Polynomials ▪ Evaluating Polynomial Expressions

A LOOK INTO MATH ▷

If you have ever exercised strenuously and then taken your pulse immediately afterward, you may have discovered that your pulse slowed quickly at first and then gradually leveled off. A typical scatterplot of this phenomenon is shown in Figure 5.4(a). These data points cannot be modeled accurately with a line, so a new expression, called a *polynomial,* is needed to model them. A graph of this new expression is shown in Figure 5.4(b) and discussed in Exercise 79. (**Source:** V. Thomas, *Science and Sport.*)

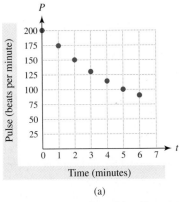

(a)

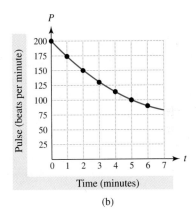

(b)

Figure 5.4 Heart Rate After Exercising

Monomials and Polynomials

A **monomial** is a number, a variable, or a product of numbers and variables raised to natural number powers. Examples of monomials include

$$-3, \quad xy^2, \quad 5a^2, \quad -z^3, \quad \text{and} \quad -\frac{1}{2}xy^3.$$

A monomial may contain more than one variable, but monomials do not contain division by variables. For example, the expression $\frac{3}{z}$ is not a monomial. If an expression contains addition or subtraction signs, it is *not* a monomial.

The **degree of a monomial** is the sum of the exponents of the variables. If the monomial has only one variable, its degree is the exponent of that variable. Remember, when a variable does not have a written exponent, the exponent is implied to be 1. A nonzero number has degree 0, and the number 0 has *undefined* degree. The number in a monomial is called the **coefficient of the monomial**. Table 5.2 contains the degree and coefficient of several monomials.

TABLE 5.2   **Properties of Monomials**

Monomial	-5	$6a^3b$	$-xy$	$7y^3$
Degree	0	4	2	3
Coefficient	-5	6	-1	7

A **polynomial** is the sum of one or more monomials. Each monomial is called a *term* of the polynomial. Addition or subtraction signs separate terms. The expression $2x^2 - 3x + 5$ is a **polynomial in one variable** with three terms. Examples of polynomials in one variable include

$$-2x, \quad 3x + 1, \quad 4y^2 - y + 7, \quad \text{and} \quad x^5 - 3x^3 + x - 7.$$

These polynomials have, 1, 2, 3, and 4 terms, respectively. A polynomial with *two terms* is called a **binomial**, and a polynomial with *three terms* is called a **trinomial**.

A polynomial can have more than one variable, as in

$$x^2y^2, \quad 2xy^2 + 5x^2y - 1, \quad \text{and} \quad a^2 + 2ab + b^2.$$

Note that all variables in a polynomial are raised to natural number powers. The **degree of a polynomial** is the degree of the term (or monomial) with highest degree.

EXAMPLE 1   Identifying properties of polynomials

Determine whether the expression is a polynomial. If it is, state how many terms and variables the polynomial contains and its degree.

(a) $7x^2 - 3x + 1$ **(b)** $5x^3 - 3x^2y^3 + xy^2 - 2y^3$ **(c)** $4x^2 + \dfrac{5}{x + 1}$

Solution
(a) The expression $7x^2 - 3x + 1$ is a polynomial with three terms and one variable. The first term $7x^2$ has degree 2 because the exponent on the variable is 2. The second term $-3x$ has degree 1 because the exponent on the variable is implied to be 1. The third term 1 has degree 0 because it is a nonzero number. The term with highest degree is $7x^2$, so the polynomial has degree **2**.
(b) The expression $5x^3 - 3x^2y^3 + xy^2 - 2y^3$ is a polynomial with four terms and two variables. The first term has degree 3 because the exponent on the variable is 3. The second term has degree 5 because the *sum* of the exponents on the variables is 5. Likewise, the third term has degree 3 and the fourth term has degree 3. The term with highest degree is $-3x^2y^3$, so the polynomial has degree $2 + 3 = 5$.
(c) The expression $4x^2 + \dfrac{5}{x + 1}$ is not a polynomial because it contains division by the polynomial $x + 1$.

Now Try Exercises 21, 23, 25

Addition of Polynomials

Suppose that we have 2 identical rectangles with length L and width W, as illustrated in Figure 5.5. Then the area of one rectangle is LW and the total area is

$$LW + LW.$$

This area is equivalent to 2 times LW, which can be expressed as $2LW$, or

$$LW + LW = 2LW.$$

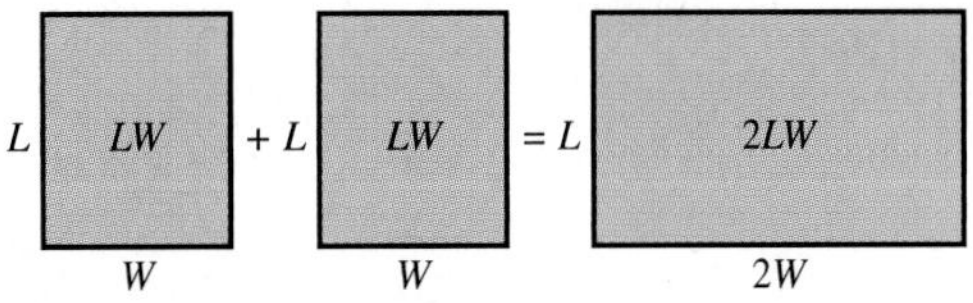

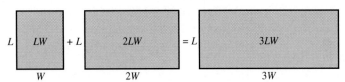

Figure 5.5 Adding $LW + LW$

If two monomials contain the same variables raised to the same powers, we call them **like terms**. We can add or subtract *like* terms but not *unlike* terms. The terms LW and $2LW$ are like terms and can be combined geometrically, as shown in Figure 5.6. If we joined one of the small rectangles with area LW and a larger rectangle with area $2LW$, then the total area is $3LW$.

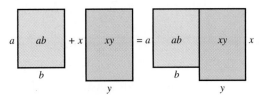

Figure 5.6 Adding $LW + 2LW$

The *distributive property* justifies combining like terms.

$$1LW + 2LW = (1 + 2)LW = 3LW$$

The rectangles shown in Figure 5.7 have areas of ab and xy. Their combined area is their sum, $ab + xy$. However, because these monomials are unlike terms, they cannot be combined into one term.

Figure 5.7 Unlike terms: $ab + xy$

EXAMPLE **2**   **Adding like terms**

State whether each pair of expressions contains like terms or unlike terms. If they are like terms, add them.
(a) $5x^2, -x^2$ **(b)** $7a^2b, 10ab^2$ **(c)** $4rt^2, \frac{1}{2}rt^2$

Solution
(a) The terms $5x^2$ and $-x^2$ have the same variable raised to the same power, so they are like terms. To add like terms add their coefficients. Note that the coefficient of $-x^2$ is -1.

$$5x^2 + (-x^2) = \left(5 + (-1)\right)x^2 \qquad \text{Distributive property}$$
$$= 4x^2 \qquad\qquad\qquad \text{Add.}$$

(b) The terms $7a^2b$ and $10ab^2$ have the same variables, but these variables are not raised to the same power. They are unlike terms and therefore cannot be added.

(c) The terms $4rt^2$ and $\frac{1}{2}rt^2$ have the same variables raised to the same powers, so they are like terms.

$$4rt^2 + \frac{1}{2}rt^2 = \left(4 + \frac{1}{2}\right)rt^2 \qquad \text{Distributive property}$$

$$= \frac{9}{2}rt^2 \qquad \text{Add.} \qquad \boxed{\text{Now Try Exercises } 33, 35, 39}$$

To add two polynomials, add like terms, as illustrated in the next example.

EXAMPLE 3 Adding polynomials

Add each pair of polynomials by combining like terms.
(a) $(3x + 4) + (-4x + 2)$
(b) $(y^2 - 2y + 1) + (3y^2 + y + 11)$
(c) $(3a^3 - 4a + 1) + (4a^3 + a^2 - 5)$

Solution
(a) $(3x + 4) + (-4x + 2) = 3x + (-4x) + 4 + 2$
$$= (3 - 4)x + (4 + 2)$$
$$= -x + 6$$

(b) $(y^2 - 2y + 1) + (3y^2 + y + 11) = y^2 + 3y^2 - 2y + y + 1 + 11$
$$= (1 + 3)y^2 + (-2 + 1)y + (1 + 11)$$
$$= 4y^2 - y + 12$$

NOTE: With practice the first two steps can be done mentally.

(c) $(3a^3 - 4a + 1) + (4a^3 + a^2 - 5) = 3a^3 + 4a^3 + a^2 - 4a + 1 - 5$
$$= (3 + 4)a^3 + a^2 - 4a - 4$$
$$= 7a^3 + a^2 - 4a - 4$$

$\boxed{\text{Now Try Exercises } 41, 43, 47}$

Polynomials can also be added vertically, as demonstrated in the next example.

EXAMPLE 4 Adding polynomials vertically

Simplify $(3x^2 - 3x + 5) + (-x^2 + x - 6)$.

Solution
Write the polynomials in a vertical format and then add each column of like terms.

$$\begin{array}{r} 3x^2 - 3x + 5 \\ -x^2 + x - 6 \\ \hline 2x^2 - 2x - 1 \end{array} \quad \text{Add.}$$

Regardless of the method used, the same answer should be obtained. However, adding vertically requires that *like terms be placed in the same column*.　$\boxed{\text{Now Try Exercise } 53}$

Subtraction of Polynomials

To subtract one integer from another, add the first integer with the *additive inverse* or *opposite* of the second integer. For example, $3 - 5$ is evaluated as follows.

$$3 - 5 = 3 + (-5) \qquad \text{Add the opposite.}$$
$$= -2 \qquad \text{Simplify.}$$

Similarly, to subtract one polynomial from another, add the first polynomial and the *opposite* of the second polynomial. To find the opposite of a polynomial, simply negate each term. Table 5.3 lists some polynomials and their opposites.

CRITICAL THINKING

What is the result when a polynomial and its opposite are added?

TABLE 5.3   Opposites of Polynomials

Polynomial	Opposite
$2x - 4$	$-2x + 4$
$-x^2 - 2x + 9$	$x^2 + 2x - 9$
$6x^3 - 12$	$-6x^3 + 12$
$-3x^4 - 2x^2 - 8x + 3$	$3x^4 + 2x^2 + 8x - 3$

EXAMPLE 5   **Subtracting polynomials**

Simplify each expression.
(a) $(3x - 4) - (5x + 1)$
(b) $(5x^2 + 2x - 3) - (6x^2 - 7x + 9)$
(c) $(6x^3 + x^2) - (-3x^3 - 9)$

Solution
(a) To subtract $(5x + 1)$ from $(3x - 4)$, we add the opposite of $(5x + 1)$, or $(-5x - 1)$.

$$(3x - 4) - (5x + 1) = (3x - 4) + (-5x - 1)$$
$$= (3 - 5)x + (-4 - 1)$$
$$= -2x - 5$$

(b) The opposite of $(6x^2 - 7x + 9)$ is $(-6x^2 + 7x - 9)$.

$$(5x^2 + 2x - 3) - (6x^2 - 7x + 9) = (5x^2 + 2x - 3) + (-6x^2 + 7x - 9)$$
$$= (5 - 6)x^2 + (2 + 7)x + (-3 - 9)$$
$$= -x^2 + 9x - 12$$

(c) The opposite of $(-3x^3 - 9)$ is $(3x^3 + 9)$.

$$(6x^3 + x^2) - (-3x^3 - 9) = (6x^3 + x^2) + (3x^3 + 9)$$
$$= (6 + 3)x^3 + x^2 + 9$$
$$= 9x^3 + x^2 + 9$$

Now Try Exercises 63, 65, 69

NOTE: Some students prefer to subtract one polynomial from another by noting that a subtraction sign in front of parentheses changes the signs of all of the terms within the parentheses. For example, part (a) of the previous example could be worked as follows.

$$(3x - 4) - (5x + 1) = 3x - 4 - 5x - 1$$
$$= (3 - 5)x + (-4 - 1)$$
$$= -2x - 5$$

EXAMPLE 6   Subtracting polynomials vertically

Simplify $(5x^2 - 2x + 7) - (-3x^2 + 3)$.

Solution
To subtract one polynomial from another vertically, simply add the first polynomial and the opposite of the second polynomial. No x-term occurs in the second polynomial, so insert $0x$.

$$5x^2 - 2x + 7$$
$$\underline{3x^2 + 0x - 3} \qquad \text{Opposite of } -3x^2 + 3 \text{ is } 3x^2 - 3 \text{ or } 3x^2 + 0x - 3.$$
$$8x^2 - 2x + 4 \qquad \text{Add like terms in each column.} \qquad \boxed{\text{Now Try Exercise 77}}$$

Evaluating Polynomial Expressions

Frequently, monomials and polynomials represent formulas that may be evaluated. We illustrate such applications in the next two examples.

EXAMPLE 7   Writing and evaluating a monomial

Write the monomial that represents the volume of the box having a square bottom, as shown in Figure 5.8. Find the volume of the box if $x = 3$ feet and $y = 2$ feet.

CRITICAL THINKING

Write an expression that gives the volume of six identical cubes having sides of length L.

Figure 5.8

Solution
The volume of a box is found by multiplying the length, width, and height together. Because the length and width are both x and the height is y, the monomial xxy represents the volume of the box. This can be written x^2y. To calculate the volume let $x = 3$ and $y = 2$ in the monomial x^2y.

$$x^2y = 3^2 \cdot 2 = 9 \cdot 2 = 18 \text{ cubic feet} \qquad \boxed{\text{Now Try Exercise 81}}$$

EXAMPLE 8   Modeling sales of personal computers

Worldwide sales of personal computers have increased dramatically in recent years, as illustrated in Figure 5.9. The polynomial

$$0.7868x^2 + 12x + 79.5$$

approximates the number of computers sold in millions, where $x = 0$ corresponds to 1997, $x = 1$ to 1998, and so on. Estimate the number of personal computers sold in 2002 by using both the graph and the polynomial.   (*Source:* International Data Corporation.)

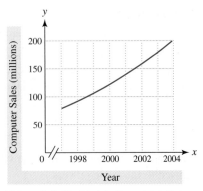

Figure 5.9 Worldwide Computer Sales

Solution

From the graph shown in Figure 5.10, it appears that personal computer sales were slightly more than 150 million, or about 160 million, in 2002. The year 2002 corresponds to $x = 5$ in the given polynomial, so substitute 5 for x and evaluate the resulting expression.

$$0.7868x^2 + 12x + 79.5 = 0.7868(5)^2 + 12(5) + 79.5$$

$$\approx 159 \text{ million}$$

The graph and the polynomial give similiar results.

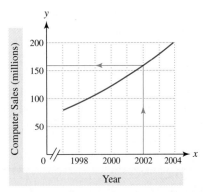

Figure 5.10 Worldwide Computer Sales   Now Try Exercise 80

5.2 PUTTING IT ALL TOGETHER

In this section we discussed monomials and polynomials, including how to add, subtract, and evaluate them. The following table summarizes several important concepts related to these topics.

Concept	Explanation	Examples
Monomial	A number, variable, or product of numbers and variables raised to natural number powers Degree is the sum of the exponents. Coefficient is the number in a monomial.	$4x^2y$ Degree: 3, coefficient: 4 $-6x^2$ Degree: 2, coefficient: -6 $-a^4$ Degree: 4, coefficient: -1 x Degree: 1, coefficient: 1 -8 Degree: 0, coefficient: -8
Polynomial	A sum of one or more monomials	$4x^2 + 8xy^2 + 3y^2$ Trinomial $-9x^4 + 100$ Binomial $-3x^2y^3$ Monomial
Like Terms	Monomials containing the same variables raised to the same powers	$10x$ and $-2x$, $4x^2$ and $3x^2$ $5ab^2$ and $-ab^2$, $5z$ and $\frac{1}{2}z$
Addition of Polynomials	To add polynomials combine like terms by applying the distributive property.	$(x^2 + 3x + 1) + (2x^2 - 2x + 7)$ $= (1 + 2)x^2 + (3 - 2)x + (1 + 7)$ $= 3x^2 + x + 8$ $3xy + 5xy = (3 + 5)xy = 8xy$
Opposite of a Polynomial	To obtain the opposite of a polynomial negate each term.	*Polynomial* *Opposite* $-2x^2 + x - 6$ $2x^2 - x + 6$ $a^2 - b^2$ $-a^2 + b^2$ $-3x - 18$ $3x + 18$
Subtraction of Polynomials	To subtract one polynomial from another, add the first polynomial to the opposite of the second polynomial.	$(x^2 + 3x) - (2x^2 - 5x)$ $= (x^2 + 3x) + (-2x^2 + 5x)$ $= (1 - 2)x^2 + (3 + 5)x$ $= -x^2 + 8x$
Evaluating a Polynomial	To evaluate a polynomial in x, substitute a value for x in the expression and simplify.	To evaluate the polynomial $3x^2 - 2x + 1$ for $x = 2$, substitute 2 for x and simplify. $3(2)^2 - 2(2) + 1 = 9$

5.2 Exercises

MyMathLab

Math XL
PRACTICE   WATCH   DOWNLOAD   READ   REVIEW

CONCEPTS

1. A _____ is a number, a variable, or a product of numbers and variables raised to a natural number power.

2. The coefficient of $-3xy$ is _____.

3. A _____ is a monomial or a sum of monomials.

4. The _____ of a monomial is the sum of the exponents of the variables.

5. A polynomial with two terms is called a _____.

6. A polynomial with three terms is called a _____.

7. $4x^3 + x^2$ has _____ terms and its degree is _____.

8. To add two polynomials, combine _____ terms.

9. To subtract two polynomials, add the first polynomial to the _____ of the second polynomial.

10. Polynomials can be added horizontally or _____.

PROPERTIES OF POLYNOMIALS

Exercises 11–18: Identify the degree and coefficient of the monomial.

11. $3x^2$

12. y

13. $-ab$

14. $2xy$

15. $-5rt$

16. $8x^2y^5$

17. -6

18. $\frac{1}{2}$

Exercises 19–30: Determine whether the expression is a polynomial. If it is, state how many terms and variables the polynomial contains. Then state its degree.

19. $-x$

20. $7z$

21. $4x^2 - 5x + 9$

22. $x^3 - 9$

23. $x + \dfrac{1}{x}$

24. $\dfrac{5}{xy + 1}$

25. $3x^2y - xy^3$

26. $a^3 + 3a^2b + 3ab^2$

27. $3x^{-2}y^{-3}$

28. $5^2a^3b^4$

29. -2^3a^4bc

30. $-7y^{-1}z^{-3}$

Exercises 31–40: State whether the given pair of expressions are like terms. If they are like terms, add them.

31. $5x, -4x$

32. $x^2, 8x^2$

33. $x^3, -6x^3$

34. $4xy, -9xy$

35. $9x, -xy$

36. $5x^2y, -3xy^2$

37. ab, ba

38. $rt^2, -2t^2r$

39. $7xy^2, -3xy^2$

40. a, b

ADDITION OF POLYNOMIALS

Exercises 41–52: Add the polynomials.

41. $(3x + 5) + (-4x + 4)$

42. $(-x + 5) + (2x - 5)$

43. $(3x^2 + 4x + 1) + (x^2 + 4x - 6)$

44. $(-x^2 - x + 7) + (2x^2 + 3x - 1)$

45. $(a^3 - 6) + (4a^3 + 7)$

46. $(2b^4 - 3b^2) + (-3b^4 + b^2)$

47. $(y^3 + 3y^2 - 5) + (3y^3 + 4y - 4)$

48. $(4z^4 + z^2 - 10) + (-z^4 + 4z - 5)$

49. $(-xy + 5) + (5xy - 4)$

50. $(2a^2 + b^2) + (3a^2 - 5b^2)$

51. $(a^3b^2 + a^2b^3) + (a^2b^3 - a^3b^2)$

52. $(a^2 + ab + b^2) + (a^2 - ab + b^2)$

Exercises 53–56: Add the polynomials vertically.

53. $\begin{array}{r} 4x^2 - 2x + 1 \\ \underline{5x^2 + 3x - 7} \end{array}$

54. $\begin{array}{r} 8x^2 + 3x + 5 \\ \underline{-x^2 - 3x - 9} \end{array}$

55. $\begin{array}{r} -x^2 + x \\ \underline{2x^2 - 8x - 1} \end{array}$

56. $\begin{array}{r} a^3 - 3a^2b + 3ab^2 - b^3 \\ \underline{a^3 + 3a^2b + 3ab^2 + b^3} \end{array}$

SUBTRACTION OF POLYNOMIALS

Exercises 57–62: Write the opposite of the polynomial.

57. $5x^2$

58. $17x + 12$

59. $3a^2 - a + 4$

60. $-b^3 + 3b$

61. $-2t^2 - 3t + 4$

62. $7t^2 + t - 10$

Exercises 63–74: Subtract the polynomials.

63. $(3x + 1) - (-x + 3)$

64. $(-2x + 5) - (x + 7)$

65. $(-x^2 + 6x + 8) - (2x^2 + x - 2)$

66. $(2y^2 + 3y - 2) - (y^2 - y - 4)$

67. $(a^2 - 2a) - (4a^2 + 3a)$

68. $(7b^3 + 3b) - (-3b^3 - b)$

69. $(z^3 - 2z^2 - z) - (4z^2 + 5z + 1)$

70. $(3z^4 - z) - (-z^4 + 4z^2 - 5)$

71. $(4xy + x^2y^2) - (xy - x^2y^2)$

72. $(a^2 + b^2) - (-a^2 + b^2)$

73. $(ab^2) - (ab^2 + a^3b)$

74. $(x^2 + 3xy + 4y^2) - (x^2 - xy + 4y^2)$

Exercises 75–78: (Refer to Example 6.) Subtract the poly-nomials vertically.

75. $(x^2 + 2x - 3) - (2x^2 + 7x + 1)$

76. $(5x^2 - 9x - 1) - (x^2 - x + 3)$

77. $(3x^3 - 2x) - (5x^3 + 4x + 2)$

78. $(a^2 + 3ab + 2b^2) - (a^2 - 3ab + 2b^2)$

APPLICATIONS

79. *Exercise and Heart Rate*   The polynomial given by $1.6t^2 - 28t + 200$ calculates the heart rate shown in Figure 5.4(b) where t represents the elapsed time in minutes since exercise stopped.
   (a) What is the heart rate when the athlete first stops exercising?
   (b) What is the heart rate after 5 minutes?
   (c) Describe what happens to the heart rate after exercise stops.

80. *Cellular Phone Subscribers*   In the early years of cellular phone technology—from 1986 through 1991—the number of subscribers in millions could be modeled by the polynomial $0.163x^2 - 0.146x + 0.205$, where $x = 1$ corresponds to 1986, $x = 2$ to 1987, and so on. The graph illustrates this growth.

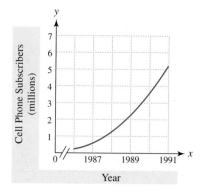

   (a) Use the graph to estimate the number of cellular phone subscribers in 1990.
   (b) Use the polynomial to estimate the number of cellular phone subscribers in 1990.
   (c) Do your answers from parts (a) and (b) agree?

81. *Areas of Squares*   Write a monomial that equals the sum of the areas of the squares. Then calculate this sum for $z = 10$ inches.

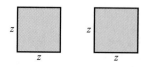

82. *Areas of Rectangles*   Find a monomial that equals the sum of the areas of the three rectangles. Find this sum for $a = 5$ yards and $b = 3$ yards.

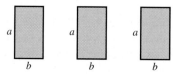

83. *Area of a Figure*   Find a polynomial that equals the area of the figure. Calculate its area for $x = 6$ feet.

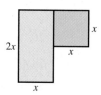

84. *Area of a Rectangle*   Write a polynomial that gives the area of the rectangle. Calculate its area for $x = 3$ feet.

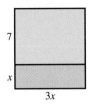

85. *Areas of Circles*   Write a polynomial that gives the sum of the areas of two circles, one with radius x and the other with radius y. Find this sum for $x = 2$ feet and $y = 3$ feet. Leave your answer in terms of π.

86. *Squares and Circles*   Write a polynomial that gives the sum of the areas of a square having sides of length x and a circle having diameter x. Approximate this sum to the nearest hundredth of a square foot for $x = 6$ feet.

87. *World Population*   The table lists actual and projected world population P in billions for selected years t.

t	1974	1987	1999	2012
P	4	5	6	7

Source: U.S. Census Bureau.

(a) Find the slope of each line segment connecting consecutive data points in the table. Can these data be modeled with a line? Explain.

(b) Does the polynomial $0.077t - 148$ give good estimates for the world population in year t? Explain how you decided.

88. *Price of a Stamp*   The table lists the price P of a first-class postage stamp for selected years t.

t	1963	1975	1987	2002	2007
P	5¢	13¢	25¢	37¢	41¢

Source: U.S. Postal Service.

(a) Does the polynomial $0.835t - 1635$ model the data in the table exactly?

(b) Does it give approximations that are within 1.5¢ of the actual values?

WRITING ABOUT MATHEMATICS

89. Explain what the terms monomial, binomial, trinomial, and polynomial mean. Give an example of each.

90. Explain how to determine the degree of a polynomial having one variable. Give an example.

91. Explain how to obtain the opposite of a polynomial. Give an example.

92. Explain how to subtract two polynomials. Give an example.

CHECKING BASIC CONCEPTS
SECTIONS 5.1 AND 5.2

1. Evaluate each expression.
   (a) -5^2 **(b)** $3^2 - 2^3$

2. Simplify each expression.
   (a) $10^3 \cdot 10^5$ **(b)** $(3x^2)(-4x^5)$
   (c) $(a^3b)^2$ **(d)** $\left(\dfrac{x}{z^3}\right)^4$

3. State the number of terms and variables in the polynomial $5x^3y - 2x^2y + 5$. What is its degree?

4. A box has a rectangular bottom twice as long as it is wide.

(a) If the bottom has width W and the box has height H, write a monomial that gives the volume of the box.

(b) Find the volume of the box for $W = 12$ inches and $H = 10$ inches.

5. Simplify each expression.
   (a) $(2a^2 + 3a - 1) + (a^2 - 3a + 7)$
   (b) $(4z^3 + 5z) - (2z^3 - 2z + 8)$
   (c) $(x^2 + 2xy + y^2) - (x^2 - 2xy + y^2)$

6. Evaluate $5x^2 - 7x$ for $x = -2$.

5.3 MULTIPLICATION OF POLYNOMIALS

Multiplying Monomials ▪ Review of the Distributive Properties ▪
Multiplying Monomials and Polynomials ▪ Multiplying Polynomials

A LOOK INTO MATH ▷ The study of polynomials dates back to Babylonian civilization in about 1800–1600 B.C. Many eighteenth-century mathematicians devoted their entire careers to the study of polynomials. Polynomials still play an important role in mathematics, often being used to approximate unknown quantities. In this section we discuss the basics of multiplying polynomials. (***Source:*** *Historical Topics for the Mathematics Classroom, Thirty-first Yearbook,* NCTM.)

Multiplying Monomials

A monomial is a number, a variable, or a product of numbers and variables raised to natural number powers. To multiply monomials, we often use the product rule for exponents.

ЄXAMPLE 1   **Multiplying monomials**

Multiply.
(a) $-5x^2 \cdot 4x^3$ **(b)** $(7xy^4)(x^3y^2)$

Solution

(a) $\quad -5x^2 \cdot 4x^3 = (-5)(4)x^2x^3 \qquad$ Commutative property

$\qquad\qquad\qquad = -20x^{2+3} \qquad\qquad$ The product rule

$\qquad\qquad\qquad = -20x^5 \qquad\qquad\quad$ Simplify.

(b) $(7xy^4)(x^3y^2) = 7xx^3y^4y^2 \qquad$ Commutative property

$\qquad\qquad\qquad = 7x^{1+3}y^{4+2} \qquad$ The product rule

$\qquad\qquad\qquad = 7x^4y^6 \qquad\qquad$ Simplify.   Now Try Exercises 11, 15

Review of the Distributive Properties

Distributive properties are used frequently for multiplying monomials and polynomials. For all real numbers a, b, and c,

$$a(b + c) = ab + ac \quad \text{and}$$
$$a(b - c) = ab - ac.$$

The first distributive property above can be visualized geometrically. For example,

$$3(x + 2) = 3x + 6$$

is illustrated in Figure 5.11. The dimensions of the large rectangle are 3 by $x + 2$, and its area is $3(x + 2)$. The areas of the two small rectangles, $3x$ and 6, equal the area of the large rectangle. Therefore $3(x + 2) = 3x + 6$.

In the next example we use the distributive properties to multiply expressions.

Figure 5.11 Area: $3x + 6$

ЄXAMPLE 2   **Using distributive properties**

Multiply.
(a) $2(3x + 4)$ **(b)** $(3x^2 + 4)5$ **(c)** $-x(3x - 6)$

Solution

(a) $2(3x + 4) = 2 \cdot 3x + 2 \cdot 4 = 6x + 8$

(b) $(3x^2 + 4)5 = 3x^2 \cdot 5 + 4 \cdot 5 = 15x^2 + 20$

(c) $-x(3x - 6) = -x \cdot 3x + x \cdot 6 = -3x^2 + 6x$

Now Try Exercises 17, 21, 23

Multiplying Monomials and Polynomials

A monomial consists of one term, whereas a polynomial consists of one or more terms separated by $+$ or $-$ signs. To multiply a monomial by a polynomial, we apply the distributive properties and the product rule.

EXAMPLE 3 Multiplying monomials and polynomials

Multiply.
(a) $9x(2x^2 - 3)$ (b) $(5x - 8)x^2$
(c) $-7(2x^2 - 4x + 6)$ (d) $4x^3(x^4 + 9x^2 - 8)$

Solution

(a)
$$9x(2x^2 - 3) = 9x \cdot 2x^2 - 9x \cdot 3 \qquad \text{Distributive property}$$
$$= 18x^3 - 27x \qquad \text{The product rule}$$

(b)
$$(5x - 8)x^2 = 5x \cdot x^2 - 8 \cdot x^2 \qquad \text{Distributive property}$$
$$= 5x^3 - 8x^2 \qquad \text{The product rule}$$

(c)
$$-7(2x^2 - 4x + 6) = -7 \cdot 2x^2 + 7 \cdot 4x - 7 \cdot 6 \qquad \text{Distributive property}$$
$$= -14x^2 + 28x - 42 \qquad \text{Simplify.}$$

(d)
$$4x^3(x^4 + 9x^2 - 8) = 4x^3 \cdot x^4 + 4x^3 \cdot 9x^2 - 4x^3 \cdot 8 \qquad \text{Distributive property}$$
$$= 4x^7 + 36x^5 - 32x^3 \qquad \text{The product rule}$$

Now Try Exercises 25, 27, 29, 31

We can also multiply monomials and polynomials that contain more than one variable.

EXAMPLE 4 Multiplying monomials and polynomials

Multiply.
(a) $2xy(7x^2y^3 - 1)$ (b) $-ab(a^2 - b^2)$

Solution

(a)
$$2xy(7x^2y^3 - 1) = 2xy \cdot 7x^2y^3 - 2xy \cdot 1 \qquad \text{Distributive property}$$
$$= 14xx^2yy^3 - 2xy \qquad \text{Commutative property}$$
$$= 14x^3y^4 - 2xy \qquad \text{The product rule}$$

(b)
$$-ab(a^2 - b^2) = -ab \cdot a^2 + ab \cdot b^2 \qquad \text{Distributive property}$$
$$= -aa^2b + abb^2 \qquad \text{Commutative property}$$
$$= -a^3b + ab^3 \qquad \text{The product rule}$$

Now Try Exercises 33, 37

Multiplying Polynomials

Monomials, binomials, and trinomials are examples of polynomials. Recall that a monomial has one term, a binomial has two terms, and a trinomial has three terms. In the next example we multiply two binomials, using both geometric and symbolic techniques.

EXAMPLE 5 Multiplying binomials

Multiply $(x + 4)(x + 2)$
(a) geometrically and   **(b)** symbolically.

Solution

(a) To multiply $(x + 4)(x + 2)$ geometrically, draw a rectangle $x + 4$ long and $x + 2$ wide, as shown in Figure 5.12(a). The area of this rectangle equals length times width, or $(x + 4)(x + 2)$. The large rectangle can be divided into four smaller rectangles, which have areas of x^2, $4x$, $2x$, and 8, as shown in Figure 5.12(b). Thus

$$(x + 4)(x + 2) = x^2 + 4x + 2x + 8$$
$$= x^2 + 6x + 8.$$

(b) To multiply $(x + 4)(x + 2)$ symbolically, apply the distributive property.

$$(x + 4)(x + 2) = (x + 4)(x) + (x + 4)(2)$$
$$= x \cdot x + 4 \cdot x + x \cdot 2 + 4 \cdot 2$$
$$= x^2 + 4x + 2x + 8$$
$$= x^2 + 6x + 8$$

Now Try Exercise 41

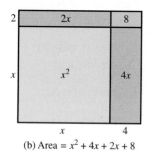

$x + 2$

$x + 4$
(a) Area $= (x + 4)(x + 2)$

2      $2x$ 8

x x^2 $4x$

x 4
(b) Area $= x^2 + 4x + 2x + 8$

Figure 5.12

The distributive properties used in part (b) of the previous example show that if we want to multiply $(x + 4)$ by $(x + 2)$, we should multiply every term in $x + 4$ by every term in $x + 2$.

$$(x + 4)(x + 2) = x^2 + 2x + 4x + 8$$
$$= x^2 + 6x + 8$$

NOTE: This process of multiplying binomials is sometimes called *FOIL*. This acronym may be used to remind us to multiply the first terms (F), outside terms (O), inside terms (I), and last terms (L).

Multiply the *First terms* to obtain x^2. $(x + 4)(x + 2)$

Multiply the *Outside terms* to obtain $2x$. $(x + 4)(x + 2)$

Multiply the *Inside terms* to obtain $4x$. $(x + 4)(x + 2)$

Multiply the *Last terms* to obtain 8. $(x + 4)(x + 2)$

The following statement summarizes how to multiply two polynomials in general.

MULTIPLYING POLYNOMIALS

The product of two polynomials may be found by multiplying every term in the first polynomial by every term in the second polynomial and then combining like terms.

EXAMPLE 6

Multiplying binomials

Multiply. Draw arrows to show how each term is found.
(a) $(3x + 2)(x + 1)$ **(b)** $(1 - x)(1 + 2x)$ **(c)** $(4x - 3)(x^2 - 2x)$

Solution

(a) $(3x + 2)(x + 1) = 3x \cdot x + 3x \cdot 1 + 2 \cdot x + 2 \cdot 1$
$= 3x^2 + 3x + 2x + 2$
$= 3x^2 + 5x + 2$

(b) $(1 - x)(1 + 2x) = 1 \cdot 1 + 1 \cdot 2x - x \cdot 1 - x \cdot 2x$
$= 1 + 2x - x - 2x^2$
$= 1 + x - 2x^2$

(c) $(4x - 3)(x^2 - 2x) = 4x \cdot x^2 - 4x \cdot 2x - 3 \cdot x^2 + 3 \cdot 2x$
$= 4x^3 - 8x^2 - 3x^2 + 6x$
$= 4x^3 - 11x^2 + 6x$ Now Try Exercises 53, 55, 61

EXAMPLE 7

Multiplying polynomials

Multiply.
(a) $(2x + 3)(x^2 + x - 1)$ **(b)** $(a - b)(a^2 + ab + b^2)$
(c) $(x^4 + 2x^2 - 5)(x^2 + 1)$

Solution

(a) Multiply every term in $(2x + 3)$ by every term in $(x^2 + x - 1)$.

$(2x + 3)(x^2 + x - 1) = 2x \cdot x^2 + 2x \cdot x - 2x \cdot 1 + 3 \cdot x^2 + 3 \cdot x - 3 \cdot 1$
$= 2x^3 + 2x^2 - 2x + 3x^2 + 3x - 3$
$= 2x^3 + 5x^2 + x - 3$

(b) $(a - b)(a^2 + ab + b^2) = a \cdot a^2 + a \cdot ab + a \cdot b^2 - b \cdot a^2 - b \cdot ab - b \cdot b^2$
$= a^3 + a^2b + ab^2 - a^2b - ab^2 - b^3$
$= a^3 - b^3$

(c) $(x^4 + 2x^2 - 5)(x^2 + 1) = x^4 \cdot x^2 + x^4 \cdot 1 + 2x^2 \cdot x^2 + 2x^2 \cdot 1 - 5 \cdot x^2 - 5 \cdot 1$
$= x^6 + x^4 + 2x^4 + 2x^2 - 5x^2 - 5$
$= x^6 + 3x^4 - 3x^2 - 5$ Now Try Exercises 65, 69, 71

Polynomials can be multiplied vertically in a manner similar to multiplication of real numbers. For example, multiplication of 123 times 12 is performed as follows.

$$
\begin{array}{r}
1\ 2\ 3 \\
\times\ 1\ 2 \\
\hline
2\ 4\ 6 \\
1\ 2\ 3 \\
\hline
1\ 4\ 7\ 6
\end{array}
$$

A similar method can be used to multiply polynomials vertically.

EXAMPLE 8   Multiplying polynomials vertically

Multiply $2x^2 - 4x + 1$ and $x + 3$.

Solution

Write the polynomials vertically. Then multiply every term in the first polynomial by each term in the second polynomial. Arrange the results so that *like terms are in the same column.*

$$
\begin{array}{r}
2x^2 - 4x + 1 \\
x + 3 \\
\hline
6x^2 - 12x + 3 \\
2x^3 - 4x^2 + x \\
\hline
2x^3 + 2x^2 - 11x + 3
\end{array}
$$

Multiply top row by 3.
Multiply top row by x.
Add each column.   **Now Try Exercise 73**

MAKING CONNECTIONS

Vertical and Horizontal Formats

Whether you decide to add, subtract, or multiply polynomials vertically or horizontally, remember that the same answer is obtained either way.

EXAMPLE 9   Finding the volume of a box

A box has a width 3 inches less than its height and a length 4 inches more than its height.
(a) If h represents the height of the box, write a polynomial that represents the volume of the box.
(b) Use this polynomial to calculate the volume of the box if $h = 10$ inches.

Solution

(a) If h is the height, then $h - 3$ is the width and $h + 4$ is the length, as illustrated in Figure 5.13. Its volume equals the product of these three expressions.

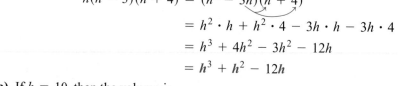

$$h(h - 3)(h + 4) = (h^2 - 3h)(h + 4)$$
$$= h^2 \cdot h + h^2 \cdot 4 - 3h \cdot h - 3h \cdot 4$$
$$= h^3 + 4h^2 - 3h^2 - 12h$$
$$= h^3 + h^2 - 12h$$

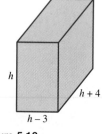

Figure 5.13

(b) If $h = 10$, then the volume is

$$10^3 + 10^2 - 12(10) = 1000 + 100 - 120 = 980 \text{ cubic inches.}$$

Now Try Exercise 81

5.3 PUTTING IT ALL TOGETHER

The following table summarizes multiplication of polynomials.

Concept	Explanation	Examples
Distributive Properties	For all real numbers a, b, and c, $$a(b + c) = ab + ac \quad \text{and}$$ $$a(b - c) = ab - ac.$$	$5(x + 3) = 5x + 15$, $3(x - 6) = 3x - 18$, and $-2x(3 - 5x^3) = -6x + 10x^4$

Concept	Explanation	Examples
Multiplying Polynomials	The product of two polynomials may be found by multiplying every term in the first polynomial by every term in the second polynomial and then combining like terms.	$3x(5x^2 + 2x - 7)$ $= 3x \cdot 5x^2 + 3x \cdot 2x - 3x \cdot 7$ $= 15x^3 + 6x^2 - 21x$ $(x + 2)(7x - 3)$ $= x \cdot 7x - x \cdot 3 + 2 \cdot 7x - 2 \cdot 3$ $= 7x^2 - 3x + 14x - 6$ $= 7x^2 + 11x - 6$

5.3 Exercises

CONCEPTS

1. The equation $x^2 \cdot x^3 = x^5$ illustrates what rule of exponents?

2. The equation $3(x - 2) = 3x - 6$ illustrates what property?

3. A monomial has _____ term(s), a binomial has _____ term(s), and a trinomial has _____ term(s).

4. The product of two polynomials may be found by multiplying every _____ in the first polynomial by every _____ in the second polynomial and then combining like terms.

5. Are $(x + y)(x + y)$ and $x^2 + y^2$ equivalent?

6. Polynomials can be multiplied horizontally or _____.

MULTIPLICATION OF MONOMIALS

Exercises 7–16: Multiply.

7. $x^2 \cdot x^5$

8. $-a \cdot a^5$

9. $-3a \cdot 4a$

10. $7x \cdot 5x$

11. $4x^3 \cdot 5x^2$

12. $6b^6 \cdot 3b^5$

13. $xy^2 \cdot 4xy$

14. $3ab \cdot ab^2$

15. $(-3xy^2)(4x^2y)$

16. $(-r^2t^2)(-r^3t)$

MULTIPLICATION OF MONOMIALS AND POLYNOMIALS

Exercises 17–38: Multiply and simplify the expression.

17. $3(x + 4)$

18. $-7(4x - 1)$

19. $-5(9x + 1)$

20. $10(1 - 6x)$

21. $(4 - z)z$

22. $3z(1 - 5z)$

23. $-y(5 + 3y)$

24. $(2y - 8)2y$

25. $3x(5x^2 - 4)$

26. $-6x(2x^3 + 1)$

27. $(6x - 6)x^2$

28. $(1 - 2x^2)3x^2$

29. $-8(4t^2 + t + 1)$

30. $7(3t^2 - 2t - 5)$

31. $n^2(-5n^2 + n - 2)$

32. $6n^3(2 - 4n + n^2)$

33. $xy(x + y)$

34. $ab(2a - 3b)$

35. $x^2(x^2y - xy^2)$

36. $2y^2(xy - 5)$

37. $-ab(a^3 - 2b^3)$

38. $5rt(r^2 + 2rt + t^2)$

MULTIPLICATION OF POLYNOMIALS

*Exercises 39–44: (Refer to Example 5.) Multiply the given expression **(a)** geometrically and **(b)** symbolically.*

39. $x(x + 3)$

40. $2x(x + 5)$

41. $(x + 2)(x + 2)$

42. $(x + 1)(x + 3)$

43. $(x + 3)(x + 6)$

44. $(x + 5)(x + 2)$

Exercises 45–72: Multiply and simplify the expression.

45. $(x + 3)(x + 5)$

46. $(x - 4)(x - 7)$

47. $(x - 8)(x - 9)$

48. $(x + 10)(x + 10)$

49. $(3z - 2)(2z - 5)$

50. $(z + 6)(2z - 1)$

51. $(8b - 1)(8b + 1)$

52. $(3t + 2)(3t - 2)$

53. $(10y + 7)(y - 1)$

54. $(y + 6)(2y + 7)$

55. $(5 - 3a)(1 - 2a)$

56. $(4 - a)(5 + 3a)$

57. $(1 - 3x)(1 + 3x)$

58. $(10 - x)(5 - 2x)$

59. $(x - 1)(x^2 + 1)$

60. $(x + 2)(x^2 - x)$

61. $(x^2 + 4)(4x - 3)$

62. $(3x^2 - 1)(3x^2 + 1)$

63. $(2n + 1)(n^2 + 3)$

64. $(2 - n^2)(1 + n^2)$

65. $(m + 1)(m^2 + 3m + 1)$

66. $(m - 2)(m^2 - m + 5)$

67. $(3x - 2)(2x^2 - x + 4)$

68. $(5x + 4)(x^2 - 3x + 2)$

69. $(x + 1)(x^2 - x + 1)$

70. $(x - 2)(x^2 + 4x + 4)$

71. $(4b^2 + 3b + 7)(b^2 + 3)$

72. $(-3a^2 - 2a + 1)(3a^2 - 3)$

Exercises 73–78: Multiply the polynomials vertically.

73. $(x + 2)(x^2 - 3x + 1)$

74. $(2y - 3)(3y^2 - 2y - 2)$

75. $(a - 2)(a^2 + 2a + 4)$

76. $(b - 3)(b^2 + 3b + 9)$

77. $(3x^2 - x + 1)(2x^2 + 1)$

78. $(2x^2 - 3x - 5)(2x^2 + 3)$

79. Thinking Generally   If a polynomial with m terms and a polynomial with n terms are multiplied, how many terms are there in the product before like terms are combined?

80. Thinking Generally   When a polynomial with m terms is multiplied by a second polynomial, the product contains k terms before like terms are combined. How many terms does the second polynomial contain?

APPLICATIONS

81. *Volume of a Box*   (Refer to Example 9.) A box has a width 4 inches less than its height and a length 2 inches more than its height.
   (a) If h is the height of the box, write a polynomial that represents the volume of the box.
   (b) Use this polynomial to calculate the volume for $h = 25$ inches.

82. *Surface Area of a Box*   Use the drawing of the box to write a polynomial that represents each of the following.

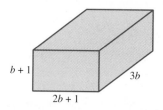

   (a) The area of its bottom
   (b) The area of its front
   (c) The area of its right side
   (d) The total area of its six sides

83. *Perimeter of a Pen*   A rectangular pen for a pet has a perimeter of 100 feet. If one side of the pen has length x, then its area is given by $x(50 - x)$.
   (a) Multiply this expression.
   (b) Evaluate the expression obtained in part (a) for $x = 25$.

84. *Rectangular Garden*   A rectangular garden has a perimeter of 500 feet.
   (a) If one side of the garden has length x, then write a polynomial expression that gives its area. Multiply this expression completely.
   (b) Evaluate the expression for $x = 50$ and interpret your answer.

85. *Surface Area of a Cube*   Write a polynomial that represents the total area of the six sides of the cube having edges with length $x + 1$.

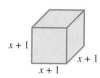

86. *Surface Area of a Sphere*   The surface area of a sphere with radius r is $4\pi r^2$. Write a polynomial that gives the surface area of a sphere with radius $x + 2$. Leave your answer in terms of π.

87. *Toy Rocket*   A toy rocket is shot straight up into the air. Its height h in feet above the ground after t seconds is represented by the expression $t(64 - 16t)$.
   (a) Multiply this expression.
   (b) Evaluate the expression obtained in part (a) and the given expression for $t = 2$.
   (c) Are your answers in part (b) the same? Should they be the same?

88. *Toy Rocket on the Moon*   (Refer to the preceding exercise.) If the same toy rocket were flown on the moon, then its height h in feet after t seconds would be $t(64 - \frac{5}{2}t)$.
   (a) Multiply this expression.
   (b) Evaluate the expression obtained in part (a) and the given expression for $t = 2$. Did the rocket go higher on the moon?

WRITING ABOUT MATHEMATICS

89. Explain how the acronym FOIL relates to multiplying two binomials, such as $x + 3$ and $2x + 1$.

90. Does the FOIL method work for multiplying a binomial and a trinomial? Explain.

91. Explain in words how to multiply any two polynomials. Give an example.

92. Give two properties of real numbers that are used for multiplying $3x(5x^2 - 3x + 2)$. Explain your answer.

GROUP ACTIVITY
WORKING WITH REAL DATA

Directions: Form a group of 2 to 4 people. Select someone to record the group's responses for this activity. All members of the group should work cooperatively to answer the questions. If your instructor asks for your results, each member of the group should be prepared to respond.

Biology   Some types of worms have a remarkable ability to live without moisture. The following table from one study shows the number of worms W surviving after x days without moisture.

x (days)	0	20	40	80	120	160
W (worms)	50	48	45	36	20	3

Source: D. Brown and P. Rothery, *Models in Biology.*

(a) Use the equation $W = -0.0014x^2 - 0.076x + 50$ to find W for each x-value in the table.
(b) Discuss how well this equation approximates the data.
(c) Use this equation to estimate the number of worms on day 60 and on day 180. Which answer is most accurate? Explain.

5.4 SPECIAL PRODUCTS

Product of a Sum and Difference ▪ Squaring Binomials ▪ Cubing Binomials

Polynomials are often used to approximate real-world phenomena in applications. Polynomials have played an important role in the development of everyday products such as computers, cell phones, and automobiles. Even digital images in computers and interest calculations at a bank make use of polynomials. In this section we discuss how to multiply some special types of binomials.

Product of a Sum and Difference

Products of the form $(a + b)(a - b)$ occur frequently in mathematics. Other examples include

$$(x + y)(x - y) \quad \text{and} \quad (2r + 3t)(2r - 3t).$$

These products can always be multiplied by using the techniques discussed in Section 5.3. However, there is a faster way to multiply these special products.

$$
\begin{aligned}
(a + b)(a - b) &= a \cdot a - a \cdot b + b \cdot a - b \cdot b \\
&= a^2 - ab + ba - b^2 \\
&= a^2 - b^2
\end{aligned}
$$

In words, the product of a sum of two numbers and their difference equals the difference of their squares. We generalize this method as follows.

PRODUCT OF A SUM AND DIFFERENCE

For any real numbers a and b,

$$(a + b)(a - b) = a^2 - b^2.$$

EXAMPLE 1   Finding products of sums and differences

Multiply.
(a) $(x + y)(x - y)$ **(b)** $(z - 2)(z + 2)$
(c) $(2r + 3t)(2r - 3t)$ **(d)** $(5m^2 - 4n^2)(5m^2 + 4n^2)$

Solution
(a) If we let $a = x$ and $b = y$, then we can apply the rule

$$(a + b)(a - b) = a^2 - b^2.$$

Thus

$$
\begin{aligned}
(x + y)(x - y) &= (x)^2 - (y)^2 \\
&= x^2 - y^2.
\end{aligned}
$$

(b) Because the expressions $(z + 2)(z - 2)$ and $(z - 2)(z + 2)$ are equal by the commutative property, we can apply the formula for the product of a sum and difference.

$$
\begin{aligned}
(z - 2)(z + 2) &= (z)^2 - (2)^2 \\
&= z^2 - 4
\end{aligned}
$$

(c) Let $a = 2r$ and $b = 3t$. Then the product can be evaluated as follows.

$$\begin{aligned}(2r + 3t)(2r - 3t) &= (2r)^2 - (3t)^2 \\ &= 4r^2 - 9t^2\end{aligned}$$

(d) $(5m^2 - 4n^2)(5m^2 + 4n^2) = (5m^2)^2 - (4n^2)^2 = 25m^4 - 16n^4$

Now Try Exercises 11, 17, 23

The next example demonstrates how to multiply mentally some products of numbers.

EXAMPLE 2

Finding a product

Use the product of a sum and difference to find $22 \cdot 18$.

Solution

Because $22 = 20 + 2$ and $18 = 20 - 2$, rewrite and evaluate $22 \cdot 18$ as follows.

$$\begin{aligned}22 \cdot 18 &= (20 + 2)(20 - 2) && \text{Product of a sum and difference} \\ &= 20^2 - 2^2 && (a + b)(a - b) = a^2 - b^2 \\ &= 400 - 4 && \text{Evaluate exponents.} \\ &= 396 && \text{Subtract.}\end{aligned}$$

Now Try Exercise 27

Squaring Binomials

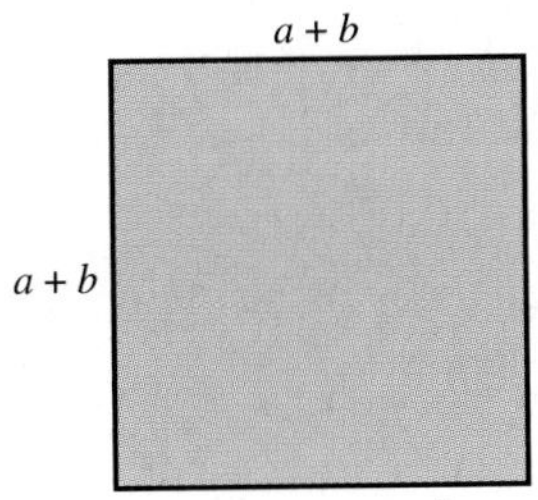

Area = $(a + b)^2$

Figure 5.14

Because each side of the square shown in Figure 5.14 has length $(a + b)$, its area equals

$$(a + b)(a + b),$$

which can be written as $(a + b)^2$. We can multiply this expression as follows.

$$\begin{aligned}(a + b)^2 &= (a + b)(a + b) \\ &= a^2 + ab + ba + b^2 \\ &= a^2 + 2ab + b^2\end{aligned}$$

This result is illustrated geometrically in Figure 5.15, where the area of the large square is $(a + b)^2$. This area can also be found by adding the areas of the four small rectangles.

$$a^2 + ab + ba + b^2 = a^2 + 2ab + b^2$$

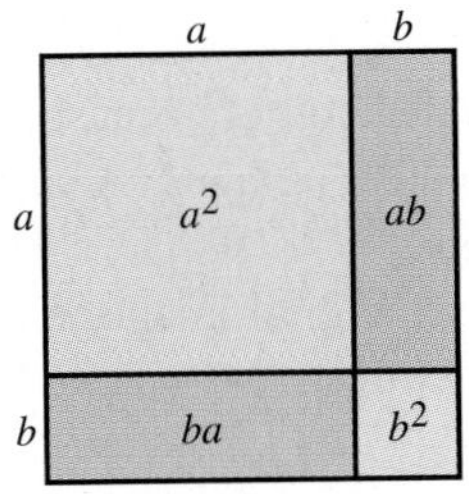

$(a + b)^2 = a^2 + 2ab + b^2$

Figure 5.15

The geometric and symbolic results are the same. Note that to obtain the middle term, $2ab$, we can multiply the two terms in the binomial and *double* the result.

A similar product that is also the square of a binomial can be calculated as

$$\begin{aligned}(a - b)^2 &= (a - b)(a - b) \\ &= a^2 - ab - ba + b^2 \\ &= a^2 - 2ab + b^2.\end{aligned}$$

The results are summarized as follows.

SQUARING A BINOMIAL

For any real numbers a and b,

$$(a + b)^2 = a^2 + 2ab + b^2 \quad \text{and}$$
$$(a - b)^2 = a^2 - 2ab + b^2.$$

That is, the square of a binomial equals the square of the first term, plus (or minus) twice the product of the two terms, plus the square of the last term.

NOTE: $(a + b)^2 \neq a^2 + b^2$. Do not forget the middle term when squaring a binomial.

EXAMPLE 3   Squaring a binomial

Multiply.
(a) $(x + 3)^2$ **(b)** $(2x - 5)^2$ **(c)** $(1 - 5y)^2$ **(d)** $(7a^2 + 3b)^2$

Solution
(a) If we let $a = x$ and $b = 3$, then we can apply the formula
$$(a + b)^2 = a^2 + 2ab + b^2.$$
Thus
$$(x + 3)^2 = (x)^2 + 2(x)(3) + (3)^2$$
$$= x^2 + 6x + 9.$$
(b) Applying the formula $(a - b)^2 = a^2 - 2ab + b^2$ with $a = 2x$ and $b = 5$ gives
$$(2x - 5)^2 = (2x)^2 - 2(2x)(5) + (5)^2$$
$$= 4x^2 - 20x + 25.$$
(c) $(1 - 5y)^2 = (1)^2 - 2(1)(5y) + (5y)^2 = 1 - 10y + 25y^2$
(d) $(7a^2 + 3b)^2 = (7a^2)^2 + 2(7a^2)(3b) + (3b)^2 = 49a^4 + 42a^2b + 9b^2$

Now Try Exercises 33, 37, 41, 47

MAKING CONNECTIONS

Multiplying Binomial and Special Products

If you forget these special products, you can still multiply polynomials by using earlier techniques. For example, the binomial in Example 3(b) can be multiplied as
$$(2x - 5)^2 = (2x - 5)(2x - 5)$$
$$= 2x \cdot 2x - 2x \cdot 5 - 5 \cdot 2x + 5 \cdot 5$$
$$= 4x^2 - 10x - 10x + 25$$
$$= 4x^2 - 20x + 25.$$

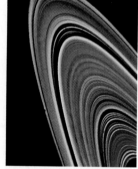

Figure 5.16 Digital Picture

▶ **REAL-WORLD CONNECTION** NASA first developed digital pictures because they were easy to transmit through space and because they provided clear images. A digital image supplied by NASA is shown in Figure 5.16.

Today, digital cameras are readily available, and the Internet uses digital images exclusively. The next example shows how polynomials relate to digital pictures.

EXAMPLE 4   Calculating the size of a digital picture

A digital picture comprises tiny square units called *pixels*. Shading individual pixels creates a picture. A simplified version of a digital picture of the letter T is shown in Figure 5.17. The dimensions of this picture are 3 pixels by 3 pixels with a 1-pixel border.
(a) Suppose that a square digital picture, including its border, is $x + 2$ pixels by $x + 2$ pixels. Find a polynomial that gives the total number of pixels in the picture, including the border.
(b) Let $x = 3$ and evaluate the polynomial. Does it agree with Figure 5.17?

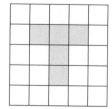

Figure 5.17

Solution

(a) The total number of pixels equals $(x + 2)$ times $(x + 2)$, or $(x + 2)^2$.

$$(x + 2)^2 = x^2 + 4x + 4$$

(b) For $x = 3$, the polynomial evaluates to $3^2 + 4 \cdot 3 + 4 = 25$, the total number of pixels. This result agrees with Figure 5.17, which has a total of $5 \cdot 5 = 25$ pixels with a 3×3–pixel picture of the letter T inside. Now Try Exercise 90

Cubing Binomials

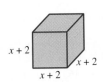

Figure 5.18
Volume $= (x + 2)^3$

To calculate the volume of the cube shown in Figure 5.18, we find the product of its length, width, and height. Because all sides have the same measure, its volume is $(x + 2)^3$. That is, the volume equals the **cube** of $x + 2$.

To multiply the expression $(x + 2)^3$, we proceed as follows.

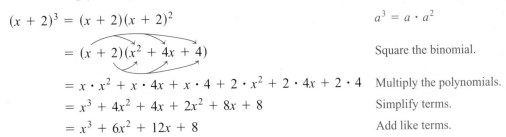

$$(x + 2)^3 = (x + 2)(x + 2)^2 \qquad a^3 = a \cdot a^2$$

$$= (x + 2)(x^2 + 4x + 4) \qquad \text{Square the binomial.}$$

$$= x \cdot x^2 + x \cdot 4x + x \cdot 4 + 2 \cdot x^2 + 2 \cdot 4x + 2 \cdot 4 \qquad \text{Multiply the polynomials.}$$

$$= x^3 + 4x^2 + 4x + 2x^2 + 8x + 8 \qquad \text{Simplify terms.}$$

$$= x^3 + 6x^2 + 12x + 8 \qquad \text{Add like terms.}$$

EXAMPLE 5 Cubing a binomial

Multiply $(2z - 3)^3$.

CRITICAL THINKING

Suppose that a student is convinced that the expressions

$(x + y)^3$ and $x^3 + y^3$

are equal. How could you convince the student otherwise?

Solution

$$(2z - 3)^3 = (2z - 3)(2z - 3)^2 \qquad a^3 = a \cdot a^2$$

$$= (2z - 3)(4z^2 - 12z + 9) \qquad \text{Square the binomial.}$$

$$= 8z^3 - 24z^2 + 18z - 12z^2 + 36z - 27 \qquad \text{Multiply the polynomials.}$$

$$= 8z^3 - 36z^2 + 54z - 27 \qquad \text{Add like terms.}$$

NOTE: $(2z - 3)^3 \neq (2z)^3 - (3)^3 = 8z^3 - 27$. Now Try Exercise 55

EXAMPLE 6 Calculating interest

If a savings account pays x percent annual interest, where x is expressed as a decimal, then after 3 years a sum of money will grow by a factor of $(1 + x)^3$.

(a) Multiply this expression.

(b) Evaluate the expression for $x = 0.10$ (or 10%), and interpret the result.

Solution

(a)
$$(1 + x)^3 = (1 + x)(1 + x)^2 \qquad a^3 = a \cdot a^2$$

$$= (1 + x)(1 + 2x + x^2) \qquad \text{Square the binomial.}$$

$$= 1 + 2x + x^2 + x + 2x^2 + x^3 \qquad \text{Multiply the polynomials.}$$

$$= 1 + 3x + 3x^2 + x^3 \qquad \text{Add like terms.}$$

(b) Let $x = 0.1$ in the expression $1 + 3x + 3x^2 + x^3$. Then

$$1 + 3(0.1) + 3(0.1)^2 + (0.1)^3 = 1.331.$$

The sum of money will increase by a factor of 1.331. For example, if $1000 are deposited in this account, it will grow to $1331 after 3 years.   **Now Try Exercise** 85

5.4 PUTTING IT ALL TOGETHER

The following table summarizes some special products of polynomials.

Concept	Explanation	Examples
Product of a Sum and Difference	For any real numbers x and y, $$(x + y)(x - y) = x^2 - y^2.$$	$(x + 6)(x - 6) = x^2 - 36$, $(2x - 3)(2x + 3) = 4x^2 - 9$, and $(x^2 + y^2)(x^2 - y^2) = x^4 - y^4$
Squaring a Binomial	For all real numbers x and y, $$(x + y)^2 = x^2 + 2xy + y^2$$ and $$(x - y)^2 = x^2 - 2xy + y^2.$$	$(x + 4)^2 = x^2 + 8x + 16$, $(5x - 2)^2 = 25x^2 - 20x + 4$, $(1 - 7x)^2 = 1 - 14x + 49x^2$, and $(x^2 + y^2)^2 = x^4 + 2x^2y^2 + y^4$
Cubing a Binomial	Multiply the binomial by its square.	$(x + 3)^3$ $= (x + 3)(x + 3)^2$ $= (x + 3)(x^2 + 6x + 9)$ $= x^3 + 6x^2 + 9x + 3x^2 + 18x + 27$ $= x^3 + 9x^2 + 27x + 27$

5.4 Exercises

CONCEPTS

1. $(a + b)(a - b) = $ ____

2. Are the expressions $(3z - 1)(3z + 1)$ and $9z^2 - 1$ equal for all real numbers z? Explain.

3. $(a + b)^2 = $ ____

4. $(a - b)^2 = $ ____

5. Are the expressions $(x + y)^2$ and $x^2 + y^2$ equal for all real numbers x and y? Explain.

6. Are the expressions $(r - t)^2$ and $r^2 - t^2$ equal for all real numbers r and t? Explain.

7. Are the expressions $(z + 5)^3$ and $z^3 + 5^3$ equal for all real numbers z? Explain.

8. $(a + b)^3 = (a + b) \cdot $ ____

PRODUCT OF A SUM AND DIFFERENCE

Exercises 9–26: Multiply.

9. $(x - 3)(x + 3)$

10. $(x + 6)(x - 6)$

11. $(x + 1)(x - 1)$

12. $(x - 7)(x + 7)$

13. $(4x - 1)(4x + 1)$

14. $(10x + 3)(10x - 3)$

15. $(1 + 2a)(1 - 2a)$ **16.** $(4 - 9b)(4 + 9b)$

17. $(2x + 3y)(2x - 3y)$ **18.** $(5r - 6t)(5r + 6t)$

19. $(ab - 5)(ab + 5)$ **20.** $(2xy + 7)(2xy - 7)$

21. $(ab + 4)(ab - 4)$ **22.** $(2 - 3yz)(2 + 3yz)$

23. $(a^2 - b^2)(a^2 + b^2)$ **24.** $(3x^2 + y^2)(3x^2 - y^2)$

25. $(x^3 - y^3)(x^3 + y^3)$

26. $(2a^4 + b^4)(2a^4 - b^4)$

Exercises 27–32: (Refer to Example 2.) Use the product of a sum and a difference to evaluate the expression.

27. $101 \cdot 99$ **28.** $52 \cdot 48$

29. $23 \cdot 17$ **30.** $29 \cdot 31$

31. $90 \cdot 110$ **32.** $38 \cdot 42$

SQUARING BINOMIALS

Exercises 33–48: Multiply.

33. $(x + 1)^2$ **34.** $(a + 3)^2$

35. $(a - 2)^2$ **36.** $(x - 7)^2$

37. $(2x + 3)^2$ **38.** $(7x - 2)^2$

39. $(3b + 5)^2$ **40.** $(7t + 10)^2$

41. $\left(\frac{3}{4}a - 4\right)^2$ **42.** $\left(\frac{1}{5}a + 1\right)^2$

43. $(1 - b)^2$ **44.** $(1 - 4a)^2$

45. $(5 + y^3)^2$ **46.** $(9 - 5x^2)^2$

47. $(a^2 + b)^2$ **48.** $(x^3 - y^3)^2$

CUBING BINOMIALS

Exercises 49–58: Multiply.

49. $(a + 1)^3$ **50.** $(b + 4)^3$

51. $(x - 2)^3$ **52.** $(y - 7)^3$

53. $(2x + 1)^3$ **54.** $(4z + 3)^3$

55. $(6u - 1)^3$ **56.** $(5v + 3)^3$

57. $t(t + 2)^3$ **58.** $a(2a - b)^3$

MULTIPLICATION OF POLYNOMIALS

Exercises 59–76: Multiply, using any appropriate method.

59. $4(5x + 9)$ **60.** $(2x + 1)(3x - 5)$

61. $(x - 5)(x + 7)$ **62.** $(x + 10)(x + 10)$

63. $(3x - 5)^2$ **64.** $(x - 3)(x + 9)$

65. $(5x + 3)(5x + 4)$ **66.** $-x^3(x^2 - x + 1)$

67. $(4b - 5)(4b + 5)$ **68.** $(x + 5)^3$

69. $-5x(4x^2 - 7x + 2)$ **70.** $(4x^2 - 5)(4x^2 + 5)$

71. $(4 - a)^3$ **72.** $2x(x - 3)^3$

73. $x(x + 3)^2$ **74.** $(x - 1)^2(x + 1)$

75. $(x + 2)(x - 2)(x + 1)(x - 1)$

76. $(x - y)(x + y)(x^2 + y^2)$

77. Thinking Generally Multiply $(a^n + b^n)(a^n - b^n)$.

78. Thinking Generally Multiply $(a^n + b^n)^2$.

APPLICATIONS

Exercises 79–82: Do each part and verify that your answers are the same.
 (a) Find the area of the large square by multiplying its length and width.
 (b) Find the sum of the areas of the smaller rectangles inside the large square.

79.
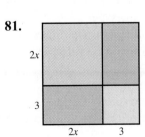

80.
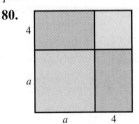

81.
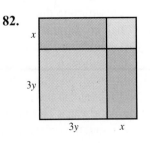

82.
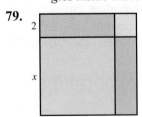

Exercises 83 and 84: Find a polynomial that represents the following.
  (a) *The outside surface area given by the six sides of the cube*
  (b) *The volume of the cube*

83.
$x + 5$
$x + 5$
$x + 5$

84.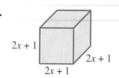
$2x + 1$
$2x + 1$
$2x + 1$

85. *Compound Interest* (Refer to Example 6.) If a sum of money is deposited in a savings account that is paying x percent annual interest (expressed as a decimal), then this sum of money increases by a factor of $(1 + x)^2$ after 2 years.
  (a) Multiply this expression.
  (b) Evaluate the polynomial expression found in part (a) for an annual interest rate of 10%, or $x = 0.10$, and interpret the answer.

86. *Compound Interest* If a sum of money is deposited in a savings account that is paying x percent annual interest, then this sum of money increases by a factor of $\left(1 + \frac{1}{100}x\right)^3$ after 3 years.
  (a) Multiply this expression.
  (b) Evaluate the polynomial expression found in part (a) for an annual interest rate of 8%, or $x = 8$, and interpret the answer.

87. *Probability* If there is an x percent chance of rain on each of two consecutive days, then the expression $(1 - x)^2$ gives the percent chance of no rain either day. Assume that all percentages are expressed as decimals.
  (a) Multiply this expression.
  (b) Evaluate the polynomial expression found in part (a) for a 50% chance of rain, or $x = 0.50$, and interpret the answer.

88. *Probability* If there is an x percent chance of rolling a 6 with one die, then the expression $(1 - x)^3$ gives the percent chance of not rolling a 6 with three dice. Assume that all percentages are expressed as decimals or fractions.
  (a) Multiply this expression.
  (b) Evaluate the polynomial expression found in part (a) for a 16.$\overline{6}$% chance of rolling a 6, or $x = \frac{1}{6}$, and interpret the answer.

89. *Swimming Pool* A square swimming pool has an 8-foot–wide sidewalk around it.
  (a) If the sides of the pool have length z, find a polynomial that gives the area of the sidewalk.
  (b) Evaluate the polynomial in part (a) for $z = 60$ and interpret the answer.

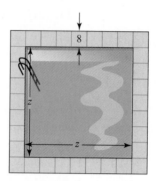
8
z
z

90. *Digital Picture* (Refer to Example 4.) Suppose that a digital picture, including its border, is $x + 2$ pixels by $x + 2$ pixels and that the actual picture inside the border is $x - 2$ pixels by $x - 2$ pixels.

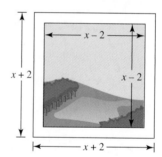
$x - 2$
$x + 2$
$x - 2$
$x + 2$

  (a) Find a polynomial that gives the number of pixels in the border.
  (b) Evaluate the polynomial in part (a) for $x = 5$.
  (c) Sketch a digital picture of the letter H with $x = 5$. Does the picture agree with the answer in part (b)?
  (d) Digital pictures typically have large values for x. If a picture has $x = 500$, find the total number of pixels in its border.

WRITING ABOUT MATHEMATICS

91. Explain why $(a + b)^2$ does not equal $a^2 + b^2$ in general for real numbers a and b.

92. Explain how to find the cube of a binomial.

CHECKING BASIC CONCEPTS
SECTIONS 5.3 AND 5.4

1. Multiply each expression.
   (a) $(-3xy^4)(5x^2y)$ (b) $-x(6 - 4x)$
   (c) $3ab(a^2 - 2ab + b^2)$

2. Multiply each expression.
   (a) $(x + 3)(4x - 3)$
   (b) $(x^2 - 1)(2x^2 + 2)$
   (c) $(x + y)(x^2 - xy + y^2)$

3. Multiply each expression.
   (a) $(5x + 2)(5x - 2)$ (b) $(x + 3)^2$
   (c) $(2 - 7x)^2$ (d) $(t + 2)^3$

4. *Surface Area* A sphere with radius r has surface area $A = 4\pi r^2$.
   (a) Find an expression for the surface area for $r = a + 7$.
   (b) Find the surface area for $a = 8$. Leave your answer in terms of π.

5. Complete each part and verify that your answers are the same.
   (a) Find the area of the large square by squaring the length of one of its sides.
   (b) Find the sum of the areas of the smaller rectangles inside the large square.

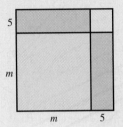

5.5 INTEGER EXPONENTS AND THE QUOTIENT RULE

Negative Integers as Exponents ▪ The Quotient Rule ▪ Other Rules for Exponents ▪ Scientific Notation

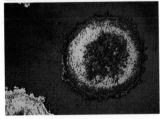

A LOOK INTO MATH ▷ Technology has brought with it the need for both very small and very large numbers. The size of an average virus (shown to the left) is 5 millionths of a centimeter, whereas the distance to the nearest star, Alpha Centauri, is 25 trillion miles. Exponents are often used to represent such numbers. In this section we discuss properties of integer exponents and some of their applications. (**Source:** C. Ronan, *The Natural History of the Universe.*)

Negative Integers as Exponents

So far we have defined exponents that are whole numbers. For example,

$$5^0 = 1 \quad \text{and} \quad 2^3 = 2 \cdot 2 \cdot 2 = 8.$$

We can also define exponents that are negative integers as follows.

NEGATIVE INTEGER EXPONENTS

Let a be a nonzero real number and n be a positive integer. Then

$$a^{-n} = \frac{1}{a^n}.$$

That is, a^{-n} is the reciprocal of a^n.

In the next example we evaluate negative exponents.

EXAMPLE 1 Evaluating negative exponents

Simplify each expression.
(a) 2^{-3} (b) 7^{-1} (c) x^{-2} (d) $(x+y)^{-8}$

Solution

(a) Because $a^{-n} = \dfrac{1}{a^n}$, $2^{-3} = \dfrac{1}{2^3} = \dfrac{1}{2 \cdot 2 \cdot 2} = \dfrac{1}{8}$.

(b) $7^{-1} = \dfrac{1}{7^1} = \dfrac{1}{7}$

(c) $x^{-2} = \dfrac{1}{x^2}$

(d) $(x+y)^{-8} = \dfrac{1}{(x+y)^8}$

Now Try Exercises 13, 25, 32(b)

TECHNOLOGY NOTE: Negative Exponents

Calculators can be used to evaluate negative exponents. The figure shows how a graphing calculator evaluates the expressions in parts (a) and (b) of Example 1.

```
2^(-3)▶Frac
              1/8
7^(-1)▶Frac
              1/7
```

The rules for exponents discussed in this chapter so far also apply to expressions having negative exponents. For example, we can apply the product rule, $a^m \cdot a^n = a^{m+n}$, as follows.

$$2^{-3} \cdot 2^2 = 2^{-3+2} = 2^{-1} = \frac{1}{2}$$

We can check this result by evaluating the expression without using the product rule.

$$2^{-3} \cdot 2^2 = \frac{1}{2^3} \cdot 2^2 = \frac{1}{8} \cdot 4 = \frac{4}{8} = \frac{1}{2}$$

EXAMPLE 2 Using the product rule with negative exponents

Evaluate each expression.
(a) $5^2 \cdot 5^{-4}$ (b) $3^{-2} \cdot 3^{-1}$

Solution

(a) $5^2 \cdot 5^{-4} = 5^{2+(-4)} = 5^{-2} = \dfrac{1}{5^2} = \dfrac{1}{25}$

(b) $3^{-2} \cdot 3^{-1} = 3^{-2+(-1)} = 3^{-3} = \dfrac{1}{3^3} = \dfrac{1}{27}$

Now Try Exercise 15

EXAMPLE 3   **Using the rules of exponents**

Simplify the expression. Write the answer using positive exponents.

(a) $x^2 \cdot x^{-5}$ **(b)** $(y^3)^{-4}$ **(c)** $(rt)^{-5}$ **(d)** $(ab)^{-3}(a^{-2}b)^3$

Solution

(a) Using the product rule, $a^m \cdot a^n = a^{m+n}$, gives

$$x^2 \cdot x^{-5} = x^{2+(-5)} = x^{-3} = \frac{1}{x^3}.$$

(b) Using the power rule, $(a^m)^n = a^{mn}$, gives

$$(y^3)^{-4} = y^{3(-4)} = y^{-12} = \frac{1}{y^{12}}.$$

(c) Using the power rule, $(ab)^n = a^n b^n$, gives

$$(rt)^{-5} = r^{-5}t^{-5} = \frac{1}{r^5} \cdot \frac{1}{t^5} = \frac{1}{r^5 t^5}.$$

This expression could also be simplified as follows.

$$(rt)^{-5} = \frac{1}{(rt)^5} = \frac{1}{r^5 t^5}$$

(d)
$$
\begin{aligned}
(ab)^{-3}(a^{-2}b)^3 &= a^{-3}b^{-3}a^{-6}b^3 \\
&= a^{-3+(-6)}b^{-3+3} \\
&= a^{-9}b^0 \\
&= \frac{1}{a^9} \cdot 1 \\
&= \frac{1}{a^9}
\end{aligned}
$$

Now Try Exercises 27, 33, 35(a)

The Quotient Rule

Consider the division problem

$$\frac{3^4}{3^2} = \frac{3 \cdot 3 \cdot 3 \cdot 3}{3 \cdot 3} = \frac{3}{3} \cdot \frac{3}{3} \cdot 3 \cdot 3 = 1 \cdot 1 \cdot 3^2 = 3^2.$$

Because there are two more 3s in the numerator than in the denominator, the result is $3^{4-2} = 3^2$. That is, to divide exponential expressions having the *same base*, subtract the exponent of the denominator from the exponent of the numerator and keep the same base. This rule is called the *quotient rule*, which we express in symbols as follows.

THE QUOTIENT RULE

For any nonzero number a and integers m and n,

$$\frac{a^m}{a^n} = a^{m-n}.$$

EXAMPLE 4   **Using the quotient rule**

Simplify each expression. Write the answer using positive exponents.

(a) $\dfrac{4^3}{4^5}$ (b) $\dfrac{6a^7}{3a^4}$ (c) $\dfrac{xy^7}{x^2y^5}$

Solution

(a) $\dfrac{4^3}{4^5} = 4^{3-5} = 4^{-2} = \dfrac{1}{4^2} = \dfrac{1}{16}$ ⌐ Subtract

(b) $\dfrac{6a^7}{3a^4} = \dfrac{6}{3} \cdot \dfrac{a^7}{a^4} = 2a^{7-4} = 2a^3$

(c) $\dfrac{xy^7}{x^2y^5} = \dfrac{x^1}{x^2} \cdot \dfrac{y^7}{y^5} = x^{1-2}y^{7-5} = x^{-1}y^2 = \dfrac{y^2}{x}$

Now Try Exercises 19(b), 37(b), 39(a)

MAKING CONNECTIONS

The Quotient Rule and Simplifying Quotients

Some quotients can be simplified mentally. Because

$$\frac{x^5}{x^3} = \frac{x \cdot x \cdot x \cdot x \cdot x}{x \cdot x \cdot x},$$

the quotient $\frac{x^5}{x^3}$ has five factors of x in the numerator and three factors of x in the denominator. There are two more factors of x in the numerator than in the denominator, $5 - 3 = 2$, so this expression simplifies to x^2. Similarly,

$$\frac{x^3}{x^5} = \frac{x \cdot x \cdot x}{x \cdot x \cdot x \cdot x \cdot x}$$

has two more factors of x in the denominator than in the numerator. This expression simplifies to $\frac{1}{x^2}$. Use this technique to simplify the expressions

$$\frac{z^7}{z^4}, \quad \frac{a^5}{a^8}, \quad \text{and} \quad \frac{x^6y^2}{x^3y^7}.$$

Other Rules for Exponents

Other rules can be used to simplify expressions with negative exponents.

QUOTIENTS AND NEGATIVE EXPONENTS

The following three rules hold for any nonzero real numbers a and b and positive integers m and n.

1. $\dfrac{1}{a^{-n}} = a^n$ **2.** $\dfrac{a^{-n}}{b^{-m}} = \dfrac{b^m}{a^n}$ **3.** $\left(\dfrac{a}{b}\right)^{-n} = \left(\dfrac{b}{a}\right)^n$

We demonstrate the validity of these rules as follows.

1. $\dfrac{1}{a^{-n}} = \dfrac{1}{\dfrac{1}{a^n}} = 1 \cdot \dfrac{a^n}{1} = a^n$

2. $\dfrac{a^{-n}}{b^{-m}} = \dfrac{\dfrac{1}{a^n}}{\dfrac{1}{b^m}} = \dfrac{1}{a^n} \cdot \dfrac{b^m}{1} = \dfrac{b^m}{a^n}$

3. $\left(\dfrac{a}{b}\right)^{-n} = \dfrac{a^{-n}}{b^{-n}} = \dfrac{\dfrac{1}{a^n}}{\dfrac{1}{b^n}} = \dfrac{1}{a^n} \cdot \dfrac{b^n}{1} = \dfrac{b^n}{a^n} = \left(\dfrac{b}{a}\right)^n$

In the next example we apply these rules.

EXAMPLE 5   **Working with quotients and negative exponents**

Simplify each expression. Write the answer using positive exponents.

(a) $\dfrac{1}{2^{-5}}$ **(b)** $\dfrac{3^{-3}}{4^{-2}}$ **(c)** $\dfrac{5x^{-4}y^2}{10x^2y^{-4}}$ **(d)** $\left(\dfrac{2}{z^2}\right)^{-4}$

Solution

(a) $\dfrac{1}{2^{-5}} = 2^5 = 2 \cdot 2 \cdot 2 \cdot 2 \cdot 2 = 32$

(b) $\dfrac{3^{-3}}{4^{-2}} = \dfrac{4^2}{3^3} = \dfrac{16}{27}$

(c) $\dfrac{5x^{-4}y^2}{10x^2y^{-4}} = \dfrac{y^2y^4}{2x^2x^4} = \dfrac{y^6}{2x^6}$

(d) $\left(\dfrac{2}{z^2}\right)^{-4} = \left(\dfrac{z^2}{2}\right)^4 = \dfrac{z^8}{2^4} = \dfrac{z^8}{16}$

Now Try Exercises   21(b), 23, 41(a), 47

Scientific Notation

Powers of 10 are important because they are used in science to express numbers that are either very small or very large in absolute value. Table 5.4 on the next page lists some powers of 10. Note that if the power of 10 decreases by 1, the result decreases by a factor of $\frac{1}{10}$, or equivalently, the decimal point is moved one place to the left. Table 5.5 on the next page shows the values of some important powers of 10.

TECHNOLOGY NOTE:
Powers of 10

Calculators make use of scientific notation, as illustrated in the accompanying figure. The letter E denotes a power of 10. That is,

$2.5\text{E}13 = 2.5 \times 10^{13}$ and
$5\text{E}{-}6 = 5 \times 10^{-6}$.

NOTE: The calculator has been set in *scientific mode*.

TABLE 5.4 Powers of 10

Power of 10	Value
10^3	1000
10^2	100
10^1	10
10^0	1
10^{-1}	$\frac{1}{10} = 0.1$
10^{-2}	$\frac{1}{100} = 0.01$
10^{-3}	$\frac{1}{1000} = 0.001$

TABLE 5.5 Important Powers of 10

Number	Value
10^3	Thousand
10^6	Million
10^9	Billion
10^{12}	Trillion
10^{-1}	Tenth
10^{-2}	Hundredth
10^{-3}	Thousandth
10^{-6}	Millionth

Recall that numbers written in decimal notation are sometimes said to be in *standard form*. Decimal numbers that are either very large or very small in absolute value can be expressed in *scientific notation*.

▶ **REAL-WORLD CONNECTION** The distance to the nearest star, Alpha Centauri, is 25 trillion miles. This distance can be written in scientific notation as 2.5×10^{13} because

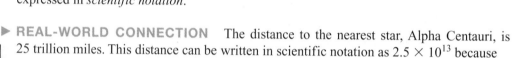

$$25,000,000,000,000 = 2.5 \times 10^{13}.$$
$$\underbrace{\qquad\qquad\qquad}_{\text{13 decimal places}}$$

The 10^{13} indicates that the decimal point in 2.5 should be moved **13** places to the **right**.

A typical virus is about 5 millionths of a centimeter in diameter, which can be written in scientific notation as 5×10^{-6} because

$$0.000005 = 5 \times 10^{-6} \text{ cm.}$$
$$\underbrace{\qquad\qquad}_{\text{6 decimal places}}$$

The 10^{-6} indicates that the decimal point in 5 should be moved **6** places to the **left**.

The following definition provides a more complete explanation of scientific notation.

SCIENTIFIC NOTATION

A real number a is in **scientific notation** when a is written in the form $b \times 10^n$, where $1 \le |b| < 10$ and n is an integer.

EXAMPLE 6 Converting scientific notation to standard form

Write each number in standard form.
(a) 5.23×10^4 **(b)** 8.1×10^{-3} **(c)** 6×10^{-2}

Solution
(a) The positive exponent 4 indicates that the decimal point in 5.23 is to be moved 4 places to the *right*.

$$5.23 \times 10^4 = 5.\underset{1\ 2\ 3\ 4}{2\ 3\ 0\ 0.} = 52,300$$

(b) A negative exponent -3 indicates that the decimal point in 8.1 is to be moved 3 places to the *left*.

$$8.1 \times 10^{-3} = 0.\underset{\underset{1\ 2\ 3}{\smallsmile}}{008}.1 = 0.0081$$

(c) $6 \times 10^{-2} = 0.\underset{\underset{1\ 2}{\smallsmile}}{06}. = 0.06$

Now Try Exercises 59, 61

The following steps can be used for writing a positive number a in scientific notation.

WRITING A POSITIVE NUMBER IN SCIENTIFIC NOTATION

For a positive, rational number a expressed as a decimal, if $1 \leq a < 10$, then $a = a \times 10^0$. Otherwise, use the following process to write a in scientific notation.

1. Move the decimal point in a until it becomes a number b such that $1 \leq b < 10$.
2. Count the number of places that the decimal point was moved. Let this positive integer be n.
3. If the decimal point was moved to the *left*, then $a = b \times 10^n$. If the decimal point was moved to the *right*, then $a = b \times 10^{-n}$.

NOTE: The scientific notation for a negative number a is the opposite of the scientific notation of $|a|$. For example, $450 = 4.5 \times 10^2$ and $-450 = -4.5 \times 10^2$.

EXAMPLE 7 Writing a number in scientific notation

Write each number in scientific notation.
(a) 281,000,000   (U.S. population in 2000)
(b) 0.001   (Approximate time in seconds for sound to travel one foot)

Solution
(a) Move the assumed decimal point in 281,000,000 eight places to the *left* to obtain 2.81.

$$2.8\underset{\underset{1\ 2\ 3\ 4\ 5\ 6\ 7\ 8}{\smallsmile}}{1\ 0\ 0\ 0\ 0\ 0\ 0}0.$$

The scientific notation for 281,000,000 is 2.81×10^8.
(b) Move the decimal point in 0.001 three places to the *right* to obtain 1.

$$0.\underset{\underset{1\ 2\ 3}{\smallsmile}}{001}.$$

The scientific notation for 0.001 is 1×10^{-3}.

Now Try Exercises 73, 77

MAKING CONNECTIONS

Scientific Notation

When a positive number a is expressed in scientific notation, a negative exponent on 10 indicates that $a < 1$, thus a is relatively small, and a positive exponent on 10 indicates that $a \geq 10$, thus a is relatively large. This can be helpful when converting from scientific notation to standard notation or vice versa. For example, to write the number 3.4×10^7 in standard form, we move the decimal point in 3.4 seven places. Because the exponent on 10 is positive, the resulting number should be relatively large. Moving the decimal point to the right results in 34,000,000. To express the number 0.00087 in scientific notation, we move the decimal point four places to obtain 8.7. Because the number 0.00087 is relatively small, the exponent on 10 will be negative. The resulting scientific notation is 8.7×10^{-4}.

Numbers in scientific notation can be multiplied by applying properties of real numbers and properties of exponents.

$$(6 \times 10^4) \cdot (3 \times 10^3) = (6 \cdot 3) \times (10^4 \cdot 10^3) \qquad \text{Properties of real numbers}$$
$$= 18 \times 10^7 \qquad \text{Product rule}$$
$$= 1.8 \times 10^8 \qquad \text{Scientific notation}$$

Division can also be performed with scientific notation.

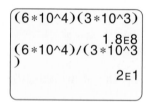

$$\frac{6 \times 10^4}{3 \times 10^3} = \frac{6}{3} \times \frac{10^4}{10^3} \qquad \text{Property of fractions}$$
$$= 2 \times 10^1 \qquad \text{Quotient rule}$$

Figure 5.19

These results are supported in Figure 5.19, where the calculator is in scientific mode. In the next example we show how to use scientific notation in an application.

EXAMPLE **8**   Analyzing the cost of advertising

In 2005, a total of $\$2.46 \times 10^{11}$ was spent on advertising in the United States. At that time the population of the United States was 2.98×10^8. Determine how much was spent per person on advertising.

Solution
To determine the amount spent per person, divide $\$2.46 \times 10^{11}$ by 2.98×10^8.

$$\frac{2.46 \times 10^{11}}{2.98 \times 10^8} = \frac{2.46}{2.98} \times 10^{11-8} \approx 0.826 \times 10^3 = 826$$

In 2005, about $826 was spent on advertising for every person in the United States.

Now Try Exercise **93**

CRITICAL THINKING

Estimate the number of seconds that you have been alive. Write your answer in scientific notation.

5.5 PUTTING IT ALL TOGETHER

In this section we discussed negative integer exponents and scientific notation. The following table summarizes several important concepts related to these topics. Assume that a and b are nonzero real numbers and that m and n are integers.

Concept	Explanation	Examples
Negative Integer Exponents	$a^{-n} = \dfrac{1}{a^n}$	$2^{-4} = \dfrac{1}{2^4} = \dfrac{1}{16}, \quad a^{-8} = \dfrac{1}{a^8},$ and $(xy)^{-2} = \dfrac{1}{(xy)^2} = \dfrac{1}{x^2 y^2}$
Quotient Rule	$\dfrac{a^m}{a^n} = a^{m-n}$	$\dfrac{7^2}{7^4} = 7^{2-4} = 7^{-2} = \dfrac{1}{7^2} = \dfrac{1}{49}$ and $\dfrac{x^6}{x^3} = x^{6-3} = x^3$
Quotients and Negative Integer Exponents	1. $\dfrac{1}{a^{-n}} = a^n$ 2. $\dfrac{a^{-n}}{b^{-m}} = \dfrac{b^m}{a^n}$ 3. $\left(\dfrac{a}{b}\right)^{-n} = \left(\dfrac{b}{a}\right)^n$	1. $\dfrac{1}{5^{-2}} = 5^2 = 25$ 2. $\dfrac{x^{-4}}{y^{-2}} = \dfrac{y^2}{x^4}$ 3. $\left(\dfrac{2}{3}\right)^{-3} = \left(\dfrac{3}{2}\right)^3 = \dfrac{3^3}{2^3} = \dfrac{27}{8}$
Scientific Notation	Write a as $b \times 10^n$, where $1 \leq |b| < 10$ and n is an integer.	$23{,}500 = 2.35 \times 10^4,$ $0.0056 = 5.6 \times 10^{-3},$ and $1000 = 1 \times 10^3$

5.5 Exercises

MyMathLab

PRACTICE　WATCH　DOWNLOAD　READ　REVIEW

CONCEPTS

1. $a^{-n} = $ _____

2. $2^{-3} = $ _____

3. $\dfrac{1}{a^{-n}} = $ _____

4. $\dfrac{1}{2^{-3}} = $ _____

5. $\dfrac{a^m}{a^n} = $ _____

6. $\dfrac{2^5}{2^2} = $ _____

7. $\dfrac{a^{-n}}{b^{-m}} = $ _____

8. $\dfrac{2^{-3}}{5^{-2}} = $ _____

9. $\left(\dfrac{a}{b}\right)^{-n} = $ _____

10. $\left(\dfrac{3}{5}\right)^{-2} = $ _____

11. To write a number a in scientific notation as $b \times 10^n$, the number b must satisfy _____.

12. Are 0.005 and 5×10^3 equivalent numbers?

NEGATIVE EXPONENTS

Exercises 13–24: Simplify the expression.

13. (a) 4^{-1} **(b)** $\left(\dfrac{1}{3}\right)^{-2}$

14. (a) 6^{-2} **(b)** 2.5^{-1}

15. (a) $2^3 \cdot 2^{-2}$ **(b)** $10^4 \cdot 10^{-2}$

16. (a) $3^{-4} \cdot 3^2$ **(b)** $10^{-1} \cdot 10^{-2}$

17. (a) $3^{-2} \cdot 3^{-1} \cdot 3^{-1}$ **(b)** $(2^3)^{-1}$

18. (a) $2^{-3} \cdot 2^5 \cdot 2^{-4}$ **(b)** $(3^{-2})^{-2}$

19. (a) $(3^2 4^3)^{-1}$ **(b)** $\dfrac{4^5}{4^2}$

20. (a) $(2^{-2} 3^2)^{-2}$ **(b)** $\dfrac{5^5}{5^3}$

21. (a) $\dfrac{1^9}{1^7}$ **(b)** $\dfrac{1}{4^{-3}}$

22. (a) $\dfrac{-6^4}{6}$ **(b)** $\dfrac{1}{6^{-2}}$

23. (a) $\dfrac{5^{-2}}{5^{-4}}$ **(b)** $\left(\dfrac{2}{7}\right)^{-2}$

24. (a) $\dfrac{7^{-3}}{7^{-1}}$ **(b)** $\left(\dfrac{3}{4}\right)^{-3}$

Exercises 25–48: Simplify the expression. Write the answer using positive exponents.

25. (a) x^{-1} **(b)** a^{-4}

26. (a) y^{-2} **(b)** z^{-7}

27. (a) $x^{-2} \cdot x^{-1} \cdot x$ **(b)** $a^{-5} \cdot a^{-2} \cdot a^{-1}$

28. (a) $y^{-3} \cdot y^4 \cdot y^{-5}$ **(b)** $b^5 \cdot b^{-3} \cdot b^{-6}$

29. (a) $x^2 y^{-3} x^{-5} y^6$ **(b)** $(xy)^{-3}$

30. (a) $a^{-2} b^{-6} b^3 a^{-1}$ **(b)** $(ab)^{-1}$

31. (a) $(2t)^{-4}$ **(b)** $(x + 1)^{-7}$

32. (a) $(8c)^{-2}$ **(b)** $(a + b)^{-9}$

33. (a) $(a^{-2})^{-4}$ **(b)** $(rt^3)^{-2}$

34. (a) $(4x^3)^{-3}$ **(b)** $(xy^{-3})^{-2}$

35. (a) $(ab)^2 (a^2)^{-3}$ **(b)** $\dfrac{x^4}{x^2}$

36. (a) $(x^3)^{-2}(xy)^4 y^{-5}$ **(b)** $\dfrac{y^9}{y^5}$

37. (a) $\dfrac{a^{10}}{a^{-3}}$ **(b)** $\dfrac{4z}{2z^4}$

38. (a) $\dfrac{b^5}{b^{-2}}$ **(b)** $\dfrac{12x^2}{24x^7}$

39. (a) $\dfrac{-4xy^5}{6x^3 y^2}$ **(b)** $\dfrac{x^{-4}}{x^{-1}}$

40. (a) $\dfrac{12a^6 b^2}{8ab^3}$ **(b)** $\dfrac{y^{-2}}{y^{-7}}$

41. (a) $\dfrac{10b^{-4}}{5b^{-5}}$ **(b)** $\left(\dfrac{a}{b}\right)^3$

42. (a) $\dfrac{8a^{-2}}{2a^{-3}}$ **(b)** $\left(\dfrac{2x}{y}\right)^5$

43. (a) $\dfrac{1}{y^{-5}}$ **(b)** $\dfrac{4}{2t^{-3}}$

44. (a) $\dfrac{1}{z^{-6}}$ **(b)** $\dfrac{5}{10b^{-5}}$

45. (a) $\dfrac{1}{(xy)^{-2}}$ **(b)** $\dfrac{1}{(a^2 b)^{-3}}$

46. (a) $\dfrac{1}{(ab)^{-1}}$ **(b)** $\dfrac{1}{(rt^4)^{-2}}$

47. (a) $\left(\dfrac{a}{b}\right)^{-2}$ **(b)** $\left(\dfrac{u}{4v}\right)^{-1}$

48. (a) $\left(\dfrac{2x}{y}\right)^{-3}$ **(b)** $\left(\dfrac{5u}{3v}\right)^{-2}$

49. Thinking Generally   For positive integers m and n show that $\dfrac{a^n}{a^m} = \dfrac{1}{a^{m-n}}$.

50. Thinking Generally   For positive integers m and n show that $\dfrac{a^{-n}}{a^{-m}} = a^{m-n}$.

SCIENTIFIC NOTATION

Exercises 51–56: (Refer to Table 5.5.) Write the value of the power of 10 in words.

51. 10^3 **52.** 10^6

53. 10^9 **54.** 10^{-1}

55. 10^{-2} **56.** 10^{-6}

Exercises 57–68: Write the expression in standard form.

57. 2×10^3

58. 5×10^2

59. 4.5×10^4

60. 7.1×10^6

61. 8×10^{-3}

62. 9×10^{-1}

63. 4.56×10^{-4}

64. 9.4×10^{-2}

65. 3.9×10^7

66. 5.27×10^6

67. -5×10^5

68. -9.5×10^3

Exercises 69–80: Write the number in scientific notation.

69. 2000

70. 11,000

71. 567

72. 9300

73. 12,000,000

74. 600,000

75. 0.004

76. 0.0008

77. 0.000895

78. 0.0123

79. −0.05

80. −0.934

Exercises 81–88: Evaluate the expression. Write the answer in standard form.

81. $(5 \times 10^3)(3 \times 10^2)$

82. $(2.1 \times 10^2)(2 \times 10^4)$

83. $(-3 \times 10^{-3})(5 \times 10^2)$

84. $(4 \times 10^2)(1 \times 10^3)(5 \times 10^{-4})$

85. $\dfrac{4 \times 10^5}{2 \times 10^2}$

86. $\dfrac{9 \times 10^2}{3 \times 10^6}$

87. $\dfrac{8 \times 10^{-6}}{4 \times 10^{-3}}$

88. $\dfrac{6.3 \times 10^2}{2 \times 10^{-3}}$

APPLICATIONS

89. *Light-year* The distance that light travels in 1 year is called a *light-year.* Light travels at 1.86×10^5 miles per second, and there are about 3.15×10^7 seconds in 1 year.
  (a) Estimate the number of miles in 1 light-year.
  (b) Except for the sun, Alpha Centauri is the nearest star, and its distance is 4.27 light-years from Earth. Estimate its distance in miles. Write your answer in scientific notation.

90. *Milky Way* It takes 2×10^8 years for the sun to make one orbit around the Milky Way galaxy. Write this number in standard form.

91. *Speed of the Sun* (Refer to the two previous exercises.) Assume that the sun's orbit in the Milky Way galaxy is circular with a diameter of 10^5 light-years. Estimate how many miles the sun travels in 1 year.

92. *Distance to the Moon* The moon is about 240,000 miles from Earth.
  (a) Write this number in scientific notation.
  (b) If a rocket traveled at 4×10^4 miles per hour, how long would it take for it to reach the moon?

93. *Gross Domestic Product* The gross domestic product (GDP) is the total national output of goods and services valued at market prices *within* the United States. The GDP of the United States in 2005 was $12,460,000,000,000. (**Source:** Bureau of Economic Analysis.)
  (a) Write this number in scientific notation.
  (b) In 2005, the U.S. population was 2.98×10^8. On average, how many dollars of goods and services were produced by each individual?

94. *Average Family Net Worth*   A family refers to a group of two or more people related by birth, marriage, or adoption who reside together. In 2000, the average family net worth was $280,000, and there were about 7.2×10^7 families. Calculate the total family net worth in the United States in 2000.   (***Source:*** U.S. Census Bureau.)

WRITING ABOUT MATHEMATICS

95. Explain what a negative exponent is and how it is different from a positive exponent. Give an example.

96. Explain why scientific notation is helpful for writing some numbers.

GROUP ACTIVITY
WORKING WITH REAL DATA

Directions: Form a group of 2 to 4 people. Select a person to record the group's responses for this activity. All members of the group should work cooperatively to answer the questions. If your instructor asks for your results, each member of the group should be prepared to respond.

Water in a Lake   East Battle Lake in Minnesota covers an area of about 1950 acres or 8.5×10^7 square feet, and its average depth is about 3.2×10^1 feet.

(a) Estimate the cubic feet of water in the lake. (*Hint:* volume = area $\times$ average depth.)

(b) One cubic foot of water equals about 7.5 gallons. How many gallons of water are in this lake?

(c) The population of the United States is about 3.02×10^8, and the average American uses about 1.5×10^2 gallons of water per day. Could this lake supply the American population with water for 1 day?

5.6   DIVISION OF POLYNOMIALS

Division by a Monomial ■ Division by a Polynomial

A LOOK INTO MATH ▷

The study of polynomials has occupied the minds of mathematicians for centuries. During the sixteenth century, Girolamo Cardano and other Italian mathematicians discovered how to solve higher degree polynomial equations. In this section we demonstrate how to divide polynomials. Division is often needed to factor polynomials and to solve polynomial equations.   (***Source:*** H. Eves, *An Introduction to the History of Mathematics.*)

Division by a Monomial

To add two fractions with like denominators, we use the property

$$\frac{a}{d} + \frac{b}{d} = \frac{a+b}{d}.$$

Girolamo Cardano (1501–1576)

For example, $\frac{1}{7} + \frac{3}{7} = \frac{1+3}{7} = \frac{4}{7}$.

To divide a polynomial by a monomial we use the same property, only in reverse. That is,

$$\frac{a+b}{d} = \frac{a}{d} + \frac{b}{d}.$$

Note that each term in the numerator is divided by the monomial in the denominator. The next example shows how to divide a polynomial by a monomial.

EXAMPLE 1   Dividing a polynomial by a monomial

Divide.

(a) $\dfrac{a^5 + a^3}{a^2}$ (b) $\dfrac{5x^4 + 10x}{10x}$ (c) $\dfrac{3y^2 + 2y - 12}{6y}$

Solution

(a) $\dfrac{a^5 + a^3}{a^2} = \dfrac{a^5}{a^2} + \dfrac{a^3}{a^2} = a^{5-2} + a^{3-2} = a^3 + a$

(b) $\dfrac{5x^4 + 10x}{10x} = \dfrac{5x^4}{10x} + \dfrac{10x}{10x} = \dfrac{5}{10} \cdot \dfrac{x^4}{x} + \dfrac{10x}{10x} = \dfrac{1}{2}x^3 + 1$

(c) $\dfrac{3y^2 + 2y - 12}{6y} = \dfrac{3y^2}{6y} + \dfrac{2y}{6y} - \dfrac{12}{6y} = \dfrac{1}{2}y + \dfrac{1}{3} - \dfrac{2}{y}$ Now Try Exercises 15, 17

MAKING CONNECTIONS

Division and Simplification

A common mistake made when dividing expressions is to "cancel" incorrectly. Note in Example 1(b) that

$$\frac{5x^4 + 10x}{10x} \neq 5x^4 + \frac{10x}{10x}.$$

The monomial must be divided into *every* term in the numerator.

When dividing two natural numbers, we can check our work by multiplying. For example, $\frac{10}{5} = 2$, and we can check this result by finding the product $5 \cdot 2 = 10$. Similarly, to check

$$\frac{a^5 + a^3}{a^2} = a^3 + a,$$

we can multiply a^2 and $a^3 + a$.

$a^2(a^3 + a) = a^2 \cdot a^3 + a^2 \cdot a$ Distributive property

$ = a^5 + a^3$ It checks.

EXAMPLE 2   Dividing and checking

Divide the expression $\frac{8x^3 - 4x^2 + 6x}{2x^2}$ and then check the result.

Solution
Be sure to divide $2x^2$ into *every* term in the numerator.

$$\frac{8x^3 - 4x^2 + 6x}{2x^2} = \frac{8x^3}{2x^2} - \frac{4x^2}{2x^2} + \frac{6x}{2x^2} = 4x - 2 + \frac{3}{x}$$

Check:

$$2x^2\left(4x - 2 + \frac{3}{x}\right) = 2x^2 \cdot 4x - 2x^2 \cdot 2 + 2x^2 \cdot \frac{3}{x}$$

$$= 8x^3 - 4x^2 + 6x$$ Now Try Exercise 13

EXAMPLE 3   Finding the length of a rectangle

The rectangle in Figure 5.20 has an area $A = x^2 + 2x$ and width x. Write an expression for its length L in terms of x.

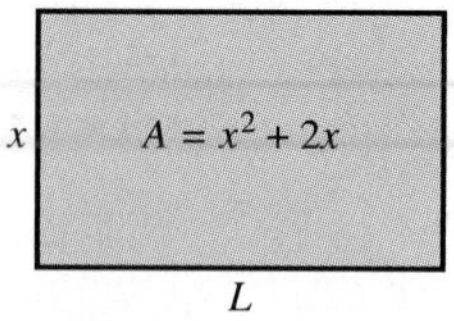

Figure 5.20

Solution
The area A of a rectangle equals length L times width W, or $A = LW$. Solving for L gives

$$L = \frac{A}{W}.$$

Thus to find the length of the given rectangle, divide the area by the width.

$$L = \frac{x^2 + 2x}{x} = \frac{x^2}{x} + \frac{2x}{x} = x + 2$$

The length of the rectangle is $x + 2$. The answer checks because $x(x + 2) = x^2 + 2x$.

Now Try Exercise 43

Division by a Polynomial

To enable you to understand division by a polynomial better, we first need to review some terminology related to long division of natural numbers. To compute $271 \div 4$, we complete long division as follows.

Quotient ⟶ 67 R 3 ⟵ Remainder
Divisor ⟶ 4)271 ⟵ Dividend
          24
          ──
          31
          28
          ──
          3

To check this result, we find the product of the quotient and divisor and then add the remainder. Because $67 \cdot 4 + 3 = 271$, the answer checks. The quotient and remainder can also be expressed as $67\frac{3}{4}$. Division of polynomials is similar to long division of natural numbers.

EXAMPLE 4   Dividing polynomials

Divide $\frac{6x^2 + 13x + 3}{3x + 2}$ and check.

Solution
Begin by dividing the first term of $3x + 2$ into the first term of $6x^2 + 13x + 3$. That is, divide $3x$ into $6x^2$ to obtain $2x$. Then find the product of $2x$ and $3x + 2$, or $6x^2 + 4x$, place it below $6x^2 + 13x$, and subtract. Bring down the 3.

$$\begin{array}{r} 2x \\ 3x + 2 \overline{)6x^2 + 13x + 3} \\ \underline{6x^2 + 4x} \\ 9x + 3 \end{array}$$

$\dfrac{6x^2}{3x} = 2x$

$2x(3x + 2) = 6x^2 + 4x$

Subtract: $13x - 4x = 9x$. Bring down the 3.

In the next step, divide $3x$ into the first term of $9x + 3$ to obtain 3. Then find the product of 3 and $3x + 2$, or $9x + 6$, place it below $9x + 3$, and subtract.

$$
\begin{array}{r}
2x + 3 \\
3x + 2 \overline{) 6x^2 + 13x + 3} \\
\underline{6x^2 + 4x} \\
9x + 3 \\
\underline{9x + 6} \\
-3
\end{array}
\qquad
\begin{array}{l}
\dfrac{9x}{3x} = 3 \\
\\
\\
\\
3(3x + 2) = 9x + 6 \\
\text{Subtract: } 3 - 6 = -3.
\end{array}
$$

The quotient is $2x + 3$ with remainder -3. This result can also be written as

$$
2x + 3 + \frac{-3}{3x + 2}, \qquad \text{Quotient} + \frac{\text{Remainder}}{\text{Divisor}}
$$

in the same manner that 67 R 3 can be written as $67\frac{3}{4}$.

Check polynomial division by adding the remainder to the product of the divisor and the quotient. That is,

$$
(\text{Divisor})(\text{Quotient}) + \text{Remainder} = \text{Dividend}.
$$

For this example, the equation becomes

$$
\begin{aligned}
(3x + 2)(2x + 3) + (-3) &= 3x \cdot 2x + 3x \cdot 3 + 2 \cdot 2x + 2 \cdot 3 - 3 \\
&= 6x^2 + 9x + 4x + 6 - 3 \\
&= 6x^2 + 13x + 3. \qquad \text{It checks.}
\end{aligned}
$$

Now Try Exercise **23**

EXAMPLE 5   **Dividing polynomials having a missing term**

Simplify $(3x^3 + 2x - 4) \div (x - 2)$.

Solution

Because the dividend does not have an x^2-term, insert $0x^2$ as a "place holder." Then begin by dividing x into $3x^3$ to obtain $3x^2$.

$$
\begin{array}{r}
3x^2 \\
x - 2 \overline{) 3x^3 + 0x^2 + 2x - 4} \\
\underline{3x^3 - 6x^2} \\
6x^2 + 2x
\end{array}
\qquad
\begin{array}{l}
\dfrac{3x^3}{x} = 3x^2 \\
\\
3x^2(x - 2) = 3x^3 - 6x^2 \\
\text{Subtract: } 0x^2 - (-6x^2) = 6x^2. \text{ Bring down } 2x.
\end{array}
$$

In the next step, divide x into $6x^2$.

$$
\begin{array}{r}
3x^2 + 6x \\
x - 2 \overline{) 3x^3 + 0x^2 + 2x - 4} \\
\underline{3x^3 - 6x^2} \\
6x^2 + 2x \\
\underline{6x^2 - 12x} \\
14x - 4
\end{array}
\qquad
\begin{array}{l}
\dfrac{6x^2}{x} = 6x \\
\\
\\
\\
6x(x - 2) = 6x^2 - 12x \\
\text{Subtract: } 2x - (-12x) = 14x. \text{ Bring down } -4.
\end{array}
$$

Now divide x into $14x$.

$$
\begin{array}{r}
3x^2 + 6x + \mathbf{14} \\
x - 2{\overline{\smash{\big)}\,3x^3 + 0x^2 + 2x - 4}} \\
\underline{3x^3 - 6x^2} \\
6x^2 + 2x \\
\underline{6x^2 - 12x} \\
14x - 4 \\
\underline{14x - 28} \\
24
\end{array}
$$

$\dfrac{14x}{x} = 14$

$14(x - 2) = 14x - 28$

Subtract: $-4 - (-28) = 24$.

The quotient is $3x^2 + 6x + 14$ with remainder 24. This result can also be written as

$$3x^2 + 6x + 14 + \frac{24}{x - 2}.$$

Now Try Exercise 31

EXAMPLE 6   Dividing with a quadratic divisor

Divide $x^3 - 3x^2 + 3x + 2$ by $x^2 + 1$.

Solution
Begin by writing $x^2 + 1$ as $x^2 + 0x + 1$.

$$
\begin{array}{r}
x - 3 \\
x^2 + 0x + 1{\overline{\smash{\big)}\,x^3 - 3x^2 + 3x + 2}} \\
\underline{x^3 + 0x^2 + x} \\
-3x^2 + 2x + 2 \\
\underline{-3x^2 + 0x - 3} \\
2x + 5
\end{array}
$$

The quotient is $x - 3$ with remainder $2x + 5$. This result can also be written as

$$x - 3 + \frac{2x + 5}{x^2 + 1}.$$

Now Try Exercise 35

5.6 PUTTING IT ALL TOGETHER

In this section we discussed division of polynomials by monomials and other polynomials. The following table summarizes several important concepts related to division of polynomials.

Concept	Explanation	Examples
Division by a Monomial	Use the property $$\frac{a + b}{d} = \frac{a}{d} + \frac{b}{d}.$$ Be sure to divide the denominator into every term in the numerator.	$\dfrac{2x^3 + 4x}{2x^2} = \dfrac{2x^3}{2x^2} + \dfrac{4x}{2x^2} = x + \dfrac{2}{x}$ and $\dfrac{a^2 - 2a}{4a} = \dfrac{a^2}{4a} - \dfrac{2a}{4a} = \dfrac{a}{4} - \dfrac{1}{2}$

Concept	Explanation	Examples
Division by a Polynomial	Is done similarly to the way long division of natural numbers is performed	Divide $x^2 + 3x + 3$ by $x + 1$. $$\begin{array}{r} x + 2 \\ x + 1 \overline{)x^2 + 3x + 3} \\ \underline{x^2 + x} \\ 2x + 3 \\ \underline{2x + 2} \\ 1 \end{array}$$ The quotient is $x + 2$ with remainder 1, which can be expressed as $$x + 2 + \frac{1}{x + 1}.$$
Checking a Result	Dividend = (Divisor)(Quotient) + Remainder	When $x^2 + 3x + 3$ is divided by $x + 1$, the quotient is $x + 2$ with remainder 1. Thus $$(x + 1)(x + 2) + 1 = x^2 + 3x + 3,$$ and the answer checks.

5.6 Exercises

PRACTICE   WATCH   DOWNLOAD   READ   REVIEW

CONCEPTS

1. $\dfrac{a + b}{d} =$ _____

2. $\dfrac{a + b - c}{d} =$ _____

3. When dividing a polynomial by a monomial, the monomial must be divided into every _____ of the polynomial.

4. Are the expressions $\dfrac{5x^2 + 2x}{2x}$ and $5x^2 + 1$ equal?

5. Are the expressions $\dfrac{5x^2 + 2x}{2x}$ and $\dfrac{5x^2}{2x}$ equal?

6. Because $\dfrac{37}{9} = 4$ with remainder 1, it follows that $37 =$ _____ · _____ + _____.

7. Because $2x^3 - x + 5$ divided by $x + 1$ equals $2x^2 - 2x + 1$ with remainder 4, it follows that $2x^3 - x + 5 =$ _____ · _____ + _____.

8. When dividing $2x^3 + 3x - 1$ by $x - 1$, insert _____ into the dividend as a "place holder" for the missing x^2-term.

DIVISION BY A MONOMIAL

Exercises 9–14: Divide and check.

9. $\dfrac{6x^2}{3x}$

10. $\dfrac{-5x^2}{10x^4}$

11. $\dfrac{z^4 + z^3}{z}$

12. $\dfrac{t^3 - t}{t}$

13. $\dfrac{a^5 - 6a^3}{2a^3}$

14. $\dfrac{b^4 - 4b}{4b^2}$

Exercises 15–22: Divide.

15. $\dfrac{4x - 7x^4}{x^2}$

16. $\dfrac{1 + 6x^4}{3x^3}$

17. $\dfrac{9x^4 - 3x + 6}{3x}$

18. $\dfrac{y^3 - 4y + 6}{y}$

19. $\dfrac{12y^4 - 3y^2 + 6y}{3y^2}$ **20.** $\dfrac{2x^2 - 6x + 9}{12x}$

21. $\dfrac{15m^4 - 10m^3 + 20m^2}{5m^2}$ **22.** $\dfrac{n^8 - 8n^6 + 4n^4}{2n^5}$

Exercises 23–28: Divide and check.

23. $\dfrac{2x^2 - 3x + 1}{x - 2}$ **24.** $\dfrac{4x^2 - x + 3}{x + 2}$

25. $\dfrac{x^2 + 2x + 1}{x + 1}$ **26.** $\dfrac{4x^2 - 4x + 1}{2x - 1}$

27. $\dfrac{x^3 - x^2 + x - 2}{x - 1}$ **28.** $\dfrac{2x^3 + 3x^2 + 3x - 1}{2x + 1}$

Exercises 29–40: Divide.

29. $\dfrac{4x^3 - 3x^2 + 7x + 3}{4x + 1}$ **30.** $\dfrac{10x^3 - x^2 - 17x - 7}{5x + 2}$

31. $\dfrac{x^3 - x + 2}{x - 2}$ **32.** $\dfrac{6x^3 + 8x^2 + 4}{3x + 4}$

33. $(3x^3 + 2) \div (x - 1)$

34. $(-3x^3 + 8x^2 + x) \div (3x + 4)$

35. $(x^3 + 3x^2 + 1) \div (x^2 + 1)$

36. $(x^4 - x^3 + x^2 - x + 1) \div (x^2 - 1)$

37. $\dfrac{x^3 + 1}{x^2 - x + 1}$ **38.** $\dfrac{4x^3 + 3x + 2}{2x^2 - x + 1}$

39. $\dfrac{x^3 + 8}{x + 2}$ **40.** $\dfrac{x^4 - 16}{x - 2}$

41. Thinking Generally   If the quotient in a polynomial division problem is an integer, what must be true about the degrees of the dividend and divisor?

42. Thinking Generally   If the quotient in a polynomial division problem is a polynomial of degree 1, what must be true about the degrees of the dividend and divisor?

APPLICATIONS

43. *Area of a Rectangle*   The area A of a rectangle is $8x^2$, and one of its sides has length $2x$. Find an expression for the length L of the other side.

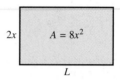

44. *Area of a Rectangle*   The area A of a rectangle is $x^2 - 1$, and one of its sides has length $x + 1$. Find the width W of the other side.

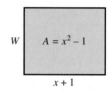

45. *Volume of a Box*   The volume V of a box is $2x^3 + 4x^2$, and the area of its bottom is $2x^2$. Find the height of the box in terms of x. Make a possible sketch of the box, and label the length of each side.

46. *Area of a Triangle*   A triangle has height h and area $A = 2h^2 - 4h$. Find its base b in terms of h. Make a possible sketch of the triangle, and label the height and base. (*Hint:* $A = \frac{1}{2}bh$.)

WRITING ABOUT MATHEMATICS

47. Suppose that one polynomial is divided into another polynomial and the remainder is 0. What does the product of the divisor and quotient equal? Explain.

48. A student simplifies the expression $\dfrac{4x^3 - 1}{4x^2}$ to $x - 1$. Explain the student's error.

CHECKING BASIC CONCEPTS
SECTIONS 5.5 AND 5.6

1. Simplify each expression. Write the result with positive exponents.

   (a) 9^{-2} (b) $\dfrac{3x^{-3}}{6x^4}$ (c) $(4ab^{-4})^{-2}$

2. Simplify each expression. Write the result with positive exponents.

   (a) $\dfrac{1}{z^{-5}}$ (b) $\dfrac{x^{-3}}{y^{-6}}$ (c) $\left(\dfrac{3}{x^2}\right)^{-3}$

3. Write each expression in scientific notation.

   (a) 45,000   (b) 0.000234   (c) 0.01

4. Write each expression in standard form.

   (a) 4.71×10^4 (b) 6×10^{-3}

5. Simplify $\dfrac{25a^4 - 15a^3}{5a^3}$.

6. Divide $3x^2 - x - 4$ by $x - 1$. State the quotient and remainder.

7. *Distance to the Sun*   The distance to the sun is approximately 93 million miles.
   (a) Write this distance in scientific notation.
   (b) Light travels at 1.86×10^5 miles per second. How long does it take for the sun's light to reach Earth?

CHAPTER 5   SUMMARY

SECTION 5.1 ■ RULES FOR EXPONENTS

Bases and Exponents   The expression b^n has base b, exponent n, and equals the expression $\underbrace{b \cdot b \cdot b \cdot \cdots \cdot b}_{n \text{ times}}$, when n is a natural number.

Example: 2^3 has base 2, exponent 3, and equals $2 \cdot 2 \cdot 2 = 8$.

Evaluating Expressions   When evaluating expressions, evaluate exponents *before* performing addition, subtraction, multiplication, division, or negation. In general, operations within parentheses should be evaluated *before* using the order of operations.

1. Evaluate exponents.
2. Perform negation.
3. Do multiplication and division from left to right.
4. Do addition and subtraction from left to right.

Example: $-3^2 + 3 \cdot 4 = -9 + 3 \cdot 4 = -9 + 12 = 3$

Zero Exponents   For any nonzero number b, $b^0 = 1$. Note that 0^0 is undefined.

Examples: $5^0 = 1$ and $\left(\dfrac{x}{y}\right)^0 = 1$, where x and y are nonzero.

Product Rule   For any real number a and natural numbers m and n,

$$a^m \cdot a^n = a^{m+n}.$$

Examples: $3^4 \cdot 3^2 = 3^6$ and $x^3 x^2 x^4 = x^9$

Power Rules   For any real numbers a and b and natural numbers m and n,

$$(a^m)^n = a^{mn}, \quad (ab)^n = a^n b^n, \quad \text{and} \quad \left(\frac{a}{b}\right)^n = \frac{a^n}{b^n}, b \neq 0.$$

Examples: $(x^2)^3 = x^6, \quad (3x)^4 = 3^4 x^4 = 81x^4, \quad \text{and} \quad \left(\frac{2}{y}\right)^3 = \frac{2^3}{y^3} = \frac{8}{y^3}$

SECTION 5.2 ■ ADDITION AND SUBTRACTION OF POLYNOMIALS

Terms Related to Polynomials

Monomial	A number, variable, or product of numbers and variables raised to natural number powers
Degree of a Monomial	Sum of the exponents of the variables
Coefficient of a Monomial	The number in a monomial

Example: The monomial $-3x^2 y^3$ has degree 5 and coefficient -3.

Polynomial	A monomial or a sum of monomials
Term of a Polynomial	Each monomial is a term of the polynomial.
Binomial	A polynomial with two terms
Trinomial	A polynomial with three terms
Degree of a Polynomial	The degree of the term with highest degree
Opposite of a Polynomial	The opposite is found by negating each term.

Example: $2x^3 - 4x + 5$ is a trinomial with degree 3. Its opposite is $-2x^3 + 4x - 5$.

Like Terms	Two monomials with the same variables raised to the same powers

Examples: $3xy^2$ and $-xy^2$ are like terms.

$5x^3$ and $3x^3$ are like. terms.

$5x^2$ and $5x$ are unlike terms.

Addition of Polynomials   Combine like terms, using the distributive property.

Example: $(2x^2 - 4x) + (-x^2 - x) = (2 - 1)x^2 + (-4 - 1)x$

$$= x^2 - 5x$$

Subtraction of Polynomials   Add the first polynomial to the opposite of the second polynomial.

Example: $(4x^4 - 5x) - (7x^4 + 6x) = (4x^4 - 5x) + (-7x^4 - 6x)$

$$= (4 - 7)x^4 + (-5 - 6)x$$

$$= -3x^4 - 11x$$

SECTION 5.3 ■ MULTIPLICATION OF POLYNOMIALS

Multiplication of Monomials   Use the commutative property and the product rule.

Examples: $-2x^3 \cdot 3x^2 = -2 \cdot 3 \cdot x^3 \cdot x^2 = -6x^5$

$(2xy^2)(3x^2y^3) = 2 \cdot 3 \cdot x \cdot x^2 \cdot y^2 \cdot y^3 = 6x^3y^5$

└── Assumed exponent of 1

Distributive Properties

$$a(b + c) = ab + ac \quad \text{and} \quad a(b - c) = ab - ac$$

Examples: $4x(3x + 6) = 4x \cdot 3x + 4x \cdot 6 = 12x^2 + 24x$

$ab(a^2 - b^2) = ab \cdot a^2 - ab \cdot b^2 = a^3b - ab^3$

Multiplication of Monomials and Polynomials   Apply the distributive properties. Be sure to multiply every term in the polynomial by the monomial.

Example: $-2x^2(4x^2 - 5x - 3) = -8x^4 + 10x^3 + 6x^2$

Multiplication of Polynomials   The product of two polynomials may be found by multiplying every term in the first polynomial by every term in the second polynomial. Be sure to combine like terms.

Examples: $(x + 3)(2x - 5) = 2x^2 - 5x + 6x - 15$

$= 2x^2 + x - 15$

$(2x + 1)(x^2 - 5x + 2) = 2x^3 - 10x^2 + 4x + x^2 - 5x + 2$

$= 2x^3 - 9x^2 - x + 2$

SECTION 5.4 ■ SPECIAL PRODUCTS

Product of a Sum and Difference

$$(a + b)(a - b) = a^2 - b^2$$

Examples: $(x + 4)(x - 4) = x^2 - 16$

$(2r - 3t)(2r + 3t) = (2r)^2 - (3t)^2 = 4r^2 - 9t^2$

Squaring Binomials

$$(a + b)^2 = a^2 + 2ab + b^2 \quad \text{and} \quad (a - b)^2 = a^2 - 2ab + b^2$$

Examples: $(2x + 1)^2 = (2x)^2 + 2(2x)1 + 1^2 = 4x^2 + 4x + 1$

$(z^2 - 2)^2 = (z^2)^2 - 2z^2(2) + 2^2 = z^4 - 4z^2 + 4$

Cubing Binomials   To multiply $(a + b)^3$ write it as $(a + b)(a + b)^2$.

Example: $(x + 4)^3 = (x + 4)(x + 4)^2$

$= (x + 4)(x^2 + 8x + 16)$ Square the binomial.

$= x^3 + 8x^2 + 16x + 4x^2 + 32x + 64$ Distributive property

$= x^3 + 12x^2 + 48x + 64$ Combine like terms.

SECTION 5.5 ■ INTEGER EXPONENTS AND THE QUOTIENT RULE

Negative Integers as Exponents   For any nonzero real number a and positive integer n,

$$a^{-n} = \frac{1}{a^n}.$$

Examples: $5^{-2} = \frac{1}{5^2}$ and $x^{-4} = \frac{1}{x^4}$

The Quotient Rule   For any nonzero real number a and integers m and n,

$$\frac{a^m}{a^n} = a^{m-n}.$$

Examples: $\dfrac{6^4}{6^2} = 6^{4-2} = 6^2 = 36$ and $\dfrac{xy^3}{x^4y^2} = x^{1-4}y^{3-2} = x^{-3}y^1 = \dfrac{y}{x^3}$

Other Rules   For any nonzero real numbers a and b and positive integers m and n,

$$\frac{1}{a^{-n}} = a^n, \quad \frac{a^{-n}}{b^{-m}} = \frac{b^m}{a^n}, \quad \text{and} \quad \left(\frac{a}{b}\right)^{-n} = \left(\frac{b}{a}\right)^n.$$

Examples: $\dfrac{1}{4^{-3}} = 4^3$, $\dfrac{x^{-3}}{y^{-2}} = \dfrac{y^2}{x^3}$,   and $\left(\dfrac{4}{5}\right)^{-2} = \left(\dfrac{5}{4}\right)^2$

Scientific Notation   A real number a written as $b \times 10^n$, where $1 \le |b| < 10$ and n is an integer

Examples: $2.34 \times 10^3 = 2340$ Move the decimal point 3 places to the right.
$2.34 \times 10^{-3} = 0.00234$ Move the decimal point 3 places to the left.

SECTION 5.6 ■ DIVISION OF POLYNOMIALS

Division of a Polynomial by a Monomial   Divide the monomial into *every* term of the polynomial.

Example: $\dfrac{5x^3 - 10x^2 + 15x}{5x} = \dfrac{5x^3}{5x} - \dfrac{10x^2}{5x} + \dfrac{15x}{5x} = x^2 - 2x + 3$

Division of a Polynomial by a Polynomial   Division of polynomials is performed similarly to long division of natural numbers.

Example: Divide $2x^3 + 4x^2 - 3x + 1$ by $x + 1$.

$$
\begin{array}{r}
2x^2 + 2x - 5 \\
x + 1 \overline{)\,2x^3 + 4x^2 - 3x + 1} \\
\underline{2x^3 + 2x^2} \\
2x^2 - 3x \\
\underline{2x^2 + 2x} \\
-5x + 1 \\
\underline{-5x - 5} \\
6
\end{array}
$$

The quotient is $2x^2 + 2x - 5$ with remainder 6, which can be written as

$$2x^2 + 2x - 5 + \frac{6}{x + 1}.$$

CHAPTER 5 REVIEW EXERCISES

SECTION 5.1

Exercises 1–6: Evaluate the expression.

1. 5^3

2. -3^4

3. $4(-2)^0$

4. $3 + 3^2 - 3^0$

5. $\dfrac{-5^2}{5}$

6. $\left(\dfrac{-5}{5}\right)^2$

Exercises 7–22: Simplify the expression.

7. $6^2 \cdot 6^3$

8. $10^5 \cdot 10^7$

9. $z^4 \cdot z^5$

10. $y^2 \cdot y \cdot y^3$

11. $5x^2 \cdot 6x^7$

12. $(ab^3)(a^3b)$

13. $(2^5)^2$

14. $(m^4)^5$

15. $(ab)^3$

16. $(x^2y^3)^4$

17. $(xy)^3(x^2y^4)^2$

18. $(a^2b^9)^0$

19. $(r - t)^4(r - t)^5$

20. $(a + b)^2(a + b)^4$

21. $\left(\dfrac{3}{x - y}\right)^2$

22. $\left(\dfrac{x + y}{2}\right)^3$

SECTION 5.2

Exercises 23 and 24: Identify the degree and coefficient of the monomial.

23. $6x^7$

24. $-x^2y^3$

Exercises 25–28: Determine whether the expression is a polynomial. If it is, state how many terms and variables the polynomial contains. Then state its degree.

25. $8y$

26. $8x^3 - 3x^2 + x - 5$

27. $a^2 + 2ab + b^2$

28. $\dfrac{1}{xy}$

29. Add the polynomials vertically.

$$3x^2 + 4x + 8$$
$$\underline{2x^2 - 5x - 5}$$

30. Write the opposite of $6x^2 - 3x - 7$.

Exercises 31–36: Simplify.

31. $(4x - 3) + (-x + 7)$

32. $(3x^2 - 1) - (5x^2 + 12)$

33. $(x^2 + 5x + 6) - (3x^2 - 4x + 1)$

34. $(a^3 + 4a^2) + (a^3 - 5a^2 + 7a)$

35. $(xy + y^2) + (4y^2 - 4xy)$

36. $(7x^2 + 2xy + y^2) - (7x^2 - 2xy + y^2)$

SECTION 5.3

Exercises 37–50: Multiply and simplify.

37. $-x^2 \cdot x^3$

38. $-(r^2t^3)(rt)$

39. $-3(2t - 5)$

40. $2y(1 - 6y)$

41. $6x^3(3x^2 + 5x)$

42. $-x(x^2 - 2x + 9)$

43. $-ab(a^2 - 2ab + b^2)$

44. $(a - 2)(a + 5)$

45. $(8x - 3)(x + 2)$

46. $(2x - 1)(1 - x)$

47. $(y^2 + 1)(2y + 1)$

48. $(y^2 - 1)(2y^2 + 1)$

49. $(z + 1)(z^2 - z + 1)$

50. $(4z - 3)(z^2 - 3z + 1)$

Exercises 51 and 52: Multiply the expression
  (a) geometrically and
  (b) symbolically.

51. $z(z + 1)$

52. $2x(x + 2)$

SECTION 5.4

Exercises 53–68: Multiply.

53. $(z + 2)(z - 2)$

54. $(5z - 9)(5z + 9)$

55. $(1 - 3y)(1 + 3y)$

56. $(5x + 4y)(5x - 4y)$

57. $(rt + 1)(rt - 1)$

58. $(2m^2 - n^2)(2m^2 + n^2)$

59. $(x + 1)^2$ **60.** $(4x + 3)^2$

61. $(y - 3)^2$ **62.** $(2y - 5)^2$

63. $(4 + a)^2$ **64.** $(4 - a)^2$

65. $(x^2 + y^2)^2$ **66.** $(xy - 2)^2$

67. $(z + 5)^3$ **68.** $(2z - 1)^3$

Exercises 69 and 70: Use the product of sum and a difference to evaluate the expression.

69. $59 \cdot 61$ **70.** $22 \cdot 18$

SECTION 5.5

Exercises 71–76: Simplify the expression.

71. 9^{-1} **72.** 3^{-2}

73. $4^3 \cdot 4^{-2}$ **74.** $10^{-6} \cdot 10^3$

75. $\dfrac{1}{6^{-2}}$ **76.** $\dfrac{5^7}{5^9}$

Exercises 77–92: Simplify the expression. Write the answer using positive exponents.

77. z^{-2} **78.** y^{-4}

79. $a^{-4} \cdot a^2$ **80.** $x^2 \cdot x^{-5} \cdot x$

81. $(2t)^{-2}$ **82.** $(ab^2)^{-3}$

83. $(xy)^{-2}(x^{-2}y)^{-1}$ **84.** $\dfrac{x^6}{x^2}$

85. $\dfrac{4x}{2x^4}$ **86.** $\dfrac{20x^5y^3}{30xy^6}$

87. $\left(\dfrac{a}{b}\right)^5$ **88.** $\dfrac{4}{t^{-4}}$

89. $\left(\dfrac{x}{3}\right)^{-3}$ **90.** $\dfrac{2}{(ab)^{-1}}$

91. $\left(\dfrac{x}{y}\right)^{-2}$ **92.** $\left(\dfrac{3u}{2v}\right)^{-1}$

Exercises 93–96: Write the expression in standard form.

93. 6×10^2 **94.** 5.24×10^4

95. 3.7×10^{-3} **96.** 6.234×10^{-2}

Exercises 97–100: Write the number in scientific notation.

97. 10,000   **98.** 56,100,000

99. 0.000054   **100.** 0.001

Exercises 101 and 102: Evaluate the expression. Write the result in standard form.

101. $(4 \times 10^2)(6 \times 10^4)$ **102.** $\dfrac{8 \times 10^3}{4 \times 10^4}$

SECTION 5.6

Exercises 103–110: Divide and check.

103. $\dfrac{5x^2 + 3x}{3x}$ **104.** $\dfrac{6b^4 - 4b^2 + 2}{2b^2}$

105. $\dfrac{3x^2 - x + 2}{x - 1}$ **106.** $\dfrac{9x^2 - 6x - 2}{3x + 2}$

107. $\dfrac{4x^3 - 11x^2 - 7x - 1}{4x + 1}$

108. $\dfrac{2x^3 - x^2 - 1}{2x - 1}$ **109.** $\dfrac{x^3 - x^2 - x + 1}{x^2 + 1}$

110. $\dfrac{x^4 + 3x^3 + 8x^2 + 7x + 5}{x^2 + x + 1}$

APPLICATIONS

111. *Heart Rate*   An athlete starts running and continues for 10 seconds. The polynomial $t^2 + 60$ calculates the heart rate of the athlete in beats per minute t seconds after beginning the run, where $t \le 10$.
 (a) What is the athlete's heart rate when the athlete first starts to run?
 (b) What is the athlete's heart rate after 10 seconds?
 (c) What happens to the athlete's heart rate while the athlete is running?

112. *Areas of Rectangles*   Find a monomial equal to the sum of the areas of the rectangles. Calculate this sum for $x = 3$ feet and $y = 4$ feet.

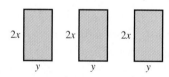

113. *Area of a Rectangle*   Write a polynomial that gives the area of the rectangle. Calculate its area for $z = 6$ inches.

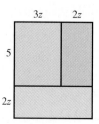

114. *Area of a Square* Find the area of the square whose sides have length x^2y.

115. *Compound Interest* If P dollars are deposited in an account that pays 6% annual interest, then the amount of money after 3 years is given by $P(1 + 0.06)^3$. Find this amount when $P = \$700$.

116. *Volume of a Sphere* The expression for the volume of a sphere with radius r is $\frac{4}{3}\pi r^3$. Find a polynomial that gives the volume of a sphere with radius $x + 2$. Leave your answer in terms of π.

117. *Height Reached by a Baseball* A baseball is hit straight up. Its height h in feet above the ground after t seconds is given by $t(96 - 16t)$.
  (a) Multiply this expression.
  (b) Evaluate both the expression in part (a) and the given expression for $t = 2$. Interpret the result.

118. *Rectangular Building* A rectangular building has a perimeter of 1200 feet.
  (a) If one side of the building has length L, write a polynomial expression that gives its area. (Be sure to multiply your expression.)
  (b) Evaluate the expression in part (a) for $L = 50$ and interpret the answer.

119. *Geometry* Complete each part and verify that your answers are equal.

  (a) Find the area of the large square by multiplying its length and width.
  (b) Find the sum of the areas of the smaller rectangles inside the large square.

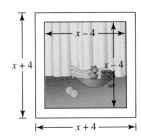

120. *Digital Picture* A digital picture, including its border, is $x + 4$ pixels by $x + 4$ pixels, and the actual picture inside the border is $x - 4$ pixels by $x - 4$ pixels.

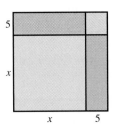

  (a) Find a polynomial that gives the number of pixels in the border.
  (b) Let $x = 100$ and evaluate the polynomial.

121. *Federal Debt* In 1990, the federal debt held by the public was $\$2.19$ trillion, and the population of the United States was 249 million. Use scientific notation to approximate the national debt per person. (*Source:* U.S. Department of the Treasury.)

122. *Alcohol Consumption* In 2004, about 233 million people in the United States were aged 14 or older. They consumed, on average, 2.23 gallons of alcohol per person. Use scientific notation to estimate the total number of gallons of alcohol consumed by this age group. (*Source:* Department of Health and Human Services.)

CHAPTER 5   TEST   Pass the Test   Video solutions to all test exercises

1. Is $5x^2 - 3xy - 7y^3$ a polynomial? If it is, state how many terms and variables the polynomial contains. Then state its degree.

2. Write the opposite of $-x^3 + 4x - 8$.

Exercises 3–6: Simplify.

3. $(-3x + 4) + (7x + 2)$

4. $(y^3 - 2y + 6) - (4y^3 + 5)$

5. $(5x^2 - x + 3) - (4x^2 - 2x + 10)$

6. $(a^3 + 5ab) + (3a^3 - 3ab)$

7. Evaluate each expression by hand.
    (a) $-4^2 + 10$ **(b)** 8^{-2} **(c)** $\dfrac{1}{2^{-3}}$
    (d) $-3x^0$

Exercises 8–15: Write the given expression with positive exponents.

8. $6y^4 \cdot 4y^7$ **9.** $(a^2b^3)^2(ab^2)$

10. $x^7 \cdot x^{-3}$ **11.** $(a^{-1}b^2)^{-3}$

12. $ab(a^2 - b^2)$ **13.** $\left(\dfrac{3a^2}{2b^{-3}}\right)^{-2}$

14. $\dfrac{12xy^4}{6x^2y}$ **15.** $\left(\dfrac{2}{a + b}\right)^4$

Exercises 16–21: Multiply and simplify.

16. $3x^2(4x^3 - 6x + 1)$ **17.** $(z - 3)(2z + 4)$

18. $(7y^2 - 3)(7y^2 + 3)$ **19.** $(3x - 2)^2$

20. $(m + 3)^3$

21. $(y + 2)(y^2 - 2y + 3)$

22. Evaluate $78 \cdot 82$ using the product of a sum and a difference.

23. Write 6.1×10^{-3} in standard form.

24. Write 5410 in scientific notation.

Exercises 25 and 26: Divide.

25. $\dfrac{9x^3 - 6x^2 + 3x}{3x^2}$ **26.** $\dfrac{x^3 + x^2 - x + 1}{x + 2}$

27. *Concert Tickets* Tickets for a concert are sold for $20 each.
    (a) Write a polynomial that gives the revenue from selling t tickets.
    (b) Putting on the concert costs management $2000 to hire the band plus $2 for each ticket sold. What is the total cost of the concert if t tickets are sold?
    (c) Subtract the polynomial that you found in part (b) from the polynomial that you found in part (a). What does this polynomial represent?

28. *Areas of Rectangles* Find a polynomial representing the sum of the areas of two identical rectangles that have width $2x$ and length $3x$. Calculate this sum for $x = 10$ feet.

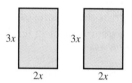

29. *Volume of a Box* Write a polynomial that represents the volume of the box. Be sure to multiply your answer completely.

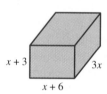

30. *Height Reached by a Golf Ball* When a golf ball is hit into the air its height in feet above the ground after t seconds is given by $t(88 - 16t)$.
    (a) Multiply this expression.
    (b) Evaluate the expression in part (a) for $t = 3$. Interpret the result.

CHAPTER 5   EXTENDED AND DISCOVERY EXERCISES

Exercises 1–6: Arithmetic and Scientific Notation The product $(4 \times 10^3) \times (2 \times 10^2)$ can be evaluated as

$$(4 \times 2) \times (10^3 \times 10^2) = 8 \times 10^5,$$

and the quotient $(4 \times 10^3) \div (2 \times 10^2)$ *can be evaluated as*

$$\frac{4 \times 10^3}{2 \times 10^2} = \frac{4}{2} \times \frac{10^3}{10^2} = 2 \times 10^1.$$

How would you evaluate $(4 \times 10^3) + (2 \times 10^2)$? *How would you evaluate* $(4 \times 10^3) - (2 \times 10^2)$? *Make a conjecture as to how numbers in scientific notation should be added and subtracted. Try your method on these problems and then check your answers with a calculator set in scientific mode. Does your method work?*

1. $(4 \times 10^3) + (3 \times 10^3)$

2. $(5 \times 10^{-2}) - (2 \times 10^{-2})$

3. $(1.2 \times 10^4) - (3 \times 10^3)$

4. $(2 \times 10^2) + (6 \times 10^1)$

5. $(2 \times 10^{-1}) + (4 \times 10^{-2})$

6. $(2 \times 10^{-3}) - (5 \times 10^{-2})$

Exercises 7 and 8: **Constructing a Box**   *A box is constructed from a rectangular piece of metal by cutting squares from the corners and folding up the sides. The square, cutout corners are x inches by x inches.*

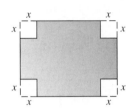 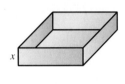

7. Suppose that the dimensions of the metal piece are 20 inches by 30 inches.
   (a) Write a polynomial that gives the volume of the box.
   (b) Find the volume of the box for $x = 4$ inches.

8. Suppose that the metal piece is square with sides of length 25 inches.
   (a) Write a polynomial expression that gives the outside surface area of the box. (Assume that the box does not have a top.)
   (b) Find this area for $x = 3$ inches.

Exercises 9–12: **Calculators and Polynomials**   *A graphing calculator can be used to help determine whether two polynomial expressions in one variable are equal. For example, suppose that a student believes that $(x + 2)^2$ and $x^2 + 4$ are equal. Then the first two calculator tables shown demonstrate that the two expressions are not equal except for $x = 0$.*

X	Y1
-3	1
-2	0
-1	1
0	4
1	9
2	16
3	25
$Y_1 \blacksquare (X+2)^2$	

X	Y1
-3	13
-2	8
-1	5
0	4
1	5
2	8
3	13
$Y_1 \blacksquare X^2 + 4$	

The next two calculator tables support the fact that $(x + 1)^2$ and $x^2 + 2x + 1$ are equal for all x.

X	Y1
-3	4
-2	1
-1	0
0	1
1	4
2	9
3	16
$Y_1 \blacksquare (X+1)^2$	

X	Y1
-3	4
-2	1
-1	0
0	1
1	4
2	9
3	16
$Y_1 \blacksquare X^2 + 2X + 1$	

Use a graphing calculator to determine whether the first expression is equal to the second expression. If the expressions are not equal, multiply the first expression and simplify it.

9. $3x(4 - 5x),\ 12x - 5x$

10. $(x - 1)^2,\ x^2 - 1$

11. $(x - 1)(x^2 + x + 1),\ x^3 - 1$

12. $(x - 2)^3,\ x^3 - 8$

CHAPTERS 1–5   CUMULATIVE REVIEW EXERCISES

Exercises 1 and 2: Evaluate each expression by hand.

1. **(a)** $18 - 2 \cdot 5$ **(b)** $42 \div 7 + 2$

2. **(a)** $21 - (-8)$ **(b)** $-\frac{7}{3} \div \left(-\frac{14}{9}\right)$

Exercises 3 and 4: Solve the equation. Note that these equations may have no solutions, one solution, or infinitely many solutions.

3. **(a)** $(x - 3) + x = 4 + x$
   (b) $2(5x - 4) = 1 + 10x$

4. **(a)** $2 + 6x = 2(3x + 1)$
   (b) $11x - 9 = -31$

5. Find the average speed of a car that travels 306 miles in 4 hours 30 minutes.

6. Find three consecutive numbers whose sum is -114.

7. Write each value as a fraction in lowest terms.
   (a) 42%   **(b)** 0.076

8. Solve the formula $Wx = 4x + Y$ for x.

9. Graph the equation $4x - 5y = 20$.

10. Sketch a line with slope $-\frac{2}{3}$ that passes through the point $(1, 1)$.

11. Write the slope–intercept form for the line shown.

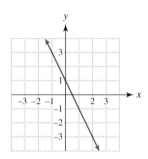

12. Find the x- and y-intercepts for the graph of the equation $2y = 3x - 6$.

Exercises 13 and 14: Write the slope–intercept form of a line that satisfies the given information.

13. Parallel to $3x - 6y = 7$, passing through $(2, -3)$

14. Passing through $(-2, -5)$ and $(1, 4)$

Exercises 15–18: Solve the system of equations. Note that these systems may have no solutions, one solution, or infinitely many solutions.

15. $4x + 3y = -6$
    $8x + 6y = 12$

16. $x - 3y = 5$
    $3x + y = 5$

17. $x + 4y = -8$
    $-3x - 12y = 24$

18. $x - 5y = 30$
    $2x + y = -6$

Exercises 19 and 20: Shade the solution set for the system of inequalities.

19. $x + y < 3$
    $y \geq x + 2$

20. $x - 2y > 4$
    $3x + y < 6$

21. Simplify the expression.
  (a) $3x^2 \cdot 5x^3$ **(b)** $(x^3y)^2(x^4y^5)$

22. Simplify.
  (a) $(5x^2 - 3x + 4) - (3x^2 - 2x + 1)$
  (b) $(7a^3 - 4a^2 - 5) + (5a^3 + 4a^2 + a)$

23. Multiply and simplify.
  (a) $(2x + 3)(x - 7)$ **(b)** $(y + 3)(y^2 - 3y - 1)$
  (c) $(4x + 7)(4x - 7)$ **(d)** $(5a + 3)^2$

24. Simplify the expression. Write the answer using positive exponents.
  (a) $x^{-5} \cdot x^3 \cdot x$ **(b)** $\left(\dfrac{2}{x^3}\right)^{-3}$
  (c) $\dfrac{3x^2y^{-1}}{6x^{-2}y}$ **(d)** $(xy^{-2})^3(x^{-2}y)^{-2}$

25. Write $24{,}000{,}000{,}000$ in scientific notation.

26. Write 4.71×10^{-7} in standard form.

27. Divide.
  (a) $\dfrac{8x^3 - 2x}{2x}$ **(b)** $\dfrac{2x^2 + x - 14}{x + 3}$

28. *Price Decrease*   If the price of a computer is reduced from \$1200 to \$900, find the percent change.

29. *Mixing an Acid Solution*   How many milliliters of a 3% acid solution should be added to 400 milliliters of a 6% acid solution to dilute it to a 5% acid solution?

30. *Surface Area of a Box*   Use the drawing of the box to write a polynomial that represents the area of each of the following.

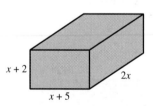

  (a) The bottom
  (b) The front
  (c) The right side
  (d) All six sides

CHAPTER 6

Factoring Polynomials and Solving Equations

Tamagawa University

Have you ever noticed that cars are designed so that their exteriors are curved and smooth? This characteristic is especially true for solar cars. In fact, a side view of a solar car often resembles the cross section of an airplane wing, as illustrated in the accompanying figure. The reason for this design is to reduce drag from air resistance. Reducing drag increases fuel efficiency and gives the car a smoother ride in the wind.

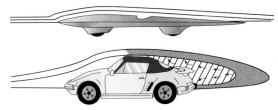

Views of a Solar Car and a Standard Car

Mathematics plays an important role in the design of cars. Cubic polynomials called *cubic splines* are used extensively by engineers to obtain a smooth shape for new cars. Although the topic of cubic splines is covered in advanced mathematics and engineering courses, in this chapter we introduce many concepts necessary for understanding polynomials and polynomial equations, which are used in the design of cars.

What we see depends mainly on what we look for.

—John Lubbock

Source: R. Burden and J. Faires. *Numerical Analysis.*

6.1 INTRODUCTION TO FACTORING

Common Factors ▪ Factoring by Grouping

Polynomials are frequently used in applications to approximate data and to model such things as changes in the weather, sales of a new video game, and the growth of young children. As a result, scientists and mathematicians commonly solve equations involving polynomials. One way to solve these equations is to use **factoring**. When factoring a polynomial, we usually write it as a *product* of lower degree polynomials. For example, because $x(x - 1) = x^2 - x$, we say that the polynomial $x^2 - x$ can be *factored* as $x(x - 1)$. Note that $x^2 - x$ is a polynomial of degree 2, whereas x and $x - 1$ are both degree 1 polynomials. In this section we introduce two basic methods of factoring polynomials.

Common Factors

When factoring a polynomial, we first look for factors that are common to each term. By applying a distributive property we can often write a polynomial as a product. For example, each term in the polynomial $8x^2 + 6x$ has a factor of $2x$ because

$$8x^2 = 2x \cdot 4x \quad \text{and} \quad 6x = 2x \cdot 3.$$

Therefore by the distributive property,

$$8x^2 + 6x = 2x(4x + 3).$$

Thus the product $2x(4x + 3)$ equals $8x^2 + 6x$. We check this result by multiplying.

$$2x(4x + 3) = 2x \cdot 4x + 2x \cdot 3$$
$$= 8x^2 + 6x \qquad \text{It checks.}$$

This factorization is shown visually in Figure 6.1, where possible dimensions for a rectangle with an area of $8x^2 + 6x$ are $2x$ by $4x + 3$.

$$
\begin{array}{c}
2x \begin{array}{|c|c|} \hline 8x^2 & 6x \\ \hline \end{array} \\
\;\; 4x \qquad\;\; 3
\end{array}
$$

Figure 6.1 $8x^2 + 6x = 2x(4x + 3)$

The expressions $8x^2 + 6x$ and $2x(4x + 3)$ are equal for all values of x. Figure 6.2 also illustrates this fact with a partial table of values for each expression.

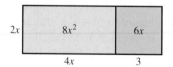

X	Y₁
-3	54
-2	20
-1	2
0	0
1	14
2	44
3	90

Y₁▤8X^2+6X

(a)

X	Y₁
-3	54
-2	20
-1	2
0	0
1	14
2	44
3	90

Y₁▤2X(4X+3)

(b)

Figure 6.2 $8x^2 + 6x = 2x(4x + 3)$

EXAMPLE 1   **Factoring an expression**

Factor the expression and sketch a rectangle that illustrates the factorization.
(a) $10x + 6$ **(b)** $6x^2 + 15x$

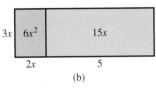

(a)

(b)

Figure 6.3

Solution

(a) Each term in the polynomial $10x + 6$ has a factor of 2 because
$$10x = 2 \cdot 5x \quad \text{and} \quad 6 = 2 \cdot 3.$$
By the distributive property, $10x + 6 = 2(5x + 3)$. This factorization is illustrated visually in Figure 6.3(a).

(b) Each term in the polynomial $6x^2 + 15x$ has a factor of $3x$ because
$$6x^2 = 3x \cdot 2x \quad \text{and} \quad 15x = 3x \cdot 5.$$
By the distributive property, $6x^2 + 15x = 3x(2x + 5)$. This factorization is illustrated visually in Figure 6.3(b).   Now Try Exercises 9, 11

EXAMPLE 2   **Finding common factors**

Factor.
(a) $15x^2 + 10x$ **(b)** $6y^3 - 2y^2$ **(c)** $3z^3 + 9z^2 - 6z$ **(d)** $2x^2y^2 + 4xy^3$

Solution

(a) In the expression $15x^2 + 10x$, the terms $15x^2$ and $10x$ both contain a common factor of $5x$ because
$$15x^2 = 5x \cdot 3x \quad \text{and} \quad 10x = 5x \cdot 2.$$
Therefore this polynomial can be factored as
$$15x^2 + 10x = 5x(3x + 2).$$

(b) In the expression $6y^3 - 2y^2$, the terms $6y^3$ and $2y^2$ both contain a common factor of $2y^2$ because
$$6y^3 = 2y^2 \cdot 3y \quad \text{and} \quad 2y^2 = 2y^2 \cdot 1.$$
Therefore this polynomial can be factored as
$$6y^3 - 2y^2 = 2y^2(3y - 1).$$

(c) In the expression $3z^3 + 9z^2 - 6z$, the terms $3z^3$, $9z^2$, and $6z$ all contain a common factor of $3z$ because
$$3z^3 = 3z \cdot z^2, \quad 9z^2 = 3z \cdot 3z, \quad \text{and} \quad 6z = 3z \cdot 2.$$
Therefore this polynomial can be factored as
$$3z^3 + 9z^2 - 6z = 3z(z^2 + 3z - 2).$$

(d) In the expression $2x^2y^2 + 4xy^3$, the terms $2x^2y^2$ and $4xy^3$ both contain a common factor of $2xy^2$ because
$$2x^2y^2 = 2xy^2 \cdot x \quad \text{and} \quad 4xy^3 = 2xy^2 \cdot 2y.$$
Thus $2x^2y^2 + 4xy^3 = 2xy^2(x + 2y)$.   Now Try Exercises 13, 15, 17, 19

MAKING CONNECTIONS

Checking Common Factors with Multiplication

When factoring we can check our work by multiplying. For example, if we are uncertain whether the equation

$$6y^3 - 2y^2 = 2y^2(3y - 1)$$

is correct, we can apply the distributive property to the right side of the above equation to obtain

$$2y^2(3y - 1) = 2y^2 \cdot 3y - 2y^2 \cdot 1$$
$$= 6y^3 - 2y^2. \qquad \text{It checks.}$$

In most situations we factor out the *greatest common factor* (GCF). For example, the polynomial $12b^3 + 8b^2$ has a common factor of $2b$. We could factor this polynomial as

$$12b^3 + 8b^2 = 2b(6b^2 + 4b).$$

However, we can factor out $4b^2$ instead.

$$12b^3 + 8b^2 = 4b^2(3b + 2)$$

Because $4b^2$ is the common factor with the greatest (integer) coefficient and highest degree, we say that $4b^2$ is the **greatest common factor** of $12b^3 + 8b^2$. In Examples 1 and 2 we factored out the greatest common factor for each expression.

When finding the greatest common factor for a polynomial, it is often helpful first to completely factor each term of the polynomial. To completely factor a term, write its coefficient as the product of prime numbers and write any powers of variables as repeated multiplication. For example, the complete factorization of $24x^4$ is $2 \cdot 2 \cdot 2 \cdot 3 \cdot x \cdot x \cdot x \cdot x$ and the complete factorization of $18xy^3$ is $2 \cdot 3 \cdot 3 \cdot x \cdot y \cdot y \cdot y$. The next example shows how to find the greatest common factor for a polynomial by using the complete factorization of each term.

EXAMPLE 3   Finding the greatest common factor

Find the greatest common factor for each expression. Then factor the expression.
(a) $9x^2 + 6x$ **(b)** $4z^4 + 8z^2$ **(c)** $8a^2b^3 - 16a^3b^2$

Solution
(a) Because

$$9x^2 = 3 \cdot 3 \cdot x \cdot x \quad \text{and}$$
$$6x = 3 \cdot 2 \cdot x,$$

both terms have common factors of 3 and x. The GCF is the product of these two factors, or $3 \cdot x = \mathbf{3x}$. Thus the expression $9x^2 + 6x$ can be factored as $\mathbf{3x}(3x + 2)$. Note that the term $3x$ inside the parentheses is the product of the (black) factors of $9x^2$ that were not part of the greatest common factor and the term 2 inside the parentheses is the product of the (black) factors of $6x$ that were not part of the greatest common factor.

(b) Because

$$4z^4 = 2 \cdot 2 \cdot z \cdot z \cdot z \cdot z \quad \text{and}$$
$$8z^2 = 2 \cdot 2 \cdot 2 \cdot z \cdot z,$$

both terms have common factors of **2, 2,** z, and z. The GCF is the product of these four factors, or $2 \cdot 2 \cdot z \cdot z = \mathbf{4z^2}$. Thus the expression $4z^4 + 8z^2$ can be factored as $\mathbf{4z^2(z^2 + 2)}$.

(c) Because

$$8a^2b^3 = 2 \cdot 2 \cdot 2 \cdot a \cdot a \cdot b \cdot b \cdot b \quad \text{and}$$
$$16a^3b^2 = 2 \cdot 2 \cdot 2 \cdot 2 \cdot a \cdot a \cdot a \cdot b \cdot b,$$

both terms have common factors of **2, 2, 2,** a, a, b, and b. The GCF is the product of these seven factors, or $2 \cdot 2 \cdot 2 \cdot a \cdot a \cdot b \cdot b = \mathbf{8a^2b^2}$. Here $8a^2b^3 - 16a^3b^2$ can be factored as $\mathbf{8a^2b^2(b - 2a)}$. 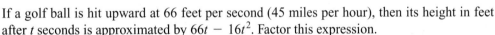 Now Try Exercises 21, 23, 35

NOTE: With practice, you may find that you can determine the GCF mentally without factoring each term as was done in Example 3.

In the next example, we factor an expression that occurs in a scientific application.

 EXAMPLE 4

Modeling the flight of a golf ball

If a golf ball is hit upward at 66 feet per second (45 miles per hour), then its height in feet after t seconds is approximated by $66t - 16t^2$. Factor this expression.

Solution
The GCF for $66t$ and $16t^2$ is $2t$ because

$$66t = 2t \cdot 33 \quad \text{and} \quad 16t^2 = 2t \cdot 8t.$$

Therefore this polynomial can be factored as

$$66t - 16t^2 = 2t(33 - 8t).$$  Now Try Exercise 77

Factoring by Grouping

Factoring by grouping is a technique that makes use of the associative and distributive properties. The next example illustrates one step in this factoring technique.

EXAMPLE 5

Factoring out binomials

Factor.
(a) $5x(x + 3) + 6(x + 3)$ **(b)** $x^2(2x - 5) - 4x(2x - 5)$

Solution
(a) Each term in the expression $5x(x + 3) + 6(x + 3)$ contains the binomial $(x + 3)$. Therefore the distributive property can be used to factor this expression.

$$5x(x + 3) + 6(x + 3) = (5x + 6)(x + 3)$$

(b) Each term in $x^2(2x - 5) - 4x(2x - 5)$ contains the binomial $(2x - 5)$. Therefore the distributive property can be used to factor this expression.

$$x^2(2x - 5) - 4x(2x - 5) = (x^2 - 4x)(2x - 5)$$
$$= x(x - 4)(2x - 5)$$

Now Try Exercises 39, 43

Now consider the polynomial

$$x^3 + x^2 + 2x + 2.$$

We can factor this polynomial by first grouping it into two binomials.

$(x^3 + x^2) + (2x + 2)$ Associative property

$x^2(x + 1) + 2(x + 1)$ Factor out common factors.

$(x^2 + 2)(x + 1)$ Factor out $(x + 1)$.

When factoring by grouping, we factor out a common factor more than once.

The first step in factoring a four-term polynomial by grouping requires the use of the associative property to write the polynomial as the *sum* of two binomials. However, this property must be applied carefully to avoid sign errors. The terms of a polynomial are separated by addition $(+)$ or subtraction $(-)$ symbols. There are three of these symbols in a four-term polynomial. The next two examples illustrate that the middle arithmetic symbol in a four-term polynomial determines how the associative property is applied.

EXAMPLE 6   **Factoring by grouping when the middle symbol is $(+)$**

Factor each polynomial.
(a) $2x^3 - 4x^2 + 3x - 6$ **(b)** $3x + 3y + ax + ay$

Solution
(a) Use the associative property to write the polynomial as the *sum* of two binomials.

$2x^3 - 4x^2 + 3x - 6 = (2x^3 - 4x^2) + (3x - 6)$ Associative property

$= 2x^2(x - 2) + 3(x - 2)$ Factor out common factors.

$= (2x^2 + 3)(x - 2)$ Factor out $(x - 2)$.

(b) Group the polynomial into the *sum* of two binomials.

$3x + 3y + ax + ay = (3x + 3y) + (ax + ay)$ Associative property

$= 3(x + y) + a(x + y)$ Factor out common factors.

$= (3 + a)(x + y)$ Factor out $(x + y)$.

Now Try Exercises 45, 63

EXAMPLE 7   **Factoring by grouping when the middle symbol is $(-)$**

Factor each polynomial.
(a) $3y^3 - y^2 - 9y + 3$ **(b)** $z^3 + 4z^2 - 5z - 20$

Solution

(a) Begin by changing the middle subtraction to addition by adding the opposite of $9y$. Then apply the associative property to write the result as the *sum* of two binomials.

$$
\begin{aligned}
3y^3 - y^2 - 9y + 3 &= 3y^3 - y^2 + (-9y) + 3 && \text{Add the opposite of } 9y. \\
&= (3y^3 - y^2) + (-9y + 3) && \text{Associative property} \\
&= y^2(3y - 1) - 3(3y - 1) && \text{Factor out common factors.} \\
&= (y^2 - 3)(3y - 1) && \text{Factor out } (3y - 1).
\end{aligned}
$$

Note that in the third step, -3 was factored from the second binomial.

(b) Begin by changing the middle subtraction to addition by adding the opposite of $5z$. Then apply the associative property to write the result as the *sum* of two binomials.

$$
\begin{aligned}
z^3 + 4z^2 - 5z - 20 &= z^3 + 4z^2 + (-5z) - 20 && \text{Add the opposite of } 5z. \\
&= (z^3 + 4z^2) + (-5z - 20) && \text{Associative property} \\
&= z^2(z + 4) - 5(z + 4) && \text{Factor out common factors.} \\
&= (z^2 - 5)(z + 4) && \text{Factor out } (z + 4).
\end{aligned}
$$

Now Try Exercises **53, 59**

When factoring some polynomials, it may be necessary to factor out the greatest common factor before completing other factoring techniques such as factoring by grouping. In the next example the GCF is factored out before grouping is applied.

EXAMPLE 8 Factoring out the GCF before grouping

Completely factor each polynomial.
(a) $6x^3 - 12x^2 - 3x + 6$ **(b)** $2x^5 - 8x^4 + 6x^3 - 24x^2$

Solution

(a) The GCF of $6x^3 - 12x^2 - 3x + 6$ is 3, so factor out 3 before factoring the remaining polynomial by grouping.

$$
\begin{aligned}
6x^3 - 12x^2 - 3x + 6 &= 3(2x^3 - 4x^2 - x + 2) && \text{Factor out the GCF.} \\
&= 3\big(2x^3 - 4x^2 + (-x) + 2\big) && \text{Add the opposite of } x. \\
&= 3\big((2x^3 - 4x^2) + (-x + 2)\big) && \text{Associative property} \\
&= 3\big(2x^2(x - 2) - 1(x - 2)\big) && \text{Factor out common factors.} \\
&= 3(2x^2 - 1)(x - 2) && \text{Factor out } (x - 2).
\end{aligned}
$$

(b) The GCF of $2x^5 - 8x^4 + 6x^3 - 24x^2$ is $2x^2$, so factor out $2x^2$ before factoring the remaining polynomial by grouping.

$$
\begin{aligned}
2x^5 - 8x^4 + 6x^3 - 24x^2 &= 2x^2(x^3 - 4x^2 + 3x - 12) && \text{Factor out the GCF.} \\
&= 2x^2\big((x^3 - 4x^2) + (3x - 12)\big) && \text{Associative property} \\
&= 2x^2\big(x^2(x - 4) + 3(x - 4)\big) && \text{Factor out common factors.} \\
&= 2x^2(x^2 + 3)(x - 4) && \text{Factor out } (x - 4).
\end{aligned}
$$

Now Try Exercises **65, 69**

6.1 PUTTING IT ALL TOGETHER

In this section we introduced basic concepts used to factor polynomials. They are summarized in the following table.

Concept	Explanation	Examples
Common Factor	Factor out a monomial common to each term in a polynomial.	$6z^2 - 6z = 6z(z - 1)$ $4y^3 - 6y^2 = 2y^2(2y - 3)$ $5x^3 - 10x^2 + 15x = 5x(x^2 - 2x + 3)$ $2a^3b^3 - 4a^2b^3 = 2a^2b^3(a - 2)$
Greatest Common Factor (GCF)	The common factor with the greatest (integer) coefficient and highest degree	The GCF of $10x^4 + 15x^2$ is $5x^2$. Common factors include $1, 5, x, 5x, x^2$, and $5x^2$. However, $5x^2$ is *the greatest common factor*.
Factoring by Grouping	Factoring by grouping is a method that can be used to factor *four terms* into a product of two binomials. It makes use of the associative and distributive properties.	$2x^3 + 3x^2 + 2x + 3$ $= (2x^3 + 3x^2) + (2x + 3)$ $= x^2(2x + 3) + 1(2x + 3)$ $= (x^2 + 1)(2x + 3)$ $4x^3 - 24x^2 - 3x + 18$ $= 4x^3 - 24x^2 + (-3x) + 18$ $= (4x^3 - 24x^2) + (-3x + 18)$ $= 4x^2(x - 6) - 3(x - 6)$ $= (4x^2 - 3)(x - 6)$

6.1 Exercises

MyMathLab   Math XL   PRACTICE   WATCH   DOWNLOAD   READ   REVIEW

CONCEPTS

1. When you write a polynomial as a product of two or more polynomials, it is called _____.

2. A common factor in the expression $ab + ac$ is _____.

3. When factoring, we can check our work by _____.

4. To completely factor a term, write the coefficient as the product of _____ numbers and write any powers of variables as repeated _____.

5. The _____ of a polynomial is the common factor with the greatest coefficient and highest degree.

6. Factoring by _____ is a method that can be used to factor four terms into a product of two binomials by using the associative and distributive properties.

7. Identify four common factors of $2x^2 + 4x$.

8. Identify the greatest common factor (GCF) of the expression $2x^2 + 4x$.

COMMON FACTORS

Exercises 9–12: Factor the expression. Then make a sketch of a rectangle that illustrates this factorization.

9. $2x + 4$

10. $6 + 3x$

11. $z^2 + 4z$

12. $2z^2 + 10z$

Exercises 13–20: Factor the expression.

13. $3x^2 + 9x$

14. $10y^2 + 2y$

15. $4y^3 - 2y^2$

16. $6x^4 + 9x^2$

17. $2z^3 + 8z^2 - 4z$

18. $5x^4 - 15x^3 - 10x^2$

19. $6x^2y - 3xy^2$

20. $7x^3y^3 + 14x^2y^2$

Exercises 21–38: Identify the greatest common factor. Then factor the expression.

21. $6x - 18x^2$

22. $16x^2 - 24x^3$

23. $8y^3 - 12y^2$

24. $12y^3 - 8y^2 + 4y$

25. $6z^3 + 3z^2 + 9z$

26. $16z^3 - 24z^2 - 36z$

27. $x^4 - 5x^3 - 4x^2$

28. $2x^4 + 8x^2$

29. $5y^5 + 10y^4 - 15y^3 + 10y^2$

30. $7y^4 - 14y^3 - 21y^2 + 7y$

31. $xy + xz$

32. $ab - bc$

33. $ab^2 - a^2b$

34. $4x^2y + 6xy^2$

35. $5x^2y^4 + 10x^3y^3$

36. $3r^3t^3 - 6r^4t^2$

37. $a^2b + ab^2 + ab$

38. $6ab^2 - 9ab + 12b^2$

FACTORING BY GROUPING

Exercises 39–44: Factor.

39. $x(x + 1) - 2(x + 1)$

40. $5x(3x - 2) + 2(3x - 2)$

41. $(z + 5)z + (z + 5)4$

42. $3y^2(y - 2) + 5(y - 2)$

43. $4x^3(x - 5) - 2x(x - 5)$

44. $8x^2(x + 3) + (x + 3)$

Exercises 45–64: Factor by grouping.

45. $x^3 + 2x^2 + 3x + 6$

46. $x^3 + 6x^2 + x + 6$

47. $2y^3 + y^2 + 2y + 1$

48. $4y^3 + 10y^2 + 2y + 5$

49. $2z^3 - 6z^2 + 5z - 15$

50. $15z^3 - 5z^2 + 6z - 2$

51. $4t^3 - 20t^2 + 3t - 15$

52. $4t^3 - 12t^2 + 3t - 9$

53. $9r^3 + 6r^2 - 6r - 4$

54. $3r^3 + 12r^2 - 2r - 8$

55. $7x^3 + 21x^2 - 2x - 6$

56. $6x^3 + 3x^2 - 10x - 5$

57. $2y^3 - 7y^2 - 4y + 14$

58. $y^3 - 5y^2 - 3y + 15$

59. $z^3 - 4z^2 - 7z + 28$

60. $12z^3 - 18z^2 - 10z + 15$

61. $2x^4 - 3x^3 + 4x - 6$

62. $x^4 + x^3 + 5x + 5$

63. $ax + bx + ay + by$

64. $ax - bx + ay - by$

Exercises 65–74: Completely factor the polynomial.

65. $3x^3 + 6x^2 + 3x + 6$

66. $5x^3 - 5x^2 + 5x - 5$

67. $6y^4 - 24y^3 - 2y^2 + 8y$

68. $6x^4 - 12x^3 + 3x^2 - 6x$

69. $x^5 + 2x^4 - 3x^3 - 6x^2$

70. $y^6 + 3y^5 - 2y^4 - 6y^3$

71. $4x^5 + 2x^4 - 12x^3 - 6x^2$

72. $18y^5 + 27y^4 + 12y^3 + 18y^2$

73. $x^3y + x^2y^2 - 2x^2y - 2xy^2$

74. $6x^3y - 3x^2y^2 + 18x^2y - 9xy^2$

75. Thinking Generally Factor a from $ax^2 + bx + c$.

76. Thinking Generally Factor c from $ax^2 + bx + c$.

APPLICATIONS

77. *Flight of a Golf Ball* The height of a golf ball in feet after t seconds is given by $80t - 16t^2$.
 (a) Identify the greatest common factor.
 (b) Factor this expression.

78. *Flight of a Golf Ball* Repeat the previous exercise if the height of a golf ball in feet after t seconds is given by $128t - 16t^2$.

79. *Volume of a Box* A box is constructed by cutting out square corners of a rectangular piece of cardboard and folding up the sides. If the cutout corners have sides with length x, then the volume of the box is given by the polynomial $4x^3 - 60x^2 + 200x$.

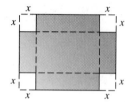 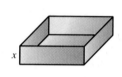

(a) Find the volume of the box when $x = 3$ inches.
(b) Factor out the greatest common factor for this expression.

80. *Volume of a Box* (Refer to the preceding exercise.) A box is constructed from a square piece of metal that is 20 inches on a side.
  (a) If the square corners of length x are cut out, write a polynomial that gives the volume of the box.
  (b) Evaluate the polynomial when $x = 4$ inches.
  (c) Factor out the greatest common factor for this polynomial expression.

WRITING ABOUT MATHEMATICS

81. Use an example to explain the difference between a common factor and the greatest common factor.

82. Use an example to explain how to factor a polynomial by grouping. What two properties of real numbers did you use?

6.2 FACTORING TRINOMIALS I ($x^2 + bx + c$)

Review of the FOIL Method ▪ Factoring Trinomials Having a Leading Coefficient of 1

A LOOK INTO MATH ▷ In Section 6.1 we discussed two types of factoring: common factors and grouping. In this section we introduce methods for factoring certain types of trinomials. These techniques are frequently used in mathematics classes that you might take in the future. We use factoring later in this chapter to solve equations.

Review of the FOIL Method

A **trinomial** is a polynomial that has three terms. We begin by reviewing products of binomials that result in trinomials.

$$(x + 2)(x + 3) = x \cdot x + x \cdot 3 + 2 \cdot x + 2 \cdot 3$$
$$= x^2 + 5x + 6$$

Note that the first term, x^2, in the trinomial results from multiplying the *first* terms of each binomial. The middle term, $5x$, results from adding the product of the *outside* terms and the product of the *inside* terms. Finally the last term, 6, results from multiplying the *last* terms of each binomial. We discussed this method of multiplying binomials, called FOIL, in Section 5.3 and illustrate it as follows.

$$(x + 2)(x + 3) = x^2 + 5x + 6$$

F L
$I \rightarrow 2x$
$O \rightarrow +3x$
$5x \longleftarrow$ The middle term checks.

Factoring Trinomials Having a Leading Coefficient of 1

Any trinomial of degree 2 in the variable x can be written in *standard form* as $ax^2 + bx + c$, where a, b, and c are constants. The constant a is called the **leading coefficient** of the trinomial. In this section we focus on trinomials where $a = 1$ and b and c are integers.

Recall that the binomials $(x + m)$ and $(x + n)$ are multiplied as follows.

$$(x + m)(x + n) = x^2 + nx + mx + mn$$
$$= x^2 + (m + n)x + mn$$

Note that the coefficient of the x-term is the sum of m and n and that the constant (or third) term is the product of m and n. Thus to factor a trinomial in the form $x^2 + bx + c$, we start by finding two numbers, m and n, such that when they are multiplied $m \cdot n = c$ and when they are added $m + n = b$. We illustrate this statement with $x^2 + 6x + 8$.

Standard Form	*Example*
$ax^2 + bx + c$	$x^2 + 6x + 8$
$m \cdot n = c$	$m \cdot n = 8$
$m + n = b$	$m + n = 6$

To determine possible values for m and n, we list factor pairs for 8 and search for a pair whose sum is 6, as in Table 6.1.

Because $2 \cdot 4 = 8$ and $2 + 4 = 6$, we let $m = 2$ and $n = 4$. We then factor the given trinomial as

$$x^2 + 6x + 8 = (x + 2)(x + 4).$$

TABLE 6.1

Factor Pairs for 8

Factors	1, 8	2, 4
Sum	9	6

Note that, if you can find this factor pair mentally, making a table is not necessary.

We check the result by multiplying the two binomials.

$$(x + 2)(x + 4) = x^2 + 6x + 8$$

The middle term checks.

FACTORING $x^2 + bx + c$

To factor the trinomial $x^2 + bx + c$, find two numbers m and n that satisfy

$$m \cdot n = c \quad \text{and} \quad m + n = b.$$

Then $x^2 + bx + c = (x + m)(x + n)$.

EXAMPLE 1 Factoring a trinomial having only positive coefficients

Factor each trinomial.
(a) $x^2 + 7x + 12$ **(b)** $x^2 + 13x + 30$ **(c)** $z^2 + 9z + 20$

Solution
(a) To factor $x^2 + 7x + 12$ we need to find a factor pair for 12 whose sum is 7. To do so we make Table 6.2 as shown on the next page.

TABLE 6.2   **Factor Pairs for 12**

Factors	1, 12	2, 6	3, 4
Sum	13	8	7

The required factor pair is **3** and **4** because $3 \cdot 4 = 12$ and $3 + 4 = 7$. Therefore the given trinomial can be factored as

$$x^2 + 7x + 12 = (x + 3)(x + 4).$$

(b) To factor $x^2 + 13x + 30$ we need to find a factor pair for 30 whose sum is 13. The required pair is **3** and **10**. Thus

$$x^2 + 13x + 30 = (x + 3)(x + 10).$$

(c) To factor $z^2 + 9z + 20$ we need to find a factor pair for 20 whose sum is 9. The required pair is 4 and 5. Thus

$$z^2 + 9z + 20 = (z + 4)(z + 5).$$

Now Try Exercises 21, 23, 25

In the next example, the coefficients of the middle terms are negative.

EXAMPLE 2   **Factoring trinomials having a negative middle coefficient**

Factor each trinomial.
(a) $x^2 - 7x + 10$ **(b)** $x^2 - 8x + 15$ **(c)** $y^2 - 9y + 18$

Solution
(a) To factor $x^2 - 7x + 10$ we need to find a factor pair for 10 whose sum equals -7. To have a positive product *and* a negative sum, *both* numbers must be negative, as shown in Table 6.3.

TABLE 6.3   **Factor Pairs for 10**

Factors	$-1, -10$	$-2, -5$
Sum	-11	-7

The required pair is -2 and -5 because $-2 \cdot (-5) = 10$ and $-2 + (-5) = -7$. Therefore the given trinomial can be factored as

$$x^2 - 7x + 10 = (x - 2)(x - 5).$$

(b) To factor $x^2 - 8x + 15$ we need to find a factor pair for 15 whose sum is -8. The required pair is -3 and -5. Thus

$$x^2 - 8x + 15 = (x - 3)(x - 5).$$

(c) To factor $y^2 - 9y + 18$ we need to find a factor pair for 18 whose sum is -9. The required pair is -3 and -6. Thus

$$y^2 - 9y + 18 = (y - 3)(y - 6).$$

Now Try Exercises 29, 31, 33

In Examples 1 and 2 the coefficient of the last term was always positive. In the next example, this coefficient is negative and the coefficient of the middle term is either positive or negative.

EXAMPLE 3 **Factoring trinomials having a negative constant term**

Factor each trinomial.
(a) $x^2 - 3x - 4$ **(b)** $x^2 + 7x - 8$ **(c)** $t^2 - 2t - 24$

Solution
(a) To factor $x^2 - 3x - 4$ we need to find a factor pair for -4 whose sum is -3. To have a negative product one factor must be positive and the other factor must be negative, as shown in Table 6.4.

TABLE 6.4   **Factor Pairs for -4**

Factors	$-1, 4$	$1, -4$	$-2, 2$
Sum	3	-3	0

The required pair is **1** and -4 because $1 \cdot (-4) = -4$ and $1 + (-4) = -3$. Therefore the given trinomial can be factored as
$$x^2 - 3x - 4 = (x + 1)(x - 4),$$
which can be checked by multiplying $(x + 1)(x - 4)$.
(b) To factor $x^2 + 7x - 8$ we need to find a factor pair for -8 whose sum is 7. The required pair is -1 and **8**. Thus
$$x^2 + 7x - 8 = (x - 1)(x + 8).$$
(c) To factor $t^2 - 2t - 24$ we need to find a factor pair for -24 whose sum is -2. The required pair is -6 and 4. Thus
$$t^2 - 2t - 24 = (t - 6)(t + 4). \quad \text{Now Try Exercises } 39, 55$$

A polynomial with integer coefficients that cannot be factored by using integer coefficients is called a **prime** polynomial. The next example illustrates that some trinomials of the form $x^2 + bx + c$ cannot be factored into the product of two binomials.

EXAMPLE 4 **Discovering that a trinomial is prime**

Factor each trinomial.
(a) $x^2 + 9x + 12$ **(b)** $x^2 + 5x - 4$

Solution
(a) To factor $x^2 + 9x + 12$, we need to find a factor pair for 12 whose sum is 9. Table 6.5 reveals that no such factor pair exists.

TABLE 6.5   **Factor Pairs for 12**

Factors	1, 12	2, 6	3, 4
Sum	13	8	7

The trinomial $x^2 + 9x + 12$ is prime.
(b) At first glance it may appear that the required factor pair is 4 and 1 because $4 \cdot 1 = 4$ and $4 + 1 = 5$. However, it is important to pay close attention to the signs of the coefficients. To factor $x^2 + 5x - 4$, we need to find a factor pair for -4 whose sum is 5. No such factor pair exists. The trinomial $x^2 + 5x - 4$ is prime. Now Try Exercises 19, 43

When factoring some trinomials, it may be necessary to factor out the greatest common factor before attempting to factor the trinomial into the product of two binomials. The next example illustrates this process.

EXAMPLE 5 Factoring out the GCF before factoring further

Factor each trinomial completely.
(a) $7x^2 + 35x + 42$ **(b)** $2x^4 - 4x^3 - 6x^2$

Solution
(a) Because the GCF of $7x^2 + 35x + 42$ is 7, factor out 7 before factoring the remaining trinomial.

$$7x^2 + 35x + 42 = 7(x^2 + 5x + 6)$$

To factor $x^2 + 5x + 6$, we need to find a factor pair for 6 whose sum is 5. The required pair is 2 and 3. Thus

$$7x^2 + 35x + 42 = 7(x + 2)(x + 3).$$

CRITICAL THINKING

A cube has a surface area of $6x^2 + 24x + 24$. What is the length of each side?

(b) Because the GCF of $2x^4 - 4x^3 - 6x^2$ is $2x^2$, factor out $2x^2$ before factoring the remaining trinomial.

$$2x^4 - 4x^3 - 6x^2 = 2x^2(x^2 - 2x - 3)$$

To factor $x^2 - 2x - 3$, we need to find a factor pair for -3 whose sum is -2. The required pair is -3 and 1. Thus

$$2x^4 - 4x^3 - 6x^2 = 2x^2(x - 3)(x + 1)$$

Now Try Exercises 65, 69

MAKING CONNECTIONS

The Signs in the Binomial Factors

If a trinomial of the form $x^2 + bx + c$ can be factored, the signs of the coefficients in the trinomial can be used to determine the signs in the binomial factors. If c is positive, the binomial factors must have the same signs. If c is negative, the binomial factors must have opposite signs. If b and c represent positive numbers, this can be summarized as follows.

Form of the Trinomial	Signs in the Binomial Factors
$x^2 + bx + c$	$(\ +\ )(\ +\ )$
$x^2 - bx + c$	$(\ -\ )(\ -\ )$
$x^2 + bx - c$	$(\ -\ )(\ +\ )$
$x^2 - bx - c$	$(\ -\ )(\ +\ )$

EXAMPLE 6 Finding the dimensions of a rectangle

Find one possibility for the dimensions of a rectangle that has an area of $x^2 + 6x + 5$.

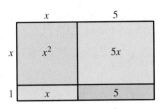

Figure 6.4 Area = $x^2 + 6x + 5$

Solution

The area of a rectangle equals length times width. If we can factor $x^2 + 6x + 5$, then the factors can represent its length and width. Because

$$x^2 + 6x + 5 = (x + 1)(x + 5),$$

one possibility for the rectangle's dimensions is width $x + 1$ and length $x + 5$, as illustrated in Figure 6.4.

Now Try Exercise 83

6.2 PUTTING IT ALL TOGETHER

In this section we discussed factoring trinomials of the form $x^2 + bx + c$. The following table summarizes these techniques.

Concept	Explanation	Examples
Factoring Trinomials of the Form $x^2 + bx + c$	Find two numbers m and n that satisfy $mn = c$ and $m + n = b$. Then $$x^2 + bx + c = (x + m)(x + n).$$	$x^2 + 9x + 20 = (x + 4)(x + 5)$ because $4 \cdot 5 = 20$ and $4 + 5 = 9$. $x^2 + x - 6 = (x - 2)(x + 3)$ because $-2 \cdot 3 = -6$ and $-2 + 3 = 1$. $x^2 - 8x + 12 = (x - 6)(x - 2)$ because $-6 \cdot (-2) = 12$ and $-6 + (-2) = -8$.

6.2 Exercises

CONCEPTS

1. In the trinomial $x^2 + bx + c$, the leading coefficient is _____.

2. Multiply $(x + m)(x + n)$. What is the coefficient of the x-term? What is the constant term?

3. To factor $x^2 + bx + c$, start by finding two numbers m and n that satisfy _____ = c and _____ = b.

4. A trinomial with integer coefficients that cannot be factored using integer coefficients is _____.

5. List all positive integer pairs that have a product of 12.

6. List all negative integer pairs that have a product of 30.

Exercises 7–14: Find the integer pair that has the given product and sum.

7. Product: 28   Sum: 11

8. Product: 35   Sum: 12

9. Product: −30   Sum: −7

10. Product: −100   Sum: 21

11. Product: −50   Sum: 5

12. Product: −15   Sum: −2

13. Product: 28   Sum: −11

14. Product: 80   Sum: −42

FACTORING TRINOMIALS

Exercises 15–60: Factor the trinomial. If the trinomial cannot be factored, write "prime."

15. $x^2 + 3x + 2$
16. $x^2 + 5x + 4$

17. $y^2 + 4y + 4$
18. $y^2 + 8y + 7$

19. $z^2 + 3z + 7$
20. $z^2 + 4z + 5$

21. $x^2 + 8x + 15$
22. $x^2 + 9x + 14$

23. $m^2 + 13m + 36$
24. $m^2 + 15m + 36$

25. $n^2 + 20n + 100$
26. $n^2 + 52n + 100$

27. $x^2 - 6x + 5$
28. $x^2 - 6x + 8$

29. $y^2 - 7y + 12$
30. $y^2 - 12y + 27$

31. $z^2 - 13z + 40$
32. $z^2 - 15z + 54$

33. $a^2 - 16a + 63$
34. $a^2 - 82a + 81$

35. $y^2 - 6y + 10$
36. $y^2 - 2y + 3$

37. $b^2 - 30b + 125$
38. $b^2 - 19b + 90$

39. $x^2 + 13x - 90$
40. $x^2 + 15x - 100$

41. $m^2 + 4m - 45$
42. $m^2 + 4m - 60$

43. $a^2 + 16a - 63$
44. $a^2 + 13a - 42$

45. $n^2 + 10n - 200$
46. $n^2 + 2n - 120$

47. $x^2 + 22x - 23$
48. $x^2 + 18x - 19$

49. $a^2 + 4a - 32$
50. $a^2 + 9a - 36$

51. $b^2 - b - 20$
52. $b^2 - b - 12$

53. $m^2 - 14m - 22$
54. $m^2 - 11m - 24$

55. $x^2 - x - 72$
56. $x^2 - 2x - 80$

57. $y^2 - 15y - 34$
58. $y^2 - 10y - 39$

59. $z^2 - 5z - 66$
60. $z^2 - 6z - 55$

Exercises 61–70: Factor the trinomial completely.

61. $5x^2 - 10x - 40$
62. $2x^2 + 8x - 10$

63. $y^3 - 7y^2 + 10y$
64. $z^3 + 9z^2 + 20z$

65. $3a^3 + 21a^2 + 18a$
66. $5b^3 - 5b^2 - 60b$

67. $2x^3 - 6x^2 + 8x$
68. $4y^3 - 20y^2 + 32y$

69. $2m^4 - 10m^3 - 28m^2$
70. $6n^4 - 18n^3 + 12n^2$

Exercises 71–78: Factor each trinomial.

71. $5 + 6x + x^2$
    (*Hint:* Write the expression in standard form.)

72. $8 + 6x + x^2$
73. $3 - 4x + x^2$

74. $10 - 7x + x^2$

75. $12 + 4x - x^2$
    (*Hint:* Write $(m - x)(n + x)$ and find m and n.)

76. $28 + 3x - x^2$
77. $32 - 4x - x^2$

78. $40 - 3x - x^2$

79. Thinking Generally   Factor the trinomial expression $x^2 + (k + 1)x + k$.

80. Thinking Generally   Factor the trinomial expression $x^2 + (k - 2)x - 2k$.

GEOMETRY

81. A square has an area of $x^2 + 2x + 1$. Find the length of a side. Make a sketch of the square.

82. A square has an area of $x^2 + 6x + 9$. Find the length of a side. Make a sketch of the square.

83. A rectangle has an area of $x^2 + 3x + 2$. Find one possibility for its width and length. Make a sketch of the rectangle.

84. A rectangle has an area of $x^2 + 9x + 8$. Find one possibility for its width and length. Make a sketch of the rectangle.

85. A cube has a surface area of $6x^2 + 12x + 6$. Find the length of a side. (*Hint:* First factor out the GCF.)

86. A cube has a surface area of $6x^2 + 36x + 54$. Find the length of a side.

87. Write a polynomial in factored form that represents the total area of the figure.

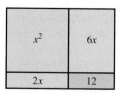

88. Write a polynomial in factored form that represents the total area of the figure.

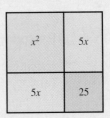

89. Explain how to determine whether a trinomial has been factored correctly. Give an example.

90. Factoring $x^2 + bx + c$ involves finding two integers m and n such that $mn = c$ and $m + n = b$. When you are finding m and n, is it better first to determine values of m and n so that the product is c or the sum is b? Explain your reasoning.

CHECKING BASIC CONCEPTS
SECTIONS 6.1 AND 6.2

1. What is the greatest common factor for the expression $8x^3 - 12x^2 + 24x$?

2. Factor $12z^3 - 18z^2$.

3. Factor each expression completely.
 (a) $6y(y - 2) + 5(y - 2)$
 (b) $2x^3 + x^2 + 10x + 5$
 (c) $4z^3 - 12z^2 + 4z - 12$

4. Factor each trinomial completely.
 (a) $x^2 + 6x + 8$ **(b)** $x^2 - x - 42$

 (c) $a^2 + 3a - 5$ **(d)** $4a^3 + 20a^2 + 24a$

5. Write a polynomial in factored form that represents the total area of the figure.

6.3 FACTORING TRINOMIALS II ($ax^2 + bx + c$)

Factoring Trinomials by Grouping ▪ Factoring with FOIL in Reverse

A LOOK INTO MATH ▷

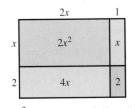

$2x^2 + 5x + 2 = (x + 2)(2x + 1)$

Figure 6.5

The sum of the areas of the four small rectangles shown in Figure 6.5 is $2x^2 + 5x + 2$. Note that the length of the large rectangle is $2x + 1$ and that its width is $x + 2$. One way to determine these dimensions is to factor $2x^2 + 5x + 2$. However, this trinomial has a leading coefficient of 2. Thus far we have factored only trinomials that have a leading coefficient of 1. In this section we discuss two methods used to factor trinomials in the form $ax^2 + bx + c$, where $a \neq 1$. In Example 1(a), we demonstrate how to factor the trinomial $2x^2 + 5x + 2$ as $(x + 2)(2x + 1)$.

Factoring Trinomials by Grouping

To factor the polynomial given by $x^2 + 6x + 5$ we find two numbers, m and n, such that $mn = 5$ and $m + n = 6$. For this trinomial we let $m = 1$ and $n = 5$, which gives

$$x^2 + 6x + 5 = (x + 1)(x + 5).$$

To factor the polynomial $2x^2 + 7x + 3$, which has a leading coefficient of 2, we must find two numbers m and n such that $mn = 2 \cdot 3 = 6$ and $m + n = 7$. One solution is

$m = 1$ and $n = 6$. Using grouping, we can now factor this trinomial by writing $7x$ as the sum $1x + 6x$.

$$
\begin{aligned}
2x^2 + 7x + 3 &= 2x^2 + \overbrace{x + 6x}^{7x} + 3 & &\text{Write } 7x \text{ as } x + 6x. \\
&= (2x^2 + x) + (6x + 3) & &\text{Associative property} \\
&= x(2x + 1) + 3(2x + 1) & &\text{Factor out common factors.} \\
&= (x + 3)(2x + 1) & &\text{Factor out } (2x + 1).
\end{aligned}
$$

This technique of factoring trinomials by grouping is summarized as follows.

FACTORING $ax^2 + bx + c$ BY GROUPING

To factor $ax^2 + bx + c$ perform the following steps. (Assume that a, b, and c are integers and have no common factors.)

1. Find two numbers, m and n, such that $mn = ac$ and $m + n = b$.
2. Write the trinomial as $ax^2 + mx + nx + c$.
3. Use grouping to factor this expression into two binomials.

EXAMPLE **1**   Factoring $ax^2 + bx + c$ by grouping

Factor each trinomial.
(a) $2x^2 + 5x + 2$ **(b)** $3z^2 + z - 2$ **(c)** $10t^2 - 11t + 3$

Solution

(a) To factor $2x^2 + 5x + 2$ we need to find m and n that satisfy $mn = 2 \cdot 2 = 4$ and $m + n = 5$. Table 6.6 shows that two such numbers are $m = 1$ and $n = 4$.

TABLE 6.6

Factor Pairs for 4

Factors	1, 4	2, 2
Sum	5	4

$$
\begin{aligned}
2x^2 + 5x + 2 &= 2x^2 + \overbrace{x + 4x}^{5x} + 2 & &\text{Write } 5x \text{ as } x + 4x. \\
&= (2x^2 + x) + (4x + 2) & &\text{Associative property} \\
&= x(2x + 1) + 2(2x + 1) & &\text{Factor out common factors.} \\
&= (x + 2)(2x + 1) & &\text{Factor out } (2x + 1).
\end{aligned}
$$

(b) To factor $3z^2 + 1z - 2$ we need to find m and n that satisfy $mn = 3 \cdot (-2) = -6$ and $m + n = 1$. Two such numbers are $m = 3$ and $n = -2$.

$$
\begin{aligned}
3z^2 + z - 2 &= 3z^2 + 3z - 2z - 2 & &\text{Write } z \text{ as } 3z - 2z. \\
&= (3z^2 + 3z) + (-2z - 2) & &\text{Associative property} \\
&= 3z(z + 1) - 2(z + 1) & &\text{Factor out common factors.} \\
&= (3z - 2)(z + 1) & &\text{Factor out } (z + 1).
\end{aligned}
$$

(c) To factor $10t^2 - 11t + 3$ we need to find m and n that satisfy $mn = 10 \cdot 3 = 30$ and $m + n = -11$. Two such numbers are $m = -5$ and $n = -6$.

$$
\begin{aligned}
10t^2 - 11t + 3 &= 10t^2 - 5t - 6t + 3 & &\text{Write } -11t \text{ as } -5t - 6t. \\
&= (10t^2 - 5t) + (-6t + 3) & &\text{Associative property} \\
&= 5t(2t - 1) - 3(2t - 1) & &\text{Factor out common factors.} \\
&= (5t - 3)(2t - 1) & &\text{Factor out } (2t - 1).
\end{aligned}
$$

Now Try Exercises **11, 21, 33**

Different Ways to Factor by Grouping

In Example 1(c) we could have written $-11t$ as $-6t - 5t$, rather than $-5t - 6t$. Then the factoring could have been written as

$$10t^2 - 11t + 3 = 10t^2 - 6t - 5t + 3$$
$$= (10t^2 - 6t) + (-5t + 3)$$
$$= 2t(5t - 3) - 1(5t - 3)$$
$$= (2t - 1)(5t - 3),$$

which gives the same result.

In the previous section we showed that some trinomials of the form $x^2 + bx + c$ are prime and cannot be factored into the product of two binomials with integer coefficients. The next example illustrates that some trinomials of the form $ax^2 + bx + c$, with $a \neq 1$, may also be prime.

EXAMPLE 2   Discovering that a trinomial is prime

Factor each trinomial.
(a) $3x^2 + 5x + 4$ **(b)** $2x^2 - 8x - 3$

Solution
(a) To factor $3x^2 + 5x + 4$, we need to find integers m and n such that $mn = 3 \cdot 4 = 12$ and $m + n = 5$. Because the middle term is positive, we consider only positive factors of 12. Table 6.7 reveals that no such integers exist.

TABLE 6.7   **Factor Pairs for 12**

Factors	1, 12	2, 6	3, 4
Sum	13	8	7

The trinomial $3x^2 + 5x + 4$ is prime.

(b) To factor $2x^2 - 8x - 3$, we must find integers m and n such that $mn = 2 \cdot (-3) = -6$ and $m + n = -8$. Table 6.8 reveals that no such integers exist.

TABLE 6.8   **Factor Pairs for −6**

Factors	−1, 6	−2, 3	2, −3	1, −6
Sum	5	1	−1	−5

The trinomial $2x^2 - 8x - 3$ is prime.   Now Try Exercises 13, 23

Although some trinomials may look as though they can be factored using the process discussed in Example 1, it is important to remember to first factor out the greatest common factor whenever possible. In the next example we factor out the GCF before factoring the trinomial further.

EXAMPLE 3   Factoring out the GCF before factoring further

Factor each trinomial completely.
(a) $15x^2 - 50x - 40$ **(b)** $4x^3 - 22x^2 + 30x$

Solution

(a) Because the GCF of $15x^2 - 50x - 40$ is 5, factor out 5 before factoring the remaining trinomial.

$$15x^2 - 50x - 40 = 5(3x^2 - 10x - 8)$$

To factor $3x^2 - 10x - 8$, we need numbers m and n such that $mn = 3 \cdot (-8) = -24$ and $m + n = -10$. Two such numbers are -12 and 2.

$$
\begin{aligned}
15x^2 - 50x - 40 &= 5(3x^2 - 12x + 2x - 8) && \text{Write } -10x \text{ as } -12x + 2x. \\
&= 5\big((3x^2 - 12x) + (2x - 8)\big) && \text{Associative property} \\
&= 5\big(3x(x - 4) + 2(x - 4)\big) && \text{Factor out common factors.} \\
&= 5(3x + 2)(x - 4) && \text{Factor out } (x - 4).
\end{aligned}
$$

(b) Because the GCF of $4x^3 - 22x^2 + 30x$ is $2x$, factor out $2x$ before factoring the remaining trinomial.

$$4x^3 - 22x^2 + 30x = 2x(2x^2 - 11x + 15)$$

To factor the trinomial $2x^2 - 11x + 15$, we need to find numbers m and n such that $mn = 2 \cdot 15 = 30$ and $m + n = -11$. Two such numbers are -6 and -5.

$$
\begin{aligned}
4x^3 - 22x^2 + 30x &= 2x(2x^2 - 6x - 5x + 15) && \text{Write } -11x \text{ as } -6x - 5x. \\
&= 2x\big((2x^2 - 6x) + (-5x + 15)\big) && \text{Associative property} \\
&= 2x\big(2x(x - 3) - 5(x - 3)\big) && \text{Factor out common factors.} \\
&= 2x(2x - 5)(x - 3) && \text{Factor out } (x - 3).
\end{aligned}
$$

Now Try Exercises 51, 55

Factoring with FOIL in Reverse

Rather than factoring a trinomial by grouping, we can use FOIL in reverse to determine its binomial factors. For example, the factors of $2x^2 + 5x + 2$ are two binomials:

$$2x^2 + 5x + 2 = (\underline{} + \underline{})(\underline{} + \underline{}),$$

where the expressions to be placed in the four blanks are yet to be found. By the FOIL method, we know that the product of the first terms of these two binomials is $2x^2$. Because $2x^2 = 2x \cdot x$, we can write

$$2x^2 + 5x + 2 = (\underline{2x} + \underline{})(\underline{x} + \underline{}).$$

By FOIL, the product of the last terms in these binomials must be 2. Because $2 = 1 \cdot 2$, we could put 1 and 2 in the blanks. However, we must be sure to place them correctly so that the product of the outside terms plus the product of the inside terms is $5x$.

$$(2x + 1)(x + 2) = 2x^2 + 5x + 2$$

$1x$
$+4x$
$\overline{5x}$ ←——Middle term checks.

If we had interchanged the 1 and 2, we would have obtained an incorrect result.

$$(2x + 2)(x + 1) = 2x^2 + 4x + 2$$

Middle term is *not* 5x.

In the next example we factor expressions of the form $ax^2 + bx + c$, where $a \neq 1$. We use FOIL in reverse and *trial and error* to find the correct factors.

EXAMPLE 4

Factoring the form $ax^2 + bx + c$

Factor each trinomial.
(a) $3x^2 + 5x + 2$ **(b)** $6x^2 + 7x - 3$ **(c)** $6 + 4y^2 - 11y$

Solution
(a) To factor $3x^2 + 5x + 2$ we start by finding factors of $3x^2$, which include $3x$ and x, so we write

$$3x^2 + 5x + 2 = (3x + \underline{\quad})(x + \underline{\quad}).$$

The factors of the last term 2 are **1** and **2**. (Because the middle term is positive, we do not consider -1 and -2.) If we place values of 1 and 2 in the binomial as follows, the middle term becomes $7x$ rather than $5x$.

$$(3x + 1)(x + 2) = 3x^2 + 7x + 2$$

Middle term is *not* 5x.

By reversing the positions of 1 and 2, we obtain a correct factorization.

$$(3x + 2)(x + 1) = 3x^2 + 5x + 2$$

Middle term checks.

(b) To factor $6x^2 + 7x - 3$, we start by finding factors of $6x^2$, which include $2x$ and $3x$ or $6x$ and x. Factors of the last term, -3, include -1 and 3 or 1 and -3. The following two factorizations give incorrect results.

$$(2x + 1)(3x - 3) = 6x^2 - 3x - 3 \qquad (6x + 1)(x - 3) = 6x^2 - 17x - 3$$

Middle term is *not* 7x.   Middle term is *not* 7x.

To obtain a middle term of $7x$ we use the following arrangement.

$$(3x - 1)(2x + 3) = 6x^2 + 7x - 3$$

Middle term checks.

To find the correct factorization we often need to try more than once.

(c) To factor $6 + 4y^2 - 11y$, we start by writing the trinomial in standard form as $4y^2 - 11y + 6$. Then we find possible factors of the first term, $4y^2$, which include $2y$ and $2y$ or $4y$ and y. Factors of the last term, 6, include either -1 and -6 or -2 and -3. (Because the middle term is negative, we do not use the positive factors of 1 and 6 or 2 and 3.) To obtain a middle term of $-11y$ we use the following arrangement.

$$(4y - 3)\ (y - 2) = 4y^2 - 11y + 6$$

$$-3y$$
$$-8y$$
$$-11y \longleftarrow \text{Middle term checks.}$$

Now Try Exercises 15, 29, 65

MAKING CONNECTIONS

The Signs in the Binomial Factors

Let a, b, and c represent positive integers. If a trinomial of the form $ax^2 + bx + c$ can be factored, the signs in the binomial factors can be summarized as follows.

Form of the Trinomial	Signs in the Binomial Factors
$ax^2 + bx + c$	$(\ +\ )(\ +\ )$
$ax^2 - bx + c$	$(\ -\ )(\ -\ )$
$ax^2 + bx - c$	$(\ -\ )(\ +\ )$
$ax^2 - bx - c$	$(\ -\ )(\ +\ )$

In the next example we demonstrate how to factor trinomials having a negative leading coefficient. This task is sometimes accomplished by first factoring out -1.

EXAMPLE 5 **Factoring trinomials having a negative leading coefficient**

Factor each trinomial.
(a) $-6x^2 + 17x - 5$ (b) $1 - x - 2x^2$

Solution
(a) One way to factor $-6x^2 + 17x - 5$ is to start by factoring out -1. Then we can apply FOIL in reverse.

Opposite of $-6x^2 + 17x - 5$

$$-6x^2 + 17x - 5 = -1(6x^2 - 17x + 5) \qquad \text{Factor out } -1.$$
$$= -1(3x - 1)(2x - 5) \qquad \text{Factor the trinomial.}$$
$$= -(3x - 1)(2x - 5) \qquad \text{Rewrite.}$$

(b) To factor the trinomial $1 - x - 2x^2$, write it in standard form, factor out -1, and then apply FOIL in reverse.

$$1 - x - 2x^2 = -2x^2 - x + 1 \qquad \text{Standard form}$$
$$= -1(2x^2 + x - 1) \qquad \text{Factor out } -1.$$
$$= -(2x - 1)(x + 1) \qquad \text{Factor the trinomial.}$$

Now Try Exercises 67, 71

6.3 PUTTING IT ALL TOGETHER

In this section we discussed factoring trinomials in the form $ax^2 + bx + c$ by using grouping and FOIL in reverse. The following table summarizes these two methods.

Concept	Explanation	Examples
Factoring Trinomials by Grouping	To factor $ax^2 + bx + c$, find two numbers, m and n, such that $mn = ac$ and $m + n = b$. Then write $$ax^2 + mx + nx + c.$$ Use grouping to factor this expression into two binomials.	For $3x^2 + 10x + 8$ let $m = 6$ and $n = 4$ because $mn = 24$ and $m + n = 10$. $$\begin{aligned} 3x^2 + 10x + 8 &= 3x^2 + 6x + 4x + 8 \\ &= (3x^2 + 6x) + (4x + 8) \\ &= 3x(x + 2) + 4(x + 2) \\ &= (3x + 4)(x + 2) \end{aligned}$$
Factoring Trinomials by Using FOIL in Reverse	To factor $ax^2 + bx + c$, find factors of ax^2 and of c. Choose and arrange these factors in two binomials so that the middle term is bx.	For $3x^2 + 10x + 8$ the factors of $3x^2$ are $3x$ and x. The positive factors of 8 are either 2 and 4, or 1 and 8. A middle term of $10x$ can be obtained as follows. $$\begin{array}{c} (3x + 4)\ (x + 2) = 3x^2 + 10x + 8 \\ \ \llcorner 4x \lrcorner \\ \llcorner\!\!-\!\!+6x\!\!-\!\!\lrcorner \\ 10x \longleftarrow \text{Middle term checks.} \end{array}$$
Choosing Signs When Factoring $ax^2 + bx + c$	For $ax^2 + bx + c$ with $a > 0$, 1. If $c > 0$ and $b > 0$, use $(\_\_ + \_\_)(\_\_ + \_\_)$. 2. If $c > 0$ and $b < 0$, use $(\_\_ - \_\_)(\_\_ - \_\_)$. 3. If $c < 0$, use $(\_\_ - \_\_)(\_\_ + \_\_)$ or $(\_\_ + \_\_)(\_\_ - \_\_)$.	1. $2x^2 + 7x + 3 = (2x + 1)(x + 3)$ $(c = 3, b = 7)$ 2. $2x^2 - 7x + 3 = (2x - 1)(x - 3)$ $(c = 3, b = -7)$ 3. $2x^2 + 5x - 3 = (2x - 1)(x + 3)$ $(c = -3, b = 5)$ $2x^2 - 5x - 3 = (2x + 1)(x - 3)$ $(c = -3, b = -5)$

6.3 Exercises

MyMathLab PRACTICE WATCH DOWNLOAD READ REVIEW

CONCEPTS

1. To factor the polynomial $ax^2 + bx + c$ by grouping, you first find two numbers, m and n, such that $mn = $ _____ and $m + n = $ _____.

2. To factor the polynomial $ax^2 + bx + c$ with FOIL in reverse, you first find possible factors for _____ and for _____.

3. If $3x^2 + 5x + 2$ is factored, the result is the product $(3x \_\_ 2)(x \_\_ 1)$.

4. If $3x^2 - x - 2$ is factored, the result is the product $(3x \_\_ 2)(x \_\_ 1)$.

5. If $3x^2 - 5x + 2$ is factored, the result is the product $(3x \_\_ 2)(x \_\_ 1)$.

6. If $3x^2 + x - 2$ is factored, the result is the product $(3x \_\_ 2)(x \_\_ 1)$.

7. If $4x^2 + 11x + 6$ is factored, the result is the product $(4x + \_\_)(\_\_ + 2)$.

8. If $4x^2 - 5x - 6$ is factored, the result is the product $(x - \_\_)(\_\_ + 3)$.

9. If $4x^2 + 4x - 3$ is factored, the result is the product $(2x - \_\_)(\_\_ + 3)$.

10. If $4x^2 - 8x + 3$ is factored, the result is the product $(2x - \_\_)(\_\_ - 3)$.

FACTORING TRINOMIALS

Exercises 11–50: Factor the trinomial. If the trinomial cannot be factored, write "prime."

11. $2x^2 + 7x + 3$

12. $2x^2 + 3x + 1$

13. $3y^2 + 2y + 4$

14. $2y^2 + 5y + 1$

15. $3x^2 + 4x + 1$

16. $3x^2 + 10x + 3$

17. $6x^2 + 11x + 3$

18. $6x^2 + 17x + 5$

19. $5x^2 - 11x + 2$

20. $7x^2 - 8x + 1$

21. $2y^2 - 7y + 5$

22. $2y^2 - 11y + 12$

23. $3m^2 - 11m - 6$

24. $5m^2 - 7m - 2$

25. $7z^2 - 37z + 10$

26. $3z^2 - 11z + 6$

27. $3t^2 - 7t - 6$

28. $8t^2 - 6t - 9$

29. $15r^2 + r - 6$

30. $12r^2 + r - 6$

31. $24m^2 - 23m - 12$

32. $24m^2 + 29m - 4$

33. $25x^2 + 5x - 2$

34. $30x^2 + 7x - 2$

35. $6x^2 + 11x - 2$

36. $12x^2 + 28x - 5$

37. $15y^2 - 7y + 2$

38. $14y^2 - 5y + 1$

39. $21n^2 + 4n - 1$

40. $21n^2 + 10n + 1$

41. $14y^2 + 23y + 3$

42. $28y^2 + 25y + 3$

43. $28z^2 - 25z + 3$

44. $15z^2 - 19z + 6$

45. $30x^2 - 29x + 6$

46. $50x^2 - 55x + 12$

47. $20a^2 + 18a - 5$

48. $40a^2 + 21a - 2$

49. $18t^2 + 23t - 6$

50. $33t^2 + 7t - 10$

Exercises 51–60: Factor the trinomial completely.

51. $12a^2 + 12a - 9$

52. $21b^2 - 14b - 56$

53. $12y^3 - 11y^2 + 2y$

54. $10z^3 + 19z^2 + 6z$

55. $24x^3 - 30x^2 + 9x$

56. $8y^3 - 16y^2 + 6y$

57. $8x^4 - 6x^3 + 2x^2$

58. $10y^3 + 15y^2 - 5y$

59. $28x^4 + 56x^3 + 21x^2$

60. $20y^4 + 42y^3 - 20y^2$

61. **Thinking Generally** Factor the trinomial expression $3x^2 + (3k + 1)x + k$.

62. **Thinking Generally** Factor the trinomial expression $3x^2 + (3k - 2)x - 2k$.

Exercises 63–72: Factor.

63. $2 + 15x + 7x^2$

64. $3 + 16x + 5x^2$

65. $2 - 5x + 2x^2$

66. $5 - 6x + x^2$

67. $3 - 2x - 8x^2$

68. $5 - 3x - 2x^2$

69. $-2x^2 - 7x + 15$

70. $-5x^2 - 19x + 4$

71. $-5x^2 + 14x + 3$

72. $-6x^2 + 17x + 14$

73. A rectangle has an area of $6x^2 + 7x + 2$. Find possible dimensions for this rectangle. Make a sketch of the rectangle.

74. A rectangle has an area of $2x^2 + 5x + 3$. Find possible dimensions for the rectangle. Make a sketch of the rectangle.

75. Write a polynomial in factored form that represents the total area of the figure.

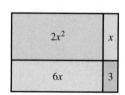

76. Write a polynomial in factored form that represents the total area of the figure.

WRITING ABOUT MATHEMATICS

77. Explain how the sign of the third term in the trinomial $ax^2 + bx + c$ affects how it is factored.

78. Explain the steps to be used to factor $ax^2 + bx + c$ by grouping.

GROUP ACTIVITY
WORKING WITH REAL DATA

Directions: Form a group of 2 to 4 people. Select someone to record the group's responses for this activity. All members of the group should work cooperatively to answer the questions. If your instructor asks for your results, each member of the group should be prepared to respond.

AIDS Cases From 1993 to 2003 the cumulative number N of AIDS cases in thousands can be approximated by $N = -2x^2 + 76x + 430$, where $x = 0$ corresponds to the year 1993.

Year	1993	1995	1997	1999	2001	2003
Cases	422	565	677	762	844	930

Source: U.S. Department of Health and Human Services.

(a) Use the equation to find N for each year in the table.
(b) Discuss how well this equation approximates the data.
(c) Rewrite the equation with the right side completely factored.
(d) Use your equation from part (c) to find N for each year in the table. Do your answers agree with those found in part (a)?

6.4 SPECIAL TYPES OF FACTORING

Difference of Two Squares ▪ Perfect Square Trinomials ▪ Sum and Difference of Two Cubes

A LOOK INTO MATH ▷ Some polynomials can be factored by using special methods. These methods are used to solve equations and simplify expressions. In this section we discuss some of these methods.

Difference of Two Squares

In Section 5.4 we showed that

$$(a - b)(a + b) = a^2 - b^2.$$

We can use this equation to factor a difference of two squares. For example, to factor the expression $x^2 - 25$ we can write it in the form $a^2 - b^2$, where $a = x$ and $b = 5$. Then the equation

$$a^2 - b^2 = (a - b)(a + b)$$

becomes

$$x^2 - 25 = (x - 5)(x + 5).$$

DIFFERENCE OF TWO SQUARES

For any real numbers a and b,

$$a^2 - b^2 = (a - b)(a + b).$$

In the next example we apply this method to other expressions.

EXAMPLE 1   **Factoring the difference of two squares**

Factor each difference of two squares.
(a) $x^2 - 36$ **(b)** $4x^2 - 9$ **(c)** $100 - 16t^2$ **(d)** $49y^2 - 64z^2$

Solution

(a) Substitute x for a and 6 for b. Then the equation

$$a^2 - b^2 = (a - b)(a + b)$$

becomes

$$x^2 - 6^2 = (x - 6)(x + 6).$$

(b) The expression $4x^2 - 9$ can be written as $(2x)^2 - 3^2$. Thus

$$4x^2 - 9 = (2x - 3)(2x + 3).$$

(c) The expression $100 - 16t^2$ can be written as $(10)^2 - (4t)^2$. Thus

$$100 - 16t^2 = (10 - 4t)(10 + 4t).$$

(d) The expression $49y^2 - 64z^2$ can be written as $(7y)^2 - (8z)^2$. Thus

$$49y^2 - 64z^2 = (7y - 8z)(7y + 8z).$$

Now Try Exercises 15, 17, 23, 27

MAKING CONNECTIONS

Sum of Squares versus Difference of Squares

The sum of two squares, $a^2 + b^2$, cannot be factored by using real numbers. However, the difference of two squares, $a^2 - b^2$, can be factored. For example, $x^2 + 4$ cannot be factored, but $x^2 - 4$ can be factored as $(x - 2)(x + 2)$.

Perfect Square Trinomials

In Section 5.4 we also showed how to compute $(a + b)^2$ and $(a - b)^2$:

$$(a + b)^2 = a^2 + 2ab + b^2 \quad \text{and}$$
$$(a - b)^2 = a^2 - 2ab + b^2.$$

The expressions $a^2 + 2ab + b^2$ and $a^2 - 2ab + b^2$ are called **perfect square trinomials.** If we can recognize a perfect square trinomial, we can use the following formulas to factor it.

PERFECT SQUARE TRINOMIALS

For any real numbers a and b,

$$a^2 + 2ab + b^2 = (a + b)^2 \quad \text{and}$$
$$a^2 - 2ab + b^2 = (a - b)^2.$$

When factoring a trinomial as a perfect square trinomial, we must first verify that the middle term is correct. This technique is demonstrated in the next example.

EXAMPLE 2 **Factoring perfect square trinomials**

If possible, factor each trinomial as a perfect square trinomial.
(a) $x^2 + 10x + 25$ **(b)** $4x^2 - 4x + 1$
(c) $9z^2 + 18z + 4$ **(d)** $x^2 - 4xy + 4y^2$

Solution
(a) To factor $x^2 + 10x + 25$ we let $a^2 = x^2$ and $b^2 = 5^2$ so that $a = x$ and $b = 5$. To be a perfect square trinomial, the middle term must be $2ab$. That is, the middle term must be *twice* the product of x and 5. So

$$2ab = 2 \cdot x \cdot 5 = 10x,$$

which is the middle term in the given expression. Thus

$$a^2 + 2ab + b^2 = (a + b)^2 \quad \text{becomes} \quad x^2 + 10x + 25 = (x + 5)^2.$$

(b) To factor $4x^2 - 4x + 1$ we let $a^2 = (2x)^2$ and $b^2 = 1^2$ so that $a = 2x$ and $b = 1$. The middle term must be $2ab$. So

$$2ab = 2 \cdot 2x \cdot 1 = 4x,$$

which is the middle term in the given expression, except for the subtraction sign. Thus

$$a^2 - 2ab + b^2 = (a - b)^2 \quad \text{becomes} \quad 4x^2 - 4x + 1 = (2x - 1)^2.$$

(c) To factor $9z^2 + 18z + 4$ we let $a^2 = (3z)^2$ and $b^2 = 2^2$ so that $a = 3z$ and $b = 2$. The middle term must be twice the product of $3z$ and 2. So

$$2ab = 2 \cdot 3z \cdot 2 = 12z,$$

which is *not* the middle term of $18z$. We cannot factor this expression as a perfect square trinomial.

(d) To factor $x^2 - 4xy + 4y^2$ we let $a^2 = x^2$ and $b^2 = (2y)^2$ so that $a = x$ and $b = 2y$. The middle term must be twice the product of x and $2y$. So

$$2ab = 2 \cdot x \cdot 2y = 4xy,$$

which is the middle term, except for the subtraction sign. Thus

$$a^2 - 2ab + b^2 = (a - b)^2 \quad \text{becomes} \quad x^2 - 4xy + 4y^2 = (x - 2y)^2.$$

Now Try Exercises 35, 39, 41, 49

Special Factoring and General Techniques

If you do not recognize a polynomial as the difference of two squares or a perfect square trinomial, you can still factor the polynomial by using the methods discussed in earlier sections.

Sum and Difference of Two Cubes

The sum or difference of two cubes may be factored—a result of the two equations

$$(a + b)(a^2 - ab + b^2) = a^3 + b^3 \quad \text{and}$$
$$(a - b)(a^2 + ab + b^2) = a^3 - b^3.$$

These equations can be verified by multiplying the left side to obtain the right side. For example, multiplying the polynomials on the left side of the first equation results in

$$(a + b)(a^2 - ab + b^2) = a \cdot a^2 - a \cdot ab + a \cdot b^2 + b \cdot a^2 - b \cdot ab + b \cdot b^2$$
$$= a^3 - a^2b + ab^2 + a^2b - ab^2 + b^3$$
$$= a^3 + b^3.$$

SUM AND DIFFERENCE OF TWO CUBES

For any real numbers a and b,

Opposite Signs

$$a^3 + b^3 = (a + b)(a^2 - ab + b^2) \quad \text{and}$$
$$a^3 - b^3 = (a - b)(a^2 + ab + b^2).$$

Opposite Signs

Any binomial whose terms can be expressed as cubes can be factored as a sum or difference of cubes. We demonstrate this method in the next example.

EXAMPLE 3   Factoring the sum and difference of two cubes

Factor each polynomial.
(a) $z^3 + 8$ (b) $x^3 - 27$ (c) $8x^3 - 1$

Solution
(a) To factor $z^3 + 8$ we let $a^3 = z^3$ and $b^3 = 2^3$ so that $a = z$ and $b = 2$. Then

$$a^3 + b^3 = (a + b)(a^2 - ab + b^2)$$

becomes

$$z^3 + 2^3 = (z + 2)(z^2 - z \cdot 2 + 2^2)$$
$$= (z + 2)(z^2 - 2z + 4).$$

(b) To factor $x^3 - 27$ we let $a^3 = x^3$ and $b^3 = 3^3$ so that $a = x$ and $b = 3$. Then

$$a^3 - b^3 = (a - b)(a^2 + ab + b^2)$$

becomes

$$x^3 - 3^3 = (x - 3)(x^2 + x \cdot 3 + 3^2)$$
$$= (x - 3)(x^2 + 3x + 9).$$

(c) To factor $8x^3 - 1$ we let $a^3 = (2x)^3$ and $b^3 = 1^3$ so that $a = 2x$ and $b = 1$. Then

$$(2x)^3 - 1^3 = (2x - 1)((2x)^2 + 2x \cdot 1 + 1^2)$$
$$= (2x - 1)(4x^2 + 2x + 1). \quad \text{Now Try Exercises } 55, 63$$

In this section we have discussed special methods for factoring polynomials that can be identified as the difference of two squares, perfect square trinomials, the sum of two cubes, or the difference of two cubes. The next example demonstrates how to recognize and factor such polynomials.

EXAMPLE 4 **Recognizing polynomials that can be factored with special methods**

Factor each polynomial.
(a) $8x^3 + 27$ **(b)** $4y^2 - 20y + 25$ **(c)** $9z^2 - 64$

Solution
(a) Because this polynomial has only two terms, it cannot be factored as a perfect square trinomial. Since it is not a difference, it cannot be factored as the difference of two squares. We will try to factor the polynomial as the sum of two cubes. To factor $8x^3 + 27$, we note that $8x^3 = (2x)^3$ and $27 = 3^3$. Then

$$8x^3 + 27 = (2x)^3 + 3^3$$
$$= (2x + 3)((2x)^2 - 2x \cdot 3 + 3^2)$$
$$= (2x + 3)(4x^2 - 6x + 9).$$

(b) Because this polynomial has three terms, we will try to factor it as a perfect square trinomial. To factor $4y^2 - 20y + 25$, we let $a^2 = (2y)^2$ and $b^2 = 5^2$ so that $a = 2y$ and $b = 5$. The middle term must be twice the product of $2y$ and 5 so that $2ab = 2 \cdot 2y \cdot 5 = 20y$. This is the correct middle term except for the subtraction sign. Thus

$$a^2 - 2ab + b^2 = (a - b)^2 \quad \text{becomes} \quad 4y^2 - 20y + 25 = (2y - 5)^2.$$

(c) Because this polynomial is a difference of two terms that appear to be square terms, we will try to factor the polynomial as the difference of two squares. The expression $9z^2 - 64$ can be written as $(3z)^2 - 8^2$. Thus

$$9z^2 - 64 = (3z - 8)(3z + 8).$$

Now Try Exercises 29, 45, 65

When using the special factoring methods discussed in this section, it is important to remember to first factor out the greatest common factor whenever possible. In the next example we factor out the GCF before factoring further.

EXAMPLE 5 Factoring out the GCF before factoring further

Factor each polynomial completely.
(a) $27x^3 + 72x^2 + 48x$ (b) $18a^3 - 8ab^2$

Solution

(a) Because the GCF of $27x^3 + 72x^2 + 48x$ is $3x$, factor out $3x$ before factoring the remaining trinomial.

$$27x^3 + 72x^2 + 48x = 3x(9x^2 + 24x + 16)$$

The expression $9x^2 + 24x + 16$ is a perfect square trinomial with $a^2 = (3x)^2$ and $b^2 = 4^2$ so that $a = 3x$ and $b = 4$. Thus

$$a^2 + 2ab + b^2 = (a + b)^2 \quad \text{becomes} \quad 9x^2 + 24x + 16 = (3x + 4)^2.$$

As a result, $27x^3 + 72x^2 + 48x = 3x(3x + 4)^2$.

(b) Because the GCF of $18a^3 - 8ab^2$ is $2a$, factor out $2a$ before factoring the remaining polynomial.

$$18a^3 - 8ab^2 = 2a(9a^2 - 4b^2)$$

The expression $9a^2 - 4b^2$ is the difference of two squares and can be written as $(3a)^2 - (2b)^2$. Thus

$$18a^3 - 8ab^2 = 2a(3a - 2b)(3a + 2b).$$

Now Try Exercises 69, 77

6.4 PUTTING IT ALL TOGETHER

In this section we discussed some special types of factoring, which are summarized in the following table.

Factoring	Explanation	Examples
Difference of Two Squares	$a^2 - b^2 = (a - b)(a + b)$ **NOTE:** The *sum* of two squares, $a^2 + b^2$, cannot be factored by using real numbers.	$x^2 - 49 = (x - 7)(x + 7)$ $81 - z^2 = (9 - z)(9 + z)$ $4r^2 - 25t^2 = (2r - 5t)(2r + 5t)$ $16a^2 + b^2$ cannot be factored.
Perfect Square Trinomial	$a^2 + 2ab + b^2 = (a + b)^2$ $a^2 - 2ab + b^2 = (a - b)^2$ Be sure to verify that the given middle term equals $2ab$ before factoring.	$m^2 + 2m + 1 = (m + 1)^2$ $25y^2 - 30y + 9 = (5y - 3)^2$ $36r^2 + 12rt + t^2 = (6r + t)^2$ $x^2 + 5x + 4$ is *not* a perfect square trinomial because $2ab = 2 \cdot x \cdot 2 = 4x \neq 5x.$

Factoring	Explanation	Examples
Sum and Difference of Two Cubes	$a^3 + b^3 = (a + b)(a^2 - ab + b^2)$ $a^3 - b^3 = (a - b)(a^2 + ab + b^2)$	$y^3 + 27 = (y + 3)(y^2 - y \cdot 3 + 3^2)$ $= (y + 3)(y^2 - 3y + 9)$ $27r^3 - 64t^3$ $= (3r - 4t)((3r)^2 + 3r \cdot 4t + (4t)^2)$ $= (3r - 4t)(9r^2 + 12rt + 16t^2)$

6.4 Exercises

CONCEPTS

1. $a^2 - b^2 =$ _____

2. The expression $a^2 + b^2$ (can/cannot) be factored by using real numbers.

3. If the expression $36x^2 - 49y^2$ is written in the form $a^2 - b^2$, then $a =$ _____ and $b =$ _____.

4. $a^2 + 2ab + b^2 =$ _____

5. $a^2 - 2ab + b^2 =$ _____

6. $x^2 +$ _____ $+ 9$ is a perfect square trinomial.

7. $4r^2 -$ _____ $+ 25t^2$ is a perfect square trinomial.

8. $a^3 + b^3 =$ _____

9. $a^3 - b^3 =$ _____

10. If the expression $8x^3 + 27y^3$ is written in the form $a^3 + b^3$, then $a =$ _____ and $b =$ _____.

11. $y^3 - 8 = (y \underline{} 2)(y^2 \underline{} 2y + 4)$

12. $64z^3 + 27 = (4z \underline{} 3)(16z^2 \underline{} 12z + 9)$

FACTORING THE DIFFERENCE OF TWO SQUARES

Exercises 13–30: Factor.

13. $x^2 - 1$

14. $x^2 - 16$

15. $z^2 - 100$

16. $z^2 - 81$

17. $4y^2 - 1$

18. $9y^2 - 16$

19. $36z^2 - 25$

20. $49z^2 - 64$

21. $9 - x^2$

22. $25 - x^2$

23. $1 - 9y^2$

24. $49 - 16y^2$

25. $4a^2 - 9b^2$

26. $16a^2 - b^2$

27. $36m^2 - 25n^2$

28. $49m^2 - 100n^2$

29. $81r^2 - 49t^2$

30. $625r^2 - 121t^2$

FACTORING PERFECT SQUARE TRINOMIALS

Exercises 31–52: Factor as a perfect square trinomial whenever possible.

31. $x^2 + 8x + 16$

32. $x^2 + 4x + 4$

33. $z^2 + 12z + 25$

34. $z^2 - 18z + 36$

35. $x^2 - 6x + 9$

36. $x^2 - 10x + 25$

37. $9y^2 + 6y + 1$

38. $16y^2 + 8y + 1$

39. $4z^2 - 4z + 1$

40. $25z^2 - 12z + 1$

41. $9t^2 + 16t + 4$

42. $4t^2 + 12t + 9$

43. $9x^2 + 30x + 25$

44. $25x^2 + 60x + 36$

45. $4a^2 - 36a + 81$

46. $9a^2 - 60a + 100$

47. $x^2 + 2xy + y^2$

48. $x^2 - 6xy + 9y^2$

49. $r^2 - 10rt + 25t^2$

50. $15r^2 + 10rt + t^2$

51. $4y^2 - 10yz + 9z^2$

52. $25y^2 - 20yz + 4z^2$

FACTORING SUMS AND DIFFERENCES OF TWO CUBES

Exercises 53–66: Factor.

53. $z^3 + 1$

54. $z^3 + 8$

55. $x^3 + 64$

56. $x^3 + 125$

57. $y^3 - 8$

58. $y^3 - 27$

59. $n^3 - 1$

60. $n^3 - 64$

61. $8x^3 + 1$

62. $27x^3 - 1$

63. $m^3 - 64n^3$

64. $m^3 + 8n^3$

65. $8x^3 + 125y^3$

66. $27x^3 + 64y^3$

GENERAL FACTORING USING SPECIAL METHODS

Exercises 67–84: Factor the expression completely.

67. $4x^2 - 16$

68. $12x^2 - 60x + 75$

69. $2y^2 - 28y + 98$

70. $y^3 - 9y$

71. $5z^3 + 40$

72. $4z^3 + 36z^2 + 100z$

73. $x^3y - xy^3$

74. $8m^3 - 8$

75. $2m^3 - 10m^2 + 18m$

76. $2a^3b - 18ab^3$

77. $700x^4 - 63x^2y^2$

78. $135r^3 - 5t^3$

79. $16a^3 + 2b^3$

80. $192x^2y^2 - 3y^4$

81. $4b^4 + 24b^3 + 36b^2$

82. $2y^4 + 24y^3 + 72y^2$

83. $500r^3 - 32t^3$

84. $8r^3 - 64t^3$

GEOMETRY

85. A square has an area of $4x^2 + 12x + 9$. Find the length of a side. Make a sketch of the square.

86. A square has an area of $9x^2 + 30x + 25$. Find the length of a side. Make a sketch of the square.

WRITING ABOUT MATHEMATICS

87. Explain how factoring $x^3 + y^3$ is different from factoring $x^3 - y^3$.

88. Using the techniques discussed in this section, can you factor the expression $4x^2 + 9y^2$ into two binomials? Explain your reasoning.

CHECKING BASIC CONCEPTS
SECTIONS 6.3 AND 6.4

1. Factor each trinomial.
   (a) $2x^2 - 5x - 12$ **(b)** $6x^2 + 17x - 14$

2. Factor completely when possible.
   (a) $3y^2 + 4y - 2$ **(b)** $6y^3 - 10y^2 - 4y$

3. Write a polynomial in factored form that represents the total area of the figure.

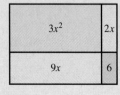

4. Factor each polynomial.
   (a) $z^2 - 64$ **(b)** $9r^2 - 4t^2$

5. Factor each trinomial.
   (a) $x^2 + 12x + 36$ **(b)** $9a^2 - 12ab + 4b^2$

6. Factor.
   (a) $m^3 - 27$ **(b)** $125n^3 + 27$

7. Factor completely.
   (a) $16x^2 - 4$ **(b)** $3y^4 + 24y$

6.5 SUMMARY OF FACTORING

Guidelines for Factoring Polynomials ▪ Factoring Polynomials

A LOOK INTO MATH ▷ A well-known line from the movie *Forrest Gump* states that "Life is like a box of choco-lates. You never know what you're gonna get." When we factor a polynomial, the techniques needed to complete the factoring are usually not spelled out as part of the problem— "You never know what you're gonna get." In this section we discuss general guidelines that can be used to factor polynomials.

Guidelines for Factoring Polynomials

The following guidelines can be used to factor polynomials in general.

FACTORING POLYNOMIALS

STEP 1: Factor out the greatest common factor, if possible.

STEP 2: A. If the polynomial has *four terms*, try factoring by grouping.
B. If the polynomial is a *binomial*, try one of the following.

1. $a^2 - b^2 = (a - b)(a + b)$ Difference of two squares
2. $a^3 - b^3 = (a - b)(a^2 + ab + b^2)$ Difference of two cubes
3. $a^3 + b^3 = (a + b)(a^2 - ab + b^2)$ Sum of two cubes

C. If the polynomial is a *trinomial*, check for a perfect square.

1. $a^2 + 2ab + b^2 = (a + b)^2$ Perfect square trinomial
2. $a^2 - 2ab + b^2 = (a - b)^2$ Perfect square trinomial

Otherwise, apply FOIL in reverse or apply grouping, as described in Sections 6.2 and 6.3.

STEP 3: Check to make sure that the polynomial is *completely* factored.

NOTE: Always perform Step 1 first. Factoring out the greatest common factor usually makes it easier to factor the resulting polynomial. After a polynomial has been factored, remember to perform Step 3 so that you are sure the given polynomial is completely factored.

Factoring Polynomials

In the first example, we apply Step 1 to a polynomial with a common factor.

EXAMPLE 1 Factoring out a common factor

Factor $5x^3 - 10x^2 + 15x$.

Solution
STEP 1: The greatest common factor is **5x**.

$$5x^3 - 10x^2 + 15x = 5x(x^2 - 2x + 3)$$

STEP 2C: The trinomial $x^2 - 2x + 3$ is prime and cannot be factored further.

STEP 3: The completely factored polynomial is $5x(x^2 - 2x + 3)$.   Now Try Exercise **9**

When factoring polynomials completely, it is often necessary to apply more than one factoring technique. In several of the next examples we factor polynomials that require more than one method of factoring.

EXAMPLE 2   **Factoring a difference of squares**

Factor $3x^4 - 48x^2$.

Solution
STEP 1: The greatest common factor is $3x^2$.

$$3x^4 - 48x^2 = 3x^2(x^2 - 16)$$

STEP 2B: The binomial $x^2 - 16$ can be factored as a difference of squares.

$$3x^2(x^2 - 16) = 3x^2(x - 4)(x + 4)$$

STEP 3: The completely factored polynomial is $3x^2(x - 4)(x + 4)$.   Now Try Exercise 37

EXAMPLE 3   **Factoring a perfect square trinomial**

Factor $36y^3 - 24y^2 + 4y$.

Solution
STEP 1: The greatest common factor is $4y$.

$$36y^3 - 24y^2 + 4y = 4y(9y^2 - 6y + 1)$$

STEP 2C: We can factor $9y^2 - 6y + 1$ as a perfect square trinomial.

$$4y(9y^2 - 6y + 1) = 4y(3y - 1)(3y - 1)$$

STEP 3: The completely factored polynomial is $4y(3y - 1)^2$.   Now Try Exercise 39

MAKING CONNECTIONS

Factoring Polynomials

We can often determine how a polynomial should be factored by considering the number of terms in the polynomial. This is summarized as follows.

Type of Polynomial	*Factoring Technique*
4-term Polynomial	Grouping
Trinomial	Perfect square trinomial
	FOIL in reverse or grouping
Binomial	Difference of squares
	Sum or difference of cubes

EXAMPLE 4   Factoring a sum of cubes

Factor $27z^3 + 64$.

Solution

STEP 1: There are no common factors.

STEP 2B: The binomial $27z^3 + 64$ can be written as $(3z)^3 + 4^3$ and can be factored as a sum of cubes.

$$27z^3 + 64 = (3z + 4)\left((3z)^2 - 3z \cdot 4 + 4^2\right)$$
$$= (3z + 4)(9z^2 - 12z + 16)$$

NOTE: The trinomial $9z^2 - 12z + 16$ cannot be factored further.

STEP 3: The completely factored polynomial is $(3z + 4)(9z^2 - 12z + 16)$.

Now Try Exercise 29

EXAMPLE 5   Factoring a trinomial

Factor $14x^4 + 7x^3 - 42x^2$.

Solution

STEP 1: The greatest common factor is $7x^2$.

$$14x^4 + 7x^3 - 42x^2 = 7x^2(2x^2 + x - 6)$$

STEP 2C: We can factor $2x^2 + x - 6$ using FOIL in reverse.

$$7x^2(2x^2 + x - 6) = 7x^2(2x - 3)(x + 2)$$

STEP 3: The completely factored polynomial is $7x^2(2x - 3)(x + 2)$.

Now Try Exercise 33

EXAMPLE 6   Factoring by grouping

Factor $15x^3 + 10x^2 - 60x - 40$.

Solution

STEP 1: The greatest common factor is 5.

$$15x^3 + 10x^2 - 60x - 40 = 5(3x^3 + 2x^2 - 12x - 8)$$

STEP 2A: Because the resulting polynomial has four terms, we apply grouping.

$$5(3x^3 + 2x^2 - 12x - 8) = 5\left((3x^3 + 2x^2) + (-12x - 8)\right) \quad \text{Associative property}$$
$$= 5\left(x^2(3x + 2) - 4(3x + 2)\right) \quad \text{Factor out common factors.}$$
$$= 5(x^2 - 4)(3x + 2) \quad \text{Factor out } (3x + 2).$$

STEP 2B: The binomial $x^2 - 4$ can now be factored as a difference of squares.

$$5(x^2 - 4)(3x + 2) = 5(x - 2)(x + 2)(3x + 2)$$

STEP 3: The completely factored polynomial is $5(x - 2)(x + 2)(3x + 2)$.

Now Try Exercise 31

EXAMPLE 7   **Factoring a polynomial having two variables**

Factor $18x^3y - 8xy^3$.

Solution

STEP 1: The greatest common factor is $2xy$.

$$18x^3y - 8xy^3 = 2xy(9x^2 - 4y^2)$$

STEP 2B: The binomial $9x^2 - 4y^2$ can be written as $(3x)^2 - (2y)^2$ and can be factored as a difference of squares.

$$9x^2 - 4y^2 = (3x - 2y)(3x + 2y)$$

STEP 3: The completely factored polynomial is $2xy(3x - 2y)(3x + 2y)$.

Now Try Exercise 53

EXAMPLE 8   **Applying several techniques**

Factor $3x^5 - 3x^3 - 24x^2 + 24$.

Solution

STEP 1: The greatest common factor is 3.

$$3x^5 - 3x^3 - 24x^2 + 24 = 3(x^5 - x^3 - 8x^2 + 8)$$

STEP 2A: The resulting four-term polynomial can be factored by grouping.

$$
\begin{aligned}
3(x^5 - x^3 - 8x^2 + 8) &= 3\big((x^5 - x^3) + (-8x^2 + 8)\big) && \text{Associative property}\\
&= 3\big(x^3(x^2 - 1) - 8(x^2 - 1)\big) && \text{Factor out common factors.}\\
&= 3(x^3 - 8)(x^2 - 1) && \text{Factor out } x^2 - 1.
\end{aligned}
$$

STEP 2B: Both binomials in this expression can be factored further. The binomial $x^3 - 8$ can be factored as a difference of cubes and the binomial $x^2 - 1$ can be factored as a difference of squares.

$$3(x^3 - 8)(x^2 - 1) = 3(x - 2)(x^2 + 2x + 4)(x - 1)(x + 1)$$

NOTE: The trinomial $x^2 + 2x + 4$ cannot be factored further.

STEP 3: The completely factored polynomial is $3(x - 2)(x^2 + 2x + 4)(x - 1)(x + 1)$.

Now Try Exercise 43

6.5   PUTTING IT ALL TOGETHER

The following table provides a summary of factoring rules that may be helpful when applying the guidelines for factoring polynomials presented in this section.

Concept	Explanation	Examples
Greatest Common Factor	Factor out the greatest common factor, or monomial, that occurs in each term.	$2x^2 - 4x + 10 = 2(x^2 - 2x + 5)$ $3x^3 + 6x = 3x(x^2 + 2)$ $7xy - x^2y = xy(7 - x)$

Concept	Explanation	Examples
Factoring by Grouping	Use the associative and distributive properties to factor a polynomial with four terms.	$\begin{aligned} x^3 - 3x^2 + 2x - 6 &= (x^3 - 3x^2) + (2x - 6) \\ &= x^2(x - 3) + 2(x - 3) \\ &= (x^2 + 2)(x - 3) \end{aligned}$
Factoring Binomials	Use the difference of squares, the difference of cubes, or the sum of cubes.	$\begin{aligned} 9x^2 - 4 &= (3x - 2)(3x + 2) \\ x^3 - 27 &= (x - 3)(x^2 + 3x + 9) \\ x^3 + 27 &= (x + 3)(x^2 - 3x + 9) \end{aligned}$
Factoring Trinomials	Use FOIL in reverse or grouping.	$x^2 + 5x - 6 = (x + 6)(x - 1)$ Check middle term: $-x + 6x = 5x$. $\begin{aligned} 4x^2 + 4x - 3 &= (4x^2 - 2x) + (6x - 3) \\ &= 2x(2x - 1) + 3(2x - 1) \\ &= (2x + 3)(2x - 1) \end{aligned}$

6.5 Exercises

MyMathLab   Math XL PRACTICE   WATCH   DOWNLOAD   READ   REVIEW

CONCEPTS

1. What do the letters GCF mean?

2. A good first step for factoring polynomials is to factor out the _____.

3. If a polynomial has four terms, what factoring method might be appropriate?

4. Can $x^2 + 1$ be factored? Explain.

5. Can $x^3 + 1$ be factored? Explain.

6. The last step for factoring is to be sure the polynomial is _____ factored.

WARM UP

Exercises 7–22: Factor completely, if possible.

7. $4x - 2$
8. $x^2 + 3x$
9. $2y^2 - 4y + 4$
10. $5y^2 - 25y + 10$
11. $z^2 - 4$
12. $9z^2 - 25$
13. $a^3 + 8$
14. $8a^3 - 1$
15. $4b^2 - 12b + 9$
16. $b^2 + 4b + 4$
17. $m^2 + 9$
18. $4m^2 + 49$
19. $x^3 - x^2 + 5x - 5$
20. $3x^3 + 6x^2 + x + 2$
21. $y^2 - 5y + 4$
22. $y^2 - 3y - 10$

GENERAL FACTORING

Exercises 23–60: Factor completely.

23. $x^3 + 4x^2 - 9x - 36$
24. $6x^2 - 19x + 15$
25. $8a^3 - 64$
26. $ab^2 - 4a$
27. $12x^4 - 18x^3 + 4x^2 - 6x$
28. $3x^2y + 24xy + 48y$
29. $54t^4 + 16t$
30. $3t^3 + 18t^2 - 48t$
31. $2r^3 + 6r^2 - 2r - 6$
32. $3r^4 + 3r^3 - 24r - 24$
33. $6z^4 - 21z^3 - 45z^2$
34. $3x^4y + 24xy^4$
35. $12b^4 - 10b^3 + 2b^2$
36. $6a^4b + 4a^3b + 18a^2b + 12ab$
37. $6y^2z - 24z^3$
38. $6y^3z - 48z^4$

39. $3x^2y - 30xy + 75y$ **40.** $8x^3 + y^3$

41. $27m^3 - 8n^3$ **42.** $45m^3 - 69m^2 + 12m$

43. $3x^5 - 12x^3 - 3x^2 + 12$

44. $8x^3 - 8$ **45.** $5a^2 - 27a - 18$

46. $2a^2 - 6ab + 3a - 9b$

47. $3rt^2 + 33rt + 90r$ **48.** $9t^2 + 24t + 16$

49. $9b^3 + 6b^2 + 12b + 8$ **50.** $5b^3 - 55b^2 - 60b$

51. $6n^3 + 2n^2 - 10n$ **52.** $7n^4 + 28n^3 - 63n^2$

53. $4x^2 - 36y^2$ **54.** $64x^2 - 25y^2$

55. $2a^3 - 16a^2 + 32a$ **56.** $24a^3 + 72a^2 + 54a$

57. $32xy^3 + 4x$ **58.** $24x^3 - 4x^2 - 160x$

59. $8b^4 + 24b^3 - 2b^2 - 6b$

60. $3z^3 - 6z^2 - 27z + 54$

GEOMETRY

61. *Dimensions of a Square*   If three identical squares have a total area of $27x^2 + 18x + 3$, find a possible length of one side of one of the squares.

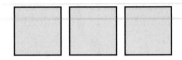

62. *Dimensions of a Cube*   If three identical cubes have a total volume of $3x^3 + 18x^2 + 36x + 24$, find a possible length of one side of one of the cubes.

WRITING ABOUT MATHEMATICS

63. Explain how the number of terms in a polynomial can help determine what method should be used to factor it.

64. Describe a method for determining whether a polynomial has been factored correctly.

6.6 SOLVING EQUATIONS BY FACTORING I (QUADRATICS)

The Zero-Product Property ■ Solving Quadratic Equations ■ Applications

A LOOK INTO MATH ▷

If a golf ball is hit upward at 132 feet per second, or 90 miles per hour, then its height h in feet above the ground after t seconds is given by $h = 132t - 16t^2$. The expression $132t - 16t^2$ is an example of a *quadratic polynomial*. To determine the elapsed time between when the ball is hit and when it strikes the ground (or when $h = 0$) we solve the *quadratic equation*

$$132t - 16t^2 = 0.$$

One method for solving this equation is by factoring. (See Example 4.) In this section we discuss how to use factoring to solve a variety of equations.

The Zero-Product Property

To solve equations we often use the **zero-product property**, which states that if the product of two numbers is 0, then at least one of the numbers must be 0.

ZERO-PRODUCT PROPERTY

For all real numbers a and b, if $ab = 0$, then $a = 0$ or $b = 0$ (or both).

NOTE: The zero-product property works only for 0. If $ab = 1$, then it does *not* follow that $a = 1$ or $b = 1$. For example, $a = \frac{1}{3}$ and $b = 3$ satisfy the equation $ab = 1$.

After factoring an expression, we can use the zero-product property to solve an equation. The left side of the equation

$$3t^2 - 9t = 0$$

may be factored to obtain

$$3t(t - 3) = 0.$$

Note that the product of $3t$ and $t - 3$ is 0. By the zero-product property, either

$$3t = 0 \quad \text{or} \quad t - 3 = 0.$$

Solving each equation for t results in

$$t = 0 \quad \text{or} \quad t = 3.$$

These values can be checked by substituting them into the given equation $3t^2 - 9t = 0$.

$$3(0)^2 - 9(0) = 0 \qquad \text{Let } t = 0. \text{ It checks.}$$
$$3(3)^2 - 9(3) = 0 \qquad \text{Let } t = 3. \text{ It checks.}$$

The t–values of 0 and 3 are called **zeros** of the polynomial $3t^2 - 9t$, because when either is substituted in this polynomial, the result is 0.

EXAMPLE 1   Applying the zero-product property

Solve each equation.
(a) $x(x - 1) = 0$
(b) $2z^2 = 0$
(c) $(t + 3)(t + 2) = 0$
(d) $x(x - 2)(2x + 1) = 0$

Solution
(a) By the zero-product property, $x(x - 1) = 0$ when $x = 0$ or $x - 1 = 0$. The solutions are 0 and 1.
(b) $2z^2 = 2 \cdot z \cdot z$ and $2 \neq 0$, so $2z^2 = 0$ when $z = 0$.
(c) $(t + 3)(t + 2) = 0$ implies that $t + 3 = 0$ or $t + 2 = 0$. The solutions to the equation are -3 and -2.
(d) We apply the zero-product property to $x(x - 2)(2x + 1) = 0$. Thus $x = 0$ or $x - 2 = 0$ or $2x + 1 = 0$. The solutions are $-\frac{1}{2}$, 0, and 2.

Now Try Exercises 12, 13, 17, 21

Solving Quadratic Equations

Any **quadratic polynomial** in the variable x can be written as $ax^2 + bx + c$ with $a \neq 0$. Any **quadratic equation** in the variable x can be written as $ax^2 + bx + c = 0$ with $a \neq 0$. This form of quadratic equation is called the **standard form** of a quadratic equation. For example, $x^2 + 2x - 3$ is a quadratic polynomial and $x^2 + 2x - 3 = 0$ is a quadratic equation written in standard form. When a quadratic polynomial in the variable x is in standard form and we read it from left to right, the terms contain descending powers of x. In other words, the first term contains x^2, the second term contains x, and the third term is a constant (the exponent on x is 0).

To solve a quadratic equation we often use factoring and the zero-product property. This method is summarized by the following steps. Although it is not necessary to label each step in the solution to a quadratic equation, it is important to keep these steps in mind.

SOLVING QUADRATIC EQUATIONS

To solve a quadratic equation by factoring, follow these steps.

STEP 1: If necessary, write the equation in standard form as $ax^2 + bx + c = 0$.

STEP 2: Factor the left side of the equation using any method.

STEP 3: Apply the zero-product property.

STEP 4: Solve each of the resulting equations. Check any solutions.

EXAMPLE 2   Solving equations by factoring

Solve each quadratic equation. Check your answers.
(a) $x^2 + 2x = 0$ (b) $y^2 = 16$ (c) $z^2 - 3z + 2 = 0$ (d) $2x^2 = 5 - 9x$

Solution

(a) Because $x^2 + 2x = 0$ is in standard form, we begin by factoring out the GCF of x.

$$x^2 + 2x = 0 \qquad \text{Given equation}$$
$$x(x + 2) = 0 \qquad \text{Factor out } x. \text{ (Step 2)}$$
$$x = 0 \quad \text{or} \quad x + 2 = 0 \qquad \text{Zero-product property (Step 3)}$$
$$x = 0 \quad \text{or} \quad x = -2 \qquad \text{Solve for } x. \text{ (Step 4)}$$

To check these values, substitute -2 and 0 for x in the given equation.

$$(-2)^2 + 2(-2) \overset{?}{=} 0 \qquad (0)^2 + 2(0) \overset{?}{=} 0 \quad \text{Substitute } -2 \text{ and } 0.$$
$$0 = 0 \qquad\qquad\qquad 0 = 0 \quad \text{Both answers check.}$$

Therefore the solutions are -2 and 0.

(b) To write $y^2 = 16$ in standard form we begin by subtracting 16 from each side to obtain 0 on the right side.

$$y^2 = 16 \qquad \text{Given equation}$$
$$y^2 - 16 = 0 \qquad \text{Subtract 16. (Step 1)}$$
$$(y - 4)(y + 4) = 0 \qquad \text{Difference of squares (Step 2)}$$
$$y - 4 = 0 \quad \text{or} \quad y + 4 = 0 \qquad \text{Zero-product property (Step 3)}$$
$$y = 4 \quad \text{or} \quad y = -4 \qquad \text{Solve for } y. \text{ (Step 4)}$$

To check these values, substitute -4 and 4 for y in the given equation.

$$(-4)^2 \overset{?}{=} 16 \qquad (4)^2 \overset{?}{=} 16 \qquad \text{Substitute } -4 \text{ and } 4.$$
$$16 = 16 \qquad 16 = 16 \qquad \text{Both answers check.}$$

The solutions are -4 and 4.

(c) We begin by factoring the left side of the equation, $z^2 - 3z + 2$.

$$z^2 - 3z + 2 = 0 \qquad \text{Given equation}$$
$$(z - 1)(z - 2) = 0 \qquad \text{Factor. (Step 2)}$$
$$z - 1 = 0 \quad \text{or} \quad z - 2 = 0 \qquad \text{Zero-product property (Step 3)}$$
$$z = 1 \quad \text{or} \quad z = 2 \qquad \text{Solve for } z. \text{ (Step 4)}$$

To check these values, substitute 1 and 2 for z in the given equation.

$$1^2 - 3(1) + 2 \stackrel{?}{=} 0 \qquad 2^2 - 3(2) + 2 \stackrel{?}{=} 0 \qquad \text{Substitute 1 and 2.}$$
$$0 = 0 \qquad\qquad\qquad 0 = 0 \qquad \text{Both answers check.}$$

The solutions are 1 and 2.

(d) We write $2x^2 = 5 - 9x$ in standard form by adding -5 and $9x$ to each side.

$$2x^2 = 5 - 9x \qquad\qquad \text{Given equation}$$
$$2x^2 + 9x - 5 = 0 \qquad\qquad \text{Add } -5 \text{ and } 9x. \text{ (Step 1)}$$
$$(2x - 1)(x + 5) = 0 \qquad\qquad \text{Factor. (Step 2)}$$
$$2x - 1 = 0 \quad \text{or} \quad x + 5 = 0 \qquad \text{Zero-product property (Step 3)}$$
$$x = \frac{1}{2} \quad \text{or} \quad x = -5 \qquad \text{Solve for } x. \text{ (Step 4)}$$

To check these values, substitute -5 and $\frac{1}{2}$ for x in the given equation.

$$2(-5)^2 \stackrel{?}{=} 5 - 9(-5) \qquad 2\left(\frac{1}{2}\right)^2 \stackrel{?}{=} 5 - 9\left(\frac{1}{2}\right) \qquad \text{Substitute } -5 \text{ and } \tfrac{1}{2}.$$
$$50 = 50 \qquad\qquad\qquad \frac{1}{2} = \frac{1}{2} \qquad \text{Both answers check.}$$

The solutions are -5 and $\frac{1}{2}$.

Now Try Exercises 25, 33, 41, 47

EXAMPLE 3   Solving an equation by factoring

Solve $6x^2 - x = 12$.

Solution

We cannot solve the equation $6x^2 - x = 12$ by factoring out the common factor of x in $6x^2 - x$ and setting each factor equal to 12. Instead we apply the zero-product property by first writing the given equation in the standard form: $ax^2 + bx + c = 0$.

$$6x^2 - x = 12 \qquad\qquad \text{Given equation}$$
$$6x^2 - x - 12 = 0 \qquad\qquad \text{Subtract 12. (Step 1)}$$
$$(2x - 3)(3x + 4) = 0 \qquad\qquad \text{Factor. (Step 2)}$$
$$2x - 3 = 0 \quad \text{or} \quad 3x + 4 = 0 \qquad \text{Zero-product property (Step 3)}$$
$$x = \frac{3}{2} \quad \text{or} \quad x = -\frac{4}{3} \qquad \text{Solve for } x. \text{ (Step 4)}$$

The solutions are $-\frac{4}{3}$ and $\frac{3}{2}$.

Now Try Exercise 49

MAKING CONNECTIONS

Equations and Expressions

The words "equation" and "expression" occur frequently in mathematics. However, they are *not* interchangeable. We often want to *solve* equations to find the values of the variable that make the equation true. We *factor* and *simplify* expressions. An equation is a statement that two expressions are equal. For example, $2x^2 - 3 = 5x + 1$ is an equation where $2x^2 - 3$ and $5x + 1$ are each expressions

Applications

▶ **REAL-WORLD CONNECTION**   To solve application problems we often need to solve equations. The next example illustrates how to solve the application presented in the introduction to this section.

EXAMPLE 4   Modeling the flight of a golf ball

If a golf ball is hit upward at 132 feet per second, or 90 miles per hour, then its height h in feet after t seconds is $h = 132t - 16t^2$. After how long does the golf ball strike the ground?

Solution
The golf ball strikes the ground when its height is 0.

$$132t - 16t^2 = 0 \qquad \text{Let } h = 0.$$
$$4t(33 - 4t) = 0 \qquad \text{Factor out } 4t.$$
$$4t = 0 \quad \text{or} \quad 33 - 4t = 0 \qquad \text{Zero-product property}$$
$$t = 0 \quad \text{or} \quad -4t = -33 \qquad \text{Divide by 4; subtract 33.}$$
$$t = 0 \quad \text{or} \quad t = \frac{33}{4} \qquad \text{Solve for } t.$$

The ball strikes the ground after $\frac{33}{4} = 8.25$ seconds. The solution of 0 is not used in this problem because it corresponds to the time when the ball is hit.   Now Try Exercise 65(a)

▶ **REAL-WORLD CONNECTION**   When you try to stop a car, the greater the speed the greater is the stopping distance. In fact, if you drive twice as fast, the braking distance will be about four times as much. And if you drive three times faster, the braking distance will be about nine times as much.

EXAMPLE 5   Modeling braking distance

The braking distance D in feet required to stop a car traveling at x miles per hour on dry, level pavement can be approximated by $D = \frac{1}{11}x^2$.   (**Source:** L. Haefner.)
(a) Calculate the braking distance for a car traveling 70 miles per hour.
(b) If the braking distance is 44 feet, calculate the speed of the car.
(c) If you have a calculator available, use it to solve part (b) numerically with a table of values.

Solution
(a) If $x = 70$, then $D = \frac{1}{11}(70)^2 = \frac{4900}{11} \approx 445$ feet.
(b) *Symbolic Solution*   Let $D = 44$ in the given equation and solve.

$$\frac{1}{11}x^2 = 44 \qquad \text{Let } D = 44.$$

$$\frac{1}{11}x^2 - 44 = 0 \qquad \text{Subtract 44.}$$

$$x^2 - 484 = 0 \qquad \text{Multiply by 11.}$$

$$(x - 22)(x + 22) = 0 \qquad \text{Difference of two squares}$$
$$\qquad \qquad \qquad \qquad \qquad \text{\textbf{Note:} } 22^2 = 484$$

$$x - 22 = 0 \quad \text{or} \quad x + 22 = 0 \qquad \text{Zero-product property}$$
$$x = 22 \quad \text{or} \quad x = -22 \qquad \text{Solve for } x.$$

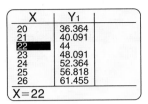

X	Y₁
20	36.364
21	40.091
22	44
23	48.091
24	52.364
25	56.818
26	61.455

X=22

Figure 6.6

The car is traveling at (approximately) 22 miles per hour. (Note that $x = -22$ has no physical meaning in this problem.)

(c) Numerical Solution Let $Y_1 = X^2/11$ and make a table of values. Scroll through the table, as shown in Figure 6.6, to where $x = 22$ when $y_1 = 44$. Thus the solution is 22 miles per hour.   Now Try Exercise 67

▶ **REAL-WORLD CONNECTION** Quadratic equations sometimes are used in applications involving rectangular shapes. The next application involves finding the dimensions of a digital photograph.

EXAMPLE 6 Finding the dimensions of a digital photograph

A small digital photograph is 20 pixels longer than it is wide, as illustrated in Figure 6.7. It has a total of 2400 pixels. Find the dimensions of this photograph.

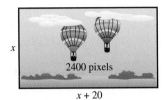

2400 pixels

x

$x + 20$

Figure 6.7 Dimensions of a Photograph

Solution
From Figure 6.7 the rectangular photograph has an area of 2400 pixels.

$$x(x + 20) = 2400 \qquad \text{Area} = \text{width} \times \text{length}$$
$$x^2 + 20x = 2400 \qquad \text{Distributive property}$$
$$x^2 + 20x - 2400 = 0 \qquad \text{Subtract 2400.}$$
$$(x - 40)(x + 60) = 0 \qquad \text{Factor.}$$
$$x - 40 = 0 \quad \text{or} \quad x + 60 = 0 \qquad \text{Zero-product property}$$
$$x = 40 \quad \text{or} \quad x = -60 \qquad \text{Solve for } x.$$

CRITICAL THINKING

Are the two solutions to $x(2x + 1) = 1$ found by letting $x = 1$ or $2x + 1 = 1$? Explain your reasoning. What are the solutions to this equation?

The valid solution is 40. Thus the dimensions of the photograph are 40 pixels by $40 + 20 = 60$ pixels.   Now Try Exercise 71

6.6 PUTTING IT ALL TOGETHER

In this section we discussed solving quadratic equations by using factoring and the zero-product property. These topics are summarized in the following table.

Concept	Explanation	Examples
Zero-Product Property	If the product of two or more expressions is 0, then at least one of the expressions must equal 0.	$ab = 0$ implies that $a = 0$ or $b = 0$. $x(x + 1)$ implies that $x = 0$ or $x + 1 = 0$. $z(z - 1)(z + 2) = 0$ implies that $z = 0$ or $z - 1 = 0$ or $z + 2 = 0$.

continued on next page

continued from previous page

Concept	Explanation	Examples
Solving Quadratic Equations by Factoring	1. Write the equation as $$ax^2 + bx + c = 0.$$ 2. Factor the left side of this equation. 3. Apply the zero-product property. 4. Solve each resulting equation. Check any solutions.	$2x^2 + 11x = 6$ $2x^2 + 11x - 6 = 0$ Step 1 $(2x - 1)(x + 6) = 0$ Step 2 $2x - 1 = 0$ or $x + 6 = 0$ Step 3 $x = \dfrac{1}{2}$ or $x = -6$ Step 4

6.6 Exercises

MyMathLab   Math XL PRACTICE   WATCH   DOWNLOAD   READ   REVIEW

CONCEPTS

1. If $ab = 0$, then either $a =$ _____ or $b =$ _____.

2. Can the zero-product property be used to state that if $(x - 1)(x - 2) = 3$, then either $x - 1 = 3$ or $x - 2 = 3$? Explain your answer.

3. If $2x(x + 6) = 0$, then either _____ or _____.

4. What is a good first step when you are solving the equation $4x^2 + 1 = 4x$ by factoring?

5. What is the next step when you are solving the equation $(x + 5)(x - 4) = 0$?

6. Factoring is an important method for _____ equations.

7. Any quadratic equation in the variable x can be written in standard form as _____.

8. Standard form for $x^2 + 1 = 6x$ is _____.

9. Because $2(4) - 8 = 0$, the value 4 is called a _____ of the polynomial $2x - 8$.

10. What is the zero of the polynomial $3x - 6$?

ZERO-PRODUCT PROPERTY

Exercises 11–22: Solve the equation.

11. $xy = 0$

12. $m^2 = 0$

13. $2x(x + 8) = 0$

14. $x(x + 10) = 0$

15. $(y - 1)(y - 2) = 0$

16. $(y + 4)(y - 3) = 0$

17. $(2z - 1)(4z - 3) = 0$

18. $(6z + 5)(z - 7) = 0$

19. $(1 - 3n)(3 - 7n) = 0$

20. $(5 - n)(5 + n) = 0$

21. $x(x - 5)(x - 8) = 0$

22. $x(x + 1)(x - 6) = 0$

SOLVING QUADRATIC EQUATIONS

Exercises 23–58: Solve and check.

23. $x^2 - x = 0$

24. $2x^2 + 4x = 0$

25. $z^2 - 5z = 0$

26. $6z^2 - 3z = 0$

27. $10y^2 + 15y = 0$

28. $2y^2 + 3y = 0$

29. $x^2 - 1 = 0$

30. $x^2 - 9 = 0$

31. $4n^2 - 1 = 0$

32. $9n^2 - 4 = 0$

33. $z^2 + 3z + 2 = 0$

34. $z^2 - 2z - 3 = 0$

35. $x^2 - 12x + 35 = 0$

36. $x^2 - x - 20 = 0$

37. $2b^2 + 3b - 2 = 0$

38. $3b^2 + b - 2 = 0$

39. $6y^2 + 19y + 10 = 0$

40. $4y^2 - 25y - 21 = 0$

41. $x^2 = 25$

42. $x^2 = 81$

43. $t^2 = 5t$

44. $10t^2 = -5t$

45. $3m^2 = -9m$

46. $4m^2 = 9$

47. $x^2 = 5x + 6$

48. $2x^2 + 3x = 14$

49. $12z^2 + 11z = 15$

50. $12z^2 = 5 - 4z$

51. $t(t + 1) = 2$

52. $t(t - 7) = -12$

53. $x(2x + 5) = 3$

54. $x(3x + 2) = 5$

55. $12x^2 + 12x = -3$

56. $20x^2 - 8 = 6x$

57. $30y^2 + 50y + 20 = 0$

58. $30y^2 - 25y + 5 = 0$

GEOMETRY

59. *Dimensions of a Square*   A square has an area of 144 square feet. Find the length of a side.

60. *Dimensions of a Cube*   A cube has a surface area of 96 square feet. Find the length of a side.

61. *Radius of a Circle*   The numerical difference between the area and the circumference of a circle is 8π. Find the radius of the circle. (*Hint:* First factor out π in your equation.)

62. *Dimensions of a Rectangle*   A rectangle is 5 feet longer than it is wide and has an area of 126 square feet. What are its dimensions?

Exercises 63 and 64: Pythagorean Theorem   Suppose that a right triangle has legs a and b and hypotenuse c, as illustrated in the figure. Then these values satisfy $a^2 + b^2 = c^2$.

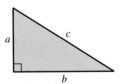

Use the Pythagorean Theorem as described above to find the value of x in the figure.

63.

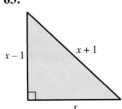

64.

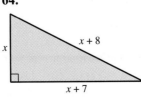

APPLICATIONS

65. *Flight of a Golf Ball*   (Refer to Example 4.) The height h in feet of a golf ball after t seconds is given by $h = 96t - 16t^2$.

    (a) How long did it take for the ball to hit the ground?

    (b) Make a table of h for $t = 0, 1, 2, \ldots, 6$. After how many seconds did the golf ball reach its maximum height?

66. *Flight of a Baseball*   The height h in feet of a baseball after t seconds is given by $h = -16t^2 + 88t + 3$. At what values of t is the height of the baseball 75 feet?

67. *Braking Distance*   (Refer to Example 5.) The braking distance D in feet required to stop a car traveling x miles per hour on dry, level pavement can be approximated by $D = \frac{1}{11}x^2$.

    (a) Calculate the braking distance for 30 miles per hour and 60 miles per hour. How do your answers compare?

    (b) If the braking distance is 33 feet, estimate the speed of the car.

    (c) If you have a calculator, use it to solve part (b) numerically. Do your answers agree?

68. *Braking Distance*   The braking distance D in feet required to stop a car traveling x miles per hour on wet, level pavement is approximated by $D = \frac{1}{9}x^2$.

    (a) Calculate the braking distance for 36 miles per hour and 72 miles per hour. How do your answers compare?

    (b) If the braking distance is 49 feet, estimate the speed of the car.

    (c) If you have a calculator, use it to solve part (b) numerically. Do your answers agree?

69. *Women in the Workforce*   The number of women W in the workforce in millions can be estimated by the equation $W = \frac{19}{3125}x^2 + \frac{11}{2}$, where $x = 0$ corresponds to 1900, $x = 10$ to 1910, and so on until $x = 100$ corresponds to 2000. (*Source:* U.S. Census Bureau.)

(a) How many women were in the workforce in 1930 and in 2000?

(b) Use a table of values to estimate the year in which 45 million women were in the workforce.

70. *AIDS Deaths*   The cumulative number of AIDS deaths D from 1993 to 2003 can be modeled by

$$D = -1808x^2 + 71{,}154x + 359{,}262,$$

where $x = 1$ corresponds to 1993, $x = 2$ to 1994, and so on until $x = 11$ corresponds to 2003. (*Source:* U.S. Department of Health and Human Services.)

(a) Estimate the cumulative number of AIDS deaths in 2002.

(b) Use a table of values to estimate the year in which the cumulative number of AIDS deaths reached 721,000.

71. *Digital Photographs*   (Refer to Example 6.) A digital photograph is 10 pixels longer than it is wide and has a total area of 2000 pixels. Find the dimensions of this rectangular photograph.

72. *Dimensions of a Building*   The rectangular floor of a shed has a length 4 feet longer than its width, and its area is 140 square feet. Let x be the width of the floor.

(a) Write a quadratic equation whose solution gives the width of the floor.

(b) Solve this equation.

WRITING ABOUT MATHEMATICS

73. List four steps for solving a quadratic equation by factoring.

74. Explain why factoring is important.

CHECKING BASIC CONCEPTS

SECTIONS 6.5 AND 6.6

1. Factor out the greatest common factor.
 (a) $9a^2 - 18a + 27$ (b) $7xy^2 + 28x$

2. Factor completely.
 (a) $6z^4 - 28z^3 + 16z^2$ (b) $2r^2t^2 - 18r^2$

3. Factor completely.
 (a) $36x^3 - 48x^2 + 16x$
 (b) $24b^3 - 81$

4. Solve each quadratic equation.
 (a) $4y^2 - 6y = 0$ (b) $5z^2 + 2z = 3$

5. Solve $x^2 + 2x - 3 = 0$ symbolically and numerically with a table of values.

6. If a golf ball is hit upward at 60 miles per hour, then its height h in feet after t seconds is given by $h = 88t - 16t^2$. Use factoring to determine when the golf ball strikes the ground.

6.7 SOLVING EQUATIONS BY FACTORING II (HIGHER DEGREE)

Polynomials Having Common Factors ■ Special Types of Polynomials

A LOOK INTO MATH ▷

In this section we discuss factoring polynomials having higher degree. Polynomials of degree 2 or higher are often used in applications. For example, the polynomial

$$0.0013x^3 - 0.085x^2 + 1.6x + 12$$

models natural gas consumption in the United States (in trillions of cubic feet) where $x = 0$ corresponds to 1960, $x = 10$ to 1970, and so on until $x = 40$ corresponds to 2000, as shown in Figure 6.8. In Exercises 75–78 we discuss further the consumption of natural gas. (**Source:** Department of Energy.)

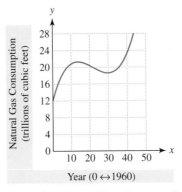

Figure 6.8 Natural Gas Consumption

Polynomials Having Common Factors

The first step in factoring a polynomial is to factor out the greatest common factor (GCF). For example, to factor $x^3 - x$ we start by factoring out the GCF, which is x. The resulting expression can be factored as the difference of two squares.

$$x^3 - x = x(x^2 - 1) \qquad \text{Factor out } x.$$
$$= x(x - 1)(x + 1) \qquad \text{Difference of two squares}$$

EXAMPLE 1   **Factoring trinomials with common factors**

Factor each trinomial completely.
(a) $-4x^2 + 28x - 40$ **(b)** $10x^3 + 28x^2 - 6x$

Solution
(a) Each term in $-4x^2 + 28x - 40$ has a factor of -4, so we start by factoring out -4. (In this example, we factor out the *negative* of the GCF to obtain a positive leading coefficient.)

$$-4x^2 + 28x - 40 = -4(x^2 - 7x + 10) \qquad \text{Factor out } -4.$$
$$= -4(x - 5)(x - 2) \qquad \text{Factor the trinomial.}$$

(b) We start by factoring out the GCF for $10x^3 + 28x^2 - 6x$, which is $2x$.

$$10x^3 + 28x^2 - 6x = 2x(5x^2 + 14x - 3) \qquad \text{Factor out } 2x.$$
$$= 2x(5x - 1)(x + 3) \qquad \text{Factor the trinomial.}$$

Now Try Exercises 11, 19

Many equations involving higher degree polynomials can be solved using the four-step process discussed in Section 6.6. It is important to remember that the zero-product property applies to *all* factors that contain the variable. In the next example we apply the zero-product property to three factors, resulting in three solutions.

EXAMPLE 2   **Solving polynomial equations**

Solve each equation.
(a) $x^3 - x^2 - 6x = 0$ **(b)** $4x^4 + 10x^3 = 6x^2$

Solution

(a) We start by factoring out the GCF, which is x.

$$x^3 - x^2 - 6x = 0 \qquad \text{Given equation}$$
$$x(x^2 - x - 6) = 0 \qquad \text{Factor out } x.$$
$$x(x - 3)(x + 2) = 0 \qquad \text{Factor the trinomial.}$$
$$x = 0 \quad \text{or} \quad x - 3 = 0 \quad \text{or} \quad x + 2 = 0 \qquad \text{Zero-product property}$$
$$x = 0 \quad \text{or} \quad x = 3 \quad \text{or} \quad x = -2 \qquad \text{Solve for } x.$$

The solutions are $-2, 0,$ and 3.

(b) We start by subtracting $6x^2$ from each side to obtain 0 on one side of the equation.

$$4x^4 + 10x^3 = 6x^2 \qquad \text{Given equation}$$
$$4x^4 + 10x^3 - 6x^2 = 0 \qquad \text{Subtract } 6x^2.$$
$$2x^2(2x^2 + 5x - 3) = 0 \qquad \text{Factor out the GCF, } 2x^2.$$
$$2x^2(2x - 1)(x + 3) = 0 \qquad \text{Factor the trinomial.}$$
$$2x \cdot x = 0 \quad \text{or} \quad 2x - 1 = 0 \quad \text{or} \quad x + 3 = 0 \qquad \text{Zero-product property}$$
$$x = 0 \quad \text{or} \quad x = \frac{1}{2} \quad \text{or} \quad x = -3 \qquad \text{Solve for } x.$$

The solutions are $-3, 0,$ and $\frac{1}{2}$.

Now Try Exercises 55, 61

▶ **REAL-WORLD CONNECTION**   The corners of a square piece of metal are cut out to form a box, as shown in Figure 6.9. This square piece of metal has sides with length 10 inches, and the cutout corners are squares with length x. The outside surface area A of this box, including the bottom and the sides but *not* the top, is $A = 100 - 4x^2$. (See Critical Thinking in the margin on the next page.)

CRITICAL THINKING

Calculate the outside surface area A of the box shown in Figure 6.9 two different ways.

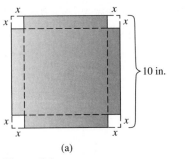

Figure 6.9

EXAMPLE 3   **Finding a dimension of a box**

Find the value of x in Figure 6.9 if its outside surface area A is 84 square inches.

Solution
Let $A = 84$ in the equation $A = 100 - 4x^2$ and solve for x.

$$100 - 4x^2 = 84 \qquad A = 84$$
$$16 - 4x^2 = 0 \qquad \text{Subtract 84.}$$
$$4(4 - x^2) = 0 \qquad \text{Factor out 4.}$$
$$4(2 - x)(2 + x) = 0 \qquad \text{Difference of squares}$$
$$2 - x = 0 \quad \text{or} \quad 2 + x = 0 \qquad \text{Zero-product property}$$
$$x = 2 \quad \text{or} \quad x = -2 \qquad \text{Solve for } x.$$

If squares measuring 2 inches on a side are cut out of each corner, the surface area of the box is 84 square inches. (Note that $x = -2$ has no physical meaning in this problem.)

Now Try Exercise 73

Special Types of Polynomials

Some types of polynomials of higher degree can be factored by using methods that we have already presented.

EXAMPLE 4   **Factoring higher degree polynomials**

Factor each polynomial completely.
(a) $x^4 - 16$ **(b)** $y^4 + 5y^2 + 4$ **(c)** $x^4 + 2x^2y^2 + y^4$ **(d)** $r^4 - t^4$

Solution
(a) We view this polynomial as the difference of two squares, where $x^4 = (x^2)^2$. Then we factor twice.

$$x^4 - 16 = (x^2)^2 - 4^2 \qquad \text{Rewrite.}$$
$$= (x^2 - 4)(x^2 + 4) \qquad \text{Difference of squares}$$
$$= (x - 2)(x + 2)(x^2 + 4) \qquad \text{Difference of squares}$$

Note that $x^2 + 4$ does not factor.

(b) Because $a^2 + 5a + 4 = (a + 1)(a + 4)$, we let $a = y^2$ and then factor the given trinomial.

$$y^4 + 5y^2 + 4 = (y^2)^2 + 5(y^2) + 4$$
$$= (y^2 + 1)(y^2 + 4)$$

Note that neither $y^2 + 1$ nor $y^2 + 4$ can be factored further.

(c) Because $a^2 + 2ab + b^2 = (a + b)^2$, we let $a = x^2$ and $b = y^2$ and then factor the given trinomial.

$$x^4 + 2x^2y^2 + y^4 = (x^2)^2 + 2x^2y^2 + (y^2)^2$$
$$= (x^2 + y^2)^2$$

(d) Because $a^2 - b^2 = (a - b)(a + b)$, we let $a = r^2$ and $b = t^2$ and then factor the given binomial.

$$r^4 - t^4 = (r^2)^2 - (t^2)^2$$
$$= (r^2 - t^2)(r^2 + t^2)$$
$$= (r - t)(r + t)(r^2 + t^2)$$

Note that $r^2 + t^2$ cannot be factored. **Now Try Exercises** 35, 41, 43

EXAMPLE 5 Solving an equation

Solve $x^5 - 81x = 0$.

Solution
We start by factoring out the common factor of x.

$$x^5 - 81x = 0 \qquad \text{Given equation}$$
$$x(x^4 - 81) = 0 \qquad \text{Factor out } x.$$
$$x(x^2 - 9)(x^2 + 9) = 0 \qquad \text{Difference of two squares}$$
$$x(x - 3)(x + 3)(x^2 + 9) = 0 \qquad \text{Difference of two squares}$$
$$x = 0 \ \text{ or } \ x - 3 = 0 \ \text{ or } \ x + 3 = 0 \ \text{ or } \ x^2 + 9 = 0 \qquad \text{Zero-product property}$$
$$x = 0 \ \text{ or } \ x = 3 \ \text{ or } \ x = -3 \qquad \text{Solve for } x.$$

Note that $x^2 + 9 = 0$ has no real number solutions because the square of a number plus 9 is never 0. The solutions are $-3, 0,$ and 3. **Now Try Exercise** 59

6.7 PUTTING IT ALL TOGETHER

The following table summarizes factoring higher degree polynomials and solving higher degree polynomial equations.

Concept	Explanation	Examples
Common Factors	A first step when factoring polynomials is to factor out the GCF.	$x^3 - 4x^2 - 5x = x(x^2 - 4x - 5)$ $= x(x - 5)(x + 1)$ $4x^4 - 16x^2 = 4x^2(x^2 - 4)$ $= 4x^2(x - 2)(x + 2)$

Concept	Explanation	Examples
Factoring Trinomials	Some higher degree trinomials can be factored by using the same methods that we used to factor quadratic trinomials. **NOTE:** $a^2 + b^2$ cannot be factored.	$x^4 + 6x^2 + 5 = (x^2 + 5)(x^2 + 1)$ $4y^4 - 25 = (2y^2 - 5)(2y^2 + 5)$ $2z^4 - 32 = 2(z^4 - 16)$ $\qquad = 2(z^2 - 4)(z^2 + 4)$ $\qquad = 2(z - 2)(z + 2)(z^2 + 4)$
Solving Equations by Factoring	Use factoring and the zero-product property to solve polynomial equations.	$3x^3 - 12x^2 + 9x = 0$ $3x(x^2 - 4x + 3) = 0$ $3x(x - 3)(x - 1) = 0$ $3x = 0 \ $ or $\ x - 3 = 0 \ $ or $\ x - 1 = 0$ $x = 0 \ $ or $\ x = 3 \ $ or $\ x = 1$ The solutions are 0, 1, and 3.

6.7 Exercises

CONCEPTS

1. When you are factoring polynomials, a good first step is to factor out the _____.

2. When you are solving an equation by factoring, the _____ property is used.

3. The zero-product property applies to all _____ containing the variable.

4. Because $a^2 - 2ab + b^2 = (a - b)^2$, it follows that $x^4 - 2x^2y^2 + y^4 =$_____.

5. Because $x^2 + 3x + 2 = (x + 1)(x + 2)$, it follows that $z^4 + 3z^2 + 2 =$ _____.

6. Is $x^4 - 1 = (x^2 - 1)(x^2 + 1)$ factored completely? Explain.

7. When you are solving the equation $x^3 = x$, what is a good first step?

8. When you are solving $x^4 - x^2 = 0$, what is a good first step?

FACTORING POLYNOMIALS

Exercises 9–20: Factor the polynomial completely.

9. $5x^2 - 5x - 30$

10. $3x^2 - 15x + 12$

11. $-4y^2 - 32y - 48$

12. $-7y^2 + 14y + 21$

13. $-20z^2 - 110z - 50$

14. $-12z^2 - 54z + 30$

15. $60 - 64t - 28t^2$

16. $18 - 45t - 27t^2$

17. $r^3 - r$

18. $r^3 + 2r^2 - 3r$

19. $3x^3 + 3x^2 - 18x$

20. $6x^3 - 26x^2 - 20x$

Exercises 21–46: Factor completely.

21. $72z^3 + 12z^2 - 24z$

22. $6z^3 - 4z^2 - 42z$

23. $x^4 - 4x^2$

24. $4x^4 - 36x^2$

25. $t^4 + t^3 - 2t^2$

26. $t^4 + 5t^3 - 24t^2$

27. $x^4 - 5x^2 + 6$

28. $x^4 - 3x^2 - 10$

29. $2x^4 + 7x^2 + 3$

30. $3x^4 - 8x^2 + 5$

31. $y^4 + 6y^2 + 9$

32. $y^4 - 10y^2 + 25$

33. $x^4 - 9$

34. $x^4 - 25$

35. $x^4 - 81$

36. $4x^4 - 64$

37. $z^5 + 2z^4 + z^3$

38. $6z^5 - 47z^4 + 35z^3$

39. $2x^2 + xy - y^2$

40. $2x^2 + 5xy + 2y^2$

41. $a^4 - 2a^2b^2 + b^4$　　**42.** $a^3 + 2a^2b + ab^2$

43. $x^3 - xy^2$　　**44.** $2x^2y - 2y^3$

45. $4x^3 + 4x^2y + xy^2$　　**46.** $x^2y - 6xy^2 + 9y^3$

SOLVING EQUATIONS

Exercises 47–50: Do the following.

47. (a) Factor $x^3 - 4x$.
　　(b) Solve $x^3 - 4x = 0$.

48. (a) Factor $4x^3 - 16x$.
　　(b) Solve $4x^3 - 16x = 0$.

49. (a) Factor $2y^3 - 6y^2 - 36y$.
　　(b) Solve $2y^3 - 6y^2 - 36y = 0$.

50. (a) Factor $z^4 - 13z^2 + 36$.
　　(b) Solve $z^4 - 13z^2 + 36 = 0$.

Exercises 51–72: Solve.

51. $3x^2 + 33x + 72 = 0$　　**52.** $4x^2 - 16x - 20 = 0$

53. $25x^2 = 50x + 75$　　**54.** $10x^2 = 20x + 80$

55. $y^3 - 3y^2 - 4y = 0$　　**56.** $y^3 - 3y^2 + 2y = 0$

57. $3z^3 + 6z^2 = 72z$　　**58.** $4z^3 = 4z^2 + 24z$

59. $x^4 - 36x^2 = 0$　　**60.** $4x^4 = 100x^2$

61. $r^4 + 6r^3 = 7r^2$　　**62.** $r^4 + 30r^2 = 11r^3$

63. $x^4 - 13x^2 = -36$　　**64.** $x^4 - 17x^2 + 16 = 0$

65. $x^4 + 1 = 2x^2$　　**66.** $x^4 - 8x^2 + 16 = 0$

67. $a^4 = 81$　　**68.** $b^3 = -8$

69. $x^3 - 2x^2 - x + 2 = 0$

70. $x^3 - x^2 + 4x - 4 = 0$

71. $x^3 - 5x^2 + x - 5 = 0$

72. $3x^3 + 2x^2 - 27x - 18 = 0$

APPLICATIONS

73. *Dimensions of a Box*　(Refer to Example 3.) A box is made from a rectangular piece of metal with length 20 inches and width 15 inches. The box has no top.

(a) What are the limitations on the size of x? Explain. your answer.

(b) Write an expression that gives the outside surface area of the box. (*Hint:* Consider the size of the metal sheet and how much was cut out.)

(c) If the outside surface area of the box is 275 square inches, find x.

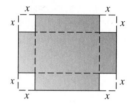 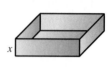

74. *Dimensions of a Box*　Refer to the previous exercise.
(a) Find a polynomial that gives the volume of the box for a given x.
(b) Factor your polynomial completely.
(c) What are the zeros of your polynomial? What do they represent in this problem?

Exercises 75–78: U.S. Natural Gas Consumption　(*Refer to the introduction to this section.*) *The polynomial*

$$0.0013x^3 - 0.085x^2 + 1.6x + 12$$

models natural gas consumption in trillions of cubic feet, where $x = 0$ *corresponds to 1960,* $x = 1$ *to 1961, and so on.*

75. How much natural gas was consumed in 1990?

76. In which year (between 1970 and 1990) was natural gas consumption about 20.4 trillion cubic feet?

77. Explain any difficulties encountered when you try to solve the equation

$$0.0013x^3 - 0.085x^2 + 1.6x + 12 = 23.2.$$

78. How might you solve this equation without factoring? If you were to find the solution to this equation, what would it represent?

WRITING ABOUT MATHEMATICS

79. Compare factoring the polynomial $x^2 + 6x + 5$ with factoring the polynomial $z^4 + 6z^2 + 5$.

80. Suppose that a polynomial can be factored. Explain how its factors can be used to find the zeros of the polynomial. Give an example.

CHECKING BASIC CONCEPTS
SECTION 6.7

1. Factor the trinomial completely.
   (a) $3x^2 - 6x - 24$ (b) $-10y^2 + 5y + 5$

2. Factor the binomial completely.
   (a) $z^4 - 25$ (b) $7t^4 - 7$

3. Factor.
   (a) $x^4 - 8x^2 + 16$ (b) $2y^3 + 17y^2 - 30y$

4. Solve $t^4 + t^3 = 12t^2$.

5. Solve $x^3 - 3x^2 + 2x - 6 = 0$.

CHAPTER 6 SUMMARY

SECTION 6.1 ■ INTRODUCTION TO FACTORING

Terms Related to Factoring Polynomials

Factoring	Writing a polynomial as a product, usually of lower degree polynomials
Common Factor	An expression that is a factor of each term in a polynomial

Example: Some common factors of $4x^4 + 8x^2$ are $2x$, x^2, and $4x^2$.

Greatest Common Factor (GCF)	The common factor with the greatest (integer) coefficient and the highest degree

Example: The GCF of $4x^4 + 8x^2$ is $4x^2$.
$$4x^4 + 8x^2 = 4x^2(x^2 + 2)$$

Factoring by Grouping	Used to factor a four-term polynomial into two binomials

Example:
$$x^3 + 5x^2 + 3x + 15 = x^2(x + 5) + 3(x + 5)$$
$$= (x^2 + 3)(x + 5)$$

SECTION 6.2 ■ FACTORING TRINOMIALS I ($x^2 + bx + c$)

Review of the FOIL Method   A method used for multiplying two binomials

First Terms	$(2x + 3)(5x + 4)$:	$2x \cdot 5x = 10x^2$
Outside Terms	$(2x + 3)(5x + 4)$:	$2x \cdot 4 = 8x$
Inside Terms	$(2x + 3)(5x + 4)$:	$3 \cdot 5x = 15x$
Last Terms	$(2x + 3)(5x + 4)$:	$3 \cdot 4 = 12$

The product is the sum of these four terms:
$$(2x + 3)(5x + 4) = 10x^2 + 8x + 15x + 12 = 10x^2 + 23x + 12.$$

Factoring Trinomials Having a Leading Coefficient of 1   To factor the trinomial $x^2 + bx + c$, find two numbers, m and n, that satisfy

$$m \cdot n = c \quad \text{and} \quad m + n = b.$$

Then $x^2 + bx + c = (x + m)(x + n)$.

Example: Because $-3 \cdot 5 = -15$ and $-3 + 5 = 2$,

$$x^2 + 2x - 15 = (x - 3)(x + 5).$$

SECTION 6.3 ▪ FACTORING TRINOMIALS II ($ax^2 + bx + c$)

Factoring Trinomials by Grouping   To factor $ax^2 + bx + c$, perform the following steps. (Assume that a, b, and c are integers and have no factor in common.)

1. Find two numbers, m and n, such that $mn = ac$ and $m + n = b$.
2. Write the trinomial as $ax^2 + mx + nx + c$.
3. Use grouping to factor this expression into two binomials.

Example: To factor $3x^2 + 10x - 8$, find two numbers whose product is -24 and whose sum is 10. These two numbers are $m = 12$ and $n = -2$, so write $10x$ as $12x - 2x$.

$$\begin{aligned} 3x^2 + 10x - 8 &= 3x^2 + 12x - 2x - 8 \\ &= (3x^2 + 12x) + (-2x - 8) \\ &= 3x(x + 4) - 2(x + 4) \\ &= (3x - 2)(x + 4) \end{aligned}$$

Factoring with FOIL in Reverse   Use trial and error and FOIL in reverse to find the factors of a trinomial.

Example: To factor $3x^2 + 10x - 8$, first find factors of $3x^2$.

$$(3x + \underline{\quad})(x + \underline{\quad})$$

Then place factors of -8 so that the resulting middle term is $10x$.

$$(3x + \underline{-2}) \cdot (x + \underline{4})$$
$$-2x$$
$$12x$$
$$10x \longleftarrow \text{Middle term checks.}$$

SECTION 6.4 ▪ SPECIAL TYPES OF FACTORING

Difference of Two Squares

$$a^2 - b^2 = (a - b)(a + b)$$

Examples: $x^2 - 16 = (x - 4)(x + 4)$ $(a = x, b = 4)$
$4r^2 - 9t^2 = (2r - 3t)(2r + 3t)$ $(a = 2r, b = 3t)$

Perfect Square Trinomials

$$a^2 + 2ab + b^2 = (a + b)^2 \quad \text{and} \quad a^2 - 2ab + b^2 = (a - b)^2$$

Examples: $4x^2 + 4x + 1 = (2x)^2 + 2(2x)1 + 1^2 = (2x + 1)^2$ $(a = 2x, b = 1)$
$x^2 - 10x + 25 = x^2 - 2 \cdot x \cdot 5 + 5^2 = (x - 5)^2$ $(a = x, b = 5)$

Sums and Differences of Two Cubes

$$a^3 + b^3 = (a + b)(a^2 - ab + b^2) \quad \text{and} \quad a^3 - b^3 = (a - b)(a^2 + ab + b^2)$$

Examples: $x^3 + 8 = (x + 2)(x^2 - 2x + 4)$ $(a = x, b = 2)$
$27x^3 - 1 = (3x - 1)(9x^2 + 3x + 1)$ $(a = 3x, b = 1)$

SECTION 6.5 ■ SUMMARY OF FACTORING

Guidelines for Factoring Polynomials   The following guidelines can be used to factor polynomials in general.

STEP 1: Factor out the greatest common factor, if possible.

STEP 2: A. If the polynomial has *four terms*, try factoring by grouping.
   B. If the polynomial is a *binomial,* try one of the following.

   1. $a^2 - b^2 = (a - b)(a + b)$ Difference of two squares
   2. $a^3 - b^3 = (a - b)(a^2 + ab + b^2)$ Difference of two cubes
   3. $a^3 + b^3 = (a + b)(a^2 - ab + b^2)$ Sum of two cubes

   C. If the polynomial is a *trinomial*, check for a perfect square.

   1. $a^2 + 2ab + b^2 = (a + b)^2$ Perfect square trinomial
   2. $a^2 - 2ab + b^2 = (a - b)^2$ Perfect square trinomial

   Otherwise, apply FOIL in reverse or apply grouping, as described in Sections 6.2 and 6.3.

STEP 3: Check to make sure that the polynomial is *completely* factored.

Examples: $12x^3 - 12x^2 + 3x = 3x(4x^2 - 4x + 1) = 3x(2x - 1)^2$ Steps 1, 2C, and 3
$9x^3 - 6x^2 + 18x - 12 = 3(3x^3 - 2x^2 + 6x - 4) = 3(x^2 + 2)(3x - 2)$ Steps 1, 2A, and 3
$16x^3 - 100x = 4x(4x^2 - 25) = 4x(2x - 5)(2x + 5)$ Steps 1, 2B, and 3

SECTION 6.6 ■ SOLVING EQUATIONS BY FACTORING I (QUADRATICS)

Zero-Product Property

For any real numbers a and b, if $ab = 0$, then $a = 0$ or $b = 0$ (or both).
The zero-product property is used to solve equations.

Examples: $xy = 0$ implies that $x = 0$ or $y = 0$.

$(x + 5)(x - 3) = 0$ implies $x + 5 = 0$ or $x - 3 = 0$.

Zero of a Polynomial   A number a is a zero of a polynomial if the result is 0 when a is substituted in that polynomial.

Example: The number -2 is a zero of $x^2 - 4$ because $(-2)^2 - 4 = 0$.

Solving Quadratic Equations by Factoring   To solve a quadratic equation by factoring, follow these steps.

STEP 1: If necessary, use algebra to write the equation as $ax^2 + bx + c = 0$.

STEP 2: Factor the left side of the equation using any method.

STEP 3: Apply the zero-product property.

STEP 4: Solve each of the resulting equations. Check any solutions.

Example:
$$x^2 + 7x = 8 \quad \text{Given equation}$$
$$x^2 + 7x - 8 = 0 \quad \text{Step 1}$$
$$(x + 8)(x - 1) = 0 \quad \text{Step 2}$$
$$x + 8 = 0 \quad \text{or} \quad x - 1 = 0 \quad \text{Step 3}$$
$$x = -8 \quad \text{or} \quad x = 1 \quad \text{Step 4}$$

SECTION 6.7 ■ SOLVING EQUATIONS BY FACTORING II (HIGHER DEGREE)

Factoring Polynomials of Higher Degree   The distributive property and the techniques for factoring quadratic polynomials can also be applied to polynomials of higher degree.

Examples: $10r^3 + 15r = 5r(2r^2 + 3)$ (To check, multiply the right side.)

Because $2x^2 + x - 1 = (x + 1)(2x - 1)$,
it follows that $2z^4 + z^2 - 1 = (z^2 + 1)(2z^2 - 1)$.

Solving Equations by Factoring   Use algebra to obtain 0 on one side of the equation. Factor the other side and apply the zero-product property.

Example:
$$x^3 = 4x$$
$$x^3 - 4x = 0 \quad \text{Subtract } 4x.$$
$$x(x^2 - 4) = 0 \quad \text{Factor out the GCF, } x.$$
$$x(x - 2)(x + 2) = 0 \quad \text{Difference of squares}$$
$$x = 0 \quad \text{or} \quad x - 2 = 0 \quad \text{or} \quad x + 2 = 0 \quad \text{Zero-product property}$$
$$x = 0 \quad \text{or} \quad x = 2 \quad \text{or} \quad x = -2 \quad \text{Solve for } x.$$

CHAPTER 6   REVIEW EXERCISES

SECTION 6.1

Exercises 1–4: Identify the greatest common factor for the expression and then factor the expression.

1. $8z^3 - 4z^2$

2. $6x^4 + 3x^3 - 12x^2$

3. $9xy + 15yz^2$

4. $a^2b^3 + a^3b^2$

Exercises 5–12: Use grouping to factor the given polynomial completely.

5. $x(x + 2) - 3(x + 2)$

6. $y^2(x - 5) + 3y(x - 5)$ **7.** $z^3 - 2z^2 + 5z - 10$

8. $t^3 + t^2 + 8t + 8$ **9.** $x^3 - 3x^2 + 6x - 18$

10. $ax + bx - ay - by$ **11.** $x^5 + 3x^4 - 2x^3 - 6x^2$

12. $2y^4 + 6y^3 + 2y^2 + 6y$

SECTION 6.2

Exercises 13–16: Find an integer pair that has the given product and sum.

13. Product: 20   Sum: 9

14. Product: -21 Sum: 4

15. Product: 36   Sum: -13

16. Product: -100 Sum: -21

Exercises 17–26: Factor the trinomial completely.

17. $x^2 - x - 12$ **18.** $x^2 + 10x + 24$

19. $x^2 + 6x - 16$ **20.** $x^2 - x - 42$

21. $x^2 + 2x - 3$ **22.** $x^2 + 22x + 120$

23. $2x^3 + 6x^2 - 20x$ **24.** $x^4 - 3x^3 - 28x^2$

25. $10 - 7x + x^2$ **26.** $24 + 2x - x^2$

SECTION 6.3

Exercises 27–36: Factor the trinomial completely.

27. $9x^2 + 3x - 2$ **28.** $2x^2 + 3x - 5$

29. $3x^2 + 14x + 15$ **30.** $35x^2 - 2x - 1$

31. $24x^2 - 7x - 5$ **32.** $4x^2 + 33x - 27$

33. $12x^3 + 48x^2 + 21x$ **34.** $8x^4 + 14x^3 - 30x^2$

35. $12 - 5x - 2x^2$ **36.** $1 + 3x - 10x^2$

SECTION 6.4

Exercises 37–50: Factor completely.

37. $z^2 - 4$ **38.** $9z^2 - 64$

39. $36 - y^2$ **40.** $100a^2 - 81b^2$

41. $x^2 + 14x + 49$ **42.** $x^2 - 10x + 25$

43. $4x^2 - 12x + 9$ **44.** $9x^2 + 48x + 64$

45. $8t^3 - 1$ **46.** $27r^3 + 8t^3$

47. $2x^3 - 50x$ **48.** $24x^3 + 81$

49. $2x^3 + 28x^2 + 98x$ **50.** $2x^4 - 128x$

SECTION 6.5

Exercises 51–58: Factor completely.

51. $9y^2 - 6y + 6$ **52.** $yz^2 - 9y$

53. $x^4 + 7x^3 - 4x^2 - 28x$ **54.** $12x^3 + 36x^2 + 27x$

55. $3ab^3 - 24a$ **56.** $5x^3 + 20x$

57. $24x^3 - 6xy^2$ **58.** $x^3y + 27y$

SECTION 6.6

Exercises 59–70: Solve the equation.

59. $mn = 0$ **60.** $y^2 = 0$

61. $(4x - 3)(x + 9) = 0$

62. $(1 - 4x)(6 + 5x) = 0$

63. $z(z - 1)(z - 2) = 0$ **64.** $z^2 - 7z = 0$

65. $y^2 - 64 = 0$ **66.** $y^2 + 9y + 14 = 0$

67. $x^2 = x + 6$ **68.** $10x^2 + 11x = 6$

69. $t(t - 14) = 72$ **70.** $t(2t - 1) = 10$

SECTION 6.7

Exercises 71–80: Factor completely.

71. $5x^2 - 15x - 50$ **72.** $-3x^2 - 6x + 45$

73. $y^3 - 4y$ **74.** $3y^3 + 6y^2 - 9y$

75. $2z^4 + 14z^3 + 20z^2$ **76.** $8z^4 - 32z^2$

77. $x^4 - 6x^2 + 9$ **78.** $2x^4 - 15x^2 - 27$

79. $a^2 + 10ab + 25b^2$ **80.** $x^3 - xy^2$

Exercises 81–88: Solve.

81. $16x^2 - 72x - 40 = 0$

82. $2x^3 - 11x^2 + 15x = 0$ **83.** $t^3 = 25t$

84. $t^4 - 7t^3 + 12t^2 = 0$ **85.** $z^4 + 16 = 8z^2$

86. $z^4 - 256 = 0$ **87.** $y^3 = -64$

88. $y^3 - y^2 - y + 1 = 0$

APPLICATIONS

89. A square has area $9x^2 + 42x + 49$. Find the length of a side. Make a sketch of the square.

90. A rectangle has area $x^2 + 6x + 5$. Find possible dimensions for the rectangle. Make a sketch of the rectangle.

91. A cube has surface area $6x^2 + 12x + 6$. Find the length of a side.

92. Write a polynomial in factored form that represents the total area of the rectangle.

x^2	$3x$
x	3

93. Write a polynomial in factored form that represents the total area of the rectangle.

$2x^2$	$3x$
$12x$	18

94. *Radius of a Circle* The area and the circumference of a circle are numerically equal. Find the radius of the circle.

95. *Dimensions of a Shed* The floor of a rectangular shed is 7 feet longer than it is wide and has an area of 120 square feet. What are its dimensions?

96. *Flight of a Ball* A ball is hit upward. Its height h in feet after t seconds is given by $h = -16t^2 + 80t + 4$. At what times is the ball 100 feet in the air?

97. *Stopping Distance* The distance D in feet that it takes to stop a car traveling x miles per hour on wet, level pavement can be approximated by $D = \frac{1}{9}x^2 + \frac{11}{3}x$.
  (a) Estimate the distance required for the car to stop when it is traveling 45 miles per hour.
  (b) If the stopping distance is 80 feet, what is the speed of the car?
  (c) If you have a calculator, use it to solve part (b) numerically with a table of values. Do your answers agree?

98. *Revenue* A company makes tops for the boxes of pickup trucks. The total revenue R in dollars from selling the tops for p dollars each is given by $R = p(200 - p)$, where $p \le 200$.
  (a) Find R when $p = \$100$.
  (b) Find p when $R = \$7500$.
  (c) If you have a calculator, use it to solve part (b) numerically with a table of values. Do your answers agree?

99. *Airline Passengers* The number N of worldwide airline passengers in millions from 1950 to 2000 is approximated by

$$N = 0.68y^2 + 3.8y + 24,$$

where $y = 0$ corresponds to 1950, $y = 1$ to 1951, and so on until $y = 50$ corresponds to 2000.
  (a) Estimate the number of airline passengers in 1970.
  (b) Use a table of values to estimate the year in which the number of airline passengers reached 544 million.

100. *Digital Photographs* A digital photograph is 30 pixels longer than it is wide and has a total area of 4000 pixels. Find the dimensions of this photograph.

101. *Dimensions of a Box* A box is made from a rectangular piece of metal with length 50 inches and width 40 inches by cutting out square corners of length x and folding up the sides.
  (a) Write an expression that gives the surface area of the inside of the box.
  (b) If the surface area of the box is 1900 square inches, find x.

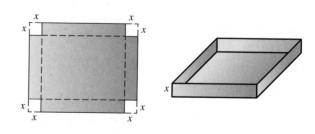

CHAPTER 6   TEST   🎓 Pass the Test   Video solutions to all test exercises

Exercises 1 and 2: Identify the greatest common factor for the expression. Then factor the expression.

1. $4x^2y - 20xy^2 + 12xy$ **2.** $9a^3b^2 + 3a^2b^2$

Exercises 3 and 4: Factor by grouping.

3. $ay + by + az + bz$ **4.** $3x^3 + x^2 - 15x - 5$

Exercises 5–8: Factor the trinomial.

5. $y^2 + 4y - 12$ **6.** $4x^2 + 20x + 25$

7. $4z^2 - 19z + 12$ **8.** $21 - 17t + 2t^2$

Exercises 9–14: Factor completely.

9. $6x^3 + 3x^2 - 3x$ **10.** $2z^4 - 12z^2 - 54$

11. $36y^3 - 100y$ **12.** $7x^4 + 56x$

13. $16a^4 + 24a^3 + 9a^2$ **14.** $2b^4 - 32$

Exercises 15–20: Solve the equation.

15. $x^2 - 16 = 0$ **16.** $y^2 = y + 20$

17. $9z^2 + 16 = 24z$ **18.** $x(x - 5) = 66$

19. $y^3 = 9y$ **20.** $x^4 - 5x^2 + 4 = 0$

21. A square has area $9x^2 + 30x + 25$. Find the length of a side in terms of x.

22. Write a polynomial in factored form that represents the total area of the rectangle.

23. *Braking Distance*   The braking distance D in feet required for a car traveling at x miles per hour to stop on dry, level pavement can be modeled by $D = \frac{1}{11}x^2$.
  (a) Calculate the distance required for the car to stop when it is traveling 55 miles per hour.
  (b) If the braking distance is 99 feet, estimate the speed of the car.

24. *Flight of a Ball*   A ball is thrown upward. Its height h in feet after t seconds is given by $h = -16t^2 + 48t + 4$. At what times is the ball 36 feet in the air?

CHAPTER 6   EXTENDED AND DISCOVERY EXERCISES

Exercises 1–6: Difference of Two Squares   The difference of two squares can be factored by using

$$a^2 - b^2 = (a - b)(a + b).$$

This equation can also be used in some situations where an expression may not appear to be the difference of two squares. For example, because $(\sqrt{3})^2 = 3$, $x^2 - 3$ can be written and then factored as

$$x^2 - 3 = x^2 - (\sqrt{3})^2$$
$$= (x - \sqrt{3})(x + \sqrt{3}).$$

Use this concept to factor the following expressions as the difference of two squares.

1. $x^2 - 5$ **2.** $y^2 - 7$

3. $3z^2 - 25$ **4.** $7t^2 - 2$

5. $x - 4$ for $x \geq 0$ (*Hint:* $(\sqrt{x})^2 = x$.)

6. $x - 7$ for $x \geq 0$

Exercises 7–12: Solving Equations   (Refer to Exercises 1–6.) Solve the equation by factoring it as the difference of two squares.

7. $x^2 - 3 = 0$ **8.** $y^2 - 7 = 0$

9. $3x^2 - 25 = 0$ **10.** $7x^2 - 11 = 0$

11. $x^4 - 9 = 0$ **12.** $x^4 - 25 = 0$

CHAPTERS 1–6   CUMULATIVE REVIEW EXERCISES

1. Write 144 as a product of prime numbers.

2. Evaluate $-2x + 3y$ for $x = -2$ and $y = 4$.

Exercises 3 and 4: Evaluate by hand and then simplify to lowest terms.

3. $\frac{3}{5} \cdot \frac{15}{21}$ **4.** $\frac{4}{5} - \frac{1}{10}$

Exercises 5 and 6: Evaluate by hand.

5. $26 - 3 \cdot 6 \div 2$ **6.** $-2^2 + \frac{3+2}{8+2}$

7. Solve $4t - 7 = 25$ and check.

8. Complete the table. Then use the table to solve the equation $2x + 3 = 5$.

x	-2	-1	0	1	2
$2x + 3$					

9. Translate the sentence "Triple a number decreased by 5 equals the number decreased by 7" into an equation using the variable n. Then solve the equation.

10. Convert 5.7% to fraction and decimal notation.

11. Convert 0.123 to a percentage.

12. Solve $P = 2W + 2L$ for W.

13. Solve $5 - 3z < -1$.

14. Make a scatterplot having the following five points: $(-2, 3), (-1, 2), (0, -1), (1, 1)$, and $(2, 2)$.

Exercises 15 and 16: Graph the given equation. Determine any intercepts.

15. $y = 3x - 2$ **16.** $y = -2$

17. Identify the x-intercept and the y-intercept. Then write the slope–intercept form of the line.

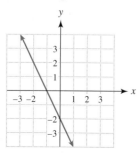

Exercises 18 and 19: Find the slope–intercept form for the line satisfying the given conditions.

18. Perpendicular to $2x - 3y = -6$ and passing through the point $(1, 2)$

19. Passing through the points $(-2, 1)$ and $(1, 5)$

20. The graphs of two equations are shown. Use the graphs to identify the solution to the system of equations. Then check your answer.

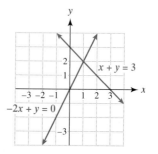

Exercises 21 and 22: Solve the system of equations.

21. $\begin{array}{r} y = -1 \\ 2x + y = 1 \end{array}$ **22.** $\begin{array}{r} 5x + y = -5 \\ -x + 2y = 12 \end{array}$

Exercises 23 and 24: The graphs of two linear equations are shown.

(a) State the number of solutions to the system of equations.

(b) Is the system consistent or inconsistent? If the system is consistent, state whether the equations are dependent or independent.

23.   **24.**

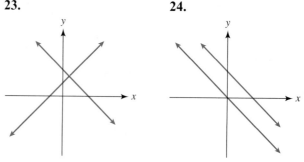

Exercises 25 and 26: Shade the solution set to the given inequality.

25. $x \le 2$ **26.** $2x + 3y \ge 6$

Exercises 27 and 28: Shade the solution set to the system of inequalities.

27. $x > -1$
$\quad y < x$

28. $2x - y \le 4$
$\quad x + 2y \ge 2$

Exercises 29–34: Simplify the expression.

29. -2^4

30. $(xy)^0$

31. $(xy)^4(x^3y^{-4})^2$

32. $7x^3(-2x^2 + 3x)$

33. $(7x - 2)(3x + 5)$

34. $(x - 7)^2$

Exercises 35–38: Simplify and write the expression using positive exponents.

35. $a^{-4} \cdot a^2$

36. $(2t^3)^{-2}$

37. $(xy)^{-3}(x^{-1}y^2)^{-1}$

38. $\left(\dfrac{2x}{y^{-2}}\right)^5$

39. Write 6.23×10^{-3} in standard form.

40. Write 543,000 in scientific notation.

Exercises 41 and 42: Divide and check.

41. $\dfrac{6x^3 + 12x^2}{3x}$

42. $\dfrac{3x^3 - x + 1}{x^2 + 1}$

Exercises 43 and 44: Factor by grouping.

43. $2y^2(x + 2) - 5(x + 2)$ **44.** $t^3 + 6t^2 + t + 6$

Exercises 45–52: Factor completely.

45. $x^2 + 3x - 28$

46. $6y^2 + y - 12$

47. $25x^2 - 4y^2$

48. $64x^2 - 16x + 1$

49. $27t^3 - 8$

50. $-4x^2 + 4x + 24$

51. $x^4 - 12x^2 + 27$

52. $x^3y - x^2y^2$

Exercises 53–56: Solve the equation.

53. $y^4 = 25y^2$

54. $8z^2 + 8z - 16 = 0$

55. $4z^3 = 49z$

56. $x^4 - 18x^2 + 81 = 0$

APPLICATIONS

57. *Shoveling the Driveway* Two people are shoveling snow from a driveway. The first person shovels 10 square feet per minute, while the second person shovels 8 square feet per minute.

(a) Write and simplify an expression that gives the total square feet that the two people shovel in x minutes.

(b) How many minutes would it take for them to clear a driveway with an area of 900 square feet?

58. *Running Distance* A person runs the following distances over three days: $1\frac{3}{4}$ miles, $2\frac{1}{2}$ miles, and $2\frac{2}{3}$ miles. How far does the person run altogether?

59. *Cost of a Car* A 7% sales tax on a car amounted to $1470. Find the cost of the car.

60. *Burning Calories* An athlete can burn 12 calories per minute while cross-country skiing and 9 calories per minute while running at 5 miles per hour. If the athlete burns 615 calories in 60 minutes, how long is spent on each activity?

61. *Renting a Car* A rental car costs $20 plus $0.25 per mile driven.

(a) Write an equation that gives the cost C of driving the car x miles.

(b) Determine the number of miles that the car is driven if the rental cost is $100.

62. *Triangle* In an isosceles triangle, the measures of the two smaller angles are equal and their sum is 20° more than the larger angle.

(a) Let x be the measure of one of the two smaller angles and y be the measure of the larger angle. Write a system of linear equations whose solution gives the measures of these angles.

(b) Solve your system.

63. *Radios and CD Players* A business manufactures at least three radios for each CD player. The total number of radios and CD players cannot exceed 90 per week. Shade the region that represents numbers of radios R and CD players P that can be produced each week within these restrictions.

64. *Areas of Rectangles* Find a monomial equal to the sum of the areas of the rectangles. Calculate this sum for $x = 2$ yards and $y = 3$ yards.

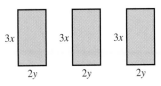

65. *Volume of a Cube* Find an expression for the volume of a cube whose sides have length $2xy^2$.

66. *Area of a Square*   Complete each part and verify that your answers are the same.
   (a) Find the area of the large square by multiplying its length and width
   (b) Find the sum of the areas of the four smaller rectangles inside the large square.

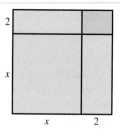

67. *Area of a Square*   A square has area $x^2 + 12x + 36$. Find the length of a side.

68. *Flight of a Golf Ball*   A golf ball is hit upward. Its height h in feet after t seconds is given by the formula $h = 64t - 16t^2$.
   (a) How long does it take for the ball to hit the ground?
   (b) At what times is the ball 48 feet in the air?

Rational Expressions

One of the most significant problems facing the U.S. transportation system is chronic highway congestion. According to a newly released "Highway Statistics," Americans drove about 3 trillion miles in 2005, nearly a 25% increase over 1995. There were 241.2 million vehicles registered in the United States in 2005. Our ability to keep traffic moving smoothly and safely is key to keeping our economy strong, and traffic congestion costs motorists more than $80 billion annually in wasted time and fuel. People spend billions of hours stuck in traffic.

Traffic congestion is subject to a nonlinear effect. If the amount of traffic doubles on a highway, the wait in traffic may more than double. At times, only a slight increase in the traffic rate can result in a dramatic increase in the time spent waiting. (See Example 6 in Section 7.1.) Mathematics can describe this effect by using rational expressions, which are used frequently to diagnose and solve traffic and highway engineering problems. In this chapter we introduce rational expressions and some of their applications.

> There is nothing wrong with making mistakes. Just don't respond with encores.
>
> —ANONYMOUS

Sources: Philip J. Longman, "American Gridlock." *U.S. News & World Report*, May 28, 2001; U.S. Department of Transportation, 2006.

7.1 INTRODUCTION TO RATIONAL EXPRESSIONS

Basic Concepts ▪ Simplifying Rational Expressions ▪ Applications

A LOOK INTO MATH ▷

Have you ever been moving smoothly in traffic and noticed how it can come to a halt all of a sudden? Mathematics shows that if the number of cars on a road increases even slightly, then the movement of traffic can slow dramatically. To understand this process better, we need to discuss rational expressions.

Basic Concepts

In Chapters 5 and 6 we discussed polynomials. Examples of polynomials include

$$3, \quad 2x, \quad x^2 + 4, \quad \text{and} \quad x^3 - 1.$$

In this chapter we discuss *rational expressions*. Rational expressions can be written as quotients (fractions) of two polynomials. Examples of rational expressions include

$$\frac{3}{2x}, \quad \frac{2x}{x^2 + 4}, \quad \frac{x^2 + 4}{3}, \quad \text{and} \quad \frac{x^3 - 1}{x^2 + 4}.$$

> **RATIONAL EXPRESSION**
>
> A rational expression can be written as $\frac{P}{Q}$, where P and Q are polynomials. A rational expression is defined whenever $Q \neq 0$.

We can evaluate polynomials for different values of a variable. For example, for $x = 2$ the polynomial $x^2 - 3x + 1$ evaluates to

$$(2)^2 - 3(2) + 1 = -1.$$

Rational expressions can be evaluated similarly.

EXAMPLE 1   Evaluating rational expressions

If possible, evaluate each expression for the given value of the variable.

(a) $\dfrac{1}{x + 1}$ $x = 2$ **(b)** $\dfrac{y^2}{2y - 1}$ $y = -4$

(c) $\dfrac{5z + 8}{z^2 - 2z + 1}$ $z = 1$ **(d)** $\dfrac{2 - x}{x - 2}$ $x = -3$

Solution

(a) If $x = 2$, then $\frac{1}{x + 1}$ becomes $\frac{1}{2 + 1} = \frac{1}{3}$.

(b) If $y = -4$, then $\frac{y^2}{2y - 1}$ becomes $\frac{(-4)^2}{2(-4) - 1} = -\frac{16}{9}$.

(c) If $z = 1$, then $\frac{5z + 8}{z^2 - 2z + 1}$ becomes $\frac{5(1) + 8}{1^2 - 2(1) + 1}$, or $\frac{13}{0}$, which is undefined because division by 0 is not possible.

(d) If $x = -3$, then $\frac{2 - x}{x - 2}$ becomes $\frac{2 - (-3)}{-3 - 2} = \frac{5}{-5} = -1$.

Now Try Exercises 11, 15, 17, 21

Division by 0 is undefined. As a result, rational expressions are different from polynomials because they are undefined whenever their denominators are 0. For example, the expression in Example 1(c) is undefined when $z = 1$.

EXAMPLE 2 Determining when a rational expression is undefined

Find all values of the variable for which each expression is undefined.

(a) $\dfrac{1}{x}$ **(b)** $\dfrac{4t}{t - 3}$ **(c)** $\dfrac{1 - 6r}{r^2 - 4}$ **(d)** $\dfrac{4}{x^2 + 1}$

Solution

(a) A rational expression is undefined when its denominator is 0. Thus $\frac{1}{x}$ is undefined when $x = 0$.

(b) The expression $\frac{4t}{t - 3}$ is undefined when its denominator, $t - 3$, is 0, or when $t = 3$.

(c) The expression $\frac{1 - 6r}{r^2 - 4}$ is undefined when its denominator, $r^2 - 4$, is 0. Here

$$r^2 - 4 = (r - 2)(r + 2) = 0$$

implies that the denominator is 0 when $r = -2$ or $r = 2$.

(d) In the expression $\frac{4}{x^2 + 1}$ the denominator, $x^2 + 1$, is never 0 because any real number squared plus 1 is always greater than or equal to 1. Thus this rational expression is defined for all real numbers x. Now Try Exercises **29, 31, 35, 37**

Simplifying Rational Expressions

In Chapter 1, we used the *basic principle of fractions*,

$$\frac{a \cdot c}{b \cdot c} = \frac{a}{b}.$$

When we use this basic principle, the fraction $\frac{8}{12}$, for example, simplifies to

$$\frac{8}{12} = \frac{2 \cdot 4}{3 \cdot 4} = \frac{2}{3}.$$

EXAMPLE 3 Simplifying fractions

Simplify each fraction by applying the basic principle of fractions.

(a) $\dfrac{5}{10}$ **(b)** $-\dfrac{36}{48}$

Solution

(a) $\dfrac{5}{10} = \dfrac{1 \cdot 5}{2 \cdot 5} = \dfrac{1}{2}$ **(b)** $-\dfrac{36}{48} = -\dfrac{3 \cdot 12}{4 \cdot 12} = -\dfrac{3}{4}$ Now Try Exercises **43, 47**

We can also apply this basic principle to rational expressions. For example,

$$\frac{x(x - 1)}{4(x - 1)} = \frac{x}{4},$$

provided that $x \neq 1$.

NOTE: This simplification is not valid when $x = 1$ because the expression is undefined for this x-value. When simplifying rational expressions, *we assume that values of the variable making a rational expression undefined are excluded.*

BASIC PRINCIPLE OF RATIONAL EXPRESSIONS

The following property can be used to simplify rational expressions, where P, Q, and R are polynomials.

$$\frac{P \cdot R}{Q \cdot R} = \frac{P}{Q} \qquad Q \text{ and } R \text{ are nonzero.}$$

NOTE: $\frac{P \cdot R}{Q \cdot R} = \frac{P}{Q} \cdot \frac{R}{R} = \frac{P}{Q} \cdot 1 = \frac{P}{Q}$, provided that $Q \neq 0$ and $R \neq 0$.

Like fractions, rational expressions can be written in *lowest terms*. For example, the rational expression $\frac{x^2 - 1}{x^2 + 2x + 1}$ can be written in lowest terms by factoring the numerator and the denominator and then applying the basic principle of rational expressions.

$$\frac{x^2 - 1}{x^2 + 2x + 1} = \frac{(x - 1)(x + 1)}{(x + 1)(x + 1)} \qquad \text{Factor the numerator and the denominator.}$$

$$= \frac{x - 1}{x + 1} \qquad \text{Apply } \frac{PR}{QR} = \frac{P}{Q} \text{ with } R = x + 1.$$

Because the basic principle of rational expressions cannot be applied further to $\frac{x-1}{x+1}$, we say that this expression is written in **lowest terms**.

EXAMPLE 4 Simplifying rational expressions

Simplify each expression.

(a) $\dfrac{8y}{4y^2}$ (b) $\dfrac{2x + 6}{3x + 9}$ (c) $\dfrac{(z + 1)(z - 5)}{(z - 5)(z + 3)}$ (d) $\dfrac{x^2 - 9}{2x^2 + 7x + 3}$

Solution

(a) Factor out the greatest common factor, $4y$, in the numerator and the denominator. Then apply the basic principle of rational expressions.

$$\frac{8y}{4y^2} = \frac{2 \cdot 4y}{y \cdot 4y} = \frac{2}{y}$$

(b) Use the distributive property and then the basic principle of rational expressions.

$$\frac{2x + 6}{3x + 9} = \frac{2(x + 3)}{3(x + 3)} = \frac{2}{3}$$

(c) Use the commutative property to rewrite the denominator of the expression.

$$\frac{(z + 1)(z - 5)}{(z - 5)(z + 3)} = \frac{(z + 1)(z - 5)}{(z + 3)(z - 5)} = \frac{z + 1}{z + 3}$$

(d) Start by factoring the numerator and the denominator.

$$\frac{x^2 - 9}{2x^2 + 7x + 3} = \frac{(x - 3)(x + 3)}{(2x + 1)(x + 3)} = \frac{x - 3}{2x + 1}$$

Now Try Exercises 55, 59, 65, 83

MAKING CONNECTIONS

Expressions and Equations

Expressions and equations are different concepts. An expression does not contain an equals sign, whereas an equation is a statement that two expressions are equal and always contains an equals sign. For example,

$$\frac{x}{x+4} \quad \text{and} \quad \frac{2}{x}$$

are two rational *expressions*, and

$$\frac{x}{x+4} = \frac{2}{x}$$

is a rational *equation*. In this section we evaluate and simplify rational expressions. Later, we will *solve* rational equations by finding x-values that make an equation true.

A negative sign can be placed in a fraction in a number of ways. For example,

$$-\frac{5}{7} = \frac{-5}{7} = \frac{5}{-7}$$

illustrates three fractions that are equal. This property can also be applied to rational expressions, as demonstrated in the next example.

EXAMPLE 5 Distributing a negative sign

Simplify each expression.

(a) $\dfrac{-x-6}{2x+12}$ (b) $\dfrac{10-z}{z-10}$ (c) $-\dfrac{5-x}{x-5}$

Solution

(a) Factor -1 out of the numerator and 2 out of the denominator.

$$\frac{-x-6}{2x+12} = \frac{-1(x+6)}{2(x+6)} = -\frac{1}{2}$$

(b) Factor -1 out of the numerator.

$$\frac{10-z}{z-10} = \frac{-1(-10+z)}{z-10} = \frac{-1(z-10)}{z-10} = -1$$

(c) Rewrite the expression with the negative sign in the numerator and then apply the distributive property. Be sure to include parentheses around the numerator.

$$-\frac{5-x}{x-5} = \frac{-(5-x)}{x-5} = \frac{-5+x}{x-5} = \frac{x-5}{x-5} = 1$$

The same answer can be obtained by distributing the negative sign over the denominator.

$$-\frac{5-x}{x-5} = \frac{5-x}{-(x-5)} = \frac{5-x}{-x+5} = \frac{5-x}{5-x} = 1$$

Now Try Exercises 67, 71, 75

The result for Example 5(b) becomes more obvious if we substitute a number for z. For example, if we let $z = 6$, then

$$\frac{10 - z}{z - 10} = \frac{10 - 6}{6 - 10} = \frac{4}{-4} = -1.$$

MAKING CONNECTIONS

Negative Signs and Rational Expressions

In general, $(b - a)$ equals $-1(a - b)$. Thus if $a \neq b$, then $\frac{b - a}{a - b} = -1$.

Applications

▶ **REAL-WORLD CONNECTION**   The next example is based on the introduction to this chapter and illustrates modeling traffic flow with a rational expression.

EXAMPLE 6   Modeling traffic flow

Suppose that 10 cars per minute can pass through a construction zone. If traffic arrives *randomly* at an average rate of x cars per minute, the average time T in minutes spent waiting in line and passing through the construction zone is given by

$$T = \frac{1}{10 - x},$$

where $x < 10$.   (**Source:** N. Garber and L. Hoel, *Traffic and Highway Engineering.*)

(a) Complete Table 7.1 by finding T for each value of x.

TABLE 7.1   **Waiting in Traffic**

x (cars/minute)	5	7	9	9.5	9.9	9.99
T (minutes)						

(b) Interpret the results.

Solution

(a) When $x = 5$ cars per minute, then $T = \frac{1}{10 - 5} = \frac{1}{5}$ minute. Other values are found similarly, as shown in Table 7.2.

TABLE 7.2   **Waiting in Traffic**

x (cars/minute)	5	7	9	9.5	9.9	9.99
T (minutes)	$\frac{1}{5}$	$\frac{1}{3}$	1	2	10	100

(b) As the average traffic rate increases from 9 cars per minute to 9.9 cars per minute, the time needed to pass through the construction zone increases from 1 minute to 10 minutes. As x nears 10 cars per minute, a small increase in x increases the waiting time dramatically.

Now Try Exercise 101

This nonlinear effect for traffic congestion in Example 6 is shown in Figure 7.1, where points from Table 7.2 have been plotted and a curve passing through them has been sketched. A vertical dashed line was also sketched at $x = 10$. This dashed line is called a **vertical asymptote** and indicates that the rational expression is undefined at this value of x. Near the vertical asymptote, the waiting time T increases dramatically for small increases in x. The graph of T does not intersect or cross this vertical asymptote.

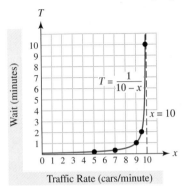

Figure 7.1

TECHNOLOGY NOTE: Making Tables

Table 7.2 can also be created with a graphing calculator by using the Ask feature, as illustrated in the following displays.

```
TABLE SETUP
 TblStart=5
 ΔTbl=1
 Indpnt:  Auto  Ask
 Depend:  Auto  Ask
```

```
  X   | Y₁
  5   | .2
  7   | .33333
  9   | 1
  9.5 | 2
  9.9 | 10
  9.99| 100

Y₁=1/(10−X)
```

AN APPLICATION INVOLVING PROBABILITY (OPTIONAL)   If 10 marbles, one blue and nine red, are placed in a jar, then the *probability*, or *likelihood*, of picking the blue marble at random is 1 *chance* in 10, or $\frac{1}{10}$. The probability of drawing a red marble at random is 9 chances in 10, or $\frac{9}{10}$. **Probability** is a real number between 0 and 1. A probability of 0, or 0%, indicates that an event is impossible, whereas a probability of 1, or 100%, indicates that an event is certain. Rational expressions are often used to describe probability.

EXAMPLE 7   Calculating probability

Suppose that n balls, numbered 1 to n, are placed in a container and only one ball has the winning number.
(a) What is the probability of drawing the winning ball at random?
(b) Calculate this probability for $n = 100$, 1000, and 10,000.
(c) What happens to the probability of drawing the winning ball as the number of balls increases?

Solution
(a) There is 1 chance in n of drawing the winning ball, so the probability is $\frac{1}{n}$.
(b) For $n = 100$, 1000, and 10,000, the probabilities are $\frac{1}{100}$, $\frac{1}{1000}$, and $\frac{1}{10,000}$.
(c) As the number of balls increases, the probability of picking the winning ball decreases.

Now Try Exercise 105

7.1 PUTTING IT ALL TOGETHER

In this section we discussed some basic concepts of rational expressions. They are summarized in the following table.

Concept	Explanation	Examples
Rational Expression	An expression of the form $\frac{P}{Q}$, where P and Q are polynomials with $Q \neq 0$	$\dfrac{1}{x}, \dfrac{x-3}{2x^2-1}, \dfrac{2x+9}{5x}$, and $\dfrac{x^2+3x-5}{1}$
Undefined Values for Rational Expressions	If the *denominator* of a rational expression is 0, then the rational expression is undefined for that value of the variable.	$\dfrac{1}{x-3}$ is undefined when $x = 3$. $\dfrac{5y}{y^2-1}$ is undefined when $y = 1$ or when $y = -1$.
Basic Principle of Rational Expressions	Factor the numerator and the denominator completely. Then apply $$\frac{P \cdot R}{Q \cdot R} = \frac{P}{Q}.$$	$\dfrac{4xy^2}{6xy^3} = \dfrac{2(2xy^2)}{3y(2xy^2)} = \dfrac{2}{3y}$ $\dfrac{4x(x-4)}{(x+1)(x-4)} = \dfrac{4x}{x+1}$

7.1 Exercises

MyMathLab   Math XP PRACTICE   WATCH   DOWNLOAD   READ   REVIEW

CONCEPTS

1. A rational expression can be written as _____, where P and Q are _____ with $Q \neq 0$.

2. Is $\frac{x}{2x^2+1}$ a rational expression? Why or why not?

3. A rational expression is undefined whenever _____.

4. The rational expression $\frac{1}{x-a}$ is undefined whenever $x =$ _____.

5. The basic principle of fractions states that a fraction can be simplified by using $\frac{a \cdot c}{b \cdot c} =$ _____.

6. The basic principle of rational expressions can be used to simplify _____ expressions.

7. The fraction $\frac{18}{60}$ simplifies to _____.

8. The expression $\frac{3(x+1)}{2x(x+1)}$ simplifies to _____.

9. The expression $\frac{x-a}{a-x}$ simplifies to _____.

10. The expression $-\frac{x-1}{x+1}$ simplifies to $\frac{?}{x+1}$.

EVALUATING RATIONAL EXPRESSIONS

Exercises 11–24: If possible, evaluate the expression for the given value of the variable.

11. $\dfrac{3}{x}$ $x = -7$

12. $\dfrac{3}{x+3}$ $x = 0$

13. $-\dfrac{x}{x-5}$ $x = -4$

14. $-\dfrac{4x}{5x+1}$ $x = 1$

15. $\dfrac{y+1}{y^2}$ $y = -2$

16. $\dfrac{3y-1}{y^2+1}$ $y = -1$

17. $\dfrac{7z}{z^2-4}$ $z = -2$

18. $\dfrac{5}{z^2-3z+2}$ $z = -1$

19. $\dfrac{5}{3t+6}$ $t = -2$

20. $\dfrac{4t}{2t+5}$ $t = -\dfrac{5}{2}$

21. $\dfrac{4-x}{x-4}$ $x = -2$

22. $\dfrac{x-7}{7-x}$ $x = 4$

23. $-\dfrac{6-x}{x-6}$ $x = 0$

24. $\dfrac{8-2x}{2x-8}$ $x = -5$

Exercises 25–28: Complete the table for the given expression. If a value is undefined, place a dash in the table.

25.

x	-2	-1	0	1	2
$\dfrac{x}{x+1}$					

26.

x	-2	-1	0	1	2
$\dfrac{2x}{3x-1}$					

27.

x	-2	-1	0	1	2
$\dfrac{3x}{2x^2+1}$					

28.

x	-2	-1	0	1	2
$\dfrac{2x-1}{x^2-1}$					

Exercises 29–42: Find any values of the variable that make the expression undefined.

29. $-\dfrac{8}{x}$ **30.** $\dfrac{7}{x+1}$

31. $\dfrac{4}{z-3}$ **32.** $\dfrac{7-z}{z-7}$

33. $\dfrac{4y}{5y+4}$ **34.** $\dfrac{3+y}{3y-7}$

35. $\dfrac{5t+2}{t^2+1}$ **36.** $\dfrac{8t}{t^2+25}$

37. $\dfrac{8x}{x^2-25}$ **38.** $\dfrac{x+4}{x^2-36}$

39. $\dfrac{x^2+3x+2}{x^2+5x+6}$ **40.** $\dfrac{2x-1}{x^2-7x+10}$

41. $\dfrac{8z^2+z+1}{2z^2-7z+5}$ **42.** $\dfrac{4n^2+17n-15}{3n^2-8n+4}$

SIMPLIFYING RATIONAL EXPRESSIONS

Exercises 43–50: Simplify the fraction to lowest terms.

43. $\dfrac{12}{18}$ **44.** $\dfrac{24}{32}$

45. $\dfrac{24}{48}$ **46.** $-\dfrac{22}{33}$

47. $-\dfrac{6}{15}$ **48.** $\dfrac{8}{22}$

49. $-\dfrac{25}{75}$ **50.** $-\dfrac{36}{42}$

Exercises 51–54: First simplify the fraction, then simplify the rational expression.

51. (a) $\dfrac{8}{16}$ (b) $\dfrac{x+2}{2x+4}$

52. (a) $\dfrac{6}{9}$ (b) $\dfrac{2x+6}{3x+9}$

53. (a) $\dfrac{7-3}{3-7}$ (b) $\dfrac{7-x}{x-7}$

54. (a) $-\dfrac{8-5}{5-8}$ (b) $-\dfrac{x-5}{5-x}$

Exercises 55–92: Simplify the expression.

55. $\dfrac{5x^4}{10x^6}$

56. $\dfrac{6y^2}{9y}$

57. $\dfrac{8xy^3}{6x^2y^2}$

58. $\dfrac{36x^2y^5}{6x^5y}$

59. $\dfrac{x + 4}{2x + 8}$

60. $\dfrac{5x - 10}{x - 2}$

61. $\dfrac{3z - 9}{5z - 15}$

62. $\dfrac{4z + 8}{10 + 5z}$

63. $\dfrac{(x + 1)(x - 1)}{(x + 6)(x - 1)}$

64. $\dfrac{(2x + 1)(x + 9)}{(4x - 3)(x + 9)}$

65. $\dfrac{(5y + 3)(2y - 1)}{(2y - 1)(y + 2)}$

66. $\dfrac{(4y - 1)(5y + 7)}{(5y + 7)(1 - 4y)}$

67. $\dfrac{x - 7}{7 - x}$

68. $\dfrac{5 - x}{x - 5}$

69. $\dfrac{a - b}{b - a}$

70. $\dfrac{2t - 3r}{3r - 2t}$

71. $\dfrac{-6 - x}{18 + 3x}$

72. $\dfrac{-2x - 6}{x + 3}$

73. $\dfrac{x + 1}{-2x - 2}$

74. $\dfrac{3x + 21}{-7 - x}$

75. $-\dfrac{9 - x}{x - 9}$

76. $-\dfrac{4 - 2x}{x - 2}$

77. $\dfrac{(3x + 5)(x - 1)}{(3x - 5)(1 - x)}$

78. $\dfrac{(2 - x)(x - 2)}{(x - 2)(2 - x)}$

79. $\dfrac{n^2 - n}{n^2 - 5n}$

80. $\dfrac{3n^2 - 4n}{n^2 + 4n}$

81. $\dfrac{x^2 - 3x}{6x - 18}$

82. $\dfrac{4x^2 + 16x}{5x^2 + 20x}$

83. $\dfrac{z^2 - 3z + 2}{z^2 - 4z + 3}$

84. $\dfrac{z^2 - 3z - 10}{z^2 - 2z - 8}$

85. $\dfrac{2x^2 + 7x - 4}{6x^2 + x - 2}$

86. $\dfrac{5x^2 + 3x - 2}{5x^2 + 13x - 6}$

87. $\dfrac{x - 3}{3x^2 - 11x + 6}$

88. $\dfrac{2x - 1}{4x^2 + 6x - 4}$

89. $-\dfrac{a - 9}{9 - a}$

90. $-\dfrac{-b - 6}{b + 6}$

91. $\dfrac{-2x - 1}{4x + 2}$

92. $\dfrac{4x + 3}{-8x - 6}$

Exercises 93–100: Do the following.
  (a) *Decide whether you are given an expression or an equation.*
  (b) *If you are given an expression, simplify it. If you are given an equation, solve it.*

93. $x + 1 = 7$

94. $x^2 - 4 = 0$

95. $\dfrac{x}{x(x + 1)}$

96. $\dfrac{x - 2}{(x - 2)(x - 8)}$

97. $\dfrac{x^2 - 4}{x + 2}$

98. $\dfrac{x - 4}{8 - 2x}$

99. $\dfrac{x}{2(1 + 3)} = 1$

100. $\dfrac{x}{2} + 2 = \dfrac{8}{2}$

APPLICATIONS

101. *Modeling Traffic Flow*   (Refer to Example 6.) Five vehicles per minute can pass through a construction zone. If the traffic arrives randomly at an average rate of x vehicles per minute, the average time T in minutes spent waiting in line and passing through the construction zone is given by $T = \dfrac{1}{5 - x}$ for $x < 5$.   (*Source:* N. Garber.)
  (a) Evaluate T for $x = 3$ and interpret the result.
  (b) Complete the table and interpret the results.

x	2	4	4.5	4.9	4.99
T					

102. *Standing in Line*   A worker at a poolside store can serve 20 customers per hour. If children arrive randomly at an average rate of x per hour, then the average number of children N waiting in line is given by $N = \dfrac{x^2}{400 - 20x}$ for $x < 20$.   (*Source:* N. Garber.)
  (a) Complete the table.

x	5	10	18	19	20
N					

  (b) Compare the number of children waiting in line if the average rate increases from 18 to 19 children per hour.

103. *Probability* Suppose that a coin is flipped. What is the probability that a head appears?

104. *Probability* A die shows the numbers 1, 2, 3, 4, 5, and 6. If each number has an equal chance of appearing on any given roll, what is the probability that a 2 or 4 appears?

105. *Probability* (Refer to Example 7.) Suppose that there are *n* balls in a container and that three balls have a winning number. If a ball is drawn randomly, do each of the following.
  (a) Write a rational expression that gives the probability of drawing a winning ball.
  (b) Write a rational expression that gives the probability of not drawing a winning ball. Evaluate your expression for $n = 100$ and interpret the result.

106. *Surface Area of a Cylinder* If a cylindrical container has a volume of π cubic feet, then its surface area *S* in square feet (excluding the top and bottom) is given by $S = \frac{2\pi}{r}$, where *r* is the radius of the cylinder.

  (a) Calculate *S* when $r = \frac{1}{2}$ foot.
  (b) What happens to this surface area when *r* becomes large? Sketch this situation.

(c) What happens to the surface area when *r* becomes small (nearly 0)? Sketch this situation.

107. *Distance and Time* A car is traveling at 60 miles per hour.
  (a) How long does it take the car to travel 360 miles?
  (b) Write a rational expression that gives the time that it takes the car to travel *M* miles.

108. *Distance and Time* A bicyclist rides uphill at 10 miles per hour for 5 miles and then rides downhill at 20 miles per hour for 5 miles. What is the bicyclist's average speed? (*Hint:* Average speed equals distance divided by time.)

Exercises 109 and 110: Traffic Flow (Refer to Example 6.)
The figure shows a graph of the waiting time T in minutes at a construction zone when cars are arriving randomly at an average rate of x cars per minute.
  (a) Give the equation of the vertical asymptote.
  (b) Explain how the graph relates to traffic flow.

109.

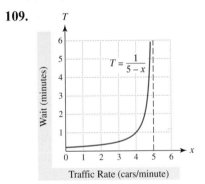

Traffic Rate (cars/minute)

110.

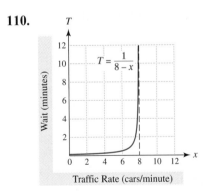

Traffic Rate (cars/minute)

WRITING ABOUT MATHEMATICS

111. Explain what a rational expression is. When is a rational expression undefined?

112. Does the rational expression $\frac{5x + 2}{10x + 4}$ equal $\frac{5x}{10x} + \frac{2}{4}$? Explain your answer.

GROUP ACTIVITY
WORKING WITH REAL DATA

Directions: Form a group of 2 to 4 people. Select someone to record the group's responses for this activity. All members of the group should work cooperatively to answer the questions. If your instructor asks for your results, each member of the group should be prepared to respond.

Students Per Computer   In the early years of microcomputers, school districts could not afford to buy a computer for every student. As the price of computers decreased, more and more school districts have been able to attain this goal. The following table lists numbers of students per computer during these early years.

Year	1983	1985	1987	1989
Students/Computer	125	50	32	22

Year	1991	1993	1995	1997
Students/Computer	18	14	10	6

Source: Quality Education Data, Inc.

(a) Make a scatterplot of the data. Would a straight line model the data accurately? Explain.

(b) Discuss how well the formula

$$S = \frac{125}{1 + 0.7(y - 1983)}, \quad y \geq 1983$$

models these data, where S represents the students per computer and y represents the year.

(c) In what year does the formula reveal that there were about 17 students per computer?

7.2   MULTIPLICATION AND DIVISION OF RATIONAL EXPRESSIONS

Review of Multiplication and Division of Fractions ▪
Multiplication of Rational Expressions ▪ Division of Rational Expressions

Stopping distance for a car can vary depending on the road conditions. If the road is slippery, it takes farther to stop. Also, it takes farther to stop if the car is traveling downhill. Rational expressions are frequently used by highway engineers to estimate the stopping distance of a car on slippery surfaces or on hills.

In previous chapters we reviewed how to add, subtract, multiply, and divide real numbers and polynomials. In this section we show how to multiply and divide rational expressions; in the next section we discuss addition and subtraction of rational expressions.

Review of Multiplication and Division of Fractions

To multiply two fractions we use the property

$$\frac{a}{b} \cdot \frac{c}{d} = \frac{ac}{bd}.$$

In the next example we review multiplication of fractions. (See Section 1.2.)

EXAMPLE 1   Multiplying fractions

Multiply and simplify your answers to lowest terms.

(a) $\frac{3}{7} \cdot \frac{4}{5}$ (b) $2 \cdot \frac{3}{4}$ (c) $\frac{4}{21} \cdot \frac{7}{8}$

Solution

(a) $\frac{3}{7} \cdot \frac{4}{5} = \frac{12}{35}$ (b) $2 \cdot \frac{3}{4} = \frac{2}{1} \cdot \frac{3}{4} = \frac{6}{4} = \frac{3}{2}$

(c) $\frac{4}{21} \cdot \frac{7}{8} = \frac{7 \cdot 4}{21 \cdot 8} = \frac{1}{3} \cdot \frac{1}{2} = \frac{1}{6}$ Now Try Exercises 9, 11, 13

To divide two fractions we "invert and multiply." That is, we change a division problem to a multiplication problem by using

$$\frac{a}{b} \div \frac{c}{d} = \frac{a}{b} \cdot \frac{d}{c}.$$

EXAMPLE 2   Dividing fractions

Divide and simplify your answers to lowest terms.

(a) $\frac{1}{3} \div \frac{5}{7}$ (b) $\frac{4}{5} \div 8$ (c) $\frac{8}{9} \div \frac{10}{3}$

Solution

(a) $\frac{1}{3} \div \frac{5}{7} = \frac{1}{3} \cdot \frac{7}{5} = \frac{7}{15}$

(b) $\frac{4}{5} \div 8 = \frac{4}{5} \cdot \frac{1}{8} = \frac{4}{40} = \frac{1}{10}$

(c) $\frac{8}{9} \div \frac{10}{3} = \frac{8}{9} \cdot \frac{3}{10} = \frac{24}{90} = \frac{4 \cdot 6}{15 \cdot 6} = \frac{4}{15}$ Now Try Exercises 17, 19, 21

Multiplication of Rational Expressions

Multiplying rational expressions is similar to multiplying fractions.

PRODUCTS OF RATIONAL EXPRESSIONS

To multiply two rational expressions multiply the numerators and multiply the denominators. That is,

$$\frac{A}{B} \cdot \frac{C}{D} = \frac{AC}{BD},$$

where B and D are nonzero.

EXAMPLE 3   Multiplying rational expressions

Multiply and simplify to lowest terms. Leave your answers in factored form.

(a) $\frac{3}{x} \cdot \frac{2x - 5}{x - 1}$ (b) $\frac{x - 1}{4x} \cdot \frac{x + 3}{x - 1}$

(c) $\frac{x^2 - 4}{x + 3} \cdot \frac{x + 3}{x + 2}$ (d) $\frac{4}{x^2 + 3x + 2} \cdot \frac{x^2 + 2x + 1}{8}$

Solution

(a) $\dfrac{3}{x} \cdot \dfrac{2x-5}{x-1} = \dfrac{3(2x-5)}{x(x-1)}$

Multiply the numerators and the denominators.

(b) $\dfrac{x-1}{4x} \cdot \dfrac{x+3}{x-1} = \dfrac{(x-1)(x+3)}{4x(x-1)}$

Multiply the numerators and the denominators.

$= \dfrac{(x+3)(x-1)}{4x(x-1)}$

Commutative property

$= \dfrac{x+3}{4x}$

Simplify.

(c) $\dfrac{x^2-4}{x+3} \cdot \dfrac{x+3}{x+2} = \dfrac{(x-2)(x+2)}{x+3} \cdot \dfrac{x+3}{x+2}$

Factor.

$= \dfrac{(x-2)(x+2)(x+3)}{(x+3)(x+2)}$

Multiply the numerators and the denominators.

$= \dfrac{(x-2)(x+2)(x+3)}{(x+2)(x+3)}$

Commutative property

$= x - 2$

Simplify.

(d) $\dfrac{4}{x^2+3x+2} \cdot \dfrac{x^2+2x+1}{8} = \dfrac{4(x^2+2x+1)}{8(x^2+3x+2)}$

Multiply the numerators and the denominators.

$= \dfrac{4(x+1)(x+1)}{8(x+2)(x+1)}$

Factor.

$= \dfrac{x+1}{2(x+2)}$

Simplify.

Now Try Exercises 33, 35, 45, 49

▶ **REAL-WORLD CONNECTION**   The next example illustrates a rational expression that is used in highway design.

EXAMPLE 4   **Estimating stopping distance**

If a car is traveling at 60 miles per hour on a slippery road, then its stopping distance D in feet can be calculated by

$$D = \frac{3600}{30} \cdot \frac{1}{x},$$

where x is the coefficient of friction between the tires and the road and $0 < x \le 1$. The more slippery the road is, the smaller the value of x.   (**Source:** L. Haefner, *Introduction to Transportation Systems*.)
(a) Multiply and simplify the formula for D.
(b) Compare the stopping distance on an icy road with $x = 0.1$ and on dry pavement with $x = 0.4$.

Solution
(a) Because

$$\frac{3600}{30} \cdot \frac{1}{x} = \frac{3600}{30x} = \frac{120 \cdot 30}{x \cdot 30} = \frac{120}{x},$$

it follows that $D = \frac{120}{x}$.

(b) When $x = 0.1$, $D = \frac{120}{0.1} = 1200$ feet, and when $x = 0.4$, $D = \frac{120}{0.4} = 300$ feet. When a car is moving at 60 miles per hour, an icy road can increase the stopping distance by 900 feet, or by a factor of 4.   Now Try Exercise 77

Division of Rational Expressions

Dividing rational expressions is similar to dividing fractions.

> ### QUOTIENTS OF RATIONAL EXPRESSIONS
>
> To divide two rational expressions multiply by the reciprocal of the divisor. That is,
>
> $$\frac{A}{B} \div \frac{C}{D} = \frac{A}{B} \cdot \frac{D}{C},$$
>
> where B, C, and D are nonzero.

EXAMPLE 5   Dividing rational expressions

Divide and simplify to lowest terms.

(a) $\dfrac{5}{2x} \div \dfrac{10}{x-4}$ **(b)** $\dfrac{x^2-9}{x^2+4} \div (x-3)$ **(c)** $\dfrac{x^2-x}{x^2-x-2} \div \dfrac{x}{x-2}$

Solution

(a)
$$\frac{5}{2x} \div \frac{10}{x-4} = \frac{5}{2x} \cdot \frac{x-4}{10}$$ Invert and multiply.
$$= \frac{5(x-4)}{20x}$$ Multiply the numerators and the denominators.
$$= \frac{x-4}{4x}$$ Simplify. Note that $\frac{5}{20} = \frac{1}{4}$.

(b)
$$\frac{x^2-9}{x^2+4} \div (x-3) = \frac{x^2-9}{x^2+4} \cdot \frac{1}{x-3}$$ Invert and multiply.
$$= \frac{x^2-9}{(x^2+4)(x-3)}$$ Multiply the numerators and the denominators.
$$= \frac{(x+3)(x-3)}{(x^2+4)(x-3)}$$ Factor the numerator.
$$= \frac{x+3}{x^2+4}$$ Simplify.

(c)
$$\frac{x^2-x}{x^2-x-2} \div \frac{x}{x-2} = \frac{x^2-x}{x^2-x-2} \cdot \frac{x-2}{x}$$ Invert and multiply.
$$= \frac{(x^2-x)(x-2)}{(x^2-x-2)x}$$ Multiply the numerators and the denominators.
$$= \frac{x(x-1)(x-2)}{x(x+1)(x-2)}$$ Factor numerator and denominator.
$$= \frac{x-1}{x+1}$$ Simplify.

Now Try Exercises 53, 61, 69

7.2 PUTTING IT ALL TOGETHER

In this section we showed how to multiply and divide rational expressions. The following table summarizes these methods.

Concept	Explanation	Examples
Basic Principle of Rational Expressions	$\frac{PR}{QR} = \frac{P}{Q}$, where Q and R are nonzero polynomials. Apply this principle to simplify rational expressions to lowest terms.	$\frac{(x + 2)(2x - 3)}{(x - 1)(2x - 3)} = \frac{x + 2}{x - 1}$
Multiplication of Rational Expressions	Multiply the numerators and multiply the denominators: $$\frac{A}{B} \cdot \frac{C}{D} = \frac{AC}{BD}.$$ Then simplify the result to lowest terms.	$\frac{4x}{x - 1} \cdot \frac{x - 1}{x + 1} = \frac{4x(x - 1)}{(x - 1)(x + 1)}$ $= \frac{4x}{x + 1}$
Division of Rational Expressions	Multiply the first expression by the reciprocal of the second expression: $$\frac{A}{B} \div \frac{C}{D} = \frac{A}{B} \cdot \frac{D}{C}.$$ Then simplify the result to lowest terms.	$\frac{x + 1}{x - 3} \div \frac{x + 1}{x - 5} = \frac{x + 1}{x - 3} \cdot \frac{x - 5}{x + 1}$ $= \frac{(x - 5)(x + 1)}{(x - 3)(x + 1)}$ $= \frac{x - 5}{x - 3}$

7.2 Exercises

MyMathLab   Math XL PRACTICE   WATCH   DOWNLOAD   READ   REVIEW

CONCEPTS

1. $\frac{2}{3} \cdot \frac{5}{7} =$ _____

2. $\frac{2}{3} \div \frac{5}{7} =$ _____

3. $\frac{A}{B} \cdot \frac{C}{D} =$ _____

4. $\frac{A}{B} \div \frac{C}{D} =$ _____

5. Simplify $\frac{(x + 7)(x + 2)}{(x + 1)(x + 2)}$.

6. $\frac{AC}{BC} =$ _____

7. Does $\frac{x + 2}{x}$ equal $\frac{x}{x} + 2$?

8. To divide $\frac{1}{x}$ by $\frac{y}{z}$ multiply _____ by _____.

REVIEW OF FRACTIONS

Exercises 9–24: Simplify to lowest terms.

9. $\frac{1}{2} \cdot \frac{4}{5}$

10. $\frac{6}{7} \cdot \frac{7}{18}$

11. $\frac{3}{7} \cdot 4$

12. $5 \cdot \frac{4}{5}$

13. $\frac{5}{4} \cdot \frac{8}{15}$

14. $\frac{3}{10} \cdot \frac{5}{9}$

15. $\frac{1}{3} \cdot \frac{2}{3} \cdot \frac{9}{11}$

16. $\frac{2}{5} \cdot \frac{10}{11} \cdot \frac{1}{4}$

17. $\frac{2}{3} \div \frac{1}{6}$

18. $\frac{5}{7} \div \frac{5}{8}$

19. $\frac{8}{9} \div \frac{5}{3}$

20. $\frac{7}{3} \div 6$

21. $8 \div \frac{4}{5}$

22. $\frac{1}{2} \div \frac{5}{4} \div \frac{2}{5}$

23. $\frac{4}{5} \div \frac{2}{3} \div \frac{1}{2}$

24. $\frac{7}{20} \div \frac{14}{5}$

MULTIPLYING RATIONAL EXPRESSIONS

Exercises 25–32: Simplify the expression.

25. $\dfrac{x+5}{x+5}$

26. $\dfrac{2x-3}{2x-3}$

27. $\dfrac{(z+1)(z+2)}{(z+4)(z+2)}$

28. $\dfrac{(2z-7)(z+5)}{(3z+5)(z+5)}$

29. $\dfrac{8y(y+7)}{12y(y+7)}$

30. $\dfrac{6(y+1)}{12(y+1)}$

31. $\dfrac{x(x+2)(x+3)}{x(x-2)(x+3)}$

32. $\dfrac{2(x+1)(x-1)}{4(x+1)(x-1)}$

Exercises 33–52: Multiply and simplify to lowest terms. Leave your answers in factored form.

33. $\dfrac{8}{x}\cdot\dfrac{x+1}{x}$

34. $\dfrac{7}{2x}\cdot\dfrac{x}{x-1}$

35. $\dfrac{8+x}{x}\cdot\dfrac{x-3}{x+8}$

36. $\dfrac{5x^2+x}{2x-1}\cdot\dfrac{1}{x}$

37. $\dfrac{z+3}{z+4}\cdot\dfrac{z+4}{z-7}$

38. $\dfrac{2z+1}{3z}\cdot\dfrac{3z}{z+2}$

39. $\dfrac{5x+1}{3x+2}\cdot\dfrac{3x+2}{5x+1}$

40. $\dfrac{x+1}{x+3}\cdot\dfrac{x+3}{x+1}$

41. $\dfrac{(t+1)^2}{t+2}\cdot\dfrac{(t+2)^2}{t+1}$

42. $\dfrac{(t-1)^2}{(t+5)^2}\cdot\dfrac{t+5}{t-1}$

43. $\dfrac{x^2}{x^2+4}\cdot\dfrac{x+4}{x}$

44. $\dfrac{x-1}{x^2}\cdot\dfrac{x^2}{x^2+1}$

45. $\dfrac{z^2-1}{z^2-4}\cdot\dfrac{z-2}{z+1}$

46. $\dfrac{z^2-9}{z-5}\cdot\dfrac{z-5}{z+3}$

47. $\dfrac{y^2-2y}{y^2-1}\cdot\dfrac{y+1}{y-2}$

48. $\dfrac{y^2-4y}{y+1}\cdot\dfrac{y+1}{y-4}$

49. $\dfrac{2x^2-x-3}{3x^2-8x-3}\cdot\dfrac{3x+1}{2x-3}$

50. $\dfrac{6x^2+11x-2}{3x^2+11x-4}\cdot\dfrac{3x-1}{6x-1}$

51. $\dfrac{(x-3)^3}{x^2-2x+1}\cdot\dfrac{x-1}{(x-3)^2}$

52. $\dfrac{x^2+4x+4}{x^2-2x+1}\cdot\dfrac{(x-1)^2}{(x+2)^2}$

Exercises 53–74: Divide and simplify to lowest terms. Leave your answers in factored form.

53. $\dfrac{2}{x}\div\dfrac{2x+3}{x}$

54. $\dfrac{6}{2x}\div\dfrac{x+2}{2x}$

55. $\dfrac{x-2}{3x}\div\dfrac{2-x}{6x}$

56. $\dfrac{x+1}{2x-1}\div\dfrac{x+1}{x}$

57. $\dfrac{z+2}{z+1}\div\dfrac{z+2}{z-1}$

58. $\dfrac{z+7}{z-4}\div\dfrac{z+7}{z-4}$

59. $\dfrac{3y+4}{2y+1}\div\dfrac{3y+4}{y+2}$

60. $\dfrac{y+5}{y-2}\div\dfrac{y}{y+3}$

61. $\dfrac{t^2-1}{t^2+1}\div\dfrac{t+1}{4}$

62. $\dfrac{4}{2t^3}\div\dfrac{8}{t^2}$

63. $\dfrac{y^2-9}{y^2-25}\div\dfrac{y+3}{y+5}$

64. $\dfrac{y+1}{y-4}\div\dfrac{y^2-1}{y^2-16}$

65. $\dfrac{2x^2-4x}{2x-1}\div\dfrac{x-2}{2x-1}$

66. $\dfrac{x-4}{x^2+x}\div\dfrac{5}{x+1}$

67. $\dfrac{2z^2-5z-3}{z^2+z-20}\div\dfrac{z-3}{z-4}$

68. $\dfrac{z^2+12z+27}{z^2-5z-14}\div\dfrac{z+3}{z+2}$

69. $\dfrac{t^2-1}{t^2+5t-6}\div(t+1)$

70. $\dfrac{t^2-2t-3}{t^2-5t-6}\div(t-3)$

71. $\dfrac{a-b}{a+b}\div\dfrac{a-b}{2a+3b}$

72. $\dfrac{x^3-y^3}{x^2-y^2}\div\dfrac{x^2+xy+y^2}{x-y}$

73. $\dfrac{x-y}{x^2+2xy+y^2}\div\dfrac{1}{(x+y)^2}$

74. $\dfrac{a^2-b^2}{4a^2-9b^2}\div\dfrac{a-b}{2a+3b}$

75. Thinking Generally Simplify $\dfrac{a-b}{b-c}\cdot\dfrac{c-b}{b-a}$.

76. Thinking Generally Simplify $\dfrac{a-b}{b-c}\div\dfrac{b-a}{a-b}$.

APPLICATIONS

77. *Stopping on Slippery Roads*   (Refer to Example 4.) If a car is traveling at 30 miles per hour on a slippery road, then its stopping distance D in feet can be calculated by

$$D = \frac{900}{30} \cdot \frac{1}{x},$$

where x is the coefficient of friction between the tires and the road and $0 < x \le 1$. (*Source:* L. Haefner.)
(a) Multiply and simplify the formula for D.
(b) Compare the stopping distance on an icy road with $x = 0.1$ and on dry pavement with $x = 0.4$.

78. *Stopping on Hills*   If a car is traveling at 50 miles per hour on a hill with wet pavement, then its stopping distance D is given by

$$D = \frac{2500}{30} \cdot \frac{1}{x + 0.3},$$

where x equals the slope of the hill.   (*Source:* L. Haefner.)
(a) Multiply and simplify the formula for D.
(b) Compare the stopping distance for an uphill slope of $x = 0.1$ to a downhill slope of $x = -0.1$.

79. *Probability*   Suppose that one jar holds n balls and that a second jar holds $n + 1$ balls. Each jar contains one winning ball.

(a) The probability, or chance, of drawing the winning ball from the first jar and *not* drawing it from the second jar is

$$\frac{1}{n} \cdot \frac{n}{n + 1}.$$

Simplify this expression.
(b) Find this probability for $n = 99$.

80. *U.S. AIDS Cases*   The cumulative number of AIDS cases C in the United States from 1982 to 1994 can be modeled by $C = 3200x^2 + 1586$, and the cumulative number of AIDS deaths D from 1982 to 1994 can be modeled by $D = 1900x^2 + 619$. In these equations $x = 0$ corresponds to 1982, $x = 1$ to 1983, and so on until $x = 12$ corresponds to 1994.   (*Source:* U.S. Department of Health.)
(a) Write the rational expression $\frac{D}{C}$ in terms of x.
(b) Evaluate your expression for $x = 4, 7$, and 10. Round your answers to the nearest thousandth. Interpret the results.
(c) Explain what the rational expression $\frac{D}{C}$ represents.

WRITING ABOUT MATHEMATICS

81. Explain how to multiply two rational expressions.

82. Explain how to divide two rational expressions.

CHECKING BASIC CONCEPTS
SECTIONS 7.1 AND 7.2

1. If possible, evaluate the expression $\frac{3}{x^2 - 1}$ for $x = -1$ and $x = 3$.

2. Simplify to lowest terms.
(a) $\frac{6x^3y^2}{15x^2y^3}$ **(b)** $\frac{5x - 15}{x - 3}$ **(c)** $\frac{x^2 - x - 6}{x^2 + x - 12}$

3. Multiply and simplify to lowest terms.
(a) $\frac{4}{3x} \cdot \frac{2x}{6}$ **(b)** $\frac{2x + 4}{x^2 - 1} \cdot \frac{x + 1}{x + 2}$

4. Divide and simplify to lowest terms.
(a) $\frac{7}{3z^2} \div \frac{14}{5z^3}$ **(b)** $\frac{x^2 + x}{x - 3} \div \frac{x}{x - 3}$

5. *Waiting in Line*   Customers are waiting in line at a department store. They arrive randomly at an average rate of x per minute. If the clerk can wait on 2 customers per minute, then the average time in minutes spent waiting in line is given by $T = \frac{1}{2 - x}$ for $x < 2$.   (*Source:* N. Garber, *Traffic and Highway Engineering*.)
(a) Complete the table.

x	0.5	1.0	1.5	1.9
T				

(b) What happens to the waiting time as x increases but remains less than 2?

7.3 ADDITION AND SUBTRACTION WITH LIKE DENOMINATORS

Review of Addition and Subtraction of Fractions ■ Rational Expressions Having Like Denominators

When companies manufacture a large number of items, quality control is important. For example, suppose that a company makes batteries. Because it is not practical to check every battery to make sure that it works properly, inspectors often check a random sample of batteries. By using mathematics and rational expressions, this technique helps determine the likelihood that all the batteries are good. (See Example 6.)

In this section we discuss methods for adding and subtracting rational expressions having like denominators. These methods are similar to the ones used to add and subtract fractions having like denominators.

Review of Addition and Subtraction of Fractions

In Section 1.2 we demonstrated how the property

$$\frac{a}{c} + \frac{b}{c} = \frac{a+b}{c}$$

can be used to add fractions having like denominators. For example,

$$\frac{3}{7} + \frac{2}{7} = \frac{3+2}{7} = \frac{5}{7}.$$

To subtract two fractions having like denominators the property

$$\frac{a}{c} - \frac{b}{c} = \frac{a-b}{c}$$

is used. For example,

$$\frac{2}{5} - \frac{4}{5} = \frac{2-4}{5} = -\frac{2}{5}.$$

EXAMPLE 1   Adding and subtracting fractions having like denominators

Simplify each expression to lowest terms.

(a) $\frac{3}{8} + \frac{4}{8}$ **(b)** $\frac{5}{9} + \frac{1}{9}$ **(c)** $\frac{12}{5} - \frac{7}{5}$ **(d)** $\frac{23}{20} - \frac{13}{20}$

Solution

(a) $\frac{3}{8} + \frac{4}{8} = \frac{3+4}{8} = \frac{7}{8}$

(b) $\frac{5}{9} + \frac{1}{9} = \frac{5+1}{9} = \frac{6}{9} = \frac{2}{3}$

(c) $\frac{12}{5} - \frac{7}{5} = \frac{12-7}{5} = \frac{5}{5} = 1$

(d) $\frac{23}{20} - \frac{13}{20} = \frac{23-13}{20} = \frac{10}{20} = \frac{1}{2}$

Now Try Exercises 15, 19

TECHNOLOGY NOTE: **Arithmetic of Fractions**

Many calculators have the capability to perform addition and subtraction of fractions, as illustrated in the following figures. Compare these results with those from Example 1.

```
(3/8)+(4/8)▶Frac
                7/8
(5/9)+(1/9)▶Frac
                2/3
```

```
(12/5)-(7/5)▶Fra
c
                 1
(23/20)-(13/20)▶
Frac
               1/2
```

Rational Expressions Having Like Denominators

Addition and subtraction of rational expressions having like denominators are similar to addition and subtraction of fractions. The following property can be used to add two rational expressions having like denominators.

SUMS OF RATIONAL EXPRESSIONS

To add two rational expressions having like denominators, add their numerators. Keep the same denominator.

$$\frac{A}{C} + \frac{B}{C} = \frac{A + B}{C} \qquad C \text{ is nonzero.}$$

When we add rational expressions with like denominators, we add the numerators. Then we combine like terms and simplify the resulting expression by applying the basic principle of rational expressions. For example, we can add $\frac{2x}{x+1}$ and $\frac{1-x}{x+1}$ as follows.

$$
\begin{aligned}
\frac{2x}{x+1} + \frac{1-x}{x+1} &= \frac{2x + 1 - x}{x+1} \qquad &\text{Add the numerators.}\\
&= \frac{2x - x + 1}{x+1} \qquad &\text{Commutative property}\\
&= \frac{x+1}{x+1} \qquad &\text{Combine like terms.}\\
&= 1 \qquad &\text{Simplify.}
\end{aligned}
$$

It is important to understand that the expressions $\frac{2x}{x+1} + \frac{1-x}{x+1}$ and 1 are *equivalent expressions*. That is, they are equal for *every* value of x except -1, for which the first expression is undefined.

In the next example, we add rational expressions with like denominators and simplify the result to lowest terms.

EXAMPLE 2   **Adding rational expressions having like denominators**

Add and simplify to lowest terms.

(a) $\dfrac{3}{b} + \dfrac{2}{b}$

(b) $\dfrac{z}{z+2} + \dfrac{2}{z+2}$

(c) $\dfrac{x-1}{x^2+x} + \dfrac{1}{x^2+x}$

(d) $\dfrac{t^2+t}{t-1} + \dfrac{1-3t}{t-1}$

Solution

(a) $\dfrac{3}{b} + \dfrac{2}{b} = \dfrac{3+2}{b} = \dfrac{5}{b}$ Add the numerators.

(b) $\dfrac{z}{z+2} + \dfrac{2}{z+2} = \dfrac{z+2}{z+2}$ Add the numerators.

$= 1$ Simplify.

(c) $\dfrac{x-1}{x^2+x} + \dfrac{1}{x^2+x} = \dfrac{x-1+1}{x^2+x}$ Add the numerators.

$= \dfrac{x}{x(x+1)}$ Factor the denominator.

$= \dfrac{1}{x+1}$ Simplify.

(d) $\dfrac{t^2+t}{t-1} + \dfrac{1-3t}{t-1} = \dfrac{t^2+t+1-3t}{t-1}$ Add the numerators.

$= \dfrac{t^2-2t+1}{t-1}$ Combine like terms.

$= \dfrac{(t-1)(t-1)}{t-1}$ Factor the numerator.

$= t-1$ Simplify.

Now Try Exercises 25, 32, 39, 41

EXAMPLE 3   **Adding rational expressions having two variables**

Add and simplify to lowest terms.

(a) $\dfrac{4}{xy} + \dfrac{5}{xy}$ (b) $\dfrac{a}{a^2-b^2} + \dfrac{b}{a^2-b^2}$ (c) $\dfrac{1}{x-y} + \dfrac{-1}{y-x}$

Solution

(a) $\dfrac{4}{xy} + \dfrac{5}{xy} = \dfrac{4+5}{xy} = \dfrac{9}{xy}$ Add the numerators.

(b) $\dfrac{a}{a^2-b^2} + \dfrac{b}{a^2-b^2} = \dfrac{a+b}{a^2-b^2}$ Add the numerators.

$= \dfrac{a+b}{(a-b)(a+b)}$ Factor the denominator.

$= \dfrac{1}{a-b}$ Simplify.

(c) First write $\dfrac{1}{x-y} + \dfrac{-1}{y-x}$ with a common denominator. Note that, if we multiply the second term by 1, written in the form $\dfrac{-1}{-1}$, it becomes

$$\dfrac{-1}{y-x} \cdot \dfrac{-1}{-1} = \dfrac{(-1)(-1)}{(y-x)(-1)} = \dfrac{1}{-y+x} = \dfrac{1}{x-y}.$$

Thus the given sum can be simplified as follows.

$$\frac{1}{x - y} + \frac{-1}{y - x} = \frac{1}{x - y} + \frac{1}{x - y} \qquad \text{Rewrite the second term.}$$

$$= \frac{2}{x - y} \qquad \text{Add the numerators.}$$

Now Try Exercises 57, 59, 61

Next we consider subtraction of rational expressions having like denominators.

DIFFERENCES OF RATIONAL EXPRESSIONS

To subtract two rational expressions having like denominators, subtract their numerators. Keep the same denominator.

$$\frac{A}{C} - \frac{B}{C} = \frac{A - B}{C} \qquad C \text{ is nonzero.}$$

Subtraction of rational expressions with like denominators is similar to addition except that instead of adding numerators, we subtract them. For example, the expressions $\frac{3x}{x - 4}$ and $\frac{2x}{x - 4}$ have like denominators and can be subtracted as follows.

$$\frac{3x}{x - 4} - \frac{2x}{x - 4} = \frac{3x - 2x}{x - 4} \qquad \text{Subtract the numerators.}$$

$$= \frac{x}{x - 4} \qquad \text{Combine like terms.}$$

In the next example, we subtract rational expressions having like denominators and simplify the result to lowest terms.

EXAMPLE 4   Subtracting rational expressions having like denominators

Subtract and simplify to lowest terms.

(a) $\dfrac{a + 1}{a} - \dfrac{1}{a}$ **(b)** $\dfrac{2y}{3y - 1} - \dfrac{3y}{3y - 1}$ **(c)** $\dfrac{1 + x}{2x^2 + 5x - 3} - \dfrac{-2}{2x^2 + 5x - 3}$

Solution

(a) $\dfrac{a + 1}{a} - \dfrac{1}{a} = \dfrac{a + 1 - 1}{a} = \dfrac{a}{a} = 1$

(b) $\dfrac{2y}{3y - 1} - \dfrac{3y}{3y - 1} = \dfrac{2y - 3y}{3y - 1} = \dfrac{-y}{3y - 1}$ or $-\dfrac{y}{3y - 1}$

(c) $\dfrac{1 + x}{2x^2 + 5x - 3} - \dfrac{-2}{2x^2 + 5x - 3} = \dfrac{1 + x - (-2)}{2x^2 + 5x - 3}$

$$= \frac{x + 3}{(2x - 1)(x + 3)}$$

$$= \frac{1}{2x - 1}$$

Now Try Exercises 27, 33, 51

If the numerator of the second fraction in a difference has more than one term, it is important to put parentheses around the second numerator.

$$\frac{x+1}{2x+1} - \frac{3-x}{2x+1} = \frac{x+1-(3-x)}{2x+1}$$ Subtract the numerators; insert parentheses.

$$= \frac{x+1-3-(-x)}{2x+1}$$ Distributive property

$$= \frac{x+1-3+x}{2x+1}$$ Double negative property

$$= \frac{2x-2}{2x+1}$$ Combine like terms.

NOTE: If parentheses were not inserted in the previous calculation, the numerator would be

$$x + 1 - 3 - x = -2,$$

which would give an incorrect result.

EXAMPLE 5   Subtracting rational expressions having like denominators

Subtract and simplify to lowest terms.

(a) $\dfrac{2x}{x+1} - \dfrac{x-1}{x+1}$ (b) $\dfrac{x+y}{3y} - \dfrac{x-y}{3y}$

Solution

(a) $\dfrac{2x}{x+1} - \dfrac{x-1}{x+1} = \dfrac{2x-(x-1)}{x+1}$ Subtract the numerators.

$$= \frac{2x-x+1}{x+1}$$ Distributive property

$$= \frac{x+1}{x+1}$$ Simplify the numerator.

$$= 1$$ Simplify.

(b) $\dfrac{x+y}{3y} - \dfrac{x-y}{3y} = \dfrac{x+y-(x-y)}{3y}$ Subtract the numerators.

$$= \frac{x+y-x+y}{3y}$$ Distributive property

$$= \frac{2y}{3y}$$ Simplify the numerator.

$$= \frac{2}{3}$$ Simplify.

Now Try Exercises 37, 63

▶ **REAL-WORLD CONNECTION**   The introduction to this section discusses how quality control uses rational expressions. The next example illustrates one way this can occur.

EXAMPLE 6   Analyzing quality control

A container holds a mixture of size A and size AA batteries. In this container, there is a total of n batteries, including 2 defective A batteries and 4 defective AA batteries. If a battery is picked at random by a quality control inspector, then the probability, or chance, of one of the defective batteries being chosen is given by the expression $\frac{2}{n} + \frac{4}{n}$.

(a) Simplify this expression.   (b) Interpret the result.

Solution

(a) Because the denominators are the same, we simply add the numerators.

$$\frac{2}{n} + \frac{4}{n} = \frac{2+4}{n} = \frac{6}{n}$$

(b) There are 6 in n chances that a defective battery is chosen.   Now Try Exercise 73

7.3 PUTTING IT ALL TOGETHER

In this section we discussed how to add and subtract rational expressions having like denominators. After adding or subtracting rational expressions, be sure to factor the numerator and the denominator completely. Then simplify the result to lowest terms by applying the basic principle of rational expressions. The following table summarizes this discussion, where A, B, and C represent polynomials and C is nonzero.

Concept	Explanation	Examples
Addition of Rational Expressions	$\dfrac{A}{C} + \dfrac{B}{C} = \dfrac{A+B}{C}$	$\dfrac{x}{x+1} + \dfrac{1-x}{x+1} = \dfrac{x+1-x}{x+1} = \dfrac{1}{x+1}$ $\dfrac{2x}{x^2-1} + \dfrac{x}{x^2-1} = \dfrac{2x+x}{x^2-1} = \dfrac{3x}{x^2-1}$
Subtraction of Rational Expressions	$\dfrac{A}{C} - \dfrac{B}{C} = \dfrac{A-B}{C}$ If B consists of more than one term, put parentheses around B and apply the distributive property.	$\dfrac{2x}{x^2-4} - \dfrac{x+2}{x^2-4} = \dfrac{2x-(x+2)}{x^2-4}$ $= \dfrac{2x-x-2}{x^2-4}$ $= \dfrac{x-2}{(x+2)(x-2)}$ $= \dfrac{1}{x+2}$

7.3 Exercises

MyMathLab
PRACTICE   WATCH   DOWNLOAD   READ   REVIEW

CONCEPTS

1. To add two rational expressions having like denominators, ——— their ———. The ——— do not change.

2. To subtract two rational expressions having like denominators, ——— their ———. The ——— do not change.

3. $\frac{2}{5} + \frac{1}{5} = $ ———

4. $\frac{6}{7} - \frac{2}{7} = $ ———

5. $\frac{A}{C} + \frac{B}{C} = $ ———

6. $\frac{A}{C} - \frac{B}{C} = $ ———

7. $5 - (x + 1) = $ ———

8. $\frac{5}{x - 1} - \frac{x + 1}{x - 1} = $ ———

9. $3x - (2x - 5) = $ ———

10. $\frac{3x}{7} - \frac{2x - 5}{7} = $ ———

ADDITION AND SUBTRACTION OF FRACTIONS

Exercises 11–24: Simplify to lowest terms.

11. $\frac{1}{2} + \frac{1}{2}$

12. $\frac{3}{7} + \frac{2}{7}$

13. $\frac{4}{5} + \frac{2}{5}$

14. $\frac{3}{11} + \frac{5}{11}$

15. $\frac{1}{6} + \frac{5}{6}$

16. $\frac{3}{10} + \frac{5}{10}$

17. $\frac{4}{7} - \frac{1}{7}$

18. $\frac{5}{13} - \frac{7}{13}$

19. $\frac{7}{8} - \frac{3}{8}$

20. $\frac{9}{16} - \frac{5}{16}$

21. $\frac{11}{12} - \frac{5}{12}$

22. $\frac{7}{24} - \frac{3}{24}$

23. $\frac{7}{15} + \frac{4}{15} - \frac{1}{15}$

24. $\frac{11}{36} - \frac{5}{36} + \frac{1}{36}$

ADDITION AND SUBTRACTION OF RATIONAL EXPRESSIONS

Exercises 25–70: Simplify to lowest terms.

25. $\frac{2}{x} + \frac{1}{x}$

26. $\frac{9}{x} - \frac{7}{x}$

27. $\frac{7 + 2x}{4x} - \frac{7}{4x}$

28. $\frac{x - 1}{5x} + \frac{2x + 1}{5x}$

29. $\frac{y + 3}{y - 3} + \frac{2y - 12}{y - 3}$

30. $\frac{5 - y}{y + 2} + \frac{y}{y + 2}$

31. $\frac{x}{x - 3} - \frac{3}{x - 3}$

32. $\frac{2x}{2x + 1} + \frac{1}{2x + 1}$

33. $\frac{5z}{4z + 3} - \frac{z}{4z + 3}$

34. $\frac{z}{2z + 1} - \frac{1 - z}{2z + 1}$

35. $\frac{t + 5}{t + 6} + \frac{t + 7}{t + 6}$

36. $\frac{t + 1}{t - 4} + \frac{1 - t}{t - 4}$

37. $\frac{5x}{2x + 3} - \frac{3x - 3}{2x + 3}$

38. $\frac{x}{5 - x} - \frac{2x - 5}{5 - x}$

39. $\frac{x - 4}{x^2 - x} + \frac{4}{x^2 - x}$

40. $\frac{2x - 2}{4x^2 - 1} + \frac{1}{4x^2 - 1}$

41. $\frac{z^2 - 1}{z - 2} + \frac{3 - 3z}{z - 2}$

42. $\frac{x^2 + 2}{x + 1} + \frac{3x}{x + 1}$

43. $\frac{x^2 + 4x - 1}{4x + 2} - \frac{x^2 - 4x - 5}{4x + 2}$

44. $\frac{2x^2 - x + 5}{x^2 - 9} - \frac{x^2 - x + 14}{x^2 - 9}$

45. $\frac{3y}{5} + \frac{2y - 5}{5}$

46. $\frac{3y - 22}{11} + \frac{8y}{11}$

47. $\frac{x + y}{4} + \frac{x - y}{4}$

48. $\frac{x + y}{4} - \frac{x - y}{4}$

49. $\frac{z^2 + 4}{z - 2} - \frac{4z}{z - 2}$

50. $\frac{z^2 + 2z}{z + 1} + \frac{1}{z + 1}$

51. $\frac{2x^2 - 5x}{2x + 1} - \frac{3}{2x + 1}$

52. $\frac{2x^2}{x + 2} + \frac{9x + 10}{x + 2}$

53. $\frac{3n}{2n^2 - n + 5} + \frac{4n}{2n^2 - n + 5}$

54. $\frac{n}{n^2 + n + 1} - \frac{1}{n^2 + n + 1}$

55. $\frac{1}{x + 3} + \frac{2}{x + 3} + \frac{3}{x + 3}$

56. $\frac{x}{2x - 5} - \frac{1}{2x - 5} + \frac{2x + 1}{2x - 5}$

57. $\dfrac{8}{ab} + \dfrac{1}{ab}$

58. $\dfrac{6}{xy} + \dfrac{9}{xy}$

59. $\dfrac{x}{(x+y)^2} + \dfrac{y}{(x+y)^2}$

60. $\dfrac{x-2y}{x^2-y^2} + \dfrac{y}{x^2-y^2}$

61. $\dfrac{8}{a-b} + \dfrac{-8}{b-a}$

62. $\dfrac{6}{x-y} + \dfrac{6}{y-x}$

63. $\dfrac{a+b}{4a} - \dfrac{a-b}{4a}$

64. $\dfrac{x-y}{5x} - \dfrac{x+y}{5x}$

65. $\dfrac{x}{x+y} + \dfrac{y}{x+y}$

66. $\dfrac{x}{x^2-y^2} - \dfrac{y}{x^2-y^2}$

67. $\dfrac{a^2}{a+b} - \dfrac{b^2}{a+b}$

68. $\dfrac{a^2}{a+b} + \dfrac{2ab+b^2}{a+b}$

69. $\dfrac{4x^2}{2x+3y} - \dfrac{9y^2}{2x+3y}$

70. $\dfrac{x^3}{x^2+xy+y^2} - \dfrac{y^3}{x^2+xy+y^2}$

71. Thinking Generally If $\dfrac{2}{3+x} + \dfrac{3}{3+x}$ equals $\dfrac{5}{10}$, what must be true about x?

72. Thinking Generally If $\dfrac{8}{6+x} - \dfrac{4}{3+y}$ equals 0, what must be true about x and y?

APPLICATIONS

73. *Quality Control*   (Refer to Example 6.) A container holds a total of $n + 1$ batteries. In this container, there are 6 defective B batteries, 5 defective C batteries, and 3 defective D batteries. If a battery is chosen at random by a quality control inspector, the probability, or chance, of one of the defective batteries being chosen is

$$\frac{6}{n+1} + \frac{5}{n+1} + \frac{3}{n+1}.$$

(a) Simplify this expression.
(b) Evaluate the simplified expression for $n = 99$ and interpret the result.

74. *Intensity of a Light Bulb*   The farther a person is from a light bulb, the less intense its light is. The equation $I = \dfrac{19}{4d^2}$ approximates the intensity of light from a 60-watt light bulb at a distance of d meters, where I is measured in watts per square meter. (**Source:** R. Weidner.)

(a) Find I for $d = 2$ meters and interpret the result.
(b) The intensity of light from a 100-watt light bulb is about $I = \dfrac{32}{4d^2}$. Find an expression for the sum of the intensities of light from a 100-watt bulb and a 60-watt bulb.

WRITING ABOUT MATHEMATICS

75. Explain how to add two rational expressions having like denominators. Give an example.

76. Explain how to subtract two rational expressions having like denominators. Give an example.

7.4 ADDITION AND SUBTRACTION WITH UNLIKE DENOMINATORS

Finding Least Common Multiples ■ Review of Fractions Having Unlike Denominators ■ Rational Expressions Having Unlike Denominators

A LOOK INTO MATH ▷

The sum and difference of rational expressions frequently occur in the design of electrical devices, such as cell phones and televisions. In Section 7.3, we added and subtracted rational expressions having like denominators. Although the denominators of rational expressions are often unlike, rational expressions can still be added and subtracted after a common denominator is found. One way to find the least common denominator for a sum or difference of rational expressions is to find the least common multiple of the denominators. In the following subsection we show how to find the least common multiple.

Finding Least Common Multiples

▶ **REAL-WORLD CONNECTION**   Two friends work part-time at a store. The first person works every sixth day, and the second person works every eighth day. If they both work today, how many days will pass before they work on the same day again?

We can answer this question by listing the days that each person works.

First person: 6,   12,   18,   **24**,   30,   36,   42,   **48**,   54

Second person: 8,   16,   **24**,   32,   40,   **48**,   56,   64

After 24 days, the two friends work on the same day. The next time is after 48 days. The numbers 24 and 48 are *common multiples* of 6 and 8. (Find another.) However, 24 is the *least* common multiple (LCM) of 6 and 8.

Another way to find the least common multiple of 6 and 8 is to factor each number into prime numbers.

$$6 = 2 \cdot 3 \quad \text{and} \quad 8 = 2 \cdot 2 \cdot 2$$

To find the least common multiple, first list each factor the greatest number of times that it occurs in either factorization. Then find the product of these numbers. For this example, the factor 2 occurs three times in the factorization of 8 and only once in the factorization of 6, so list 2 three times. The factor 3 appears only once in the factorization of 6 so list it once:

$$2, \ 2, \ 2, \ 3.$$

The least common multiple is their product: $2 \cdot 2 \cdot 2 \cdot 3 = 24$.

This same procedure can also be used to find the least common multiple for two or more polynomials.

FINDING THE LEAST COMMON MULTIPLE

The least common multiple (LCM) of two or more polynomials can be found as follows.

STEP 1: Factor each polynomial completely.

STEP 2: List each factor the greatest number of times that it occurs in any factorization.

STEP 3: Find the product of this list of factors. The result is the LCM.

EXAMPLE 1   Finding least common multiples

Find the least common multiple of each pair of expressions.
(a) $2x, 5x^2$ **(b)** $x^2 - x, x - 1$
(c) $x + 2, x - 3$ **(d)** $x^2 + 2x + 1, x^2 + 3x + 2$

Solution
(a) STEP 1: Factor $2x$ and $5x^2$ completely.

$$2x = 2 \cdot x \quad \text{and} \quad 5x^2 = 5 \cdot x \cdot x$$

STEP 2: In either factorization, the factor 2 occurs at most once, the factor 5 occurs at most once, and the factor x occurs at most twice. The list of factors is: **2, 5, x, x.**

STEP 3: The LCM equals the product

$$2 \cdot 5 \cdot x \cdot x = 10x^2.$$

(b) STEP 1: Factor $x^2 - x$ and $x - 1$ completely. Note that $x - 1$ cannot be factored.

$$x^2 - x = x(x - 1) \quad \text{and} \quad x - 1 = x - 1$$

STEP 2: Both factors, x and $x - 1$, occur at most once in either factorization. The list of factors is x, $(x - 1)$.

STEP 3: The LCM is the product $x(x - 1)$, or $x^2 - x$.

(c) STEP 1: Neither $x + 2$ nor $x - 3$ can be factored.

STEP 2: The list of factors is $(x + 2)$, $(x - 3)$.

STEP 3: The LCM is the product $(x + 2)(x - 3)$, or $x^2 - x - 6$.

(d) STEP 1: Factor $x^2 + 2x + 1$ and $x^2 + 3x + 2$ completely.

$$x^2 + 2x + 1 = (x + 1)(x + 1) \quad \text{and} \quad x^2 + 3x + 2 = (x + 1)(x + 2)$$

STEP 2: In either factorization, the factor $(x + 1)$ occurs at most twice and the factor $(x + 2)$ occurs at most once. The list is $(x + 1)$, $(x + 1)$, $(x + 2)$.

STEP 3: The LCM is the product $(x + 1)^2(x + 2)$.   Now Try Exercises 17, 21, 33, 37

Review of Fractions Having Unlike Denominators

Before we can find the sum $\frac{1}{2} + \frac{1}{3}$ by hand, we need to rewrite these fractions by using their least common denominator. The least common denominator (LCD) of $\frac{1}{2}$ and $\frac{1}{3}$ corresponds to the least common multiple of 2 and 3, which is 6. As a result, we rewrite these fractions as

$$\frac{1}{2} \cdot \frac{3}{3} = \frac{3}{6} \quad \text{and} \quad \frac{1}{3} \cdot \frac{2}{2} = \frac{2}{6}.$$

Their sum is $\frac{1}{2} + \frac{1}{3} = \frac{3}{6} + \frac{2}{6} = \frac{5}{6}$.

EXAMPLE 2   **Adding and subtracting fractions having unlike denominators**

Simplify each expression.

(a) $\frac{3}{10} + \frac{4}{15}$ **(b)** $\frac{7}{8} - \frac{1}{6}$

Solution

(a) The LCD for $\frac{3}{10}$ and $\frac{4}{15}$ equals the LCM of 10 and 15, which is 30. We rewrite these fractions as

$$\frac{3}{10} \cdot \frac{3}{3} = \frac{9}{30} \quad \text{and} \quad \frac{4}{15} \cdot \frac{2}{2} = \frac{8}{30}.$$

Their sum is

$$\frac{3}{10} + \frac{4}{15} = \frac{9}{30} + \frac{8}{30} = \frac{17}{30}.$$

(b) The LCD for $\frac{7}{8}$ and $\frac{1}{6}$ is the LCM of 8 and 6, which is 24. We rewrite these fractions as

$$\frac{7}{8} \cdot \frac{3}{3} = \frac{21}{24} \quad \text{and} \quad \frac{1}{6} \cdot \frac{4}{4} = \frac{4}{24}.$$

Their difference is

$$\frac{7}{8} - \frac{1}{6} = \frac{21}{24} - \frac{4}{24} = \frac{17}{24}.$$ Now Try Exercises 51, 53

Rational Expressions Having Unlike Denominators

The first step in adding or subtracting rational expressions having unlike denominators is to rewrite each expression by using the least common denominator. Then the sum or difference can be found by using the techniques discussed in Section 7.3.

NOTE: The LCD of two or more rational expressions equals the LCM of their denominators.

To add or subtract rational expressions, we often rewrite a rational expression with a different denominator. This technique is demonstrated in the next example and is used in future examples.

EXAMPLE 3   **Rewriting rational expressions**

Rewrite each rational expression so it has the given denominator D.

(a) $\dfrac{3}{2x}$, $D = 8x^2$ **(b)** $\dfrac{1}{x + 1}$, $D = x^2 - 1$

Solution

(a) We need to write $\dfrac{3}{2x}$ so that it is equivalent to $\dfrac{?}{8x^2}$. Because $8x^2 = 2x \cdot 4x$, we can multiply $\dfrac{3}{2x}$ by 1 in the form $\dfrac{4x}{4x}$ as follows.

$$\frac{3}{2x} \cdot \frac{4x}{4x} = \frac{12x}{8x^2} \qquad \text{Multiply rational expressions.}$$

(b) We must write $\dfrac{1}{x + 1}$ so that it is equivalent to $\dfrac{?}{x^2 - 1}$. Because $x^2 - 1 = (x + 1)(x - 1)$, we can multiply $\dfrac{1}{x + 1}$ by 1 in the form $\dfrac{x - 1}{x - 1}$ as follows.

$$\frac{1}{x + 1} \cdot \frac{x - 1}{x - 1} = \frac{x - 1}{x^2 - 1} \qquad \text{Multiply rational expressions.}$$

Now Try Exercises 43, 45

EXAMPLE 4   **Adding rational expressions having unlike denominators**

Find each sum and leave your answer in factored form.

(a) $\dfrac{5}{8y} + \dfrac{7}{4y^2}$ **(b)** $\dfrac{1}{x - 1} + \dfrac{1}{x + 1}$ **(c)** $\dfrac{x}{x^2 + 2x + 1} + \dfrac{1}{x + 1}$

Solution

(a) First find the LCM for $8y$ and $4y^2$.

$$8y = 2 \cdot 2 \cdot 2 \cdot y \quad \text{and} \quad 4y^2 = 2 \cdot 2 \cdot y \cdot y$$

Thus the LCM is $2 \cdot 2 \cdot 2 \cdot y \cdot y = 8y^2$. Because

$$8y^2 = 8y \cdot y \quad \text{and} \quad 8y^2 = 4y^2 \cdot 2,$$

we multiply the first expression by $\dfrac{y}{y}$ and the second expression by $\dfrac{2}{2}$.

$$\frac{5}{8y} + \frac{7}{4y^2} = \frac{5}{8y} \cdot \frac{y}{y} + \frac{7}{4y^2} \cdot \frac{2}{2} \qquad \text{Rewrite by using the LCD.}$$

$$= \frac{5y}{8y^2} + \frac{14}{8y^2} \qquad \text{Multiply the fractions.}$$

$$= \frac{5y + 14}{8y^2} \qquad \text{Add the numerators.}$$

(b) The LCM for $x - 1$ and $x + 1$ is their product, $(x - 1)(x + 1)$.

$$\frac{1}{x - 1} + \frac{1}{x + 1} = \frac{1}{x - 1} \cdot \frac{x + 1}{x + 1} + \frac{1}{x + 1} \cdot \frac{x - 1}{x - 1}$$ Rewrite by using the LCD.

$$= \frac{x + 1}{(x - 1)(x + 1)} + \frac{x - 1}{(x + 1)(x - 1)}$$ Multiply the fractions.

$$= \frac{x + 1 + x - 1}{(x - 1)(x + 1)}$$ Add the numerators.

$$= \frac{2x}{(x - 1)(x + 1)}$$ Simplify the numerator.

(c) First find the LCM for $x^2 + 2x + 1$ and $x + 1$. Because

$$x^2 + 2x + 1 = (x + 1)(x + 1),$$

their LCM is $(x + 1)(x + 1) = (x + 1)^2$.

$$\frac{x}{x^2 + 2x + 1} + \frac{1}{x + 1} = \frac{x}{(x + 1)^2} + \frac{1}{x + 1} \cdot \frac{x + 1}{x + 1}$$ Rewrite by using the LCD.

$$= \frac{x}{(x + 1)^2} + \frac{x + 1}{(x + 1)^2}$$ Multiply the fractions.

$$= \frac{2x + 1}{(x + 1)^2}$$ Add the numerators.

Now Try Exercises 59, 71, 77

Subtraction of rational expressions is performed in a manner similar to addition and is illustrated in the next example.

EXAMPLE 5 Subtracting rational expressions having unlike denominators

Simplify each expression. Write your answer in lowest terms and leave it in factored form.

(a) $\dfrac{5}{z} - \dfrac{z}{z - 1}$

(b) $\dfrac{5}{x + 1} - \dfrac{1}{x^2 - 1}$

(c) $\dfrac{x}{x^2 - 2x} - \dfrac{1}{x^2 + 2x}$

(d) $\dfrac{3}{x - 1} - \dfrac{3}{x^2 - x} + \dfrac{5}{x}$

Solution

(a) The LCD is $z(z - 1)$.

$$\frac{5}{z} - \frac{z}{z - 1} = \frac{5}{z} \cdot \frac{z - 1}{z - 1} - \frac{z}{z - 1} \cdot \frac{z}{z}$$ Rewrite by using the LCD.

$$= \frac{5(z - 1)}{z(z - 1)} - \frac{z^2}{z(z - 1)}$$ Multiply the fractions.

$$= \frac{5(z - 1) - z^2}{z(z - 1)}$$ Add the numerators.

$$= \frac{-z^2 + 5z - 5}{z(z - 1)}$$ Simplify the numerator.

(b) The LCD is $(x - 1)(x + 1)$. Note that $x^2 - 1 = (x - 1)(x + 1)$.

$$\frac{5}{x + 1} - \frac{1}{x^2 - 1} = \frac{5}{x + 1} \cdot \frac{x - 1}{x - 1} - \frac{1}{(x - 1)(x + 1)}$$ Rewrite by using the LCD.

$$= \frac{5(x - 1) - 1}{(x - 1)(x + 1)}$$ Subtract the numerators.

$$= \frac{5x - 5 - 1}{(x - 1)(x + 1)}$$ Distributive property

$$= \frac{5x - 6}{(x - 1)(x + 1)}$$ Simplify the numerator.

(c) Start by factoring each denominator. Because

$$x^2 - 2x = x(x - 2) \quad \text{and} \quad x^2 + 2x = x(x + 2),$$

the LCD is $x(x - 2)(x + 2)$.

$$\frac{x}{x^2 - 2x} - \frac{1}{x^2 + 2x} = \frac{x}{x(x - 2)} \cdot \frac{x + 2}{x + 2} - \frac{1}{x(x + 2)} \cdot \frac{x - 2}{x - 2}$$ Rewrite by using the LCD.

$$= \frac{x(x + 2)}{x(x - 2)(x + 2)} - \frac{x - 2}{x(x - 2)(x + 2)}$$ Multiply the rational expressions.

$$= \frac{x(x + 2) - (x - 2)}{x(x - 2)(x + 2)}$$ Subtract the numerators.

$$= \frac{x^2 + 2x - x + 2}{x(x - 2)(x + 2)}$$ Distributive property

$$= \frac{x^2 + x + 2}{x(x - 2)(x + 2)}$$ Simplify the numerator.

(d) The given expression contains three rational expressions. Begin by finding the LCM of the three denominators: $x - 1, x^2 - x,$ and x. Because $x^2 - x = x(x - 1)$, the LCM is $x(x - 1)$.

$$\frac{3}{x - 1} - \frac{3}{x^2 - x} + \frac{5}{x} = \frac{3}{x - 1} \cdot \frac{x}{x} - \frac{3}{x(x - 1)} + \frac{5}{x} \cdot \frac{x - 1}{x - 1}$$ Rewrite by using the LCD.

$$= \frac{3x}{x(x - 1)} - \frac{3}{x(x - 1)} + \frac{5(x - 1)}{x(x - 1)}$$ Multiply the expressions.

$$= \frac{3x - 3 + 5x - 5}{x(x - 1)}$$ Combine the expressions.

$$= \frac{8x - 8}{x(x - 1)}$$ Simplify the numerator.

$$= \frac{8(x - 1)}{x(x - 1)}$$ Factor the numerator.

$$= \frac{8}{x}$$ Simplify to lowest terms.

Now Try Exercises 69, 83, 85, 87

CRITICAL THINKING

Find the reciprocal of the sum $x + \frac{1}{x}$.

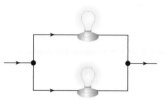

Figure 7.2

▶ **REAL-WORLD CONNECTION**   Sums of rational expressions occur in applications involving electricity. The flow of electricity through a wire can be compared to the flow of water through a hose. Voltage is the force "pushing" the electricity and corresponds to water pressure in a hose. Resistance is the opposition to the flow of electricity and corresponds to the diameter of a hose. More resistance results in less flow of electricity. An ordinary light bulb is an example of a resistor in an electrical circuit. Resistance is often measured in units called *ohms*. For example, a standard 60-watt light bulb has a resistance of about 200 ohms.

Suppose that two light bulbs are wired in parallel so that electricity can flow through either light bulb, as illustrated in Figure 7.2. If the individual resistances of the light bulbs are R and S, then the total resistance of the circuit is given by the *reciprocal* of the sum

$$\frac{1}{R} + \frac{1}{S}.$$

(**Source:** R. Weidner and R. Sells, *Elementary Classical Physics, Vol. 2.*)

EXAMPLE 6   **Modeling electrical resistance**

Add $\frac{1}{R} + \frac{1}{S}$, and then find the reciprocal of the result.

Solution
The LCD for $\frac{1}{R}$ and $\frac{1}{S}$ is RS.

$$\frac{1}{R} + \frac{1}{S} = \frac{1}{R} \cdot \frac{S}{S} + \frac{1}{S} \cdot \frac{R}{R} \qquad \text{Rewrite by using the LCD.}$$

$$= \frac{S}{RS} + \frac{R}{RS} \qquad \text{Multiply the fractions.}$$

$$= \frac{S + R}{RS} \qquad \text{Add the numerators.}$$

In general, the reciprocal of $\frac{a}{b}$ is $\frac{b}{a}$, so the reciprocal of $\frac{S + R}{RS}$ is $\frac{RS}{S + R}$. This final expression can be used to find the total resistance of the circuit.   Now Try Exercise 103

7.4   PUTTING IT ALL TOGETHER

In this section we discussed how to add and subtract rational expressions having unlike denominators. The following table summarizes this discussion.

Concept	Explanation	Examples
Least Common Multiple (LCM)	1. Factor each polynomial completely. 2. List each factor the greatest number of times that it occurs in any factorization. 3. The LCM is the product of this list.	1. $x^2 - 2x = x(x - 2)$ $\quad x^2 - 6x + 8 = (x - 2)(x - 4)$ 2. $x, (x - 2), (x - 4)$ 3. LCM $= x(x - 2)(x - 4)$

Concept	Explanation	Examples
Least Common Denominator (LCD)	The LCD of two or more rational expressions equals the LCM of their denominators.	The LCD of $\dfrac{2}{x^2 - 2x}$ and $\dfrac{3}{x^2 - 6x + 8}$ is $x(x - 2)(x - 4)$ because the LCM of $x^2 - 2x$ and $x^2 - 6x + 8$ is $x(x - 2)(x - 4)$, as shown in the preceding numbered list.
Addition and Subtraction of Rational Expressions Having Unlike Denominators	First rewrite each expression by using the LCD. Then add or subtract the expressions. Finally, write your answer in lowest terms.	The LCM of x and $x + 2$ is $x(x + 2)$. $$\frac{2}{x} + \frac{5}{x + 2} = \frac{2}{x} \cdot \frac{x + 2}{x + 2} + \frac{5}{x + 2} \cdot \frac{x}{x}$$ $$= \frac{2x + 4}{x(x + 2)} + \frac{5x}{x(x + 2)}$$ $$= \frac{7x + 4}{x(x + 2)}$$

7.4 Exercises

CONCEPTS

1. Give a common multiple of 6 and 9 that is not the *least* common multiple.

2. The LCM of x and y is _____.

3. To rewrite $\frac{3}{4}$ having denominator 12, multiply $\frac{3}{4}$ by the fraction _____.

4. To rewrite $\frac{4}{x - 1}$ having denominator $x^2 - 1$, multiply $\frac{4}{x - 1}$ by the rational expression _____.

5. What is the LCD for $\frac{1}{4}$ and $\frac{5}{6}$?

6. What is the LCD for $\frac{1}{2x}$ and $\frac{5}{6x^2}$?

LEAST COMMON MULTIPLES

Exercises 7–14: Find the least common multiple.

7. 4, 6

8. 6, 9

9. 2, 3

10. 5, 4

11. 10, 15

12. 8, 12

13. 24, 36

14. 32, 40

Exercises 15–38: Find the least common multiple. Leave your answer in factored form.

15. $4x, 6x$

16. $6x, 9x$

17. $5x, 10x^2$

18. $4x^2, 12x$

19. $x, x + 1$

20. $4x, x - 1$

21. $2x + 1, x + 3$

22. $5x + 3, x + 9$

23. $4x^2, 9x^3$

24. $12x^3, 15x^5$

25. $x^2 - x, x^2 + x$

26. $x^2 + 2x, x^2$

27. $(x + 1)^2, x + 1$

28. $(x - 8)^2, (x - 8)(x + 1)$

29. $(2x - 1)^3, (2x - 1)(x + 3)$

30. $(x - 4)(x + 4), (x - 4)(x + 3)$

31. $4x^2 - 1, 2x + 1$

32. $x^2 + 4x + 3, x + 3$

33. $x^2 - 1, x + 1$

34. $x^2 - 4, x - 2$

35. $x^2 + 4, 4x$ **36.** $4x^2 + x, x$

37. $2x^2 + 7x + 6, x^2 + 5x + 6$

38. $x^2 - 3x + 2, x^2 + 2x - 3$

ADDITION AND SUBTRACTION OF RATIONAL EXPRESSIONS

Exercises 39–50: Rewrite the rational expression so it has the given denominator D. For example, if the fraction $\frac{1}{4}$ is written with denominator $D = 8$, it becomes $\frac{2}{8}$.

39. $\frac{1}{3}$, $D = 9$ **40.** $\frac{3}{4}$, $D = 24$

41. $\frac{5}{7}$, $D = 21$ **42.** $\frac{4}{5}$, $D = 30$

43. $\frac{1}{4x}$, $D = 8x^3$ **44.** $\frac{5}{3x}$, $D = 9x^2$

45. $\frac{1}{x + 2}$, $D = x^2 - 4$ **46.** $\frac{3}{x - 3}$, $D = x^2 - 9$

47. $\frac{1}{x + 1}$, $D = x^2 + x$ **48.** $\frac{3}{x - 3}$, $D = x^2 - 3x$

49. $\frac{2x}{x + 1}$, $D = x^2 + 2x + 1$

50. $\frac{x}{2x - 1}$, $D = 2x^2 + 11x - 6$

Exercises 51–102: Simplify the expression. Write your answer in lowest terms and leave it in factored form.

51. $\frac{4}{5} + \frac{1}{2}$ **52.** $\frac{3}{8} + \frac{1}{4}$

53. $\frac{5}{9} - \frac{1}{3}$ **54.** $\frac{7}{10} - \frac{3}{15}$

55. $\frac{4}{25} + \frac{2}{5}$ **56.** $\frac{7}{9} - \frac{1}{3}$

57. $\frac{1}{5} + \frac{3}{4} - \frac{1}{2}$ **58.** $\frac{6}{7} - \frac{8}{9} + \frac{2}{3}$

59. $\frac{1}{3x} + \frac{3}{4x}$ **60.** $\frac{4}{2x^2} + \frac{7}{3x}$

61. $\frac{5}{z^2} - \frac{7}{z^3}$ **62.** $\frac{8}{z} - \frac{3}{2z}$

63. $\frac{1}{x} - \frac{1}{y}$ **64.** $\frac{1}{xy} - \frac{4}{y}$

65. $\frac{a}{b} + \frac{b}{a}$ **66.** $\frac{3}{x} - \frac{4}{y}$

67. $\frac{1}{2x + 4} + \frac{3}{x + 2}$ **68.** $\frac{1}{5x - 10} - \frac{x}{x - 2}$

69. $\frac{2}{t - 2} - \frac{1}{t}$ **70.** $\frac{7}{2t} + \frac{1}{t + 5}$

71. $\frac{5}{n - 1} + \frac{n}{n + 1}$ **72.** $\frac{4n}{3n - 2} + \frac{n}{n + 1}$

73. $\frac{3}{x - 3} + \frac{6}{3 - x}$ **74.** $\frac{x}{x - 8} + \frac{x}{8 - x}$

75. $\frac{1}{5k - 1} + \frac{1}{1 - 5k}$ **76.** $\frac{4}{4 - 3k} + \frac{3k}{3k - 4}$

77. $\frac{2x}{(x - 1)^2} + \frac{4}{x - 1}$ **78.** $\frac{5}{(x + 5)} - \frac{x}{(x + 5)^2}$

79. $\frac{2y}{y(2y - 1)} + \frac{1}{2y - 1}$ **80.** $\frac{5y}{y(y + 1)} - \frac{5}{y + 1}$

81. $\frac{1}{x + 2} - \frac{1}{x^2 + 2x}$ **82.** $\frac{1}{x - 3} - \frac{2}{x^2 - 3x}$

83. $\frac{3}{x - 2} - \frac{1}{x^2 - 4}$ **84.** $\frac{x}{9 - x^2} - \frac{1}{3 - x}$

85. $\frac{2}{x^2 - 3x} - \frac{1}{x^2 + 3x}$ **86.** $\frac{3}{x^2 + 4x} - \frac{2}{x^2 - 4x}$

87. $\frac{1}{x - 2} - \frac{1}{x + 2} + \frac{1}{x}$ **88.** $\frac{1}{x^2} - \frac{2}{x} + \frac{2}{x - 1}$

89. $\frac{x}{x^2 + 4x + 4} + \frac{1}{x + 2}$

90. $\frac{1}{x^2 - 3x - 4} - \frac{1}{x + 1}$

91. $\frac{x}{(x + 1)(x + 2)} - \frac{1}{(x + 2)(x + 3)}$

92. $\frac{2x}{(x - 1)(x - 2)} - \frac{5}{x - 2}$

93. $\frac{1}{a + b} - \frac{1}{a - b}$

94. $\frac{x}{x^2 - y^2} - \frac{1}{x + y}$

95. $\frac{r}{r - t} + \frac{t}{t - r} - 1$

96. $\frac{1}{x} + \frac{2}{x^2 - 2x} + \frac{5}{x - 2}$

97. $\dfrac{1}{2a} + \dfrac{1}{3a} + \dfrac{1}{4a}$

98. $\dfrac{1}{b} + \dfrac{1}{b+1} + \dfrac{1}{b+2}$

99. $\dfrac{2}{x-y} + \dfrac{3}{y-x} + \dfrac{1}{x-y}$

100. $\dfrac{a}{a-b} + \dfrac{b}{b-a} + \dfrac{3}{b}$

101. $\dfrac{3}{x-3} - \dfrac{3}{x^2-3x} - \dfrac{6}{x(x-3)}$

102. $\dfrac{3}{2a-4} + \dfrac{5}{2a} - \dfrac{3}{a^2-2a}$

APPLICATIONS

103. *Electricity*   In Example 6, we showed that the expressions

$$\frac{1}{R} + \frac{1}{S} \quad \text{and} \quad \frac{S+R}{RS}$$

are equivalent. Evaluate both expressions by using $R = 120$ and $S = 200$. Are your answers the same?

104. *Intensity of a Light Bulb*   The formula $I = \frac{32}{4d^2}$ approximates the intensity of light from a 100-watt light bulb at a distance of d meters, where I is in watts per square meter. For light from a 40-watt bulb the equation for its intensity becomes $I = \frac{16}{5d^2}$. (*Source:* R. Weidner.)

   (a) Find an expression for the sum of the intensities of light from the two light bulbs.

   (b) Find the combined intensity of their light at $d = 5$ meters.

105. *Photography*   A lens in a camera has a focal length, which is important for focusing the camera. If an object is at a distance D from the lens that has a focal length F, then to be in focus the distance S between the lens and the film should satisfy the equation

$$\frac{1}{S} = \frac{1}{F} - \frac{1}{D},$$

as illustrated in the figure. Write the difference $\frac{1}{F} - \frac{1}{D}$ as one term.

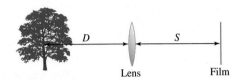

106. *Geometry*   Find the sum of the areas of the two rectangles shown in the figure. Write your answer in factored form.

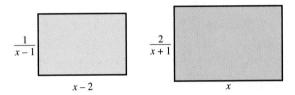

WRITING ABOUT MATHEMATICS

107. Explain how to find the least common multiple of two polynomials.

108. Explain how to subtract two rational expressions having unlike denominators.

CHECKING BASIC CONCEPTS
SECTIONS 7.3 AND 7.4

1. Simplify each expression.

   (a) $\dfrac{x}{x+2} + \dfrac{2}{x+2}$ **(b)** $\dfrac{2}{3x} - \dfrac{x}{3x}$

   (c) $\dfrac{z^2+z}{z+2} + \dfrac{z}{z+2}$

2. Find the least common multiple of each pair of expressions.

   (a) $3x, 5x$ **(b)** $4x, x^2 + x$

   (c) $x+1, x-1$

3. Simplify each expression.

   (a) $\dfrac{1}{x+1} + \dfrac{5}{x}$ **(b)** $\dfrac{5}{x-3} + \dfrac{1}{3-x}$

   (c) $\dfrac{-4}{4x+2} - \dfrac{x+2}{2x+1}$

4. Simplify the expression $\dfrac{a}{a-b} - \dfrac{b}{a+b}$.

7.5   COMPLEX FRACTIONS

Basic Concepts ▪ Simplifying Complex Fractions

If two and a half pizzas are cut so that each piece equals one-eighth of a pizza, then there are 20 pieces of pizza. This problem can be written as

$$\frac{2\frac{1}{2}}{\frac{1}{8}} \quad \text{or} \quad \frac{2 + \frac{1}{2}}{\frac{1}{8}}.$$

These expressions are examples of *complex* fractions. Typically, we want to rewrite a complex fraction as a standard fraction in the form $\frac{a}{b}$. In this section we show how to simplify complex fractions.

Basic Concepts

A **complex fraction** is a rational expression that contains fractions in its numerator, denominator, or both. Examples of complex fractions include

$$\frac{1 + \frac{1}{2}}{1 - \frac{1}{3}}, \quad \frac{5z}{\frac{7}{z} - \frac{4}{2z}}, \quad \text{and} \quad \frac{\frac{2x}{3} + \frac{1}{x}}{x - \frac{1}{x + 1}}.$$

The expression $\frac{a}{b} \div \frac{c}{d}$ can be written as the complex fraction

$$\frac{\frac{a}{b}}{\frac{c}{d}}.$$

Because $\frac{a}{b} \div \frac{c}{d} = \frac{a}{b} \cdot \frac{d}{c}$, we can simplify some complex fractions by multiplying the numerator and the reciprocal of the denominator. This "invert and multiply" strategy is summarized by the following.

> ### SIMPLIFYING BASIC COMPLEX FRACTIONS
>
> For any real numbers a, b, c, and d,
>
> $$\frac{\frac{a}{b}}{\frac{c}{d}} = \frac{a}{b} \cdot \frac{d}{c},$$
>
> where b, c, and d are nonzero.

Simplifying Complex Fractions

There are two basic methods of simplifying a complex fraction. The first is to simplify both the numerator and the denominator and then divide the resulting two fractions. The second is to multiply the numerator and denominator by the least common denominator of all fractions within the complex fraction.

SIMPLIFYING THE NUMERATOR AND DENOMINATOR   The following steps outline Method I for simplifying complex fractions.

SIMPLIFYING COMPLEX FRACTIONS (Method I)

To simplify a complex fraction, perform the following steps.

STEP 1: Write the numerator as a single fraction; write the denominator as a single fraction.

STEP 2: Divide the denominator into the numerator by multiplying the numerator and the reciprocal of the denominator.

STEP 3: Simplify the result to lowest terms.

Sometimes a complex fraction already has both its numerator and denominator written as single fractions. As a result, we can start with Step 2. This situation is shown in Example 1.

EXAMPLE 1   Simplifying basic complex fractions

Simplify each complex fraction.

(a) $\dfrac{\frac{3}{4}}{\frac{9}{8}}$ **(b)** $\dfrac{3\frac{3}{4}}{1\frac{1}{2}}$ **(c)** $\dfrac{\frac{x}{4}}{\frac{3}{2y}}$ **(d)** $\dfrac{\frac{(x-1)^2}{4}}{\frac{x-1}{8}}$

Solution

(a) We can skip Step 1. To simplify $\frac{3}{4} \div \frac{9}{8}$, multiply $\frac{3}{4}$ by the reciprocal of $\frac{9}{8}$, or $\frac{8}{9}$.

$$\frac{\frac{3}{4}}{\frac{9}{8}} = \frac{3}{4} \cdot \frac{8}{9} \quad \text{Invert and multiply (Step 2).}$$

$$= \frac{24}{36} \quad \text{Multiply the fractions.}$$

$$= \frac{2}{3} \quad \text{Simplify (Step 3).}$$

(b) Start by writing $3\frac{3}{4}$ and $1\frac{1}{2}$ as improper fractions.

$$\frac{3\frac{3}{4}}{1\frac{1}{2}} = \frac{\frac{15}{4}}{\frac{3}{2}} \quad \text{Write as improper fractions (Step 1).}$$

$$= \frac{15}{4} \cdot \frac{2}{3} \quad \text{Invert and multiply (Step 2).}$$

$$= \frac{5}{2} \quad \text{Multiply and simplify (Step 3).}$$

(c) We can skip Step 1. To simplify $\frac{x}{4} \div \frac{3}{2y}$, multiply $\frac{x}{4}$ by the reciprocal of $\frac{3}{2y}$, or $\frac{2y}{3}$.

$$\frac{\dfrac{x}{4}}{\dfrac{3}{2y}} = \frac{x}{4} \cdot \frac{2y}{3} \qquad \text{Invert and multiply (Step 2).}$$

$$= \frac{2xy}{12} \qquad \text{Multiply the fractions.}$$

$$= \frac{xy}{6} \qquad \text{Simplify (Step 3).}$$

(d) We can skip Step 1. To simplify $\frac{(x-1)^2}{4} \div \frac{x-1}{8}$, multiply $\frac{(x-1)^2}{4}$ by the reciprocal of $\frac{x-1}{8}$, or $\frac{8}{x-1}$.

$$\frac{\dfrac{(x-1)^2}{4}}{\dfrac{x-1}{8}} = \frac{(x-1)^2}{4} \cdot \frac{8}{x-1} \qquad \text{Invert and multiply (Step 2).}$$

$$= \frac{8(x-1)(x-1)}{4(x-1)} \qquad \text{Multiply the fractions.}$$

$$= 2(x-1) \qquad \text{Simplify (Step 3).}$$

Now Try Exercises 15, 17, 21, 31

EXAMPLE 2   Simplifying complex fractions

Simplify. Write your answer in lowest terms.

(a) $\dfrac{\dfrac{1}{a} + \dfrac{1}{b}}{\dfrac{1}{a} - \dfrac{1}{b}}$ **(b)** $\dfrac{x - \dfrac{4}{x}}{x + \dfrac{4}{x}}$ **(c)** $\dfrac{\dfrac{1}{x} + \dfrac{2}{x-1}}{\dfrac{2}{x} - \dfrac{1}{x-1}}$ **(d)** $\dfrac{\dfrac{1}{x} + \dfrac{1}{y}}{\dfrac{1}{x^2} - \dfrac{1}{y^2}}$

Solution
(a) In Step 1, write the numerator as one fraction by using the LCD, ab.

$$\frac{1}{a} + \frac{1}{b} = \frac{b}{ab} + \frac{a}{ab} = \frac{b+a}{ab}$$

Write the denominator as one fraction by using the LCD, ab.

$$\frac{1}{a} - \frac{1}{b} = \frac{b}{ab} - \frac{a}{ab} = \frac{b-a}{ab}$$

Finally, use these results to simplify the given complex fraction.

$$\frac{\dfrac{1}{a}+\dfrac{1}{b}}{\dfrac{1}{a}-\dfrac{1}{b}} = \frac{\dfrac{b+a}{ab}}{\dfrac{b-a}{ab}}$$ Simplify the numerator and the denominator (Step 1).

$$= \frac{b+a}{ab} \cdot \frac{ab}{b-a}$$ Invert and multiply (Step 2).

$$= \frac{ab(b+a)}{ab(b-a)}$$ Multiply.

$$= \frac{b+a}{b-a}$$ Simplify (Step 3).

(b) The LCD for both the numerator and the denominator is x.

$$\frac{x-\dfrac{4}{x}}{x+\dfrac{4}{x}} = \frac{\dfrac{x^2}{x}-\dfrac{4}{x}}{\dfrac{x^2}{x}+\dfrac{4}{x}}$$ Write by using the LCD (Step 1).

$$= \frac{\dfrac{x^2-4}{x}}{\dfrac{x^2+4}{x}}$$ Combine terms.

$$= \frac{x^2-4}{x} \cdot \frac{x}{x^2+4}$$ Invert and multiply (Step 2).

$$= \frac{x(x^2-4)}{x(x^2+4)}$$ Multiply.

$$= \frac{x^2-4}{x^2+4}$$ Simplify (Step 3).

(c) The LCD for the numerator and the denominator is $x(x-1)$.

$$\frac{\dfrac{1}{x}+\dfrac{2}{x-1}}{\dfrac{2}{x}-\dfrac{1}{x-1}} = \frac{\dfrac{x-1}{x(x-1)}+\dfrac{2x}{x(x-1)}}{\dfrac{2(x-1)}{x(x-1)}-\dfrac{x}{x(x-1)}}$$ Write by using the LCD (Step 1).

$$= \frac{\dfrac{x-1+2x}{x(x-1)}}{\dfrac{2(x-1)-x}{x(x-1)}}$$ Combine terms.

$$= \frac{\dfrac{3x-1}{x(x-1)}}{\dfrac{x-2}{x(x-1)}}$$ Simplify.

$$= \frac{3x-1}{x(x-1)} \cdot \frac{x(x-1)}{x-2}$$ Invert and multiply (Step 2).

$$= \frac{3x-1}{x-2}$$ Multiply and simplify (Step 3).

(d) The LCD for the numerator is xy, and the LCD for the denominator is x^2y^2.

$$\frac{\dfrac{1}{x}+\dfrac{1}{y}}{\dfrac{1}{x^2}-\dfrac{1}{y^2}} = \frac{\dfrac{y}{xy}+\dfrac{x}{xy}}{\dfrac{y^2}{x^2y^2}-\dfrac{x^2}{x^2y^2}} \qquad \text{Write by using the LCD (Step 1).}$$

$$= \frac{\dfrac{y+x}{xy}}{\dfrac{y^2-x^2}{x^2y^2}} \qquad \text{Combine terms.}$$

$$= \frac{y+x}{xy}\cdot\frac{x^2y^2}{y^2-x^2} \qquad \text{Invert and multiply (Step 2).}$$

$$= \frac{x^2y^2(y+x)}{xy(y^2-x^2)} \qquad \text{Multiply.}$$

$$= \frac{x^2y^2(y+x)}{xy(y-x)(y+x)} \qquad \text{Factor the denominator (Step 3).}$$

$$= \frac{xy}{y-x} \qquad \text{Simplify.}$$

Now Try Exercises **33, 39, 43, 47**

MULTIPLYING BY THE LCD Method II for simplifying a complex fraction is to multiply the numerator and denominator by the least common denominator of all the fractions in *both* the numerator and the denominator.

The following steps outline Method II for simplifying complex fractions.

SIMPLIFYING COMPLEX FRACTIONS (Method II)

To simplify a complex fraction, perform the following steps.

STEP 1: Find the LCD of all fractions within the complex fraction.

STEP 2: Multiply the numerator and the denominator of the complex fraction by the LCD.

STEP 3: Simplify the result to lowest terms.

We can use this method to simplify the complex fraction in Example 2(a). Because the LCD for the numerator and the denominator is ab, we multiply the complex fraction by 1, expressed in the form $\frac{ab}{ab}$. This method is equivalent to multiplying the numerator and the denominator by ab.

$$\frac{\dfrac{1}{a} + \dfrac{1}{b}}{\dfrac{1}{a} - \dfrac{1}{b}} = \frac{\left(\dfrac{1}{a} + \dfrac{1}{b}\right)ab}{\left(\dfrac{1}{a} - \dfrac{1}{b}\right)ab}$$ Multiply by $\frac{ab}{ab} = 1$.

$$= \frac{\dfrac{ab}{a} + \dfrac{ab}{b}}{\dfrac{ab}{a} - \dfrac{ab}{b}}$$ Distributive property

$$= \frac{b + a}{b - a}$$ Simplify.

EXAMPLE 3   **Simplifying complex fractions**

Simplify.

(a) $\dfrac{2z}{\dfrac{4}{z} + \dfrac{3}{z}}$ (b) $\dfrac{\dfrac{1}{x - 3}}{\dfrac{1}{x} + \dfrac{3}{x - 3}}$ (c) $\dfrac{\dfrac{1}{a} - \dfrac{1}{b}}{\dfrac{1}{2b^2} - \dfrac{1}{2a^2}}$

Solution

(a) The LCD for the numerator *and* the denominator is z, so multiply each by z (Step 1).

$$\frac{2z}{\dfrac{4}{z} + \dfrac{3}{z}} = \frac{(2z)z}{\left(\dfrac{4}{z} + \dfrac{3}{z}\right)z}$$ Multiply by $\frac{z}{z}$ (Step 2).

$$= \frac{2z^2}{\dfrac{4z}{z} + \dfrac{3z}{z}}$$ Distributive property

$$= \frac{2z^2}{4 + 3}$$ Simplify (Step 3).

$$= \frac{2z^2}{7}$$ Add.

(b) The LCD is the product $x(x - 3)$ (Step 1).

$$\frac{\dfrac{1}{x - 3}}{\dfrac{1}{x} + \dfrac{3}{x - 3}} = \frac{\dfrac{1}{x - 3}}{\left(\dfrac{1}{x} + \dfrac{3}{x - 3}\right)} \cdot \frac{x(x - 3)}{x(x - 3)}$$ Multiply by 1 (Step 2).

$$= \frac{\dfrac{x(x - 3)}{x - 3}}{\dfrac{x(x - 3)}{x} + \dfrac{3x(x - 3)}{x - 3}}$$ Distributive property

$$= \frac{x}{(x - 3) + 3x}$$ Simplify the fractions (Step 3).

$$= \frac{x}{4x - 3}$$ Simplify the denominator.

(c) The LCD for the numerator *and* the denominator is $2a^2b^2$ (Step 1).

$$\dfrac{\dfrac{1}{a} - \dfrac{1}{b}}{\dfrac{1}{2b^2} - \dfrac{1}{2a^2}} = \dfrac{\dfrac{1}{a} - \dfrac{1}{b}}{\dfrac{1}{2b^2} - \dfrac{1}{2a^2}} \cdot \dfrac{2a^2b^2}{2a^2b^2}$$ Multiply by 1 (Step 2).

$$= \dfrac{\left(\dfrac{1}{a} - \dfrac{1}{b}\right)2a^2b^2}{\left(\dfrac{1}{2b^2} - \dfrac{1}{2a^2}\right)2a^2b^2}$$ Multiply the fractions.

$$= \dfrac{\dfrac{2a^2b^2}{a} - \dfrac{2a^2b^2}{b}}{\dfrac{2a^2b^2}{2b^2} - \dfrac{2a^2b^2}{2a^2}}$$ Distributive property

$$= \dfrac{2ab^2 - 2a^2b}{a^2 - b^2}$$ Simplify the fractions (Step 3).

$$= \dfrac{2ab(b - a)}{(a - b)(a + b)}$$ Factor.

$$= -\dfrac{2ab}{a + b}$$ Simplify.

Now Try Exercises 35, 37, 51

CRITICAL THINKING

Are the expressions $\dfrac{\dfrac{a}{b}}{\dfrac{a}{b} + 1}$

and $\dfrac{1}{1 + 1}$ equal? Explain.

Are the expressions $\dfrac{\dfrac{a}{b} + 1}{\dfrac{a}{b}}$

and $1 + \dfrac{b}{a}$ equal? Explain.

7.5 PUTTING IT ALL TOGETHER

In this section we discussed how to simplify complex fractions to the form $\frac{a}{b}$. The following table summarizes this discussion.

Concept	Explanation	Examples
Complex Fraction	A rational expression that contains fractions in its numerator, denominator, or both	$\dfrac{3 + \dfrac{1}{x + 1}}{3 - \dfrac{1}{x + 1}}$ and $\dfrac{\dfrac{x}{y} - \dfrac{y}{x}}{\dfrac{x}{y} + \dfrac{y}{x}}$
Simplifying Basic Complex Fractions	$\dfrac{\dfrac{a}{b}}{\dfrac{c}{d}} = \dfrac{a}{b} \cdot \dfrac{d}{c}$	$\dfrac{\dfrac{2}{x}}{\dfrac{4}{x - 1}} = \dfrac{2}{x} \cdot \dfrac{x - 1}{4} = \dfrac{x - 1}{2x}$

Concept	Explanation	Examples
Method I: Simplifying the Numerator and Denominator First	Combine the terms in the numerator, combine the terms in the denominator, and then invert and multiply.	$$\dfrac{\dfrac{1}{x}+\dfrac{3}{x}}{\dfrac{5}{y}-\dfrac{4}{y}}=\dfrac{\dfrac{4}{x}}{\dfrac{1}{y}}=\dfrac{4}{x}\cdot\dfrac{y}{1}=\dfrac{4y}{x}$$
Method II: Multiplying the Numerator and Denominator by the LCD	Multiply the numerator *and* the denominator by the LCD of *all* fractions within the expression.	$$\dfrac{\dfrac{2}{x}+\dfrac{1}{y}}{\dfrac{4}{y}-\dfrac{1}{x}}=\dfrac{\left(\dfrac{2}{x}+\dfrac{1}{y}\right)xy}{\left(\dfrac{4}{y}-\dfrac{1}{x}\right)xy}$$ $$=\dfrac{2y+x}{4x-y}$$ Note that the LCD is xy.

7.5 Exercises

MyMathLab

Math XL
PRACTICE   WATCH   DOWNLOAD   READ   REVIEW

CONCEPTS

1. $\dfrac{\frac{1}{2}}{\frac{3}{4}}=$ _____

2. $\dfrac{\frac{a}{b}}{\frac{c}{d}}=$ _____

3. A complex fraction is a rational expression that contains _____ in its numerator, denominator, or both.

4. Write "the quantity x plus one half divided by the quantity x minus one half" as a complex fraction.

5. What operation does the fraction bar represent?

6. Write the expression $\frac{x}{2}\div\frac{1}{x-1}$ as a complex fraction.

7. Write the expression $\frac{a}{b}\div\frac{c}{d}$ as a complex fraction.

8. What is the LCD for $\frac{1}{x+2}$ and $\frac{1}{x}$?

SIMPLIFYING COMPLEX FRACTIONS

Exercises 9–14: For the complex fraction, determine the LCD of all the fractions appearing in both the numerator and the denominator.

9. $\dfrac{\dfrac{x}{5}-\dfrac{1}{6}}{\dfrac{2}{15}-3x}$

10. $\dfrac{\dfrac{1}{2}-\dfrac{1}{x}}{\dfrac{1}{2}+\dfrac{1}{x}}$

11. $\dfrac{\dfrac{2}{x+1}-x}{\dfrac{2}{x-1}+x}$

12. $\dfrac{\dfrac{1}{4x}-\dfrac{4}{x}}{\dfrac{1}{2x}+\dfrac{1}{3x}}$

13. $\dfrac{\dfrac{1}{2x-1}-\dfrac{1}{2x+1}}{\dfrac{x+1}{x}}$

14. $\dfrac{\dfrac{1}{4x^2}-\dfrac{1}{2x^3}}{\dfrac{1}{x-1}+\dfrac{1}{x-1}}$

Exercises 15–54: Simplify the complex fraction.

15. $\dfrac{\frac{2}{3}}{\frac{5}{6}}$

16. $\dfrac{\frac{8}{9}}{\frac{5}{4}}$

17. $\dfrac{2\frac{1}{2}}{1\frac{3}{4}}$

18. $\dfrac{1\frac{2}{3}}{3\frac{1}{2}}$

19. $\dfrac{1\frac{1}{2}}{2\frac{1}{3}}$

20. $\dfrac{2\frac{1}{5}}{2\frac{1}{10}}$

21. $\dfrac{\frac{r}{t}}{\frac{2r}{t}}$

22. $\dfrac{\frac{8}{p}}{\frac{4}{p}}$

23. $\dfrac{\dfrac{6}{x}}{\dfrac{2}{y}}$

24. $\dfrac{\dfrac{3}{14x}}{\dfrac{6}{7x}}$

25. $\dfrac{\dfrac{6}{m-2}}{\dfrac{2}{m-2}}$

26. $\dfrac{\dfrac{3}{n+1}}{\dfrac{6}{n+1}}$

27. $\dfrac{\dfrac{p+1}{p}}{\dfrac{p+2}{p}}$

28. $\dfrac{\dfrac{2p}{2p+5}}{\dfrac{1}{4p+10}}$

29. $\dfrac{\dfrac{5}{z^2-1}}{\dfrac{z}{z^2-1}}$

30. $\dfrac{\dfrac{z}{z-2}}{\dfrac{z}{z-2}}$

31. $\dfrac{\dfrac{y}{y^2-9}}{\dfrac{1}{y+3}}$

32. $\dfrac{\dfrac{2y}{2y-1}}{\dfrac{1}{4y^2-1}}$

33. $\dfrac{x-\dfrac{1}{x}}{x+\dfrac{1}{x}}$

34. $\dfrac{4-\dfrac{1}{x}}{4+\dfrac{1}{x}}$

35. $\dfrac{x}{\dfrac{2}{x}+\dfrac{1}{x}}$

36. $\dfrac{5x}{1+\dfrac{1}{x}}$

37. $\dfrac{\dfrac{3}{x+1}}{\dfrac{4}{x+1}-\dfrac{1}{x+1}}$

38. $\dfrac{\dfrac{5}{2x-3}-\dfrac{4}{2x-3}}{\dfrac{7}{2x-3}+\dfrac{8}{2x-3}}$

39. $\dfrac{\dfrac{1}{m^2n}+\dfrac{1}{mn^2}}{\dfrac{1}{m^2n}-\dfrac{1}{mn^2}}$.

40. $\dfrac{\dfrac{3}{x-1}-\dfrac{2}{x}}{\dfrac{3}{x-1}+\dfrac{2}{x}}$

41. $\dfrac{\dfrac{1}{2x}+\dfrac{1}{y}}{\dfrac{1}{y}-\dfrac{1}{2x}}$

42. $\dfrac{\dfrac{3}{x}-\dfrac{2}{y}}{\dfrac{3}{x}+\dfrac{2}{y}}$

43. $\dfrac{\dfrac{1}{ab}+\dfrac{1}{a}}{\dfrac{1}{ab}-\dfrac{1}{b}}$

44. $\dfrac{\dfrac{1}{a}+\dfrac{2}{3b}}{\dfrac{1}{a}-\dfrac{5}{2b}}$

45. $\dfrac{\dfrac{2}{q}-\dfrac{1}{q+1}}{\dfrac{1}{q+1}}$

46. $\dfrac{\dfrac{5}{p}+\dfrac{4}{p-5}}{\dfrac{5}{p}-\dfrac{5}{p-5}}$

47. $\dfrac{\dfrac{1}{x+1}+\dfrac{1}{x+2}}{\dfrac{1}{x+1}-\dfrac{1}{x+2}}$

48. $\dfrac{\dfrac{1}{x-3}-\dfrac{1}{x+3}}{1-\dfrac{1}{x^2-9}}$

49. $\dfrac{\dfrac{1}{2x-1}-\dfrac{1}{2x+1}}{\dfrac{x+1}{x}}$

50. $\dfrac{\dfrac{1}{4x^2}-\dfrac{1}{x^3}}{\dfrac{1}{x-1}+\dfrac{1}{x-1}}$

51. $\dfrac{\dfrac{1}{ab^2}-\dfrac{1}{a^2b}}{\dfrac{1}{b}-\dfrac{1}{a}}$

52. $\dfrac{\dfrac{1}{x^2}-\dfrac{1}{y^2}}{\dfrac{1}{x}-\dfrac{1}{y}}$

53. $\dfrac{1}{a^{-1}+b^{-1}}$

54. $\dfrac{a^2-b^2}{a^{-2}-b^{-2}}$

APPLICATIONS

55. *Annuity*   If P dollars are deposited every 2 weeks in an account paying an annual interest rate r expressed as a decimal, then the amount A in the account after 2 years can be approximated by

$$\left(P\left(1+\frac{r}{26}\right)^{52}-P\right)\div\frac{r}{26}.$$

Write this expression as a complex fraction.

56. *Annuity*   (Continuation of the preceding exercise) Use a calculator to evaluate the expression when $r = 0.026$ (2.6%) and $P = \$100$. Interpret the result.

57. *Resistance in Electricity* Light bulbs are often wired so that electricity can flow through either bulb, as illustrated in the accompanying figure.

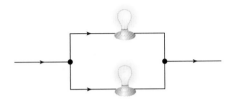

In this way, if one bulb burns out, the other bulb still works. If two light bulbs have resistances T and S, their combined resistance R is

$$R = \frac{1}{\frac{1}{T} + \frac{1}{S}}.$$

Simplify this formula.

58. *Resistance in Electricity* (Refer to the preceding exercise.) Evaluate the formula

$$R = \frac{1}{\frac{1}{T} + \frac{1}{S}}$$

when $T = 100$ and $S = 200$.

WRITING ABOUT MATHEMATICS

59. A student simplifies a complex fraction as shown below. Explain the student's mistake and how you would simplify the complex fraction correctly.

$$\frac{\frac{1}{x} + 1}{\frac{1}{x}} \overset{?}{=} \frac{\frac{1}{x}}{\frac{1}{x}} + 1 = 2$$

60. Explain one method for simplifying a complex fraction.

7.6 RATIONAL EQUATIONS AND FORMULAS

Solving Rational Equations ▪ Rational Expressions and Equations ▪ Graphical and Numerical Solutions ▪ Solving a Formula for a Variable ▪ Applications

A LOOK INTO MATH ▷

In Section 7.1 we demonstrated that, if cars arrive randomly at a construction site at an average rate of x cars per minute and if 10 cars per minute can pass through the site, then the average time T that a driver spends waiting in line and passing through the construction site is

$$T = \frac{1}{10 - x},$$

where T is in minutes and $x < 10$. If the highway department wants to limit the average wait for a car to $\frac{1}{2}$ minute or less, then mathematics can be used to determine the corresponding value of x. However, before solving this problem, we discuss rational equations. (***Source:*** N. Garber and L. Hoel, *Traffic and Highway Engineering*.)

Solving Rational Equations

▶ **REAL-WORLD CONNECTION** If an equation contains one or more rational expressions, it is called a **rational equation**. Rational equations occur in mathematics whenever a rational expression is set equal to a constant. For example, the rational expression

$$\frac{1}{10 - x}$$

from the introduction can be used to estimate the average time that drivers wait to get through a construction site. If the wait is $\frac{1}{2}$ minute, we determine x by solving the *rational equation*

$$\frac{1}{10 - x} = \frac{1}{2}.$$

To solve this equation, we multiply each side by the LCD: $2(10 - x)$.

$$\frac{2(10 - x)}{10 - x} = \frac{2(10 - x)}{2} \qquad \text{Multiply by the LCD.}$$

$$2 = 10 - x \qquad \text{Simplify.}$$

$$x = 8 \qquad \text{Add } x; \text{ subtract 2.}$$

The average wait is $\frac{1}{2}$ minute when cars arrive randomly at an average rate of 8 cars per minute. In general, if a rational equation is in the form

$$\frac{a}{b} = \frac{c}{d},$$

we can multiply each side of this equation by the common denominator bd to obtain

$$\frac{a(bd)}{b} = \frac{c(bd)}{d},$$

which simplifies to $ad = cb$. This technique can be used to solve some types of basic rational equations.

SOLVING BASIC RATIONAL EQUATIONS

The equations

$$\frac{a}{b} \diagdown\hspace{-1.2em}\diagup \frac{c}{d} \qquad \text{and} \qquad ad = bc$$

are equivalent, provided that b and d are nonzero. Note that converting the first equation to the second equation is sometimes called *cross multiplying*.

EXAMPLE 1   **Solving rational equations**

Solve each equation.

(a) $\dfrac{5}{3} = \dfrac{4}{x}$ **(b)** $\dfrac{x + 1}{5} = \dfrac{3x}{2}$ **(c)** $\dfrac{4}{3x - 4} = x$ **(d)** $\dfrac{1}{x} + \dfrac{2}{x} = \dfrac{3}{7}$

Solution

(a)

$$\frac{5}{3} = \frac{4}{x} \qquad \text{Given equation}$$

$$5x = 12 \qquad \text{Cross multiply.}$$

$$x = \frac{12}{5} \qquad \text{Divide by 5.}$$

The solution is $\frac{12}{5}$.

(b)

$$\frac{x + 1}{5} = \frac{3x}{2}$$ Given equation

$$2(x + 1) = 15x$$ Cross multiply.

$$2x + 2 = 15x$$ Distributive property

$$-13x = -2$$ Subtract 2 and $15x$.

$$x = \frac{2}{13}$$ Divide by -13.

The solution is $\frac{2}{13}$.

(c)

$$\frac{4}{3x - 4} = \frac{x}{1}$$ Write x as $\frac{x}{1}$.

$$x(3x - 4) = 4 \cdot 1$$ Cross multiply.

$$3x^2 - 4x = 4$$ Distributive property

$$3x^2 - 4x - 4 = 0$$ Subtract 4.

$$(3x + 2)(x - 2) = 0$$ Factor.

$$3x + 2 = 0 \quad \text{or} \quad x - 2 = 0$$ Zero-product property

$$x = -\frac{2}{3} \quad \text{or} \quad x = 2$$ Solve each equation.

The solutions are $-\frac{2}{3}$ and 2.

(d)

$$\frac{1}{x} + \frac{2}{x} = \frac{3}{7}$$ Given equation

$$\frac{3}{x} = \frac{3}{7}$$ Add the rational expressions.

$$3x = 21$$ Cross multiply.

$$x = 7$$ Divide by 3.

The solution is 7.

Now Try Exercises 9, 13, 33, 43

Another technique for solving rational equations is to multiply each side by the least common denominator. Unlike cross multiplying, this technique can always be used.

EXAMPLE 2   **Multiplying by the LCD**

Solve each equation. Check your answer.

(a) $\dfrac{1}{x - 1} - \dfrac{1}{x} = \dfrac{1}{9x}$ **(b)** $\dfrac{1}{x - 1} + \dfrac{1}{x + 1} = \dfrac{12}{x^2 - 1}$

Solution

(a) Start by multiplying each term by the LCD, $9x(x - 1)$.

$$\frac{1}{x - 1} - \frac{1}{x} = \frac{1}{9x}$$ Given equation

$$\frac{9x(x - 1)}{x - 1} - \frac{9x(x - 1)}{x} = \frac{9x(x - 1)}{9x}$$ Multiply each term by the LCD.

$$9x - 9(x - 1) = x - 1$$ Simplify each rational expression.

$$9 = x - 1$$ Distributive property

$$x = 10$$ Add 1 and rewrite.

Check:

$$\frac{1}{10-1} - \frac{1}{10} \overset{?}{=} \frac{1}{9(10)}$$ Substitute 10 for x.

$$\frac{10}{90} - \frac{9}{90} \overset{?}{=} \frac{1}{90}$$ LCD $= 90$

$$\frac{1}{90} = \frac{1}{90}$$ The answer checks.

(b) Start by multiplying each term by the LCD, $x^2 - 1 = (x-1)(x+1)$.

$$\frac{1}{x-1} + \frac{1}{x+1} = \frac{12}{x^2-1}$$ Given equation

$$\frac{(x-1)(x+1)}{x-1} + \frac{(x-1)(x+1)}{x+1} = \frac{12(x-1)(x+1)}{(x-1)(x+1)}$$ Multiply each term by the LCD.

$$(x+1) + (x-1) = 12$$ Simplify each rational expression. Combine like terms.

$$2x = 12$$

$$x = 6$$ Divide by 2.

Check:

$$\frac{1}{6-1} + \frac{1}{6+1} \overset{?}{=} \frac{12}{6^2-1}$$ Substitute 6 for x.

$$\frac{1}{5} + \frac{1}{7} \overset{?}{=} \frac{12}{35}$$ Simplify.

$$\frac{7}{35} + \frac{5}{35} \overset{?}{=} \frac{12}{35}$$ The LCD is 35.

$$\frac{12}{35} = \frac{12}{35}$$ The answer checks.

Now Try Exercises 51, 57

In Example 2 we checked our answers. Although the answer checked in both cases, it may not always check. When multiplying a rational equation by the LCD, it is possible to obtain *extraneous solutions* that do not satisfy the *given* equation. This situation is demonstrated in Example 3. Before solving Example 3, we present a step-by-step strategy for solving rational equations.

STEPS FOR SOLVING A RATIONAL EQUATION

STEP 1: Find the LCD of the terms in the equation.

STEP 2: Multiply each side of the equation by the LCD.

STEP 3: Simplify each term.

STEP 4: Solve the resulting equation.

STEP 5: Check each answer in the *given* equation. Reject any value that makes a denominator equal 0.

NOTE: If the rational equation is in the form $\frac{a}{b} = \frac{c}{d}$, or "fraction equals fraction," cross multiplying to obtain $ad = bc$ may be helpful. However, be sure to check your answers.

EXAMPLE 3   Solving an equation having an extraneous solution

If possible, solve $\dfrac{1}{x-2} + \dfrac{1}{x+2} = \dfrac{4}{x^2-4}$.

Solution

The LCD is $(x - 2)(x + 2) = x^2 - 4$ (Step 1).

$$\frac{1}{x - 2} + \frac{1}{x + 2} = \frac{4}{x^2 - 4} \qquad \text{Given equation}$$

$$\frac{(x - 2)(x + 2)}{x - 2} + \frac{(x - 2)(x + 2)}{x + 2} = \frac{4(x^2 - 4)}{x^2 - 4} \qquad \text{Multiply by the LCD (Step 2).}$$

$$(x + 2) + (x - 2) = 4 \qquad \text{Simplify each term (Step 3).}$$

$$2x = 4 \qquad \text{Combine like terms (Step 4).}$$

$$x = 2 \qquad \text{Divide by 2.}$$

Check: $\qquad \dfrac{1}{2 - 2} + \dfrac{1}{2 + 2} \stackrel{?}{=} \dfrac{4}{2^2 - 4} \qquad$ Substitute 2 for x (Step 5).

Note that both sides of the equation are undefined because it is not possible to divide by 0. Therefore 2 is not a solution; rather, it is an *extraneous solution*. That is, there are no solutions to the given equation. Now Try Exercise 55

Rational Expressions and Equations

Rational expressions and rational equations are not the same concepts. For example,

$$\frac{3}{x} \quad \text{and} \quad \frac{x}{x^2 - x}$$

are both rational expressions. Neither one contains an equals sign. We often simplify or evaluate rational expressions. By factoring the denominator and applying the basic principle of rational expressions, we can simplify the second expression.

$$\frac{x}{x^2 - x} = \frac{x}{x(x - 1)} \qquad \text{Factor the denominator.}$$

$$= \frac{1}{x - 1} \qquad \text{Simplify.}$$

Thus the expressions $\frac{x}{x^2 - x}$ and $\frac{1}{x - 1}$ are equal for every value of x (except 0 and 1). Expressions can also be evaluated. For example, if we replace x with 3, the expression $\frac{3}{x}$ equals $\frac{3}{3}$, or 1.

When two expressions are set equal, an equation is formed that is typically true for only a limited number of x-values. For example, $\frac{3}{x} = \frac{1}{x - 1}$ is a rational equation that is true for only one x-value.

$$\frac{3}{x} = \frac{1}{x - 1} \qquad \text{Given equation}$$

$$3(x - 1) = x \cdot 1 \qquad \text{Cross multiply.}$$

$$3x - 3 = x \qquad \text{Distributive property}$$

$$2x = 3 \qquad \text{Add 3; subtract } x.$$

$$x = \frac{3}{2} \qquad \text{Divide each side by 2.}$$

The only solution is $\frac{3}{2}$. (Check this solution.) If we replace x with any value other than $\frac{3}{2}$, the equation $\frac{3}{x} = \frac{1}{x - 1}$ is a false statement.

EXAMPLE 4   Identifying expressions and equations

Determine whether you are given an expression or an equation. If it is an expression, simplify it and then evaluate it for $x = 5$. If it is an equation, solve it.

(a) $\dfrac{x - 1}{x + 1} = \dfrac{x}{x + 3}$ (b) $\dfrac{x^2 - 2x}{x - 1} + \dfrac{1}{x - 1}$

Solution
(a) There is an equals sign, so it is an equation.

$$\dfrac{x - 1}{x + 1} = \dfrac{x}{x + 3} \qquad \text{Given equation}$$

$$(x - 1)(x + 3) = x(x + 1) \qquad \text{Cross multiply.}$$

$$x^2 + 2x - 3 = x^2 + x \qquad \text{Multiply.}$$

$$2x - 3 = x \qquad \text{Subtract } x^2.$$

$$x = 3 \qquad \text{Subtract } x; \text{ add } 3.$$

Check: $\dfrac{3 - 1}{3 + 1} \overset{?}{=} \dfrac{3}{3 + 3}$ Replace x with 3 in given equation.

$\dfrac{1}{2} = \dfrac{1}{2}$ The answer checks.

The solution to the equation is 3.

(b) There is no equals sign, so it is an expression. The common denominator is $x - 1$, so add the numerators.

$$\dfrac{x^2 - 2x}{x - 1} + \dfrac{1}{x - 1} = \dfrac{x^2 - 2x + 1}{x - 1} \qquad \text{Add the numerators.}$$

$$= \dfrac{(x - 1)(x - 1)}{x - 1} \qquad \text{Factor the numerator.}$$

$$= x - 1 \qquad \text{Simplify.}$$

The given expression simplifies to $x - 1$. When x equals 5, the expression evaluates to $5 - 1$, or 4.

Now Try Exercises 65, 69

MAKING CONNECTIONS

Expressions versus Equations

1. If a problem does not contain an equals sign, you are probably adding, subtracting, multiplying, dividing, or otherwise simplifying an expression. Your answer will be an expression, not a value for x.

2. If the problem has an equals sign, it is an equation to be solved. Your answer will be a value (or values) for x that makes the equation a true statement. One strategy for solving a rational equation is to multiply each side of the equation by the LCD. Be sure to check all your answers.

Graphical and Numerical Solutions

Like other types of equations, rational equations can also be solved graphically and numerically. Graphs of rational expressions are not lines. They are typically curves that can be graphed by plotting several points and then sketching a graph or by using a graphing calculator.

EXAMPLE 5   **Solving a rational equation graphically and numerically**

Solve $\frac{2}{x} = x + 1$ graphically and numerically.

Solution

Graphical Solution   To solve $\frac{2}{x} = x + 1$, graph $y_1 = \frac{2}{x}$ and $y_2 = x + 1$. The graph of y_2 is a line with slope 1 and y-intercept 1. To graph y_1 make a table of values, as shown in Table 7.3. Then plot the points and connect them with a smooth curve. Note that y_1 is undefined for $x = 0$. Generally, it is a good idea to plot at least three points on each side of an asymptote. These points were plotted and the curves sketched in Figure 7.3. Note that the graphs of y_1 and y_2 intersect at $(-2, -1)$ and $(1, 2)$. The solutions to the equation are -2 and 1. Check these solutions.

TABLE 7.3

x	$\frac{2}{x}$
-3	$-\frac{2}{3}$
-2	-1
-1	-2
0	—
1	2
2	1
3	$\frac{2}{3}$

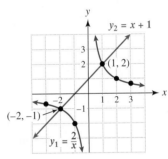

Figure 7.3

Numerical Solution   Make a table of values for $y_1 = \frac{2}{x}$ and $y_2 = x + 1$, as shown in Table 7.4. Note that $\frac{2}{x} = x + 1$ when $x = -2$ or $x = 1$.

TABLE 7.4

x	-3	-2	-1	0	1	2	3
$y_1 = \frac{2}{x}$	$-\frac{2}{3}$	-1	-2	—	2	1	$\frac{2}{3}$
$y_2 = x + 1$	-2	-1	0	1	2	3	4

Now Try Exercise 77

CRITICAL THINKING

If you solve Example 5 symbolically, what will be the result? Verify your answer.

A graphing calculator can be used efficiently to create Figure 7.3 and Table 7.4, as illustrated in Figure 7.4 on the next page.

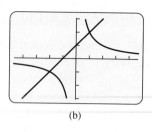

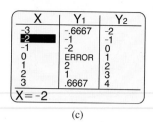

(a)   (b)   (c)

Figure 7.4

Solving a Formula for a Variable

▶ **REAL-WORLD CONNECTION**   Formulas in applications often involve both the use of rational expressions and the need to solve an equation for a variable. This idea is illustrated in the next two examples.

EXAMPLE 6   Finding time in a distance problem

If a person travels at a speed, or rate, r for time t, then the distance d traveled is $d = rt$.
(a) How far does a person travel in 2 hours when traveling at 60 miles per hour?
(b) Solve the formula $d = rt$ for t.
(c) How long does it take a person to go 250 miles when traveling at 40 miles per hour?

Solution
(a) $d = rt = 60 \cdot 2 = 120$ miles
(b) To solve $d = rt$ for t divide each side by r to obtain

$$\frac{d}{r} = \frac{rt}{r} \quad \text{or} \quad \frac{d}{r} = t.$$

(c) $t = \dfrac{d}{r} = \dfrac{250}{40} = \dfrac{25}{4} = 6.25$ hours, or 6 hours and 15 minutes.   Now Try Exercise 89

EXAMPLE 7   Solving a formula for a variable

Solve each equation for the specified variable.
(a) $A = \frac{bh}{2}$ for b **(b)** $P = \frac{nRT}{V}$ for V **(c)** $S = 2\pi rh + \pi r^2$ for h

Solution
(a) First, multiply each side by 2.

$$A = \frac{bh}{2} \qquad \text{Given equation}$$

$$2A = bh \qquad \text{Multiply each side by 2.}$$

$$\frac{2A}{h} = b \qquad \text{Divide each side by } h.$$

(b) Begin by multiplying each side by V.

$$P = \frac{nRT}{V} \qquad \text{Given equation}$$

$$PV = nRT \qquad \text{Multiply each side by } V.$$

$$V = \frac{nRT}{P} \qquad \text{Divide each side by } P.$$

(c) Begin by subtracting πr^2 from each side.

$$S = 2\pi rh + \pi r^2 \qquad \text{Given equation}$$

$$S - \pi r^2 = 2\pi rh \qquad \text{Subtract } \pi r^2 \text{ from each side.}$$

$$\frac{S - \pi r^2}{2\pi r} = h \qquad \text{Divide each side by } 2\pi r.$$

Now Try Exercises 91, 93, 94

Applications

▶ **REAL-WORLD CONNECTION**   Rational equations sometimes occur in time and rate problems, as demonstrated in the next two examples.

EXAMPLE 8   **Mowing a lawn**

Two people are mowing a large lawn. One person has a riding mower, and the other person has a push mower. The person with the riding mower can cut the lawn alone in 4 hours, and the person with the push mower can cut the lawn alone in 9 hours. How long does it take for them, working together, to cut the lawn?

Solution
The first person can cut the entire lawn in 4 hours, so this person can cut $\frac{1}{4}$ of the lawn in 1 hour, $\frac{2}{4}$ of the lawn in 2 hours, and in general, $\frac{t}{4}$ of the lawn in t hours. The second person can cut the lawn in 9 hours so (using similar reasoning) this person can cut $\frac{t}{9}$ of the lawn in t hours. Together they can cut

$$\frac{t}{4} + \frac{t}{9}$$

of the lawn in t hours. The job is complete when the fraction of the lawn cut reaches 1. To find out how long this task takes, solve the equation

$$\frac{t}{4} + \frac{t}{9} = 1.$$

Begin by multiplying each side by the LCD, or 36.

$$\frac{t}{4} + \frac{t}{9} = 1 \qquad \text{Equation to be solved}$$

$$\frac{36t}{4} + \frac{36t}{9} = 36(1) \qquad \text{Multiply each term by 36.}$$

$$9t + 4t = 36 \qquad \text{Simplify.}$$

$$13t = 36 \qquad \text{Combine like terms.}$$

$$t = \frac{36}{13} \qquad \text{Divide each side by 13.}$$

Working together they can cut the lawn in $\frac{36}{13} \approx 2.8$ hours.   Now Try Exercise 105

CRITICAL THINKING

If one person can mow a lawn in x hours and another person can mow it in y hours, how long does it take them, working together, to mow the lawn?

EXAMPLE 9   **Solving a distance problem**

Suppose that the winner of a 600-mile car race finishes 12 minutes ahead of the second-place finisher. If the winner averages 5 miles per hour faster than the second racer, find the average speed of each racer.

Solution

Here we apply the four-step method for solving an application problem.

STEP 1: *Identify any variables.*

x: speed of slower car in miles per hour

$x + 5$:   speed of faster car in miles per hour

STEP 2: *Write an equation.* To determine the time required for each car to finish the race we use the equation $t = \frac{d}{r}$. Because the race is 600 miles, the time for the slower car is $\frac{600}{x}$ and the time for the faster car is $\frac{600}{x+5}$. The difference between these times is 12 minutes, or $\frac{12}{60} = \frac{1}{5}$ hour. Thus to determine x, we solve

$$\frac{600}{x} - \frac{600}{x+5} = \frac{1}{5}.$$

NOTE: Speeds are in miles per *hour*, so time must be in hours not minutes.

STEP 3: *Solve the equation.* Start by multiplying by the LCD, $5x(x+5)$.

$$\frac{600}{x} - \frac{600}{x+5} = \frac{1}{5} \qquad \text{Equation to be solved}$$

$$\frac{600 \cdot 5x(x+5)}{x} - \frac{600 \cdot 5x(x+5)}{x+5} = \frac{1 \cdot 5x(x+5)}{5} \qquad \text{Multiply each term by the LCD.}$$

$$3000(x+5) - 3000x = x(x+5) \qquad \text{Simplify.}$$

$$3000x + 15{,}000 - 3000x = x^2 + 5x \qquad \text{Distributive property}$$

$$15{,}000 = x^2 + 5x \qquad \text{Combine like terms.}$$

$$x^2 + 5x - 15{,}000 = 0 \qquad \text{Rewrite the equation.}$$

$$(x-120)(x+125) = 0 \qquad \text{Factor.}$$

$$x - 120 = 0 \quad \text{or} \quad x + 125 = 0 \qquad \text{Zero-product property}$$

$$x = 120 \quad \text{or} \quad x = -125 \qquad \text{Solve.}$$

The slower car travels at 120 miles per hour, and the faster car travels 5 miles per hour faster, or 125 miles per hour. (The solution -125 has no meaning in this problem.)

STEP 4: *Check your answer.* The slower car travels 600 miles at 120 miles per hour, which takes $\frac{600}{120} = 5$ hours. The faster car travels 600 miles at 125 miles per hour, which requires $\frac{600}{125} = 4.8$ hours. The difference between their times is $5 - 4.8 = 0.2$ hour, or $0.2 \times 60 = 12$ minutes, so the answer checks.   Now Try Exercise 107

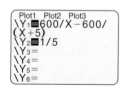

TECHNOLOGY NOTE: Solving Rational Equations Numerically

Tables can be used to find solutions to rational equations. The displays in the margin show the positive solution of 120 from Example 9. Note the use of parentheses for entering the formula for Y_1.

7.6 PUTTING IT ALL TOGETHER

In this section we discussed two methods of solving rational equations. The following table summarizes these methods.

Concept	Explanation	Examples
Solving the Equation $\dfrac{a}{b} = \dfrac{c}{d}$	Cross multiply to obtain $ad = bc$.	$\dfrac{5}{2x} = \dfrac{3}{6}$ is equivalent to $6x = 30$. $\dfrac{x}{2} = \dfrac{8}{x}$ is equivalent to $x^2 = 16$. (Cross multiplication works only for equations having *one* rational expression on each side: "fraction equals fraction.")
Multiplying by the LCD	**STEP 1:** Find the LCD. **STEP 2:** Multiply each term by the LCD. **STEP 3:** Simplify each term. **STEP 4:** Solve the resulting equation. **STEP 5:** Check each possible solution.	Solve $\dfrac{1}{x} - \dfrac{2}{3x} = \dfrac{5}{6}$. **STEP 1:** The LCD is $6x$. **STEP 2:** $\dfrac{1(6x)}{x} - \dfrac{2(6x)}{3x} = \dfrac{5(6x)}{6}$ **STEP 3:** $6 - 4 = 5x$ **STEP 4:** $\dfrac{2}{5} = x$ **STEP 5:** $\dfrac{1}{\frac{2}{5}} - \dfrac{2}{3\left(\frac{2}{5}\right)} = \dfrac{5}{2} - \dfrac{5}{3} = \dfrac{5}{6}$ It checks.

7.6 Exercises

MyMathLab

Math XL PRACTICE   WATCH   DOWNLOAD   READ   REVIEW

CONCEPTS

1. If an equation contains one or more rational expressions, it is called a _____ equation.

2. Give an example of a rational expression and an example of a rational equation.

3. The equation $\dfrac{a}{b} = \dfrac{c}{d}$ is equivalent to _____, provided that _____ and _____ are nonzero.

4. Are the equations $\dfrac{2}{x-1} = 5$ and $5(x-1) = 2$ equivalent provided that $x \neq 1$?

5. One way to solve the equation $\dfrac{5}{3x} + \dfrac{3}{4x} = 1$ is to multiply each side by the LCD, or _____.

6. To solve the equation $T = \dfrac{R}{SV}$ for V, multiply each side by the variable _____ and then divide each side by the variable _____.

SOLVING RATIONAL EQUATIONS

Exercises 7–64: Solve and check your answer.

7. $\dfrac{x}{2} = \dfrac{3}{4}$

8. $\dfrac{2x}{3} = \dfrac{2}{5}$

9. $\dfrac{3}{z} = \dfrac{6}{5}$

10. $\dfrac{2}{7} = \dfrac{1}{z}$

11. $\dfrac{12}{7} = \dfrac{2}{t}$

12. $\dfrac{10}{t} = \dfrac{5}{7}$

13. $\dfrac{3y}{4} = \dfrac{7y}{2}$

14. $\dfrac{y}{6} = \dfrac{5y}{3}$

15. $\dfrac{2}{3} = \dfrac{1}{2x + 1}$

16. $\dfrac{1}{x + 4} = \dfrac{3}{5}$

17. $\dfrac{5}{2x} = \dfrac{8}{x + 2}$

18. $\dfrac{1}{x - 1} = \dfrac{5}{3x}$

19. $\dfrac{1}{z - 1} = \dfrac{2}{z + 1}$

20. $\dfrac{4}{z + 3} = \dfrac{2}{z - 2}$

21. $\dfrac{3}{n + 5} = \dfrac{2}{n - 5}$

22. $\dfrac{4}{3n + 2} = \dfrac{1}{n - 1}$

23. $\dfrac{m}{m - 1} = \dfrac{5}{4}$

24. $\dfrac{5m}{2m - 1} = \dfrac{3}{2}$

25. $\dfrac{5x}{5 - x} = \dfrac{1}{3}$

26. $\dfrac{x + 2}{3x} = \dfrac{4}{3}$

27. $\dfrac{6}{5 - 2x} = 2$

28. $\dfrac{x + 1}{x} = 6$

29. $\dfrac{2x}{2x + 1} = \dfrac{-1}{2x + 1}$

30. $\dfrac{x}{x - 4} = \dfrac{4}{x - 4}$

31. $\dfrac{1}{1 - x} = \dfrac{3}{1 + x}$

32. $\dfrac{2x}{1 - 2x} = \dfrac{1}{2}$

33. $\dfrac{1}{z + 2} = -z$

34. $\dfrac{1}{z - 2} = \dfrac{z}{3}$

35. $\dfrac{-1}{2x + 5} = \dfrac{x}{3}$

36. $\dfrac{x}{2} = \dfrac{1}{3x + 5}$

37. $\dfrac{x}{2} + \dfrac{x}{4} = 3$

38. $\dfrac{x}{4} - \dfrac{x}{3} = 1$

39. $\dfrac{3x}{4} - \dfrac{x}{2} = 1$

40. $\dfrac{2x}{3} + \dfrac{x}{3} = 6$

41. $\dfrac{4}{t + 1} + \dfrac{1}{t + 1} = -1$

42. $\dfrac{2}{t - 5} - \dfrac{5}{t - 5} = 3$

43. $\dfrac{1}{x} + \dfrac{2}{x} = \dfrac{1}{2}$

44. $\dfrac{1}{2x} - \dfrac{2}{x} = -3$

45. $\dfrac{2}{x - 1} + 1 = \dfrac{4}{x^2 - 1}$

46. $\dfrac{1}{x} + 2 = \dfrac{1}{x^2 + x}$

47. $\dfrac{1}{x + 2} = \dfrac{4}{4 - x^2} - 1$

48. $\dfrac{1}{x - 3} + 1 = \dfrac{6}{x^2 - 9}$

49. $\dfrac{5}{4z} - \dfrac{2}{3z} = 1$

50. $\dfrac{3}{z + 1} - \dfrac{1}{z + 1} = 2$

51. $\dfrac{4}{y - 1} + \dfrac{1}{y} = \dfrac{6}{5}$

52. $\dfrac{6}{y + 1} + \dfrac{6}{y} = 5$

53. $\dfrac{1}{2x} - \dfrac{1}{x + 3} = 0$

54. $\dfrac{2}{x} - \dfrac{6}{2x - 1} = -1$

55. $\dfrac{1}{x - 1} + \dfrac{1}{x + 1} = \dfrac{2}{x^2 - 1}$

56. $\dfrac{1}{2x + 1} + \dfrac{1}{2x - 1} = \dfrac{2}{4x^2 - 1}$

57. $\dfrac{1}{x - 2} + \dfrac{1}{x + 2} = \dfrac{6}{x^2 - 4}$

58. $\dfrac{2}{x + 3} - \dfrac{1}{x - 3} = \dfrac{1}{x^2 - 9}$

59. $\dfrac{1}{p + 1} + \dfrac{1}{p + 2} = \dfrac{1}{p^2 + 3p + 2}$

60. $\dfrac{1}{p - 1} - \dfrac{1}{p + 3} = \dfrac{1}{p^2 + 2p - 3}$

61. $\dfrac{1}{x - 2} + \dfrac{3}{2x - 4} = \dfrac{6}{3x - 6}$

62. $\dfrac{4}{x + 1} - \dfrac{4}{2x + 2} = \dfrac{1}{(x + 1)^2}$

63. $\dfrac{1}{r^2 - r - 2} + \dfrac{2}{r^2 - 2r} = \dfrac{1}{r^2 + r}$

64. $\dfrac{3}{r^2 - 1} + \dfrac{1}{r^2 + r} = \dfrac{3}{r^2 - r}$

*Exercises 65–72: **Expressions and Equations**   (Refer to Example 4.) Determine whether you are given an expression or an equation. If it is an expression, simplify (if possible) and evaluate it for $x = 2$. If it is an equation, solve it.*

65. $\dfrac{1}{x} - \dfrac{1 - x}{x}$

66. $\dfrac{1}{x} - x = 0$

67. $\dfrac{1}{2x} - \dfrac{1}{4x} = \dfrac{1}{8}$

68. $\dfrac{1}{2x} - \dfrac{1}{4x}$

69. $\dfrac{x + 1}{x - 1} = \dfrac{2x - 3}{2x - 5}$

70. $\dfrac{2x - 1}{4x + 1} = \dfrac{x + 1}{2x - 1}$

71. $\dfrac{4x + 4}{x + 2} + \dfrac{x^2}{x + 2}$

72. $\dfrac{x^2 - 2}{x - 2} - \dfrac{x}{x - 2}$

GRAPHICAL AND NUMERICAL SOLUTIONS

Exercises 73–76: Use the graph to solve the given equation. Check your answers.

73. $\dfrac{1}{x} = 4x$

74. $\dfrac{x}{x-1} = x$

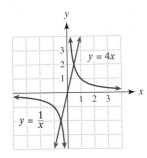

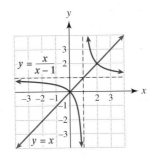

75. $\dfrac{x-1}{x} = -2x$

76. $\dfrac{3}{x^2-1} = 1$

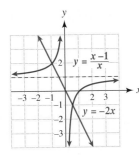

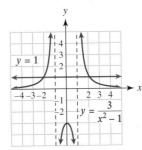

Exercises 77–84: (Refer to Example 5.) Solve the equation
  (a) *graphically and*
  (b) *numerically.*

77. $\dfrac{3}{x} = x + 2$

78. $-\dfrac{2}{x} = 1 - x$

79. $\dfrac{3x}{2} = \dfrac{1}{2}x - 1$

80. $\dfrac{x}{3} = 2 - \dfrac{2}{3}x$

81. $\dfrac{3}{x-1} = 3$

82. $\dfrac{2}{x+5} = 1$

83. $\dfrac{4}{x^2} = 1$

84. $\dfrac{-18}{x^2} = -2$

Exercises 85–88: Solve the rational equation graphically to the nearest thousandth.

85. $\dfrac{1}{\pi x - 2} + \dfrac{\sqrt{2}}{2.1x} = 1.3$

86. $\dfrac{6}{x^2-4} + \dfrac{\pi}{2x} = \dfrac{3}{2}$

87. $\dfrac{1}{x^3} - \dfrac{\pi}{4} = 0$

88. $\dfrac{1}{2\pi x^2} - \dfrac{x}{2} = 1$

SOLVING AN EQUATION FOR A VARIABLE

Exercises 89–100: (Refer to Examples 6 and 7.) Solve the equation for the specified variable.

89. $m = \dfrac{F}{a}$ for a

90. $m = \dfrac{2K}{v^2}$ for K

91. $I = \dfrac{V}{R+r}$ for r

92. $\dfrac{1}{T} = \dfrac{r}{R-r}$ for R

93. $h = \dfrac{2A}{b}$ for b

94. $h = \dfrac{2A}{b_1+b_2}$ for b_1

95. $\dfrac{3}{k} = \dfrac{z}{z+5}$ for z

96. $\dfrac{5}{r} = \dfrac{t+r}{t}$ for t

97. $T = \dfrac{ab}{a+b}$ for b

98. $A = \dfrac{2b}{a-b}$ for b

99. $\dfrac{3}{k} = \dfrac{1}{x} - \dfrac{2}{y}$ for x

100. $\dfrac{1}{R} = \dfrac{1}{R_1} + \dfrac{1}{R_2}$ for R_1

APPLICATIONS

101. *Waiting in Line* (Refer to the introduction for this section.) Solve the equation

$$\frac{1}{10-x} = 1$$

to determine the traffic rate x in cars per minute corresponding to an average waiting time of 1 minute.

102. *Waiting in Line* At a post office customers arrive randomly at an average rate of x people per minute. The clerk can wait on 4 customers per minute. The average time T in minutes spent waiting in line is given by

$$T = \frac{1}{4-x},$$

where $x < 4$.
  (a) Evaluate T for $x = 3, 3.9$, and 3.99. What happens to the waiting time as the arrival rate nears 4 people per minute?
  (b) Find x when the waiting time is 5 minutes.

103. *Shoveling a Sidewalk* It takes an older employee 4 hours to shovel the snow from a sidewalk, but a younger employee can shovel the same sidewalk in 3 hours. How long will it take them to clear the walk if they work together?

104. *Pumping Water* One pump can empty a pool in 5 days, whereas a second pump can empty the pool in 7 days. How long will it take the two pumps, working together, to empty the pool?

105. *Painting a House* One painter can paint a house in 8 days, yet a more experienced painter can paint the house in 4 days. How long will it take the two painters, working together, to paint the house?

106. *Working Together* Suppose that one person can mow a lawn in x hours and that a second person can mow the same lawn in y hours.
 (a) Show that it takes $\frac{xy}{x+y}$ hours for the two people, working together, to mow the lawn.
 (b) Use this expression to solve Example 8.

107. *Bicycle Race* The winner of a 6-mile bicycle race finishes 2 minutes ahead of a teammate and travels, on average, 2 miles per hour faster than the teammate. Find the average speed of each racer.

108. *Freeway Travel* Two drivers travel 150 miles on a freeway and then stop at a wayside rest area. The first driver travels 5 miles per hour faster and arrives $\frac{1}{7}$ hour ahead of the second. Find the average speed of each car.

109. *Braking Distance* If a car is traveling *downhill* at 30 miles per hour on wet pavement, then the braking distance B in feet for this car is given by

$$B = \frac{30}{0.3 + m},$$

where $m < 0$ is the slope of the hill. (*Source:* L. Haefner, *Introduction to Transportation Systems.*)
 (a) Find the braking distance for $m = -0.05$ and interpret the result.
 (b) Find m if the braking distance is 150 feet.

110. *Slippery Roads* If a car is traveling at 30 miles per hour on a level road, then its braking distance in feet is $\frac{30}{x}$, where x is the coefficient of friction between the road and the tires. The variable x is positive and satisfies $x \leq 1$. The closer the value of x is to 0, the more slippery is the road. (*Source:* L. Haefner.)
 (a) Evaluate the expression for $x = 1, 0.5$, and 0.1. Interpret the results.
 (b) Find x for a braking distance of 150 feet.

111. *River Current* A boat can travel 36 miles upstream in the same time that it can travel 54 miles downstream. If the speed of the current is 3 miles per hour, find the speed of the boat without a current.

112. *Airplane Speed* An airplane can travel 380 miles into the wind in the same time that it can travel 420 miles with the wind. If the wind speed is 10 miles per hour, find the speed of the airplane without any wind.

113. *Airplane Speed* An airplane can travel 450 miles into the wind in the same time that it can travel 750 miles with the wind. If the wind speed is 50 miles per hour, find the speed of the airplane without any wind.

114. *River Current* A boat can travel 114 miles upstream in the same time that it can travel 186 miles downstream. If the speed of the current is 6 miles per hour, find the speed of the boat without a current.

115. *Running and Walking* An athlete runs 10 miles and then walks home. The trip home takes 1 hour longer than it took to run that distance. If the athlete runs 5 miles per hour faster than she walks, what are her average running and walking speeds?

116. *Speed Limit* A person drives 390 miles on a stretch of road. Half the distance is driven traveling 5 miles per hour below the speed limit, and half the distance is driven traveling 5 miles per hour above the speed limit. If the time spent traveling at the slower speed exceeds the time spent traveling at the faster speed by 24 minutes, find the speed limit.

117. *Unknown Number* Find a number n that makes the expression $\frac{2+n}{8-n}$ equal 1.

118. *Unknown Number* Find a number n that makes the expression $\frac{2n+4}{4-3n}$ equal $-\frac{9}{11}$.

119. *Unknown Number* Find a number n that makes the expression $\frac{4+6n}{4-4n}$ equal $-\frac{49}{31}$.

120. *Highway Curves* To make a highway curve safe, highway engineers often bank it, as shown in the figure. If a curve is designed for a speed of 50 miles per hour and is banked with positive slope m, then a minimum radius R in feet for the curve is given by

$$R = \frac{2500}{15m + 2}.$$

(*Source:* N. Garber and L. Hoel, *Traffic and Highway Engineering.*)

(a) Find R for $m = 0.1$. Interpret the result.
(b) If $R = 500$, find m. Interpret the result.

WRITING ABOUT MATHEMATICS

121. Do all rational equations have solutions? Explain.

122. Why is it important to check your answer when solving rational equations? Explain.

CHECKING BASIC CONCEPTS
SECTIONS 7.5 AND 7.6

1. Simplify each complex fraction.

(a) $\dfrac{\dfrac{x}{3}}{\dfrac{2x}{5}}$
(b) $\dfrac{\dfrac{2}{2x} - \dfrac{1}{3x}}{6x}$

(c) $\dfrac{\dfrac{1}{a} - \dfrac{1}{b}}{\dfrac{1}{a} + \dfrac{1}{b}}$
(d) $\dfrac{\dfrac{1}{r^2} - \dfrac{1}{t^2}}{\dfrac{2}{r} - \dfrac{2}{t}}$

2. Solve each equation. Check your answer.

(a) $\dfrac{1}{2x} = \dfrac{3}{x + 1}$
(b) $\dfrac{x}{2x + 3} = \dfrac{4}{5}$

3. Solve each equation. Check your answer.

(a) $\dfrac{1}{2x} + \dfrac{3}{2x} = 1$
(b) $\dfrac{3}{x + 1} - \dfrac{2}{x} = -2$

4. Solve the equation and check your answer.

$$\dfrac{1}{x - 1} = \dfrac{2}{x^2 - 1} - \dfrac{1}{2}$$

5. Solve each equation for the specified variable.

(a) $\dfrac{ax}{2} - 3y = b$ for x

(b) $\dfrac{1}{2m - 1} = \dfrac{k}{m}$ for m

6. *Braking Distance* If a car is traveling *uphill* at 60 miles per hour on wet pavement, then the braking distance D in feet for this car is given by the rational expression

$$D = \frac{120}{0.3 + m},$$

where $m > 0$ is the slope of the hill. (*Source:* L. Haefner, *Introduction to Transportation Systems.*)

(a) Find D for $m = 0.1$ and interpret the result.
(b) Find the slope of the road if D is 200 feet.

7.7 PROPORTIONS AND VARIATION

Proportions ■ Direct Variation ■ Inverse Variation

Proportions frequently are used to solve applications. The following are a few examples.

- If someone earns $120 per day, then that person can earn $600 in 5 days.
- If a car goes 280 miles on 10 gallons of gas, then it can go 560 miles on 20 gallons of gas.
- If a person walks 1 mile in 20 minutes, then that person can walk $\frac{1}{2}$ mile in 10 minutes.

In this section we discuss several applications of proportions.

Proportions

▶ **REAL-WORLD CONNECTION**   A **ratio** is a comparison of two quantities, expressed as a quotient. For example, a math class might have 7 boys for every 8 girls. Thus the boy–girl ratio in this class is *7 to 8*, or $\frac{7}{8}$. In mathematics, ratios are typically expressed as fractions.

Ratios and proportions are sometimes used to find how much space remains for music on a compact disc (CD). A 700-megabyte CD can store about 80 minutes of music. Suppose that some music has been recorded on the CD and that 256 megabytes are still available. Using ratios and proportions, we can find how many more minutes of music could be recorded. A **proportion** is a statement that two *ratios* are equal.   (**Source:** Maxell Corporation.)

Let x represent the number of minutes available on a CD. Then **80** minutes are to **700** megabytes as x minutes are to **256** megabytes. By setting the *ratios* $\frac{80}{700}$ and $\frac{x}{256}$ equal to each other, we obtain the *proportion*

$$\frac{80}{700} = \frac{x}{256}. \qquad \frac{\text{Minutes}}{\text{Megabytes}} = \frac{\text{Minutes}}{\text{Megabytes}}$$

Solving this equation for x gives

$$700x = 80(256) \qquad \text{Cross multiply.}$$

$$x = \frac{80 \cdot 256}{700} \approx 29.3 \text{ minutes.} \qquad \text{Divide by 700.}$$

About 29 minutes are available to record on the CD.

MAKING CONNECTIONS

Proportions and Fractional Parts

We could have solved the preceding problem by noting that the fraction of the CD still available for recording music is $\frac{256}{700}$. So $\frac{256}{700}$ of 80 minutes is

$$\frac{256}{700} \cdot 80 \approx 29.3 \text{ minutes.}$$

EXAMPLE 1   Calculating the water content in snow

Six inches of light, fluffy snow are equivalent to about half an inch of rain in terms of water content. If 15 inches of this type of snow fall, estimate the water content.

Solution

Let x be the equivalent amount of rain. Then **6** inches of snow are to $\frac{1}{2}$ inch of rain as **15** inches of snow are to x inches of rain, which can be written as the proportion

$$\frac{6}{\frac{1}{2}} = \frac{15}{x}. \qquad\qquad \frac{\text{Snow}}{\text{Rain}} = \frac{\text{Snow}}{\text{Rain}}$$

Solving this equation gives

$$6x = \frac{15}{2} \quad\text{or}\quad x = \frac{15}{12} = 1.25.$$

Thus 15 inches of light, fluffy snow are equivalent to about 1.25 inches of rain.

Now Try Exercise **57**

Proportions frequently occur in geometry when we work with similar figures. Two triangles are similar if the measures of their corresponding angles are equal. Corresponding sides of similar triangles are proportional. Figure 7.5 shows two right triangles that are similar because each has angles of 30°, 60°, and 90°.

We can find the length of side x by using proportions. Side x is to 16 as 5.5 is to 8, which can be written as the proportion

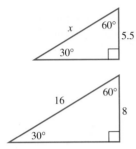

Figure 7.5

$$\frac{x}{16} = \frac{5.5}{8}. \qquad\qquad \frac{\text{Hypotenuse}}{\text{Hypotenuse}} = \frac{\text{Shorter leg}}{\text{Shorter leg}}$$

Solving yields the equation

$$8x = 5.5(16) \qquad\qquad \text{Cross multiply.}$$
$$x = 11. \qquad\qquad \text{Divide by 8.}$$

NOTE: Proportions can be set up in different ways and still produce the correct result. For example, we could say that x is to 5.5 in the smaller triangle as 16 is to 8 in the larger triangle.

$$\frac{x}{5.5} = \frac{16}{8} \qquad\qquad \frac{\text{Hypotenuse}}{\text{Shorter leg}} = \frac{\text{Hypotenuse}}{\text{Shorter leg}}$$

Solving, we obtain $8x = 5.5(16)$, or $x = 11$, which is the same answer.

EXAMPLE **2** | Calculating the height of a tree

A 6-foot-tall person casts a 4-foot-long shadow. If a nearby tree casts a 36-foot-long shadow, estimate the height of the tree. See Figure 7.6.

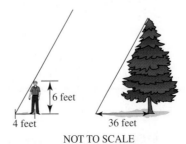

NOT TO SCALE

Figure 7.6

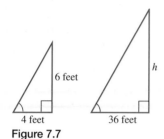

Figure 7.7

Solution

The triangles shown in Figure 7.7 are similar because the measures of the corresponding angles are equal. Therefore their sides are proportional. Let h be the height of the tree.

$$\frac{h}{6} = \frac{36}{4} \qquad \frac{\text{Height}}{\text{Height}} = \frac{\text{Shadow length}}{\text{Shadow length}}$$

$$4h = 6(36) \qquad \text{Cross multiply.}$$

$$h = \frac{6(36)}{4} \qquad \text{Divide each side by 4.}$$

$$h = 54 \qquad \text{Simplify.}$$

The tree is 54 feet tall.

Now Try Exercise 56

Direct Variation

▶ **REAL-WORLD CONNECTION**   If your wage is $12 per hour, the amount you earn is proportional to the number of hours that you work. If you work H hours, your total pay P satisfies the equation

$$\frac{P}{H} = \frac{12}{1}, \qquad \frac{\text{Pay}}{\text{Hours}}$$

or, equivalently,

$$P = 12H.$$

We say that your pay P is *directly proportional* to the number of hours H worked. The constant of proportionality is 12.

DIRECT VARIATION

Let x and y denote two quantities. Then y is **directly proportional** to x, or y **varies directly** with x, if there is a nonzero number k such that

$$y = kx.$$

The number k is called the **constant of proportionality**, or the **constant of variation**.

The following 4-step process can be helpful when solving a variation application. This process is used to solve Examples 3 and 4 and can also be used to solve other types of variation problems.

SOLVING A VARIATION APPLICATION

When solving a variation problem, the following steps can be used.

STEP 1: Write the general equation for the type of variation problem that you are solving.

STEP 2: Substitute given values in this equation so the constant of variation k is the only unknown value in the equation. Solve for k.

STEP 3: Substitute the value of k in the general equation in Step 1.

STEP 4: Use this equation to find the requested quantity.

EXAMPLE 3 Solving a direct variation problem

Let y be directly proportional to x, or vary directly with x. Suppose $y = 7$ when $x = 5$. Find y when $x = 11$.

Solution
STEP 1: The general equation for direct variation is $y = kx$.

STEP 2: Substitute 7 for y and **5** for x in $y = kx$. Solve for k.

$$7 = k(5) \qquad \text{Let } y = 7 \text{ and } x = 5.$$

$$\frac{7}{5} = k \qquad \text{Divide each side by 5.}$$

STEP 3: Replace k with $\frac{7}{5}$ in the equation $y = kx$ to obtain $y = \frac{7}{5}x$.

STEP 4: To find y, let $x = 11$. Then $y = \frac{7}{5}(11) = \frac{77}{5} = 15.4$.   Now Try Exercise 31

EXAMPLE 4 Solving a direct variation application

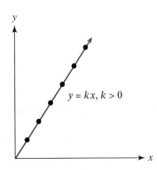

The strength of a beam of wood varies directly with its width. A beam that is 2.5 inches wide can support 800 pounds. How much weight can a similar beam support if its width is 3.2 inches?

Solution
STEP 1: The general equation for direct variation is $y = kx$, where y is the weight and x is the width.

STEP 2: Substitute **800** for y and **2.5** for x in $y = kx$. Solve for k.

$$800 = k(2.5) \qquad \text{Let } y = 800 \text{ and } x = 2.5.$$

$$\frac{800}{2.5} = k \qquad \text{Divide each side by 2.5.}$$

$$k = 320 \qquad \text{Simplify; rewrite the equation.}$$

STEP 3: Replace k with **320** in $y = kx$ to obtain $y = 320x$.

STEP 4: Find the weight y by substituting **3.2** for x in $y = 320x$.

$$y = 320(3.2) = 1024 \text{ pounds} \qquad \text{Now Try Exercise 59}$$

   The graph of $y = kx$ is a line passing through the origin, as illustrated in Figure 7.8. Sometimes data in a scatterplot indicate that two quantities are directly proportional. The constant of proportionality k corresponds to the slope of the graph, which can be negative.

$y = kx, k > 0$

Figure 7.8 Direct Variation

EXAMPLE **5**   Modeling college tuition

Table 7.5 lists the tuition for taking various numbers of credits.
(a) A scatterplot of the data is shown in Figure 7.9. Could the data be modeled using a line?

TABLE 7.5

Credits	Tuition
3	$189
5	$315
8	$504
11	$693
17	$1071

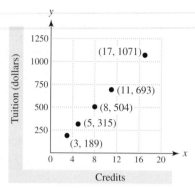

Figure 7.9

(b) Explain why tuition is directly proportional to the number of credits taken.
(c) Find the constant of proportionality. Interpret your result.
(d) Predict the cost of taking 16 credits.

Solution

(a) The data are linear and suggest a line passing through the origin.
(b) Because the data can be modeled by a line passing through the origin, tuition is directly proportional to the number of credits taken. Hence doubling the credits will double the tuition and tripling the credits will triple the tuition.
(c) The slope of the line equals the constant of proportionality k. If we use the first and last data points $(3, 189)$ and $(17, 1071)$, the slope is

$$k = \frac{1071 - 189}{17 - 3} = 63.$$

That is, tuition is $63 per credit. If we graph the line $y = 63x$, it models the data, as shown in Figure 7.10. This graph can also be created with a graphing calculator.

(d) If y represents tuition and x represents the credits taken, 16 credits would cost

$$y = 63(16) = \$1008.$$

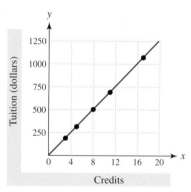

Figure 7.10

Now Try Exercise **63**

Ratios and the Constant of Proportionality

The constant of proportionality in Example 5 can also be found by calculating the ratios $\frac{y}{x}$, where y is the tuition and x is the credits taken. Note that each ratio in the table is 63 because the equation $y = 63x$ is equivalent to the equation $\frac{y}{x} = 63$.

x	3	5	8	11	17
y	189	315	504	693	1071
$\frac{y}{x}$	63	63	63	63	63

Inverse Variation

▶ REAL-WORLD CONNECTION   When two quantities vary inversely, an increase in one quantity results in a decrease in the second quantity. For example, at 30 miles per hour a car travels 120 miles in 4 hours, whereas at 60 miles per hour the car travels 120 miles in 2 hours. Doubling the speed (or rate) decreases the travel time by half. Distance equals rate times time, so $d = rt$. Thus

$$120 = rt, \quad \text{or equivalently,} \quad t = \frac{120}{r}.$$

We say that the time t to travel 120 miles is *inversely proportional* to the speed or rate r. The constant of proportionality or constant of variation is 120.

INVERSE VARIATION

Let x and y denote two quantities. Then y is **inversely proportional** to x, or y **varies inversely** with x, if there is a nonzero number k such that

$$y = \frac{k}{x}.$$

The data shown in Figure 7.11 represent inverse variation and are modeled by $y = \frac{k}{x}$. Note that, as x increases, y decreases. We assume k is positive.

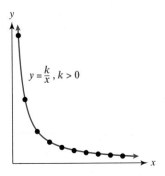

$y = \frac{k}{x}, k > 0$

Figure 7.11 Inverse Variation

EXAMPLE 6   Solving an inverse variation problem

Let y be inversely proportional to x, or vary inversely with x. Suppose $y = 5$ when $x = 6$. Find y when $x = 21$.

Solution

STEP 1: The general equation for inverse variation is $y = \frac{k}{x}$.

STEP 2: Because $y = 5$ when $x = 6$, substitute 5 for y and 6 for x in $y = \frac{k}{x}$. Solve for k.

$$5 = \frac{k}{6} \qquad \text{Let } y = 5 \text{ and } x = 6.$$

$$30 = k \qquad \text{Multiply each side by 6.}$$

STEP 3: Replace k with 30 in the equation $y = \frac{k}{x}$ to obtain $y = \frac{30}{x}$.

STEP 4: To find y, let $x = 21$. Then $y = \frac{30}{21} = \frac{10}{7}$.   Now Try Exercise 37

▶ **REAL-WORLD CONNECTION**   A wrench is commonly used to loosen a nut on a bolt. See Figure 7.12. If the nut is difficult to loosen, a wrench with a longer handle is often helpful.

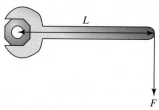

Figure 7.12

EXAMPLE 7   Illustrating inverse variation with a wrench

TABLE 7.6

L (inches)	F (pounds)
6	12
8	9
12	6
18	4
24	3

Table 7.6 lists the force F necessary to loosen a particular nut with wrenches of different lengths L.

(a) Make a scatterplot of the data and discuss the graph. Are the data linear?

(b) Explain why the force F is inversely proportional to the handle length L. Find k so that $F = \frac{k}{L}$ models the data.

(c) Predict the force needed to loosen the nut with a 15-inch wrench.

Solution

(a) The scatterplot shown in Figure 7.13 reveals that the data are nonlinear. As the length L of the wrench increases, the force F necessary to loosen the nut decreases.

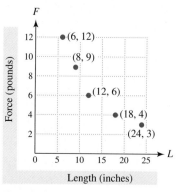

Figure 7.13

(b) If F is inversely proportional to L, then $F = \frac{k}{L}$, or $FL = k$. That is, the product of F and L equals the constant of proportionality k. In Table 7.6, the product of F and L always equals 72 for each data point. Thus F is inversely proportional to L with constant of proportionality $k = 72$.

(c) If $L = 15$, then $F = \frac{72}{15} = 4.8$. A wrench with a 15-inch handle requires a force of 4.8 pounds to loosen the nut.   Now Try Exercise 67

TECHNOLOGY NOTE: Scatterplots and Graphs

A graphing calculator can be used to create scatterplots and graphs. A scatterplot of the data in Table 7.6 is shown in the first figure. In the second figure the data and the equation $y = \frac{72}{x}$ are graphed. Note that each tick mark represents 5 units.

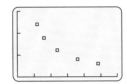 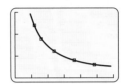

Example 7 demonstrates that the product of the variables in an inverse variation equation equals a constant. This result is true in general. If we multiply each side of $y = \frac{k}{x}$ by x, we obtain $xy = k$. That is, the product of the variables in inverse variation equals the constant of variation k. In the next example, we use this fact to determine if a table of data represents inverse variation.

EXAMPLE 8   Analyzing data

Determine whether the data in each table represent direct variation, inverse variation, or neither.

(a)

x	4	5	10	20
y	40	32	16	8

(b)

x	2	5	9	11
y	18	45	81	99

(c)

x	2	4	6	8
y	8	20	30	56

Solution

(a) As x increases, y decreases. Because $xy = 160$ for each data point in the table, the equation $y = \frac{160}{x}$ models the data. The data represent inverse variation.

(b) As $\frac{y}{x} = 9$ for each data point in the table, the equation $y = 9x$ models the data in the table. These data represent direct variation.

(c) Neither the product xy nor the ratio $\frac{y}{x}$ are constant for the data in the table. Therefore these data represent neither direct variation nor inverse variation.

Now Try Exercises 43, 45, 47

7.7 PUTTING IT ALL TOGETHER

In this section we introduced some basic concepts of proportion and variation. They are summarized in the following table.

Concept	Explanation	Examples
Proportion	A statement that two ratios are equal	$\dfrac{8}{17} = \dfrac{49}{x}$ and $\dfrac{x}{6} = \dfrac{3}{14}$
Direct Variation	Two quantities x and y vary according to the equation $y = kx$, where k is a nonzero constant. The constant of proportionality (or variation) is k.	$y = 4x$ or $\dfrac{y}{x} = 4$ <table><tr><td>x</td><td>1</td><td>2</td><td>4</td></tr><tr><td>y</td><td>4</td><td>8</td><td>16</td></tr></table> Note that if x doubles, then y also doubles.
Inverse Variation	Two quantities x and y vary according to the equation $y = \frac{k}{x}$, where k is a nonzero constant. The constant of proportionality (or variation) is k.	$y = \dfrac{3}{x}$ or $xy = 3$ <table><tr><td>x</td><td>1</td><td>3</td><td>6</td></tr><tr><td>y</td><td>3</td><td>1</td><td>$\frac{1}{2}$</td></tr></table> Note that if x doubles, then y decreases by half.

7.7 Exercises

MyMathLab   Math XL PRACTICE   WATCH   DOWNLOAD   READ   REVIEW

CONCEPTS

1. What is a proportion?

2. If 5 is to 6 as x is to 7, write a proportion that allows you to find x.

3. Suppose that y is directly proportional to x. If x doubles, what happens to y?

4. Suppose that y is inversely proportional to x. If x doubles, what happens to y?

5. If y varies directly with x, then $\frac{y}{x}$ equals a _____.

6. If y varies inversely with x, then xy equals a _____.

7. Would the food bill B generally vary directly or inversely with the number of people N being fed? Explain your reasoning.

8. Would the time T needed to paint a building vary directly or inversely with the number of painters N working on the job? Explain your reasoning.

PROPORTIONS

Exercises 9–20: Solve the proportion.

9. $\dfrac{x}{24} = \dfrac{5}{8}$

10. $\dfrac{x}{5} = \dfrac{3}{7}$

11. $\dfrac{14}{x} = \dfrac{2}{3}$

12. $\dfrac{4}{9} = \dfrac{9}{x}$

13. $\dfrac{3}{16} = \dfrac{h}{256}$ **14.** $\dfrac{20}{a} = \dfrac{15}{4}$

15. $\dfrac{3}{4} = \dfrac{2x}{7}$ **16.** $\dfrac{7}{3z} = \dfrac{5}{4}$

17. $\dfrac{x}{6} = \dfrac{8}{3x}$ **18.** $\dfrac{4}{x} = \dfrac{4x}{9}$

19. $\dfrac{x}{7} = \dfrac{7}{4x}$ **20.** $\dfrac{2}{3x} = \dfrac{27x}{8}$

21. Thinking Generally Solve $\dfrac{a}{b} = \dfrac{c}{d}$ for b.

22. Thinking Generally Solve $\dfrac{a+b}{c^2} = \dfrac{1}{2}$ for b.

Exercises 23–30: Do the following.
  (a) *Write a proportion that models the situation.*
  (b) *Solve the proportion for x.*

23. 5 is to 8, as 9 is to x

24. x is to 11, as 7 is to 4

25. A triangle has sides of 4, 7, and 10. In a similar triangle the shortest side is 8 and the longest side is x.

26. A rectangle has sides of 5 and 12. In a similar rectangle the longer side is 10 and the shorter side is x.

27. If you earn \$98 in 7 hours, then you can earn x dollars in 11 hours.

28. If 14 gallons of gasoline contain 1.4 gallons of ethanol, then 22 gallons of gasoline contain x gallons of ethanol.

29. If 3 cassette tapes can record 180 minutes of music, then 7 cassette tapes can record x minutes.

30. If a gas pump fills a 25-gallon tank in 6 minutes, it can fill a 14-gallon tank in x minutes.

VARIATION

*Exercises 31–36: **Direct Variation** Suppose that y is directly proportional to x.*
  (a) *Use the given information to find the constant of proportionality k.*
  (b) *Then use y = kx to find y for x = 6.*

31. $y = 4$ when $x = 2$ **32.** $y = 5$ when $x = 10$

33. $y = 3$ when $x = 2$ **34.** $y = 11$ when $x = 55$

35. $y = -60$ when $x = 8$

36. $y = -17$ when $x = 68$

*Exercises 37–42: **Inverse Variation** Suppose that y is inversely proportional to x.*
  (a) *Use the given information to find the constant of proportionality k.*
  (b) *Then use $y = \frac{k}{x}$ to find y for x = 8.*

37. $y = 6$ when $x = 4$ **38.** $y = 2$ when $x = 24$

39. $y = 80$ when $x = \frac{1}{2}$ **40.** $y = \frac{1}{4}$ when $x = 32$

41. $y = 20$ when $x = 20$ **42.** $y = \frac{8}{3}$ when $x = 12$

Exercises 43–48: (Refer to Example 8.)
  (a) *Determine whether the data represent direct variation, inverse variation, or neither.*
  (b) *If the data represent either direct or inverse variation, find an equation that models the data.*
  (c) *Graph the equation and the data when possible.*

43.

x	2	3	4	5
y	3	4.5	6	7.5

44.

x	10	20	30	40
y	12	6	5	4

45.

x	3	6	9	12
y	12	6	4	3

46.

x	2	6	10	14
y	105	35	21	15

47.

x	4	6	12	20
y	10	20	30	40

48.

x	1	5	9	15
y	6	30	54	90

Exercises 49–54: Use the graph to determine whether the data represent direct variation, inverse variation, or neither. Find the constant of variation whenever possible.

49.
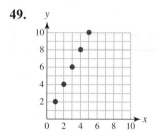

50.
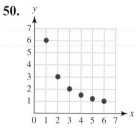

51.
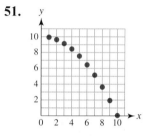

52.
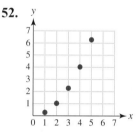

53.
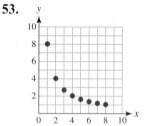

54.
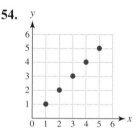

APPLICATIONS

55. *Recording Music*   A 750-megabyte CD can record 85 minutes of music. How many minutes can be recorded on 420 megabytes?

56. *Height of a Tree*   (Refer to Example 2.) A 5-foot person casts an 8-foot shadow, and a nearby tree casts a 30-foot shadow. Estimate the height of the tree.

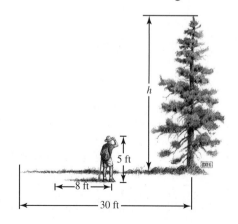

57. *Water Content in Snow*   (Refer to Example 1.) Eight inches of heavy, wet snow are equivalent to an inch of rain. Estimate the water content in 13 inches of heavy, wet snow.

58. *Wages*   If a person working for an hourly wage earns $143 in 13 hours, how much will that person earn in 15 hours?

59. *Strength of a Beam*   (Refer to Example 4.) The strength of a metal beam varies directly with its width. A beam that is 6.2 inches wide can support 2800 pounds. How much weight can a similar beam support if it is 4.7 inches wide?

60. *Strength of a Beam*   The strength of a wood beam varies inversely with its length. A beam that is 33 feet long can support 1200 pounds. How much weight can a similar beam support if it is 23 feet long?

61. *Making Fudge*   If $2\frac{2}{3}$ cups of sugar can make 14 pieces of fudge, how much sugar is needed to make 49 pieces of fudge?

62. *Making Coffee*   If 6 tablespoons of coffee grounds make 10 cups of coffee, how many tablespoons of coffee grounds are needed to make 35 cups of coffee?

63. *Rolling Resistance of Cars*   If you were to try to push a car, you would experience *rolling resistance*. This resistance equals the force necessary to keep the car moving slowly in neutral gear. The following table shows the rolling resistance R for passenger cars of different gross weights W.   (**Source:** N. Garber and L. Hoel, *Traffic and Highway Engineering.*)

W (pounds)	2000	2500	3000	3500
R (pounds)	24	30	36	42

(a) Do the data represent direct or inverse variation? Explain.
(b) Find an equation that models the data. Graph the equation with the data.
(c) Estimate the rolling resistance of a 3200-pound car.

64. *Transportation Costs*   The use of a particular toll bridge varies inversely according to the toll. If the toll is $0.50, then 8000 vehicles use the bridge. Estimate the number of users if the toll is $0.80.   (**Source:** N. Garber.)

65. *Flow of Water*   The gallons of water G flowing in 1 minute through a hose with a cross-sectional area A are shown in the table.

A (square inch)	0.2	0.3	0.4	0.5
G (gallons)	5.4	8.1	10.8	13.5

(a) Do the data represent direct or inverse variation? Explain.

(b) Find an equation that models the data. Graph the equation with the data.

(c) Interpret the constant of variation k.

66. *Hooke's Law*   The table shows the distance D that a spring stretches when a weight W is hung on it.

W (pounds)	2	6	9	15
D (inches)	1.6	4.8	7.2	12

(a) Do the data represent direct or inverse variation? Explain.

(b) Find an equation that models the data.

(c) How far will the spring stretch if an 11-pound weight is hung on it, as depicted in the figure?

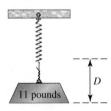

67. *Tightening Lug Nuts*   (Refer to Example 7.) When a tire is mounted on a car, the lug nuts should not be over-tightened. The following table shows the maximum force used with wrenches of different lengths.

L (inches)	8	10	16
F (pounds)	150	120	75

Source: Tires Plus.

(a) Model the data, using the equation $F = \frac{k}{L}$.

(b) How much force should be used with a wrench 15 inches long?

68. *Ozone and UV Radiation*   Ozone in the atmosphere filters out approximately 90% of the harmful ultraviolet (UV) rays from the sun. Depletion of the ozone layer increases the amount of UV radiation reaching Earth's surface. An increase in UV radiation is associated with skin cancer. The following graph shows the percentage increase y in UV radiation for a decrease in the ozone layer of x percent.   (*Source:* R. Turner, D. Pearce, and I. Bateman, *Environmental Economics*.)

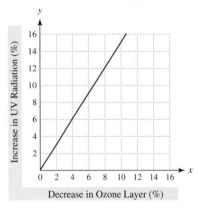

(a) Does this graph represent direct or inverse variation?

(b) Find an equation for the line in the graph.

(c) Estimate the percentage increase in UV radiation if the ozone layer decreases by 5%.

69. *Air Temperature and Altitude*   In the first 6 miles of Earth's atmosphere, air cools as the altitude increases. The following graph shows the temperature change y in degrees Fahrenheit at an altitude of x miles. (*Source:* A. Miller and R. Anthes, *Meteorology*.)

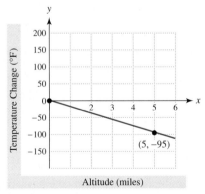

(a) Does this graph represent direct variation or inverse variation?

(b) Find an equation that models the data in the graph.

(c) Is the constant of proportionality k positive or negative? Interpret k.

(d) Find the change in air temperature 2.5 miles high.

70. *Cost of Tuition*　(Refer to Example 5.) The cost of tuition is directly proportional to the number of credits taken. If 6 credits cost $483, find the cost of 13 credits. What does the constant of proportionality represent?

71. *Electrical Resistance*　The electrical resistance of a wire is directly proportional to its length. If a 30-foot-long wire has a resistance of 3 ohms, find the resistance of an 18-foot-long wire.

72. *Resistance and Current*　The current that flows through an electrical circuit is inversely proportional

to the resistance. When the resistance R is 180 ohms, the current I is 0.6 amp. Find the current when the resistance is 54 ohms.

WRITING ABOUT MATHEMATICS

73. Explain what it means for a quantity y to be directly proportional to a quantity x.

74. Explain what it means for a quantity y to be inversely proportional to a quantity x.

CHECKING BASIC CONCEPTS
SECTION 7.7

1. Solve each proportion.

  (a) $\dfrac{x}{9} = \dfrac{2}{5}$　**(b)** $\dfrac{4}{3} = \dfrac{5}{b}$

2. Write a proportion that models each situation. Then solve it.
  (a) 4 is to 6 as 8 is to x
  (b) If 2 compact discs can record 148 minutes of music, then 5 compact discs can record x minutes of music.

3. Suppose that y is inversely proportional to x. If $y = 4$ when $x = 15$, find the constant of proportionality k. Find y when $x = 10$.

4. Decide whether the data in the table represent direct or inverse variation. Explain your reasoning. Find the constant of variation.

  (a)

x	2	4	6	8
y	3	6	9	12

  (b)

x	2	4	6	8
y	12	6	4	3

5. *Wages*　If a person working for an hourly wage earns $272 in 17 hours, how much will the person earn in 10 hours?

CHAPTER 7　SUMMARY
SECTION 7.1 ■ INTRODUCTION TO RATIONAL EXPRESSIONS

Rational Expression
A rational expression can be written in the form $\dfrac{P}{Q}$, where P and Q are polynomials, and is defined whenever $Q \neq 0$.

Example:　$\dfrac{x^2}{x-5}$ is a rational expression that is defined for all real numbers except $x = 5$.

Simplifying a Rational Expression　To simplify a rational expression, factor the numerator and the denominator. Then apply the basic principle of rational expressions,

$$\frac{PR}{QR} = \frac{P}{Q}.$$

Examples:　$\dfrac{x^2-4}{x^2-3x+2} = \dfrac{(x+2)(x-2)}{(x-1)(x-2)} = \dfrac{x+2}{x-1},$　$\dfrac{x-3}{3-x} = -\dfrac{x-3}{x-3} = -1$

SECTION 7.2 ■ MULTIPLICATION AND DIVISION OF RATIONAL EXPRESSIONS

Multiplying Rational Expressions   To multiply two rational expressions, multiply the numerators and multiply the denominators.

$$\frac{A}{B} \cdot \frac{C}{D} = \frac{AC}{BD}$$

Example: $\dfrac{3}{x - 1} \cdot \dfrac{4}{x + 1} = \dfrac{12}{(x - 1)(x + 1)}$

Dividing Rational Expressions   To divide two rational expressions, multiply by the reciprocal of the divisor.

$$\frac{A}{B} \div \frac{C}{D} = \frac{A}{B} \cdot \frac{D}{C} = \frac{AD}{BC}$$

Example: $\dfrac{x + 2}{x^2 + 3x} \div \dfrac{x + 2}{x} = \dfrac{x + 2}{x(x + 3)} \cdot \dfrac{x}{x + 2} = \dfrac{x(x + 2)}{x(x + 3)(x + 2)} = \dfrac{1}{x + 3}$

SECTION 7.3 ■ ADDITION AND SUBTRACTION WITH LIKE DENOMINATORS

Addition of Rational Expressions Having Like Denominators   To add two rational expressions having like denominators, add their numerators. Keep the same denominator.

$$\frac{A}{C} + \frac{B}{C} = \frac{A + B}{C}$$

Example: $\dfrac{5x}{x + 4} + \dfrac{1}{x + 4} = \dfrac{5x + 1}{x + 4}$

Subtraction of Rational Expressions Having Like Denominators   To subtract two rational expressions having like denominators, subtract their numerators. Keep the same denominator.

$$\frac{A}{C} - \frac{B}{C} = \frac{A - B}{C}$$

Example: $\dfrac{6}{2x + 1} - \dfrac{2}{2x + 1} = \dfrac{4}{2x + 1}$

SECTION 7.4 ■ ADDITION AND SUBTRACTION WITH UNLIKE DENOMINATORS

Finding Least Common Multiples   The least common multiple (LCM) of two or more polynomials can be found as follows.

STEP 1: Factor each polynomial completely.

STEP 2: List each factor the greatest number of times that it occurs in any factorization.

STEP 3: Find the product of this list of factors. It is the LCM.

Example: $4x^2(x + 1) = 2 \cdot 2 \cdot x \cdot x \cdot (x + 1)$
$2x(x^2 - 1) = 2 \cdot x \cdot (x + 1) \cdot (x - 1)$

Listing each factor the greatest number of times and multiplying gives

$$2 \cdot 2 \cdot x \cdot x \cdot (x + 1)(x - 1) = 4x^2(x^2 - 1),$$

which is the LCM of $4x^2(x + 1)$ and $2x(x^2 - 1)$.

Finding the Least Common Denominator   The least common denominator (LCD) is the least common multiple (LCM) of the denominators.

Example: From the preceding example, the LCD for $\frac{1}{4x^2(x + 1)}$ and $\frac{1}{2x(x^2 - 1)}$ is $4x^2(x^2 - 1)$.

Addition and Subtraction of Rational Expressions Having Unlike Denominators   First write each rational expression by using the LCD. Then add or subtract the resulting rational expressions.

Example: $\dfrac{1}{x - 1} - \dfrac{1}{x} = \dfrac{x}{x(x - 1)} - \dfrac{x - 1}{x(x - 1)} = \dfrac{x - (x - 1)}{x(x - 1)} = \dfrac{1}{x(x - 1)}$

Note that the LCD is $x(x - 1)$.

SECTION 7.5 ■ COMPLEX FRACTIONS

Complex Fractions   A complex fraction is a rational expression that contains fractions in its numerator, denominator, or both. The following equation can be used to simplify some complex fractions.

$$\frac{\dfrac{a}{b}}{\dfrac{c}{d}} = \frac{a}{b} \cdot \frac{d}{c}$$

Example: $\dfrac{\dfrac{x}{3}}{\dfrac{x}{x - 1}} = \dfrac{x}{3} \cdot \dfrac{x - 1}{x} = \dfrac{x(x - 1)}{3x} = \dfrac{x - 1}{3}$

Simplifying Complex Fractions

Method I   Combine terms in the numerator, combine terms in the denominator, and simplify the resulting expression.

Method II   Multiply the numerator and denominator by the LCD for both and simplify the resulting expression.

Example: *Method I* $\dfrac{\dfrac{1}{a} - \dfrac{1}{b}}{\dfrac{1}{a} + \dfrac{1}{b}} = \dfrac{\dfrac{b - a}{ab}}{\dfrac{b + a}{ab}} = \dfrac{b - a}{ab} \cdot \dfrac{ab}{b + a} = \dfrac{b - a}{b + a}$

Method II   The least common denominator is *ab*.

$$\frac{\frac{1}{a}-\frac{1}{b}}{\frac{1}{a}+\frac{1}{b}} = \frac{\left(\frac{1}{a}-\frac{1}{b}\right)ab}{\left(\frac{1}{a}+\frac{1}{b}\right)ab} = \frac{\frac{ab}{a}-\frac{ab}{b}}{\frac{ab}{a}+\frac{ab}{b}} = \frac{b-a}{b+a}$$

SECTION 7.6 ■ RATIONAL EQUATIONS AND FORMULAS

Solving Rational Equations   One way to solve the equation $\frac{a}{b}=\frac{c}{d}$ is to cross multiply to obtain $ad=bc$. (Check each answer.) A general way to solve rational equations is to follow these steps.

STEP 1: Find the LCD of the terms in the equation.

STEP 2: Multiply each side of the equation by the LCD.

STEP 3: Simplify each term.

STEP 4: Solve the resulting equation.

STEP 5: Check each answer in the *given* equation. Reject any value that makes a denominator equal 0.

Examples: $\frac{1}{2x}=\frac{2}{x+3}$ implies that $x+3=4x$, or $x=1$. This answer checks.

To solve $\frac{5}{x}-\frac{1}{3x}=\frac{7}{3}$ multiply each term by the LCD, $3x$:

$$\frac{5(3x)}{x}-\frac{3x}{3x}=\frac{7(3x)}{3},$$

which simplifies to $15-1=7x$, or $x=2$. This answer checks.

Solving for a Variable   Many formulas contain more than one variable. To solve for a particular variable, use the rules of algebra to isolate the variable.

Example: To solve $S=\frac{2\pi}{r}$ for r, multiply each side by r to obtain $Sr=2\pi$ and then divide each side by S to obtain $r=\frac{2\pi}{S}$.

SECTION 7.7 ■ PROPORTIONS AND VARIATION

Proportions   A proportion is a statement that two ratios are equal.

Example: $\frac{5}{x}=\frac{4}{7}$

Similar Triangles   Two triangles are similar if the measures of their corresponding angles are equal. Corresponding sides of similar triangles are proportional.

Example: A right triangle has legs of lengths 3 and 4. A similar right triangle has a shorter leg with length 6. Its longer leg can be found by solving the proportion $\frac{3}{6}=\frac{4}{x}$ to obtain $x=8$.

Direct Variation   A quantity y is *directly proportional* to a quantity x, or y *varies directly* with x, if there is a nonzero constant k such that $y=kx$. The number k is called the *constant of proportionality* or the *constant of variation*.

Example: If y varies directly with x, then the ratios $\frac{y}{x}$ always equal k. The following data satisfy $\frac{y}{x} = 4$, so the constant of variation is 4. Thus $y = 4x$.

x	1	2	3	4
y	4	8	12	16

Inverse Variation   A quantity y is *inversely proportional* to a quantity x, or y *varies inversely* with x, if there is a nonzero constant k such that $y = \frac{k}{x}$.

Example: If y varies inversely with x, then the products xy always equal k. The following data satisfy $xy = 12$, so the constant of variation is 12. Thus $y = \frac{12}{x}$.

x	1	2	4	6
y	12	6	3	2

CHAPTER 7   REVIEW EXERCISES

SECTION 7.1

Exercises 1–4: If possible, evaluate the expression for the given value of x.

1. $\dfrac{3}{x-3}$ $x = -2$

2. $\dfrac{4x}{5 - x^2}$ $x = 3$

3. $\dfrac{-x}{7-x}$ $x = 7$

4. $\dfrac{4x}{x^2 - 3x + 2}$ $x = 2$

5. Complete the table for the rational expression. If a value is undefined, place a dash in the table.

x	-2	-1	0	1	2
$\frac{3x}{x-1}$					

6. Find the x-values that make $\dfrac{8}{x^2 - 4}$ undefined.

Exercises 7–12: Simplify to lowest terms.

7. $\dfrac{25x^3y^4}{15x^5y}$

8. $\dfrac{x^2 - 36}{x + 6}$

9. $\dfrac{x-9}{9-x}$

10. $\dfrac{x^2 - 5x}{5x}$

11. $\dfrac{2x^2 + 5x - 3}{2x^2 + x - 1}$

12. $\dfrac{3x^2 + 10x - 8}{3x^2 + x - 2}$

SECTION 7.2

Exercises 13–16: Multiply and write in lowest terms.

13. $\dfrac{x-3}{x+1} \cdot \dfrac{2x+2}{x-3}$

14. $\dfrac{2x+5}{(x+5)(x-1)} \cdot \dfrac{x-1}{2x+5}$

15. $\dfrac{z+3}{z-4} \cdot \dfrac{z-4}{(z+3)^2}$

16. $\dfrac{x^2}{x^2 - 4} \cdot \dfrac{x+2}{x}$

Exercises 17–22: Divide and write in lowest terms.

17. $\dfrac{x+1}{2x} \div \dfrac{3x+3}{5x}$

18. $\dfrac{4}{x^3} \div \dfrac{x+1}{2x^2}$

19. $\dfrac{x-5}{x+2} \div \dfrac{2x-10}{x+2}$

20. $\dfrac{x^2 - 6x + 5}{x^2 - 25} \div \dfrac{x-1}{x+5}$

21. $\dfrac{x^2 - y^2}{x+y} \div \dfrac{x-y}{x+y}$

22. $\dfrac{a^3 - b^3}{a+b} \div \dfrac{a-b}{2a+2b}$

SECTION 7.3

Exercises 23–30: Add or subtract and write in lowest terms.

23. $\dfrac{2}{x+10} + \dfrac{8}{x+10}$

24. $\dfrac{9}{x-1} - \dfrac{8}{x-1}$

25. $\dfrac{x+2y}{2x} + \dfrac{x-2y}{2x}$

26. $\dfrac{x}{x+3} + \dfrac{3}{x+3}$

27. $\dfrac{x}{x^2-1} - \dfrac{1}{x^2-1}$

28. $\dfrac{2x}{x^2-25} + \dfrac{10}{x^2-25}$

29. $\dfrac{3}{xy} - \dfrac{1}{xy}$

30. $\dfrac{x+y}{2y} + \dfrac{x-y}{2y}$

SECTION 7.4

Exercises 31–36: Find the least common multiple for the expressions. Leave your answer in factored form.

31. $3x, 5x$

32. $5x^2, 10x$

33. $x, x-5$

34. $10x^2, x^2-x$

35. $x^2-1, (x+1)^2$

36. x^2-4x, x^2-16

Exercises 37–42: Rewrite the rational expression by using the given denominator D.

37. $\dfrac{3}{8}, D=24$

38. $\dfrac{4}{3x}, D=12x$

39. $\dfrac{3x}{x-2}, D=x^2-4$

40. $\dfrac{2}{x+1}, D=x^2+x$

41. $\dfrac{3}{5x}, D=5x^2-5x$

42. $\dfrac{2x}{2x-3}, D=2x^2+x-6$

Exercises 43–54: Simplify the expression.

43. $\dfrac{5}{8} + \dfrac{1}{6}$

44. $\dfrac{3}{4x} + \dfrac{1}{x}$

45. $\dfrac{5}{9x} - \dfrac{2}{3x}$

46. $\dfrac{7}{x-1} - \dfrac{3}{x}$

47. $\dfrac{1}{x+1} + \dfrac{1}{x-1}$

48. $\dfrac{4}{3x^2} - \dfrac{3}{2x}$

49. $\dfrac{1+x}{3x} - \dfrac{3}{2x}$

50. $\dfrac{x}{x^2-1} - \dfrac{1}{x-1}$

51. $\dfrac{2}{x-y} - \dfrac{3}{x+y}$

52. $\dfrac{2}{x} - \dfrac{1}{2x} + \dfrac{2}{3x}$

53. $\dfrac{3}{2y} + \dfrac{1}{2x}$

54. $\dfrac{x}{y-x} + \dfrac{y}{x-y}$

SECTION 7.5

Exercises 55–64: Simplify the complex fraction.

55. $\dfrac{\frac{3}{4}}{\frac{7}{11}}$

56. $\dfrac{\frac{x}{5}}{\frac{2x}{7}}$

57. $\dfrac{\frac{m}{n}}{\frac{2m}{n^2}}$

58. $\dfrac{\frac{3}{p-1}}{\frac{1}{p+1}}$

59. $\dfrac{\frac{3}{m-1}}{\frac{2m-2}{m+1}}$

60. $\dfrac{\frac{2}{2n+1}}{\frac{8}{2n-1}}$

61. $\dfrac{\frac{1}{2x} - \frac{1}{3x}}{\frac{2}{3x} - \frac{1}{6x}}$

62. $\dfrac{\frac{2}{xy} - \frac{1}{y}}{\frac{2}{xy} + \frac{1}{y}}$

63. $\dfrac{\frac{1}{x} - \frac{1}{x+1}}{\frac{x}{x+1}}$

64. $\dfrac{\frac{2}{x-1} - \frac{1}{x+1}}{\frac{1}{x^2-1}}$

SECTION 7.6

Exercises 65–70: Solve and check your answer.

65. $\dfrac{x}{5} = \dfrac{4}{7}$

66. $\dfrac{4}{x} = \dfrac{3}{2}$

67. $\dfrac{3}{z+1} = \dfrac{1}{2z}$

68. $\dfrac{x+2}{x} = \dfrac{3}{5}$

69. $\dfrac{1}{x+1} = \dfrac{2}{x-2}$

70. $\dfrac{x}{3} = \dfrac{-1}{x+4}$

Exercises 71–84: If possible, solve. Check your answer.

71. $\dfrac{1}{5x} + \dfrac{3}{5x} = \dfrac{1}{5}$ **72.** $\dfrac{1}{x-1} + \dfrac{2x}{x-1} = 1$

73. $\dfrac{1}{x} + \dfrac{2}{3x} = \dfrac{1}{3}$ **74.** $\dfrac{1}{x+3} + \dfrac{2x}{x+3} = \dfrac{3}{2}$

75. $\dfrac{5}{x} - \dfrac{3}{x+1} = \dfrac{1}{2}$ **76.** $\dfrac{1}{x-1} - \dfrac{1}{x+1} = \dfrac{1}{4}$

77. $\dfrac{4}{p} - \dfrac{5}{p+2} = 0$

78. $\dfrac{1}{x-3} - \dfrac{1}{x+3} = \dfrac{1}{x^2-9}$

79. $\dfrac{1}{x+1} = \dfrac{-x}{x+1}$

80. $\dfrac{2}{x} = \dfrac{2}{x^2+x} - 4$

81. $\dfrac{2}{x^2-2x} + \dfrac{1}{x^2-4} = \dfrac{1}{x^2+2x}$

82. $\dfrac{3}{x^2-3x} - \dfrac{1}{x^2-9} = \dfrac{1}{x^2+3x}$

83. $\dfrac{1}{x^2} - \dfrac{5}{x^2+4x} = \dfrac{1}{x^2+4x}$

84. $\dfrac{5}{x^2-1} - \dfrac{1}{x^2+2x+1} = \dfrac{3}{x^2-1}$

Exercises 85 and 86: Solve for the specified variable.

85. $\dfrac{1}{a} + \dfrac{2}{b} = \dfrac{3}{c}$ for b **86.** $y = \dfrac{x}{x-1}$ for x

SECTION 7.7

Exercises 87 and 88: Solve the proportion.

87. $\dfrac{x}{6} = \dfrac{1}{3}$ **88.** $\dfrac{5}{x} = \dfrac{7}{3}$

*Exercises 89 and 90: **Proportions**   Do the following.*
(a) Write a proportion that models the situation.
(b) Solve the proportion for x.

89. A rectangle has sides of 6 and 13. In a similar rectangle the longer side is 20 and the shorter side is x.

90. If you earn \$341 in 11 hours, then you can earn x dollars in 8 hours.

*Exercises 91 and 92: **Direct Variation**   Suppose that y is directly proportional to x.*
(a) Use the given information to find the constant of proportionality k.
(b) Then use y = kx to find y for x = 5.

91. $y = 8$ when $x = 2$ **92.** $y = 21$ when $x = 7$

*Exercises 93 and 94: **Inverse Variation**   Suppose that y is inversely proportional to x.*
(a) Use the given information to find the constant of proportionality k.
(b) Then use $y = \dfrac{k}{x}$ to find y for x = 5.

93. $y = 2.5$ when $x = 4$ **94.** $y = 7$ when $x = 3$

Exercises 95 and 96: Do the following.
(a) Determine whether the data represent direct or inverse variation.
(b) Find an equation that models the data.
(c) Graph the data and your equation.

95.

x	2	3	4	5
y	30	20	15	12

96.

x	2	4	6	8
y	6	12	18	24

Exercises 97 and 98: Use the graph to determine whether the data represent direct or inverse variation. Find the constant of variation.

97.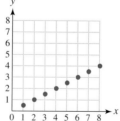

98.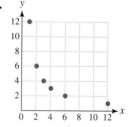

APPLICATIONS

99. *Modeling Traffic Flow*   Fifteen vehicles per minute can pass through an intersection. If vehicles arrive randomly at an average rate of x per minute, the average waiting time T in minutes is given by $T = \dfrac{1}{15-x}$ for $x < 15$.   (***Source:*** N. Garber, *Traffic and Highway Engineering*.)

(a) Find T when $x = 10$ and interpret the result.

(b) Complete the table.

x	5	10	13	14	14.9
T					

(c) What happens to the waiting time as the traffic rate x approaches 15 vehicles per minute?

100. *Distance and Time*   A car traveled at 50 miles per hour for 150 miles and then traveled at 75 miles per hour for 150 miles. What was the car's average speed?

101. *Emptying a Swimming Pool*   A large pump can empty a swimming pool in 100 hours, whereas a small pump can empty the pool in 160 hours. How long will it take to empty the pool if both pumps are used?

102. *Jogging*   Two athletes jog 10 miles. One of the athletes jogs 2 miles per hour faster and finishes 10 minutes ahead of the other athlete. Find the average speed of each athlete.

103. *River Current*   A boat can travel 16 miles upstream in the same time that it can travel 48 miles downstream. If the speed of the current is 4 miles per hour, find the boat's speed.

104. *Height of a Tree*   A 5-foot-tall person has a 6-foot-long shadow, and a nearby tree has a 40-foot-long shadow. Estimate the tree's height.

105. *Transportation Costs*   Use of a toll road varies inversely with the toll. If the toll is $0.25, then 400 vehicles use the road. Estimate the number of users for a toll of $0.50.

106. *Water Content in Snow*   Twenty inches of extremely dry, powdery snow are equivalent to an inch of rain. Estimate the water content in 32 inches of this type of snow.

107. *Tightening a Bolt*   The torque exerted on a nut by a wrench is inversely proportional to the length of the wrench's handle. Suppose that a 12-inch wrench can be used to tighten a nut by using 30 pounds of force. How much force is necessary to tighten the same nut by using a 10-inch wrench?

108. *Cost of Carpet*   The cost of carpet is directly proportional to the amount of carpet purchased. If 17 square yards cost $612, find the cost of 13 square yards.

109. *Cold Water Survival*   When a person falls through the ice on a lake, the cold water removes body heat 25 times faster than air at the same temperature. To survive, a person has between 30 and 90 seconds to get out of the water. How long could a person remain in air at the same temperature?

110. *Strength of a Beam*   The strength of a wood beam varies inversely with its length. A beam that is 18 feet long can support 900 pounds. How much weight can a similar beam support if its length is 21 feet?

CHAPTER 7   TEST   Pass the Test   Video solutions to all test exercises

1. Evaluate the expression $\frac{3x}{2x-1}$ for $x = 3$.

2. Find any x-value that makes $\frac{x-1}{x+2}$ undefined.

Exercises 3 and 4: Simplify the expression.

3. $\frac{x^2 - 25}{x - 5}$

4. $\frac{3x^2 - 15x}{3x}$

Exercises 5–12: Simplify the expression. Write your answer in lowest terms.

5. $\frac{x-2}{x+4} \cdot \frac{3x+12}{x-2}$

6. $\frac{z+1}{z+3} \cdot \frac{2z+6}{z+1}$

7. $\frac{x+1}{5x} \div \frac{2x+2}{x-1}$

8. $\frac{2}{x^2} \div \frac{x+3}{3x}$

9. $\dfrac{x}{x+4} + \dfrac{3x+1}{x+4}$

10. $\dfrac{4t+1}{2t-3} - \dfrac{3t-6}{2t-3}$

11. $\dfrac{1}{y^2+y} - \dfrac{y-1}{y^2-y}$

12. $\dfrac{1}{xy} + \dfrac{x}{y} - \dfrac{1}{y^2}$

Exercises 13 and 14: Simplify the complex fraction.

13. $\dfrac{\dfrac{a}{3b}}{\dfrac{5a}{b^2}}$

14. $\dfrac{1+\dfrac{1}{p-1}}{1-\dfrac{1}{p-1}}$

Exercises 15–22: Solve the equation and check your answer.

15. $\dfrac{2}{7} = \dfrac{5}{x}$

16. $\dfrac{x+3}{2x} = 1$

17. $\dfrac{1}{2x} + \dfrac{2}{5x} = \dfrac{9}{10}$

18. $\dfrac{1}{x-1} + \dfrac{2}{x+2} = \dfrac{3}{2}$

19. $\dfrac{1}{x^2-1} - \dfrac{4}{x+1} = \dfrac{3}{x-1}$

20. $\dfrac{1}{x^2-4x} + \dfrac{2}{x^2-16} = \dfrac{2}{x^2+4x}$

21. $\dfrac{x}{2x-1} = \dfrac{1-x}{2x-1}$

22. $\dfrac{x}{x-5} + \dfrac{x}{x+5} = \dfrac{10x}{x^2-25}$

Exercises 23 and 24: Solve the equation for the specified variable.

23. $y = \dfrac{2}{3x-5}$ for x

24. $\dfrac{a+b}{ab} = 1$ for b

25. Suppose that y is directly proportional to x.
   (a) If $y = 14$ when $x = 4$, find k so that $y = kx$.
   (b) Then use $y = kx$ to find y for $x = 6$.

26. Use the table to determine whether y varies directly or inversely with x. Find the constant of variation.

x	2	4	8	16
y	16	8	4	2

27. *Emptying a Swimming Pool*   It takes a large pump 40 hours to empty a swimming pool, whereas a small pump can empty the pool in 60 hours. How long will it take to empty the pool if both pumps are used?

28. *Height of a Building*   A 5-foot-tall post has a 4-foot-long shadow, and a nearby building has a 54-foot-long shadow. Estimate the height of the building.

29. *Standing in Line*   A department store clerk can wait on 30 customers per hour. If people arrive randomly at an average rate of x per hour, then the average number of customers N waiting in line is given by $N = \dfrac{x^2}{900-30x}$, for $x < 30$. Evaluate the expression for $x = 24$ and interpret your result.

CHAPTER 7   EXTENDED AND DISCOVERY EXERCISES

1. *Graph of a Rational Function*   A car wash can clean 15 cars per hour. If cars arrive randomly at an average rate of x per hour, the average number N of cars waiting in line is given by

$$N = \dfrac{x^2}{225-15x},$$

where $x < 15$.   (*Source:* N. Garber and L. Hoel, *Traffic and Highway Engineering.*)

   (a) Complete the table.

x	3	9	12	13	14
N					

   (b) For what value of x is N undefined?
   (c) Plot the points from the table. Then graph $x = 15$ as a vertical dashed line.
   (d) Sketch a graph of N that passes through these points. Do not allow your graph to cross the vertical, dashed line, called an asymptote.
   (e) Use the graph to explain why a small increase in x can sometimes lead to a long wait.
   (f) Explain what happens over a long period of time if the arrival rate x exceeds 15 cars per hour. (*Hint:* The formula is not valid for $x \geq 15$.)

Exercises 2–6: Graphing Rational Functions   *Complete the following.*

(a) *Use the given equation to complete the table of values for y.*

x	−4	−3	−2	−1	0	1	2	3	4
y									

(b) *Determine any x-value that will make the expression undefined.*

(c) *Sketch a dashed, vertical line (asymptote) in the xy-plane at any undefined values of x.*

(d) *Plot the points from the table.*

(e) *Sketch a graph of the equation. Do not let your graph cross the vertical dashed line.*

2. $y = \dfrac{1}{x - 1}$ **3.** $y = \dfrac{1}{x + 1}$

4. $y = \dfrac{4}{x^2 + 1}$ **5.** $y = \dfrac{x}{x + 1}$

6. $y = \dfrac{x}{x - 1}$

CHAPTERS 1–7 CUMULATIVE REVIEW EXERCISES

1. Evaluate $\pi r^2 h$ when $r = 2$ and $h = 6$.

2. Translate the phrase "two less than twice a number" into an algebraic expression using the variable x.

3. Find the reciprocal of $-\frac{8}{5}$.

Exercises 4 and 5: Evaluate and simplify to lowest terms.

4. $\frac{1}{2} \div \frac{5}{4}$ **5.** $\frac{5}{8} + \frac{1}{8}$

6. Simplify $\dfrac{4x}{9y} \div \dfrac{6x}{3y}$.

Exercises 7 and 8: Simplify the expression.

7. $-2 + 7x + 4 - 5x$

8. $-4(4 - y) + (5 - 3y)$

9. Solve $-2x + 11 = 13$ and check the solution.

10. Convert 0.045 to a percentage.

11. Solve $V = 6LW$ for W.

12. Solve $-3x + 1 \geq x$.

Exercises 13 and 14: Graph the equation. Determine any intercepts.

13. $2x - 3y = 6$ **14.** $x = 1$

15. Sketch a line with slope $m = 3$ passing through $(-2, -1)$. Write its slope–intercept form.

16. The table lists points located on a line. Write the slope–intercept form of the line.

x	−2	−1	0	1
y	−5	−3	−1	1

Exercises 17 and 18: Find the slope–intercept form for the line satisfying the given conditions.

17. Parallel to $y = -\frac{2}{3}x + 1$ passing through $(2, -1)$

18. Passing through $(-1, 2)$ and $(2, 4)$

19. A fish population is initially 2000 and is increasing at 200 per month. Write an equation that gives the number of fish N after x months.

20. Determine which ordered pair is a solution to the system of equations: $(2, -6)$ or $(1, -2)$.

$$4x + y = 2$$
$$x - 4y = 9$$

21. A table listing solutions for the equations $y = 2x$ and $y = 3 - x$ is given. Identify the solution to both equations.

x	−1	0	1	2
y = 2x	−2	0	2	4
y = 3 − x	4	3	2	1

22. Solve the system of equations.
$$-3r - t = 2$$
$$2r + t = -4$$

Exercises 23 and 24: Determine if the system of linear equations has no solutions, one solution, or infinitely many solutions.

23. $2x - y = 5$
$ -2x + y = -5$

24. $4x - 6y = 12$
$ -6x + 9y = 18$

25. Solve the system of equations symbolically, graphically, and numerically.
$$-2x + y = 0$$
$$x + y = 3$$

26. Use the graph to identify one solution to $2x + y < 1$. *Answers may vary.* Then shade the solution set.

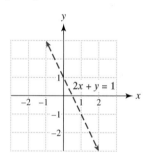

Exercises 27–32: Simplify the expression.

27. $(5x^2 - 3) + (-x^2 + 4)$

28. $3z^2 \cdot 5z^6$

29. $(ab)^3$

30. $(2y - 3)(5y + 2)$

31. $(x^2 - y^2)^2$

32. $(2t + 5)^2$

Exercises 33–36: Simplify and write the expression using positive exponents.

33. $2^{-4} \cdot 2^5$

34. $\dfrac{1}{3^{-2}}$

35. $(3x^2)^{-3}$

36. $\dfrac{4x^2}{2x^4}$

37. Write 0.00123 in scientific notation.

38. Divide and check: $\dfrac{2x^2 - x + 3}{x - 1}$.

Exercises 39 and 40: Identify the greatest common factor for each expression. Then factor the expression.

39. $20z^3 - 15z^2$

40. $12x^2y + 15xy^2$

Exercises 41–46: Factor completely.

41. $6 + 13x - 5x^2$

42. $9z^2 - 4$

43. $t^2 + 16t + 64$

44. $x^3 - 16x$

45. $x^3 + 2x^2 - 99x$

46. $a^2 + 6ab + 9b^2$

Exercises 47–50: Solve the equation.

47. $(2x - 7)(x + 5) = 0$

48. $2x^2 - 4x = 0$

49. $y^2 + 5y - 14 = 0$

50. $x^3 = 4x$

51. Evaluate $\dfrac{1}{x - 2}$ at $x = 3$. For what value of x is the expression undefined?

52. Simplify $\dfrac{x^2 + 2x + 1}{x^2 - 1}$ to lowest terms.

Exercises 53–56: Simplify completely.

53. $\dfrac{x}{x + 1} + \dfrac{1}{x + 1}$

54. $\dfrac{1}{x - 1} - \dfrac{1}{x^2 - 1}$

55. $\dfrac{3x}{4y} \cdot \dfrac{y}{9x^2}$

56. $\dfrac{x}{x^2 - 4} \div \dfrac{2x}{x - 2}$

57. Simplify $\dfrac{1 + \frac{2}{x}}{1 - \frac{2}{x}}$.

Exercises 58–60: Solve.

58. $\dfrac{5}{x} = \dfrac{7}{8}$

59. $\dfrac{4}{3x} - \dfrac{3}{4x} = 1$

60. $\dfrac{1}{x - 1} + \dfrac{2}{x + 2} = \dfrac{3}{2}$

61. Solve $z = 3x - 2y$ for x.

62. Suppose that y is directly proportional to x and that $y = 7$ when $x = 14$. Find y when $x = 11$.

63. Suppose that y is inversely proportional to x and that $y = 8$ when $x = 20$. Find y when $x = 2$.

64. Determine whether the data in the table represent direct or inverse variation. Find an equation that models the data.

x	1	2	4	10
y	20	10	5	2

APPLICATIONS

65. *Shoveling the Driveway*   Two people are shoveling snow from a driveway. The first person shovels 12 square feet per minute, while the second person shovels 9 square feet per minute.

(a) Write a simplified expression that gives the total square feet that the two people shovel in x minutes.

(b) How long would it take them to clear a driveway with 1890 square feet?

66. *Burning Calories* An athlete can burn 10 calories per minute while running and 4 calories per minute while walking. If the athlete burns 450 calories in 60 minutes, how long is spent on each activity?

67. *Head Start* In 1980 there were 376 thousand children participating in the federal Head Start program. In 2000 this number increased to 858 thousand students. (*Source:* Department of Health and Human Services.)

(a) Find the slope of the line passing through the points (1980, 376) and (2000, 858).

(b) Interpret the slope as a rate of change.

(c) If trends continue, estimate the number of participants in 2010.

68. *Triangle* In an isosceles triangle, the measures of the two smaller angles are equal and their sum is 32° more than the largest angle.

(a) Let x be the measure of one of the two smaller angles and y be the measure of the largest angle. Write a system of linear equations whose solution gives the measures of these angles.

(b) Solve your system.

69. *Energy* The U.S. consumption of natural gas G in trillions of cubic feet from 1990 to 2000 can be approximated by

$$G = -0.036x^2 + 0.76x + 18.7,$$

where $x = 0$ corresponds to 1990, $x = 1$ to 1991, and so on. (*Source:* Department of Energy.)

(a) Evaluate this polynomial when $x = 5$ and interpret the answer.

(b) In what year was the natural gas consumption 22.7 trillion cubic feet?

70. *Strength of a Beam* The strength of a wood beam varies inversely with its length. A beam that is 10 feet long can support 1100 pounds. How much weight can a similar beam support if it is 22 feet long?

CHAPTER 8

Radical Expressions

California is frequently hit by severe rainstorms that cause mudslides, blocked highways, and flooding. Ashy, mountainous slopes left behind by wildfires are especially susceptible to mudslides, or debris flows. Since 1905, debris flows have occurred during at least 14 rainy seasons and have been responsible for both property damage and loss of life. More than 100 Californians have been killed by debris flows during the past 30 years.

To combat the damage caused by these rainstorms, many cities have constructed open channels. These open channels are capable of carrying large volumes of rainwater to the ocean in a short time. The proper design of these channels is essential and requires the use of mathematics. For example, one important characteristic of an open channel is its slope m. A steeper slope results in runoff flowing faster through the channel than a shallow slope, thereby reducing the possibility of a devastating mudslide. The rate R at which water flows through an open channel is directly proportional to the square root of its slope. That is, $R = k\sqrt{m}$, where k is a constant determined by the shape of the channel.

> *The struggle is what teaches us.*
>
> —SUE GRAFTON

In this chapter we discuss open channels (see Section 8.5, Exercises 75–78) and other applications that use formulas having square roots and cube roots.

Sources: California Geological Survey; N. Garber and L. Hoel, *Traffic and Highway Design.*

8.1 INTRODUCTION TO RADICAL EXPRESSIONS

Square Roots ▪ Cube Roots ▪ The Pythagorean Theorem ▪ The Distance Formula ▪ Graphing (Optional)

A LOOK INTO MATH ▷

Did you know that smaller animals tend to have faster heart rates than larger animals? For example, the heart rate of a dog is typically faster than the heart rate of an elephant. To estimate the average heart rate of an animal, we can evaluate square roots. In Example 3 we use a formula having a square root to estimate heart rates of animals.

Square Roots

The number 10 is a square root of 100 because $10 \cdot 10 = 100$, or $10^2 = 100$. The number -10 is also a square root of 100 because $(-10)(-10) = 100$, or $(-10)^2 = 100$. The number 10 is called the **principal**, or **positive**, **square root** of 100 and is denoted $\sqrt{100}$. The **negative square root** of 100 is -10 and is denoted $-\sqrt{100}$. To denote both the positive and negative square roots of 100 we write $\pm\sqrt{100}$, where the symbol $\pm$ is read "plus or minus." In the radical expression $\sqrt{100}$, the symbol $\sqrt{}$ is called the **radical sign** and the number 100 is called the **radicand**. A definition of a square root is the following.

> **SQUARE ROOT**
>
> The number b is a **square root** of a if $b^2 = a$.

Every *positive* number has one positive square root and one negative square root. Square roots of some integers do not equal integers, as shown in the next example.

EXAMPLE 1   Finding square roots

Find the square roots of each number. Approximate your answer to three decimal places when appropriate.
(a) 36   (b) 5

```
√(5)
        2.236067977
```

Figure 8.1

Solution
(a) The square roots of 36 are $\pm\sqrt{36}$, or ± 6, because $6^2 = 36$ and $(-6)^2 = 36$.
(b) Because $2 \cdot 2 = 4$ and $3 \cdot 3 = 9$, the *positive* square root of 5 is between 2 and 3. Using a calculator, we get $\sqrt{5} \approx 2.236$, as shown in Figure 8.1. Thus the square roots of 5 are $\pm\sqrt{5}$, or ± 2.236 rounded to three decimal places.   Now Try Exercises 13, 23

EXAMPLE 2   Evaluating principal square roots

Evaluate each square root.
(a) $\sqrt{81}$ (b) $\sqrt{49}$ (c) $\sqrt{\frac{1}{4}}$

Solution
(a) Because $9 \cdot 9 = 81$, the principal, or *positive*, square root of 81 is $\sqrt{81} = 9$.
(b) Because $7 \cdot 7 = 49$, the principal, or *positive*, square root of 49 is $\sqrt{49} = 7$.
(c) Because $\frac{1}{2} \cdot \frac{1}{2} = \frac{1}{4}$, the principal, or *positive*, square root of $\frac{1}{4}$ is $\sqrt{\frac{1}{4}} = \frac{1}{2}$.
Now Try Exercises 27, 29, 31

▶ **REAL-WORLD CONNECTION**   In the next example we discuss the application from the introduction to this section.

EXAMPLE 3   Estimating heart rates of animals

According to one model, the rate at which an animal's heart beats varies with its weight. Smaller animals tend to have faster heart rates, whereas larger animals tend to have slower heart rates. The average heart rate R in beats per minute of an animal with weight W in pounds is given by the formula $R = \dfrac{885}{\sqrt{W}}$.   (*Source:* C. Pennycuick, *Newton Rules Biology.*)

(a) Estimate the heart rate for a 64-pound dog.
(b) Estimate the heart rate for a 1600-pound elephant.

Solution
(a) For a **64**-pound dog, the average heart rate is $\dfrac{885}{\sqrt{64}} = \dfrac{885}{8} \approx 111$ beats per minute.

(b) For a **1600**-pound elephant, the average heart rate is $\dfrac{885}{\sqrt{1600}} = \dfrac{885}{40} \approx 22$ beats per minute.

Now Try Exercise 109

If a whole number a has an integer square root, then a is a **perfect square**. For example, 25 is a perfect square because $\sqrt{25} = 5$. If a whole number is not a perfect square, then its square root is *irrational*. For example, $\sqrt{5} \approx 2.236$, so $\sqrt{5}$ is an irrational number. The expressions $\sqrt{7}$ and $\sqrt{10}$ are also irrational numbers, whereas $\sqrt{25}$ and $\sqrt{36}$ are rational numbers.

Square roots of a negative number are not real numbers, so $\sqrt{-5}$ is not a real number because there is no real number a such that $a^2 = -5$.

EXAMPLE 4   Identifying numbers

State whether each expression is one or more of the following: a real number, a rational number, an irrational number, or none.
(a) $\sqrt{16}$ (b) $\sqrt{-144}$ (c) $\sqrt{18}$

Solution
(a) Because $\sqrt{16} = 4, \sqrt{16}$ is a real number and a rational number.
(b) Because the square root of a negative number is not a real number, $\sqrt{-144}$ is not a real number, a rational number, nor an irrational number.
(c) Because $\sqrt{18} \approx 4.243$ is not a positive integer, $\sqrt{18}$ is a real number and an irrational number.

Now Try Exercises 37, 41, 45

Cube Roots

Another common radical expression is the *cube root* of a number a, denoted $\sqrt[3]{a}$. For example, $\sqrt[3]{8} = 2$ because $2^3 = 2 \cdot 2 \cdot 2 = 8$.

CUBE ROOT

The number b is a **cube root** of a if $b^3 = a$.

EXAMPLE **5**   Finding cube roots

Find the cube root of each expression. Approximate the result to three decimal places when appropriate.

(a) $\sqrt[3]{27}$ **(b)** $\sqrt[3]{-64}$ **(c)** $\sqrt[3]{15}$

```
³√(15)
        2.466212074
2.466^3
         14.9961307
```

Figure 8.2

Solution

(a) $\sqrt[3]{27} = 3$ because $3^3 = 3 \cdot 3 \cdot 3 = 27$.

(b) $\sqrt[3]{-64} = -4$ because $(-4)^3 = (-4)(-4)(-4) = -64$.

(c) Because $2^3 = 8$ and $3^3 = 27$, $\sqrt[3]{15}$ is between 2 and 3. Figure 8.2 shows that $\sqrt[3]{15} \approx 2.466$, rounded to three decimal places and that $2.466^3 \approx 15$.

Now Try Exercises 53, 55, 65

MAKING CONNECTIONS

Square Roots and Cube Roots

Although the square root of a negative number is not a real number, the cube root of a negative number is a negative real number. Every real number has one real cube root, whereas every *positive* real number has two real square roots.

The Pythagorean Theorem

One of the most famous theorems in mathematics is the **Pythagorean theorem**. It states that, if a right triangle has legs a and b with hypotenuse c (see Figure 8.3), then

$$a^2 + b^2 = c^2.$$

Figure 8.3 $a^2 + b^2 = c^2$

If the legs of a right triangle are $a = 3$ and $b = 4$, the hypotenuse is $c = 5$ because

$$3^2 + 4^2 = 5^2, \quad \text{or} \quad 9 + 16 = 25.$$

Given the lengths of two sides of a *right* triangle, the Pythagorean theorem can always be used to find the length of the missing side.

NOTE: If the lengths of the sides of a triangle are a, b, and c and if they satisfy $a^2 + b^2 = c^2$, then the triangle is a *right* triangle. For example, if the sides of a triangle are 5, 12, and 13, then it must be a right triangle because $5^2 + 12^2 = 13^2$, or $25 + 144 = 169$.

EXAMPLE **6**   Finding sides of right triangles

A right triangle has legs a and b with hypotenuse c. Find the length of the missing side.

(a) $a = 5$ feet, $b = 12$ feet   **(b)** $a = 60$ inches, $c = 61$ inches

Solution

(a) Substitute $a = 5$ and $b = 12$ in the Pythagorean theorem.

$c^2 = a^2 + b^2$	Pythagorean theorem
$c^2 = 5^2 + 12^2$	Substitute $a = 5$ and $b = 12$.
$c^2 = 25 + 144$	Square the numbers.
$c^2 = 169$	Add.
$c = 13$	Take the principal square root.

Because length is always positive, take the principal, or positive, square root. The length of the hypotenuse is 13 feet.

(b) Rather than finding the hypotenuse we must find the missing side b. Substitute $a = 60$ and $c = 61$ in the Pythagorean theorem.

$$c^2 = a^2 + b^2 \qquad \text{Pythagorean theorem}$$
$$61^2 = 60^2 + b^2 \qquad \text{Substitute } a = 60 \text{ and } c = 61.$$
$$3721 = 3600 + b^2 \qquad \text{Square the numbers.}$$
$$b^2 = 121 \qquad \text{Solve for } b^2; \text{ rewrite the equation.}$$
$$b = 11 \qquad \text{Take the principal square root.}$$

The length of leg b is 11 inches.

Now Try Exercises 73, 79

▶ **REAL-WORLD CONNECTION** In the next example we apply the Pythagorean theorem to calculate the diagonal length of a television set.

EXAMPLE 7

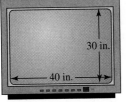

Figure 8.4

Finding the diagonal length of a television set

A flat, rectangular screen of a television set is 40 inches wide and 30 inches high, as illustrated in Figure 8.4. Find the length of the diagonal for this television set. Why is it called a 50-inch television set?

Solution

Let $a = 40$ and $b = 30$. Then the diagonal corresponds to the hypotenuse of a right triangle with legs of 40 inches and 30 inches.

$$c^2 = a^2 + b^2 \qquad \text{Pythagorean theorem}$$
$$c^2 = 40^2 + 30^2 \qquad \text{Substitute } a = 40 \text{ and } b = 30.$$
$$c^2 = 2500 \qquad \text{Square the numbers and add.}$$
$$c = 50 \qquad \text{Take the principal square root.}$$

The diagonal is 50 inches in length, which is why the set is called a 50-inch television.

Now Try Exercise 111

The Distance Formula

The Pythagorean theorem can be used to find the distance between two points in the xy-plane. Suppose that a line segment has endpoints (x_1, y_1) and (x_2, y_2), as illustrated in Figure 8.5. The lengths of the legs of the right triangle are $x_2 - x_1$ and $y_2 - y_1$. The distance d is the hypotenuse of the right triangle. Applying the Pythagorean theorem, we have

$$d^2 = (x_2 - x_1)^2 + (y_2 - y_1)^2.$$

Distance is *nonnegative*, so we let d be the principal square root and obtain

$$d = \sqrt{(x_2 - x_1)^2 + (y_2 - y_1)^2}.$$

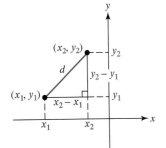

Figure 8.5

DISTANCE FORMULA

The **distance** d between the points (x_1, y_1) and (x_2, y_2) in the xy-plane is

$$d = \sqrt{(x_2 - x_1)^2 + (y_2 - y_1)^2}.$$

EXAMPLE 8   Finding distance between two points

Find the exact length of the line segment shown in Figure 8.6. Then approximate this value to two decimal places.

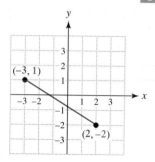

Figure 8.6

Solution

The endpoints of the line segment shown in Figure 8.6 are $(-3, 1)$ and $(2, -2)$. In the distance formula, we let $(x_1, y_1) = (-3, 1)$ and $(x_2, y_2) = (2, -2)$.

$$d = \sqrt{(x_2 - x_1)^2 + (y_2 - y_1)^2}$$ Distance formula

$$= \sqrt{(2 - (-3))^2 + (-2 - 1)^2}$$ Let $x_2 = 2$, $x_1 = -3$, $y_2 = -2$, and $y_1 = 1$.

$$= \sqrt{(5)^2 + (-3)^2}$$ Subtract.

$$= \sqrt{34}$$ Simplify.

$$\approx 5.83$$ Approximate.

The *exact* length is $\sqrt{34}$, and the *approximate* length, to two decimal places, is 5.83.

Now Try Exercise 89

NOTE: In Example 8, the order in which the points are chosen does not matter. If we let $(x_1, y_1) = (2, -2)$ and $(x_2, y_2) = (-3, 1)$, we get the same length. That is,

$$d = \sqrt{(-3 - 2)^2 + (1 - (-2))^2}$$

$$= \sqrt{(-5)^2 + (3)^2}$$

$$= \sqrt{34}.$$

EXAMPLE 9   Finding distance between two cars

Two cars stop at an intersection. Then the first car travels north at 50 miles per hour and the second car travels east at 60 miles per hour. To the nearest mile, find the distance between the cars after 30 minutes.

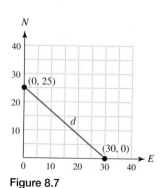

Figure 8.7

Solution

After 30 minutes, or half an hour, the first car is $(0.5)(50) = 25$ miles north of the intersection and the second car is $(0.5)(60) = 30$ miles east of the intersection. If we locate the intersection at the origin with coordinates $(0, 0)$, then the first car is located at $(0, 25)$, the second car is located at $(30, 0)$, and the distance between them is d, as shown in Figure 8.7. In the distance formula, we let $(x_1, y_1) = (0, 25)$ and $(x_2, y_2) = (30, 0)$.

$$d = \sqrt{(x_2 - x_1)^2 + (y_2 - y_1)^2}$$ Distance formula

$$= \sqrt{(30 - 0)^2 + (0 - 25)^2}$$ Let $x_2 = 30$, $x_1 = 0$, $y_2 = 0$, and $y_1 = 25$.

$$= \sqrt{(30)^2 + (-25)^2}$$ Subtract.

$$= \sqrt{1525}$$ Simplify.

$$\approx 39.1$$ Approximate.

To the nearest mile, the cars are 39 miles apart after 30 minutes.   Now Try Exercise 117

Graphing (Optional)

Like other equations, we can graph equations containing radical expressions. In the next example, we graph $y = \sqrt{x}$. To sketch this graph, we first make a table of convenient values. Then we plot these points and sketch a smooth curve through them.

EXAMPLE 10   Graphing a square root

Make a table of values for $y = \sqrt{x}$. Then sketch a graph.

Solution
In Table 8.1 we list (x, y) ordered pairs for the graph of $y = \sqrt{x}$. For convenience, we choose x-values that are perfect squares. Note that $\sqrt{x}$ is undefined whenever x is negative. A graph of Table 8.1 is shown in Figure 8.8.

TABLE 8.1

x	$\sqrt{x}$
0	0
1	1
4	2
9	3

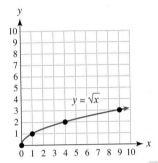

Figure 8.8

Now Try Exercise 99

8.1 PUTTING IT ALL TOGETHER

The following table summarizes several important topics from this section.

Concept	Explanation	Examples
Square Root	The number b is a square root of the number a if $b^2 = a$. The square root of a negative number is not a real number.	The square roots of 16 are ± 4. The square roots of 7 are $\pm\sqrt{7}$.
Principal Square Root: $\sqrt{a}$	The principal square root is the positive square root.	$\sqrt{25} = 5, \sqrt{4} = 2, \sqrt{169} = 13$, and $\sqrt{3} \approx 1.732$ (to three decimal places)
Cube Root: $\sqrt[3]{a}$	The number b is a cube root of the number a if $b^3 = a$. *Every* real number has one real cube root.	The cube root of 125 is $\sqrt[3]{125} = 5$. The cube root of -27 is $\sqrt[3]{-27} = -3$.

continued on next page

continued from previous page

Concept	Explanation	Examples
Pythagorean Theorem	The lengths of the sides of a right triangle satisfy $a^2 + b^2 = c^2$. 	If the legs of a right triangle are 7 inches and 24 inches, then the hypotenuse c is given by $c^2 = 7^2 + 24^2$, or $c = \sqrt{625} = 25$ inches.
Distance Formula	The distance d between (x_1, y_1) and (x_2, y_2) can be found by using $d = \sqrt{(x_2 - x_1)^2 + (y_2 - y_1)^2}$.	The distance between $(0, 2)$ and $(3, 6)$ is $d = \sqrt{(3 - 0)^2 + (6 - 2)^2}$ $= \sqrt{25}$ $= 5$.

8.1 Exercises

MyMathLab Math XL PRACTICE   WATCH   DOWNLOAD   READ   REVIEW

CONCEPTS

1. The number b is a(n) _____ of a if $b^2 = a$.

2. The square roots of 9 are _____ and _____.

3. In the expression $\sqrt{12}$, the number 12 is called the _____.

4. The principal square root of 4 is _____.

5. $\sqrt{9} =$ _____   **6.** $\sqrt[3]{8} =$ _____

7. Does $\sqrt{6}$ represent a rational or an irrational number?

8. The number b is a(n) _____ of a if $b^3 = a$.

9. The cube root of 64 is _____.

10. If a right triangle has legs a and b with hypotenuse c, then the Pythagorean theorem states that _____.

11. If a right triangle has legs with lengths 3 and 4, then its hypotenuse has length _____.

12. The distance d between the points (x_1, y_1) and (x_2, y_2) is $d =$ _____.

SQUARE ROOTS

Exercises 13–26: Find the square roots of the given number. Approximate the results to three decimal places when appropriate.

13. 4   **14.** 16

15. 49   **16.** 64

17. 121   **18.** 144

19. 400   **20.** 225

21. 196   **22.** 169

23. 8   **24.** 14

25. 24   **26.** 72

Exercises 27–36: Evaluate the square root.

27. $\sqrt{16}$ **28.** $\sqrt{1}$

29. $\sqrt{4}$ **30.** $\sqrt{169}$

31. $\sqrt{\frac{1}{9}}$ **32.** $\sqrt{\frac{1}{16}}$

33. $\sqrt{\frac{9}{4}}$ **34.** $\sqrt{\frac{25}{4}}$

35. $\sqrt{0.04}$ **36.** $\sqrt{0.36}$

Exercises 37–46: State whether the given expression is one or more or none of the following: a real number, a rational number, or an irrational number.

37. $\sqrt{100}$ **38.** $\sqrt{81}$

39. $\sqrt{2500}$ **40.** $\sqrt{1600}$

41. $\sqrt{6}$ **42.** $\sqrt{28}$

43. $\sqrt{150}$ **44.** $\sqrt{99}$

45. $\sqrt{-16}$ **46.** $\sqrt{-120}$

Exercises 47–52: Evaluate the square root. Approximate your answer to two decimal places when appropriate.

47. $\sqrt{144}$ **48.** $\sqrt{64}$

49. $-\sqrt{9}$ **50.** $-\sqrt{25}$

51. $\sqrt{10}$ **52.** $-\sqrt{45}$

CUBE ROOTS

Exercises 53–70: Find the cube root. Approximate your answer to three decimal places when appropriate.

53. $\sqrt[3]{8}$ **54.** $\sqrt[3]{-27}$

55. $\sqrt[3]{-125}$ **56.** $\sqrt[3]{64}$

57. $-\sqrt[3]{27}$ **58.** $\sqrt[3]{-64}$

59. $-\sqrt[3]{-1}$ **60.** $-\sqrt[3]{-8}$

61. $\sqrt[3]{1000}$ **62.** $\sqrt[3]{1,000,000}$

63. $\sqrt[3]{-343}$ **64.** $\sqrt[3]{-729}$

65. $\sqrt[3]{5}$ **66.** $\sqrt[3]{9}$

67. $\sqrt[3]{-16}$ **68.** $\sqrt[3]{-36}$

69. $\sqrt[3]{-9}$ **70.** $\sqrt[3]{15}$

71. **Thinking Generally** Evaluate $\sqrt{a^2}$, if $a > 0$.

72. **Thinking Generally** Evaluate $\sqrt[3]{a^3}$.

PYTHAGOREAN THEOREM

Exercises 73–80: A right triangle has legs a and b with hypotenuse c. Find the exact length of the missing side.

73. $a = 4$ inches, $b = 3$ inches

74. $a = 8$ feet, $b = 6$ feet

75. $a = 5$ meters, $b = 12$ meters

76. $a = 7$ miles, $b = 24$ miles

77. $c = 3$ miles, $b = 2$ miles

78. $c = 5$ feet, $b = 3$ feet

79. $a = 6$ feet, $c = 10$ feet

80. $a = 8$ meters, $c = 17$ meters

Exercises 81–86: Let a triangle have sides a and b with longest side c. Use the equation $a^2 + b^2 = c^2$ to determine whether the triangle is a right triangle.

81. $a = 5, b = 12, c = 13$

82. $a = 8, b = 12, c = 15$

83. $a = 6, b = 4, c = 8$

84. $a = 5, b = 10, c = 12$

85. $a = 60, b = 11, c = 61$

86. $a = 15, b = 8, c = 17$

THE DISTANCE FORMULA

Exercises 87–90: (Refer to Example 8.) Find the exact length of the line segment shown in the figure. Approximate this length to two decimal places when appropriate.

87.

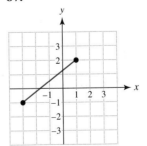

88.

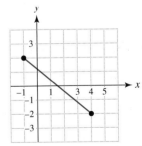

89.

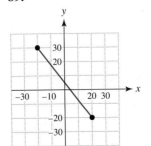

90.

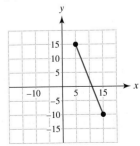

Exercises 91–98: Find the exact distance between the pair of points. Approximate this distance to three decimal places when appropriate.

91. $(1, 3), (4, 7)$

92. $(-2, 2), (3, 14)$

93. $(4, -4), (-3, 20)$

94. $(-10, -10), (1, 50)$

95. $(-3, 5), (2, -6)$

96. $(4, -7), (-2, 10)$

97. $(-2, -1), (0, -9)$

98. $(4, -8), (11, -12)$

GRAPHING

Exercises 99–104: (Refer to Example 10.) Graph the equation.

99. $y = 2\sqrt{x}$

100. $y = \sqrt{x - 1}$

101. $y = \sqrt{x + 1}$

102. $y = 4\sqrt{x}$

103. $y = \sqrt[3]{x}$

104. $y = 2\sqrt[3]{x}$

APPLICATIONS

Exercises 105 and 106: Find the exact perimeter.

105.

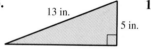

13 in.   5 in.

106.

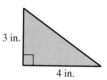

3 in.   4 in.

Exercises 107 and 108: Find the exact length of the diagonal.

107.

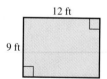

12 ft   9 ft

108.

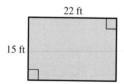

22 ft   15 ft

109. *Heart Rate*   (Refer to Example 3.) The average heart rate R in beats per minute of an animal weighing W pounds can be approximated by $R = \frac{885}{\sqrt{W}}$. Estimate the heart rate for each animal.   (*Source:* C. Pennycuick.)

(a) A 400-pound bear

(b) A 25-pound turkey

(c) A 169-pound person

110. *Stepping Frequency*   When small animals walk, they tend to take fast, short steps, whereas larger animals tend to take slower, longer steps. If an animal is h meters high at the shoulder, then the frequency F in steps per second can be estimated by

$F = \frac{0.87}{\sqrt{h}}$. The value of F is referred to as the animal's *stepping frequency*. Estimate the stepping frequency for each animal.   (*Source:* C. Pennycuick.)

h

(a) An elephant 3 meters at the shoulder

(b) A hyena 0.8 meter at the shoulder

(c) A Thomson's gazelle 0.6 meter at the shoulder

111. *Television Dimensions*   (Refer to Example 7.) A flat, rectangular television screen is 20 inches wide and 15 inches high. Find the length of the diagonal.

112. *Television Dimensions*   (Refer to Example 7.) Suppose that your friend buys a television with a 55-inch diagonal.

(a) From this information, can you determine how wide the television screen is?

(b) If the height of this screen is 27 inches, approximate the width of the screen to two decimal places.

113. *Height of a Pole*   A 40-foot guy wire is attached to the top of a telephone pole and to a stake in the ground 15 feet from the base of the pole, as shown in the figure. Find the height h of the telephone pole to the nearest tenth of a foot.

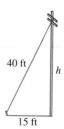

40 ft h 15 ft

114. *Flying a Kite*   Approximate, to the nearest foot, the length L of the string attached to the kite, as shown in the figure. Assume that the kite string is attached to a stake in level ground.

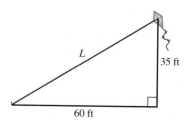

L 35 ft   60 ft

115. *Basketball Court* A basketball court is 52 feet wide and 96 feet long. Find the length of the diagonal of the court to the nearest tenth of a foot.

116. *Baseball Field* The bases in a major league baseball field are placed at the four corners of a square 90 feet on a side. Approximate, to the nearest foot, the distance d from first base to third base.

117. *Distance Between Ships* (Refer to Example 9.) Two ships pass at noon, one traveling south at 25 miles per hour and the other traveling west at 20 miles per hour. Approximate, to the nearest mile, the distance between the ships at 2:30 P.M.

118. *Distance* At 2:00 P.M. a car is 15 miles east of Smalltown, traveling north at 70 miles per hour. Approximate, to the nearest mile, the straight-line distance between the car and Smalltown at 5:00 P.M.

119. *Equilateral Triangle* The lengths of the three sides of an equilateral triangle are equal. Find the height of the triangle if its sides have length a. (*Hint:* Make a sketch.)

120. *Speed of Light* When designing computers, engineers must take into account that electricity travels at about 3×10^{10} centimeters per second.
(a) Suppose that two electronic components are located at (10, 60) and (25, 50) on an integrated circuit of a computer, where units are in centimeters. How far apart are these components?
(b) How long will it take electricity to flow in a straight line between these two components?

WRITING ABOUT MATHEMATICS

121. Explain how the figure illustrates the Pythagorean theorem.

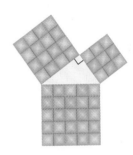

122. Suppose that you know the length of each side of a triangle. Explain how you can determine whether the triangle is a right triangle.

8.2 MULTIPLICATION AND DIVISION OF RADICAL EXPRESSIONS

The Product Rule ■ Simplifying Square Roots ■ The Quotient Rule

A LOOK INTO MATH ▷

Vehicles often leave skid marks at the scene of an accident. With the help of mathematics and radical expressions, law enforcement officers can determine the speed of a vehicle. (See Exercise 117.) In this section we show how to multiply and divide radical expressions.

The Product Rule

Let's first consider multiplication of square roots. For example, the equations

$$\sqrt{4} \cdot \sqrt{9} = 2 \cdot 3 = 6 \quad \text{and} \quad \sqrt{4 \cdot 9} = \sqrt{36} = 6$$

imply that

$$\sqrt{4} \cdot \sqrt{9} = \sqrt{4 \cdot 9}.$$

That is, the product of two square roots equals the square root of the product of their radicands.

PRODUCT RULE FOR SQUARE ROOTS

If a and b are nonnegative real numbers, then

$$\sqrt{a} \cdot \sqrt{b} = \sqrt{ab}.$$

EXAMPLE 1 Multiplying radical expressions

Multiply each expression.
(a) $\sqrt{2} \cdot \sqrt{8}$ **(b)** $\sqrt{3} \cdot \sqrt{48}$

Solution
(a) $\sqrt{2} \cdot \sqrt{8} = \sqrt{2 \cdot 8} = \sqrt{16} = 4$
(b) $\sqrt{3} \cdot \sqrt{48} = \sqrt{3 \cdot 48} = \sqrt{144} = 12$

Now Try Exercises 13, 19

Sometimes radical expressions contain variables. In the next example we simplify this type of expression.

EXAMPLE 2 Multiplying radicals containing variables

Multiply each pair of radicals. Assume that all variables are positive.
(a) $\sqrt{5} \cdot \sqrt{3xy}$ **(b)** $\sqrt{3b} \cdot \sqrt{2a}$

Solution
(a) $\sqrt{5} \cdot \sqrt{3xy} = \sqrt{5 \cdot 3xy} = \sqrt{15xy}$
(b) $\sqrt{3b} \cdot \sqrt{2a} = \sqrt{3b \cdot 2a} = \sqrt{6ab}$

Now Try Exercises 23, 25

We can also use the product rule to write a square root as the product of two square roots. Because $\sqrt{a} \cdot \sqrt{b} = \sqrt{ab}$, it follows that $\sqrt{ab} = \sqrt{a} \cdot \sqrt{b}$. For example, $21 = 3 \cdot 7$ so we can write

$$\sqrt{21} = \sqrt{3} \cdot \sqrt{7}.$$

This product may not be unique. However, when writing the square root of a natural number greater than 1 as a product, we will restrict its factors to natural numbers greater than 1.

EXAMPLE 3 Applying the product rule

Write each square root as a product of square roots.
(a) $\sqrt{35}$ **(b)** $\sqrt{3x}, x > 0$ **(c)** $\sqrt{40}$

Solution
(a) Because $35 = 5 \cdot 7$, it follows that $\sqrt{35} = \sqrt{5} \cdot \sqrt{7}$.
(b) Because $3x = 3 \cdot x$, it follows that $\sqrt{3x} = \sqrt{3} \cdot \sqrt{x}$.
(c) Because 40 has more than one pair of factors, there is more than one possibility. For example, $40 = 2 \cdot 20$ and $40 = 5 \cdot 8$, so two possibilities are

$$\sqrt{40} = \sqrt{2} \cdot \sqrt{20} \quad \text{and} \quad \sqrt{40} = \sqrt{5} \cdot \sqrt{8}.$$

Now Try Exercises 29, 33, 35

Simplifying Square Roots

A whole number a is a *perfect square* if an integer b exists such that $b^2 = a$. Thus 49 is a perfect square because $7^2 = 49$. Perfect square factors and the product rule can be used to simplify some radical expressions. For example, the largest perfect square factor of 75 is 25 because $75 = 25 \cdot 3$. Thus $\sqrt{75}$ can be simplified as

$$\sqrt{75} = \sqrt{25 \cdot 3} = \sqrt{25} \cdot \sqrt{3} = 5\sqrt{3}.$$

This procedure for simplifying square roots can be generalized as follows.

SIMPLIFYING SQUARE ROOTS

STEP 1: Determine the largest perfect square factor of the radicand.

STEP 2: Use the product rule to factor out and simplify this perfect square.

EXAMPLE 4 Simplifying square roots

Simplify each expression.
(a) $\sqrt{40}$ (b) $\sqrt{72}$ (c) $\sqrt{288}$

Solution
(a) Note that $40 = 4 \cdot 10$ and that 4 is the largest perfect square factor of 40. Thus
$$\sqrt{40} = \sqrt{4 \cdot 10} = \sqrt{4} \cdot \sqrt{10} = 2\sqrt{10}.$$

(b) The largest perfect square factor in 72 is 36 because $72 = 36 \cdot 2$. Thus
$$\sqrt{72} = \sqrt{36 \cdot 2} = \sqrt{36} \cdot \sqrt{2} = 6\sqrt{2}.$$

(c) Because $288 = 144 \cdot 2$ and $\sqrt{144} = 12$, the largest perfect square factor of 288 is 144. Thus
$$\sqrt{288} = \sqrt{144} \cdot \sqrt{2} = 12\sqrt{2}.$$

Now Try Exercises **37, 51, 57**

We can also use perfect square factors and the product rule to simplify square roots with variables in the radicand. To simplify these square roots, we often use the fact that
$$\sqrt{x^2} = x, \quad \text{whenever} \quad x \geq 0.$$

NOTE: If $x < 0$, then $\sqrt{x^2} \neq x$. For example, $\sqrt{4^2} = 4$ but $\sqrt{(-4)^2} \neq -4$.

EXAMPLE 5 Simplifying square roots containing variables

Simplify each expression. Assume that all variables are positive.
(a) $\sqrt{x^8}$ (b) $\sqrt{b^3}$ (c) $\sqrt{x^2 y^2}$

Solution
(a) Because $x^8 = x^4 \cdot x^4$, it follows that $\sqrt{x^8} = \sqrt{(x^4)^2} = x^4$.
(b) Because $b^3 = b^2 \cdot b$, it follows that $\sqrt{b^3} = \sqrt{b^2} \cdot \sqrt{b} = b\sqrt{b}$.
(c) By the product rule, $\sqrt{x^2 y^2} = \sqrt{x^2} \cdot \sqrt{y^2} = xy$.   Now Try Exercises **61, 65, 71**

In Example 6(b) and 6(c), we apply the product rule to three *nonnegative* factors. That is,

$$\sqrt{abc} = \sqrt{a} \cdot \sqrt{b} \cdot \sqrt{c}.$$

For example, $\sqrt{30} = \sqrt{2 \cdot 3 \cdot 5} = \sqrt{2} \cdot \sqrt{3} \cdot \sqrt{5}$.

EXAMPLE 6 Simplifying square roots

Simplify each expression. Assume that all variables are positive.

(a) $\sqrt{25x^2}$ **(b)** $\sqrt{45a^3}$ **(c)** $\sqrt{x^5y^2}$ **(d)** $\sqrt{2y^3} \cdot \sqrt{32y}$

Solution

(a) $\sqrt{25x^2} = \sqrt{25} \cdot \sqrt{x^2} = 5x$

(b) The largest perfect square factor of $45 = 9 \cdot 5$ is 9, and the largest perfect square factor of $a^3 = a^2 \cdot a$ is a^2. Thus

$$\sqrt{45a^3} = \sqrt{(9a^2)(5a)} = \sqrt{9} \cdot \sqrt{a^2} \cdot \sqrt{5a} = 3a\sqrt{5a}.$$

(c) The largest perfect square factor of $x^5 = x^4 \cdot x$ is x^4.

$$\sqrt{x^5y^2} = \sqrt{(x^4y^2)x} = \sqrt{x^4} \cdot \sqrt{y^2} \cdot \sqrt{x} = x^2y\sqrt{x}$$

(d) $\sqrt{2y^3} \cdot \sqrt{32y} = \sqrt{(2y^3)(32y)} = \sqrt{64y^4} = \sqrt{64} \cdot \sqrt{y^4} = 8y^2$

Now Try Exercises 63, 67, 69, 99

The Quotient Rule

Consider the equations

$$\sqrt{\frac{16}{25}} = \sqrt{\left(\frac{4}{5}\right)^2} = \frac{4}{5} \quad \text{and} \quad \frac{\sqrt{16}}{\sqrt{25}} = \frac{4}{5},$$

which imply that

$$\sqrt{\frac{16}{25}} = \frac{\sqrt{16}}{\sqrt{25}}.$$

That is, the square root of a quotient equals the quotient of the square roots. This concept can be generalized as follows.

QUOTIENT RULE FOR SQUARE ROOTS

If a and b are nonnegative real numbers with $b \neq 0$, then

$$\sqrt{\frac{a}{b}} = \frac{\sqrt{a}}{\sqrt{b}}.$$

EXAMPLE 7 Simplifying quotients

Simplify each expression. Assume that all variables are positive.

(a) $\sqrt{\dfrac{49}{64}}$ (b) $\sqrt{\dfrac{y^3}{4}}$ (c) $\sqrt{\dfrac{5x^2}{16}}$

Solution

(a) $\sqrt{\dfrac{49}{64}} = \dfrac{\sqrt{49}}{\sqrt{64}} = \dfrac{7}{8}$

(b) $\sqrt{\dfrac{y^3}{4}} = \dfrac{\sqrt{y^3}}{\sqrt{4}} = \dfrac{\sqrt{y^2}\cdot\sqrt{y}}{2} = \dfrac{y\sqrt{y}}{2}$

(c) $\sqrt{\dfrac{5x^2}{16}} = \dfrac{\sqrt{5x^2}}{\sqrt{16}} = \dfrac{\sqrt{5}\cdot\sqrt{x^2}}{4} = \dfrac{x\sqrt{5}}{4}$

Now Try Exercises 75, 93, 95

The quotient rule for square roots can sometimes be used to simplify a quotient of two square roots. For example,

$$\dfrac{\sqrt{75}}{\sqrt{3}} = \sqrt{\dfrac{75}{3}} = \sqrt{25} = 5.$$

This technique is demonstrated in the next example.

EXAMPLE 8 Simplifying quotients and products

Simplify each expression. Assume that all variables are positive.

(a) $\dfrac{\sqrt{45}}{\sqrt{15}}$ (b) $\dfrac{\sqrt{8ab}}{\sqrt{2ab}}$ (c) $\sqrt{7x}\cdot\sqrt{\dfrac{7}{x^3y^2}}$

Solution

(a) $\dfrac{\sqrt{45}}{\sqrt{15}} = \sqrt{\dfrac{45}{15}} = \sqrt{3}$

(b) $\dfrac{\sqrt{8ab}}{\sqrt{2ab}} = \sqrt{\dfrac{8ab}{2ab}} = \sqrt{4} = 2$

(c) $\sqrt{7x}\cdot\sqrt{\dfrac{7}{x^3y^2}} = \sqrt{\dfrac{49x}{x^3y^2}}$ Product rule

$= \sqrt{\dfrac{49}{x^2y^2}}$ Simplify.

$= \dfrac{\sqrt{49}}{\sqrt{x^2y^2}}$ Quotient rule

$= \dfrac{7}{xy}$ Simplify the numerator and denominator.

Now Try Exercises 83, 105, 109

CRITICAL THINKING

Does $\sqrt{x^{16}} = x^4$? Explain.

8.2 PUTTING IT ALL TOGETHER

The following table summarizes some important topics related to radical expressions. Assume that all variables are positive in each example.

Concept	Explanation	Examples
Product Rule for Square Roots	Let a and b be nonnegative real numbers. Then $$\sqrt{a} \cdot \sqrt{b} = \sqrt{ab}.$$	$\sqrt{3} \cdot \sqrt{27} = \sqrt{81} = 9$ $\sqrt{t} \cdot \sqrt{t} = \sqrt{t^2} = t$ $\sqrt{2x} \cdot \sqrt{8x} = \sqrt{16x^2} = 4x$
Simplifying Square Roots	**STEP 1:** Determine the largest perfect square factor of the radicand. **STEP 2:** Use the product rule to factor out and simplify this perfect square.	$\sqrt{8} = \sqrt{4 \cdot 2} = \sqrt{4} \cdot \sqrt{2} = 2\sqrt{2}$ $\sqrt{x^3} = \sqrt{x^2 \cdot x} = \sqrt{x^2} \cdot \sqrt{x} = x\sqrt{x}$ $\sqrt{50y^2} = \sqrt{25} \cdot \sqrt{2} \cdot \sqrt{y^2} = 5y\sqrt{2}$ $\sqrt{a^5 b^2} = \sqrt{a^4} \cdot \sqrt{a} \cdot \sqrt{b^2} = a^2 b\sqrt{a}$
Quotient Rule for Square Roots	Let a and b be nonnegative real numbers with $b \neq 0$. Then $$\sqrt{\frac{a}{b}} = \frac{\sqrt{a}}{\sqrt{b}}.$$	$\sqrt{\dfrac{5}{16}} = \dfrac{\sqrt{5}}{\sqrt{16}} = \dfrac{\sqrt{5}}{4}$ $\dfrac{\sqrt{40a^3}}{\sqrt{10a}} = \sqrt{\dfrac{40a^3}{10a}} = \sqrt{4a^2} = 2a$

8.2 Exercises

MyMathLab
PRACTICE   WATCH   DOWNLOAD   READ   REVIEW

CONCEPTS

1. $\sqrt{2} \cdot \sqrt{32} = $ _____

2. $\sqrt{a} \cdot \sqrt{b} = $ _____

3. Is 25 a perfect square?

4. The largest perfect square factor in 32 is _____.

5. $\sqrt{8} = $ _____ $\sqrt{2}$

6. If $x \geq 0$, then $\sqrt{x^2} = $ _____.

7. $\dfrac{\sqrt{32}}{\sqrt{2}} = $ _____

8. $\sqrt{\dfrac{a}{b}} = $ _____

9. $\sqrt{b^4} = $ _____

10. $\sqrt{x^3} = $ _____

MULTIPLYING SQUARE ROOTS

Exercises 11–28: Simplify the expression. Assume that all variables are positive.

11. $\sqrt{5} \cdot \sqrt{5}$

12. $\sqrt{2} \cdot \sqrt{2}$

13. $\sqrt{3} \cdot \sqrt{7}$

14. $\sqrt{11} \cdot \sqrt{5}$

15. $\sqrt{\dfrac{1}{2}} \cdot \sqrt{\dfrac{1}{2}}$

16. $\sqrt{\dfrac{2}{3}} \cdot \sqrt{\dfrac{2}{3}}$

17. $\sqrt{\dfrac{3}{5}} \cdot \sqrt{\dfrac{2}{7}}$

18. $\sqrt{\dfrac{11}{5}} \cdot \sqrt{\dfrac{1}{23}}$

19. $\sqrt{2} \cdot \sqrt{32}$

20. $\sqrt{3} \cdot \sqrt{27}$

21. $\sqrt{48} \cdot \sqrt{3}$

22. $\sqrt{45} \cdot \sqrt{5}$

23. $\sqrt{5x} \cdot \sqrt{3}$

24. $\sqrt{10} \cdot \sqrt{t}$

25. $\sqrt{a} \cdot \sqrt{5b}$

26. $\sqrt{5a} \cdot \sqrt{2b}$

27. $\sqrt{7x} \cdot \sqrt{3z}$

28. $\sqrt{x} \cdot \sqrt{6z}$

Exercises 29–36: (Refer to Example 3.) Write the square root as a product of square roots.

29. $\sqrt{33}$

30. $\sqrt{91}$

31. $\sqrt{85}$

32. $\sqrt{6}$

33. $\sqrt{7z}$

34. $\sqrt{11t}$

35. $\sqrt{30}$

36. $\sqrt{105}$

SIMPLIFYING SQUARE ROOTS

Exercises 37–44: Find the largest perfect square factor of the number.

37. 20

38. 18

39. 144

40. 50

41. 200

42. 98

43. 128

44. 63

Exercises 45–50: Complete the equation.

45. $\sqrt{75} =$ _____ $\sqrt{3}$

46. $\sqrt{32} =$ _____ $\sqrt{2}$

47. $\sqrt{24} =$ _____ $\sqrt{6}$

48. $\sqrt{45} =$ _____ $\sqrt{5}$

49. $\sqrt{28} =$ _____ $\sqrt{7}$

50. $\sqrt{54} =$ _____ $\sqrt{6}$

Exercises 51–74: Simplify the expression by factoring out the largest perfect square. Assume that all variables are positive.

51. $\sqrt{20}$

52. $\sqrt{18}$

53. $\sqrt{63}$

54. $\sqrt{128}$

55. $\sqrt{12}$

56. $\sqrt{112}$

57. $\sqrt{300}$

58. $\sqrt{72}$

59. $\sqrt{180}$

60. $\sqrt{405}$

61. $\sqrt{t^3}$

62. $\sqrt{4r^3}$

63. $\sqrt{4a^2}$

64. $\sqrt{9b^2}$

65. $\sqrt{x^4}$

66. $\sqrt{y^6}$

67. $\sqrt{8x^2}$

68. $\sqrt{27y^2}$

69. $\sqrt{125t^3}$

70. $\sqrt{20r^3}$

71. $\sqrt{x^2y^4}$

72. $\sqrt{4a^2b^2}$

73. $\sqrt{16x^2y^3}$

74. $\sqrt{36a^5b^2}$

Exercises 75–110: Simplify the expression. Assume that all variables are positive.

75. $\sqrt{\dfrac{4}{49}}$

76. $\sqrt{\dfrac{9}{64}}$

77. $\sqrt{\dfrac{16}{81}}$

78. $\sqrt{\dfrac{25}{121}}$

79. $\sqrt{\dfrac{17}{36}}$

80. $\sqrt{\dfrac{23}{144}}$

81. $\sqrt{\dfrac{0.01}{0.49}}$

82. $\sqrt{\dfrac{0.04}{0.09}}$

83. $\dfrac{\sqrt{72}}{\sqrt{2}}$

84. $\dfrac{\sqrt{27}}{\sqrt{3}}$

85. $\dfrac{\sqrt{75}}{\sqrt{3}}$

86. $\dfrac{\sqrt{245}}{\sqrt{5}}$

87. $\sqrt{5} \cdot \dfrac{\sqrt{5}}{\sqrt{9}}$

88. $\sqrt{6} \cdot \dfrac{\sqrt{6}}{\sqrt{25}}$

89. $\sqrt{\dfrac{2}{3}} \cdot \dfrac{\sqrt{2}}{\sqrt{3}}$

90. $\dfrac{\sqrt{45}}{\sqrt{3}} \cdot \sqrt{\dfrac{45}{27}}$

91. $\sqrt{\dfrac{x^2}{9}}$

92. $\sqrt{\dfrac{t^2}{16}}$

93. $\sqrt{\dfrac{2y^2}{49}}$

94. $\sqrt{\dfrac{3r^2}{81}}$

95. $\sqrt{\dfrac{20}{y^4}}$

96. $\sqrt{\dfrac{40}{t^4}}$

97. $\sqrt{2xy} \cdot \sqrt{2xy}$

98. $\sqrt{5a^2b} \cdot \sqrt{5ab}$

99. $\sqrt{2xy} \cdot \sqrt{8xy^3}$

100. $\sqrt{7t^5} \cdot \sqrt{r^2t}$

101. $\sqrt{\dfrac{2}{x^2y^2}}$

102. $\sqrt{\dfrac{5}{a^2b^2}}$

103. $\dfrac{\sqrt{75x^3}}{\sqrt{3x}}$

104. $\dfrac{\sqrt{100t^5}}{\sqrt{t^3}}$

105. $\dfrac{\sqrt{50xy}}{\sqrt{2xy}}$

106. $\dfrac{\sqrt{800rt^3}}{\sqrt{2rt}}$

107. $\dfrac{\sqrt{2x}}{\sqrt{3y}} \cdot \dfrac{\sqrt{32xy^2}}{\sqrt{3x^2y}}$

108. $\dfrac{\sqrt{6ab^2}}{\sqrt{5ab}} \cdot \dfrac{\sqrt{5ab^2}}{\sqrt{54ab}}$

109. $\sqrt{\dfrac{32}{xy}} \cdot \sqrt{\dfrac{x^3y}{2}}$

110. $\sqrt{\dfrac{rt}{24}} \cdot \sqrt{\dfrac{r^2t}{24}}$

111. Thinking Generally   Simplify $\sqrt{x^{2n}}$, where $x > 0$ and n is a natural number.

112. Thinking Generally   Simplify $\frac{\sqrt{x^{2n}}}{\sqrt{x^n}}$, where $x > 0$ and n is a natural number.

APPLICATIONS

113. *Hang Time of a Football*   A good punter can kick a football so that it stays in the air for a relatively long time. This time is called *hang time*. If a football is kicked h feet high, then its hang time T in seconds is given by $T = \sqrt{\frac{h}{4}}$.
  (a) Calculate the exact hang time for a football kicked 81 feet into the air.
  (b) Use the quotient rule to simplify the formula.

114. *Orbits and Distance*   Johannes Kepler (1571–1630) discovered a relationship between a planet's distance D from the sun and the time T it takes the planet to orbit the sun. This formula is

$$T = \sqrt{D} \cdot \sqrt{D} \cdot \sqrt{D},$$

where T is in Earth years and $D = 1$ corresponds to the average distance between Earth and the sun, or 93,000,000 miles.
  (a) If $D = 5.2$ for Jupiter, estimate the number of years required for Jupiter to orbit the sun.
  (b) Use the product rule to simplify the formula.

115. *Area and Radius*   If a circle has area A, then its radius r is given by $r = \sqrt{\frac{A}{\pi}}$.
  (a) Approximate r to two decimal places when $A = 10$ square inches.
  (b) Use the quotient rule for square roots to rewrite this formula.

116. *Squares*   If a square has area $2z^2$, then the length of a side x is given by $x = \sqrt{2z^2}$. Use the product rule for square roots to simplify this formula. Assume z is positive.

117. *Skid Marks*   Vehicles involved in accidents often leave skid marks. To determine how fast a vehicle was traveling, officials often use a test vehicle to compare skid marks on the same section of road. Suppose that a vehicle in a crash left skid marks D feet long and that a test vehicle traveling at v miles per hour leaves skid marks d feet long. Then the speed V of the vehicle involved in the crash is given by

$$V = \sqrt{\frac{v^2 D}{d}}.$$

(*Source:* N. Garber.)
  (a) Determine V if $v = 40$ mph, $D = 225$ feet, and $d = 100$ feet.
  (b) Simplify the formula by factoring out the perfect square.

118. *Perimeter*   A rectangle has width $\sqrt{2x}$ and length $\sqrt{8x}$, as illustrated in the figure.

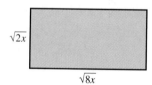

  (a) Find the exact area of the rectangle.
  (b) By what factor is the length of the rectangle greater than the width?
  (c) Let $x = 2$. Find the width, length, and area of this rectangle.

WRITING ABOUT MATHEMATICS

119. Explain how to simplify $\sqrt{x^{4n}}$, where n is a natural number and x is a positive number. (*Hint:* Start by letting $n = 1, 2, 3$ and simplify each result.)

120. A student simplifies the expression $\sqrt{64x^{16}y^9}$ to $8x^4y^3$. Identify the student's mistake. What is the correct answer?

CHECKING BASIC CONCEPTS
SECTIONS 8.1 AND 8.2

1. Evaluate each expression.
   (a) $\sqrt{81}$ (b) $\pm\sqrt{625}$

2. Approximate $\sqrt{7}$ to three decimal places.

3. Evaluate each expression.
   (a) $\sqrt[3]{8}$ (b) $-\sqrt[3]{-64}$ (c) $\sqrt[3]{-27}$

4. Find the distance between $(-3, 4)$ and $(5, -2)$.

5. Simplify each expression by factoring out the largest perfect square. Assume that all variables are positive.
   (a) $\sqrt{80}$ (b) $\sqrt{16x^3}$ (c) $\sqrt{20x^2y^2}$

6. Simplify each expression. Assume that all variables are positive.
   (a) $\sqrt{8} \cdot \sqrt{2}$ (b) $\sqrt{\dfrac{a^3}{100}}$ (c) $\sqrt{\dfrac{8}{x^3}} \cdot \sqrt{\dfrac{x}{2}}$

7. *Ladder Against a Building* The base of a 25-foot ladder is 7 feet from the wall of a building. How high up the wall does the top of the ladder rest?

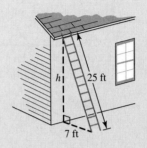

8. *Distance Between Two Ships* Two ships pass at noon, one traveling north at 20 miles per hour and the other traveling east at 25 miles per hour. Approximate, to the nearest mile, the distance between the ships at 3:30 P.M.

8.3 ADDITION AND SUBTRACTION OF RADICAL EXPRESSIONS

Addition of Radical Expressions ■ Subtraction of Radical Expressions

Sometimes open channels are constructed to protect a city from flood water. The rate R at which water flows through a channel is given by the formula $R = k\sqrt{m}$, where m is the slope of the channel and k is a constant determined by the shape of the channel. To determine how much water flows through two channels, we need to add radical expressions. In this section we learn how to add and subtract radical expressions.

Addition of Radical Expressions

We can use the distributive property to add $4x$ and $8x$ because they are *like* terms. That is,

$$4x + 8x = (4 + 8)x = 12x.$$

Similarly, we can add (or subtract) *like radicals*. Square roots with the same radicand are **like radicals**, as are cube roots with the same radicand. For example, we can add $7\sqrt{3}$ and $3\sqrt{3}$ because both contain the like radical $\sqrt{3}$.

$$7\sqrt{3} + 3\sqrt{3} = (7 + 3)\sqrt{3} = 10\sqrt{3}$$

We can also add like radicals containing cube roots, such as $7\sqrt[3]{3}$ and $3\sqrt[3]{3}$.

$$7\sqrt[3]{3} + 3\sqrt[3]{3} = (7 + 3)\sqrt[3]{3} = 10\sqrt[3]{3}$$

Sometimes two radicals that are not alike can be added by changing them to like radicals. For example, $\sqrt{27}$ and $\sqrt{3}$ are unlike radicals. However, because

$$\sqrt{27} = \sqrt{9 \cdot 3} = \sqrt{9} \cdot \sqrt{3} = 3\sqrt{3},$$

we can add $\sqrt{27}$ and $\sqrt{3}$.

$$\sqrt{27} + \sqrt{3} = 3\sqrt{3} + 1\sqrt{3} = (3 + 1)\sqrt{3} = 4\sqrt{3}$$

We cannot combine $2x + 3x^2$ because they are unlike terms. Similarly, we cannot combine $\sqrt{2} + \sqrt{3}$ because they have different radicands and thus are unlike radicals.

EXAMPLE 1   Finding like radicals

Write each pair of terms as like radicals, if possible.
(a) $\sqrt{75}, \sqrt{48}$ **(b)** $\sqrt{12}, \sqrt{20}$

Solution
(a) The expressions $\sqrt{75}$ and $\sqrt{48}$ are unlike radicals. However, they can be changed to like radicals as follows.

$$\sqrt{75} = \sqrt{25 \cdot 3} = \sqrt{25} \cdot \sqrt{3} = 5\sqrt{3}$$

and

$$\sqrt{48} = \sqrt{16 \cdot 3} = \sqrt{16} \cdot \sqrt{3} = 4\sqrt{3}$$

The expressions $5\sqrt{3}$ and $4\sqrt{3}$ are like radicals.
(b) Because

$$\sqrt{12} = \sqrt{4 \cdot 3} = \sqrt{4} \cdot \sqrt{3} = 2\sqrt{3}$$

and

$$\sqrt{20} = \sqrt{4 \cdot 5} = \sqrt{4} \cdot \sqrt{5} = 2\sqrt{5},$$

the expressions $\sqrt{12}$ and $\sqrt{20}$ *cannot* be written as like radicals.

Now Try Exercises 7, 9

In the next three examples we use these techniques to add radical expressions.

EXAMPLE 2   Adding radical expressions

Simplify each expression.
(a) $2\sqrt{7} + 6\sqrt{7}$ **(b)** $3\sqrt[3]{12} + 11\sqrt[3]{12}$
(c) $4\sqrt{48} + 5\sqrt{27}$ **(d)** $\sqrt{3} + \sqrt{27} + \sqrt{75}$

Solution
(a) Because $2\sqrt{7}$ and $6\sqrt{7}$ are like radicals, they can be combined.

$$2\sqrt{7} + 6\sqrt{7} = (2 + 6)\sqrt{7} = 8\sqrt{7}$$

(b) Because $3\sqrt[3]{12}$ and $11\sqrt[3]{12}$ are like radicals, they can be combined.

$$3\sqrt[3]{12} + 11\sqrt[3]{12} = (3 + 11)\sqrt[3]{12} = 14\sqrt[3]{12}$$

(c)
$$4\sqrt{48} + 5\sqrt{27} = 4\sqrt{16 \cdot 3} + 5\sqrt{9 \cdot 3}$$
$$= 4\sqrt{16} \cdot \sqrt{3} + 5\sqrt{9} \cdot \sqrt{3}$$
$$= 4(4\sqrt{3}) + 5(3\sqrt{3})$$
$$= 16\sqrt{3} + 15\sqrt{3}$$
$$= (16 + 15)\sqrt{3}$$
$$= 31\sqrt{3}$$

(d) $\sqrt{3} + \sqrt{27} + \sqrt{75} = \sqrt{3} + \sqrt{9 \cdot 3} + \sqrt{25 \cdot 3}$
$$= \sqrt{3} + \sqrt{9} \cdot \sqrt{3} + \sqrt{25} \cdot \sqrt{3}$$
$$= 1\sqrt{3} + 3\sqrt{3} + 5\sqrt{3}$$
$$= (1 + 3 + 5)\sqrt{3}$$
$$= 9\sqrt{3}$$

Now Try Exercises `15, 22, 23, 35`

NOTE: $\sqrt{a + b} \neq \sqrt{a} + \sqrt{b}$. For example, $\sqrt{9 + 16} \neq \sqrt{9} + \sqrt{16} = 3 + 4 = 7$. Rather, $\sqrt{9 + 16} = \sqrt{25} = 5$.

EXAMPLE **3** **Adding radical expressions containing variables**

Add each expression and simplify. Assume that all variables are positive.

(a) $\sqrt{x} + 3\sqrt{x}$ **(b)** $8\sqrt[3]{ab} + 2\sqrt[3]{ab}$ **(c)** $\sqrt{4t} + 5\sqrt{9t}$

Solution

(a) $\sqrt{x} + 3\sqrt{x} = (1 + 3)\sqrt{x} = 4\sqrt{x}$

(b) $8\sqrt[3]{ab} + 2\sqrt[3]{ab} = (8 + 2)\sqrt[3]{ab} = 10\sqrt[3]{ab}$

(c) $\sqrt{4t} + 5\sqrt{9t} = \sqrt{4} \cdot \sqrt{t} + 5\sqrt{9} \cdot \sqrt{t}$
$$= 2\sqrt{t} + 15\sqrt{t}$$
$$= (2 + 15)\sqrt{t}$$
$$= 17\sqrt{t}$$

Now Try Exercises `47, 51, 56`

EXAMPLE **4** **Adding three radical expressions**

Simplify each expression. Assume that variables are positive.

(a) $3\sqrt{50} + \sqrt{72} + \sqrt{27}$ **(b)** $\sqrt{x^3} + \sqrt{9x} + \sqrt{3x^2}$

Solution

(a) $3\sqrt{50} + \sqrt{72} + \sqrt{27} = 3\sqrt{25 \cdot 2} + \sqrt{36 \cdot 2} + \sqrt{9 \cdot 3}$
$$= 3\sqrt{25} \cdot \sqrt{2} + \sqrt{36} \cdot \sqrt{2} + \sqrt{9} \cdot \sqrt{3}$$
$$= 15\sqrt{2} + 6\sqrt{2} + 3\sqrt{3}$$
$$= 21\sqrt{2} + 3\sqrt{3}$$

(b) $\sqrt{x^3} + \sqrt{9x} + \sqrt{3x^2} = \sqrt{x^2 \cdot x} + \sqrt{9 \cdot x} + \sqrt{x^2 \cdot 3}$

$= \sqrt{x^2} \cdot \sqrt{x} + \sqrt{9} \cdot \sqrt{x} + \sqrt{x^2} \cdot \sqrt{3}$

$= x\sqrt{x} + 3\sqrt{x} + x\sqrt{3}$

$= (x + 3)\sqrt{x} + x\sqrt{3}$ Now Try Exercises 39, 43

In the next example we apply addition of radicals to find the exact perimeter of a rectangle.

EXAMPLE **5**   Finding the perimeter of a rectangle

Find the *exact* perimeter of the rectangle shown in Figure 8.9. Simplify your answer.

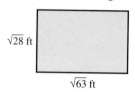

$\sqrt{28}$ ft

$\sqrt{63}$ ft

Figure 8.9

Solution

The width is $\sqrt{28}$ feet and the length is $\sqrt{63}$ feet, so the perimeter of the rectangle equals

$2\sqrt{28} + 2\sqrt{63}$ feet.

To simplify this answer, we write $\sqrt{28}$ and $\sqrt{63}$ as like radicals and then add.

$2\sqrt{28} + 2\sqrt{63} = 2\sqrt{4 \cdot 7} + 2\sqrt{9 \cdot 7}$

$= 2\sqrt{4} \cdot \sqrt{7} + 2\sqrt{9} \cdot \sqrt{7}$

$= 2(2\sqrt{7}) + 2(3\sqrt{7})$

$= 4\sqrt{7} + 6\sqrt{7}$

$= 10\sqrt{7}$ feet   Now Try Exercise 59

▶ **REAL-WORLD CONNECTION**   In the next example we apply the formula for open flood channels, which were discussed at the beginning of this section.

EXAMPLE **6**   Calculating water flow in channels

Suppose that two flood channels have flow rates R_1 and R_2 given by

$R_1 = 1000\sqrt{m_1}$ and $R_2 = 500\sqrt{m_2}$,

where R_1 and R_2 are in cubic feet per second and m_1 and m_2 are the slopes of each channel, respectively.

(a) Find $R_1 + R_2$ if $m_1 = 0.04$ and $m_2 = 0.01$.

(b) Find $R_1 + R_2$ if both channels have slope m.

Solution

(a) $R_1 + R_2 = 1000\sqrt{0.04} + 500\sqrt{0.01} = 1000(0.2) + 500(0.1) = 250$ cubic feet per second.

(b) If $m_1 = m$ and $m_2 = m$, then

$R_1 + R_2 = 1000\sqrt{m} + 500\sqrt{m} = (1000 + 500)\sqrt{m} = 1500\sqrt{m}$.

     Now Try Exercises 61, 62

Subtraction of Radical Expressions

Subtraction of radicals is similar to addition of radicals. For example, $5\sqrt{11}$ can be subtracted from $8\sqrt{11}$ as

$$8\sqrt{11} - 5\sqrt{11} = (8 - 5)\sqrt{11} = 3\sqrt{11}.$$

As with addition, we can subtract only like radicals. Sometimes two radicals can be simplified to like radicals, as demonstrated in the next two examples.

EXAMPLE 7   Subtracting radical expressions

Subtract each expression and simplify. Assume that the variable x is positive.

(a) $6\sqrt{5} - 2\sqrt{5}$ (b) $4\sqrt{50} - 3\sqrt{8}$ (c) $\sqrt{16x} - 3\sqrt{x}$

Solution

(a) $6\sqrt{5} - 2\sqrt{5} = (6 - 2)\sqrt{5} = 4\sqrt{5}$

(b) We start by changing $\sqrt{50}$ and $\sqrt{8}$ to like radicals.

$$\begin{aligned}
4\sqrt{50} - 3\sqrt{8} &= 4\sqrt{25} \cdot \sqrt{2} - 3\sqrt{4} \cdot \sqrt{2} \\
&= 4(5\sqrt{2}) - 3(2\sqrt{2}) \\
&= 20\sqrt{2} - 6\sqrt{2} \\
&= (20 - 6)\sqrt{2} \\
&= 14\sqrt{2}
\end{aligned}$$

(c) $$\begin{aligned}
\sqrt{16x} - 3\sqrt{x} &= \sqrt{16} \cdot \sqrt{x} - 3\sqrt{x} \\
&= 4\sqrt{x} - 3\sqrt{x} \\
&= (4 - 3)\sqrt{x} \\
&= \sqrt{x}
\end{aligned}$$

Now Try Exercises 25, 31, 49

EXAMPLE 8   Subtracting radical expressions

Subtract each expression and simplify. Assume that all variables are positive.

(a) $\sqrt{x^3} - \sqrt{9x}$ (b) $5\sqrt{a^2 b} - 3\sqrt{a^2 b}$ (c) $4\sqrt[3]{x} - 3\sqrt[3]{x}$

Solution

(a) $$\begin{aligned}
\sqrt{x^3} - \sqrt{9x} &= \sqrt{x^2} \cdot \sqrt{x} - \sqrt{9} \cdot \sqrt{x} \\
&= x\sqrt{x} - 3\sqrt{x} \\
&= (x - 3)\sqrt{x}
\end{aligned}$$

(b) $$\begin{aligned}
5\sqrt{a^2 b} - 3\sqrt{a^2 b} &= 5\sqrt{a^2} \cdot \sqrt{b} - 3\sqrt{a^2} \cdot \sqrt{b} \\
&= 5a\sqrt{b} - 3a\sqrt{b} \\
&= (5a - 3a)\sqrt{b} \\
&= 2a\sqrt{b}
\end{aligned}$$

(c) Because $4\sqrt[3]{x}$ and $3\sqrt[3]{x}$ are like radicals, we can subtract them.

$$4\sqrt[3]{x} - 3\sqrt[3]{x} = (4 - 3)\sqrt[3]{x} = \sqrt[3]{x}$$

Now Try Exercises 41, 45, 55

8.3 PUTTING IT ALL TOGETHER

In this section we discussed how to add and subtract radical expressions. These concepts are summarized in the following table.

Concept	Explanation	Examples
Addition of Radical Expressions	Add like radicals.	$3\sqrt{2} + 4\sqrt{2} = 7\sqrt{2}$ $\sqrt{7} + 5\sqrt{7} + 3\sqrt{7} = 9\sqrt{7}$ $4\sqrt[3]{4} + 2\sqrt[3]{4} = 6\sqrt[3]{4}$
Subtraction of Radical Expressions	Subtract like radicals.	$7\sqrt{2} - 4\sqrt{2} = 3\sqrt{2}$ $\sqrt{7} - 5\sqrt{7} - 2\sqrt{7} = -6\sqrt{7}$ $5\sqrt[3]{9} - 2\sqrt[3]{9} = 3\sqrt[3]{9}$
Addition and Subtraction of Unlike Radical Expressions	Sometimes unlike radicals can be written as like radicals.	$\sqrt{2} + \sqrt{8} = \sqrt{2} + \sqrt{4} \cdot \sqrt{2}$ $= \sqrt{2} + 2\sqrt{2}$ $= 3\sqrt{2}$

8.3 Exercises

MyMathLab   Math XL PRACTICE   WATCH   DOWNLOAD   READ   REVIEW

CONCEPTS

1. $\sqrt{2} + \sqrt{2} = $ ____

2. $3\sqrt{b} - 2\sqrt{b} = $ ____

3. $3\sqrt[3]{b} - 2\sqrt[3]{b} = $ ____

4. $3\sqrt[3]{b} + 2\sqrt[3]{b} = $ ____

5. Can you simplify $\sqrt{3} + \sqrt{12}$? Explain.

6. Can you simplify $\sqrt{3} - \sqrt{5}$? Explain.

LIKE RADICALS

Exercises 7–14: Write each pair of terms as like radicals, if possible.

7. $\sqrt{2}, \sqrt{8}$

8. $\sqrt{50}, \sqrt{8}$

9. $\sqrt{12}, \sqrt{15}$

10. $\sqrt{18}, \sqrt{27}$

11. $\sqrt{63}, \sqrt{28}$

12. $\sqrt{44}, \sqrt{99}$

13. $\sqrt{5}, \sqrt{125}$

14. $\sqrt{3}, \sqrt{27}$

OPERATIONS ON RADICAL EXPRESSIONS

Exercises 15–58: Simplify the expression. Assume that all variables are positive.

15. $2\sqrt{5} + 3\sqrt{5}$

16. $5\sqrt{6} + \sqrt{6}$

17. $\sqrt{13} + 8\sqrt{13}$

18. $5\sqrt{21} + 6\sqrt{21}$

19. $2\sqrt{8} + 3\sqrt{8}$

20. $2\sqrt{12} + 3\sqrt{12}$

21. $4\sqrt{3} + \sqrt{27} + \sqrt{18}$

22. $7\sqrt{2} + 2\sqrt{8} + \sqrt{32}$

23. $9\sqrt{72} + \sqrt{32}$

24. $10\sqrt{20} + 11\sqrt{45}$

25. $6\sqrt{15} - 2\sqrt{15}$

26. $7\sqrt{22} - \sqrt{22}$

27. $5\sqrt{75} - 7\sqrt{75}$

28. $9\sqrt{90} - 11\sqrt{90}$

29. $\sqrt{7} - 5\sqrt{28}$

30. $\sqrt{10} - \sqrt{40}$

31. $2\sqrt{54} - 2\sqrt{24}$

32. $3\sqrt{5} - \sqrt{125}$

33. $9\sqrt{50} - 3\sqrt{8} + \sqrt{2}$

34. $2\sqrt{99} - 3\sqrt{44} + 2\sqrt{11}$

35. $2\sqrt[3]{4} + 5\sqrt[3]{4}$ **36.** $7\sqrt[3]{5} + 6\sqrt[3]{5}$

37. $\sqrt[3]{20} - 2\sqrt[3]{20}$ **38.** $7\sqrt[3]{15} - 3\sqrt[3]{15}$

39. $3\sqrt{32} + \sqrt{18} + 2\sqrt{75}$

40. $5\sqrt{8} + \sqrt{20} + \sqrt{98}$

41. $3\sqrt{4x^3} - \sqrt{x}$ **42.** $\sqrt{25x} - \sqrt{4x^3}$

43. $\sqrt{4x} + \sqrt{x^3} + \sqrt{2x^2}$

44. $\sqrt{16b} + \sqrt{9b} + \sqrt{4a}$

45. $7\sqrt{a^2b} - 2\sqrt{a^2b}$

46. $\sqrt{25ab^2} - \sqrt{ab^2}$

47. $3\sqrt{t} + 2\sqrt{t}$ **48.** $\sqrt{t} + 5\sqrt{t}$

49. $\sqrt{x} - 5\sqrt{x}$ **50.** $5\sqrt{x} - 6\sqrt{x}$

51. $\sqrt{9b} + \sqrt{4b}$ **52.** $2\sqrt{25b} + 6\sqrt{36b}$

53. $\sqrt{4t^3} - 8\sqrt{t} - \sqrt{4t}$

54. $2\sqrt{8t} - 6\sqrt{2t} + 2\sqrt{9t}$

55. $4\sqrt[3]{x} - 2\sqrt[3]{x}$ **56.** $12\sqrt[3]{ab} + 5\sqrt[3]{ab}$

57. $9\sqrt[3]{x} + 2\sqrt[3]{x}$ **58.** $5\sqrt[3]{x} - 8\sqrt[3]{x}$

GEOMETRY

59. *Perimeter*   Find the exact perimeter of the triangle.

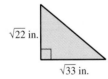

$\sqrt{22}$ in.

$\sqrt{33}$ in.

60. *Area*   Find the exact area and perimeter of the rectangle. Simplify your answer.

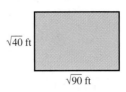

$\sqrt{40}$ ft

$\sqrt{90}$ ft

APPLICATIONS

Exercises 61 and 62: (Refer to Example 6.) Suppose that two open flood channels have flow rates R_1 and R_2 given by

$$R_1 = 1500\sqrt{m_1} \quad \text{and} \quad R_2 = 2000\sqrt{m_2},$$

where R_1 and R_2 are in cubic feet per second and m_1 and m_2 are the slopes of each channel, respectively.

61. Find $R_1 + R_2$ if $m_1 = 0.09$ and $m_2 = 0.04$.

62. Find $R_1 + R_2$ if both channels have slope m.

WRITING ABOUT MATHEMATICS

63. A student simplifies an expression *incorrectly*:

$$\sqrt{20} + \sqrt{45} \stackrel{?}{=} \sqrt{4 \cdot 5} + \sqrt{9 \cdot 5}$$
$$\stackrel{?}{=} 2\sqrt{5} + 3\sqrt{5}$$
$$\stackrel{?}{=} (2 + 3)\sqrt{5 + 5}$$
$$\stackrel{?}{=} 5\sqrt{10}.$$

Explain the error that the student made.

64. Give an example of like radicals and an example of unlike radicals.

8.4 SIMPLIFYING RADICAL EXPRESSIONS

Simplifying Products ▪ Rationalizing the Denominator

A LOOK INTO MATH ▷

An important skill in mathematics is the simplification of expressions. If expressions are complicated, then they are often difficult to work with when we are solving equations. When expressions are simpler, we can often solve these equations more easily. In this section we discuss how to simplify products and quotients of radical expressions.

Simplifying Products

Some types of radical expressions can be multiplied like binomials. For example, because

$$(a - b)(a + b) = a^2 - b^2,$$

we have

$$(2 - \sqrt{3})(2 + \sqrt{3}) = 2^2 - (\sqrt{3})^2$$
$$= 4 - 3$$
$$= 1.$$

Note that the product $(2 - \sqrt{3})(2 + \sqrt{3})$ simplifies to 1, which is less complicated.

EXAMPLE 1   Multiplying radical expressions

Multiply and simplify.

(a) $(7 + \sqrt{2})(7 - \sqrt{2})$ **(b)** $(\sqrt{x} - 7)(\sqrt{x} + 3)$, $x \geq 0$

Solution

(a) Because $(a + b)(a - b) = a^2 - b^2$, we can let $a = 7$ and $b = \sqrt{2}$. Then the given expression can be simplified as

$$(7 + \sqrt{2})(7 - \sqrt{2}) = 7^2 - (\sqrt{2})^2$$
$$= 49 - 2$$
$$= 47.$$

(b) Recall that $(y - 7)(y + 3) = y^2 - 4y - 21$. The given expression can be multiplied similarly, where $y = \sqrt{x}$.

$$(\sqrt{x} - 7)(\sqrt{x} + 3) = \sqrt{x} \cdot \sqrt{x} + 3\sqrt{x} - 7\sqrt{x} - 7 \cdot 3$$
$$= (\sqrt{x})^2 - 4\sqrt{x} - 21$$
$$= x - 4\sqrt{x} - 21$$

NOTE: $(\sqrt{x})^2 = x$ only if $x \geq 0$. If $x < 0$, then $\sqrt{x}$ is not a real number.

Now Try Exercises 9, 21

Rationalizing the Denominator

In mathematics expressions are commonly written without radicals in the denominator. Quotients containing radical expressions can appear to be different but actually be equal. For example, $\frac{1}{\sqrt{2}}$ and $\frac{\sqrt{2}}{2}$ represent the same real number even though they look different.

To show that they are equal, we can multiply $\frac{1}{\sqrt{2}}$ by 1 in the form $\frac{\sqrt{2}}{\sqrt{2}}$.

$$\frac{1}{\sqrt{2}} \cdot \frac{\sqrt{2}}{\sqrt{2}} = \frac{1 \cdot \sqrt{2}}{\sqrt{2} \cdot \sqrt{2}} = \frac{\sqrt{2}}{2}$$

NOTE: When rationalizing a denominator, we use the fact that $\sqrt{a} \cdot \sqrt{a} = a$ for any real number a, such that $a \geq 0$.

One way to standardize quotients containing radical expressions is to remove the radical expressions from the denominator by using a process called **rationalizing the denominator**. Exercise 79 at the end of this section suggests one reason why people rationalized denominators before calculators were invented. The next example demonstrates how to rationalize the denominator when it contains a single term.

EXAMPLE 2   Rationalizing the denominator

Rationalize the denominator of each expression. Assume that all variables are positive.

(a) $\dfrac{3}{\sqrt{7}}$ (b) $\dfrac{5}{2\sqrt{5}}$ (c) $-\dfrac{1}{\sqrt{t}}$ (d) $\sqrt{\dfrac{7}{x}}$

Solution

(a) The denominator contains the irrational number $\sqrt{7}$, so multiply the given expression by 1 in the form $\dfrac{\sqrt{7}}{\sqrt{7}}$.

$$\frac{3}{\sqrt{7}} = \frac{3}{\sqrt{7}} \cdot \frac{\sqrt{7}}{\sqrt{7}} = \frac{3\sqrt{7}}{7}$$

(b) $\dfrac{5}{2\sqrt{5}} = \dfrac{5}{2\sqrt{5}} \cdot \dfrac{\sqrt{5}}{\sqrt{5}} = \dfrac{5\sqrt{5}}{2 \cdot 5} = \dfrac{\sqrt{5}}{2}$

(c) $-\dfrac{1}{\sqrt{t}} = -\dfrac{1}{\sqrt{t}} \cdot \dfrac{\sqrt{t}}{\sqrt{t}} = -\dfrac{\sqrt{t}}{t}$

(d) $\sqrt{\dfrac{7}{x}} = \dfrac{\sqrt{7}}{\sqrt{x}} \cdot \dfrac{\sqrt{x}}{\sqrt{x}} = \dfrac{\sqrt{7x}}{x}$

Now Try Exercises 35, 43, 47, 49

When the denominator of a quotient is either a sum or difference and contains a square root, we multiply the numerator and denominator by the **conjugate** of the denominator to rationalize the denominator. The conjugate of a radical expression containing two terms is typically found by changing an addition symbol to a subtraction symbol or vice versa. For example, the conjugate of $4 + \sqrt{2}$ is $4 - \sqrt{2}$ and the conjugate of $\sqrt{a} - \sqrt{b}$ is $\sqrt{a} + \sqrt{b}$.

EXAMPLE 3   Writing the conjugate of radical expressions

Write the conjugate of each expression.

(a) $1 - \sqrt{2}$ (b) $\sqrt{7} + 4$ (c) $\sqrt{x} + \sqrt{y}$

Solution

(a) The conjugate of $1 - \sqrt{2}$ is $1 + \sqrt{2}$.
(b) The conjugate of $\sqrt{7} + 4$ is $\sqrt{7} - 4$.
(c) The conjugate of $\sqrt{x} + \sqrt{y}$ is $\sqrt{x} - \sqrt{y}$.

Now Try Exercises 29, 31, 33

In the next example we use the conjugate to rationalize the denominator.

EXAMPLE 4 Rationalizing the denominator

Rationalize the denominator of each expression.

(a) $\dfrac{2 + \sqrt{5}}{2 - \sqrt{5}}$ (b) $\dfrac{\sqrt{x} - \sqrt{y}}{\sqrt{x} + \sqrt{y}}$

Solution

(a) The conjugate of the denominator, $2 - \sqrt{5}$, is $2 + \sqrt{5}$, so we multiply by $\dfrac{2 + \sqrt{5}}{2 + \sqrt{5}}$.

$$\frac{2 + \sqrt{5}}{2 - \sqrt{5}} = \frac{2 + \sqrt{5}}{2 - \sqrt{5}} \cdot \frac{2 + \sqrt{5}}{2 + \sqrt{5}}$$

$$= \frac{(2 + \sqrt{5})(2 + \sqrt{5})}{(2 - \sqrt{5})(2 + \sqrt{5})}$$

$$= \frac{2 \cdot 2 + 2\sqrt{5} + 2\sqrt{5} + \sqrt{5} \cdot \sqrt{5}}{2^2 - (\sqrt{5})^2}$$

$$= \frac{4 + 4\sqrt{5} + 5}{4 - 5}$$

$$= \frac{9 + 4\sqrt{5}}{-1}$$

$$= -9 - 4\sqrt{5}$$

(b) The conjugate of the denominator is $\sqrt{x} - \sqrt{y}$, so we multiply by $\dfrac{\sqrt{x} - \sqrt{y}}{\sqrt{x} - \sqrt{y}}$.

CRITICAL THINKING

Sometimes it is necessary to rationalize the *numerator* of an expression. Use the concepts you learned about rationalizing the denominator to rationalize the numerator of $\dfrac{\sqrt{x} - \sqrt{y}}{\sqrt{x} + \sqrt{y}}$.

$$\frac{\sqrt{x} - \sqrt{y}}{\sqrt{x} + \sqrt{y}} = \frac{\sqrt{x} - \sqrt{y}}{\sqrt{x} + \sqrt{y}} \cdot \frac{\sqrt{x} - \sqrt{y}}{\sqrt{x} - \sqrt{y}}$$

$$= \frac{(\sqrt{x} - \sqrt{y})(\sqrt{x} - \sqrt{y})}{(\sqrt{x} + \sqrt{y})(\sqrt{x} - \sqrt{y})}$$

$$= \frac{\sqrt{x} \cdot \sqrt{x} - \sqrt{x} \cdot \sqrt{y} - \sqrt{x} \cdot \sqrt{y} + \sqrt{y} \cdot \sqrt{y}}{(\sqrt{x})^2 - (\sqrt{y})^2}$$

$$= \frac{x - 2\sqrt{xy} + y}{x - y}$$

Now Try Exercises **61, 69**

8.4 PUTTING IT ALL TOGETHER

In this section we discussed how to multiply radical expressions and rationalize the denominator. These concepts are summarized in the following table.

Concept	Explanation	Examples
Multiplying Radical Expressions	Can sometimes be multiplied like binomials	$(5 - \sqrt{7})(5 + \sqrt{7}) = 5^2 - (\sqrt{7})^2 = 18$ $(\sqrt{2} + 1)(\sqrt{2} - 3) = 2 - 2\sqrt{2} - 3$ $= -1 - 2\sqrt{2}$

Concept	Explanation	Examples
Rationalizing the Denominator	Multiply the expression by 1 in the form $$\frac{\sqrt{a}}{\sqrt{a}} \quad \text{or} \quad \frac{\text{conjugate}}{\text{conjugate}},$$ where "conjugate" represents the conjugate of the denominator.	$$\frac{2}{3\sqrt{7}} = \frac{2}{3\sqrt{7}} \cdot \frac{\sqrt{7}}{\sqrt{7}} = \frac{2\sqrt{7}}{21}$$ $$\frac{\sqrt{x}}{\sqrt{x}+1} = \frac{\sqrt{x}}{\sqrt{x}+1} \cdot \frac{\sqrt{x}-1}{\sqrt{x}-1}$$ $$= \frac{x - \sqrt{x}}{x-1}$$

8.4 Exercises

MyMathLab   Math XL PRACTICE   WATCH   DOWNLOAD   READ   REVIEW

CONCEPTS

1. $(a-b)(a+b) = $ _____

2. If $b \geq 0$, then $\sqrt{b} \cdot \sqrt{b} = $ _____.

3. $\left(1 - \sqrt{2}\right)\left(1 + \sqrt{2}\right) = $ _____

4. $\left(\sqrt{a} - \sqrt{b}\right)\left(\sqrt{a} + \sqrt{b}\right) = $ _____

5. The conjugate of $4 - \sqrt{11}$ is _____.

6. The conjugate of $2 + \sqrt{3}$ is _____.

7. To rationalize the denominator of $\frac{2}{\sqrt{11}}$, multiply this expression by _____.

8. To rationalize the denominator of $\frac{2}{4 - \sqrt{11}}$, multiply this expression by _____.

MULTIPLYING RADICAL EXPRESSIONS

Exercises 9–28: Multiply and simplify. Assume that all variables are positive.

9. $\left(4 - \sqrt{2}\right)\left(4 + \sqrt{2}\right)$

10. $\left(3 - \sqrt{7}\right)\left(3 + \sqrt{7}\right)$

11. $\left(8 + \sqrt{11}\right)\left(8 - \sqrt{11}\right)$

12. $\left(6 + \sqrt{2}\right)\left(6 - \sqrt{2}\right)$

13. $\left(\sqrt{5} - 7\right)\left(\sqrt{5} + 7\right)$

14. $\left(\sqrt{10} - 1\right)\left(\sqrt{10} + 1\right)$

15. $\left(\sqrt{15} + 2\right)\left(\sqrt{15} - 2\right)$

16. $\left(\sqrt{8} + 3\right)\left(\sqrt{8} - 3\right)$

17. $\left(\sqrt{x} - 5\right)\left(\sqrt{x} + 5\right)$

18. $\left(\sqrt{y} + 3\right)\left(\sqrt{y} - 3\right)$

19. $\left(7 - \sqrt{t}\right)\left(7 + \sqrt{t}\right)$

20. $\left(5 + \sqrt{t}\right)\left(5 - \sqrt{t}\right)$

21. $\left(\sqrt{y} - 1\right)\left(\sqrt{y} - 2\right)$

22. $\left(\sqrt{t} + 1\right)\left(\sqrt{t} - 2\right)$

23. $\left(3 - \sqrt{x}\right)\left(1 + \sqrt{x}\right)$

24. $\left(8 + \sqrt{z}\right)\left(2 + \sqrt{z}\right)$

25. $\left(\sqrt{x} - \sqrt{y}\right)\left(\sqrt{x} + \sqrt{y}\right)$

26. $\left(\sqrt{a} + 5\sqrt{b}\right)\left(\sqrt{a} - 5\sqrt{b}\right)$

27. $\left(\sqrt{r} - 3\sqrt{t}\right)\left(\sqrt{r} + 3\sqrt{t}\right)$

28. $\left(\sqrt{r} + 2\sqrt{t}\right)\left(\sqrt{r} - 2\sqrt{t}\right)$

RATIONALIZING THE DENOMINATOR

Exercises 29–34: Find the conjugate of the expression.

29. $\sqrt{17} - 3$ 30. $\sqrt{41} + 5$

31. $20 + \sqrt{5}$ 32. $35 - 2\sqrt{3}$

33. $\sqrt{x} - \sqrt{z}$ 34. $\sqrt{a} + 7\sqrt{b}$

Exercises 35–74: Rationalize the denominator. Assume that all variables are positive.

35. $\dfrac{2}{\sqrt{5}}$

36. $\dfrac{7}{\sqrt{7}}$

37. $\dfrac{14}{\sqrt{14}}$

38. $\dfrac{1}{\sqrt{3}}$

39. $-\dfrac{3}{\sqrt{3}}$

40. $-\dfrac{1}{\sqrt{2}}$

41. $\dfrac{5}{\sqrt{10}}$

42. $\dfrac{3}{\sqrt{18}}$

43. $-\dfrac{4}{3\sqrt{7}}$

44. $-\dfrac{9}{2\sqrt{5}}$

45. $\dfrac{3}{2\sqrt{3}}$

46. $-\dfrac{7}{5\sqrt{7}}$

47. $\dfrac{2}{\sqrt{b}}$

48. $\dfrac{4}{\sqrt{t}}$

49. $-\sqrt{\dfrac{3}{b}}$

50. $\sqrt{\dfrac{5}{2x}}$

51. $\dfrac{\sqrt{2x}}{\sqrt{3}}$

52. $\dfrac{\sqrt{5z}}{\sqrt{7}}$

53. $\dfrac{\sqrt{4x^3}}{\sqrt{36x}}$

54. $\dfrac{1}{\sqrt{9x^3}}$

55. $\dfrac{1}{\sqrt{3}-1}$

56. $\dfrac{1}{\sqrt{5}+3}$

57. $\dfrac{\sqrt{5}}{\sqrt{5}+2}$

58. $\dfrac{\sqrt{7}}{\sqrt{7}-1}$

59. $\dfrac{\sqrt{2}+1}{\sqrt{2}-1}$

60. $\dfrac{\sqrt{11}-4}{\sqrt{11}+4}$

61. $\dfrac{5-\sqrt{13}}{5+\sqrt{13}}$

62. $\dfrac{12+\sqrt{2}}{12-\sqrt{2}}$

63. $\dfrac{1}{2\sqrt{x}}$

64. $\dfrac{3}{2\sqrt{y}}$

65. $\dfrac{\sqrt{x}-2}{\sqrt{x}+2}$

66. $\dfrac{\sqrt{z}+3}{\sqrt{z}-3}$

67. $\dfrac{\sqrt{a}-2b}{\sqrt{a}+2b}$

68. $\dfrac{\sqrt{x}+2y}{\sqrt{x}-2y}$

69. $\dfrac{\sqrt{a}+\sqrt{b}}{\sqrt{a}-\sqrt{b}}$

70. $\dfrac{\sqrt{a}-\sqrt{2b}}{\sqrt{a}+\sqrt{2b}}$

71. $\dfrac{\sqrt{2x}-\sqrt{y}}{\sqrt{2x}+\sqrt{y}}$

72. $\dfrac{\sqrt{4a}+\sqrt{4b}}{\sqrt{a}-\sqrt{b}}$

73. $\dfrac{1}{\sqrt{b+1}+\sqrt{b}}$

74. $\dfrac{1}{\sqrt{z+1}-\sqrt{z}}$

GEOMETRY

75. *Perimeter and Area*   Find the *exact* perimeter and area of the rectangle shown. Simplify your answers.

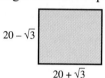

$\sqrt{91}-2$

$\sqrt{91}+2$

76. *Perimeter and Area*   Find the *exact* perimeter and area of the rectangle shown. Simplify your answers.

$20-\sqrt{3}$

$20+\sqrt{3}$

APPLICATIONS

Exercises 77 and 78: The time T in seconds for a pendulum to swing back and forth once can be calculated by

$$T = 2\pi\sqrt{\dfrac{L}{32}},$$

where L equals the length of the pendulum in feet.

77. Find T when $L = 2$ feet.

78. Rationalize the denominator in the expression for T.

WRITING ABOUT MATHEMATICS

79. Suppose that a student knows that $\sqrt{2} \approx 1.4142136$ and does not have a calculator. Which expression, $\dfrac{1}{\sqrt{2}}$ or $\dfrac{\sqrt{2}}{2}$, would be easier to evaluate by hand? Why?

80. Explain how to find the conjugate of a radical expression that has two terms. Give examples.

CHECKING BASIC CONCEPTS
SECTIONS 8.3 AND 8.4

1. Simplify each expression.
   (a) $5\sqrt{6} + 5\sqrt{6}$
   (b) $3\sqrt{75} - 2\sqrt{27}$
   (c) $\sqrt{16k} + \sqrt{25k}$

2. Find the *exact* perimeter of the rectangle shown in the figure. Simplify your answer.

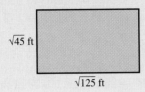

$\sqrt{45}$ ft

$\sqrt{125}$ ft

3. Multiply each expression.
   (a) $(5 + \sqrt{2})(5 - \sqrt{2})$
   (b) $(\sqrt{x} - 3)(\sqrt{x} + 5)$

4. Rationalize the denominator in each expression.
   (a) $\dfrac{5}{3\sqrt{5}}$

   (b) $\dfrac{3 + \sqrt{7}}{3 - \sqrt{7}}$

   (c) $\dfrac{1}{\sqrt{x} + \sqrt{y}}$

8.5 EQUATIONS INVOLVING RADICAL EXPRESSIONS

Solving Radical Equations ■ Solving an Equation for a Variable ■ Numerical and Graphical Methods (Optional)

A LOOK INTO MATH ▷

Because of Earth's curvature, a person can see only a limited distance to the horizon. The higher a person's elevation, the farther the person can see. The formula

$$D = 1.22\sqrt{h}$$

can be used to estimate the distance D in miles to the horizon, where h is the height in feet of the person's eyes above level ground. See Example 1. This formula has a variable as a radicand and therefore is called a **radical equation**. To determine the elevation necessary for a person to see 20 miles to the horizon, we substitute **20** for D and solve the equation

$$20 = 1.22\sqrt{h}.$$

(*Source:* F. Mannering and W. Kilareski, *Principles of Highway Engineering and Traffic Analysis.*)

In this section we discuss the basics of solving radical equations.

Solving Radical Equations

To solve an equation containing a square root, we often isolate the square root and then square each side of the equation. For example, to solve $2\sqrt{x} = 8$, we begin by dividing each side by 2 to isolate $\sqrt{x}$. Then we square each side.

$2\sqrt{x} = 8$	Given equation
$\sqrt{x} = 4$	Divide each side by 2.
$(\sqrt{x})^2 = 4^2$	Square each side.
$x = 16$	$(\sqrt{x})^2 = x$ when $x \geq 0$.

To check this answer, we substitute **16** for x in the given equation $2\sqrt{x} = 8$.

$$2\sqrt{16} \stackrel{?}{=} 8 \qquad \text{Let } x = 16.$$

$$8 = 8 \qquad \text{The answer checks.}$$

To solve radical equations having square roots, we apply the **squaring property for solving equations**.

SQUARING PROPERTY FOR SOLVING EQUATIONS

If each side of an equation is squared, then any solutions to the given equation are among the solutions to the new equation. That is, the solutions to the equation $a = b$ are among the solutions to $a^2 = b^2$.

We *must check our answers* when applying this squaring property. For example, consider the equation $2x = 2$, whose only solution is 1. If we square each side of the equation (which is not necessary to solve the equation), we obtain the new equation $4x^2 = 4$, whose solutions are 1 or -1. Here, 1 is a solution to the given equation $2x = 2$ and to the new equation, but -1 is an **extraneous solution** that satisfies the new equation but *not* the given equation.

▶ **REAL-WORLD CONNECTION**   In the next example we solve the equation presented at the beginning of this section to find out how high in feet a person's eyes must be to see 20 miles to the horizon.

EXAMPLE 1   Seeing the horizon

Solve the equation $20 = 1.22\sqrt{h}$ to find the elevation in feet necessary for a person to see 20 miles to the horizon.

Solution
To solve the equation $20 = 1.22\sqrt{h}$ for h, we begin by dividing each side by 1.22. Then we square each side of the equation.

$$20 = 1.22\sqrt{h} \qquad \text{Given equation}$$

$$\frac{20}{1.22} = \sqrt{h} \qquad \text{Divide each side by 1.22.}$$

$$\left(\frac{20}{1.22}\right)^2 = (\sqrt{h})^2 \qquad \text{Square each side.}$$

$$\left(\frac{20}{1.22}\right)^2 = h \qquad (\sqrt{h})^2 = h \text{ when } h \geq 0.$$

$$h \approx 269 \qquad \text{Approximate; rewrite.}$$

To see 20 miles to the horizon, a person needs to be elevated about 269 feet.

Check: To check this answer, we substitute **269** for h in the given equation.

$$20 \stackrel{?}{=} 1.22\sqrt{269} \qquad \text{Let } h = 269.$$

$$20 \approx 20.01 \qquad \text{It checks (approximately).}$$

Now Try Exercise **69**

EXAMPLE 2   Solving a radical equation

Solve $\sqrt{x - 1} = 2$.

Solution
Start by squaring each side of the given equation.

$$\sqrt{x - 1} = 2 \qquad \text{Given equation}$$
$$(\sqrt{x - 1})^2 = 2^2 \qquad \text{Square each side.}$$
$$x - 1 = 4 \qquad \text{Simplify.}$$
$$x = 5 \qquad \text{Add 1 to each side.}$$

Check: To be certain that **5** is a solution, we check it in the given equation.

$$\sqrt{5 - 1} \stackrel{?}{=} 2 \qquad \text{Let } x = 5.$$
$$\sqrt{4} \stackrel{?}{=} 2 \qquad \text{Subtract.}$$
$$2 = 2 \qquad \text{The answer checks.} \qquad \boxed{\text{Now Try Exercise } 11}$$

The following gives basic steps for solving a radical equation having a square root.

SOLVING AN EQUATION HAVING A SQUARE ROOT

STEP 1: Isolate the square root term on one side of the equation.

STEP 2: Apply the squaring property.

STEP 3: Solve the new equation.

STEP 4: Check your answers by substituting each result in the *given* equation.

The next example illustrates the importance of checking solutions.

EXAMPLE 3   Solving a radical equation

Solve $\sqrt{x + 2} = x$.

Solution
Because the square root is already isolated on the left side of the equation, Step 1 is not necessary. We begin by squaring each side of the given equation.

$$\sqrt{x + 2} = x \qquad \text{Given equation}$$
$$(\sqrt{x + 2})^2 = x^2 \qquad \text{Square each side (Step 2).}$$
$$x + 2 = x^2 \qquad \text{Simplify (Step 3).}$$
$$x^2 - x - 2 = 0 \qquad \text{Rewrite the equation.}$$
$$(x + 1)(x - 2) = 0 \qquad \text{Factor the trinomial.}$$
$$x = -1 \quad \text{or} \quad x = 2 \qquad \text{Solve.}$$

Check:

$$\sqrt{-1 + 2} \stackrel{?}{=} -1 \qquad \sqrt{2 + 2} \stackrel{?}{=} 2 \qquad \text{Let } x = -1 \text{ and } x = 2 \text{ (Step 4).}$$
$$\sqrt{1} \stackrel{?}{=} -1 \qquad \sqrt{4} \stackrel{?}{=} 2 \qquad \text{Add.}$$
$$1 \neq -1 \qquad 2 = 2 \qquad \text{Simplify the square root.}$$

Thus 2 is a solution to the *given* equation, but -1 is not a solution to the given equation; rather, it is an extraneous solution. $\qquad \boxed{\text{Now Try Exercise } 37}$

Sometimes the square root term must be isolated before we can apply the square root property, as illustrated in the next two examples.

EXAMPLE 4   Isolating the square root term

Solve $\sqrt{2 - x} + 1 = 5$.

Solution

Before we can square each side, we must isolate the square root term by subtracting 1 from each side.

$\sqrt{2 - x} + 1 = 5$	Given equation
$\sqrt{2 - x} = 4$	Subtract 1 from each side (Step 1).
$(\sqrt{2 - x})^2 = 4^2$	Square each side (Step 2).
$2 - x = 16$	Simplify (Step 3).
$-x = 14$	Subtract 2 from each side.
$x = -14$	Multiply each side by -1.

CRITICAL THINKING

How would you solve $2\sqrt[3]{x} = 6$? Find any solutions to this equation.

Check: To check that -14 is a solution, we substitute -14 for x in the given equation.

$\sqrt{2 - (-14)} + 1 \stackrel{?}{=} 5$	Let $x = -14$. (Step 4).
$\sqrt{16} + 1 \stackrel{?}{=} 5$	Simplify.
$4 + 1 \stackrel{?}{=} 5$	Evaluate the square root.
$5 = 5$	The answer checks. **Now Try Exercise 29**

EXAMPLE 5   Isolating the square root term

Solve $3\sqrt{x} + 6 = 3x$.

Solution

We begin by subtracting 6 from each side of the equation and then divide each side by 3 to isolate the square root.

$3\sqrt{x} + 6 = 3x$	Given equation
$3\sqrt{x} = 3x - 6$	Subtract 6 from each side (Step 1).
$\sqrt{x} = x - 2$	Divide each side by 3.
$(\sqrt{x})^2 = (x - 2)^2$	Square each side (Step 2).
$x = x^2 - 4x + 4$	Simplify (Step 3).
$0 = x^2 - 5x + 4$	Subtract x.
$0 = (x - 4)(x - 1)$	Factor.
$x = 1$ or $x = 4$	Solve for x.

Check: Let $x = 1$ and let $x = 4$ in the *given* equation (Step 4).

$$3\sqrt{1} + 6 \overset{?}{=} 3(1)$$

$$9 \neq 3 \quad \text{(Does not check)}$$

$$3\sqrt{4} + 6 \overset{?}{=} 3(4)$$

$$12 = 12 \quad \text{(Checks)}$$

The only solution is 4.

Now Try Exercise **33**

Solving an Equation for a Variable

▶ **REAL-WORLD CONNECTION**   In applications we often have to solve an equation for a variable. Sometimes these equations contain square roots. The next example illustrates this technique by using a problem from highway construction.

EXAMPLE **6**

Solving an equation for a variable

Figure 8.10

If a circular highway curve without any banking has a radius of r feet, then the speed limit S in miles per hour for the curve can be estimated by the equation $S = 1.5\sqrt{r}$. See Figure 8.10. (*Source:* N. Garber and L. Hoel, *Traffic and Highway Engineering*.)
(a) Solve this equation for r and then explain what it calculates.
(b) Find an appropriate radius for a curve that is to have a 30-mile-per-hour speed limit.

Solution
(a) Begin by dividing each side by 1.5. Then square each side of the resulting equation.

$$S = 1.5\sqrt{r} \qquad \text{Given equation}$$

$$\frac{S}{1.5} = \sqrt{r} \qquad \text{Divide each side by 1.5.}$$

$$\frac{S^2}{1.5^2} = (\sqrt{r})^2 \qquad \text{Square each side.}$$

$$r = \frac{S^2}{2.25} \qquad \text{Simplify; rewrite.}$$

This equation calculates a safe radius r for a curve with a speed limit S.
(b) If $S = 30$, then $r = \frac{30^2}{2.25} = 400$. Thus a safe radius for a curve with a 30-mile-per-hour speed limit is 400 feet.

Now Try Exercise **72**

EXAMPLE **7**

Solving an equation for a variable

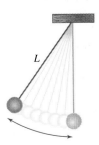

Figure 8.11

The amount of time required for a pendulum to swing back and forth once is called its *period*. The period P in seconds can be estimated with the formula

$$P = 2\pi\sqrt{\frac{L}{32}},$$

where L is the length of the pendulum in feet. See Figure 8.11.
(a) Solve this equation for L.
(b) If $P = 2$ seconds, find L.

Solution

(a) We begin by dividing each side of the formula by 2π.

$$P = 2\pi\sqrt{\dfrac{L}{32}} \qquad \text{Given formula}$$

$$\dfrac{P}{2\pi} = \sqrt{\dfrac{L}{32}} \qquad \text{Divide each side by } 2\pi.$$

$$\left(\dfrac{P}{2\pi}\right)^2 = \left(\sqrt{\dfrac{L}{32}}\right)^2 \qquad \text{Square each side.}$$

$$\dfrac{P^2}{4\pi^2} = \dfrac{L}{32} \qquad \text{Properties of exponents}$$

$$\dfrac{32P^2}{4\pi^2} = L \qquad \text{Multiply each side by 32.}$$

$$L = \dfrac{8P^2}{\pi^2} \qquad \text{Simplify; rewrite the equation.}$$

(b) If $P = 2$ seconds, then

$$L = \dfrac{8(2)^2}{\pi^2} = \dfrac{32}{\pi^2} \approx 3.24 \text{ feet.}$$

If a pendulum has a period of 2 seconds, then its length is about 3.24 feet.

Now Try Exercise 74

CRITICAL THINKING

If the length of a pendulum quadruples, what happens to the period P?

Numerical and Graphical Methods (Optional)

We can use tables and graphs to solve a radical equation. For example, suppose that we want to solve the equation

$$\sqrt{x - 1} = 2.$$

We can start by making a table of values for $y = \sqrt{x - 1}$, as shown in Table 8.2. The table reveals that $\sqrt{x - 1} = 2$ when $x = 5$. Thus 5 is the solution to the equation. (If a solution to an equation is *not* a convenient value, such as a fraction or irrational number, then finding the solution with a table of values may not be practical.)

TABLE 8.2

x	1	2	5	10
$\sqrt{x-1}$	0	1	2	3

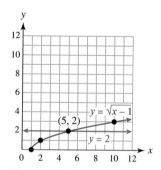

Figure 8.12

To solve this equation graphically, we start by plotting the points in Table 8.2 and connecting them with a smooth curve to obtain the graph of $y = \sqrt{x - 1}$. We find the solution to the equation by determining the x-coordinate where this graph intersects the line $y = 2$, as shown in Figure 8.12. The graphs intersect at $(5, 2)$ so the solution is 5.

TECHNOLOGY NOTE: Tables and Graphs

Graphing calculators can generate tables and graphs for solving radical equations. Calculator figures for Table 8.2 and Figure 8.12 are shown. Table 8.2 was created by using the "Ask" feature.

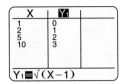
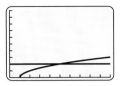

8.5 PUTTING IT ALL TOGETHER

The following table summarizes important topics from this section.

Concept	Explanation	Examples
Squaring Property for Solving Equations	The solutions to the equation $a = b$ are among the solutions to the equation $a^2 = b^2$.	$\sqrt{x+1} = 3$ $(\sqrt{x+1})^2 = 3^2$ $x + 1 = 9$ $x = 8$ *Be sure to check your answer.*
Steps for Solving an Equation Having a Square Root	**STEP 1:** Isolate the square root on one side of the given equation. **STEP 2:** Apply the squaring property. **STEP 3:** Solve the new equation. **STEP 4:** Check each answer by substituting it in the *given* equation.	Given equation: $\sqrt{2x} - 3 = 5$ **STEP 1:** $\sqrt{2x} = 8$ **STEP 2:** $2x = 64$ **STEP 3:** $x = 32$ **STEP 4:** $\sqrt{2(32)} - 3 \overset{?}{=} 5$ $\sqrt{64} - 3 \overset{?}{=} 5$ $5 = 5$ It checks.

8.5 Exercises

MyMathLab   Math XL PRACTICE   WATCH   DOWNLOAD   READ   REVIEW

CONCEPTS

1. What is a good first step in solving the equation $\sqrt{x} + 5 = 4$?

2. When you solve an equation by squaring each side, what must you do with any answers?

3. Find any solutions to $\sqrt{x} = 4$.

4. Is 8 a solution to $\sqrt{2x} - 1 = 3$?

5. Do the equations $x = 4$ and $x^2 = 16$ have the same solution set?

6. If $a = \sqrt{b}$, then $b =$ _____.

SOLVING RADICAL EQUATIONS

Exercises 7–46: Solve and check your answers.

7. $\sqrt{x} = 6$ 　　　　　　**8.** $\sqrt{z} = 3$

9. $\sqrt{t} = -4$ 　　　　　**10.** $\sqrt{r} = -9$

11. $\sqrt{x - 7} = 3$ 　　　**12.** $\sqrt{x + 10} = 8$

13. $\sqrt{1 - k} = 10$ 　　　**14.** $\sqrt{5 - k} = 5$

15. $\sqrt{2t - 3} = 4$ 　　　**16.** $\sqrt{3t + 1} = 7$

17. $\sqrt{3x + 1} = 5$ 　　　**18.** $\sqrt{8x - 2} = 6$

19. $9 - \sqrt{x} = 1$ 　　　**20.** $4 + \sqrt{2x} = 12$

21. $\sqrt{b + 1} = 2\sqrt{b}$ 　　**22.** $\sqrt{2b - 1} = \sqrt{b}$

23. $\sqrt{4z + 2} = \sqrt{3z}$

24. $\sqrt{7z + 3} = 2\sqrt{z - 1}$

25. $3\sqrt{x} = 12$ 　　　　**26.** $5\sqrt{x} = 10$

27. $-2\sqrt{x} = 4$ 　　　　**28.** $4\sqrt{x} = -36$

29. $\sqrt{x - 2} + 1 = 5$ 　　**30.** $\sqrt{4 - x} - 2 = 1$

31. $2\sqrt{x - 1} - 3 = 1$ 　　**32.** $4\sqrt{x + 1} + 1 = 5$

33. $3\sqrt{3x + 1} + x = 7x$ 　**34.** $\sqrt{2x - 1} + x = 2x$

35. $2\sqrt{x + 1} + 3 = 9$ 　　**36.** $7\sqrt{1 - 2x} - 1 = 20$

37. $\sqrt{x + 20} = x$ 　　　**38.** $\sqrt{2 - x} = x$

39. $5\sqrt{2t - 1} = 3t$ 　　　**40.** $\sqrt{3t + 4} = t$

41. $\sqrt{2x + 1} = x - 1$ 　　**42.** $\sqrt{4x + 4} = \frac{1}{2}x + 2$

43. $z + 1 = \sqrt{37 + 2z}$ 　**44.** $z - 5 = \sqrt{2z + 5}$

45. $\sqrt{(x - 1)(x + 2)} = 2$

46. $\sqrt{(x + 3)(x - 2)} = 6$

Exercises 47–56: Solve the equation for the specified variable. Assume that all variables are positive.

47. $T = \sqrt{L}$ for L 　　　**48.** $P = \sqrt{\frac{k}{5}}$ for k

49. $b = \sqrt{2a}$ for a 　　　**50.** $R = \sqrt{5b}$ for b

51. $D = 4\sqrt{t + 1}$ for t 　　**52.** $t = \sqrt{2r - 1}$ for r

53. $c = \sqrt{a^2 + b^2}$ for b 　**54.** $Z = \sqrt{R^2 + C^2}$ for R

55. $k = 4 + \sqrt{d}$ for d 　　**56.** $y = 5 - \sqrt{3t}$ for t

NUMERICAL AND GRAPHICAL SOLUTIONS

Exercises 57–64: Do the following.
　(a) *Use a table of values to solve the equation.*
　(b) *Use a graph to solve the equation.*

57. $\sqrt{x} = 2$ 　　　　　**58.** $\sqrt{x + 1} = 3$

59. $\sqrt{x - 2} = 1$ 　　　**60.** $\sqrt{x + 2} = 3$

61. $\sqrt{x + 1} = 3$ 　　　**62.** $\sqrt{x - 2} = 0$

63. $2\sqrt{x} = 2$ 　　　　**64.** $\sqrt{2x} = 2$

Exercises 65–68: Solve the equation to two decimal places either graphically or numerically.

65. $\sqrt{x - 1.1} + \pi x = 8$

66. $\sqrt{x - 1} + 0.4x = \sqrt{2x + \pi}$

67. $\sqrt{x + 3.2} + 0.1x = -x^2 + 4$

68. $\sqrt{x - 2.2} = x + 2$

APPLICATIONS

69. *Distance to the Horizon*　(Refer to Example 1.) The formula $D = 1.22\sqrt{h}$ can be used to calculate the distance in miles to the horizon for a person whose eyes are h feet above level ground.
　(a) If a person is 100 feet above level ground, how far away is the horizon?
　(b) If the horizon is 50 miles away, how far above level ground is the person?
　(c) If a person's elevation is increased by a factor of 4, how much farther away is the horizon?

70. *A Round Earth*　(Refer to the preceding exercise.) When tall sailing ships were built, many people realized that the earth was round, not flat, because a person could see land from the top of a tall mast sooner than a person standing on the deck. Suppose that the mast of a ship is 150 feet high and that the deck of the ship is 50 feet high. If the ship is traveling at 6 miles per hour, estimate how much sooner a person at the top of the mast could see land than a person standing on the deck. (Ignore the height of each person.)

71. *Right Triangle*　Suppose that a right triangle has legs a and b with hypotenuse c.
　(a) Given a and c, determine a formula for b.
　(b) Find b if $a = 7$ inches and $c = 25$ inches.

72. *Highway Curve*   (Refer to Example 6.) If a circular highway curve without any banking has a radius of r feet, then the speed limit S in miles per hour can be estimated by $S = 1.5\sqrt{r}$.   (**Source:** N. Garber.)
   (a) Use the formula to find a speed limit for a curve with a 900-foot radius.
   (b) Use the formula to find a safe radius for a curve with a 60-mile-per-hour speed limit.

73. *Skid Marks*   Vehicles involved in accidents often leave skid marks. To determine how fast a vehicle was traveling, officials often use a test vehicle to compare skid marks on the same section of road. Suppose that a vehicle in a crash left skid marks D feet long and that a test vehicle traveling at v miles per hour leaves skid marks d feet long. Then the speed V of the vehicle involved in the crash is given by

$$V = v\sqrt{\frac{D}{d}}.$$

(**Source:** N. Garber.)
   (a) Solve this formula for D.
   (b) Find D when $V = 50$ miles per hour, $v = 40$ miles per hour, and $d = 200$ feet.

74. *Pendulum*   (Refer to Example 7.) The period P in seconds of a pendulum that is L feet long is given by

$$P = 2\pi\sqrt{\frac{L}{32}}.$$

(a) If a pendulum is 5 feet long, find its period.
(b) Find L if the period is 1 second.
(c) Rewrite this formula so that it calculates the period in seconds when the length L is in inches.

Exercises 75–78: **Design of Open Channels**   *(Refer to the introduction to this chapter.) To protect cities from flooding during heavy rains, open channels are sometimes constructed to handle runoff. The rate R at which water flows through the channel can be estimated by the formula $R = k\sqrt{m}$, where m is the slope of the channel and k is a constant determined by the shape of the channel.*

75. If the slope m of an open channel increases by a factor of 4, by what factor does R increase?

76. Suppose that an open channel has slope $m = 0.04$ (or 4%) and runoff rate $R = 400$ cubic feet per second. Find k for this channel.

77. If the slope in the channel in Exercise 76 increases to $m = 0.09$, what happens to R?

78. If $k = 3400$ and $R = 340$ cubic feet per second, find the slope m.

WRITING ABOUT MATHEMATICS

79. Explain what the squaring property is used for. What must be done with any answers after the squaring property has been applied?

80. Explain the four steps for solving an equation containing a square root. Give an example.

GROUP ACTIVITY
WORKING WITH REAL DATA

Directions: Form a group of 2 to 4 people. Select someone to record the group's responses for this activity. All members of the group should work cooperatively to answer the questions. If your instructor asks for your results, each member of the group should be prepared to respond.

1. *Throwing a Stone to the Moon*   How fast would you need to throw a stone to have it reach the moon? Using square roots, we can calculate this *escape velocity*. Earth's escape velocity V_E is the *minimum velocity* necessary for an object, such as a stone or satellite, to leave Earth's gravity and continue into space. This velocity in miles per hour can be calculated with the formula

$$V_E = k\sqrt{\frac{M}{R}},$$

where $M = 1.3 \times 10^{25}$ pounds is the mass (weight) of Earth, $R = 4000$ miles is the radius of Earth, and $k = 4.4 \times 10^{-7}$ is a universal constant.   (**Source:** R. Weidner.)
   (a) Calculate V_E for Earth.
   (b) Do you think you could throw a stone that fast?

2. *Throwing a Stone to Earth*   (Refer to the preceding exercise.) Standing on the moon, how fast would you need to throw a stone to have it reach Earth? We can

continued on next page

continued from previous page

use the same formula to answer this question. Note that for the moon, $M = 1.6 \times 10^{23}$ pounds and $R = 1070$ miles.

(a) Use the formula to calculate the escape velocity V_M for the moon.

(b) How do V_E and V_M compare? Does this result agree with your intuition?

3. *Orbiting Earth and the Moon* (Refer to the two preceding exercises.) If you wanted a stone to go into a circular orbit, the necessary velocity would be less than the escape velocity, or equal to the escape velocity divided by $\sqrt{2}$.

(a) Find the velocity needed for a circular orbit around Earth.

(b) Find the velocity needed for a circular orbit around the moon.

4. *Mass of Mars* (Refer to Exercise 1.) The escape velocity for Mars is about 11,400 miles per hour, and the diameter of Mars is about 4200 miles. Use the formula $V = k\sqrt{\dfrac{M}{R}}$ to calculate the mass M of Mars.

8.6 HIGHER ROOTS AND RATIONAL EXPONENTS

Higher Roots ■ Rational Exponents

A LOOK INTO MATH ▷

Allometry is an area of biology that studies the relative sizes of different characteristics of an animal. For example, biologists have found that heavier birds tend to have wings with larger areas than lighter birds do. For some species of birds, the weight W of a bird in pounds is related to the area of its wings A in square inches by the formula

$$A = 100W^{2/3}.$$

Note that the exponent is a fraction, or rational number. Rational exponents can be used to denote square and cube roots. Powers of higher roots can also be written as rational exponents. In this section we discuss higher roots, rational exponents, and some of their applications. (*Source:* C. Pennycuick, *Newton Rules Biology.*)

Higher Roots

We can generalize square roots and cube roots to define *n*th roots of a number a. The number b is an **nth root** of a if $b^n = a$, where n is a positive integer. The **principal nth root** of a is denoted $\sqrt[n]{a}$, where n is called the **index**. For a square root the index is understood to be 2, and we usually write $\sqrt{a}$ rather than $\sqrt[2]{a}$. An odd n signifies an **odd root**, and an even n signifies an **even root**. The cube root $\sqrt[3]{a}$ is an example of an odd root, and the square root $\sqrt{a}$ is an example of an even root. When n is odd, there is exactly one real *n*th root and $\sqrt[n]{a}$ denotes that root. When n is even and a is positive, there are two real *n*th roots, one positive and one negative, and $\sqrt[n]{a}$ denotes the positive root.

NOTE: An odd root of a negative number is a negative number, but the even root of a negative number is *not* a real number and is left undefined in this section.

EXAMPLE 1   Evaluating *n*th roots

Find each root, if possible.

(a) $\sqrt[3]{-8}$ (b) $\sqrt[4]{81}$ (c) $\sqrt[5]{32}$ (d) $\sqrt[4]{-64}$

Solution

(a) $\sqrt[3]{-8} = -2$ because $(-2)^3 = (-2)(-2)(-2) = -8$.

(b) $\sqrt[4]{81} = 3$ because $3^4 = 3 \cdot 3 \cdot 3 \cdot 3 = 81$.

(c) $\sqrt[5]{32} = 2$ because $2^5 = 2 \cdot 2 \cdot 2 \cdot 2 \cdot 2 = 32$.

(d) $\sqrt[4]{-64}$ is *not* a real number because there is no real number b such that $b^4 = -64$. That is, the even root of a negative number is not a real number.

Now Try Exercises 11, 13, 15, 31

The product and quotient rules for square roots also apply to higher roots.

PRODUCT AND QUOTIENT RULES FOR RADICAL EXPRESSIONS

Let a and b be real numbers, where $\sqrt[n]{a}$ and $\sqrt[n]{b}$ are both defined. Then

$$\sqrt[n]{a} \cdot \sqrt[n]{b} = \sqrt[n]{a \cdot b} \qquad \text{Product rule}$$

and

$$\sqrt[n]{\frac{a}{b}} = \frac{\sqrt[n]{a}}{\sqrt[n]{b}}. \qquad \text{Quotient rule } (b \neq 0)$$

In the next two examples we use the product rule and the quotient rule to simplify several radical expressions.

EXAMPLE 2 **Finding products of higher roots**

Simplify each product.

(a) $\sqrt[3]{-3} \cdot \sqrt[3]{9}$ (b) $\sqrt[4]{4} \cdot \sqrt[4]{4}$ (c) $\sqrt[5]{27} \cdot \sqrt[5]{-9}$ (d) $\sqrt[6]{\frac{2}{3}} \cdot \sqrt[6]{\frac{3}{2}}$

Solution

(a) $\sqrt[3]{-3} \cdot \sqrt[3]{9} = \sqrt[3]{-3 \cdot 9} = \sqrt[3]{-27} = -3$ because $(-3)^3 = -27$.

(b) $\sqrt[4]{4} \cdot \sqrt[4]{4} = \sqrt[4]{4 \cdot 4} = \sqrt[4]{16} = 2$ because $2^4 = 16$.

(c) $\sqrt[5]{27} \cdot \sqrt[5]{-9} = \sqrt[5]{27(-9)} = \sqrt[5]{-243} = -3$ because $(-3)^5 = -243$.

(d) $\sqrt[6]{\frac{2}{3}} \cdot \sqrt[6]{\frac{3}{2}} = \sqrt[6]{\frac{2}{3} \cdot \frac{3}{2}} = \sqrt[6]{1} = 1$ because $1^6 = 1$.   Now Try Exercises 37, 39, 41, 43

EXAMPLE 3 **Finding quotients of higher roots**

Simplify each quotient.

(a) $\sqrt[3]{\frac{27}{64}}$ (b) $\sqrt[4]{\frac{14}{81}}$ (c) $\frac{\sqrt[3]{-32}}{\sqrt[3]{4}}$ (d) $\frac{\sqrt[5]{64}}{\sqrt[5]{2}}$

Solution

(a) $\sqrt[3]{\frac{27}{64}} = \frac{\sqrt[3]{27}}{\sqrt[3]{64}} = \frac{3}{4}$ because $3^3 = 27$ and $4^3 = 64$.

(b) $\sqrt[4]{\frac{14}{81}} = \frac{\sqrt[4]{14}}{\sqrt[4]{81}} = \frac{\sqrt[4]{14}}{3}$ because $3^4 = 81$.

(c) $\dfrac{\sqrt[3]{-32}}{\sqrt[3]{4}} = \sqrt[3]{-\dfrac{32}{4}} = \sqrt[3]{-8} = -2$ because $(-2)^3 = -8$.

(d) $\dfrac{\sqrt[5]{64}}{\sqrt[5]{2}} = \sqrt[5]{\dfrac{64}{2}} = \sqrt[5]{32} = 2$ because $2^5 = 32$.

Now Try Exercises 33, 35, 45, 49

Rational Exponents

When m and n are integers, the product rule states that $a^m \cdot a^n = a^{m+n}$. This rule can be extended to include exponents that are fractions. For example,

$$9^{1/2} \cdot 9^{1/2} = 9^{1/2+1/2} = 9^1 = 9.$$

If we multiply $9^{1/2}$ by itself, the result is 9. Because we know that $\sqrt{9} \cdot \sqrt{9} = 3 \cdot 3 = 9$, this discussion suggests that $9^{1/2} = \sqrt{9}$ and leads to the following definition.

THE EXPRESSION $a^{1/n}$

If n is an integer greater than 1, then

$$a^{1/n} = \sqrt[n]{a}.$$

If $a < 0$ and n is an even positive integer, then $a^{1/n}$ is not a real number.

EXAMPLE 4   Interpreting rational exponents

Write each expression in radical notation and then evaluate.
(a) $25^{1/2}$ **(b)** $27^{1/3}$ **(c)** $16^{1/4}$

Solution
(a) The exponent $\frac{1}{2}$ indicates a square root, so $25^{1/2} = \sqrt{25} = 5$.
(b) The exponent $\frac{1}{3}$ indicates a cube root, so $27^{1/3} = \sqrt[3]{27} = 3$.
(c) The exponent $\frac{1}{4}$ indicates a fourth root, so $16^{1/4} = \sqrt[4]{16} = 2$.

Now Try Exercises 51, 54, 57

Now suppose that we want to define the expression $8^{2/3}$. On the one hand, using properties of exponents, we have

$$8^{1/3} \cdot 8^{1/3} = 8^{1/3+1/3} = 8^{2/3}.$$

On the other hand, we have

$$8^{1/3} \cdot 8^{1/3} = \sqrt[3]{8} \cdot \sqrt[3]{8} = 2 \cdot 2 = 4.$$

Thus $8^{2/3} = 4$, and that value is obtained whether we interpret $8^{2/3}$ as either

$$8^{2/3} = \left(8^{1/3}\right)^2 = \left(\sqrt[3]{8}\right)^2 = 2^2 = 4$$

or

$$8^{2/3} = (8^2)^{1/3} = \sqrt[3]{8^2} = \sqrt[3]{64} = 4.$$

That is, $8^{2/3}$ indicates that either we take the **cube root** of 8 and then **square** it or we **square** 8 and then take the **cube root**. The result is 4 in both cases.

This discussion supports the following definition.

THE EXPRESSION $a^{m/n}$

If m and n are positive integers with $\frac{m}{n}$ in lowest terms, then

$$a^{m/n} = \sqrt[n]{a^m} = \left(\sqrt[n]{a}\right)^m.$$

If $a < 0$ and n is an even positive integer, then $a^{m/n}$ is not a real number.

NOTE: When evaluating $a^{m/n}$ by hand, it is often easier to take the nth root of a first and then raise it to the mth power, rather than the reverse.

EXAMPLE 5   Interpreting rational exponents

Write each expression in radical notation and then evaluate.
(a) $27^{2/3}$ **(b)** $(-32)^{3/5}$ **(c)** $16^{3/4}$

Solution
(a) The exponent $\frac{2}{3}$ indicates that we either take the cube root of 27 and then square the result or that we square 27 and then take the cube root. In either case, the result will be the same. Taking the cube root and then squaring results in

$$27^{2/3} = \left(\sqrt[3]{27}\right)^2 = 3^2 = 9.$$

(b) The exponent $\frac{3}{5}$ indicates that we either take the fifth root of -32 and then cube the result or that we cube -32 and then take the fifth root. Taking the fifth root and then cubing results in

$$(-32)^{3/5} = \left(\sqrt[5]{-32}\right)^3 = (-2)^3 = -8.$$

(c) The exponent $\frac{3}{4}$ indicates that we either take the fourth root of 16 and then cube the result or that we cube 16 and then take the fourth root. Taking the fourth root and then cubing results in

$$16^{3/4} = \left(\sqrt[4]{16}\right)^3 = (2)^3 = 8.$$

Now Try Exercises 63, 65, 67

▶ **REAL-WORLD CONNECTION**   In the next example we use rational exponents in an application that models the relationship between a bird's weight and its wing size.

EXAMPLE 6   Modeling a bird's wing size

Heavier birds tend to have larger wings than lighter birds do. For some birds the relationship between the surface area A of the bird's wings in square inches and its weight W in pounds is given by the formula $A = 100W^{2/3}$.   (**Source:** C. Pennycuick, *Newton Rules Biology*.)
(a) Use radical notation to write this formula.
(b) Use your formula to estimate A for a bird that weighs 8 pounds.

Solution

(a) The exponent $\frac{2}{3}$ in $A = 100W^{2/3}$ indicates that we take the cube root of W and then square the result or that we square W and then take the cube root. Thus in radical notation

$$A = 100\left(\sqrt[3]{W}\right)^2 \quad \text{or} \quad A = 100\sqrt[3]{W^2}.$$

(b) Let $W = 8$, substitute it in $A = 100(\sqrt[3]{W})^2$ and simplify the result.

$$A = 100\left(\sqrt[3]{8}\right)^2 = 100(2)^2 = 400 \text{ square inches}$$

Now Try Exercise 77

From properties of exponents we know that $a^{-n} = \frac{1}{a^n}$, where n is a positive integer. We now define this property for negative rational exponents.

THE EXPRESSION $a^{-m/n}$

If m and n are positive integers with $\frac{m}{n}$ in lowest terms, then

$$a^{-m/n} = \frac{1}{a^{m/n}}, \quad a \neq 0.$$

EXAMPLE 7   **Interpreting rational exponents**

Write each expression in radical notation and then evaluate.
(a) $27^{-1/3}$ **(b)** $81^{-3/4}$ **(c)** $(-8)^{-2/3}$

Solution

(a) $27^{-1/3} = \dfrac{1}{27^{1/3}} = \dfrac{1}{\sqrt[3]{27}} = \dfrac{1}{3}$

(b) $(81)^{-3/4} = \dfrac{1}{81^{3/4}} = \dfrac{1}{\left(\sqrt[4]{81}\right)^3} = \dfrac{1}{3^3} = \dfrac{1}{27}$

(c) $(-8)^{-2/3} = \dfrac{1}{(-8)^{2/3}} = \dfrac{1}{\left(\sqrt[3]{-8}\right)^2} = \dfrac{1}{(-2)^2} = \dfrac{1}{4}$

Now Try Exercises 69, 71, 73

MAKING CONNECTIONS

Rules for Radicals and Rational Exponents

The rules for radicals are a result of the properties of rational exponents.

$$\sqrt[n]{a \cdot b} = \sqrt[n]{a} \cdot \sqrt[n]{b} \quad \text{is equivalent to} \quad (a \cdot b)^{1/n} = a^{1/n} \cdot b^{1/n}.$$

$$\sqrt[n]{\frac{a}{b}} = \frac{\sqrt[n]{a}}{\sqrt[n]{b}} \quad \text{is equivalent to} \quad \left(\frac{a}{b}\right)^{1/n} = \frac{a^{1/n}}{b^{1/n}}.$$

8.6 PUTTING IT ALL TOGETHER

The following table summarizes important features of higher roots and rational exponents.

Concept	Explanation	Examples
nth Root	b is an nth root of a if $b^n = a$, where n is a positive integer. The principal nth root of a is denoted $\sqrt[n]{a}$.	$\sqrt[3]{-8} = -2$ because $(-2)^3 = -8$. $\sqrt[4]{81} = 3$ because $3^4 = 81$. $\sqrt[5]{243} = 3$ because $3^5 = 243$. $\sqrt[4]{-16}$ is undefined because $b^4 \geq 0$ for any real number b.
Product and Quotient Rules for Radical Expressions	Product rule: $$\sqrt[n]{a} \cdot \sqrt[n]{b} = \sqrt[n]{a \cdot b}$$ Quotient rule: $$\sqrt[n]{\frac{a}{b}} = \frac{\sqrt[n]{a}}{\sqrt[n]{b}}$$	$\sqrt[3]{-2} \cdot \sqrt[3]{4} = \sqrt[3]{-8} = -2$ $\sqrt[4]{3} \cdot \sqrt[4]{27} = \sqrt[4]{81} = 3$ $\dfrac{\sqrt[3]{128}}{\sqrt[3]{2}} = \sqrt[3]{\dfrac{128}{2}} = \sqrt[3]{64} = 4$ $\sqrt[4]{\dfrac{5}{16}} = \dfrac{\sqrt[4]{5}}{\sqrt[4]{16}} = \dfrac{\sqrt[4]{5}}{2}$
The Expression $a^{1/n}$	$a^{1/n} = \sqrt[n]{a}$, where n is a positive integer greater than 1.	$25^{1/2} = \sqrt{25} = 5$ $64^{1/6} = \sqrt[6]{64} = 2$ $(-64)^{1/3} = \sqrt[3]{-64} = -4$
The Expression $a^{m/n}$	$$a^{m/n} = \sqrt[n]{a^m}$$ $$= \left(\sqrt[n]{a}\right)^m,$$ where m and n are positive integers with $\frac{m}{n}$ written in lowest terms.	$8^{5/3} = \left(\sqrt[3]{8}\right)^5 = 2^5 = 32$ $81^{3/4} = \left(\sqrt[4]{81}\right)^3 = 3^3 = 27$ $(-32)^{4/5} = \left(\sqrt[5]{-32}\right)^4 = (-2)^4 = 16$
The Expression $a^{-m/n}$	$$a^{-m/n} = \frac{1}{a^{m/n}},$$ where m and n are positive integers with $\frac{m}{n}$ written in lowest terms and $a \neq 0$.	$8^{-5/3} = \dfrac{1}{8^{5/3}} = \dfrac{1}{2^5} = \dfrac{1}{32}$ $81^{-3/4} = \dfrac{1}{81^{3/4}} = \dfrac{1}{3^3} = \dfrac{1}{27}$ $(-32)^{-4/5} = \dfrac{1}{(-32)^{4/5}} = \dfrac{1}{(-2)^4} = \dfrac{1}{16}$

8.6 Exercises

MyMathLab

 Math XL PRACTICE   WATCH   DOWNLOAD   READ   REVIEW

CONCEPTS

1. For the expression $\sqrt[3]{8}$, the number 3 is the _____ and the number 8 is the _____.

2. $\sqrt[5]{32}$ is an example of an (odd/even) root.

3. $\sqrt[4]{19}$ is an example of an (odd/even) root.

4. Does $\sqrt[3]{3} = 1$? Explain your answer.

5. An odd root of a negative number is a _____.

6. An even root of a negative number is not a _____.

7. $\sqrt[n]{a} \cdot \sqrt[n]{b} =$ _____

8. $\sqrt[n]{\dfrac{a}{b}} =$ _____

9. $a^{1/n} =$ _____

10. $a^{m/n} =$ _____

SIMPLIFYING EXPRESSIONS

Exercises 11–50: Simplify the expression.

11. $\sqrt[3]{-27}$

12. $\sqrt[3]{1000}$

13. $\sqrt[5]{243}$

14. $\sqrt[5]{-1}$

15. $\sqrt[4]{16}$

16. $\sqrt[4]{81}$

17. $\sqrt[4]{10{,}000}$

18. $\sqrt[4]{625}$

19. $-\sqrt[5]{-32}$

20. $-\sqrt[5]{-243}$

21. $-\sqrt[6]{1}$

22. $\sqrt[6]{64}$

23. $\sqrt[4]{5^4}$

24. $-\sqrt[6]{7^6}$

25. $-\sqrt[4]{5^4}$

26. $\sqrt[6]{7^6}$

27. $\sqrt[3]{\dfrac{27}{8}}$

28. $\sqrt[3]{-\dfrac{8}{27}}$

29. $\sqrt[4]{\dfrac{20}{81}}$

30. $\sqrt[4]{\dfrac{3}{16}}$

31. $\sqrt[4]{-16}$

32. $\sqrt[6]{-1}$

33. $\sqrt[3]{\dfrac{8}{27}}$

34. $\sqrt[3]{\dfrac{64}{125}}$

35. $\sqrt[4]{\dfrac{7}{16}}$

36. $\sqrt[4]{\dfrac{5}{81}}$

37. $\sqrt[4]{\dfrac{4}{9}} \cdot \sqrt[4]{\dfrac{4}{9}}$

38. $\sqrt[6]{\dfrac{1}{4}} \cdot \sqrt[6]{\dfrac{1}{16}}$

39. $\sqrt[3]{16} \cdot \sqrt[3]{4}$

40. $\sqrt[3]{5} \cdot \sqrt[3]{25}$

41. $\sqrt[6]{2} \cdot \sqrt[6]{32}$

42. $\sqrt[4]{2} \cdot \sqrt[4]{8}$

43. $\sqrt[5]{16} \cdot \sqrt[5]{-2}$

44. $\sqrt[6]{4} \cdot \sqrt[6]{16}$

45. $\dfrac{\sqrt[3]{3}}{\sqrt[3]{81}}$

46. $\dfrac{\sqrt[3]{-80}}{\sqrt[3]{10}}$

47. $\dfrac{\sqrt[4]{2}}{\sqrt[4]{32}}$

48. $\dfrac{\sqrt[4]{2}}{\sqrt[4]{162}}$

49. $\dfrac{\sqrt[5]{-128}}{\sqrt[5]{4}}$

50. $\dfrac{\sqrt[5]{16}}{\sqrt[5]{0.5}}$

RATIONAL EXPONENTS

Exercises 51–76: Write the expression in radical notation and then evaluate it.

51. $36^{1/2}$

52. $100^{1/2}$

53. $27^{1/3}$

54. $8^{1/3}$

55. $(-1)^{1/3}$

56. $(-64)^{1/3}$

57. $81^{1/4}$

58. $256^{1/4}$

59. $32^{1/5}$

60. $(-1)^{1/5}$

61. $4^{3/2}$

62. $16^{5/2}$

63. $64^{2/3}$

64. $(-8)^{2/3}$

65. $16^{5/4}$

66. $81^{3/4}$

67. $(-1)^{3/5}$

68. $(-32)^{2/5}$

69. $49^{-1/2}$

70. $81^{-1/2}$

71. $(-27)^{-2/3}$

72. $(-64)^{-4/3}$

73. $625^{-3/4}$

74. $243^{-2/5}$

75. $16^{-5/4}$

76. $81^{-3/4}$

APPLICATIONS

77. **Weight and Wing Area** (Refer to Example 6.) The formula $A = 100\sqrt[3]{W^2}$ can be used to estimate the surface area in square inches of the wings of a bird that weighs W pounds. Estimate the surface area of the wings of a bird that weighs 27 pounds.

78. **Wing Span** Biologists have found that the weight W in kilograms of some types of birds and the length L in meters of their wing span are related by the formula $L = 0.91W^{1/3}$.
    (a) Use radical notation to write this formula.
    (b) Estimate the wing span of a 2-kilogram bird to two decimal places.
    (c) If a bird's weight doubles, by how much will the bird's wing span increase? Be specific.

L

79. *Radius of a Circle*   If a circle has area A, then the radius of the circle is given by

$$r = \sqrt{\frac{A}{\pi}}.$$

(a) Use rational exponents to write this formula.
(b) Find the radius of a circle that has an area of 9π square feet.

80. *Radius of a Cylinder*   If a cylinder has volume V and height h, then its radius r is given by

$$r = \sqrt{\frac{V}{\pi h}}.$$

(a) Use rational exponents to write this formula.
(b) Find the radius of a cylinder with $V = 100\pi$ cubic inches and $h = 16$ inches.

WRITING ABOUT MATHEMATICS

81. Between what two consecutive integers does $\sqrt[3]{10}$ lie? Explain your reasoning. Do not use a calculator.

82. The expression $27^{5/3}$ can be evaluated as either $\sqrt[3]{27^5}$ or $\left(\sqrt[3]{27}\right)^5$. Which of these radical expressions would you use if no calculator were available? Why?

CHECKING BASIC CONCEPTS
SECTIONS 8.5 AND 8.6

1. Solve each equation. Check your answer.
   (a) $\sqrt{3x + 4} - 2 = 2$
   (b) $\sqrt{24 - 2x} = -x$

2. Solve $S = 4\sqrt{t} - 2$ for t.

3. Evaluate each expression.
   (a) $\sqrt[3]{-27}$ **(b)** $\sqrt[4]{16}$

4. Simplify each expression.
   (a) $\sqrt[4]{9} \cdot \sqrt[4]{9}$ **(b)** $\sqrt[3]{\frac{9}{64}}$
   (c) $\dfrac{\sqrt[3]{40}}{\sqrt[3]{5}}$

5. Write each expression in radical notation and then evaluate.
   (a) $9^{1/2}$ **(b)** $16^{3/4}$ **(c)** $1000^{-1/3}$

CHAPTER 8   SUMMARY
SECTION 8.1 ■ INTRODUCTION TO RADICAL EXPRESSIONS

Square and Cube Roots

Square Root b is a square root of a if $b^2 = a$.

Principal Square Root $\sqrt{a} = b$ if $b^2 = a$ and $b \geq 0$.

   Examples: $\sqrt{25} = 5$ and $\pm\sqrt{4} = \pm 2$

Cube Root b is a cube root of a if $b^3 = a$.

   Examples: $\sqrt[3]{8} = 2$ and $\sqrt[3]{-64} = -4$

Perfect Square   A whole number a is a perfect square if an integer b exists such that $b^2 = a$.

Example: 49 is a perfect square because $7^2 = 49$.

Pythagorean Theorem   If a right triangle has legs a and b with hypotenuse c, then

$$a^2 + b^2 = c^2.$$

Example: If a right triangle has legs 3 and 4, then the hypotenuse equals

$$c = \sqrt{3^2 + 4^2} = \sqrt{9 + 16} = \sqrt{25} = 5.$$

The Distance Formula $d = \sqrt{(x_2 - x_1)^2 + (y_2 - y_1)^2}$

Example: The distance between $(-2, 1)$ and $(2, 4)$ is

$$d = \sqrt{\left(2 - (-2)\right)^2 + (4 - 1)^2} = \sqrt{16 + 9} = 5.$$

SECTION 8.2 ■ MULTIPLICATION AND DIVISION OF RADICAL EXPRESSIONS

Product Rule for Square Roots

$$\sqrt{a} \cdot \sqrt{b} = \sqrt{a \cdot b},$$

provided each expression is defined.

Example: $\sqrt{5} \cdot \sqrt{20} = \sqrt{5 \cdot 20} = \sqrt{100} = 10$

Simplifying Square Roots

STEP 1: Determine the largest perfect square factor of the radicand.

STEP 2: Use the product rule to factor out and simplify this perfect square.

Example: $\sqrt{75} = \sqrt{25 \cdot 3} = \sqrt{25} \cdot \sqrt{3} = 5\sqrt{3}$

Quotient Rule for Square Roots

$$\sqrt{\frac{a}{b}} = \frac{\sqrt{a}}{\sqrt{b}},$$

provided each expression is defined.

Examples: $\sqrt{\frac{7}{16}} = \frac{\sqrt{7}}{\sqrt{16}} = \frac{\sqrt{7}}{4}$ and $\frac{\sqrt{80}}{\sqrt{20}} = \sqrt{\frac{80}{20}} = \sqrt{4} = 2$

SECTION 8.3 ■ ADDITION AND SUBTRACTION OF RADICAL EXPRESSIONS

Addition and Subtraction   To add or subtract a radical expression, combine like radicals.

Examples: $4\sqrt{5} + 3\sqrt{5} = 7\sqrt{5}$ and $7\sqrt[3]{6} - 2\sqrt[3]{6} = 5\sqrt[3]{6}$

Like Radicals Sometimes unlike radicals can be written as like radicals so that they can be combined.

Example:
$$\sqrt{8} + \sqrt{18} = \sqrt{4 \cdot 2} + \sqrt{9 \cdot 2}$$
$$= \sqrt{4} \cdot \sqrt{2} + \sqrt{9} \cdot \sqrt{2}$$
$$= 2\sqrt{2} + 3\sqrt{2}$$
$$= 5\sqrt{2}$$

SECTION 8.4 ■ SIMPLIFYING RADICAL EXPRESSIONS

Multiplication Sometimes radical expressions can be multiplied as binomials are.

Examples:
$$(2 + \sqrt{5})(2 - \sqrt{5}) = (2)^2 - (\sqrt{5})^2 = 4 - 5 = -1$$
$$(3 + \sqrt{2})(2 - \sqrt{2}) = 6 - 3\sqrt{2} + 2\sqrt{2} - 2 = 4 - \sqrt{2}$$

Rationalizing the Denominator Multiply the numerator and denominator by 1 so that the radical no longer appears in the denominator.

Examples:
$$\frac{7}{\sqrt{7}} = \frac{7}{\sqrt{7}} \cdot \frac{\sqrt{7}}{\sqrt{7}} = \frac{7\sqrt{7}}{7} = \sqrt{7}$$
$$\frac{1}{2 + \sqrt{5}} = \frac{1}{2 + \sqrt{5}} \cdot \frac{2 - \sqrt{5}}{2 - \sqrt{5}} = \frac{2 - \sqrt{5}}{(2)^2 - (\sqrt{5})^2}$$
$$= \frac{2 - \sqrt{5}}{-1} = -2 + \sqrt{5}$$

SECTION 8.5 ■ EQUATIONS INVOLVING RADICAL EXPRESSIONS

Squaring Property for Solving Equations The solutions to $a = b$ are among the solutions to $a^2 = b^2$. (Be sure to check any answers when applying this property.)

Example: The solutions to $\sqrt{x + 2} = x$ are among the solutions to $x + 2 = x^2$.

Solving an Equation Containing a Square Root

STEP 1: Isolate the square root term on one side of the equation.

STEP 2: Apply the squaring property.

STEP 3: Solve the new equation.

STEP 4: Check your answers by substituting each result in the *given* equation.

Example:
$\sqrt{x + 2} + 1 = 5$	Given equation
$\sqrt{x + 2} = 4$	STEP 1: Subtract 1.
$x + 2 = 4^2$	STEP 2: Square each side.
$x = 14$	STEP 3: Subtract 2.

Check: $\sqrt{14 + 2} + 1 = 5$ STEP 4: Let $x = 14$; it checks.

Solving an Equation for a Variable   Equations and formulas are often solved for a variable.

Example: To solve $D = 1.22\sqrt{h}$ for h, start by dividing each side by 1.22.

$$\frac{D}{1.22} = \sqrt{h} \qquad \text{Divide each side by 1.22.}$$

$$\left(\frac{D}{1.22}\right)^2 = (\sqrt{h})^2 \qquad \text{Square each side.}$$

$$h = \frac{D^2}{1.4884} \qquad (\sqrt{h})^2 = h, h \geq 0; \text{ rewrite equation.}$$

SECTION 8.6 ■ HIGHER ROOTS AND RATIONAL EXPONENTS

nth Roots b is an nth root of a if $b^n = a$. The principal nth root of a is denoted $\sqrt[n]{a}$. Note that an even root of a negative number is not a real number.

Example: $\sqrt[3]{64} = 4$ because $4^3 = 64$.

Product and Quotient Rules for Radicals

$$\underset{\text{Product rule}}{\sqrt[n]{a} \cdot \sqrt[n]{b} = \sqrt[n]{ab}} \quad \text{and} \quad \underset{\text{Quotient rule}}{\sqrt[n]{\frac{a}{b}} = \frac{\sqrt[n]{a}}{\sqrt[n]{b}},} \quad \text{provided each expression is defined.}$$

Examples: $\sqrt[3]{2} \cdot \sqrt[3]{4} = \sqrt[3]{8} = 2$ and $\sqrt[4]{\frac{16}{81}} = \frac{\sqrt[4]{16}}{\sqrt[4]{81}} = \frac{2}{3}$

The Expression $a^{1/n}$ $a^{1/n} = \sqrt[n]{a}$

Examples: $36^{1/2} = \sqrt{36} = 6$ and $4^{1/3} = \sqrt[3]{4} \approx 1.587$

The Expression $a^{m/n}$ $a^{m/n} = \sqrt[n]{a^m}$ or $a^{m/n} = (\sqrt[n]{a})^m$

Examples: $8^{2/3} = \sqrt[3]{8^2} = 4$ and $8^{2/3} = (\sqrt[3]{8})^2 = 4$

The Expression $a^{-m/n}$ $a^{-m/n} = \frac{1}{a^{m/n}}$

Example: $8^{-2/3} = \frac{1}{8^{2/3}} = \frac{1}{4}$

CHAPTER 8   REVIEW EXERCISES

SECTION 8.1

Exercises 1–4: Find the square roots of the number. Approximate the result to three decimal places when appropriate.

1. 36

2. 400

3. 7

4. 12

Exercises 5–8: State whether the given expression is one or more or none of the following: a real number, a rational number, or an irrational number.

5. $\sqrt{81}$

6. $\sqrt{\frac{4}{9}}$

7. $\sqrt{15}$

8. $\sqrt{-5}$

Exercises 9–14: Evaluate each expression and approximate the result to two decimal places when appropriate.

9. $\sqrt{121}$

10. $\sqrt{400}$

11. $-\sqrt{49}$

12. $-\sqrt{17}$

13. $\sqrt[3]{6}$

14. $\sqrt[3]{-125}$

Exercises 15 and 16: A right triangle has legs a and b with hypotenuse c. Find the exact length of the missing side.

15. $a = 4$ inches, $b = 5$ inches

16. $a = 7$ feet, $c = 25$ feet

17. Find the *exact* length of the line segment shown. Then approximate this length to two decimal places.

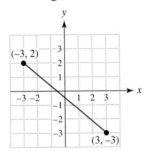

18. For the rectangle shown, find the *exact* length of the diagonal and of the perimeter. Simplify each answer.

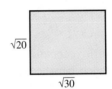

Exercises 19–22: Find the exact distance between the pair of points. Approximate this distance to three decimal places when appropriate.

19. $(2, -5), (7, 7)$

20. $(0, -3), (-1, 2)$

21. $(6, -4), (-5, -4)$

22. $(20, -30), (5, -10)$

SECTION 8.2

Exercises 23–28: Simplify the expression. Assume that all variables are positive.

23. $\sqrt{6} \cdot \sqrt{5}$

24. $\sqrt{\frac{1}{3}} \cdot \sqrt{\frac{2}{3}}$

25. $\sqrt{4} \cdot \sqrt{64}$

26. $\sqrt{2x} \cdot \sqrt{2x}$

27. $\sqrt{3t} \cdot \sqrt{27t}$

28. $\sqrt{7a} \cdot \sqrt{2b}$

Exercises 29–34: Simplify the expression by factoring out the largest perfect square. Assume that all variables are positive.

29. $\sqrt{45}$

30. $\sqrt{200}$

31. $\sqrt{t^5}$

32. $\sqrt{5b^2}$

33. $\sqrt{9x^2y}$

34. $\sqrt{49a^3b}$

Exercises 35–44: Simplify the expression. Assume that all variables are positive.

35. $\sqrt{\frac{25}{36}}$

36. $\sqrt{5} \cdot \sqrt{8}$

37. $\sqrt{\frac{3}{x}} \cdot \sqrt{\frac{27}{x}}$

38. $\dfrac{\sqrt{64}}{\sqrt{4}}$

39. $\dfrac{\sqrt{50}}{\sqrt{2}}$

40. $\dfrac{\sqrt{3r^3}}{\sqrt{r}}$

41. $\sqrt{\frac{x}{16}} \cdot \sqrt{\frac{x^2}{4}}$

42. $\sqrt{3ab^3} \cdot \sqrt{27ab}$

43. $\dfrac{\sqrt{75ab}}{\sqrt{3ab}}$

44. $\dfrac{\sqrt{8xy^3}}{\sqrt{2xy}}$

SECTION 8.3

Exercises 45–52: Simplify the expression. Assume that all variables are positive.

45. $8\sqrt{7} - 5\sqrt{7}$

46. $\sqrt{15} + 2\sqrt{15}$

47. $2\sqrt{2} + \sqrt{8}$

48. $3\sqrt{45} - 2\sqrt{20}$

49. $5\sqrt[3]{9} - 2\sqrt[3]{9}$

50. $9\sqrt[3]{10} + \sqrt[3]{10}$

51. $\sqrt{4t} + \sqrt{9t}$

52. $2\sqrt{ab} - \sqrt{4ab}$

SECTION 8.4

Exercises 53 and 54: Multiply and simplify.

53. $\left(9 - \sqrt{6}\right)\left(9 + \sqrt{6}\right)$

54. $\left(2 + \sqrt{t}\right)\left(2 - \sqrt{t}\right)$

Exercises 55–60: Rationalize the denominator. Assume that all variables are positive.

55. $\dfrac{5}{\sqrt{5}}$

56. $-\dfrac{9}{3\sqrt{11}}$

57. $-\dfrac{4}{3\sqrt{x}}$

58. $\dfrac{\sqrt{5x^3}}{\sqrt{25x}}$

59. $\dfrac{1}{\sqrt{5} - 2}$

60. $\dfrac{4 - \sqrt{b}}{4 + \sqrt{b}}$

SECTION 8.5

Exercises 61–68: Solve and check your answers.

61. $\sqrt{x} = 5$

62. $\sqrt{z} = -2$

63. $\sqrt{2 - t} = 5$

64. $\sqrt{3b + 1} = 5$

65. $\sqrt{x + 4} - 1 = 5$

66. $\sqrt{8x} = 4\sqrt{x - 4}$

67. $\frac{1}{2}\sqrt{2b + 4} + 4 = b$

68. $x - 7 = \sqrt{2x + 1}$

Exercises 69 and 70: Solve the equation for the specified variable.

69. $A = \sqrt{4P}$ for P

70. $P = 2\pi\sqrt{\dfrac{L}{g}}$ for g

Exercises 71 and 72: Do the following.
   (a) Use a table of values to solve the equation.
   (b) Use a graph to solve the equation.

71. $\sqrt{x} + 2 = 2$

72. $\sqrt{x - 1} = 4$

SECTION 8.6

Exercises 73–84: Simplify the expression.

73. $\sqrt[3]{64}$

74. $\sqrt[3]{-1000}$

75. $-\sqrt[4]{81}$

76. $\sqrt[5]{32}$

77. $\sqrt[4]{7^4}$

78. $-\sqrt[6]{\dfrac{1}{64}}$

79. $\sqrt[3]{-4} \cdot \sqrt[3]{16}$

80. $\sqrt[4]{3} \cdot \sqrt[4]{27}$

81. $\dfrac{\sqrt[3]{2}}{\sqrt[3]{16}}$

82. $\dfrac{\sqrt[4]{2}}{\sqrt[4]{32}}$

83. $\sqrt[4]{\dfrac{2}{3}} \cdot \sqrt[4]{\dfrac{8}{27}}$

84. $\sqrt[3]{\dfrac{4}{27}}$

Exercises 85–88: Write the expression in radical notation and then evaluate it.

85. $25^{3/2}$

86. $1000^{1/3}$

87. $16^{-5/4}$

88. $(-8)^{-2/3}$

APPLICATIONS

89. *Distance Between Two Ships*   Two ships pass at midnight, one traveling south at 30 miles per hour and the other traveling west at 16 miles per hour. Find the distance between the ships at 2:30 A.M.

90. *Height of a Radio Tower*   One end of a guy wire is attached to the top of a 400-foot radio tower and the other end is attached to a stake in level ground 300 feet from the base of the tower, as illustrated in the figure. Find the length L of the guy wire.

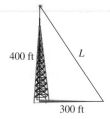

91. *Heart Rate*   The average heart rate R in beats per minute of an animal with weight W in pounds can be approximated by $R = \dfrac{885}{\sqrt{W}}$. Use this equation to estimate the weight of a dog having an average heart rate of 150 beats per minute.   (**Source:** C. Pennycuick, *Newton Rules Biology.*)

92. *Hang Time of a Football*   The length of time that a football spends in the air after it is kicked is called *hang time*. If a football is kicked h feet high, then its hang time T in seconds is given by $T = \sqrt{\dfrac{h}{4}}$. If the hang time is 4 seconds, how high did the football rise?

93. *Orbits and Distance*   The formula that relates a planet's distance D from the sun and the time T it takes to orbit the sun is

$$T = D^{3/2},$$

where T is in Earth years and $D = 1$ corresponds to the average distance between Earth and the sun, or 93,000,000 miles.
   (a) Use radical notation to write this formula.
   (b) Find T for Mars with $D = 1.52$.

94. *Skid Marks*   Vehicles involved in accidents often leave skid marks. To determine how fast a vehicle was traveling, officials often use a test vehicle to compare skid marks on the same section of road. Suppose that a vehicle in a crash left skid marks D feet long and that a test vehicle traveling at v miles per hour leaves skid marks d feet long. Then the speed V of the vehicle involved in the crash is given by

$$V = v\sqrt{\dfrac{D}{d}}.$$

Solve this formula for d.   (**Source:** N. Garber and L. Hoel, *Traffic and Highway Engineering.*)

95. *Perimeter and Area* Find the exact perimeter P and the exact area A of the rectangle shown. Simplify your answers.

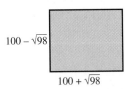

$100 - \sqrt{98}$

$100 + \sqrt{98}$

96. *Distance to the Horizon* The formula $D = 1.22\sqrt{h}$ can be used to calculate the distance in miles to the horizon for a person whose eyes are h feet above the ground. If the horizon is 10 miles away, how far above the ground are the person's eyes?

97. *Period of a Pendulum* The period P in seconds of a pendulum that is L feet long is given by

$$P = 2\pi\sqrt{\frac{L}{32}}.$$

(a) If a pendulum is 8 feet long, find its period.
(b) How long should a pendulum be to have a 2-second period?

98. *Wing Span* The weight W in pounds of some types of birds and the length L in inches of their average wing span are related by the formula $L = 27.5\sqrt[3]{W}$. (*Source:* C. Pennycuick.)

(a) Use a rational exponent to write this equation.
(b) Estimate the wing span of an 8-pound bird.

CHAPTER 8   TEST Pass the Test Video solutions to all test exercises

1. Find the square roots of 64.

2. State whether $\sqrt{8}$ is one or more or none of the following: a real number, a rational number, or an irrational number.

Exercises 3–6: Evaluate the expression. Approximate your result to two decimal places when appropriate.

3. $\sqrt{81}$

4. $\sqrt[3]{-64}$

5. $\sqrt[5]{32}$

6. $\sqrt{11}$

7. Write $25^{3/2}$ in radical notation and then evaluate it.

8. Simplify $16^{-3/4}$.

9. Let a right triangle have legs with length 5 inches and 7 inches. Find the *exact* length of the hypotenuse.

10. Find the *exact* length of the line segment shown.

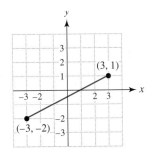

11. For the rectangle shown, find the *exact* length of the diagonal D and of the perimeter P. Simplify each answer.

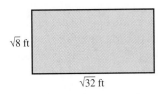

$\sqrt{8}$ ft

$\sqrt{32}$ ft

12. Find the *exact* distance between the points $(5, -4)$ and $(-2, 3)$. Then approximate your answer to three decimal places.

Exercises 13 and 14: Simplify the expression by factoring out the largest perfect square. Assume that all variables are positive.

13. $\sqrt{54}$

14. $\sqrt{9t^3}$

Exercises 15–20: Simplify the expression. Assume that all variables are positive.

15. $\sqrt{6} \cdot \sqrt{24}$

16. $\dfrac{\sqrt{90}}{\sqrt{10}}$

17. $\sqrt{3a} \cdot \sqrt{27a}$

18. $\dfrac{\sqrt{4r^5}}{\sqrt{r^3}}$

19. $\sqrt{\dfrac{x}{3}} \cdot \sqrt{\dfrac{x}{3}}$

20. $\sqrt{8ab^2} \cdot \sqrt{2a^2b}$

Exercises 21–28: Simplify the expression. Assume that all variables are positive.

21. $7\sqrt{5} + 2\sqrt{5}$ **22.** $3\sqrt{8} - \sqrt{18}$

23. $2\sqrt[3]{11} - 3\sqrt[3]{11}$ **24.** $\sqrt{25ab} + \sqrt{9ab}$

25. $\sqrt{8} + \sqrt{72} - \sqrt{27}$ **26.** $\sqrt{96} + 2\sqrt{24}$

27. $\sqrt{4b} - \sqrt{9b} + \sqrt{16b}$ **28.** $\sqrt{16x} + \sqrt{4x^3}$

Exercises 29 and 30: Rationalize the denominator of the expression. Assume that all variables are positive.

29. $\dfrac{4}{2\sqrt{3}}$ **30.** $\dfrac{2 - \sqrt{x}}{2 + \sqrt{x}}$

Exercises 31 and 32: Solve and check your answers.

31. $\sqrt{2x - 5} = 5$ **32.** $\sqrt{x + 30} = x$

33. Solve $b = \sqrt{c^2 - a^2}$ for c. Assume that $c > 0$.

34. *Distance Between Two Cars*   Two cars stop at an intersection at noon. The first car then travels north at 60 miles per hour, and the second car travels east at 40 miles per hour. Approximate, to the nearest mile, the distance between the cars at 12:45 P.M.

35. *Heart Rate*   The average heart rate R in beats per minute of an animal having weight W in pounds can be approximated by the formula $R = \dfrac{885}{\sqrt{W}}$.

(**Source:** C. Pennycuick, *Newton Rules Biology*.)

  (**a**) Use this equation to estimate the average heart rate of a 64-pound dog.

  (**b**) Estimate the weight of a cat that has an average heart rate of 177 beats per minute.

  (**c**) Use rational exponents to write the formula.

  (**d**) Solve the formula for W.

CHAPTER 8   EXTENDED AND DISCOVERY EXERCISES

Exercises 1–6: Squaring Twice   When an equation contains two or more terms with square roots, it may be necessary to square each side of the equation more than once. In these situations, isolate one of the square roots and then square each side of the equation. If a radical term remains after simplifying, repeat these steps. Use this procedure to solve the equation. Be sure to check your answers.

1. $\sqrt{2x - 1} = \sqrt{x + 1}$ **2.** $\sqrt{x} = \sqrt{x - 7} + 1$

3. $\sqrt{t - 2} = \sqrt{t + 3} - 1$

4. $\sqrt{2t - 2} + \sqrt{t} = 7$

5. $\sqrt{b + 1} - \sqrt{b - 6} = 1$

6. $\sqrt{b - 7} = \sqrt{b} - 1$

*Exercises 7–12: Power Rule   The squaring property can be extended to the **power rule for solving equations**. This rule states that the solutions to the equation $a = b$ are among the solutions to $a^n = b^n$, where n is a positive integer. For example, to solve $\sqrt[3]{x + 3} = 2$, we can cube each side of the equation.*

$$\left(\sqrt[3]{x + 3}\right)^3 = 2^3 \qquad \text{Cube each side.}$$
$$x + 3 = 8 \qquad \text{Simplify.}$$
$$x = 5 \qquad \text{Subtract 3 from each side.}$$

Use the power rule to solve the equation. Be sure to check your answers.

7. $\sqrt[3]{x} = 3$

8. $\sqrt[3]{2x} + 1 = 3$

9. $\sqrt[3]{2x - 2} + 1 = 4$

10. $\sqrt[4]{t} = 2$

11. $\sqrt[4]{4x} + 2 = 4$

12. $\sqrt[5]{5x - 3} = 2$

13. *Power Rule*   (Refer to Exercises 7–12.) Explain how you could use the power rule to solve $\sqrt[3]{3x} = \sqrt{x}$. Find any solutions.

14. *Number Puzzle*   Only one integer n between 10 and 99 is both a perfect square and a perfect cube. That is, $n = a^2$ for some integer a, and $n = b^3$ for some integer b. Find this integer.

15. *Number Puzzle*   Only one integer n between 100 and 999 is both a perfect square and a perfect cube. That is, $n = a^2$ for some integer a, and $n = b^3$ for some integer b. Find this integer.

CHAPTERS 1–8   CUMULATIVE REVIEW EXERCISES

1. Write 60 as a product of prime numbers.

2. Evaluate $45 - 3^2 \cdot 8 \div 2$.

3. Simplify $\frac{2x^2}{5y} \div \frac{x}{15y^2}$.

4. Solve $4t - 3 = 2$ and check the solution.

5. Convert 45% to fraction and decimal notation.

6. Solve $-2x + 1 \geq 4$.

Exercises 7 and 8: Graph the equation.

7. $y = 3x - 2$ 8. $y = x^2 - 1$

9. Find the slope–intercept form for the line perpendicular to $3x - 2y = 6$ passing through $(1, -1)$.

10. Solve the system of equations.

$$2x + y = 4$$
$$-x + y = 1$$

Exercises 11–14: Simplify the expression.

11. $(4x^2 - 2) - (-2x^2 + 1)$

12. $(a^3 b^5)(a^2 b)$

13. $6x(-4x^2 + 2x)$

14. $(x - 2)(x + 5)$

Exercises 15 and 16: Simplify and write the expression using positive exponents.

15. $\dfrac{1}{4^{-3}}$ 16. $(3t^2)^{-3}$

17. Write 6.5×10^4 in standard (decimal) notation.

18. Divide $\dfrac{9x^3 - 6x^2}{3x}$.

Exercises 19–22: Factor completely.

19. $x^2 + 2x - 15$ 20. $16z^2 - 9$

21. $x^3 - y^3$ 22. $x^4 + 4x^2 + 3$

Exercises 23 and 24: Solve the equation.

23. $x^2 - 5x - 14 = 0$ 24. $x^4 + 16 = 8x^2$

Exercises 25 and 26: Simplify completely.

25. $\dfrac{1}{x - 2} + \dfrac{1}{x^2 - 4}$ 26. $\dfrac{xy}{x + y} \div \dfrac{x}{x^2 - y^2}$

27. Simplify $\dfrac{\frac{1}{y} + \frac{1}{x}}{\frac{1}{y} - \frac{1}{x}}$.

28. Solve $\dfrac{1}{x - 1} + \dfrac{1}{x + 1} = \dfrac{1}{x^2 - 1}$.

29. Evaluate $\sqrt{121}$.

30. Find the exact distance between the points $(3, -4)$ and $(1, 1)$.

31. Simplify each expression completely. Assume that all variables are positive.

   (a) $\sqrt{2t} \cdot \sqrt{8t}$ (b) $\sqrt{200} + \sqrt{8}$

   (c) $\dfrac{\sqrt[3]{16x}}{\sqrt[3]{2}}$ (d) $(1 - \sqrt{2})(1 + \sqrt{2})$

   (e) $81^{3/4}$ (f) $32^{-1/5}$

32. Solve $5\sqrt{2 - x} + 1 = 11$.

33. Solve $x^2 + y^2 = 4$ for y if $y \geq 0$.

34. Simplify $\sqrt[3]{-27}$.

APPLICATIONS

35. *Wages*   Suppose that a person earns \$500 in 8 days. At the same rate, how much could the person earn in 13 days?

36. *Burning Calories*   An athlete burns 10 calories per minute on a stair climber and 9 calories per minute on a rowing machine. If the athlete burns 470 calories in 50 minutes, how long does the athlete spend on each activity?

37. *Triangle*   In a triangle, the sum of the measures of the two smaller angles equals the largest angle. The largest angle is double the smallest angle. Find the measures of each angle.

38. *Distance Between Cars*   Two cars meet at an intersection. The first car travels north at 60 miles per hour and the second car travels east at 40 miles per hour. To the nearest mile, approximate the distance between the cars after 1.5 hours.

Quadratic Equations

As society becomes more technologically advanced, personal information increasingly is being transferred over the Internet, phone lines, and through the mail. Although such transactions are fast and convenient, they provide numerous opportunities for sensitive information to fall into the wrong hands. Some personal information, such as credit card, bank account, and social security numbers, can be used by thieves to commit fraud and theft. Obtaining personal information and using it without the knowledge of the owner is called *identity theft*. The number of identity theft complaints registered with the Federal Trade Commission (FTC) has increased significantly in recent years.

The following table lists the number of identity theft complaints registered with the U.S. Trade Commission for the years 2000, 2001, 2002, and 2003. The accompanying graph reveals that the data do not lie on a straight line. A graph of a linear equation would not represent these data very well. Instead, a portion of a nonlinear graph called a parabola may be used to show the data more accurately. (See Section 9.2, Exercises 61 and 62.)

Year	2000	2001	2002	2003
Complaints	31,116	86,198	161,819	291,274

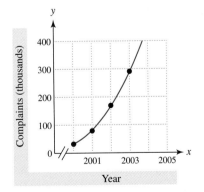

> Perseverance is not a long race; it is many short races one after another.
>
> —WALTER ELLIOTT

In this chapter we discuss parabolas and the quadratic equations associated with them. We also discuss a variety of methods for solving quadratic equations, including the square root property, completing the square, and the quadratic formula.

Source: Federal Trade Commission, *ID Theft, What's It All About?*, October 2003.

9.1 PARABOLAS

Graphing Parabolas ▪ The Graph of $y = ax^2$

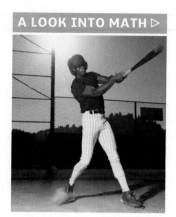

When a ball is hit into the air, it travels a curved path until it hits the ground. The graph of a linear equation (a line) could not be used to represent the height of the ball. In this section we discuss a curved graph called a parabola that can be used to represent more accurately the flight of a ball. Parabolas have applications in a wide variety of fields, such as road construction and medicine, and are important concepts in mathematics.

Graphing Parabolas

From Chapter 3 we know that the graph of $y = ax + b$ is a line that slopes upward if $a > 0$ and slopes downward if $a < 0$. In this section we discuss the graph of $y = ax^2 + bx + c$, which is a *parabola*. A parabola is either a ∪-shaped graph opening upward if $a > 0$ or a ∩-shaped graph opening downward if $a < 0$, as shown in Figure 9.1.

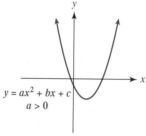

(a) Parabola opening upward

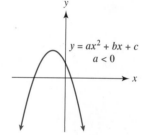

(b) Parabola opening downward

Figure 9.1

EQUATION OF A PARABOLA

The **equation of a parabola** with a vertical axis of symmetry can be written

$$y = ax^2 + bx + c,$$

where a, b, and c are constants with $a \neq 0$. If $a > 0$, the parabola opens upward; if $a < 0$, the parabola opens downward.

The graph of $y = x^2$ is a parabola that opens upward, with its *vertex* located at the origin, as shown in Figure 9.2(a). The **vertex** is the *lowest* point on the graph of a parabola that opens upward or the *highest* point on the graph of a parabola that opens downward. A parabola opening downward is shown in Figure 9.2(b). Its vertex is the point $(0, 2)$ and is the highest point on the graph. If we were to fold the xy-plane along the y-axis, the left and right sides of the graph would match. That is, the graph is symmetric with respect to the y-axis. In this case the y-axis is the **axis of symmetry** for the parabola. Figure 9.2(c) shows a parabola that opens upward with vertex $(2, -1)$ and axis of symmetry $x = 2$.

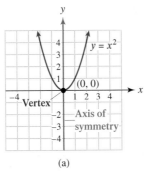

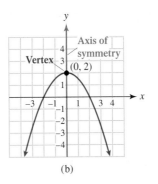

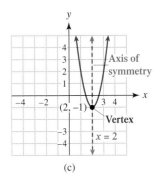

(a)　　　　　　　　　　(b)　　　　　　　　　　(c)

Figure 9.2

EXAMPLE **1** **Determining whether parabolas open upward or downward**

Determine whether each parabola opens upward or downward.
(a) $y = 2x^2 - 3x + 5$　　(b) $y = 4 + 2x - 7x^2$　　(c) $y = -(x - 2)^2$

Solution
(a) The equation is in the form $y = ax^2 + bx + c$ with $a = 2$. Because $2 > 0$, the parabola opens upward.
(b) First rewrite the equation as $y = ax^2 + bx + c$ to obtain $y = -7x^2 + 2x + 4$. Because $a = -7$ and $-7 < 0$, the parabola opens downward.
(c) To write the equation in the form $y = ax^2 + bx + c$, square the binomial and then apply the distributive property.

$$y = -(x - 2)^2 \qquad \text{Given equation}$$
$$= -(x^2 - 4x + 4) \qquad \text{Square the binomial.}$$
$$= -x^2 + 4x - 4 \qquad \text{Distributive property}$$

Because $a = -1$ and $-1 < 0$, the parabola opens downward.

Now Try Exercises 11, 13, 15

When graphing a parabola, we can use the following formula to locate the x-coordinate of the vertex.

VERTEX FORMULA

The x-coordinate of the vertex of the graph of $y = ax^2 + bx + c$, $a \neq 0$, is given by

$$x = -\frac{b}{2a}.$$

To find the y-coordinate of the vertex, substitute this x-value into the equation for the parabola.

NOTE: The equation of the axis of symmetry for $y = ax^2 + bx + c$ is $x = -\frac{b}{2a}$.

EXAMPLE 2   Finding the vertex of a parabola

Find the vertex for the graph of $y = 2x^2 - 4x + 1$.

Solution
For $y = 2x^2 - 4x + 1$, $a = 2$ and $b = -4$. The x-value of the vertex is

$$x = -\frac{b}{2a} = -\frac{(-4)}{2(2)} = 1.$$

To find the y-value of the vertex, substitute $x = 1$ in the given equation.

$$y = 2(1)^2 - 4(1) + 1 = -1.$$

Thus the vertex is located at $(1, -1)$.   Now Try Exercise 39

The vertex is an important point on the graph of a parabola. Another important value is the y-intercept. To find the y-intercept for $y = ax^2 + bx + c$, let $x = 0$.

$$y = a(0)^2 + b(0) + c = c$$

Thus the point $(0, c)$ lies on the parabola. In the next example, these concepts are used to graph the equation from Example 2.

EXAMPLE 3   Graphing a parabola

Graph $y = 2x^2 - 4x + 1$.

Solution
From Example 2 the vertex is $(1, -1)$. To find the y-intercept, let $x = 0$.

$$y = 2(0)^2 - 4(0) + 1 = 1$$

The y-intercept is 1, so the point $(0, 1)$ lies on the parabola. Table 9.1 lists the vertex and some other points on the graph of $y = 2x^2 - 4x + 1$. When plotting points for a parabola, choose at least two points on each side of the vertex. The graph is shown in Figure 9.3.

TABLE 9.1

x	y	
−1	7	
0	1	← y-intercept: 1
1	−1	← Vertex: $(1, -1)$
2	1	
3	7	

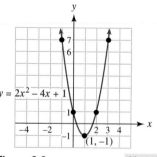

$y = 2x^2 - 4x + 1$

Figure 9.3   Now Try Exercise 49

▶ **REAL-WORLD CONNECTION**   Suppose that the parabola shown in Figure 9.4 represents a valley. If we walk from *left to right*, the valley "goes down" and then "goes up." Mathematically, we say that the graph is *decreasing* when $x \le 0$ and *increasing* when $x \ge 0$. The vertex represents the point at which the graph switches from decreasing to increasing. In Figure 9.2(b), the graph increases when $x \le 0$ and decreases when $x \ge 0$, and in Figure 9.2(c) the graph decreases when $x \le 2$ and increases when $x \ge 2$.

Decreasing: Increasing:
$x \le 0$ $x \ge 0$

Figure 9.4

NOTE: When determining where a graph is increasing and where it is decreasing, we must "walk" along the graph *from left to right*. (We read English from left to right, which might help you remember.)

EXAMPLE 4 **Graphing parabolas**

Graph each equation. Identify the vertex and the axis of symmetry. Then state where the graph is increasing and where it is decreasing.
(a) $y = x^2 - 1$ **(b)** $y = -(x + 1)^2$ **(c)** $y = x^2 + 4x + 3$

Solution

(a) Begin by making a convenient table of values (see Table 9.2). The x-coordinate of the vertex is 0 and can be found by letting $a = 1$ and $b = 0$ in $x = -\frac{b}{2a}$. Then plot the points and sketch a smooth $\cup$-shaped curve that opens upward, as shown in Figure 9.5. The lowest point on this graph is $(0, -1)$, which is the vertex. The axis of symmetry is the vertical line $x = 0$, which passes through the vertex and coincides with the y-axis. Note also the symmetry of the y-values in Table 9.2 about the vertex $(0, -1)$. This graph is decreasing when $x \leq 0$ and increasing when $x \geq 0$.

TABLE 9.2

x	$y = x^2 - 1$	
-2	3	
-1	0	
Vertex: $(0, -1) \rightarrow$ 0	-1	Equal
1	0	
2	3	

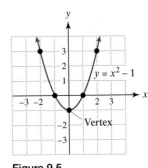

Figure 9.5

(b) Make a table of values (see Table 9.3). Plot the points and sketch a smooth $\cap$-shaped curve opening downward, as shown in Figure 9.6. The highest point on this graph is $(-1, 0)$, which is the vertex. The axis of symmetry is the vertical line $x = -1$, which passes through the vertex. This graph is increasing when $x \leq -1$ and decreasing when $x \geq -1$.

TABLE 9.3

x	$y = -(x + 1)^2$	
-3	-4	
-2	-1	
Vertex: $(-1, 0) \rightarrow$ -1	0	Equal
y-intercept: $-1 \rightarrow$ 0	-1	
1	-4	

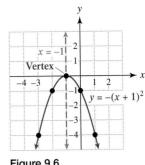

Figure 9.6

NOTE: If two points on a parabola have equal y-values, then the x-value of the vertex is midway between the x-values of these two points. For example, from Table 9.3 the points $(-3, -4)$ and $(1, -4)$ lie on the graph of $y = -(x + 1)^2$ and they both have a y-coordinate of -4. Thus the x-coordinate of the vertex is -1 because -1 is midway between -3 and 1.

(c) Make a table of values (see Table 9.4). Plot the points and sketch a smooth ∪-shaped graph opening upward, as shown in Figure 9.7. The lowest point on this graph is $(-2, -1)$, which is the vertex. The axis of symmetry is the vertical line $x = -2$, which passes through the vertex. The graph is decreasing when $x \leq -2$ and increasing when $x \geq -2$.

TABLE 9.4

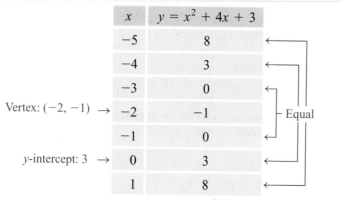

x	$y = x^2 + 4x + 3$
-5	8
-4	3
-3	0
Vertex: $(-2, -1)$ → -2	-1
-1	0
y-intercept: 3 → 0	3
1	8

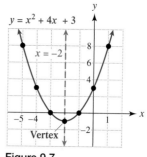

Figure 9.7

Now Try Exercises 19, 21, 23, 35

▶ **REAL-WORLD CONNECTION** In applications the vertex of a parabola can have special meaning. In the next example a parabola depicts the height reached by a baseball.

EXAMPLE 5 **Finding the height attained by a baseball**

When a baseball is hit into the air, its height y in feet after x seconds can be calculated by the equation $y = -16x^2 + 96x + 3$.
(a) Find the vertex of this parabola.
(b) Interpret the x- and y-coordinates of the vertex.

Solution
(a) For $y = -16x^2 + 96x + 3$, $a = -16$ and $b = 96$. The x-coordinate of the vertex is

$$x = -\frac{b}{2a} = -\frac{96}{2(-16)} = 3.$$

To find the y-coordinate of the vertex, let $x = 3$ in the given equation to obtain

$$y = -16(3)^2 + 96(3) + 3 = 147.$$

Thus the vertex is $(3, 147)$.
(b) Because x represents the time in seconds after the ball is hit and y represents the height of the ball in feet, a vertex of $(3, 147)$ means that after 3 seconds the height of the ball is 147 feet. Furthermore, the vertex is the highest point on the graph of a parabola that opens downward, which indicates that 147 feet is the *maximum* height attained by the baseball.

Now Try Exercise 73

The Graph of $y = ax^2$

In this subsection we discuss the graph of $y = ax^2$, where $a \neq 0$. First, we consider the case where $a > 0$ by graphing $y_1 = \frac{1}{2}x^2$, $y_2 = x^2$, and $y_3 = 2x^2$, as shown in Figure 9.8(a). Note that $a = \frac{1}{2}$, $a = 1$, and $a = 2$, respectively, and that as *a increases* the resulting parabola becomes *narrower*. The graph of $y_1 = \frac{1}{2}x^2$ is wider than the graph of $y_2 = x^2$, and the graph of $y_3 = 2x^2$ is narrower than the graph of $y_2 = x^2$. In general, the graph of $y = ax^2$ is wider than the graph of $y = x^2$ when $0 < a < 1$ and narrower than the graph of $y = x^2$ when $a > 1$. When $a > 0$, the graph of $y = ax^2$ opens upward and never lies below the x-axis.

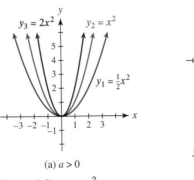

(a) $a > 0$ (b) $a < 0$

Figure 9.8 $y = ax^2$

When $a < 0$, the graph of $y = ax^2$ opens downward and never lies above the x-axis. The graphs of $y_4 = -\frac{1}{2}x^2$, $y_5 = -x^2$, and $y_6 = -2x^2$ are shown in Figure 9.8(b). The graph of $y_4 = -\frac{1}{2}x^2$ is wider than the graph of $y_5 = -x^2$, and the graph of $y_6 = -2x^2$ is narrower than the graph of $y_5 = -x^2$.

THE GRAPH OF $y = ax^2$

The graph of $y = ax^2$ is a parabola that opens upward when $a > 0$ and opens downward when $a < 0$. As the value of $|a|$ increases, the graph of $y = ax^2$ becomes narrower. The vertex is $(0, 0)$, and the axis of symmetry is the y-axis.

EXAMPLE 6   Graphing $y = ax^2$

Compare the graph of $y = -3x^2$ to the graph of $y = x^2$. Then graph both equations on the same coordinate axes.

Solution
Both graphs are parabolas. However, the graph of $y = -3x^2$ opens downward and is narrower than the graph of $y = x^2$. Their graphs are shown in Figure 9.9.

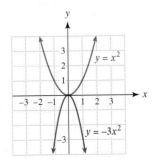

Figure 9.9   Now Try Exercise 59

9.1 PUTTING IT ALL TOGETHER

The following table summarizes some of the important topics in this section.

Concept	Explanation	Examples
Equation of a Parabola	Can be written $$y = ax^2 + bx + c, a \neq 0$$	$y = x^2 + x - 2$ and $y = -2x^2 + 4$ $(b = 0)$
Vertex Formula	The x-coordinate of the vertex for the graph of $y = ax^2 + bx + c$ with $a \neq 0$ is given by $$x = -\frac{b}{2a}.$$ The y-coordinate of the vertex is found by substituting this x-value in the equation.	If $y = -2x^2 + 8x - 7$, then $$x = -\frac{8}{2(-2)} = 2$$ and $$y = -2(2)^2 + 8(2) - 7 = 1.$$ The vertex is $(2, 1)$ and its axis of symmetry is $x = 2$. The graph opens downward because $a < 0$.
Graph of $y = ax^2 + bx + c$	Its graph is a parabola that opens upward if $a > 0$ and downward if $a < 0$. The value of $|a|$ affects the width of the parabola. The y-intercept is c.	The graph of $y = -\frac{1}{4}x^2$ opens downward and is wider than the graph of $y = x^2$, as shown in the figure. Each graph has its vertex at $(0, 0)$ and its axis of symmetry is $x = 0$.

9.1 Exercises

MyMathLab   Math XL
PRACTICE   WATCH   DOWNLOAD   READ   REVIEW

CONCEPTS

1. The graph of $y = ax^2 + bx + c$ is called a _____.

2. If a parabola opens upward, what is the lowest point on the parabola called?

3. If a parabola is symmetric with respect to the y-axis, the y-axis is called the _____.

4. The vertex of $y = x^2$ is _____.

5. Sketch a parabola that opens downward with a vertex of $(1, 2)$.

6. If $y = ax^2 + bx + c$, the x-coordinate of the vertex is given by $x =$ _____.

7. Compared to the graph of $y = x^2$, the graph of $y = 2x^2$ is _____ (wider/narrower).

8. The graph of $y = -x^2$ is similar to the graph of $y = x^2$ except that it opens _____.

9. The equation of a parabola can be written in the form $y =$ _____.

10. If a parabola opens downward, the point with the largest y-value is called the _____.

PARABOLAS

Exercises 11–18: Determine whether the parabola opens upward or downward.

11. $y = 5x^2 - 2x + 7$ 12. $y = -2x^2 + x - 6$

13. $y = 3x - 4x^2 - 7$ 14. $y = 8 - 3x + 2x^2$

15. $y = -(x + 6)^2$ 16. $y = 3 - (x + 1)^2$

17. $y = (2 - x)^2$ 18. $y = 4 - (3 - x)^2$

Exercises 19–22: Identify the vertex, axis of symmetry, and whether the parabola opens upward or downward. State where the graph is increasing and where it is decreasing.

19.

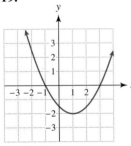

20.

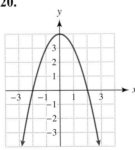

21.

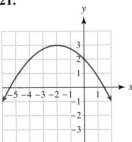

22.

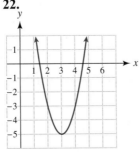

GRAPHING PARABOLAS

Exercises 23–38: Do the following.
 (a) Graph the parabola.
 (b) Identify the vertex and axis of symmetry.

23. $y = x^2 - 2$ 24. $y = x^2 - 1$

25. $y = -3x^2 + 1$ 26. $y = \frac{1}{2}x^2 + 2$

27. $y = (x - 1)^2$ 28. $y = (x + 2)^2$

29. $y = x^2 + x - 2$ 30. $y = x^2 - 2x + 2$

31. $y = 2x^2 - 3$ 32. $y = 1 - 2x^2$

33. $y = 2x - x^2$ 34. $y = x^2 + 2x - 8$

35. $y = -2x^2 + 4x - 1$ 36. $y = -\frac{1}{2}x^2 + 2x - 3$

37. $y = \frac{1}{4}x^2 - x + 5$ 38. $y = 3 - 6x - 4x^2$

Exercises 39–46: Find the vertex of the parabola.

39. $y = x^2 - 4x - 2$

40. $y = 2x^2 + 6x - 3$

41. $y = -\frac{1}{3}x^2 - 2x + 1$

42. $y = 5 - 4x + x^2$ 43. $y = 3 - 2x^2$

44. $y = \frac{1}{4}x^2 - 3x - 2$

45. $y = -0.3x^2 + 0.6x + 1.1$

46. $y = 25 - 10x + 20x^2$

Exercises 47–52: Do the following.
 (a) Find the vertex and y-intercept.
 (b) Graph the equation.

47. $y = x^2 - 4$ 48. $y = 4 - x^2$

49. $y = x^2 + 2x - 1$ 50. $y = x^2 - 2x + 1$

51. $y = -\frac{1}{2}x^2 + 2x$ 52. $y = 4x - x^2$

Exercises 53–60: Graph the parabola. Compare the graph to the graph of $y = x^2$.

53. $y = -x^2$ 54. $y = -2x^2$

55. $y = 2x^2$ 56. $y = 3x^2$

57. $y = \frac{1}{4}x^2$ 58. $y = \frac{1}{2}x^2$

59. $y = -\frac{1}{2}x^2$ 60. $y = -\frac{3}{2}x^2$

Exercises 61–72: State where the graph of the parabola is increasing and where it is decreasing.

61. $y = x^2 + 2x - 1$ **62.** $y = x^2 + 6x + 2$

63. $y = x^2 - 5x$ **64.** $y = x^2 - 3x$

65. $y = 2x^2 + 2x - 3$ **66.** $y = 3x^2 - 3x + 7$

67. $y = -x^2 + 2x + 5$ **68.** $y = -x^2 + 4x - 3$

69. $y = 4x - x^2$ **70.** $y = 6x - x^2$

71. $y = -2x^2 + x - 5$ **72.** $y = -5x^2 + 15x$

APPLICATIONS

73. *Height Reached by a Baseball* (Refer to Example 5.)
A baseball is hit into the air, and its height y in feet after x seconds is given by $y = -16x^2 + 64x + 2$.
   (a) Find the vertex of this parabola.
   (b) Interpret the vertex.

74. *Height Reached by a Golf Ball* A golf ball is hit into the air, and its height y in feet after x seconds is given by $y = -16x^2 + 128x$.
   (a) Find the vertex of this parabola.
   (b) Interpret the vertex.

75. *Maximizing Area* The rectangular pen shown in the figure requires 100 feet of fence. What dimensions give the largest area?

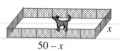

$50 - x$

76. *Maximizing Area* The rectangular pen shown in the figure requires 100 feet of fence. One side of the pen is against a building and needs no fence. What dimensions give the largest area?

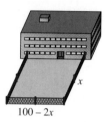

$100 - 2x$

WRITING ABOUT MATHEMATICS

77. The equation of a parabola is $y = ax^2 + bx + c$. Explain how the value of a affects the graph.

78. If the vertex of the parabola $y = ax^2 + bx + c$ with $a > 0$ is (h, k), explain where the graph is increasing and where it is decreasing.

9.2 INTRODUCTION TO QUADRATIC EQUATIONS

Basics of Quadratic Equations ▪ Graphical and Numerical Solutions ▪ The Square Root Property

A LOOK INTO MATH ▷

A taxiway used by an airplane to exit a runway often contains curves. A curve that is too sharp for the speed of the plane is a safety hazard. The equation $y = \frac{1}{2}x^2$ can be used to calculate a safe radius y for a plane that is traveling x miles per hour. We can determine the speed limit for a curve with a radius of **650** feet by solving the *quadratic equation*

$$\frac{1}{2}x^2 = 650.$$

In this section we demonstrate techniques for solving this and other quadratic equations.
(***Source:*** Federal Aviation Administration.)

Basics of Quadratic Equations

Any quadratic polynomial can be written as $ax^2 + bx + c$ with $a \neq 0$. Examples of quadratic polynomials include

$$6x^2 - 7, \quad -\frac{1}{4}x^2 + 5x, \quad \text{and} \quad x^2 + 4x - 3.$$

Setting a quadratic polynomial equal to a constant, such as 0, results in a quadratic equation. Examples of quadratic equations include

$$6x^2 - 7 = 0, \quad -\frac{1}{4}x^2 + 5x = 0, \quad \text{and} \quad x^2 + 4x - 3 = 0.$$

QUADRATIC EQUATION

A **quadratic equation** is an equation that can be written as

$$ax^2 + bx + c = 0,$$

where a, b, and c are constants with $a \neq 0$.

Solutions to the quadratic equation $ax^2 + bx + c = 0$ correspond to x-intercepts of the graph of $y = ax^2 + bx + c$. Because this graph is either ∪-shaped or ∩-shaped, it can intersect the x-axis zero, one, or two times, as illustrated in Figure 9.10. Hence a quadratic equation can have no solutions, one solution, or two solutions.

NOTE: We only consider solutions that are real numbers in this section.

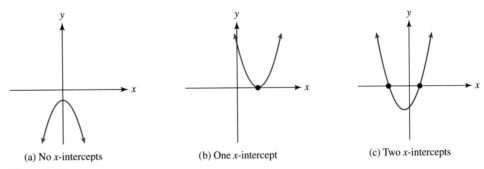

(a) No x-intercepts   (b) One x-intercept   (c) Two x-intercepts

Figure 9.10

Graphical and Numerical Solutions

In Chapter 6 we solved quadratic equations by factoring and applying the zero-product property. It is also possible to solve many quadratic equations by graphing or constructing tables. In the next example we apply all three of these techniques to quadratic equations that have no solutions, one solution, or two solutions.

EXAMPLE 1   Solving quadratic equations

Solve each quadratic equation. Support your results numerically and graphically.
(a) $2x^2 + 1 = 0$ (No real solutions)
(b) $x^2 + 4 = 4x$ (One real solution)
(c) $x^2 - 6x + 8 = 0$ (Two real solutions)

Solution

(a) *Symbolic Solution*

$$2x^2 + 1 = 0 \qquad \text{Given equation}$$
$$2x^2 = -1 \qquad \text{Subtract 1.}$$
$$x^2 = -\frac{1}{2} \qquad \text{Divide by 2.}$$

This equation has no real solutions because $x^2 \geq 0$ for all real numbers x.

Numerical and Graphical Solution The points in Table 9.5 for $y = 2x^2 + 1$ are plotted in Figure 9.11 and connected with a parabolic graph. The graph of $y = 2x^2 + 1$ has no x-intercepts, indicating that there are no real solutions. Note that the y-values in Table 9.5 are always positive.

TABLE 9.5

x	y
−2	9
−1	3
0	1
1	3
2	9

$y > 0$ for all x

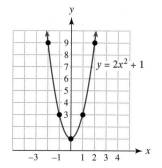

Figure 9.11 No Solutions

(b) *Symbolic Solution*

$$x^2 + 4 = 4x \qquad \text{Given equation}$$
$$x^2 - 4x + 4 = 0 \qquad \text{Subtract } 4x \text{ from each side.}$$
$$(x - 2)(x - 2) = 0 \qquad \text{Factor.}$$
$$x - 2 = 0 \quad \text{or} \quad x - 2 = 0 \qquad \text{Zero-product property}$$
$$x = 2 \qquad \text{There is one solution.}$$

Numerical and Graphical Solution Because the given equation is equivalent to $x^2 - 4x + 4 = 0$, we let $y = x^2 - 4x + 4$. The points in Table 9.6 are plotted in Figure 9.12 and connected with a parabolic graph. The graph of $y = x^2 - 4x + 4$ has one x-intercept, 2. Note that in Table 9.6, $y = 0$ when $x = 2$, indicating that the equation has one solution.

TABLE 9.6

x	y
0	4
1	1
2	**0**
3	1
4	4

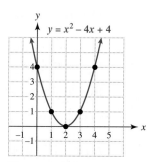

Figure 9.12 One Solution

(c) *Symbolic Solution*

$$x^2 - 6x + 8 = 0 \qquad \text{Given equation}$$

$$(x - 2)(x - 4) = 0 \qquad \text{Factor.}$$

$$x - 2 = 0 \quad \text{or} \quad x - 4 = 0 \qquad \text{Zero-product property}$$

$$x = 2 \quad \text{or} \quad x = 4 \qquad \text{There are two solutions.}$$

Numerical and Graphical Solution The points in Table 9.7 for $y = x^2 - 6x + 8$ are plotted in Figure 9.13 and connected with a parabolic graph. The graph of $y = x^2 - 6x + 8$ has two x-intercepts, 2 and 4, indicating two solutions. Note in Table 9.7 that $y = 0$ when $x = 2$ or $x = 4$.

TABLE 9.7

x	y
0	8
1	3
2	**0**
3	-1
4	**0**
5	3
6	8

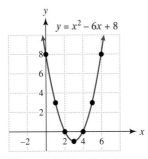

Figure 9.13 Two Solutions

Now Try Exercises 23, 29, 31

The Square Root Property

The **square root property** is a symbolic method that can be used to solve quadratic equations without x-terms. The following is an example of the square root property.

$$x^2 = 25 \quad \text{is equivalent to} \quad x = \pm 5.$$

The expression $x = \pm 5$ (read "x equals plus or minus 5") indicates that either $x = 5$ or $x = -5$. Each value is a solution to $x^2 = 25$ because $(5)^2 = 25$ and $(-5)^2 = 25$.

> ### SQUARE ROOT PROPERTY
>
> Let k be a nonnegative number. Then the solutions to the equation
>
> $$x^2 = k$$
>
> are given by $x = \pm\sqrt{k}$. If $k < 0$, then this equation has no real solutions.

EXAMPLE 2   **Using the square root property**

Solve each equation.

(a) $x^2 = 7$ **(b)** $16x^2 - 9 = 0$ **(c)** $(x - 4)^2 = 25$

Solution

(a) The given equation $x^2 = 7$ is equivalent to $x = \pm\sqrt{7}$ by the square root property. The solutions are $\sqrt{7}$ and $-\sqrt{7}$.

(b)

$$16x^2 - 9 = 0 \qquad \text{Given equation}$$

$$16x^2 = 9 \qquad \text{Add 9.}$$

$$x^2 = \frac{9}{16} \qquad \text{Divide by 16.}$$

$$x = \pm\sqrt{\frac{9}{16}} \qquad \text{Square root property}$$

$$x = \pm\frac{3}{4} \qquad \text{Simplify.}$$

The solutions are $\frac{3}{4}$ and $-\frac{3}{4}$.

(c)

$$(x - 4)^2 = 25 \qquad \text{Given equation}$$

$$(x - 4) = \pm\sqrt{25} \qquad \text{Square root property}$$

$$x - 4 = \pm 5 \qquad \text{Simplify.}$$

$$x = 4 \pm 5 \qquad \text{Add 4.}$$

$$x = 9 \quad \text{or} \quad x = -1 \qquad \text{Evaluate } 4 + 5 \text{ and } 4 - 5.$$

The solutions are 9 and -1.   `Now Try Exercises 39, 41, 45`

▶ **REAL-WORLD CONNECTION**   If an object is dropped from a height of h feet, its distance d above the ground after t seconds is given by

$$d = h - 16t^2.$$

This formula can be used to estimate the time it takes for a falling object to hit the ground.

EXAMPLE 3   **Describing a falling object**

A toy falls from a window 64 feet above the ground. Use the formula $d = h - 16t^2$ to determine how long the toy takes to hit the ground.

Solution

The height of the window is 64 feet, so let $h = 64$ and $d = 64 - 16t^2$. The toy strikes the ground when the distance d above the ground equals 0.

$$64 - 16t^2 = 0 \qquad \text{Equation to solve}$$

$$-16t^2 = -64 \qquad \text{Subtract 64.}$$

$$t^2 = \frac{-64}{-16} \qquad \text{Divide by } -16.$$

$$t^2 = 4 \qquad \text{Simplify.}$$

$$t = \pm 2 \qquad \text{Square root property}$$

Time cannot be negative in this problem, so the appropriate solution is given by $t = 2$. The toy hits the ground after 2 seconds.   `Now Try Exercise 57`

▶ **REAL-WORLD CONNECTION** In the introduction to this section we discussed how the solution to the equation

$$\frac{1}{2}x^2 = 650$$

would give a safe speed limit x for a curve with a radius of 650 feet on an airport taxiway. We solve this problem in the next example.

EXAMPLE 4   **Finding a safe speed limit**

Solve the equation $\frac{1}{2}x^2 = 650$ and interpret any solutions.

Solution
Multiply by 2 and use the square root property to solve this equation.

$\frac{1}{2}x^2 = 650$	Given equation
$x^2 = 1300$	Multiply each side by 2.
$x = \pm\sqrt{1300}$	Square root property

The solutions are $\sqrt{1300} \approx 36$ and $-\sqrt{1300} \approx -36$. The solution given by $x \approx 36$ indicates that a safe speed limit for a curve with a radius of 650 feet should be 36 miles per hour. (The negative solution has no physical meaning in this problem.)

Now Try Exercise 55

9.2 PUTTING IT ALL TOGETHER

Quadratic equations can be solved symbolically, graphically, and numerically. They have no solutions, one solution, or two solutions. We only consider solutions that are real numbers in this section. Symbolic techniques for solving quadratic equations include factoring and the square root property. We discussed factoring extensively in Chapter 6, so the following table features only the square root property.

Concept	Explanation	Examples
Quadratic Equation	Can be expressed in the form $ax^2 + bx + c = 0$, where $a \neq 0$	$3x^2 - 2x + 1 = 0$, $\quad x^2 - 5 = \frac{1}{2}$, and $\quad 7x^2 - 4 = 3x$
Square Root Property	If $k \geq 0$, the solutions to the equation $x^2 = k$ are $\pm\sqrt{k}$.	$x^2 = 100$ is equivalent to $x = \pm 10$ and $x^2 = 13$ is equivalent to $x = \pm\sqrt{13}$.

9.2 Exercises

CONCEPTS

1. Give an example of a quadratic equation. How many real solutions can a quadratic equation have?

2. The solutions to $x^2 = 9$ are _____.

3. Name two symbolic methods that can be used to solve a quadratic equation.

4. Sketch a parabola that has two x-intercepts and opens downward.

5. Sketch a parabola that has no x-intercepts and opens upward.

6. If the graph of $y = ax^2 + bx + c$ intersects the x-axis twice, how many solutions does the equation $ax^2 + bx + c = 0$ have? Explain.

Exercises 7–14: Determine whether the given equation is quadratic.

7. $x^2 - 3x + 1 = 0$

8. $2x^2 - 3 = 0$

9. $3x + 1 = 0$

10. $x^3 - 3x^2 + x = 0$

11. $-3x^2 + x = 16$

12. $x^2 - 1 = 4x$

13. $x^2 = \sqrt{x} + 1$

14. $\dfrac{1}{x - 1} = 5$

SOLVING QUADRATIC EQUATIONS

Exercises 15–18: A graph of $y = ax^2 + bx + c$ is given. Use this graph to solve $ax^2 + bx + c = 0$, if possible.

15.

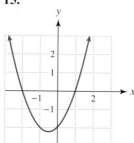

16.

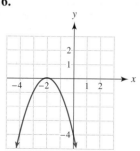

17.

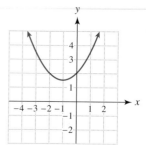

18.

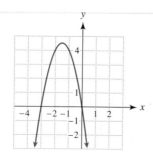

Exercises 19–22: A table of $y = ax^2 + bx + c$ is given. Use this table to solve $ax^2 + bx + c = 0$.

19.

X	Y₁
-3	6
-2	0
-1	-4
0	-6
1	-6
2	-4
3	0

Y₁ = X^2−X−6

20.

X	Y₁
-6	0
-4	-16
-2	-24
0	-24
2	-16
4	0
6	24

Y₁ = X^2+2X−24

21.

X	Y₁
-2	9
-1.5	4
-1	1
-.5	0
0	1
.5	4
1	9

Y₁ = 4X^2+4X+1

22.

X	Y₁
-4	-2.5
-3	0
-2	1.5
-1	2
0	1.5
1	0
2	-2.5

Y₁ = −X^2/2−X+3/2

Exercises 23–32: Solve the quadratic equation. Support your results numerically and graphically.

23. $x^2 - 4x - 5 = 0$

24. $x^2 - x - 6 = 0$

25. $x^2 + 2x = 3$

26. $x^2 + 4x = 5$

27. $x^2 = 9$

28. $x^2 = 4$

29. $4x^2 + 1 = 0$

30. $x^2 + 2 = 0$

31. $x^2 + 2x + 1 = 0$

32. $4x^2 + 1 = 4x$

Exercises 33–38: Solve by factoring.

33. $x^2 + 2x - 35 = 0$

34. $2x^2 - 7x + 3 = 0$

35. $6x^2 - x - 1 = 0$

36. $x^2 + 4x + 6 = -3x$

37. $2x^2 + x + 3 = 6x$

38. $6x^2 - x = 15$

Exercises 39–46: Use the square root property to solve.

39. $x^2 = 144$

40. $4x^2 - 5 = 0$

41. $4x^2 - 64 = 0$

42. $9x^2 = 7$

43. $(x + 1)^2 = 25$

44. $(x + 4)^2 = 9$

45. $(x - 1)^2 = 64$

46. $(x - 3)^2 = 0$

SOLVING EQUATIONS BY MORE THAN ONE METHOD

Exercises 47–54: Solve the quadratic equation
   (a) symbolically,
   (b) graphically, and
   (c) numerically.

47. $x^2 - 3x - 18 = 0$

48. $\frac{1}{2}x^2 + 2x - 6 = 0$

49. $x^2 - 8x + 15 = 0$

50. $2x^2 + 3 = 7x$

51. $4(x^2 + 35) = 48x$

52. $4x(2 - x) = -5$

53. $1 - 2x - 8x^2 = 0$

54. $-8x^2 + 2x + 1 = 0$

APPLICATIONS

55. *Safe Curve Speed* (Refer to Example 4.) Find a safe speed limit x for an airport taxiway curve with the given radius R.
   (a) $R = 450$ feet   **(b)** $R = 800$ feet

56. *Braking Distance* The braking distance y in feet that it takes for a car to stop on wet, level pavement can be estimated by $y = \frac{1}{9}x^2$, where x is the speed of the car in miles per hour. Find the speed associated with each braking distance. (**Source:** L. Haefner, *Introduction to Transportation Systems.*)
   (a) 25 feet   **(b)** 361 feet   **(c)** 784 feet

57. *Falling Object*   (Refer to Example 3.) How long does it take for a toy to hit the ground if it is dropped out of a window 60 feet above the ground? Does it take twice as long as it takes to fall from a window 30 feet above the ground?

58. *Distance* Two athletes start jogging at the same time. One jogs north at 6 miles per hour while the second jogs east at 8 miles per hour. After how long are the two athletes 20 miles apart?

59. *Geometry* A triangle has an area of 35 square inches, and its base is 3 inches more than its height. Find the base and height of the triangle.

60. *Trade Deficit* The U.S. trade deficit in billions of dollars for the years 1997, 1998, and 1999 can be computed by

$$y = 16x^2 - 63{,}861x + 63{,}722{,}378,$$

where x is the year. In which year was the trade deficit $164 billion?   (**Source:** Department of Commerce.)

Exercises 61 and 62: Identity Theft (Refer to the introduction to this chapter.) The number N of identity thefts in thousands x years after 2000 can be estimated by

$$N = 18.5x^2 + 30.1x + 32.6.$$

Use this formula to complete the following.

61. Estimate the number of identity thefts in 2001.

62. Estimate the year when identity thefts reached 645 thousand.

WRITING ABOUT MATHEMATICS

63. Suppose that you are asked to solve

$$ax^2 + bx + c = 0.$$

Explain how the graph of $y = ax^2 + bx + c$ can be used to find real solutions to the equation.

64. Explain why a quadratic equation could *not* have more than two solutions. (*Hint:* Consider the graph of $y = ax^2 + bx + c$.)

1. Graph each equation. Identify the vertex and axis of symmetry.
 (a) $y = x^2 - 2$
 (b) $y = x^2 - 2x - 2$

2. Compare the graph of $y_1 = 2x^2$ to the graph of $y_2 = -\frac{1}{2}x^2$.

3. State where the graph of $y = -3x^2 + 12x - 5$ is increasing and where it is decreasing.

4. Solve the quadratic equation $2x^2 - 7x + 3 = 0$ symbolically and graphically.

5. Use the square root property to solve $x^2 = 5$.

9.3 SOLVING BY COMPLETING THE SQUARE

Perfect Square Trinomials ▪ Completing the Square

A LOOK INTO MATH ▷ Some quadratic equations cannot be solved easily by using only the symbolic techniques discussed in Section 9.2. For example, although the quadratic equation $x^2 - 5x + 1 = 0$ has two real solutions, it cannot be solved by factoring or by applying the square root property. To solve it and many similar quadratic equations symbolically we can use a technique called *completing the square*. (See Example 4.)

Perfect Square Trinomials

In Chapter 6 we discussed the following formulas for perfect square trinomials:

$$a^2 + 2ab + b^2 = (a + b)^2 \quad \text{and} \quad a^2 - 2ab + b^2 = (a - b)^2.$$

An example is $4x^2 + 4x + 1 = (2x + 1)^2$, where $a = 2x$ and $b = 1$. Other examples of perfect square trinomials *with a leading coefficient of* 1 are

$$x^2 + 6x + 9, \quad x^2 - 2x + 1, \quad \text{and} \quad x^2 + 10x + 25.$$

Note that in each example the red constant term equals the square of half the blue coefficient of the x-term. That is, we can write them as

$$x^2 + 6x + \left(\frac{6}{2}\right)^2, \quad x^2 - 2x + \left(\frac{-2}{2}\right)^2, \quad \text{and} \quad x^2 + 10x + \left(\frac{10}{2}\right)^2.$$

These three trinomials can be written, respectively, in factored form as

$$(x + 3)^2, \quad (x - 1)^2, \quad \text{and} \quad (x + 5)^2.$$

In general, a perfect square trinomial with a leading coefficient of 1 can be written as

$$x^2 + bx + \left(\frac{b}{2}\right)^2,$$

where the constant term equals the square of half the coefficient of the x-term. In factored form this perfect square trinomial equals $\left(x + \frac{b}{2}\right)^2$.

EXAMPLE 1   **Determining whether a trinomial is a perfect square trinomial**

Determine whether each trinomial is a perfect square trinomial.
(a) $x^2 - 4x + 4$ **(b)** $x^2 + 14x + 64$ **(c)** $x^2 + 8x + 16$

Solution

(a) To be a perfect square trinomial the constant term **4** must equal the square of half the coefficient of the x-term. The coefficient of the x-term is -4, so we let $b = -4$. Then

$$\left(\frac{b}{2}\right)^2 = \left(\frac{-4}{2}\right)^2 = (-2)^2 = 4$$

equals the given constant term **4** and the trinomial is a perfect square trinomial.

(b) To be a perfect square trinomial the constant term **64** must equal the square of half the coefficient of the x-term. The coefficient of the x-term is 14, so we let $b = 14$. Here

$$\left(\frac{b}{2}\right)^2 = \left(\frac{14}{2}\right)^2 = 7^2 = 49 \neq 64.$$

The trinomial is not a perfect square.

(c) The coefficient of the x-term is 8, so we let $b = 8$. Then

$$\left(\frac{b}{2}\right)^2 = \left(\frac{8}{2}\right)^2 = 4^2 = 16$$

equals the given constant term **16** and the trinomial is a perfect square trinomial.

Now Try Exercises 9, 11, 13

Completing the Square

Any perfect square trinomial can be written as the square of a binomial. For example, $x^2 - 4x + 4 = (x - 2)^2$ and $x^2 + 8x + 16 = (x + 4)^2$. In general,

$$x^2 + bx + \left(\frac{b}{2}\right)^2 = \left(x + \frac{b}{2}\right)^2.$$

This general equation provides a method for solving quadratic equations by creating a perfect square trinomial on one side of the equation that can then be written as the square of a binomial. We can solve a quadratic equation in the form $x^2 + bx = d$, where b and d are constants, by adding $\left(\frac{b}{2}\right)^2$ to each side of the equation and then factoring the resulting perfect square trinomial on the left side of the equation.

In the equation $x^2 + 6x = 7$ we have $b = 6$, so we add $\left(\frac{6}{2}\right)^2 = 9$ to each side.

$x^2 + 6x = 7$	Given equation
$x^2 + 6x + 9 = 7 + 9$	Add 9 to each side.
$(x + 3)^2 = 16$	Perfect square trinomial
$x + 3 = \pm 4$	Square root property
$x = -3 \pm 4$	Add -3 to each side.
$x = 1$ or $x = -7$	Simplify $-3 + 4$ and $-3 - 4$.

The solutions are 1 and -7.

CRITICAL THINKING

Use the figure to *complete the square* for $x^2 + 6x$.

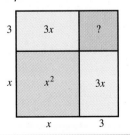

Note that the left side of the equation becomes a perfect square trinomial. We show how to create one in the next example.

ЄXAMPLЄ 2   **Creating a perfect square trinomial**

Find the term that should be added to $x^2 - 16x$ to form a perfect square trinomial.

Solution
The coefficient of the x-term is -16, so we let $b = -16$. To complete the square we divide b by 2 and then square the result.

$$\left(\frac{b}{2}\right)^2 = \left(\frac{-16}{2}\right)^2 = 64$$

If we add **64**, a perfect square trinomial is formed.

$$x^2 - 16x + 64 = (x - 8)^2 \qquad \text{Now Try Exercise } 17$$

Completing the square can be used to solve quadratic equations when a trinomial does not factor easily, as illustrated in the next three examples.

ЄXAMPLЄ 3   **Completing the square when the leading coefficient is 1**

Solve the equation $x^2 - 6x + 2 = 0$.

Solution
Start by writing the equation in the form $x^2 + bx = d$.

$$
\begin{array}{ll}
x^2 - 6x + 2 = 0 & \text{Given equation} \\
x^2 - 6x = -2 & \text{Subtract 2 from each side.} \\
x^2 - 6x + 9 = -2 + 9 & \text{Add } \left(\frac{b}{2}\right)^2 = \left(\frac{-6}{2}\right)^2 = 9. \\
(x - 3)^2 = 7 & \text{Perfect square trinomial} \\
x - 3 = \pm\sqrt{7} & \text{Square root property} \\
x = 3 \pm \sqrt{7} & \text{Add 3 to each side.}
\end{array}
$$

The solutions are $3 + \sqrt{7} \approx 5.65$ and $3 - \sqrt{7} \approx 0.35$.   Now Try Exercise 27

ЄXAMPLЄ 4   **Completing the square when the leading coefficient is 1**

Solve the equation $x^2 - 5x + 1 = 0$.

Solution
Start by writing the equation in the form $x^2 + bx = d$.

$$
\begin{array}{ll}
x^2 - 5x + 1 = 0 & \text{Given equation} \\
x^2 - 5x = -1 & \text{Subtract 1 from each side.} \\
x^2 - 5x + \dfrac{25}{4} = -1 + \dfrac{25}{4} & \text{Add } \left(\frac{b}{2}\right)^2 = \left(\frac{-5}{2}\right)^2 = \frac{25}{4}. \\
\left(x - \dfrac{5}{2}\right)^2 = \dfrac{21}{4} & \text{Perfect square trinomial} \\
x - \dfrac{5}{2} = \pm\dfrac{\sqrt{21}}{2} & \text{Square root property} \\
x = \dfrac{5 \pm \sqrt{21}}{2} & \text{Add } \frac{5}{2} \text{ to each side.}
\end{array}
$$

The solutions are $\dfrac{5 + \sqrt{21}}{2} \approx 4.79$ and $\dfrac{5 - \sqrt{21}}{2} \approx 0.21$.   Now Try Exercise 31

If the leading coefficient is a rather than 1, we can divide each side of the equation by a so that it becomes 1. For example, if we divide each side of $2x^2 + 7x = 5$ by 2, the equation becomes $x^2 + \frac{7}{2}x = \frac{5}{2}$. This second equation has a leading coefficient of 1 and can be solved by completing the square, as demonstrated in the next example.

EXAMPLE 5   **Completing the square when the leading coefficient is not 1**

Solve the equation $2x^2 + 7x = 5$.

Solution
Divide each side of the equation by 2 so that the leading coefficient of the x^2-term becomes 1.

$$2x^2 + 7x = 5 \qquad \text{Given equation}$$

$$x^2 + \frac{7}{2}x = \frac{5}{2} \qquad \text{Divide each side by 2.}$$

$$x^2 + \frac{7}{2}x + \frac{49}{16} = \frac{5}{2} + \frac{49}{16} \qquad \text{Add } \left(\frac{b}{2}\right)^2 = \left(\frac{7}{4}\right)^2 = \frac{49}{16}.$$

$$\left(x + \frac{7}{4}\right)^2 = \frac{89}{16} \qquad \text{Perfect square trinomial}$$

$$x + \frac{7}{4} = \pm\frac{\sqrt{89}}{4} \qquad \text{Square root property}$$

$$x = \frac{-7 \pm \sqrt{89}}{4} \qquad \text{Add } -\frac{7}{4} \text{ to each side.}$$

The solutions are $\frac{-7 + \sqrt{89}}{4} \approx 0.61$ and $\frac{-7 - \sqrt{89}}{4} \approx -4.11$.   Now Try Exercise 35

9.3 PUTTING IT ALL TOGETHER

The method of completing the square can be used to solve quadratic equations symbolically when the trinomial does not factor easily. After creating a perfect square trinomial on one side of the equation, we can then apply the square root property to find the solutions. The following table summarizes this technique.

Concept	Explanation	Examples
Perfect Square Trinomial	Can be expressed in the form $$x^2 + bx + \left(\frac{b}{2}\right)^2 = \left(x + \frac{b}{2}\right)^2$$	$x^2 - 2x + 1, \quad x^2 + 8x + 16, \quad$ and $\quad x^2 + 12x + 36$ $(x-1)^2 \qquad\qquad (x+4)^2 \qquad\qquad\qquad (x+6)^2$
Method of Completing the Square	To solve an equation in the form $x^2 + bx = d$, add $\left(\frac{b}{2}\right)^2$ to each side of the equation. Factor the resulting perfect square trinomial and solve for x by applying the square root property.	To solve $x^2 + 8x = 3$, add $\left(\frac{8}{2}\right)^2 = 16$ to each side. $x^2 + 8x + 16 = 3 + 16 \qquad$ Add 16 to each side. $(x+4)^2 = 19 \qquad$ Perfect square trinomial $x + 4 = \pm\sqrt{19} \qquad$ Square root property $x = -4 \pm \sqrt{19} \qquad$ Add -4 to each side. $x \approx 0.36, -8.36 \qquad$ Approximate.

9.3 Exercises

MyMathLab   PRACTICE   WATCH   DOWNLOAD   READ   REVIEW

CONCEPTS

1. $x^2 + 6x + 9 = (\underline{\hspace{1cm}})^2$

2. $x^2 - 6x + 9 = (\underline{\hspace{1cm}})^2$

3. $x^2 + bx + \left(\frac{b}{2}\right)^2 = (\underline{\hspace{1cm}})^2$

4. $x^2 - bx + \left(\frac{b}{2}\right)^2 = (\underline{\hspace{1cm}})^2$

5. In a perfect square trinomial with a leading coefficient of 1, the constant term equals the square of half the coefficient of the _____.

6. The method of _____ can be used to solve quadratic equations when the trinomial does not factor easily.

7. What should be added to $x^2 + 6x$ to form a perfect square trinomial?

8. To solve $x^2 + bx = d$ by completing the square, what value should be added to each side of the equation?

PERFECT SQUARE TRINOMIALS

Exercises 9–16: Determine whether the given trinomial is a perfect square trinomial.

9. $x^2 - 10x + 20$ **10.** $x^2 + 3x + 11$

11. $x^2 + 4x + 4$ **12.** $x^2 - x + 1$

13. $x^2 - 16x + 64$ **14.** $x^2 - 14x + 49$

15. $x^2 - 9x + \frac{81}{2}$ **16.** $x^2 - 7x + \frac{49}{4}$

Exercises 17–20: (Refer to Example 2.) Find the term that should be added to the expression to form a perfect square trinomial. Write the resulting perfect square trinomial in factored form.

17. $x^2 - 8x$ **18.** $x^2 - 5x$

19. $x^2 + 9x$ **20.** $x^2 + x$

COMPLETING THE SQUARE

Exercises 21–24: To solve by completing the square, what value should you add to each side of the equation?

21. $x^2 + 6x = -2$ **22.** $x^2 - 6x = 3$

23. $x^2 - 7x = 1$ **24.** $x^2 + 4x = 1$

Exercises 25–40: Solve by completing the square.

25. $x^2 - 2x = 15$ **26.** $x^2 - 2x - 2 = 0$

27. $x^2 + 6x - 1 = 0$ **28.** $x^2 - 16x = 5$

29. $x^2 - 3x = 5$ **30.** $x^2 + 5x = 2$

31. $x^2 - 5x + 2 = 0$ **32.** $x^2 - 9x + 7 = 0$

33. $x^2 - 4 = 2x$ **34.** $x^2 + 1 = 7x$

35. $2x^2 - 3x = 4$ **36.** $3x^2 + 6x = 5$

37. $4x^2 - 8x = 7$ **38.** $25x^2 - 20x = 1$

39. $36x^2 + 18x = -1$ **40.** $6x^2 + 4x - 1 = 0$

APPLICATIONS

41. *Geometry* The length of a rectangle is 4 inches more than its width. Find the width of the rectangle if the area is 16 square inches.

42. *Distance* An athlete is jogging west while a faster athlete is jogging south from a common starting point. At the moment when the athletes are 4 miles apart, the faster athlete has traveled 2 miles farther than the slower athlete. How far has the slower athlete traveled at this time?

WRITING ABOUT MATHEMATICS

43. Explain how to determine whether a trinomial in the form $x^2 + bx + c$ is a perfect square trinomial.

44. Explain the steps for solving $ax^2 + bx = d$, with $a \neq 1$, by completing the square.

GROUP ACTIVITY
WORKING WITH REAL DATA

Directions: Form a group of 2 to 4 people. Select someone to record the group's responses for this activity. All members of the group should work cooperatively to answer the questions. If your instructor asks for the results, each member of the group should be prepared to respond.

Minimum Wage The table shows the minimum wage for three different years.

Year (x)	1940	1968	1997
Wage (y)	$0.25	$1.60	$5.45

Source: Bureau of Labor Statistics.

The formula $y = 0.0016(x - 1940)^2 + 0.25$ can be used to estimate the minimum wage y during year x for the time period represented in the table.

(a) Use the formula to calculate the minimum wage for each year in the given table. For which year do the values for the minimum wage given by the table and the formula differ most?

(b) Estimate the minimum wage in 1976 and compare it to the actual value of $2.30.

(c) Estimate the year when the minimum wage was $1.00.

(d) Use the formula to compute the minimum wage in 2008.

9.4 THE QUADRATIC FORMULA

Solving Quadratic Equations ■ The Discriminant

A LOOK INTO MATH ▷

To estimate the stopping distance of a car, highway engineers compute two quantities. The first quantity is the *reaction distance*, which is the distance a car travels from the time a driver first recognizes a hazard until the brakes are applied. The second quantity is *braking distance*, which is the distance a car travels after a driver applies the brakes. If a car is traveling x miles per hour, highway engineers estimate the reaction distance in feet as $\frac{11}{3}x$ and the braking distance in feet as $\frac{1}{9}x^2$. The total *stopping distance d* in feet is

$$d = \frac{1}{9}x^2 + \frac{11}{3}x.$$

If a car's headlights don't illuminate the road beyond **620** feet, a safe nighttime speed limit x for the car can be determined by solving the quadratic equation

$$\frac{1}{9}x^2 + \frac{11}{3}x = 620.$$

(**Source:** L. Haefner, *Introduction to Transportation Systems*.)

In this section we learn how to solve this equation (see Example 6) with the quadratic formula.

Solving Quadratic Equations

Recall that any quadratic equation can be written in the form

$$ax^2 + bx + c = 0.$$

If we solve this equation for x in terms of a, b, and c by completing the square, we obtain the **quadratic formula**. See Exercise 79.

> ### QUADRATIC FORMULA
>
> The solutions to $ax^2 + bx + c = 0$ with $a \neq 0$ are given by
> $$x = \frac{-b \pm \sqrt{b^2 - 4ac}}{2a}.$$

NOTE: The quadratic formula can be used to solve *any* quadratic equation.

In the first example we find values for a, b, and c by writing a quadratic equation in the form $ax^2 + bx + c = 0$. These values are used in the quadratic formula.

EXAMPLE 1 Determining *a*, *b*, and *c*

Determine a, b, and c by writing each equation in the form $ax^2 + bx + c = 0$.
(a) $4x^2 - 3x + 5 = 0$ **(b)** $-2x^2 + 1 = x$ **(c)** $(x - 3)(x + 3) = 0$

Solution
(a) The equation $4x^2 - 3x + \mathbf{5} = 0$ is in the form $ax^2 + bx + c = 0$, so $a = \mathbf{4}$, $b = \mathbf{-3}$, and $c = \mathbf{5}$.
(b) Subtract x from each side of $-2x^2 + 1 = x$ to obtain $-2x^2 - x + 1 = 0$. The equation is now in the form $ax^2 + bx + c = 0$ with $a = -2$, $b = -1$, and $c = 1$.
(c) Multiply the left side of $(x - 3)(x + 3) = 0$ to obtain $x^2 - 9 = 0$. Thus $a = 1$ and $c = -9$. There is no x-term, so $b = 0$.   `Now Try Exercises 7, 9, 13`

Sometimes we can use either factoring or the quadratic formula to solve an equation. This situation is demonstrated in the next example.

EXAMPLE 2 Solving a quadratic equation having two solutions

Solve $2x^2 + 3x - 2 = 0$ by factoring and by the quadratic formula.

Solution
Factoring

$$2x^2 + 3x - 2 = 0 \qquad \text{Given equation}$$
$$(2x - 1)(x + 2) = 0 \qquad \text{Factor.}$$
$$2x - 1 = 0 \quad \text{or} \quad x + 2 = 0 \qquad \text{Zero-product property}$$
$$x = \frac{1}{2} \quad \text{or} \quad x = -2 \qquad \text{Solve for } x.$$

Quadratic Formula   Let $a = \mathbf{2}$, $b = \mathbf{3}$, and $c = \mathbf{-2}$ in the quadratic formula.

$$x = \frac{-b \pm \sqrt{b^2 - 4ac}}{2a} \qquad \text{Quadratic formula}$$
$$= \frac{-3 \pm \sqrt{3^2 - 4(2)(-2)}}{2(2)} \qquad \text{Substitute for } a, b, \text{ and } c.$$
$$= \frac{-3 \pm \sqrt{25}}{4} \qquad \text{Simplify.}$$
$$= \frac{-3 \pm 5}{4} \qquad \text{Evaluate; } \sqrt{25} = 5.$$
$$= \frac{1}{2} \quad \text{or} \quad -2 \qquad \text{Simplify.}$$

Both methods give the same solutions of $\frac{1}{2}$ and -2.   `Now Try Exercise 19`

One general strategy for solving a quadratic equation is to try factoring first. If factoring does not work, use completing the square or the quadratic formula. The next example cannot be solved by factoring, so we use the quadratic formula and support our answer graphically.

EXAMPLE 3   Solving a quadratic equation having two solutions

Solve the equation $2x^2 - 3x - 1 = 0$. Support your result graphically.

Solution

Symbolic Solution   Let $a = 2$, $b = -3$, and $c = -1$.

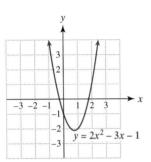

$$x = \frac{-b \pm \sqrt{b^2 - 4ac}}{2a} \qquad \text{Quadratic formula}$$

$$x = \frac{-(-3) \pm \sqrt{(-3)^2 - 4(2)(-1)}}{2(2)} \qquad \text{Substitute for } a, b, \text{ and } c.$$

$$x = \frac{3 \pm \sqrt{17}}{4} \qquad \text{Simplify.}$$

The solutions are $\frac{3 + \sqrt{17}}{4} \approx 1.78$ and $\frac{3 - \sqrt{17}}{4} \approx -0.28$.

Graphical Solution   The graph of $y = 2x^2 - 3x - 1$ is shown in Figure 9.14. The two x-intercepts correspond to the two solutions to $2x^2 - 3x - 1 = 0$. Estimating from this graph, we see that the solutions are approximately -0.25 and 1.75, which supports our symbolic solution.   Now Try Exercise 21

Figure 9.14 Two x-intercepts

MAKING CONNECTIONS

Solutions and x-intercepts

Example 3 supports the fact that the solutions to the equation $2x^2 - 3x - 1 = 0$ are equal to the x-intercepts on the graph of $y = 2x^2 - 3x - 1$. In general, the real solutions to $ax^2 + bx + c = 0$ are equal to the x-intercepts on the graph of $y = ax^2 + bx + c$.

EXAMPLE 4   Solving a quadratic equation having one solution

Solve the equation $x^2 + 4x + 4 = 0$. Support your result graphically.

Solution

Symbolic Solution   Let $a = 1$, $b = 4$, and $c = 4$.

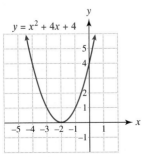

$$x = \frac{-b \pm \sqrt{b^2 - 4ac}}{2a} \qquad \text{Quadratic formula}$$

$$= \frac{-4 \pm \sqrt{4^2 - 4(1)(4)}}{2(1)} \qquad \text{Substitute for } a, b, \text{ and } c.$$

$$= \frac{-4 \pm \sqrt{0}}{2} \qquad \text{Simplify.}$$

$$= -2 \qquad \sqrt{0} = 0$$

There is one solution, -2.

Graphical Solution   The graph of $y = x^2 + 4x + 4$ is shown in Figure 9.15. The one x-intercept, -2, corresponds to the solution to $x^2 + 4x + 4 = 0$.   Now Try Exercise 23

Figure 9.15 One x-intercept

EXAMPLE 5   Recognizing a quadratic equation having no real solutions

Solve the equation $5x^2 - x + 3 = 0$. Support your result graphically.

Solution
Symbolic Solution   Let $a = 5$, $b = -1$, and $c = 3$.

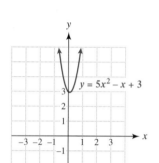

$$x = \frac{-b \pm \sqrt{b^2 - 4ac}}{2a}$$ Quadratic formula

$$= \frac{-(-1) \pm \sqrt{(-1)^2 - 4(5)(3)}}{2(5)}$$ Substitute for a, b, and c.

$$= \frac{1 \pm \sqrt{-59}}{10}$$ Simplify.

There are no real solutions to this equation because $\sqrt{-59}$ is not a real number. (Later in this chapter we discuss how to find complex solutions to quadratic equations such as this one.)

Graphical Solution   The graph of $y = 5x^2 - x + 3$ is shown in Figure 9.16. There are no x-intercepts, indicating that the equation $5x^2 - x + 3 = 0$ has no real solutions.

Figure 9.16 No x-intercepts

Now Try Exercise 25

▶ **REAL-WORLD CONNECTION**   Earlier in this section we discussed how engineers estimate safe stopping distances for automobiles. In the next example we solve the equation presented in the introduction.

EXAMPLE 6   Estimating stopping distance

If a car's headlights do not illuminate the road beyond 620 feet, estimate a safe nighttime speed limit x for the car by solving $\frac{1}{9}x^2 + \frac{11}{3}x = 620$.

Solution
Begin by subtracting 620 from each side of the equation.

$$\frac{1}{9}x^2 + \frac{11}{3}x - 620 = 0.$$ Subtract 620.

To eliminate fractions, multiply each side by the LCD, which is 9. (This step is not necessary, but it makes the problem easier to work.)

$$x^2 + 33x - 5580 = 0.$$ Multiply by 9.

Now let $a = 1$, $b = 33$, and $c = -5580$ in the quadratic formula.

$$x = \frac{-b \pm \sqrt{b^2 - 4ac}}{2a}$$ Quadratic formula

$$= \frac{-33 \pm \sqrt{33^2 - 4(1)(-5580)}}{2(1)}$$ Substitute for a, b, and c.

$$= \frac{-33 \pm \sqrt{23{,}409}}{2}$$ Simplify.

Because $\sqrt{23{,}409} = 153$, the solutions are

$$\frac{-33 + 153}{2} = 60 \quad \text{and} \quad \frac{-33 - 153}{2} = -93.$$

The negative solution has no physical meaning because negative speeds are not possible. The other solution is 60, so an appropriate speed limit is 60 miles per hour.

Now Try Exercise 69

The Discriminant

The expression $b^2 - 4ac$ in the quadratic formula is called the **discriminant**. It provides information about the number of solutions to a quadratic equation.

> ### THE DISCRIMINANT AND QUADRATIC EQUATIONS
>
> To determine the number of real solutions to the equation $ax^2 + bx + c = 0$, evaluate the discriminant $b^2 - 4ac$.
>
> 1. If $b^2 - 4ac > 0$, there are two real solutions.
> 2. If $b^2 - 4ac = 0$, there is one real solution.
> 3. If $b^2 - 4ac < 0$, there are no real solutions.

EXAMPLE 7   Using the discriminant

Use the discriminant to determine the number of solutions to $4x^2 + 25 = 20x$.

Solution

Write the equation as $4x^2 - 20x + 25 = 0$ so that $a = 4$, $b = -20$, and $c = 25$. The discriminant evaluates to

$$b^2 - 4ac = (-20)^2 - 4(4)(25) = 0.$$

Thus there is one real solution.

Now Try Exercise 53(a), (b)

We also need to be able to analyze graphs of quadratic equations, which we demonstrate in the next example.

EXAMPLE 8   Analyzing a graph of a quadratic equation

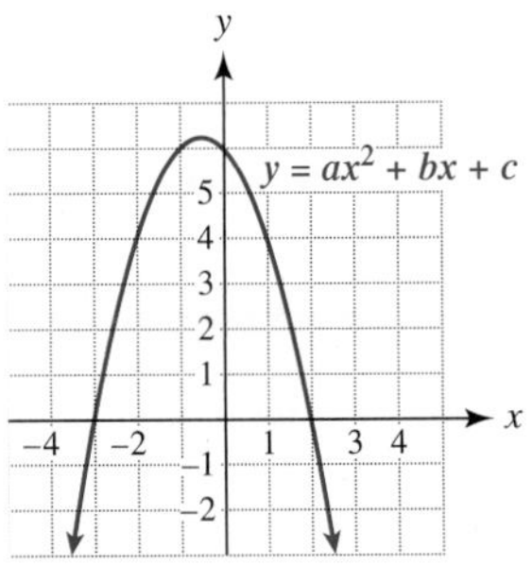

A graph of $y = ax^2 + bx + c$ is shown in Figure 9.17.
(a) State whether $a > 0$ or $a < 0$.
(b) Solve the equation $ax^2 + bx + c = 0$.
(c) Determine whether the discriminant is positive, negative, or zero.

Figure 9.17

Solution
(a) The parabola opens downward, so $a < 0$.
(b) The solutions correspond to the x-intercepts, -3 and 2.
(c) There are two real solutions, so the discriminant is positive.

Now Try Exercise 45

9.4 PUTTING IT ALL TOGETHER

Quadratic equations can be solved symbolically by using factoring, the square root property, completing the square, and the quadratic formula. Graphical and numerical methods can also be used to solve quadratic equations. In this section we discussed the quadratic formula and its discriminant, which we summarize in the following table.

Concept	Explanation	Examples
Quadratic Formula	The quadratic formula can be used to solve *any* quadratic equation written as $ax^2 + bx + c = 0$. The solutions are given by $$x = \frac{-b \pm \sqrt{b^2 - 4ac}}{2a}.$$	For the equation $$2x^2 - 3x + 1 = 0$$ with $a = 2$, $b = -3$, and $c = 1$, the solutions are $$\frac{-(-3) \pm \sqrt{(-3)^2 - 4(2)(1)}}{2(2)} = \frac{3 \pm \sqrt{1}}{4} = 1, \frac{1}{2}.$$
Discriminant	The discriminant, $b^2 - 4ac$, may be used to determine the number of solutions to $ax^2 + bx + c = 0$. 1. $b^2 - 4ac > 0$ indicates two real solutions. 2. $b^2 - 4ac = 0$ indicates one real solution. 3. $b^2 - 4ac < 0$ indicates no real solutions.	For the equation $$x^2 + 4x - 1 = 0$$ with $a = 1$, $b = 4$, and $c = -1$, the discriminant is $$b^2 - 4ac = 4^2 - 4(1)(-1) = 20 > 0,$$ indicating two real solutions.

9.4 Exercises

CONCEPTS

1. What is the quadratic formula used for?

2. How many real solutions can $ax^2 + bx + c = 0$ have?

3. What is the discriminant?

4. If the discriminant evaluates to 0, what does that indicate about the quadratic equation?

5. Name four symbolic techniques for solving a quadratic equation.

6. Does every quadratic equation have at least one real solution? Explain.

THE QUADRATIC FORMULA

Exercises 7–14: (Refer to Example 1.) Determine a, b, and c by writing the equation in the form $ax^2 + bx + c = 0$.

7. $5x^2 - 4x + 6 = 0$ 8. $-2x^2 + x - 7 = 0$

9. $3x^2 = 2x - 5$ 10. $7x^2 + 2x = 8$

11. $x^2 = x$ 12. $x(x - 1) = 0$

13. $(x - 3)(x + 4) = 0$ 14. $(2x - 1)(x + 2) = 0$

Exercises 15–20: (Refer to Example 2.) Solve the equation by factoring and by the quadratic formula.

15. $x^2 - 2x + 1 = 0$ 16. $x^2 + 8x + 16 = 0$

17. $x^2 - 2x - 3 = 0$ **18.** $x^2 - 5x + 4 = 0$

19. $2x^2 - 5x - 3 = 0$ **20.** $4x^2 + x - 3 = 0$

Exercises 21–26: Use the quadratic formula to solve the equation. Support your result graphically. If there are no real solutions, say so.

21. $2x^2 + 11x - 6 = 0$ **22.** $x^2 + 2x - 24 = 0$

23. $-x^2 + 2x - 1 = 0$ **24.** $3x^2 - x + 1 = 0$

25. $2x^2 + x + 1 = 0$ **26.** $25 + x^2 = 10x$

Exercises 27–44: Solve by using the quadratic formula. If there are no real solutions, say so.

27. $x^2 - 6x - 16 = 0$ **28.** $2x^2 - 9x + 7 = 0$

29. $4x^2 - x - 1 = 0$ **30.** $-x^2 + 2x + 1 = 0$

31. $-3x^2 + 2x - 1 = 0$ **32.** $x^2 + x + 3 = 0$

33. $36x^2 + 9 = 36x$ **34.** $4x^2 + 1.96 = 5.6x$

35. $2x(x - 3) = 2$ **36.** $x(x + 1) + x = 5$

37. $(x - 1)(x + 1) + 2 = 4x$

38. $\frac{1}{2}(x - 6) = x^2 + 1$ **39.** $\frac{1}{2}x(x + 1) = 2x^2 - \frac{3}{2}$

40. $\frac{1}{2}x^2 - \frac{1}{4}x + \frac{1}{2} = x$ **41.** $2x(x - 1) = 7$

42. $3x(x - 4) = 4$ **43.** $-3x^2 + 10x - 5 = 0$

44. $-2x^2 + 4x - 1 = 0$

THE DISCRIMINANT

Exercises 45–50: A graph of $y = ax^2 + bx + c$ is shown.
  (a) State whether $a > 0$ or $a < 0$.
  (b) Solve $ax^2 + bx + c = 0$, if possible.
  (c) Determine whether the discriminant is positive, negative, or zero.

45.

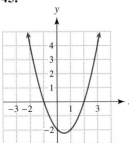

46.

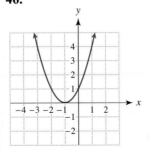

47.

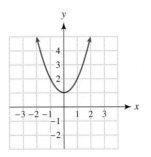

48.

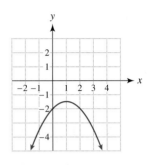

49.

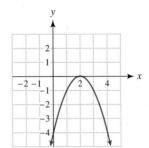

50.

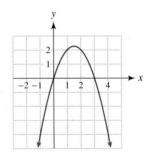

Exercises 51–58: Do the following for the given equation.
  (a) Evaluate the discriminant.
  (b) How many real solutions are there?
  (c) Support your answer for part (b) by using graphing.

51. $3x^2 + x - 2 = 0$ **52.** $5x^2 - 13x + 6 = 0$

53. $x^2 - 4x + 4 = 0$ **54.** $\frac{1}{4}x^2 + 4 = 2x$

55. $\frac{1}{2}x^2 + \frac{3}{2}x + 2 = 0$ **56.** $x - 3 = 2x^2$

57. $x(x + 3) = 3$

58. $(4x - 1)(x - 3) = -25$

Exercises 59–68: Use the quadratic formula to find any x-intercepts on the graph of the equation.

59. $y = x^2 - 2x - 1$ **60.** $y = x^2 + 3x + 1$

61. $y = -2x^2 - x + 3$ **62.** $y = -3x^2 - x + 4$

63. $y = x^2 + x + 5$ **64.** $y = 3x^2 - 2x + 5$

65. $y = x^2 + 9$ **66.** $y = x^2 + 11$

67. $y = 3x^2 + 4x - 2$ **68.** $y = 4x^2 - 2x - 3$

APPLICATIONS

Exercises 69–72: ***Estimating Stopping Distance*** *(Refer to Example 6.) Use $d = \frac{1}{9}x^2 + \frac{11}{3}x$ to find a safe speed x for the following stopping distances d.*

69. 80 feet   **70.** 210 feet

71. 390 feet   **72.** 900 feet

73. *Estimating U.S. AIDS Deaths*   The cumulative numbers in thousands of AIDS deaths from 1984 through 1994 may be estimated by

$$y = 2.39x^2 + 5.04x + 5.1,$$

where $x = 0$ corresponds to 1984, $x = 1$ to 1985, and so on until $x = 10$ corresponds to 1994. See the accompanying graph. Use the formula to estimate the year when the total number of AIDS deaths reached 200 thousand. Compare your result with that shown in the graph.

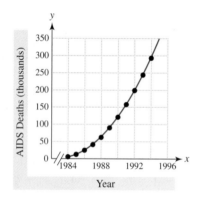

74. *Screen Dimensions*   The width of a rectangular computer screen is 2 inches more than its height. If the area of the screen is 168 square inches, find its dimensions.

75. *Area of a Garden*   There are 200 feet of fence around a rectangular garden. If the area of the garden is 2475 square feet, determine the dimensions of the garden.

76. *Height Reached by a Golf Ball*   A golf ball is hit so that its height h in feet after t seconds is given by $h = -16t^2 + 64t$. When is the golf ball 60 feet in the air?

WRITING ABOUT MATHEMATICS

77. Explain how the discriminant can be used to determine the number of real solutions to a quadratic equation.

78. If you know the value of $b^2 - 4ac$, what information does this expression give you about the graph of $y = ax^2 + bx + c$? Explain your answer.

79. Derive the quadratic formula by completing the following. Explain each step. Assume $a > 0$.
(a) Write $ax^2 + bx + c = 0$ as $x^2 + \frac{b}{a}x = -\frac{c}{a}$.
(b) Complete the square to obtain

$$\left(x + \frac{b}{2a}\right)^2 = \frac{b^2 - 4ac}{4a^2}.$$

(c) Use the square root property and solve for x.

CHECKING BASIC CONCEPTS

SECTIONS 9.3 AND 9.4

1. Determine whether the given trinomial is a perfect square trinomial.
(a) $x^2 - 6x + 12$
(b) $x^2 + 2x + 1$

2. Complete the square to solve each equation.
(a) $x^2 - 4x = -1$
(b) $x^2 - 6x + 4 = 0$

3. Use the quadratic formula to solve each equation.
(a) $2x^2 = 3x + 1$
(b) $9x^2 - 24x + 16 = 0$

4. Calculate the discriminant for each equation and give the number of real solutions.
(a) $x^2 - 5x + 5 = 0$
(b) $2x^2 - 5x + 4 = 0$
(c) $49x^2 - 56x + 16 = 0$

9.5 COMPLEX SOLUTIONS

Basic Concepts ▪ Addition, Subtraction, and Multiplication ▪ Quadratic Equations Having Complex Solutions

A common misconception is that abstract or theoretical mathematics is unimportant in today's world. Many new ideas with great practical importance were first developed as abstract concepts with no particular application in mind. For example, complex numbers, which are related to square roots of negative numbers, started as an abstract concept to solve equations. Today complex numbers are used in many sophisticated applications, such as the design of electrical circuits, ships, and airplanes.

Basic Concepts

A graph of $y = x^2 + 1$ is shown in Figure 9.18. There are no x-intercepts, so the equation $x^2 + 1 = 0$ has no real-number solutions. If we try to solve $x^2 + 1 = 0$ by subtracting 1 from each side, the result is $x^2 = -1$. Because $x^2 \geq 0$ for any real number x, there are no real solutions. However, mathematicians have invented solutions.

$$x^2 = -1$$
$$x = \pm \sqrt{-1} \qquad \text{Square root property}$$

We now define a *number* called the **imaginary unit**, denoted i.

Figure 9.18

PROPERTIES OF THE IMAGINARY UNIT i

$$i = \sqrt{-1} \quad \text{and} \quad i^2 = -1$$

By creating the number i, the solutions to the equation $x^2 + 1 = 0$ are i and $-i$. Using the real numbers and the imaginary unit i, we can define a new set of numbers called the *complex numbers*. A **complex number** can be written in **standard form**, as $a + bi$, where a and b are real numbers. The **real part** is a and the **imaginary part** is b. Every real number a is also a complex number because it can be written $a + 0i$. A complex number $a + bi$ with $b \neq 0$ is an **imaginary number**. Table 9.8 lists several complex numbers with their real and imaginary parts.

TABLE 9.8 Complex Numbers

$a + bi$	$-3 + 2i$	5	$-3i$	$-5 - 2i$	$4 + 6i$
Real Part: a	-3	5	0	-5	4
Imaginary Part: b	2	0	-3	-2	6

Figure 9.19 on the next page shows how different sets of numbers are related. Note that *the set of complex numbers contains the set of real numbers.*

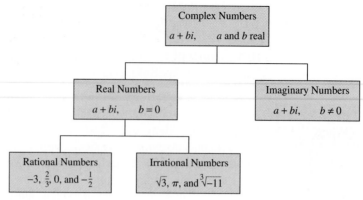

Figure 9.19

Using the imaginary unit i, we may write the square root of a negative number as a complex number. For example, $\sqrt{-5} = i\sqrt{5}$, and $\sqrt{-9} = i\sqrt{9} = 3i$.

THE EXPRESSION $\sqrt{-a}$

If $a > 0$, then $\sqrt{-a} = i\sqrt{a}$.

EXAMPLE 1 Writing the square root of a negative number

Write each square root using the imaginary unit i.

(a) $\sqrt{-4}$ **(b)** $\sqrt{-7}$ **(c)** $\sqrt{-20}$

Solution

(a) $\sqrt{-4} = i\sqrt{4} = 2i$ **(b)** $\sqrt{-7} = i\sqrt{7}$

(c) $\sqrt{-20} = i\sqrt{20} = i\sqrt{4}\sqrt{5} = 2i\sqrt{5}$

Now Try Exercises **9, 11, 15**

NOTE: Although standard form for a complex number is $a + bi$, we often write $1 + \sqrt{2}i$ or $\sqrt{7}i$ as $1 + i\sqrt{2}$ or $i\sqrt{7}$ so that it is clear that i is *not* part of the square root.

Addition, Subtraction, and Multiplication

Arithmetic operations can be defined for complex numbers.

ADDITION AND SUBTRACTION To add the complex numbers $(-3 + 2i)$ and $(2 - i)$, add the real parts and add the imaginary parts.

$$(-3 + 2i) + (2 - i) = (-3 + 2) + (2i - i)$$
$$= (-3 + 2) + (2 - 1)i$$
$$= -1 + i$$

This same process works for subtraction.

$$(6 - 3i) - (2 + 5i) = (6 - 2) + (-3i - 5i)$$
$$= (6 - 2) + (-3 - 5)i$$
$$= 4 - 8i$$

This method is summarized as follows.

SUM OR DIFFERENCE OF COMPLEX NUMBERS

Let $a + bi$ and $c + di$ be two complex numbers. Then

$$(a + bi) + (c + di) = (a + c) + (b + d)i \qquad \text{Sum}$$

and

$$(a + bi) - (c + di) = (a - c) + (b - d)i. \qquad \text{Difference}$$

EXAMPLE 2   Adding and subtracting complex numbers

Write each sum or difference in standard form.
(a) $(-7 + 2i) + (3 - 4i)$ **(b)** $3i - (5 - i)$

Solution
(a) $(-7 + 2i) + (3 - 4i) = (-7 + 3) + (2 - 4)i = -4 - 2i$
(b) $3i - (5 - i) = 3i - 5 + i = -5 + (3 + 1)i = -5 + 4i$

Now Try Exercises 19, 25

TECHNOLOGY NOTE: **Complex Numbers**

Many calculators can perform arithmetic with complex numbers. The figure shows a calculator display for the results in Example 2.

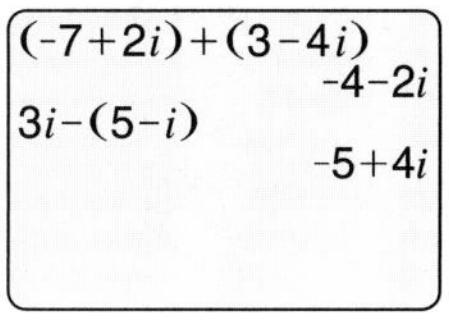

```
(-7+2i)+(3-4i)
                -4-2i
3i-(5-i)
                -5+4i
```

MULTIPLICATION   We multiply two complex numbers in the same way that we multiply binomials and then we apply the property $i^2 = -1$.

EXAMPLE 3   Multiplying complex numbers

Write each product in standard form.
(a) $(2 - 3i)(1 + 4i)$ **(b)** $(5 - 2i)(5 + 2i)$

Solution
(a) Multiply the complex numbers like binomials.

$$\begin{aligned}
(2 - 3i)(1 + 4i) &= (2)(1) + (2)(4i) - (3i)(1) - (3i)(4i) \\
&= 2 + 8i - 3i - 12i^2 \\
&= 2 + 5i - 12(-1) \\
&= 14 + 5i
\end{aligned}$$

(b) Multiply these complex numbers in the same way.

$$\begin{aligned}
(5 - 2i)(5 + 2i) &= (5)(5) + (5)(2i) - (2i)(5) - (2i)(2i) \\
&= 25 + 10i - 10i - 4i^2 \\
&= 25 - 4(-1) \\
&= 29
\end{aligned}$$

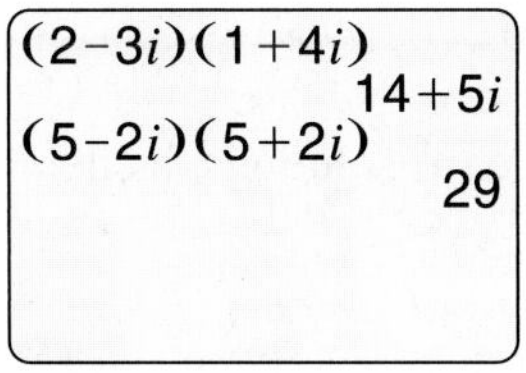

```
(2-3i)(1+4i)
                14+5i
(5-2i)(5+2i)
                   29
```

Figure 9.20

These results are supported in Figure 9.20.

Now Try Exercises 29, 30

Quadratic Equations Having Complex Solutions

The quadratic equation $ax^2 + bx + c = 0$ has no real solutions whenever the discriminant, $b^2 - 4ac$, is negative. For example, the quadratic equation $x^2 + 4 = 0$ has $a = 1$, $b = 0$, and $c = 4$. Its discriminant is

$$b^2 - 4ac = 0^2 - 4(1)(4) = -16 < 0,$$

so this equation has no real solutions. However, if we allow solutions to be complex numbers, we can solve this equation as follows.

$x^2 + 4 = 0$	Given equation
$x^2 = -4$	Subtract 4.
$x = \pm\sqrt{-4}$	Square root property
$x = \sqrt{-4}$ or $x = -\sqrt{-4}$	Meaning of $\pm$
$x = 2i$ or $x = -2i$	The expression $\sqrt{-a}$

The solutions are $\pm 2i$. We check each solution to $x^2 + 4 = 0$ as follows.

$$(2i)^2 + 4 = (2)^2 i^2 + 4 = 4(-1) + 4 = 0 \qquad \text{It checks.}$$

$$(-2i)^2 + 4 = (-2)^2 i^2 + 4 = 4(-1) + 4 = 0 \qquad \text{It checks.}$$

The following result is an extension of the square root property to include complex solutions.

THE EQUATION $x^2 + k = 0$

If $k > 0$, the solutions to $x^2 + k = 0$ are given by $x = \pm i\sqrt{k}$.

EXAMPLE 4   Solving a quadratic equation having complex solutions

Solve $x^2 + 3 = 0$.

Solution
The solutions are $\pm i\sqrt{3}$. That is, $x = i\sqrt{3}$ or $x = -i\sqrt{3}$.   Now Try Exercise 41

When $b \neq 0$, the preceding method cannot be used. Consider the quadratic equation $2x^2 + x + 3 = 0$, which has $a = 2$, $b = 1$, and $c = 3$. Its discriminant is

$$b^2 - 4ac = 1^2 - 4(2)(3) = -23 < 0.$$

This equation has no real solutions but has two complex (imaginary) solutions, as demonstrated in the next example.

EXAMPLE 5 **Solving a quadratic equation having complex solutions**

Solve $2x^2 + x + 3 = 0$. Write your answers in standard form: $a + bi$.

Solution
Because $2x^2 + x + 3 = 0$, let $a = 2$, $b = 1$, and $c = 3$.

$$x = \frac{-b \pm \sqrt{b^2 - 4ac}}{2a}$$ Quadratic formula

$$= \frac{-1 \pm \sqrt{1^2 - 4(2)(3)}}{2(2)}$$ Substitute for a, b, and c.

$$= \frac{-1 \pm \sqrt{-23}}{4}$$ Simplify.

$$= \frac{-1 \pm i\sqrt{23}}{4}$$ $\sqrt{-23} = i\sqrt{23}$

$$= -\frac{1}{4} \pm i\frac{\sqrt{23}}{4}$$ Divide each term by 4.

The solutions in standard form are $-\frac{1}{4} + i\frac{\sqrt{23}}{4}$ and $-\frac{1}{4} - i\frac{\sqrt{23}}{4}$.   **Now Try Exercise** 57

Sometimes we can use properties of radicals to simplify a solution to a quadratic equation, as demonstrated in the next example.

EXAMPLE 6 **Solving a quadratic equation having complex solutions**

Solve $\frac{3}{4}x^2 + 1 = x$. Write your answer in standard form: $a + bi$.

Solution
Begin by subtracting x from each side of the equation and then multiply by 4 to clear fractions. The resulting equation is $3x^2 - 4x + 4 = 0$. Substitute $a = 3$, $b = -4$, and $c = 4$ in the quadratic formula.

$$x = \frac{-b \pm \sqrt{b^2 - 4ac}}{2a}$$ Quadratic formula

$$= \frac{-(-4) \pm \sqrt{(-4)^2 - 4(3)(4)}}{2(3)}$$ Substitute.

$$= \frac{4 \pm \sqrt{-32}}{6}$$ Simplify.

$$= \frac{4 \pm 4i\sqrt{2}}{6}$$ $\sqrt{-32} = i\sqrt{32} = i\sqrt{16}\sqrt{2} = 4i\sqrt{2}$

$$= \frac{2}{3} \pm \frac{2}{3}i\sqrt{2}$$ Divide 6 into each term and simplify.

                                                      Now Try Exercise 47

9.5 PUTTING IT ALL TOGETHER

In this section we discussed complex numbers and quadratic equations having complex solutions. Complex numbers allow us to solve quadratic equations that could not be solved using only real numbers. The following table summarizes the important concepts in this section.

Concept	Explanation	Examples
Complex Numbers	A complex number can be expressed as $a + bi$, where a and b are real numbers. The imaginary unit i satisfies $i = \sqrt{-1}$ and $i^2 = -1$. As a result, we can write $\sqrt{-a} = i\sqrt{a}$ if $a > 0$.	$5 + 3i$ and $-3 - 7i$ $\sqrt{-13} = i\sqrt{13}$ and $\sqrt{-9} = 3i$
Addition, Subtraction, and Multiplication of Complex Numbers	To add (subtract) complex numbers, add (subtract) the real parts and then add (subtract) the imaginary parts. Multiply complex numbers in a similar manner to how *FOIL* is used to multiply binomials. Then apply the property $i^2 = -1$.	$(3 + 6i) + (-1 + 2i)$ Sum $= (3 + -1) + (6 + 2)i$ $= 2 + 8i$ $(2 - 5i) - (1 + 4i)$ Difference $= (2 - 1) + (-5 - 4)i$ $= 1 - 9i$ $(-1 + 2i)(3 + i)$ Product $= (-1)(3) + (-1)(i) + (2i)(3) + (2i)(i)$ $= -3 - i + 6i + 2i^2$ $= -3 + 5i + 2(-1)$ $= -5 + 5i$
Quadratic Formula and Complex Solutions	If the discriminant is negative, or $b^2 - 4ac < 0$, the solutions to a quadratic equation are complex numbers. If $k > 0$, the solutions to $x^2 + k = 0$ are given by $x = \pm i\sqrt{k}$.	$2x^2 - x + 3 = 0$ $x = \dfrac{-(-1) \pm \sqrt{(-1)^2 - 4(2)(3)}}{2(2)}$ $= \dfrac{1 \pm \sqrt{-23}}{4} = \dfrac{1}{4} \pm i\dfrac{\sqrt{23}}{4}$ $x^2 + 9 = 0$ is equivalent to $x = \pm 3i$ and $x^2 + 7 = 0$ is equivalent to $x = \pm i\sqrt{7}$.

9.5 Exercises

MyMathLab

PRACTICE WATCH DOWNLOAD READ REVIEW

CONCEPTS

1. Give an example of a complex number that is not a real number.

2. Can you give an example of a real number that is not a complex number? Explain.

3. $\sqrt{-1} = $ _____

4. $i^2 = $ _____

5. $\sqrt{-a} = $ _____, if $a > 0$.

6. The standard form for a complex number is _____.

7. The real part of $4 - 5i$ is _____.

8. The imaginary part of $4 - 5i$ is _____.

COMPLEX NUMBERS

Exercises 9–18: Use the imaginary unit to write the expression.

9. $\sqrt{-3}$
10. $\sqrt{-11}$
11. $\sqrt{-36}$
12. $\sqrt{-49}$
13. $\sqrt{-144}$
14. $\sqrt{-64}$
15. $\sqrt{-12}$
16. $\sqrt{-8}$
17. $\sqrt{-18}$
18. $\sqrt{-48}$

Exercises 19–38: Write the expression in standard form.

19. $(4 + 3i) + (-2 - 3i)$
20. $(2 - i) + (5 - 7i)$
21. $2i + 5i$
22. $-3i + 6i$
23. $(2 - 7i) - (1 + 2i)$
24. $(1 + 8i) - (3 + 9i)$
25. $5i - (10 - 2i)$
26. $(1 + i) - (1 - i)$
27. $4(5 - 3i)$
28. $2(-6 - i)$
29. $(-3 - 4i)(5 - 4i)$
30. $(3 + 5i)(3 - 5i)$
31. $(-4i)(5i)$
32. $(-6i)(-4i)$
33. $3i + (2 - 3i) - (1 - 5i)$
34. $4 - (5 - 7i) + (3 + 7i)$
35. $(2 + i)^2$
36. $(-1 + 2i)^2$
37. $2i(-3 + i)$
38. $5i(1 - 9i)$

39. **Thinking Generally** Evaluate $(a + bi)(a - bi)$ for any real numbers a and b.

40. **Thinking Generally** Evaluate i^{4n} for any natural number n.

COMPLEX SOLUTIONS

Exercises 41–60: Solve the equation. Write complex solutions in standard form.

41. $x^2 + 9 = 0$
42. $x^2 + 16 = 0$
43. $x^2 + 80 = 0$
44. $x^2 + 20 = 0$
45. $x^2 + \frac{1}{4} = 0$
46. $x^2 + \frac{9}{4} = 0$
47. $\frac{3}{2}x^2 + 2 = x$
48. $5x^2 + 6 = x$
49. $x^2 = -6$
50. $x^2 = -75$
51. $x^2 - 3 = 0$
52. $x^2 - 8 = 0$
53. $x^2 + 2 = 0$
54. $x^2 + 4 = 0$
55. $x^2 + 2 = x$
56. $x^2 + 2x + 3 = 0$
57. $2x^2 + 3x = -4$
58. $3x^2 - x = 1$
59. $x^2 + 1 = 4x$
60. $3x^2 + 2 = x$

WRITING ABOUT MATHEMATICS

61. A student multiplies $(2 + 3i)(4 - 5i)$ *incorrectly* to obtain $8 - 15i$. What was the student's mistake?

62. Explain what a negative discriminant indicates about the solutions to a quadratic equation.

9.6 INTRODUCTION TO FUNCTIONS

Basic Concepts ▪ Representing a Function ▪ Definition of a Function ▪
Identifying a Function

A LOOK INTO MATH ▷   In earlier chapters we showed how to use numbers to describe data. For example, instead of simply saying that it is *hot* outside, we might use the number $102°F$ to describe the temperature. We also showed that data can be modeled with formulas and graphs. Formulas and graphs are sometimes used to represent *functions*, which are important in mathematics. In this section we introduce functions and their representations.

Basic Concepts

▶ **REAL-WORLD CONNECTION**   Functions are used to calculate many important quantities. For example, suppose that a person works for $7 per hour. We could use a function f to calculate the amount of money the person earned after working x hours simply by multiplying the *input x* by 7. The result y is called the *output*. This concept is shown visually in the following diagram.

$$\text{Input } x \longrightarrow \text{Function } f \longrightarrow \text{Output } y = f(x)$$

For each valid input x, a function computes *exactly one* output y, which may be represented by the ordered pair (x, y). If the input is 5 hours, f outputs $7 \cdot 5 = \$35$; if the input is 8 hours, f outputs $7 \cdot 8 = \$56$. These results can be represented by the ordered pairs $(5, 35)$ and $(8, 56)$. Sometimes an input may not be valid. For example, if $x = -3$, there is no reasonable output because a person cannot work -3 hours.

We say that *y is a function of x* because the output y is determined by the input x. To emphasize that y is a function of x, we use the notation $y = f(x)$. The symbol $f(x)$ does not represent multiplication of a variable f and a variable x. The notation $y = f(x)$ is called *function notation*, is read "*y* equals *f* of *x*," and means that function f with input x produces output y. For example, if $x = 3$ hours, $y = f(3) = \$21$.

FUNCTION NOTATION

The notation $y = f(x)$ is called **function notation**. The **input** is x, the **output** is y, and the *name* of the function is f.

$$\underset{\text{Output}}{\overset{\text{Name}}{y = f(x)}}_{\text{Input}}$$

The expression $f(4) = 28$ is read "*f* of 4 equals 28" and indicates that f outputs 28 when the input is 4. A function computes *exactly one* output for each valid input. The letters f, g, and h are often used to denote names of functions.

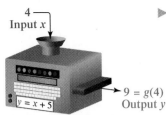

4 ─┐
Input x

$y = x + 5$

9 = g(4)
Output y

Figure 9.21 Function Machine

▶ **REAL-WORLD CONNECTION**   Functions can be used to compute a variety of quantities. For example, suppose that a boy has a sister who is exactly 5 years older than he is. If the age of the boy is x, then a function g can calculate the age of his sister by adding 5 to x. Thus $g(4) = 4 + 5 = 9$, $g(10) = 10 + 5 = 15$, and in general $g(x) = x + 5$. That is, function g adds 5 to every input x to obtain the output $y = g(x)$.

Functions can be represented by an input–output machine, as illustrated in Figure 9.21. This machine represents function g and receives input $x = 4$, adds 5 to this value, and then outputs $g(4) = 4 + 5 = 9$.

Representing a Function

▶ **REAL-WORLD CONNECTION**   Functions can be represented by words, tables, formulas, graphs, and diagrams. We begin by considering a different function f that converts yards to feet.

TABLE 9.9

x (yards)	y (feet)
1	3
2	6
3	9
4	12
5	15
6	18
7	21

VERBAL DESCRIPTION (WORDS)   To convert x yards to y feet, we must multiply x by 3. Therefore if function f computes the number of feet in x yards, a **verbal description** of f is "Multiply the input x in yards by 3 to obtain the output y in feet."

TABLE OF VALUES   A function f that converts yards to feet is shown in Table 9.9, where $y = f(x)$. Many times it is impossible to list all valid inputs x in a table. Many valid inputs, such as $x = 10$ or $x = 5.3$, are not shown in Table 9.9. Note that for each valid input x there is exactly one output y. *For a function, inputs are not listed more than once in a table.*

FORMULA   The computation performed by f to convert x yards to y feet is expressed by $y = 3x$. A formula for f is $f(x) = 3x$, where $y = f(x)$. We say that function f is *defined by* or *given by* $f(x) = 3x$. Thus $f(2) = 3 \cdot 2 = 6$.

GRAPH   A graph visually associates an x-input with a y-output. The ordered pairs

$$(1, 3), (2, 6), (3, 9), (4, 12), (5, 15), (6, 18), \text{ and } (7, 21)$$

from Table 9.9 are plotted in Figure 9.22(a). This scatterplot suggests a line for the graph of f. For each real number x there is exactly one real number y determined by $y = 3x$. If we restrict inputs to $x \geq 0$ and plot all ordered pairs $(x, 3x)$, then a line with no breaks will appear, as shown in Figure 9.22(b).

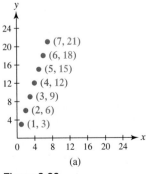

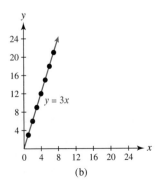

(a)   (b)

Figure 9.22

(a) Function

(b) Function

(c) Not a Function

Figure 9.23

MAKING CONNECTIONS

Functions, Points, and Graphs

If $f(a) = b$, then the point (a, b) lies on the graph of f. Conversely, if the point (a, b) lies on the graph of f, then $f(a) = b$. Thus each point on the graph of a function f can be written in the form $(a, f(a))$.

DIAGRAM   Functions may be represented by **diagrams**. Figure 9.23(a) is a diagram of a function, where an arrow is used to identify the output y associated with input x. For example, input **2** results in output **6**, which is written in function notation as $f(2) = 6$. That is, **2** yards are equivalent to **6** feet. Figure 9.23(b) shows a function f even though $f(1) = 4$ and $f(2) = 4$. Although two inputs for f have the same output, each valid input has exactly one output. In contrast, Figure 9.23(c) does not show a function because input 2 results in two different outputs, 5 and 6.

MAKING CONNECTIONS

Four Representations of a Function

Formula $f(x) = x + 1$

Table of Values

x	y
-2	-1
-1	0
0	1
1	2
2	3

Graph

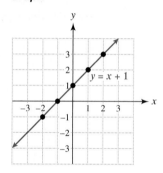

Verbal Description f adds 1 to an input x to produce an output y.

EXAMPLE **1**   Evaluating functions

Evaluate $f(x)$ at the given value of x.
(a) $f(x) = 3x - 7$ $x = -2$

(b) $f(x) = \dfrac{x}{x + 2}$ $x = 0.5$

(c) $f(x) = \sqrt{x - 1}$ $x = 10$

Solution
(a) $f(-2) = 3(-2) - 7 = -6 - 7 = -13$

(b) $f(0.5) = \dfrac{0.5}{0.5 + 2} = \dfrac{0.5}{2.5} = 0.2$

(c) $f(10) = \sqrt{10 - 1} = \sqrt{9} = 3$

Now Try Exercises **11, 13, 21**

EXAMPLE 2   Graphing a function by hand

Sketch a graph of $f(x) = x - 1$. Use the graph to evaluate $f(-2)$.

Solution
Begin by making a table of values containing at least three points. Pick convenient values of x, such as $x = -1, 0, 1$.

$$f(-1) = -1 - 1 = -2$$
$$f(0) = 0 - 1 = -1$$
$$f(1) = 1 - 1 = 0$$

Display the results, as shown in Table 9.10.

Plot the points $(-1, -2)$, $(0, -1)$, and $(1, 0)$. Then sketch a line through the points to obtain the graph of f. A graph of a line results when *infinitely* many points are plotted, as shown in Figure 9.24.

To evaluate $f(-2)$, first find $x = -2$ on the x-axis. See Figure 9.25. Then move downward to the graph of f. By moving across to the y-axis, we see that the corresponding y-value is -3. Thus $f(-2) = -3$.

TABLE 9.10

x	y
-1	-2
0	-1
1	0

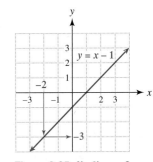

Figure 9.24

Figure 9.25 $f(-2) = -3$

Now Try Exercises 33, 43

NOTE: If a function f is given by a formula $f(x)$, then to graph f, we graph $y = f(x)$. For example, to graph $f(x) = x - 1$, simply graph $y = x - 1$.

In the next example we find a formula and then sketch a graph of a function.

EXAMPLE 3   Representing a function

Let function f square the input x and then subtract 1 to obtain the output y.
(a) Write a formula for f.
(b) Make a table of values for f. Use $x = -2, -1, 0, 1, 2$.
(c) Sketch a graph of f.

Solution
(a) *Formula*   If we square x and then subtract 1, we obtain $x^2 - 1$. Thus a formula for f is
$$f(x) = x^2 - 1.$$

TABLE 9.11

x	$f(x)$
-2	3
-1	0
0	-1
1	0
2	3

(b) *Table of Values*   Make a table of values for $f(x)$, as shown in Table 9.11. For example,
$$f(-2) = (-2)^2 - 1 = 4 - 1 = 3.$$

(c) *Graph*   To obtain a graph of $f(x) = x^2 - 1$, plot the points from Table 9.11 and then connect them with a smooth curve, as shown in Figure 9.26. Note that we need to plot enough points so that we can determine the overall shape of the graph.

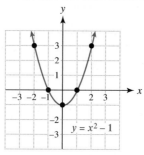

Figure 9.26

Now Try Exercises 23, 39

Definition of a Function

A function is an important concept in mathematics. *A function receives an input x and produces exactly one output y*, which can be expressed as an ordered pair:

$$(x, y).$$
Input   Output

> ### FUNCTION
>
> A **function** f is a set of ordered pairs (x, y), where each x-value corresponds to exactly one y-value.

The **domain** of f is the set of all x-values (inputs), and the **range** of f is the set of all y-values (outputs). For example, a function f that converts 1, 2, 3, and 4 yards to feet could be expressed as

$$f = \{(1, 3), (2, 6), (3, 9), (4, 12)\}.$$

The domain of f is $D = \{1, 2, 3, 4\}$, and the range of f is $R = \{3, 6, 9, 12\}$.

▶ REAL-WORLD CONNECTION   In the next example, we see how education can improve a person's chances for earning a higher income.

EXAMPLE 4   Computing average income

The function f computes the average 2004 individual income in dollars by educational attainment. This function is defined by $f(N) = 18,900$, $f(H) = 25,900$, $f(B) = 45,400$, and $f(M) = 62,300$, where N denotes no diploma, H a high school diploma, B a bachelor's degree, and M a master's degree.   (*Source:* Bureau of the Census.)
(a) Write f as a set of ordered pairs.
(b) Give the domain and range of f.
(c) Discuss the relationship between education and income.

Solution

(a) $f = \{(N, 18900), (H, 25900), (B, 45400), (M, 62300)\}$.

(b) The domain of function f is $D = \{N, H, B, M\}$, and the range of function f is $R = \{18900, 25900, 45400, 62300\}$.

(c) Education pays—on average, the greater the educational attainment, the greater the annual earnings.   Now Try Exercise **29**

The domain of a function is the set of all valid inputs. To determine the domain of a function from a formula, we must determine x-values for which the formula is defined. This concept is demonstrated in the next example.

EXAMPLE **5**   Finding domains of functions

Use $f(x)$ to find the domain of f.

(a) $f(x) = 5x$ (b) $f(x) = \dfrac{1}{x-2}$ (c) $f(x) = \sqrt{x}$

Solution

(a) Because we can always multiply a real number x by 5, $f(x) = 5x$ is defined for all real numbers. Thus the domain of f includes all real numbers.

(b) Because we cannot divide by 0, input $x = 2$ is not valid for $f(x) = \frac{1}{x-2}$. The expression for $f(x)$ is defined for all other values of x. Thus the domain of f includes all real numbers except 2, or $x \neq 2$.

(c) Because square roots of negative numbers are not real numbers, the inputs for $f(x) = \sqrt{x}$ cannot be negative. Thus the domain of f includes all nonnegative numbers, or $x \geq 0$.

Now Try Exercises **55, 59, 63**

Identifying a Function

Recall that for a function each valid input x produces exactly one output y. In the next two examples we demonstrate techniques for identifying a function.

EXAMPLE **6**   Determining whether a table of values represents a function

TABLE 9.12

x	y
1	−4
2	8
3	2
1	5
4	−6

Determine whether Table 9.12 represents a function.

Solution
The table does not represent a function because input $x = 1$ produces two outputs: −4 and 5.

Same input x

$(1, -4) \qquad (1, 5)$

Different outputs y

Now Try Exercise **77**

VERTICAL LINE TEST   To determine whether a graph represents a function, we must be convinced that it is impossible for an input x to have two or more outputs y. If two distinct points have the same x-coordinate on a graph, then the graph cannot represent a function. For example, the ordered pairs $(-1, 1)$ and $(-1, -1)$ could not lie on the graph of a function because input -1 results in *two* outputs: 1 and -1. When the points $(-1, 1)$ and $(-1, -1)$ are plotted, they lie on the same vertical line, as shown in Figure 9.27(a). A graph passing through these points intersects the vertical line twice, as illustrated in Figure 9.27(b).

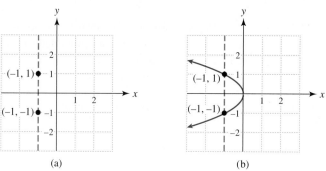

(a)   (b)

Figure 9.27

To determine whether a graph represents a function, visualize vertical lines moving across the xy-plane. If each vertical line intersects the graph *at most once*, then it is a graph of a function. This test is called the **vertical line test**. Note that the graph in Figure 9.27(b) fails the vertical line test and therefore does not represent a function.

VERTICAL LINE TEST

If every vertical line intersects a graph at no more than one point, then the graph represents a function.

EXAMPLE 7   Determining whether a graph represents a function

Determine whether the graphs shown in Figures 9.28 and 9.29 represent functions.

(a)   **(b)**

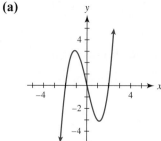

Figure 9.28

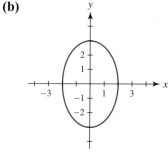

Figure 9.29

Solution

(a) Any vertical line will cross the graph at most once, as depicted in Figure 9.30. Therefore the graph *does* represent a function.

(b) The graph *does not* represent a function because there exist vertical lines that can intersect the graph twice. One such line is shown in Figure 9.31.

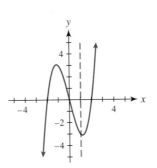

Figure 9.30 Passes Vertical Line Test

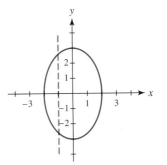

Figure 9.31 Fails Vertical Line Test

Now Try Exercises 71, 73

9.6 PUTTING IT ALL TOGETHER

One important concept in mathematics is that of a function. A function calculates exactly one output for each valid input and produces input–output ordered pairs in the form (x, y). A function typically computes something, such as area, speed, or sales tax. The following table summarizes some concepts related to functions.

Concept	Explanation	Examples
Function	A set of ordered pairs (x, y), where each x-value corresponds to exactly one y-value	$f = \{(1, 3), (2, 3), (3, 1)\}$ $f(x) = 2x$ A graph of $y = x + 2$ A table of values for $y = 4x$
Domain and Range of a Function	The domain D is the set of all valid inputs. The range R is the set of all outputs.	For $S = \{(-1, 0), (3, 4), (5, 0)\}$, $D = \{-1, 3, 5\}$ and $R = \{0, 4\}$. For $f(x) = \frac{1}{x}$ the domain includes all real numbers except 0, or $x \neq 0$.
Vertical Line Test	If every vertical line intersects a graph at no more than one point, the graph represents a function.	This graph does *not* pass this test and thus does not represent a function.

9.6 Exercises

MyMathLab
Math XL PRACTICE   WATCH   DOWNLOAD   READ   REVIEW

CONCEPTS

1. The notation $y = f(x)$ is called _____ notation.

2. The notation $y = f(x)$ is read _____.

3. If $f(x) = x^2 + 1$, then $f(2) =$ _____.

4. The set of valid inputs for a function is the _____.

5. The set of outputs for a function is the _____.

6. A function computes _____ output for each valid input.

7. What is the vertical line test used for?

8. If $f(3) = 4$, the point _____ is on the graph of f. If $(3, 6)$ is on the graph of g, then $g($ _____ $) =$ _____.

9. **Thinking Generally**   If $f(a) = b$, the point _____ is on the graph of f.

10. **Thinking Generally**   If (c, d) is on the graph of g, then $g(c) =$ _____.

REPRESENTING AND EVALUATING FUNCTIONS

Exercises 11–22: Evaluate $f(x)$ at the given values of x.

11. $f(x) = 4x - 2$ $x = -1, 0$

12. $f(x) = 5 - 3x$ $x = -4, 2$

13. $f(x) = \sqrt{x}$ $x = 0, \frac{9}{4}$

14. $f(x) = \sqrt[3]{x}$ $x = -1, 27$

15. $f(x) = x^2$ $x = -5, \frac{3}{2}$

16. $f(x) = x^3$ $x = -2, 0.1$

17. $f(x) = 3$ $x = -8, \frac{7}{3}$

18. $f(x) = 100$ $x = -\pi, \frac{1}{3}$

19. $f(x) = 5 - x^3$ $x = -2, 3$

20. $f(x) = x^2 + 5$ $x = -\frac{1}{2}, 6$

21. $f(x) = \dfrac{2}{x + 1}$ $x = -5, 4$

22. $f(x) = \dfrac{x}{x - 4}$ $x = -3, 1$

Exercises 23–28: Do the following.
   (a) *Write a formula for the function described.*
   (b) *Evaluate the function for input 10.*

23. Function I computes the number of inches in x yards.

24. Function M computes the number of miles in x feet.

25. Function A computes the area of a circle with radius r.

26. Function C computes the circumference of a circle with radius r.

27. Function A computes the square feet in x acres. (*Hint:* 43,560 square feet equal 1 acre.)

28. Function K computes the number of kilograms in x pounds. (*Hint:* 2.2 pounds equal 1 kilogram.)

Exercises 29–32: Write each function f as a set of ordered pairs. Give the domain and range of f.

29. $f(1) = 3, f(2) = -4, f(3) = 0$

30. $f(-1) = 4, f(0) = 6, f(1) = 4$

31. $f(a) = b, f(c) = d, f(e) = a, f(d) = b$

32. $f(a) = 7, f(b) = 7, f(c) = 7, f(d) = 7$

Exercises 33–42: Sketch a graph of f.

33. $f(x) = -x + 3$ 34. $f(x) = -2x + 1$

35. $f(x) = 2x$ 36. $f(x) = \frac{1}{2}x - 2$

37. $f(x) = 4 - x$ 38. $f(x) = 6 - 3x$

39. $f(x) = x^2$ 40. $f(x) = \sqrt{x}$

41. $f(x) = \sqrt{x + 1}$ 42. $f(x) = \frac{1}{2}x^2 - 1$

Exercises 43–48: Use the graph of f to evaluate the given expressions.

43. $f(0)$ and $f(2)$

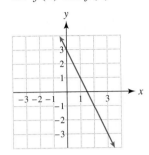

44. $f(-2)$ and $f(2)$

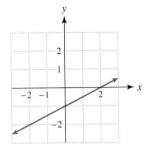

45. $f(-2)$ and $f(1)$

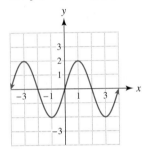

46. $f(-1)$ and $f(0)$

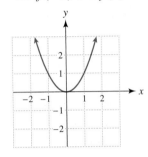

47. $f(1)$ and $f(2)$

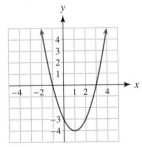

48. $f(-1)$ and $f(4)$

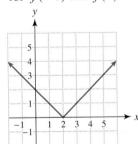

Exercises 49–52: Give a verbal description for $f(x)$.

49. $f(x) = x - \frac{1}{2}$

50. $f(x) = \frac{3}{4}x$

51. $f(x) = \frac{x}{3}$

52. $f(x) = x^2 + 1$

53. *Home Prices* The average price P of a single-family home in thousands of dollars from 1990 to 2000 can be approximated by $P(x) = 5.7(x - 1990) + 150$, where x is the year. Evaluate $P(1995)$ and interpret the result. (*Source:* U.S. Census Bureau.)

54. *Median Family Income* The median income I of a family in thousands of dollars from 2000 to 2005 can be approximated by $I(x) = 1.4(x - 2000) + 51$, where x is the year. Evaluate $P(2003)$ and interpret the result. (*Source:* U.S. Census Bureau.)

IDENTIFYING DOMAINS

Exercises 55–64: Find the domain.

55. $f(x) = 10x$

56. $f(x) = 5 - x$

57. $f(x) = x^2 - 3$

58. $f(x) = \frac{1}{2}x^2$

59. $f(x) = \frac{3}{x - 5}$

60. $f(x) = \frac{x}{x + 1}$

61. $f(x) = \frac{2x}{x^2 + 1}$

62. $f(x) = \frac{6}{1 - x}$

63. $f(x) = \sqrt{x - 1}$

64. $f(x) = \sqrt{2 - x}$

IDENTIFYING A FUNCTION

Exercises 65–68: Determine whether the diagram could represent a function.

65.

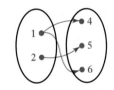

66.

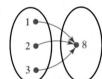

67.

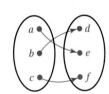

68.

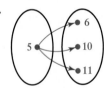

Exercises 69–76: Determine whether the graph represents a function.

69.

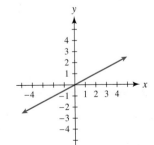

70.

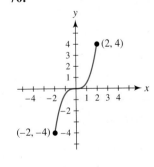

71.

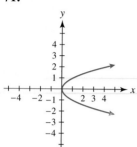

72.

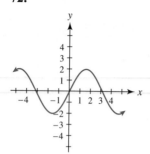

75.

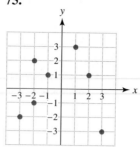

76.

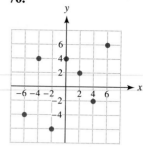

73.

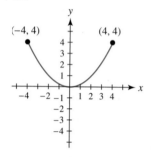

74.

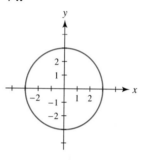

Exercises 77 and 78: Determine whether the table represents a function.

77.

x	5	10	5
y	2	1	0

78.

x	−3	−2	−1
y	10	10	10

WRITING ABOUT MATHEMATICS

79. Give an example of a function. Identify the domain of your function.

80. Explain how to evaluate a function by using a graph. Give an example.

CHECKING BASIC CONCEPTS

SECTIONS 9.5 AND 9.6

1. Use the imaginary unit i to write each expression.
   (a) $\sqrt{-64}$ **(b)** $\sqrt{-17}$

2. Simplify each expression.
   (a) $(2 - 3i) + (1 - i)$
   (b) $4i - (2 + i)$ **(c)** $(3 - 2i)(1 + i)$

3. Solve each quadratic equation.
   (a) $x^2 + 25 = 0$
   (b) $x^2 - 6x + 13 = 0$
   (c) $x^2 + x + 2 = 0$

4. Find a formula for a function f that squares the input x and then subtracts 1.

5. Use the graph to evaluate $f(0)$ and $f(2)$.

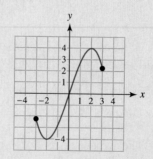

CHAPTER 9 SUMMARY

SECTION 9.1 ■ PARABOLAS

Equation of a Parabola   The equation of a parabola can be written as

$$y = ax^2 + bx + c \quad (a \neq 0).$$

Graph of $y = ax^2 + bx + c$ Its graph is $\cup$-shaped or $\cap$-shaped and is called a parabola. If $a > 0$, the parabola opens upward; if $a < 0$, the parabola opens downward. The vertex of a parabola is either the lowest point on a parabola that opens upward or the highest point on a parabola that opens downward.

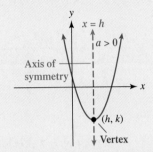
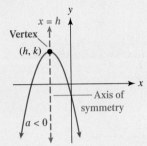

The Vertex Formula   The x-coordinate of the vertex is given by $x = -\frac{b}{2a}$.

Axis of Symmetry   The axis of symmetry is a vertical line that passes through the vertex of the graph of a parabola. If the vertex is (h, k), then the axis of symmetry is $x = h$. The parabola is symmetric with respect to this line.

The Graph of $y = ax^2$ The graph of $y = ax^2$ is a parabola that opens upward when $a > 0$ and opens downward when $a < 0$. As the value of $|a|$ increases, the graph of $y = ax^2$ becomes narrower. The vertex is $(0, 0)$, and the axis of symmetry is the y-axis.

SECTION 9.2 ■ INTRODUCTION TO QUADRATIC EQUATIONS

Quadratic Equations   Any quadratic equation can be written as $ax^2 + bx + c = 0$ and can have no real solutions, one real solution, or two real solutions. These solutions correspond to the x-intercepts on the graph of $y = ax^2 + bx + c$. These equations can be solved symbolically by factoring or by the square root property.

Example:
$$x^2 + x - 2 = 0$$
$$(x + 2)(x - 1) = 0$$
$$x = -2 \quad \text{or} \quad x = 1$$

The x-intercepts for $y = x^2 + x - 2$ are -2 and 1.

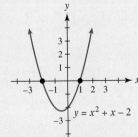

The Square Root Property   Let k be a nonnegative number. Then the solutions to the equation $x^2 = k$ are given by $x = \pm\sqrt{k}$. If $k < 0$, this equation has no real solutions.

Example: $x^2 = 16$ is equivalent to $x = \pm 4$.

SECTION 9.3 ■ SOLVING BY COMPLETING THE SQUARE

Completing the Square   To solve $x^2 + bx = d$ by completing the square add $\left(\frac{b}{2}\right)^2$ to each side of the equation.

Example:

$x^2 + 6x = 4$	Given equation
$x^2 + 6x + 9 = 4 + 9$	Add $\left(\frac{6}{2}\right)^2 = 9$ to each side.
$(x + 3)^2 = 13$	Perfect square trinomial
$x + 3 = \pm\sqrt{13}$	Square root property
$x = -3 \pm \sqrt{13}$	Add -3 to each side.
$x \approx 0.61, -6.61$	Approximate.

SECTION 9.4 ■ THE QUADRATIC FORMULA

The Quadratic Formula   The solutions to $ax^2 + bx + c = 0$ $(a \neq 0)$ are given by

$$x = \frac{-b \pm \sqrt{b^2 - 4ac}}{2a}.$$

Example: Solve $2x^2 + 3x - 1 = 0$ by letting $a = 2$, $b = 3$, and $c = -1$.

$$x = \frac{-3 \pm \sqrt{3^2 - 4(2)(-1)}}{2(2)} = \frac{-3 \pm \sqrt{17}}{4} \approx 0.28, -1.78$$

The Discriminant   The expression $b^2 - 4ac$ evaluates to a real number and is called the discriminant. If $b^2 - 4ac > 0$, there are two real solutions; if $b^2 - 4ac = 0$, there is one real solution; and if $b^2 - 4ac < 0$, there are no real solutions to $ax^2 + bx + c = 0$.

Example: For $2x^2 + 3x - 1 = 0$, the discriminant is

$$b^2 - 4ac = 3^2 - 4(2)(-1) = 17 > 0.$$

There are two real solutions to this quadratic equation.

SECTION 9.5 ■ COMPLEX SOLUTIONS

Complex Numbers

Imaginary Unit	$i = \sqrt{-1}$ and $i^2 = -1$
Standard Form	$a + bi$, where a and b are real numbers
	Examples: $4 + 3i$, $5 - 6i$, 8, and $-2i$
Real Part	The real part of $a + bi$ is a.
	Example: The real part of $3 - 2i$ is 3.
Imaginary Part	The imaginary part of $a + bi$ is b.
	Example: The imaginary part of $2 - i$ is -1.

Arithmetic Operations   Arithmetic operations are similar to arithmetic operations on binomials.

Examples: $(2 + 2i) + (3 - i) = 5 + i,$
$(1 - i) - (1 - 2i) = i,$ and
$(1 - i)(1 + i) = 1^2 - i^2 = 1 - (-1)$
$= 2$

Quadratic Formula and Complex Solutions   If the discriminant evaluates to a negative number, or $b^2 - 4ac < 0$, the solutions to a quadratic equation are complex numbers.

Square Root Property   If $k > 0$, the solutions to $x^2 + k = 0$ are given by $x = \pm i\sqrt{k}$.

SECTION 9.6 ■ INTRODUCTION TO FUNCTIONS

Function   A function is a set of ordered pairs (x, y), where each x-value corresponds to exactly one y-value. A function takes a valid input x and computes exactly one output y, forming the ordered pair (x, y).

Domain and Range of a Function   The domain D is the set of all valid inputs, or x-values, and the range R is the set of all outputs, or y-values.

Examples: $f = \{(1, 2), (2, 3), (3, 3)\}$ has $D = \{1, 2, 3\}$ and $R = \{2, 3\}$.

$f(x) = x^2$ has domain all real numbers.

Function Notation $y = f(x)$ and is read "y equals f of x."

Example: $f(x) = \frac{2x}{x - 1}$ implies that $f(3) = \frac{2 \cdot 3}{3 - 1} = \frac{6}{2} = 3$. Thus the point $(3, 3)$ is on the graph of f.

Function Representations   Functions can be given by words, tables, formulas, and graphs.

Formula $f(x) = x^2$

Table of Values   **Graph**

x	y
-2	4
-1	1
0	0
1	1
2	4

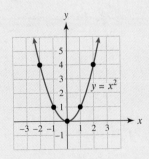

Verbal Description (Words) f squares the input x to obtain the output y.

CHAPTER 9   REVIEW EXERCISES

SECTION 9.1

Exercises 1 and 2: Identify the vertex, axis of symmetry, and whether the parabola opens upward or downward. State where the graph is increasing and where the graph is decreasing.

1.

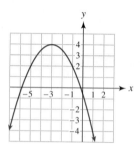

2.

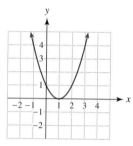

Exercises 3–6: Do the following.
  (a) Graph the parabola.
  (b) Identify the vertex and axis of symmetry.

3. $y = x^2 - 2$

4. $y = -x^2 + 4x - 3$

5. $y = -\frac{1}{2}x^2 + x + \frac{3}{2}$

6. $y = 2x^2 + 8x + 5$

Exercises 7–10: Find the vertex of the parabola.

7. $y = x^2 - 4x - 2$

8. $y = 5 - x^2$

9. $y = -\frac{1}{4}x^2 + x$

10. $y = 2 + 2x + x^2$

SECTION 9.2

Exercises 11–14: A graph of $y = ax^2 + bx + c$ is shown. Solve $ax^2 + bx + c = 0$.

11.

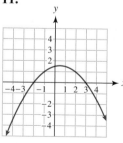

12.

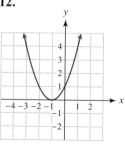

13.

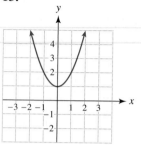

14.

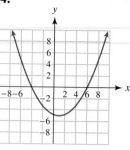

Exercises 15 and 16: A table of $y = ax^2 + bx + c$ is shown. Solve $ax^2 + bx + c = 0$.

15.

X	Y1	
-20	250	
-15	100	
-10	0	
-5	-50	
0	-50	
5	0	
10	100	
Y1■X^2+5X−50		

16.

X	Y1	
-.75	2	
-.5	0	
-.25	-1	
0	-1	
.25	0	
.5	2	
.75	5	
Y1■8X^2+2X−1		

Exercises 17–20: Solve the quadratic equation
  (a) graphically and (b) numerically.

17. $x^2 - 5x - 50 = 0$ **18.** $\frac{1}{2}x^2 + x - \frac{3}{2} = 0$

19. $\frac{1}{4}x^2 + \frac{1}{2}x = 2$ **20.** $\frac{1}{2}x + \frac{3}{4} = \frac{1}{4}x^2$

Exercises 21–24: Solve by factoring.

21. $x^2 + x - 20 = 0$ **22.** $x^2 + 11x + 24 = 0$

23. $15x^2 - 4x - 4 = 0$ **24.** $7x^2 - 25x + 12 = 0$

Exercises 25–28: Use the square root property to solve.

25. $x^2 = 100$ **26.** $3x^2 = \frac{1}{3}$

27. $4x^2 - 6 = 0$ **28.** $5x^2 = x^2 - 4$

SECTION 9.3

Exercises 29–32: Determine whether the given trinomial is a perfect square trinomial.

29. $x^2 - 8x + 8$ **30.** $x^2 - 4x + 4$

31. $x^2 + 20x + 100$ **32.** $x^2 - 2x + 2$

Exercises 33–36: Solve by completing the square.

33. $x^2 + 6x = -2$ **34.** $x^2 - 4x = 6$

35. $x^2 - 2x - 5 = 0$ **36.** $2x^2 + 6x - 1 = 0$

SECTION 9.4

Exercises 37–42: Solve by using the quadratic formula.

37. $x^2 - 9x + 18 = 0$ **38.** $x^2 - 24x + 143 = 0$

39. $6x^2 + x = 1$ **40.** $5x^2 + 1 = 5x$

41. $x(x - 8) = 5$ **42.** $2x(2 - x) = 3 - 2x$

Exercises 43–46: A graph of $y = ax^2 + bx + c$ is shown.
   (a) State whether $a > 0$ or $a < 0$.
   (b) Solve $ax^2 + bx + c = 0$.
   (c) Determine whether the discriminant is positive, negative, or zero.

43.

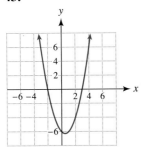

44.

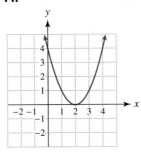

45.

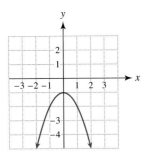

46.

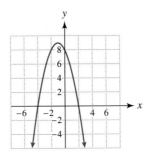

Exercises 47–50: Do the following for the equation.
   (a) Evaluate the discriminant.
   (b) How many real solutions are there?
   (c) Support your answer for part (b) graphically.

47. $2x^2 - 3x + 1 = 0$ **48.** $7x^2 + 2x - 5 = 0$

49. $3x^2 + x = -2$ **50.** $4.41x^2 + 9 = 12.6x$

SECTION 9.5

Exercises 51–54: Write the expression in standard form.

51. $(1 - 2i) + 2i$ **52.** $(1 + 3i) - (3 - i)$

53. $(1 - i)(2 + 3i)$ **54.** $(1 - i)^2(1 + i)$

Exercises 55–58: Solve the equation. Write complex solutions in standard form.

55. $x^2 + x + 5 = 0$ **56.** $2x^2 + 8 = 0$

57. $2x^2 = x - 1$ **58.** $7x^2 = 2x - 5$

SECTION 9.6

Exercises 59 and 60: Evaluate $f(x)$ for the given values of x.

59. $f(x) = 3x - 1$ $x = -2, \frac{1}{3}$

60. $f(x) = 5 - 3x^2$ $x = -3, 1$

Exercises 61 and 62: Do the following.
   (a) Write a formula for the function described.
   (b) Evaluate the function for input 5.

61. Function P computes the number of pints in q quarts.

62. Function f computes 3 less than 4 times a number x.

Exercises 63 and 64: Sketch a graph of f.

63. $f(x) = -2x$ **64.** $f(x) = x^2 - 1$

Exercises 65 and 66: Use the graph of f to evaluate the given expressions.

65. $f(0)$ and $f(-3)$ **66.** $f(-2)$ and $f(1)$

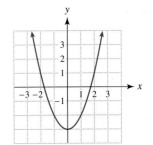

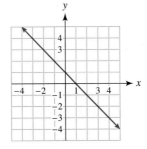

Exercises 67 and 68: Does the graph represent a function?

67.

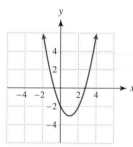

68.

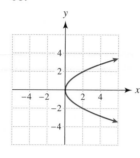

Exercises 69–72: Find the domain.

69. $f(x) = -3x + 7$

70. $f(x) = \sqrt{x}$

71. $f(x) = \frac{3}{x}$

72. $f(x) = x^2 + 2$

APPLICATIONS

73. *Braking Distance* On dry pavement a safe braking distance d in feet for a car traveling x miles per hour is $d = \frac{x^2}{12}$. For each distance d, find x. (*Source:* F. Mannering, *Principles of Highway Engineering and Traffic Control.*)

(a) $d = 144$ feet   (b) $d = 300$ feet

74. *Numbers* The product of two numbers is 143. One number is 2 more than the other.

(a) Write an equation whose solution gives the smaller number x.

(b) Solve the equation.

75. *Screen Dimensions* A square computer screen has an area of 123 square inches. Approximate its dimensions.

76. *Flying a Kite* A kite is being flown, as illustrated in the accompanying figure. If 130 feet of string have been let out, find the value of x.

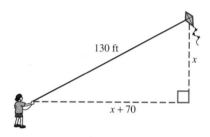

CHAPTER 9   TEST   ⬤ Pass the Test   Video solutions to all test exercises

1. Find the vertex and axis of symmetry for the graph of $y = -\frac{1}{2}x^2 + x + 1$.

2. A graph of $y = ax^2 + bx + c$ is shown. Solve $ax^2 + bx + c = 0$.

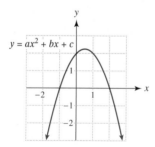

Exercises 3 and 4: Solve the quadratic equation.

3. $3x^2 + 11x - 4 = 0$ **4.** $2x^2 = 2 - 6x^2$

5. Solve $x^2 - 8x = 1$ by completing the square.

6. Solve $x(-2x + 3) = -1$ by using the quadratic formula.

7. A graph of $y = ax^2 + bx + c$ is shown.

(a) State whether $a > 0$ or $a < 0$.

(b) Solve $ax^2 + bx + c = 0$.

(c) Determine whether the discriminant is positive, negative, or zero.

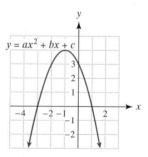

8. Complete the following for $-3x^2 + 4x - 5 = 0$.

(a) Evaluate the discriminant.

(b) How many real solutions are there?

(c) Support your answer for part (b) graphically.

Exercises 9–12: Write the expression in standard form.

9. $(-5 + i) + (7 - 20i)$ **10.** $3i - (6 - 5i)$

11. $2i(2 + 3i)$ **12.** $\left(\frac{1}{2} - i\right)\left(\frac{1}{2} + i\right)$

Exercises 13 and 14: Solve the quadratic equation.

13. $x^2 + 11 = 0$ **14.** $x^2 + 25 = 8x$

15. Evaluate $f(4)$ if $f(x) = 3x^2 - \sqrt{x}$. Give a point on the graph of f.

16. Write a formula for a function C that calculates the cost of buying x pounds of candy at \$4 per pound. Evaluate $C(5)$.

17. Sketch a graph of f.
  (a) $f(x) = -2x + 1$ **(b)** $f(x) = x^2 + 1$

18. Use the graph of f to evaluate $f(-3)$ and $f(0)$.

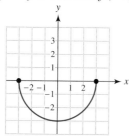

19. Determine whether the graph represents a function. Explain your reasoning.

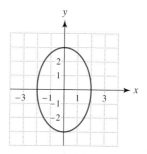

20. Find the domain of function f.
  (a) $f(x) = \frac{3}{4}x - 5$ **(b)** $f(x) = \frac{3x}{5 - x}$

21. *Braking Distance*   On wet pavement a safe braking distance d in feet for a car traveling x miles per hour is $d = \frac{x^2}{9}$. What speed corresponds to a braking distance of 250 feet?   (**Source:** F. Mannering, *Principles of Highway Engineering and Traffic Control.*)

CHAPTER 9   EXTENDED AND DISCOVERY EXERCISES

FITTING DATA WITH A PARABOLA

1. *Survival Rate of Birds*   The survival rate of sparrowhawks varies according to their age. The following table summarizes the results of one study by listing the age in years and the percentage of birds that survived the previous year. For example, 52% of sparrowhawks that reached age 6 lived to be 7 years old. (**Source:** D. Brown and P. Rothery, *Models in Biology: Mathematics, Statistics and Computing.*)

Age	1	2	3	4	5
Percent (%)	45	60	71	67	67

Age	6	7	8	9
Percent (%)	61	52	30	25

  (a) Try to explain the relationship between age and the likelihood of surviving the next year.
  (b) Make a scatterplot of the data. What type of graph might fit the data? Explain.

  (c) Graph each equation. Which graph fits (models) the data better?
  $$y_1 = -3.57x + 71.1$$
  $$y_2 = -2.07x^2 + 17.1x + 33$$

  (d) Use one of these equations to estimate the likelihood of a 5.5-year-old sparrowhawk surviving for 1 more year.

THE DISCRIMINANT

Exercises 2–5: Factoring and the Discriminant   If the discriminant of the trinomial $ax^2 + bx + c$ with integer coefficients is a perfect square, then it can be factored. For example, on the one hand, the discriminant of $6x^2 + x - 2$ is

$$1^2 - 4(6)(-2) = 49,$$

which is a perfect square ($7^2 = 49$), so we can factor the trinomial as $6x^2 + x - 2 = (2x - 1)(3x + 2)$. On the other hand, the discriminant for $x^2 + x - 1$ is

$$1^2 - 4(1)(-1) = 5,$$

which is not a perfect square, so we cannot factor this trinomial by using integers as coefficients. Similarly, if the discriminant is negative, the trinomial cannot be factored by using integer coefficients. Use the discriminant D to predict whether the trinomial can be factored. Then test your prediction.

2. $10x^2 - x - 3$

3. $4x^2 - 3x - 6$

4. $3x^2 + 2x - 2$

5. $2x^2 + x + 3$

CHAPTERS 1–9 CUMULATIVE REVIEW EXERCISES

1. Write 360 as a product of prime numbers.

2. Translate the sentence "Double a number increased by 7 equals the number decreased by 2" into an equation by using the variable n. Then solve the equation.

Exercises 3 and 4: Simplify to lowest terms.

3. $\frac{2}{3} + \frac{4}{7} \cdot \frac{21}{28}$

4. $\frac{3}{5} \div \frac{6}{5} - \frac{2}{3}$

Exercises 5 and 6: Evaluate by hand.

5. $30 - 4 \div 2 \cdot 6$

6. $\frac{3^2 - 2^3}{20 - 5 \cdot 2}$

7. Solve $2(x + 1) - 6x = x - 4$.

8. Solve $4 - 3x = -2$ graphically. Check your answer.

9. Convert 124% to fraction and decimal notation.

10. If $A = 30$ square miles and $h = 10$ miles, use the formula $A = \frac{1}{2}bh$ to find b.

11. Solve $A = \frac{h}{2}(a + b)$ for b.

12. Solve $6t - 1 < 3 - t$. Write the solution set in set-builder notation.

13. Make a scatterplot with the points $(-1, 2)$, $(1, -2)$, $(0, 3)$, $(-2, 0)$, and $(2, 3)$.

Exercises 14–16: Graph the equation and determine any intercepts.

14. $y = -\frac{1}{2}x + 2$

15. $-3x + 4y = 12$

16. $x = -2$

17. Use the graph to identify the x-intercept and the y-intercept. Then write the slope–intercept form of the line.

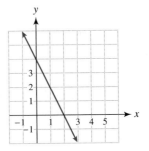

18. Write the slope–intercept form of a line that passes through the point $(-1, 3)$ with slope $m = -2$.

Exercises 19 and 20: Find the slope–intercept form of the line satisfying the given conditions.

19. Passing through the points $(-3, 5)$ and $(2, 8)$

20. Perpendicular to $x + 2y = 5$, passing through the point $(-1, 1)$

21. An insect population P is initially 4000 and increases by 500 insects per day. Write an equation in slope–intercept form that gives this population after x days.

22. Solve the system of equations. Write your answer as an ordered pair.

$$-2a + b = -5$$
$$4a - 3b = 0$$

Exercises 23 and 24: Shade the solution set.

23. $y \geq -1$

24. $3x - 2y \leq 6$

Exercises 25 and 26: Shade the solution set.

25. $y \geq x - 1$
$y \leq 2x$

26. $x - y < 3$
$2x + y \geq 6$

Exercises 27–36: Simplify and use positive exponents, when appropriate, to write the expression.

27. $(ab - 6b^2) - (4b^2 + 8ab + 2)$

28. $5 - 3^4$

29. $4(4^{-2})(3^{-1})(3^4)$

30. $\frac{2^{-4}}{4^{-2}}$

31. $(8t^{-3})(3t^2)(t^5)$

32. $2(rt)^4$

33. $(2t^3)^{-2}$

34. $(4a^2b^3)^2(2ab)^{-3}$

35. $\left(\frac{2rt^{-1}}{3r^{-2}t^3}\right)^4$

36. $\left(\frac{2a^{-1}}{ab^{-2}}\right)^{-3}$

Exercises 37–42: Multiply.

37. $2a^2(a^2 - 2a + 3)$

38. $(a + b)(a^2 - ab + b^2)$

39. $(5x + 1)(x - 7)$

40. $(y - 3)(2y + 3)$

41. $(a + b)(a - b)$

42. $(2x + 3y)^2$

43. Write 1.5×10^{-5} in standard form.

44. Write 2,130,000 in scientific notation.

Exercises 45 and 46: Divide.

45. $\frac{4x^3 - 8x^2 + 6x}{2x}$

46. $(x^4 - 9x^3 + 23x^2 - 17x + 11) \div (x - 5)$

Exercises 47–54: Factor completely.

47. $10ab^2 - 25a^3b^5$

48. $y^3 - 3y^2 + 2y - 6$

49. $6z^2 + 7z - 3$

50. $4z^2 - 9$

51. $4y^2 - 20y + 25$

52. $a^3 - 27$

53. $4z^4 - 17z^2 + 15$

54. $2a^3b + a^2b^2 - ab^3$

Exercises 55–58: Solve the equation.

55. $(x - 1)(x + 2) = 0$

56. $x^2 - 9x = 0$

57. $6y^2 - 7y = 3$

58. $x^3 = 4x$

Exercises 59 and 60: Simplify to lowest terms.

59. $\frac{x^2 - 16}{x + 4}$

60. $\frac{2x^2 - 11x - 6}{6x^2 - 5x - 4}$

61. Find the x-values that make $\frac{x - 3}{16 - x^2}$ undefined.

62. Evaluate $\frac{4x + 1}{x - 1}$ for $x = -2$.

Exercises 63–66: Simplify and write in lowest terms.

63. $\frac{x^2 - 3x + 2}{x + 7} \div \frac{x - 2}{2x + 14}$

64. $\frac{x}{2x + 3} + \frac{x + 3}{2x + 3}$

65. $\frac{5x}{x^2 - 1} - \frac{3}{x + 1}$

66. $\frac{\frac{2}{x} - \frac{2}{y}}{\frac{2}{x} + \frac{2}{y}}$

Exercises 67–69: Solve the equation.

67. $\frac{x + 2}{5} = \frac{x}{4}$

68. $\frac{1}{3x} + \frac{5}{2x} = 2$

69. $\frac{1}{x - 2} + \frac{2}{x + 2} = \frac{1}{x^2 - 4}$

70. Suppose that y is inversely proportional to x and that $y = 25$ when $x = 4$. Find y for $x = 10$.

Exercises 71 and 72: A right triangle has legs a and b with hypotenuse c. Find the exact length of the missing side.

71. $a = 8$ feet, $b = 3$ feet

72. $b = 24$ miles, $c = 25$ miles

Exercises 73 and 74: Find the exact distance between the pair of points.

73. $(-2, 3), (4, 7)$

74. $(3, -8), (-1, -8)$

Exercises 75–78: Simplify the expression. Assume that all variables are positive.

75. $\frac{\sqrt{80}}{\sqrt{5}}$

76. $\sqrt{2x^3y} \cdot \sqrt{32xy}$

77. $8\sqrt[3]{7} - 3\sqrt[3]{7}$

78. $\frac{\sqrt[4]{64}}{\sqrt[4]{4}}$

Exercises 79 and 80: Solve and check your answer.

79. $\sqrt{4 - x} = 7$

80. $x - 1 = \sqrt{2x + 1}$

Exercises 81 and 82: Find the vertex of the parabola.

81. $y = x^2 - 6x + 14$ **82.** $y = 2x^2 - 4x - 1$

Exercises 83–88: Solve the quadratic equation, using the method of your choice. Write any complex solutions in standard form.

83. $x^2 + 3x = 18$ **84.** $x^2 - 2x = 2$

85. $3x^2 - 4 = 11$ **86.** $4x^2 - 12x + 7 = 0$

87. $x^2 - 4x = -29$ **88.** $9x^2 - 6x + 17 = 0$

89. Find $f(4)$, if $f(x) = 2x^2 + \sqrt{x}$.

90. Identify the domain of each function f.
   (a) $f = \{(-3, 4), (0, 3), (2, -1)\}$
   (b) $f(x) = \dfrac{3}{x + 6}$ **(c)** $f(x) = \sqrt{x + 4}$

Exercises 91 and 92: Graph f by hand.

91. $f(x) = -2x + 1$ **92.** $f(x) = -\frac{1}{2}x^2$

APPLICATIONS

93. Motion   The table lists the distance d in miles traveled by an airplane for various elapsed times t in hours. Find an equation that represents these data.

t (hours)	2	3	4	5
d (miles)	650	975	1300	1625

94. Snowfall   By noon 4 inches of snow had fallen. For the next 6 hours snow fell at $\frac{1}{2}$ inch per hour.
   (a) Give a formula that calculates the number of inches I of snow that fell x hours past noon.
   (b) What is the slope of the graph of I?
   (c) Interpret the slope as a rate of change.
   (d) How much snow had fallen by 4 P.M.?

95. Graphical Interpretation   An athlete rides a bicycle away from home at 20 miles per hour for 1.5 hours. The athlete then turns around and rides toward home at 15 miles per hour. Sketch a graph that depicts the athlete's distance d from home after t hours.

96. Ticket Sales   The price of admission to a baseball game is $15 for children and $25 for adults. If a group of 8 people pay $170, find the number of children and the number of adults in the group.

97. Working Together   One person can shovel the snow from a sidewalk in 2 hours and another person can shovel the same sidewalk in 1.5 hours. How long will it take for them to shovel the snow from the sidewalk if they work together?

98. Flight of a Golf Ball   If a golf ball is hit upward with a velocity of 88 feet per second (60 miles per hour), then its height h in feet after t seconds can be approximated by
$$h = 88t - 16t^2.$$
   (a) What is the height of the ball after 2 seconds?
   (b) After how long does the ball strike the ground?

99. Height of a Building   A 7-foot-tall stop sign casts a 4-foot-long shadow, while a nearby building casts a 35-foot-long shadow. Find the height of the building.

100. Traffic Flow   Twenty vehicles per minute can pass through an intersection. If vehicles arrive randomly at an average rate of x per minute, the average waiting time T in minutes is $T = \frac{1}{20 - x}$ for $x < 20$.
   (**Source:** N. Garber, *Traffic and Highway Engineering.*)
   (a) Evaluate the expression for $x = 15$ and interpret the result.
   (b) Complete the table.

x	5	10	15	19	19.9
T					

   (c) What happens to the waiting time as the traffic rate approaches 20 vehicles per minute?

101. Error in Measurements   An apprentice carpenter measures the length of a board to be L feet. If the actual measurement of the board is A feet, then the relative error in this measurement is $\left| \frac{L - A}{A} \right|$. If $A = 74$ inches and the relative error is to be less than or equal to 0.005 (0.5%), what values for L are possible?

Appendix Sets

Basic Terminology

A **set** is a collection of things, and the members of a set are called **elements**. A set can be described by listing its elements between braces. For example, the set W containing the *weekdays* is

$$W = \{\text{Monday, Tuesday, Wednesday, Thursday, Friday}\}.$$

This set has 5 elements. For example, Monday *is an element of* W, which is denoted

$$\text{Monday} \in W.$$

However, Sunday *is not an element of* W, which is denoted

$$\text{Sunday} \notin W.$$

If a set contains no elements, then it is called the **empty set** or **null set**. The empty set is denoted $\varnothing$, or $\{\ \}$. For example, the set Z that contains the names of U.S. states starting with the letter Z is the empty set. That is, $Z = \varnothing$ or $Z = \{\ \}$.

NOTE: Do *not* write the empty set as $\{\varnothing\}$.

EXAMPLE 1 Listing the elements of a set

Use set notation to list the elements of each set S described.
(a) The natural numbers from 1 to 12 that are odd
(b) The days of the week that start with the letter T
(c) The last names of U.S. presidents in office during the 1990s

Solution
(a) The list of the natural numbers from 1 to 12 is

$$1, 2, 3, 4, 5, 6, 7, 8, 9, 10, 11, \text{ and } 12.$$

The set of odd natural numbers from this list is

$$S = \{1, 3, 5, 7, 9, 11\}.$$

(b) $S = \{\text{Tuesday, Thursday}\}$
(c) $S = \{\text{Bush, Clinton}\}$

Now Try Exercises 1, 3

EXAMPLE 2 Determining the elements of sets

Use $\in$ or $\notin$ to make each statement true.
(a) $5 \_\_\_\_ \{1, 2, 3, 4, 5, 6\}$
(b) $-2 \_\_\_\_ \{-4, 0, 2, 4, 6\}$
(c) $\frac{1}{2} \_\_\_\_ \{0, 0.5, 1.0, 1.5, 2.0\}$

Solution
(a) Because 5 is an element of $\{1, 2, 3, 4, 5, 6\}$, we write

$$5 \in \{1, 2, 3, 4, 5, 6\}.$$

(b) Because -2 is not an element of $\{-4, 0, 2, 4, 6\}$, we write

$$-2 \notin \{-4, 0, 2, 4, 6\}.$$

(c) Because $\frac{1}{2} = 0.5$, we write

$$\frac{1}{2} \in \{0, 0.5, 1.0, 1.5, 2.0\}.$$

Now Try Exercises **11, 13, 15**

Universal Set

When discussing sets, we assume that there is a *universal set*. The **universal set** contains all elements under consideration. For example, if the universal set U is all days of the week, then the set S containing the days that start with the letter S is

$$S = \{\text{Sunday, Saturday}\}.$$

However, if the universal set U is only the weekdays, then the set S containing the days starting with the letter S is the *empty set*, or $S = \{\ \}$.

EXAMPLE **3**   **Using different universal sets**

Determine the set O of odd integers that belong to each universal set U.
(a) $U = \{1, 2, 3, 4, 5, 6, 7, 8, 9, 10\}$
(b) $U = \{1, 6, 11, 16, 21, 26\}$
(c) $U = \{1, 2, 3, 4, \dots\}$

Solution
(a) The odd integers in U are 1, 3, 5, 7, and 9, so

$$O = \{1, 3, 5, 7, 9\}.$$

(b) The odd integers in $U = \{1, 6, 11, 16, 21, 26\}$ are 1, 11, and 21. Thus

$$O = \{1, 11, 21\}.$$

(c) The three dots in $\{1, 2, 3, 4, \dots\}$ indicate that U contains all natural numbers. Thus the set O contains all odd natural numbers, or

$$O = \{1, 3, 5, 7, 9, \dots\}.$$ Now Try Exercises **21, 23**

Subsets

If every element in a set B is contained in a set A, then we say that B is a **subset** of A, denoted $B \subseteq A$. For example, if $A = \{1, 2, 3, 4\}$ and $B = \{2, 4\}$, then $B \subseteq A$ because every element in B belongs to A. However, A is not a subset of B, denoted $A \nsubseteq B$. because the elements 1 and 3 are in A but *not* in B. The symbol $\nsubseteq$ is read "is not a subset of."

If every element in set A is in set B and every element in set B is in set A, then A and B are **equal sets**, denoted $A = B$. Note that if $A \subseteq B$ and $B \subseteq A$, then $A = B$. Why?

EXAMPLE **4**   **Determining subsets**

Let $A = \{a, b, c, d, e\}$, $B = \{b, c, d\}$, $C = \{b, e\}$, and $D = \{e, b\}$. Determine whether each statement is true or false.
(a) $A \subseteq B$ **(b)** $B \subseteq A$ **(c)** $C = D$ **(d)** $C \nsubseteq A$ **(e)** $\varnothing \subseteq B$ **(f)** $A \subseteq A$

Solution

(a) False; the elements a and e in *A* are not in *B*, so *A* is *not* a subset of *B*.

(b) True; every element in *B* is in *A*, so *B* is a subset of *A*.

(c) True; although the elements in *C* and *D* are listed in a different order, they contain exactly the same elements, so *C* and *D* are equal.

(d) False; every element in *C* is in *A*, so *C* *is* a subset of *A*.

(e) True; the empty set, or null set, is a subset of *every* set.

(f) True; every element in *A* is in *A*, so *A* is a subset of itself.

NOTE: Every set is a subset of itself.   Now Try Exercises **29, 31, 33, 35**

Venn Diagrams

Venn diagrams are often used to depict relationships among sets. A large rectangle typically represents the universal set, and subsets of the universal set are represented by regions within the universal set. In Figure A.1 the universal set *U* is represented by everything inside the large rectangle. The set *A* is represented by the red circular region within this rectangle because *A* is a subset of *U*.

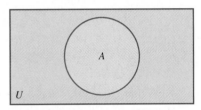

Figure A.1 $A \subseteq U$

The **complement** of a set *A*, denoted A', is the set containing all elements in the universal set that are *not* in *A*. That is, if $a \notin A$, then $a \in A'$. For example, if

$$U = \{1, 2, 3, 4, 5, 6\} \quad \text{and} \quad A = \{1, 2, 3\},$$

then

$$A' = \{4, 5, 6\}$$

because the elements 4, 5, and 6 are found in *U* but not in *A*. This situation is illustrated by the Venn diagram in Figure A.2. The red region is *A*, and the blue region is A'. Together, the red and blue regions comprise the universal set *U*.

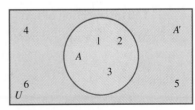

Figure A.2 *A* and A'

NOTE: Every element in *U* must be in either *A* or A', but not both.

EXAMPLE 5   Determining complements

Let the universal set U be $U = \{$red, blue, yellow, green, black, white$\}$, and let two subsets of U be $A = \{$red, blue, yellow$\}$ and $B = \{$black, white$\}$. Find each of the following.
(a) A' **(b)** B' **(c)** U'

Solution
(a) The elements in U that are not in A are in $A' = \{$green, black, white$\}$.
(b) $B' = \{$red, blue, yellow, green$\}$
(c) Because the universal set U contains every element under consideration, the complement of U, or U', is empty. That is, $U' = \{\ \}$.   Now Try Exercises 73, 75

Union and Intersection

Although we do not perform arithmetic operations, such as multiplication or division, on sets, we can find the *union or intersection* of two or more sets. The **union** of two sets A and B, denoted $A \cup B$ and read "A union B," is the set containing any element that can be found in *either* set A *or* set B. If an element is in both A and B, then this element is listed only once in $A \cup B$. For example, if

$$A = \{1, 2, 3, 4\} \quad \text{and} \quad B = \{3, 4, 5, 6\},$$

then

$$A \cup B = \{1, 2, 3, 4, 5, 6\}.$$

Note that elements 3 and 4 are in both A and B but are listed only once in $A \cup B$. A Venn diagram of this situation is shown in Figure A.3. The region that represents the union is shaded blue.

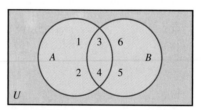

Figure A.3

The **intersection** of two sets A and B, denoted $A \cap B$ and read "A intersect B," is the set containing elements that can be found in *both* set A and set B. For example, if

$$A = \{1, 2, 3, 4\} \quad \text{and} \quad B = \{3, 4, 5, 6\},$$

then

$$A \cap B = \{3, 4\}.$$

Note that elements 3 and 4 belong to *both A and B*. This situation is illustrated by the Venn diagram in Figure A.4. The purple region containing elements 3 and 4 represents the intersection of A and B.

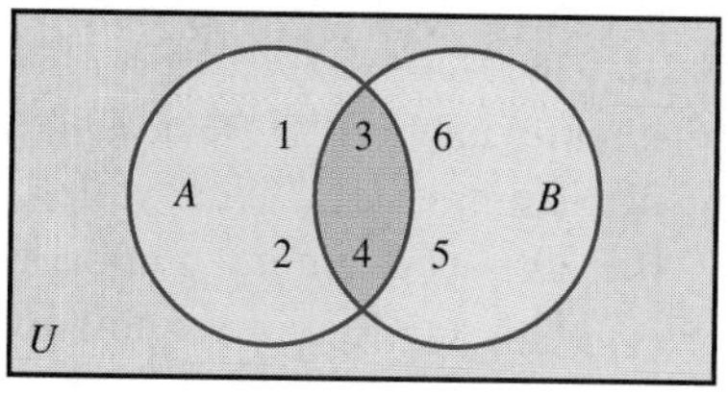

Figure A.4

EXAMPLE 6 **Finding unions and intersections of sets**

Let $A = \{a, b, c, x, z\}$, $B = \{a, x, y\}$, and $C = \{c, x, y, z\}$. Find each set.
(a) $A \cup B$ **(b)** $B \cap C$ **(c)** $A \cup C$ **(d)** $A \cap B \cap C$

Solution

(a) The union contains the elements belonging to either A *or* B or both. Thus

$$A \cup B = \{a, b, c, x, y, z\}.$$

(b) The intersection contains the elements belonging to both B *and* C. Thus

$$B \cap C = \{x, y\}.$$

(c) $A \cup C = \{a, b, c, x, y, z\}$
(d) For an element to be in the intersection of three sets, the element must belong to each set. The only element belonging to A *and* B *and* C is x. Thus $A \cap B \cap C = \{x\}$.

Now Try Exercises 63, 65, 79

EXAMPLE 7 **Using set operations**

Let $U = \{1, 2, 3, 4, 5, 6, 7, 8\}$, $A = \{2, 4, 6, 8\}$, $B = \{1, 3, 5, 6\}$, and $C = \{5, 6, 7, 8\}$. Find each set.
(a) $A \cap B$ **(b)** $B \cap B'$ **(c)** $B \cup C$ **(d)** $A \cup C'$

Solution

(a) $A \cap B = \{6\}$
(b) Because $B = \{1, 3, 5, 6\}$ and $B' = \{2, 4, 7, 8\}$, $B \cap B' = \{\ \}$.

   NOTE: The intersection of a set and its complement is always the empty set.

(c) $B \cup C = \{1, 3, 5, 6, 7, 8\}$
(d) Because $A = \{2, 4, 6, 8\}$ and $C' = \{1, 2, 3, 4\}$, $A \cup C' = \{1, 2, 3, 4, 6, 8\}$.

Now Try Exercises 64, 71, 77

Finite and Infinite Sets

Sets can either have a finite number of elements or infinitely many elements. The elements of a **finite set** can be listed explicitly, whereas the elements of an **infinite set** cannot be listed because there are infinitely many elements. For example, the set of integers S from -3 to 3 is a finite set with 7 elements because

$$S = \{-3, -2, -1, 0, 1, 2, 3\}.$$

In contrast, the set of natural numbers,

$$N = \{1, 2, 3, 4, \ldots\},$$

is an infinite set because the list continues without end.

EXAMPLE **8**   Recognizing finite and infinite sets

Identify each set as finite or infinite.
(a) The set of integers
(b) The set of natural numbers between 1 and 5, inclusive
(c) The set of rational numbers between 1 and 5, inclusive

Solution
(a) There are infinitely many integers, so the set of integers is infinite.
(b) The set of natural numbers between 1 and 5, inclusive, is $\{1, 2, 3, 4, 5\}$. Because we can list each element of this set, it is a finite set.
(c) Because there are infinitely many rational numbers (fractions) between 1 and 5, this set is an infinite set.
Now Try Exercises **81, 83**

SETS Exercises

BASIC CONCEPTS

Exercises 1–10: Use set notation to list the elements of the set described.

1. The natural numbers less than 8

2. The natural numbers greater than 4 and less than or equal to 10

3. The days of the week starting with the letter S

4. U.S. states whose names start with the letter A

5. The letters of the alphabet from A to G

6. The letters of the alphabet from D to M

7. Dogs that can fly under their own power

8. People having an income of $10 trillion in 2008

9. The even integers greater than -2 and less than 11

10. The odd integers less than 21 and greater than 12

Exercises 11–20: Insert $\in$ or $\notin$ to make the statement true.

11. 10 _____ $\{5, 10, 15, 20\}$

12. 7 _____ $\{1, 3, 5, 7, 9, 11\}$

13. 5 _____ $\{2, 4, 6, 8, 10\}$

14. -1 _____ $\{0, 1, 2, 3, 4, 5\}$

15. $\frac{1}{4}$ _____ $\{0, 0.25, 0.5, 0.75, 1\}$

16. $\frac{2}{5}$ _____ $\{0, 0.2, 0.4, 0.6, 0.8, 1\}$

17. $\frac{1}{3}$ _____ $\{0, 0.33, 0.67, 1\}$

18. $\frac{4}{2}$ _____ $\{1, 3, 4, 5, 7\}$

19. red _____ $\{$blue, green, red, yellow$\}$

20. M _____ $\{$M, T, W, R, F$\}$

Exercises 21–24: Determine the set E of even integers that belong to the given universal set U.

21. $U = \{1, 2, 3, 4, 5, 6, 7, 8, 9, 10\}$

22. $U = \{-3, -2, -1, 0, 1, 2, 3\}$

23. $U = \{1, 2, 3, 4, \ldots\}$

24. $U = \{1, 3, 5, 7, 9, 11, \ldots\}$

Exercises 25–28: Determine the set A of words starting with the letter A that belong to the given universal set U.

25. $U = \{$Apple, Orange, Pear, Apricot$\}$

26. $U = \{$Apple, Orange$\}$

27. $U = \{$Calculus, Algebra, Geometry$\}$

28. $U = \{$Calculus, Algebra, Geometry, Arithmetic$\}$

SUBSETS

Exercises 29–36: Let $A = \{1, 2, 3, 4\}$, $B = \{3, 4, 5, 6\}$, and $C = \{3, 6\}$. Determine whether the statement is true or false.

29. $A \subseteq B$
30. $B \subseteq A$
31. $C \subseteq B$
32. $\varnothing \subseteq C$
33. $C \subseteq C$
34. $A \nsubseteq B$
35. $B \nsubseteq C$
36. $C \subseteq A$

Exercises 37–44: Let $A = \{a, b, c, d, e\}$, $B = \{b, d\}$, and $C = \{a, c, e\}$. Determine whether the statement is true or false.

37. $B \subseteq A$
38. $C \subseteq A$
39. $B \nsubseteq C$
40. $C = B$
41. $C \subseteq \varnothing$
42. $C \nsubseteq B$
43. $A \subseteq A$
44. $C \nsubseteq C$

VENN DIAGRAMS

Exercises 45–52: Let A and B be two sets and U be the universal set. Sketch a Venn diagram and shade the region that illustrates the given set.

45. $A \cup B$
46. $A \cap B$
47. A'
48. B'
49. $A \cap U$
50. $B \cup U$
51. $(A \cap B)'$
52. $(A \cup B)'$

UNIONS, INTERSECTIONS, AND COMPLEMENTS

Exercises 53–62: Write the expression in terms of one set.

53. $\{1, 2, 3\} \cup \{3, 4\}$
54. $\{1, 2, 3\} \cup \{6\}$
55. $\{a, b, c\} \cap \{a, b, d\}$
56. $\{x, y, z\} \cap \{a, b, c\}$
57. $\{a, b, c\} \cup \{\ \}$
58. $\{a, b, c\} \cap \varnothing$
59. $\{a, b\} \cup \{c, b\} \cup \{d, a\}$
60. $\{1, 3, 5\} \cup \{4, 5\} \cup \{4, 5, 6\}$
61. $\{4, 5, 8, 9\} \cap \{3, 5, 8, 9\} \cap \{4, 5, 8\}$
62. $\{1, 2, 5\} \cap \{2, 1\} \cap \{1, 2, 3\}$

Exercises 63–80: Let
$$U = \{1, 2, 3, 4, 5, 6, 7, 8, 9, 10\},$$
$$A = \{1, 3, 5, 7, 9\}, B = \{2, 4, 6, 8, 10\},$$
$$C = \{3, 4, 5, 6, 7\}, \text{ and } D = \{4, 5, 6, 9\}.$$

Use set notation to list the elements in the given set.

63. $A \cup B$
64. $A \cap B$
65. $C \cap D$
66. $A \cup D$
67. $U \cap \varnothing$
68. $\varnothing \cup U$
69. $A \cup \varnothing$
70. $\varnothing \cap B$
71. $D \cap D'$
72. $A \cup A$
73. U'
74. A'
75. B'
76. D'
77. $A \cup D'$
78. $A' \cap B'$
79. $A \cap B \cap C$
80. $B \cup C \cup D$

FINITE AND INFINITE SETS

Exercises 81–86: Determine whether the given set is finite or infinite.

81. The set of whole numbers
82. The set of real numbers
83. The set of natural numbers less than 1000
84. The set of natural numbers greater than 4 and less than 20
85. The days of the week
86. The names of the students in your class

Answers to Selected Exercises

SECTION 1.1 (pp. 8–10)

1. counting **2.** zero **3.** 1 **4.** composite **5.** prime
6. composite **7.** factors **8.** formula **9.** variable
10. equals sign **11.** sum **12.** product **13.** quotient
14. difference **15.** Composite; $4 = 2 \times 2$ **17.** Neither
19. Prime **21.** Composite; $92 = 2 \times 2 \times 23$
23. Composite; $225 = 3 \times 3 \times 5 \times 5$ **25.** Prime
27. $6 = 2 \times 3$ **29.** $12 = 2 \times 2 \times 3$
31. $32 = 2 \times 2 \times 2 \times 2 \times 2$ **33.** $39 = 3 \times 13$
35. $294 = 2 \times 3 \times 7 \times 7$
37. $300 = 2 \times 2 \times 3 \times 5 \times 5$ **39.** Yes **41.** No
43. Yes **45.** Yes **47.** 10 **49.** 7 **51.** 4 **53.** 18
55. 4 **57.** 9 **59.** 8 **61.** 14 **63.** 1 **65.** 28 **67.** 7
69. 5 **71.** 18 **73.** 3 **75.** $3s$; s is the cost of soda.
77. $x + 5$; x is the number. **79.** $n + 5$; n is the number.
81. $3n$; n is the number. **83.** $p - 200$; p is the population.
85. $\frac{z}{6}$; z is the number. **87.** st; s is the speed and t is the time.
89. $\frac{x + 7}{y}$; x is one number, y is the other number.
91.

Yards (y)	1	2	3	4	5	6	7
Feet (F)	3	6	9	12	15	18	21

$F = 3y$
93. $P = 100D$ **95.** $M = 3x$; 108 mi **97.** $B = 6D$
99. $C = 12x$ **101.** 198 square feet

SECTION 1.2 (pp. 23–26)

1. $\frac{3}{4}; \frac{1}{4}$ **2.** 11; 21 **3.** 0 **4.** 1 **5.** numerators
6. denominators **7.** $\frac{1}{4}$ **8.** $\frac{a}{b}$ **9.** multiply **10.** $\frac{1}{4}$
11. $\frac{1}{a}$ **12.** $\frac{1}{5}$ **13.** $\frac{ac}{bd}$ **14.** $\frac{ad}{bc}$ **15.** $\frac{a+c}{b}$ **16.** $\frac{a-c}{b}$
17. 2 **18.** 12 **19.** 9 **20.** $\frac{6}{6}$ **21.** 4 **23.** 25 **25.** 10
27. $\frac{3}{5}$ **29.** $\frac{3}{5}$ **31.** $\frac{1}{2}$ **33.** $\frac{1}{3}$ **35.** $\frac{2}{5}$ **37.** $\frac{1}{3}$ **39.** $\frac{2}{5}$
41. $\frac{1}{4}$ **43.** $\frac{1}{6}$ **45.** $\frac{3}{20}$ **47.** 1 **49.** $\frac{3}{5}$ **51.** $\frac{12}{5}$ **53.** $\frac{3}{4}$
55. 1 **57.** $\frac{3a}{2b}$ **59.** $\frac{3}{16}$ **61.** 4 **63.** $\frac{1}{3}$
65. (a) $\frac{1}{5}$ (b) $\frac{1}{7}$ (c) $\frac{7}{4}$ (d) $\frac{8}{9}$
67. (a) 2 (b) 9 (c) $\frac{101}{12}$ (d) $\frac{17}{31}$
69. $\frac{3}{2}$ **71.** $\frac{3}{2}$ **73.** 8 **75.** $\frac{4}{3}$ **77.** $\frac{63}{8}$ **79.** 12 **81.** $\frac{3}{10}$
83. $\frac{a}{2}$ **85.** 1 **87.** (a) 1 (b) $\frac{1}{3}$ **89.** (a) 2 (b) 1
91. (a) $\frac{7}{33}$ (b) $\frac{1}{11}$ **93.** 10 **95.** 45 **97.** 15 **99.** 24
101. 12 **103.** 24 **105.** $\frac{3}{6}, \frac{4}{6}$ **107.** $\frac{28}{36}, \frac{15}{36}$ **109.** $\frac{3}{48}, \frac{28}{48}$
111. $\frac{4}{12}, \frac{9}{12}, \frac{10}{12}$ **113.** $\frac{5}{6}$ **115.** $\frac{13}{16}$ **117.** $\frac{1}{4}$ **119.** $\frac{1}{6}$
121. $\frac{59}{70}$ **123.** $\frac{13}{36}$ **125.** $\frac{17}{600}$ **127.** $\frac{9}{8}$ **129.** $4\frac{3}{4}$ ft

131. $\frac{961}{1260}$ **133.** $32\frac{5}{16}$ inches **135.** $\frac{5}{8}$ yd^2
137. $8\frac{1}{4}$ miles **139.** $\frac{6}{125}$

CHECKING BASIC CONCEPTS 1.1 & 1.2 (p. 26)

1. (a) Prime (b) Composite; $28 = 2 \times 2 \times 7$
(c) Neither (d) Composite; $180 = 2 \times 2 \times 3 \times 3 \times 5$
2. 2 **3.** 30 **4.** $x + 5$ **5.** $I = 12F$
6. (a) 3 (b) 8 **7.** (a) $\frac{5}{7}$ (b) $\frac{2}{3}$ **8.** $\frac{3}{4}$
9. (a) $\frac{1}{2}$ (b) $\frac{1}{4}$ (c) $\frac{2}{5}$ (d) $\frac{7}{12}$ **10.** $3\frac{1}{3}$ cups

SECTION 1.3 (pp. 32–33)

1. add **2.** multiply **3.** Six **4.** 2 **5.** a^6
6. base; exponent **7.** 6^2 **8.** 8^3 **9.** 4^5 **10.** x^3
11. 17; multiplication; addition
12. 2; exponents; subtraction **13.** 4; left; right
14. 2; left; right **15.** No; $2^3 = 8$, but $3^2 = 9$.
16. $5 \cdot 5$ **17.** 2^5 **19.** 3^4 **21.** $\left(\frac{1}{2}\right)^4$ **23.** a^5
25. $(x + 3)^2$ **27.** (a) 16 (b) 16 **29.** (a) 6 (b) 1
31. (a) 32 (b) 1000 **33.** (a) $\frac{4}{9}$ (b) $\frac{1}{32}$
35. (a) $\frac{8}{125}$ (b) $\frac{81}{49}$ **37.** 2^3 **39.** 5^2 **41.** 7^2 **43.** 10^3
45. $\left(\frac{1}{2}\right)^4$ **47.** $\left(\frac{2}{3}\right)^5$ **49.** 29 **51.** 4 **53.** 90 **55.** 3
57. 2 **59.** 19 **61.** 32 **63.** 125 **65.** 3 **67.** 4
69. 80 **71.** $\frac{49}{16}$ **73.** $2^3 - 8$; 0 **75.** $30 - 4 \cdot 3$; 18
77. $\frac{4^2}{2^3}$; 2 **79.** $\frac{40}{10} + 2$; 6 **81.** $100(2 + 3)$; 500
83. About 536,870,912 bytes **85.** (a) $k = 7$ (b) 32
87. (a) 8 yr (b) $80,000

SECTION 1.4 (pp. 42–44)

1. $-b$ **2.** 7 **3.** b **4.** -9 **5.** natural **6.** rational
7. 4 **8.** principal **9.** real **10.** irrational
11. rational **12.** irrational **13.** $0.\overline{27}$ **14.** 1; 4
15. $\sqrt{2}$ (answers may vary) **16.** not equal
17. approximately equal **18.** 3 **19.** 0 **20.** left
21. origin **22.** b **23.** $>$ **25.** $=$
27. (a) -9 (b) 9 **29.** (a) $-\frac{2}{3}$ (b) $\frac{2}{3}$
31. (a) -8 (b) 8 **33.** (a) $-a$ (b) a
35. 6 **37.** $\frac{1}{2}$ **39.** 0.25 **41.** 0.875 **43.** 1.5
45. 0.05 **47.** $0.\overline{6}$ **49.** $0.\overline{7}$
51. Natural, whole, integer, and rational
53. Natural, whole, integer, and rational
55. Whole, integer, and rational **57.** Rational
59. Rational **61.** Irrational
63. Natural, integer, and rational
65. Natural, integer, and rational **67.** Rational

69. (number line: (b) at -2, (a) at 0, (c) at 3)

71. (number line: (b), (a) near 0–1, (c))

73. (number line: (b) at -3, (c)(a) at 1)

75. (number line: -50 to 50)

77. (number line: (c) at -2, (b) at 2, (a) at 3)

79. 5.23 **81.** 7 **83.** 4 **85.** $\pi - 3$ **87.** $-b$
89. $<$ **91.** $>$ **93.** $>$ **95.** $<$ **97.** $>$
99. $<$ **101.** $-9, -2^3, -3, 0, 1$ **103.** $-2, -\frac{3}{2}, \frac{1}{3}, \sqrt{5}, \pi$
105. $-4^2, -\frac{17}{28}, -\frac{4}{7}, \sqrt{2}, \sqrt{7}$ **107. (a)** 16.5
(b) Answers may vary. **(c)** 17.125; answers may vary.

CHECKING BASIC CONCEPTS 1.3 & 1.4 (p. 44)

1. (a) 5^4 **(b)** 7^5 **2. (a)** 8 **(b)** 10,000 **(c)** $\frac{8}{27}$
(d) -81 **3. (a)** 4^3 **(b)** 2^6 **4. (a)** 26 **(b)** 9 **(c)** 2
(d) $\frac{1}{2}$ **(e)** 4 **(f)** 0 **5.** $5^3 \div 3$, or $\frac{5^3}{3}$ **6. (a)** 17 **(b)** $-a$
7. (a) 0.15 **(b)** 0.625 **8. (a)** Natural, integer, and rational
(b) Integer and rational **(c)** Irrational **(d)** Rational
9.
(number line: (b) at -3, (e) at -1, (a)(d) at 1, (c) at 3)

10. (a) 12 **(b)** a **11. (a)** $<$ **(b)** $<$ **(c)** $<$
12. $-7, -1.6, 0, \frac{1}{3}, \sqrt{3}, 3^2$

SECTION 1.5 (pp. 49–51)

1. two **2.** addends **3.** sum **4.** zero **5.** positive
6. negative **7.** absolute value **8.** difference
9. addition **10.** opposite; $(-b)$ **11.** addition
12. subtraction **13.** $-25; 0$ **15.** $\sqrt{21}; 0$
17. $-5.63; 0$ **19.** 4 **21.** 2 **23.** -3 **25.** 4 **27.** 2
29. -1 **31.** 10 **33.** -150 **35.** 1 **37.** -7 **39.** $\frac{1}{4}$
41. $-\frac{9}{14}$ **43.** $-\frac{5}{4}$ **45.** -1.1 **47.** 34 **49.** -1
51. 0 **53.** -3 **55.** 7 **57.** 1 **59.** $-\frac{1}{14}$ **61.** $\frac{1}{2}$
63. 2.9 **65.** -164 **67.** -11 **69.** -9 **71.** 42
73. -5 **75.** 50 **77.** 8.8 **79.** $\frac{1}{2}$ **81.** -1
83. $2 + (-5); -3$ **85.** $-5 + 7; 2$ **87.** $-(2^3); -8$
89. $-6 - 7; -13$ **91.** $6 + (-10) - 5; -9$ **93.** \$230
95. 25 yards **97.** 64,868 feet

SECTION 1.6 (pp. 59–60)

1. factors **2.** product **3.** negative **4.** positive
5. quotient **6.** dividend; divisor **7.** $\frac{1}{a}$ **8.** reciprocal
9. $-\frac{4}{3}$ **10.** reciprocal, or multiplicative inverse **11.** $-a$
12. 25 **13.** -25 **14.** $\frac{1}{b}$ **15.** positive **16.** negative
17. subtraction **18.** $5; 8$ **19.** -12 **21.** -18 **23.** 0

25. 60 **27.** $\frac{1}{4}$ **29.** -1 **31.** 200 **33.** -5000 **35.** 120
37. $-\frac{3}{2}$ **39.** 1 **41.** Negative **43.** -1 **45.** -16 **47.** 8
49. -40 **51.** -2 **53.** 10 **55.** -4 **57.** -3 **59.** -32
61. 0 **63.** 0 **65.** $-\frac{1}{22}$ **67.** $\frac{4}{15}$ **69.** $-\frac{15}{16}$ **71.** Undefined
73. -1 **75.** $-\frac{4}{3}$ **77.** 0.5 **79.** 0.1875 **81.** 3.5 **83.** $5.\overline{6}$
85. 1.4375 **87.** 0.875 **89.** $\frac{1}{4}$ **91.** $\frac{4}{25}$ **93.** $\frac{5}{8}$ **95.** $\frac{11}{16}$
97. $2.\overline{3}; \frac{7}{3}$ **99.** $2.1; \frac{21}{10}$ **101.** $-1.8\overline{3}; -\frac{11}{6}$ **103.** $0.8; \frac{4}{5}$
105. About 131 million **107.** 0.156

CHECKING BASIC CONCEPTS 1.5 & 1.6 (pp. 60–61)

1. (a) 0 **(b)** -19 **2. (a)** $\frac{8}{9}$ **(b)** -3.2
3. (a) $-1 + 5; 4$ **(b)** $4 - (-3); 7$ **4.** $144°F$
5. (a) 35 **(b)** $\frac{4}{15}$ **6. (a)** -9 **(b)** -32 **(c)** 25
7. (a) $-\frac{15}{2}$ **(b)** $\frac{15}{32}$ **8.** $-\frac{6}{7}$ **9. (a)** -5 **(b)** -5
(c) -5 **(d)** 5 **10. (a)** 0.6 **(b)** 3.875

SECTION 1.7 (pp. 69–71)

1. commutative; addition **2.** commutative; multiplication
3. associative; addition **4.** associative; multiplication
5. subtraction; division **6.** subtraction; division
7. distributive **8.** distributive **9.** b **10.** a
11. identity; addition **12.** identity; multiplication
13. $-a$ **14.** $\frac{1}{a}$ **15.** $10 + (-6)$ **17.** $6 \cdot (-5)$
19. $10 + a$ **21.** $7b$ **23.** $1 + (2 + 3)$
25. $(2 \cdot 3) \cdot 4$ **27.** $a + (5 + c)$ **29.** $x \cdot (3 \cdot 4)$
31. $a + b + c = (a + b) + c$
$\qquad\qquad = c + (a + b)$
$\qquad\qquad = c + (b + a)$
$\qquad\qquad = c + b + a$
33. 20 **35.** $ab - 8a$ **37.** $-t - z$
39. $-5 + a$ **41.** $3a + 15$ **43.** $3z - 18$
45. $a \cdot (b + c + d) = a \cdot \big((b + c) + d\big)$
$\qquad\qquad\qquad = a \cdot (b + c) + ad$
$\qquad\qquad\qquad = ab + ac + ad$
47. $11x$ **49.** $-b$ **51.** $2a$ **53.** $-14w$
55. Commutative (multiplication) **57.** Associative (addition)
59. Distributive **61.** Distributive, commutative (multiplication)
63. Distributive **65.** Associative (multiplication)
67. Distributive **69.** Identity (addition)
71. Identity (multiplication) **73.** Identity (multiplication)
75. Inverse (multiplication) **77.** Inverse (addition)
79. 30 **81.** 100 **83.** 178 **85.** 477 **87.** 79 **89.** 90
91. 816 **93.** 1 **95.** $\frac{1}{6}$ **97. (a)** 410 **(b)** 9970
(c) -6300 **(d)** $-140,000$ **99. (a)** 19,000 **(b)** $-45,100$
(c) 60,000 **(d)** $-7,900,000$ **101. (a)** 1.256 **(b)** 0.96
(c) 0.0987 **(d)** -0.0056 **(e)** 120 **(f)** 457.8
103. Commutative (addition) **105.** 19.8 miles
107. (a) $13 \times (5 \times 2)$; multiplying by 10 is easy.
(b) Associative (multiplication)

SECTION 1.8 (pp. 77–79)

1. term **2.** is not; is **3.** coefficient **4.** -1 **5.** 6
6. factors; terms **7.** like **8.** like; unlike **9.** like
10. distributive **11.** Yes; 91 **13.** Yes; -6 **15.** No
17. Yes; 1 **19.** No **21.** Yes; -9 **23.** Like **25.** Like
27. Unlike **29.** Unlike **31.** Like **33.** Unlike
35. Unlike **37.** Like **39.** $8x$ **41.** $14y$ **43.** $41a$
45. 0 **47.** Not possible **49.** Not possible **51.** $3x^2$
53. $-xy$ **55.** 0 **57.** $3x + 2$ **59.** $-2z + \frac{1}{2}$ **61.** $11y$
63. $4z - 1$ **65.** $12y - 7z$ **67.** $-3x - 4$ **69.** $-6x - 1$
71. $-\frac{1}{3}x + \frac{2}{3}$ **73.** $\frac{2}{5}x + \frac{3}{5}y + \frac{1}{5}$ **75.** $0.4x^2$ **77.** $7x^2 - 7x$
79. $2b$ **81.** $7x^3 + 2$ **83.** x **85.** 3 **87.** z **89.** $3x - 2$
91. $2z + 3$ **93.** $5x + 6x$; $11x$ **95.** $x^2 + 2x^2$; $3x^2$
97. $6x - 4x$; $2x$ **99.** (a) $1570w$ (b) 65,940 square feet
101. (a) $50x$ (b) 2400 cubic feet (c) 24 minutes

CHECKING BASIC CONCEPTS 1.7 & 1.8 (p. 79)

1. (a) $18y$ (b) $x + 10$ **2.** $20y$ **3.** (a) $5 - x$
(b) $5x - 35$ **4.** Distributive **5.** 0 **6.** (a) 60 (b) 7
(c) 368 **7.** (a) Unlike (b) Like **8.** (a) $14z$
(b) $-3y + 3$ **9.** (a) $-3y - 3$ (b) $-14y$
(c) x (d) 35 **10.** $3x + 5x$; $8x$

CHAPTER 1 REVIEW (pp. 84–87)

1. Prime **2.** Composite; $27 = 3 \times 3 \times 3$
3. Composite; $108 = 2 \times 2 \times 3 \times 3 \times 3$
4. Composite; $91 = 7 \times 13$ **5.** Neither **6.** Neither
7. 3 **8.** 5 **9.** 12 **10.** 6 **11.** 7 **12.** 7 **13.** 8
14. 6 **15.** $5c$, where c is the cost of the CD
16. $x - 5$, where x is the number **17.** $3^2 + 5$
18. $2^3 \div (3 + 1)$ **19.** $3x$, where x is the number
20. $x - 4$, where x is the number **21.** (a) $\frac{5}{8}$ (b) $\frac{3}{4}$
22. (a) $\frac{3}{4}$ (b) $\frac{3}{5}$ **23.** $\frac{5}{8}$ **24.** $\frac{2}{9}$ **25.** $\frac{3}{11}$ **26.** $\frac{12}{23}$ **27.** 2
28. $\frac{3x}{2y}$ **29.** $\frac{3}{35}$ **30.** (a) $\frac{1}{8}$ (b) 1 (c) $\frac{19}{5}$ (d) $\frac{2}{3}$ **31.** 9
32. $\frac{9}{14}$ **33.** 12 **34.** $\frac{1}{8}$ **35.** 24 **36.** 42 **37.** $\frac{1}{3}$ **38.** $\frac{1}{2}$
39. $\frac{19}{24}$ **40.** $\frac{9}{22}$ **41.** $\frac{5}{12}$ **42.** $\frac{13}{18}$ **43.** 5^6 **44.** $\left(\frac{7}{6}\right)^3$
45. 3^4 **46.** x^5 **47.** $(x + 1)^2$ **48.** $(a - 5)^3$
49. (a) 64 (b) 49 (c) 8 **50.** 5 **51.** 25 **52.** 7
53. 3 **54.** 25 **55.** 1 **56.** 2 **57.** 0 **58.** 0 **59.** 10
60. 1 **61.** 5 **62.** 19 **63.** (a) 8 (b) 3
64. (a) $-\frac{3}{7}$ (b) $-\frac{2}{5}$ **65.** (a) 0.8 (b) 0.15
66. (a) $0.\overline{5}$ (b) $0.\overline{63}$ **67.** Whole, integer, and rational
68. Rational **69.** Integer and rational **70.** Irrational
71. Irrational **72.** Rational
73.
74. (a) 5 (b) π (c) $\sqrt{2} - 1$
75. (a) $<$ (b) $>$ (c) $<$ (d) $>$

76. $-3, -\frac{2}{3}, \sqrt{3}, \pi - 1, 3$ **77.** 4 **78.** -3 **79.** 1
80. -5 **81.** 1 **82.** -2 **83.** -44 **84.** 40 **85.** $-\frac{11}{4}$
86. -7 **87.** $-\frac{2}{9}$ **88.** $-\frac{5}{4}$ **89.** $\frac{2}{7}$ **90.** $-\frac{4}{35}$ **91.** 4
92. $-\frac{3}{4}$ **93.** $3 + (-5)$; -2 **94.** $2 - (-4)$; 6 **95.** $0.\overline{7}$
96. 2.2 **97.** $\frac{3}{5}$ **98.** $\frac{3}{8}$ **99.** Commutative (multiplication)
100. Associative (addition) **101.** Distributive
102. Commutative (addition) **103.** Identity (multiplication)
104. Associative (multiplication) **105.** Distributive
106. Identity (addition) **107.** Inverse (addition)
108. Inverse (multiplication) **109.** 40 **110.** 301
111. 2475 **112.** 6580 **113.** 549.8 **114.** 43.56
115. Yes; 55 **116.** Yes; -1 **117.** No **118.** No
119. $-6x$ **120.** $15z$ **121.** $4x^2$ **122.** $3x + 1$
123. $\frac{1}{2}z + 2$ **124.** $x - 18$ **125.** $9x^2 - 6$ **126.** 0
127. 5 **128.** c **129.** $3y + 2$ **130.** $2x - 5$
131. (a) $7x$ (b) 420 square feet (c) 24 minutes
132. 16 square feet
133.

Gallons (G)	1	2	3	4	5	6
Pints (P)	8	16	24	32	40	48

$P = 8G$

134. $C = 0.25x$ **135.** $\frac{3}{20}$ **136.** $200,000 **137.** $1\frac{3}{20}$ feet
138. $15\frac{3}{8}$ miles **139.** $1101 **140.** $124°$F
141. About 129 million

CHAPTER 1 TEST (pp. 87–88)

1. (a) Prime (b) Composite; $56 = 2 \times 2 \times 2 \times 7$
2. $\frac{15}{7}$ **3.** $4^2 - 3$; 13 **4.** $\frac{3}{4}$ **5.** (a) $\frac{3}{4}$ (b) $\frac{16}{45}$ (c) $\frac{2}{7}$
(d) $\frac{15}{4}$ (e) $\frac{31}{36}$ (f) $\frac{2}{13}$ **6.** y^4 **7.** (a) 8 (b) 71
(c) -40 (d) 9 **8.** (a) Integer and rational (b) Irrational
9.
10. (a) $<$ (b) $>$ **11.** (a) -6 (b) 21
12. (a) Distributive (b) Associative (multiplication)
(c) Commutative (addition) **13.** 1734 **14.** (a) $12 - 4z$
(b) $15x - 6$ (c) $x - 19$ **15.** (a) $\frac{19}{12}x$ (b) $12\frac{2}{3}$ acres
16. $2\frac{3}{5}$ feet **17.** (a) $C = 13x$ (b) $221 **18.** $640

CHAPTER 2: LINEAR EQUATIONS AND INEQUALITIES

SECTION 2.1 (pp. 97–99)

1. true **2.** false **3.** solution **4.** solution set
5. solutions **6.** Equivalent **7.** $b + c$ **8.** subtraction
9. bc **10.** division **11.** equivalent **12.** given
13. 22 **15.** 3 **17.** -5 **19.** 8 **21.** -34 **23.** 9
25. 17 **27.** 2 **29.** 5 **31.** -15 **33.** 1963 **35.** a
37. $\frac{1}{5}$ **39.** 6 **41.** 3 **43.** 0 **45.** 7 **47.** -6

49. 3   **51.** $\frac{5}{4}$ **53.** 5   **55.** 7   **57.** $\frac{3}{2}$ **59.** a

61. (a)

Hours (x)	0	1	2	3	4	5	6
Rainfall (R)	3	3.5	4	4.5	5	5.5	6

(b) $R = 0.5x + 3$ **(c)** 4.5 inches; yes   **(d)** 4.125 inches
63. (a) $L = 300x$ **(b)** $870 = 300x$ **(c)** 2.9
65. 17.5 years   **67.** $25,000

SECTION 2.2 (pp. 108–110)

1. $ax + b = 0$ **2.** 3; 2   **3.** Exactly one   **4.** is not
5. Addition, multiplication   **6.** Distributive   **7.** LCD
8. 100   **9.** None   **10.** Infinitely many   **11.** Infinitely
many   **12.** None   **13.** Yes; $a = 3, b = -7$ **15.** Yes;
$a = \frac{1}{2}, b = 0$ **17.** No   **19.** No   **21.** Yes;
$a = 1.1, b = -0.9$ **23.** Yes; $a = 2, b = -6$ **25.** No
27.

x	0	1	2	3	4
$-3x + 7$	7	4	1	-2	-5

2

29.

x	-2	-1	0	1	2
$4 - 2x$	8	6	4	2	0

-1

31. $\frac{3}{11}$ **33.** 23   **35.** 7   **37.** $-\frac{11}{5}$ **39.** $-\frac{7}{2}$ **41.** 3
43. $\frac{9}{4}$ **45.** $-\frac{5}{2}$ **47.** $\frac{4}{7}$ **49.** $\frac{9}{8}$ **51.** 1   **53.** -0.055
55. 8   **57.** $\frac{1}{7}$ **59.** 1   **61.** $-\frac{b}{a}$ **63.** No solutions
65. One solution   **67.** Infinitely many solutions
69. No solutions   **71.** Infinitely many solutions
73. (a)

Hours (x)	0	1	2	3	4
Distance (D)	4	12	20	28	36

(b) $D = 8x + 4$ **(c)** 28 miles; yes   **(d)** 2.25 hours; the
bicyclist is 22 miles from home after 2 hours and 15 minutes.
75. 2001   **77.** 2004

CHECKING BASIC CONCEPTS 2.1 & 2.2 (p. 110)

1. (a) No   **(b)** Yes
2.

x	3	3.5	4	4.5	5
$4x - 3$	9	11	13	15	17

4

3. (a) 18   **(b)** $\frac{1}{6}$ **(c)** $2.\overline{6}$ or $\frac{8}{3}$ **(d)** 3
4. (a) One solution   **(b)** Infinitely many solutions
(c) No solutions   **5. (a)** $D = 300 - 75x$
(b) $0 = 300 - 75x$ **(c)** 4 hours

SECTION 2.3 (pp. 120–123)

1. Check your solution.   **2.** $n + 1$ and $n + 2$
3. $\frac{x}{100}$ **4.** 0.01   **5.** left   **6.** right   **7.** $\frac{P_2 - P_1}{P_1} \times 100$
8. increase; decrease   **9.** rt **10.** distance; time
11. $2 + x = 12$; 10   **13.** $\frac{x}{5} = x - 24$; 30
15. $\frac{x + 5}{2} = 7$; 9   **17.** $\frac{x}{2} = 17$; 34   **19.** 31, 32, 33
21. 34   **23.** 8   **25.** 21   **27.** 11   **29.** 7   **31.** 13
33. 96 billion kilowatt-hours   **35.** $\frac{37}{100}$; 0.37

37. $\frac{37}{25}$; 1.48   **39.** $\frac{69}{1000}$; 0.069   **41.** $\frac{1}{2000}$; 0.0005   **43.** 45%
45. 180%   **47.** 0.6%   **49.** 40%   **51.** 75%   **53.** $83.\overline{3}\%$
55. About 41.4%   **57.** $988 per credit
59. About 130.4 million   **61.** About 52,223
63. 25%; -20% **65.** $d = 8$ miles
67. $r = 20$ feet per second   **69.** $t = 5$ hours   **71.** 60 mph
73. $\frac{3}{8}$ hour   **75.** 0.6 hr at 5 mph; 0.5 hr at 8 mph
77. 3.75 hours   **79.** 30 ounces
81. $1500 at 6%; $2500 at 5%   **83.** 6 gallons   **85.** 10 g

SECTION 2.4 (pp. 133–135)

1. formula   **2.** LW **3.** $\frac{D}{G}$ **4.** $\frac{1}{2}bh$ **5.** $\frac{1}{360}$ **6.** 360
7. 180   **8.** LWH **9.** $2LW + 2WH + 2LH$ **10.** $2\pi r$
11. πr^2 **12.** $\pi r^2 h$ **13.** $\frac{x}{5}$ **14.** $\frac{1}{2}(a + b)h$ **15.** 18 ft²
17. 9 in²   **19.** $16\pi \approx 50.3$ ft²   **21.** 11 ft²   **23.** 91 in²
25. 36 in²   **27.** $8\pi \approx 25.1$ in.   **29.** 4602 ft²   **31.** 65°
33. 81°   **35.** 30°   **37.** 36°, 36°, 108°
39. $C = 12\pi \approx 37.7$ in.; $A = 36\pi \approx 113.1$ in²
41. $r = 1$ in.; $A = \pi \approx 3.14$ in²
43. $V = 2640$ in³; $S = 1208$ in²
45. $V = 2$ ft³; $S = \frac{32}{3}$ ft²   **47.** 20π in³   **49.** 600π in³
51. $\frac{147}{64}\pi \approx 7.2$ ft³   **53.** $W = \frac{A}{L}$ **55.** $h = \frac{V}{\pi r^2}$
57. $a = \frac{2A}{h} - b$ **59.** $W = \frac{V}{LH}$ **61.** $b = 2s - a - c$
63. $b = a - c$ **65.** $a = \frac{cd}{b - d}$ **67.** 15 in.   **69.** 2.98
71. 2.24   **73.** 77°F   **75.** $-40°$F   **77.** $-5°$C
79. $-20°$C   **81.** 11 mpg   **83.** 2.4 mi

CHECKING BASIC CONCEPTS 2.3 & 2.4
(pp. 135–136)

1. (a) $3x = 36$; 12   **(b)** $35 - x = 43$; -8
2. $-32, -31, -30$ **3.** 0.095   **4.** 125%
5. About 3536   **6.** 6.5 hr   **7.** $5000 at 6%; $3000 at 7%
8. 18 gal   **9.** 12 in.
10. $A = 9\pi \approx 28.3$ ft²; $C = 6\pi \approx 18.8$ ft   **11.** 30°
12. $l = \frac{A - \pi r^2}{\pi r}$

SECTION 2.5 (pp. 146–150)

1. equals sign; $<, \leq, >, \geq$ **2.** greater than; less than
3. solution   **4.** equivalent   **5.** one   **6.** infinitely many
7. number line   **8.** is not   **9.** $>$ **10.** $<$ **11.** $>$
12. No; $-4x < 8$ is equivalent to $x > -2$
13.
15.
17.

19.

21. $x < 0$ **23.** $x \le 3$ **25.** $x \ge 10$ **27.** $[6, \infty)$
29. $(-2, \infty)$ **31.** $(-\infty, 7]$ **33.** Yes   **35.** Yes   **37.** Yes
39. Yes   **41.** No   **43.** $x > -2$ **45.** $x < 1$
47.

x	1	2	3	4	5
$-2x + 6$	4	2	0	-2	-4

$x \ge 3$

49.

x	-3	-2	-1	0	1
$5 - x$	8	7	6	5	4
$x + 7$	4	5	6	7	8

$x < -1$

51. $x > 3$;
53. $y \ge -2$;
55. $z > 8$;
57. $t \le -5$;
59. $x < 5$;
61. $t \le -2$;
63. $y > -\frac{3}{20}$;
65. $z \ge -\frac{14}{3}$;

67. $\{x \mid x > 1\}$ **69.** $\{x \mid x \ge -7\}$ **71.** $\{x \mid x < 6\}$
73. $x < 7$ **75.** $x \le -\frac{4}{3}$ **77.** $x > -\frac{39}{2}$ **79.** $x \le \frac{3}{2}$
81. $x > \frac{1}{2}$ **83.** $x \le \frac{5}{4}$ **85.** $x > -\frac{8}{3}$ **87.** $x \le -\frac{3}{10}$
89. $x \le -\frac{1}{3}$ **91.** $x < 5$ **93.** $x \le -\frac{3}{2}$ **95.** $x > -\frac{9}{8}$
97. $x \le \frac{4}{7}$ **99.** $x < \frac{21}{11}$ **101.** $x > 60$ **103.** $x \ge 21$
105. $x > 40{,}000$ **107.** $x \le 70$ **109.** Less than 10 feet
111. It is less than 20 inches.   **113.** 86 or more
115. 4.5 hours   **117.** 4 days   **119. (a)** $C = 1.5x + 2000$
(b) $R = 12x$ **(c)** $P = 10.5x - 2000$ **(d)** 191 or more
compact discs   **121. (a)** After 48 minutes   **(b)** After more
than 48 minutes   **123.** Altitudes more than 4.5 miles
125. 2000 and later

CHECKING BASIC CONCEPTS 2.5 (p. 150)

1.

2. $x < 1$ **3.** Yes
4.

x	-2	-1	0	1	2
$5 - 2x$	9	7	5	3	1

; $x \ge -1$

5. (a) $x > 3$ **(b)** $x \ge -35$ **(c)** $x \le -1$
6. $x \le 12$ **7.** More than 13 inches

CHAPTER 2 REVIEW (pp. 154–157)

1. -6 **2.** 2   **3.** $\frac{9}{4}$ **4.** $-\frac{1}{2}$ **5.** 3   **6.** $-\frac{7}{3}$ **7.** $-\frac{5}{2}$
8. $-\frac{7}{2}$ **9.** Yes; $a = 5, b = -3$ **10.** Yes; $a = -4, b = 1$
11. No   **12.** No   **13.** 2   **14.** 22   **15.** $\frac{27}{5}$ **16.** 7   **17.** -7
18. $-\frac{2}{3}$ **19.** $\frac{9}{4}$ **20.** 0   **21.** $\frac{7}{8}$ **22.** $\frac{11}{4}$ **23.** No solutions
24. Infinitely many solutions   **25.** Infinitely many solutions
26. One solution
27.

x	0.5	1.0	1.5	2.0	2.5
$-2x + 3$	2	1	0	-1	-2

1.5

28.

x	-2	-1	0	1	2
$-(x + 1) + 3$	4	3	2	1	0

0

29. $6x = 72$; 12   **30.** $x + 18 = -23$; -41
31. $2x - 5 = x + 4$; 9   **32.** $x + 4 = 3x$; 2
33. 16, 17, 18, 19   **34.** $-52, -51, -50$ **35.** $\frac{17}{20}$; 0.85
36. $\frac{7}{125}$; 0.056   **37.** $\frac{3}{10{,}000}$; 0.0003   **38.** $\frac{171}{50}$; 3.42
39. 89%   **40.** 0.5%   **41.** 230%   **42.** 100%
43. $d = 24$ mi   **44.** $d = 3850$ ft
45. $r = 25$ yards per second   **46.** $t = \frac{25}{3}$ hr
47. 7.5 m^2 **48.** $36\pi \approx 113.1$ ft^2 **49.** 864 in^2, or 6 ft^2
50. 45.5 in^2 **51.** $18\pi \approx 56.5$ ft   **52.** $25\pi \approx 78.5$ in^2
53. 50°   **54.** 22.5°   **55.** $625\pi \approx 1963.5$ in^3
56. 1620 in^2, or 11.25 ft^2 **57.** 174 in^2
58. About 60.6 ft^2 **59.** $x = a - y$ **60.** $x = \frac{P - 2y}{2}$
61. $y = \frac{z}{2x}$ **62.** $b = 3S - a - c$ **63.** $b = \frac{12T - 4a}{3}$
64. $c = \frac{ab}{d - b}$ **65.** 2.82   **66.** 3.12   **67.** 59°F   **68.** 45°F
69.

70.

71.

72.

73. $x < 3$ **74.** $x \ge -1$ **75.** Yes   **76.** Yes   **77.** Yes
78. No
79.

x	0	1	2	3	4
$5 - x$	5	4	3	2	1

$x < 2$

80.

x	1	1.5	2	2.5	3
$2x - 5$	-3	-2	-1	0	1

$x \le 2.5$

81. $x > 3$ **82.** $x \ge -5$ **83.** $x \le -1$ **84.** $x < \frac{23}{3}$
85. $x \le \frac{1}{9}$ **86.** $x \ge \frac{9}{4}$ **87.** $x < 50$ **88.** $x \le 45{,}000$
89. $x \ge 16$ **90.** $x < 1995$
91. (a)

Time	12:00	1:00	2:00	3:00	4:00	5:00
Rainfall (R)	2	2.75	3.5	4.25	5	5.75

(b) $R = \frac{3}{4}x + 2$ **(c)** $\frac{23}{4} = 5\frac{3}{4}$ in.; yes   **(d)** $\frac{77}{16} = 4\frac{13}{16}$ in.
92. $2125

93. (a)

Hours (x)	1	2	3	4	5
Distance (D)	40	30	20	10	0

(b) $D = 50 - 10x$ **(c)** 20 miles; yes **(d)** 3 hours or less or from noon to 3:00 p.m. **94.** 1998 **95.** 4
96. About 82% **97.** $\frac{1}{6}$ hr, or 10 min **98.** 28 gal
99. 50 mL **100.** $500 at 3%; $800 at 5% **101.** 33 in. by 23 in. **102.** 1.8 mi **103.** 25 inches or less **104.** 74 or more
105. 6 hr **106. (a)** $C = 85x + 150,000$ **(b)** $R = 225x$
(c) $P = 140x - 150,000$ **(d)** 1071 or fewer

CHAPTER 2 TEST (pp. 157–158)

1. -6 **2.** $\frac{5}{2}$ **3.** $-\frac{16}{11}$ **4.** $\frac{25}{6}$ **5.** No solutions
6.

x	0	1	2	3	4
6 − 2x	6	4	2	0	−2

7. $x + (-7) = 6$; 13 **8.** $2x + 6 = x - 7$; -13
9. 111, 112, 113 **10.** $\frac{4}{125}$; 0.032 **11.** 34.5%
12. $37.50 **13.** 1056 ft/sec **14.** 7.5 in^2
15. $C = 30\pi \approx 94.2$ in.; $A = 225\pi \approx 706.9$ in^2
16. 30° **17.** $x = \frac{y - z}{3y}$ **18.** $x = \frac{20R - 4y}{5}$
19. $x \le 6$ **20.** $x > -\frac{1}{7}$ **21. (a)** $S = 2x + 5$ **(b)** 21 in.
(c) 17.5 in. **22.** 2000 mL **23.** 300%

CHAPTERS 1 AND 2 CUMULATIVE REVIEW (p. 159)

1. Composite; $3 \times 3 \times 5$ **2.** Prime **3.** $\frac{1}{2}$ **4.** $\frac{1}{9}$
5. $\frac{13}{24}$ **6.** $\frac{13}{15}$ **7.** Integer and rational **8.** Irrational
9. 3 **10.** 10 **11.** 15 **12.** 9 **13.** -7 **14.** $\frac{3}{7}$
15. $4x^3$ **16.** $5x + 3$ **17.** 14 **18.** -4 **19.** $\frac{1}{2}$ **20.** 1
21. Infinitely many **22.** No solutions **23.** 29, 30, 31
24. 0.047 **25.** 17% **26.** 65 mph **27.** 20
28. $25\pi \approx 78.5$ in^2 **29.** $x = \frac{a + 4}{3y}$ **30.** $x = 3A - y - z$
31. $x < 1$

32. $x \le 2$

33. $I = 36Y$ **34.** $618 **35.** 300 mL
36. $1500 at 4%; $1750 at 6%

CHAPTER 3: GRAPHING EQUATIONS

SECTION 3.1 (pp. 168–171)

1. xy-plane **2.** origin **3.** $(0, 0)$ **4.** 4 **5.** II **6.** III
7. axes **8.** x; y **9.** scatterplot **10.** line
11. $(-2, -2), (-2, 2), (0, 0), (2, 2)$
13. $(-1, 0), (0, -3), (0, 2), (2, 0)$
15. (a) I **(b)** III **17. (a)** None **(b)** I
19. (a) II **(b)** IV **21.** I and III

23.

25.

27.

29.

31.

33.

35.

37.

39. $(1940, 182), (1960, 484), (1980, 632), (2000, 430)$; in 1940 there were 182 billion cigarettes consumed in the U.S. (answers may vary slightly).

41. (a)

(b) Federal income tax receipts increased.

43. (a)

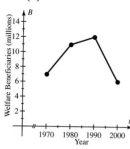

(b) The number of welfare beneficiaries increased and then decreased.

47. (a) The rate decreased. **(b)** 9 **(c)** About -73.1%

45. (a)

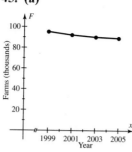

(b) The number of farms decreased.

39.

x	-2	0	2
y	3	2	1

Table values may vary.

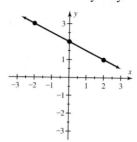

SECTION 3.2 (pp. 178–180)

1. two **2.** ordered pair **3.** solution **4.** linear **5.** is
6. solutions **7.** graph **8.** line **9.** Yes **11.** No
13. No **15.** Yes **17.** No

19.

x	-2	-1	0	1	2
y	-8	-4	0	4	8

21.

x	-8	-4	0	4	8
y	-4	0	4	8	12

23.

x	6	3	0	-3	-9
y	-2	0	2	4	8

25.

x	-3	0	3	6
y	-9	0	9	18

27.

x	-8	-4	0	4
y	-2	0	2	4

29.

x	8	6	4	2
y	-2	0	2	4

31.

x	$-\frac{1}{2}$	$-\frac{1}{4}$	0	$\frac{1}{4}$
y	-2	-1	0	1

33. They must be multiples of 5.

35.

x	-1	0	1
y	2	0	-2

Table values may vary.

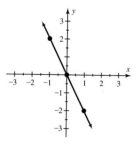

37.

x	0	1	2
y	3	2	1

Table values may vary.

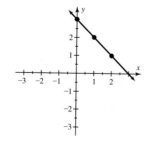

41.

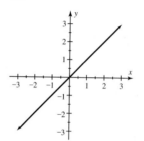

43.

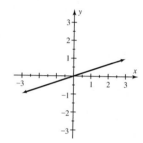

45.

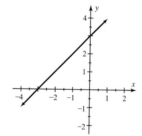

47.

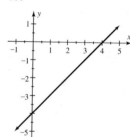

49.

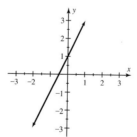

51.

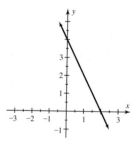

53.

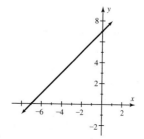

55.

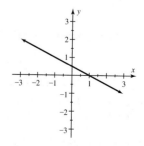

57.

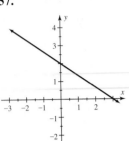

59.

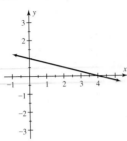

73. (a) 2.5; 91

(b)

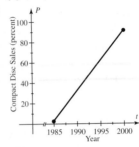

(c) 1998

CHECKING BASIC CONCEPTS 3.1 & 3.2 (pp. 180–181)

1. $(-2, 2)$, II; $(-1, -2)$, III; $(1, 3)$, I; $(3, 0)$, none

2.

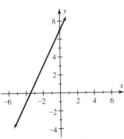

3.

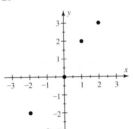

The percentage increased.

61.

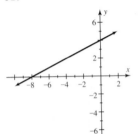

63.

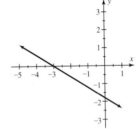

4. Yes

5.

x	-2	-1	0	1	2
y	5	3	1	-1	-3

65.

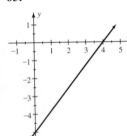

67.

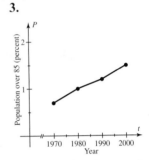

6. (a)

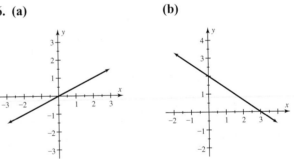

(b)

69. (a)

t	2010	2020	2030	2040	2050
P	13.2	15.0	16.7	18.5	20.3

(b) 2030

71. (a)

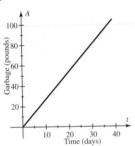

(b) About 37 days

7. (a) In 1995, receipts were $1.425 trillion; in 2001, receipts were $2.115 trillion.

(b)

(c) 1996

SECTION 3.3 (pp. 189–192)

1. Two   **2.** one; one   **3.** x-intercept   **4.** 0
5. y-intercept   **6.** 0   **7.** 3   **8.** horizontal   **9.** $y = b$
10. 3   **11.** vertical   **12.** $x = k$ **13.** 3; -2 **15.** 0; 0
17. -2 and 2; 4   **19.** 1; 1

21.

x	-2	-1	0	1	2
y	0	1	2	3	4

-2; 2

23.

x	-4	-2	0	2	4
y	-6	-4	-2	0	2

2; -2

25. x-int: 3; y-int: -2
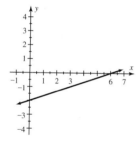

27. x-int: 5; y-int: -3
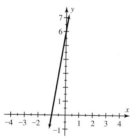

29. x-int: 6; y-int: -2
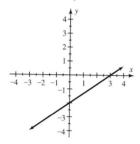

31. x-int: -1; y-int: 6
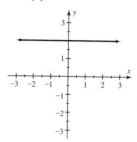

33. x-int: 7; y-int: 3
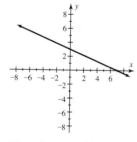

35. x-int: 4; y-int: -3
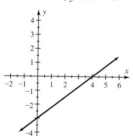

37. x-int: 4; y-int: -2
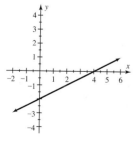

39. x-int: -4; y-int: 3
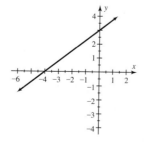

41. x-int: 3; y-int: 2
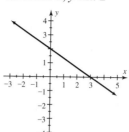

43. x-int: -2; y-int: 5
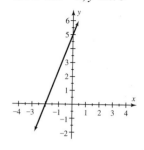

45. x-int: $\frac{C}{A}$; y-int: $\frac{C}{B}$

47. (a)
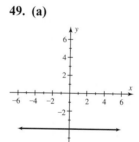

(b)
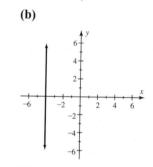

49. (a)
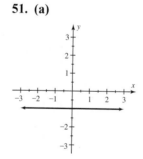

(b)
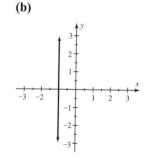

51. (a)
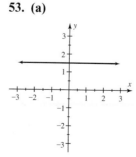

(b)
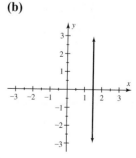

53. (a)

(b)

55. $y = 4$ **57.** $x = -1$ **59.** $y = -6$ **61.** $x = 5$

63. $y = 1$ **65.** $x = -6$ **67.** $y = 2; x = 1$
69. $y = -45; x = 20$ **71.** $y = 5; x = 0$ **73.** $x = -1$
75. $y = -\frac{5}{6}$ **77.** $x = 4$ **79.** $x = -\frac{2}{3}$ **81.** $y = 0$
83. (a) y-int: 200; x-int: 4   **(b)** The driver was initially
200 miles from home; the driver arrived home after 4 hours.
85. (a) v-int: 128; t-int: 4

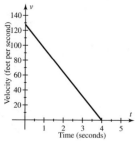

(b) The initial velocity was 128 ft/sec; the velocity after
4 seconds was 0.
87. (a) y-int: 2000; x-int: 4   **(b)** The pool initially contained
2000 gallons; the pool was empty after 4 hours.

SECTION 3.4 (pp. 202–207)

1. horizontal   **2.** vertical   **3.** rise; run   **4.** horizontal
5. vertical   **6.** $\frac{y_2 - y_1}{x_2 - x_1}$ **7.** positive   **8.** negative
9. Positive   **11.** Zero   **13.** Negative
15. 0; the rise always equals 0.
17. 1; the graph rises 1 unit for each unit of run.
19. 2; the graph rises 2 units for each unit of run.
21. Undefined; the run always equals 0.
23. $\frac{1}{2}$; the graph rises 1 unit for each 2 units of run.
25. -1; the graph falls 1 unit for each unit of run.
27. 2;   **29.** Undefined;

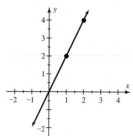 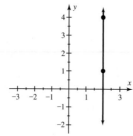

31. $-\frac{2}{3}$;   **33.** 0;

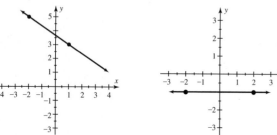

35. $\frac{5}{7}$ **37.** $-\frac{6}{7}$ **39.** 0   **41.** Undefined   **43.** $\frac{13}{20}$
45. $\frac{9}{100}$ **47.** $-\frac{5}{7}$ **49.** 49

51.   **53.**

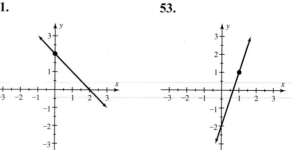

55.   **57.**

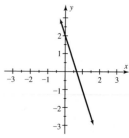

59. 2; 1; -2 **61.** -9; -1; -9
63.   **65.**

x	0	1	2	3
y	-4	-2	0	2

x	1	2	3	4
y	4	1	-2	-5

67.

x	-4	-2	0	2
y	0	3	6	9

69. (a)   **71. (a)**

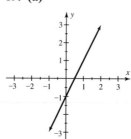 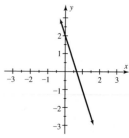

(b) 2   **(b)** -3
73. (a)   **75. (a)**

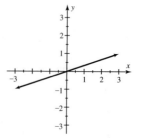 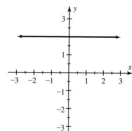

(b) $\frac{1}{3}$ **(b)** 0
77. c.   **79.** b.
81. (a) $m_1 = 1000$; $m_2 = -1000$ **(b)** $m_1 = 1000$: Water
is being added to the pool at a rate of 1000 gallons per hour.

$m_2 = -1000$: Water is being removed from the pool at a rate of 1000 gallons per hour.   **(c)** Initially the pool contained 2000 gallons of water. Over the first 3 hours, water was pumped into the pool at a rate of 1000 gallons per hour. For the next 2 hours, water was pumped out of the pool at a rate of 1000 gallons per hour.

83. (a) $m_1 = 50$; $m_2 = 0$; $m_3 = -50$ **(b)** $m_1 = 50$: The car is moving away from home at a rate of 50 mph. $m_2 = 0$: The car is not moving. $m_3 = -50$: The car is moving toward home at a rate of 50 mph.   **(c)** Initially the car is at home. Over the first 2 hours, the car travels away from home at a rate of 50 mph. Then the car is parked for 1 hour. Finally, the car travels toward home at a rate of 50 mph.

85. m^3/min

87.

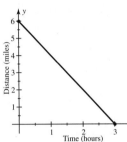

89.

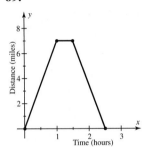

91. (a) 10   **(b)** The birth rate increased on average by 10,000 children per year.

93. (a) 2.5   **(b)** The revenue is $2.50 per screwdriver.

95. (a) $1; 2$ **(b)** $\frac{1}{50}$ **(c)** Oil should be added at a rate of 1 gallon of oil per 50 gallons of gasoline.

97. (a)

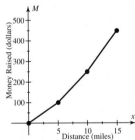

(b) $m_1 = 20$; $m_2 = 30$; $m_3 = 40$ **(c)** $m_1 = 20$: Each mile between 0 and 5 miles is worth $20 per mile. $m_2 = 30$: Each mile between 5 and 10 miles is worth $30 per mile. $m_3 = 40$: Each mile between 10 and 15 miles is worth $40 per mile.

99. (a) 1200   **(b)** Median family income increased on average by $1200 per year over this time period.   **(c)** $48,000

101. (a) Toward   **(b)** 10 ft   **(c)** 2 ft/min   **(d)** 5

CHECKING BASIC CONCEPTS 3.3 & 3.4 (pp. 207–208)

1. $-2; 3$

2.

x	-2	-1	0	1	2
y	-6	-4	-2	0	2

$1; -2$

3. (a) x-int: 6; y-int: -3 **(b)** y-int: 2

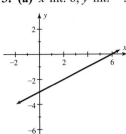

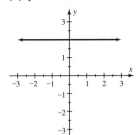

(c) x-int: -1

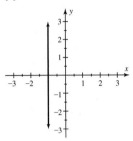

4. $y = 4$; $x = -2$ **5. (a)** $\frac{3}{4}$ **(b)** 0   **(c)** Undefined

6. $-\frac{1}{2}$

7.

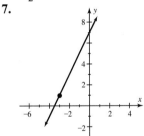

8. (a) $m_1 = 0$; $m_2 = 2$; $m_3 = -\frac{2}{3}$ **(b)** $m_1 = 0$: The depth is not changing. $m_2 = 2$: The depth increased at a rate of 2 feet per hour. $m_3 = -\frac{2}{3}$: The depth decreased at a rate of $\frac{2}{3}$ foot per hour.   **(c)** Initially the pond had a depth of 5 feet. For the first hour, there was no change in the depth of the pond. For the next hour, the depth of the pond increased at a rate of 2 feet per hour to a depth of 7 feet. Finally, the depth of the pond decreased for 3 hours at a rate of $\frac{2}{3}$ foot per hour until it was 5 feet deep.

9. (a) 28   **(b)** For counties between 450 and 600 square miles, the population increases at an average rate of 28 people per square mile.   **(c)** No. We do not know if this trend continues.

SECTION 3.5 (pp. 214–217)

1. $y = mx + b$ **2.** slope   **3.** y-intercept   **4.** horizontal
5. origin   **6.** $y = 2x + 5$ **7.** parallel   **8.** slope
9. perpendicular   **10.** reciprocals   **11.** f.
13. a.   **15.** e.   **17.** $y = x - 1$ **19.** $y = -2x + 1$

21. $y = \frac{1}{2}x - 2$ **23.** $y = -2x$ **25.** $y = \frac{3}{4}x + 2$
27. $y = x + 2$ **29.** $y = 2x - 1$

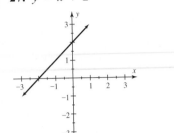

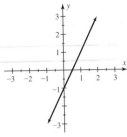

31. $y = -\frac{1}{2}x - 2$ **33.** $y = \frac{1}{3}x$

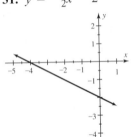

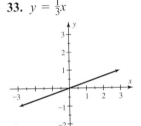

35. $y = -2x + 1$

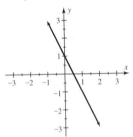

37. (a) $y = -x + 4$ (b) $-1; 4$
39. (a) $y = -2x + 4$ (b) $-2; 4$
41. (a) $y = \frac{1}{2}x + 2$ (b) $\frac{1}{2}; 2$
43. (a) $y = \frac{2}{3}x - 2$ (b) $\frac{2}{3}; -2$
45. (a) $y = \frac{1}{4}x + \frac{3}{2}$ (b) $\frac{1}{4}; \frac{3}{2}$
47. (a) $y = -\frac{1}{3}x + \frac{2}{3}$ (b) $-\frac{1}{3}; \frac{2}{3}$
49.   **51.**

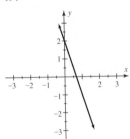

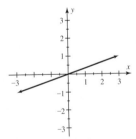

53.   **55.**

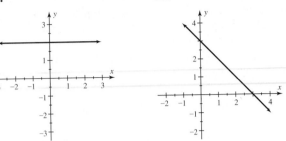

57.   **59.**

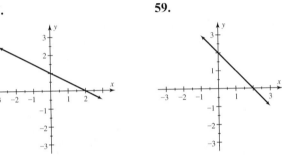

61. $y = 2x + 2$ **63.** $y = x - 2$ **65.** $y = \frac{4}{7}x + 3$
67. $y = 3x$ **69.** $y = -\frac{1}{2}x + \frac{5}{2}$ **71.** $y = 2x$
73. $y = \frac{1}{3}x + \frac{1}{3}$ **75.** (a) \$25   (b) 25 cents   (c) 25;
the fixed cost of renting the car   (d) 0.25; the cost per mile
of driving the car   **77.** (a) \$7.45   (b) $C = 0.07x + 3.95$
(c) 67 min   **79.** (a) $m = 0.35; b = 164.3$
(b) The fixed cost of owning the car for one month

SECTION 3.6 (pp. 224–227)

1. One   **2.** One   **3.** $y = mx + b$
4. $y - y_1 = m(x - x_1)$ or $y = m(x - x_1) + y_1$
5. 1; 3   **6.** distributive   **7.** Yes; every nonvertical line
has exactly one slope and one y-intercept.
8. No; it depends on the point used.
9. Yes   **11.** No   **13.** Yes   **15.** $y - 2 = \frac{3}{4}(x - 1)$
17. $y + 1 = -\frac{1}{2}(x - 3)$ **19.** $y - 1 = 4(x + 3)$
21. $y + 3 = \frac{1}{2}(x + 5)$ **23.** $y - 30 = 1.5(x - 2000)$
25. $y - 4 = \frac{7}{3}(x - 2)$ **27.** $y = \frac{3}{5}(x - 5)$
29. $y - 15 = 5(x - 1990)$ **31.** $y = 3x - 2$
33. $y = \frac{1}{3}x$ **35.** $y = \frac{2}{3}x + \frac{1}{12}$ **37.** $y = -2x + 9$
39. $y = \frac{3}{5}x - 2$ **41.** $y = -16x - 19$ **43.** $-mx_1 + y_1$
45. $y = -2x + 5$ **47.** $y = -x + 1$ **49.** $y = -\frac{1}{9}x + \frac{1}{3}$
51. $y = 2x - 7$ **53.** $y = 2x - 15$ **55.** $y = -2x - 1$
57. $y = \frac{1}{2}x - \frac{5}{2}$ **59.** (a) Toward   (b) After 1 hour the person is 250 miles from home. After 4 hours the person is 100
miles from home.   (c) 50 mph   (d) $y = -50x + 300$; the
car is traveling toward home at 50 mph.

61. (a) Entering; 300 gallons **(b)** 100; initially the tank contains 100 gallons. **(c)** $y = 50x + 100$; the amount of water is increasing at a rate of 50 gallons per minute.
(d) 8 minutes
63. (a) $-5°$F per hour **(b)** $T = -5x + 45$; the temperature is decreasing at a rate of $5°$F per hour.
(c) 9; at 9 A.M. the temperature was $0°$F.
(d)

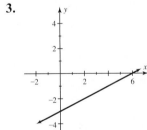

(e) At 4 A.M. the temperature was $25°$F.
65. (a) 1,130,000; 1,450,000 **(b)** 0.032; the number of inmates increased on average by 32,000 per year.
67. (a) In 2002, tuition and fees were \$4081; in 2004, tuition and fees were \$5133. **(b)** $y - 4081 = 526(x - 2002)$ or $y - 5133 = 526(x - 2004)$; tuition and fees increased on average by \$526/yr. **(c)** About \$5700; \$5659
69. (a) $y = 4.9x - 9780.6$ **(b)** 68.4 million

CHECKING BASIC CONCEPTS 3.5 & 3.6 (pp. 227–228)

1. $y = -3x + 1$ **2.** $y = \frac{4}{5}x - 4$; $\frac{4}{5}$; -4
3.

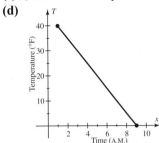

4. (a) $y = 3x - 2$ **(b)** $y = -\frac{3}{2}x$ **(c)** $y = -\frac{7}{3}x - \frac{5}{3}$
5. $y - 3 = -2(x + 1)$ **6.** $y = -2x + 1$
7. $y = 2x + 3$ **8. (a)** $y = -12x + 48$ **(b)** 12 mph
(c) 4 P.M. **(d)** 48 miles **9. (a)** 5 inches **(b)** 2 inches per hour **(c)** 5; total inches of snow that fell before noon
(d) 2; the rate of snowfall was 2 inches per hour.

SECTION 3.7 (pp. 233–235)

1. abstraction **2.** predict **3.** linear **4.** exact
5. approximate **6.** constant **7.** constant rate of change
8. initial amount **9.** f. **11.** a. **13.** e. **15.** Yes
17. No **19.** No **21.** Exact; $y = 2x - 2$
23. Approximate; $y = 2x + 2$ **25.** Approximate; $y = 2$
27. $y = 5x + 40$ **29.** $y = -20x - 5$ **31.** $y = 8$

33. (a) Yes **(b)**

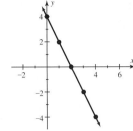

(c) $y = -2x + 4$
35. (a) No **(b)**

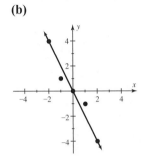

(c) $y = -2x$
37. (a) Yes **(b)**

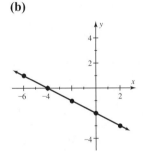

(c) $y = -\frac{1}{2}x - 2$
39. (a) No **(b)**

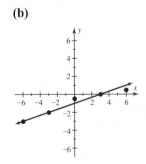

(c) $y = \frac{1}{3}x - 1$
41. $g = 5t + 200$, where g represents gallons of water and t represents time in minutes.
43. $d = 6t + 5$, where d represents distance in miles and t represents time in hours.
45. $p = 8t + 200$, where p represents total pay in dollars and t represents time in hours.
47. $r = t + 5$, where r represents total number of roofs shingled and t represents time in days.

49. (a) $-\frac{2}{45}$ acre per year   **(b)** $A = -\frac{2}{45}t + 5$

51. (a) $y = 8x - 15{,}864$ **(b)** About 208,000

53. (a)   **(b)**

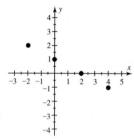
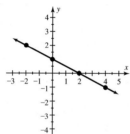

(c) 20; the mileage is 20 miles per gallon.

(d) $y = 20x$ **(e)** 140 miles

CHECKING BASIC CONCEPTS 3.7 (p. 236)

1. Yes   **2.** Approximate; $y = x - 1$

3. (a) $y = 10x + 50$ **(b)** $y = -2x + 200$

4. (a) Yes   **(b)**

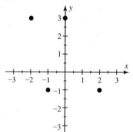
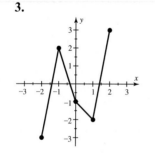

(c) $y = -\frac{1}{2}x + 1$

5. (a) $T = 0.075x$ **(b)** $4.5°F$

CHAPTER 3 REVIEW (pp. 241–246)

1. $(-2, 0)$: none; $(-1, 2)$: II; $(0, 0)$: none; $(1, -2)$: IV; $(1, 3)$: I

2.   **3.**

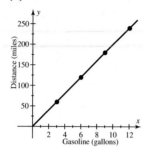

4.

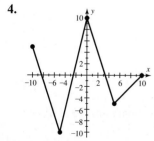

5. Yes   **6.** No   **7.** No   **8.** Yes

9.

x	-2	-1	0	1	2
y	6	3	0	-3	-6

10.

x	4	3	2.5	2	1
y	-3	-1	0	1	3

11.

x	-2	0	2	4
y	-4	2	8	14

12.

x	1	2	3	4
y	6	5	4	3

13.

x	-0.5	0	0.5	1
y	-1	0	1	2

14.

x	-1	-3	-5	-7
y	1	2	3	4

15.   **16.**

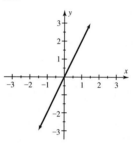
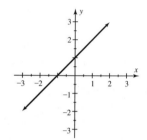

17.   **18.**

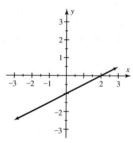
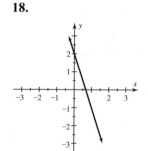

19.   **20.**

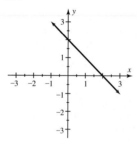
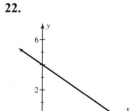

21.   **22.**

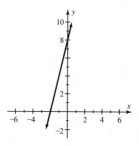

23. 3; −2   **24.** −2 and 2; −4

25.

x	−2	−1	0	1	2
y	4	3	2	1	0

2; 2

26.

x	−4	−2	0	2	4
y	−4	−3	−2	−1	0

4; −2

27. x-int: 3; y-int: −2

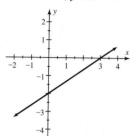

28. x-int: 1; y-int: −5

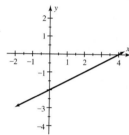

29. x-int: 4; y-int: −2

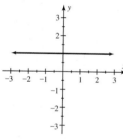

30. x-int: 2; y-int: 3

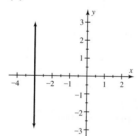

31. (a)

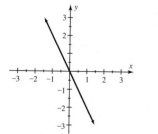

(b)

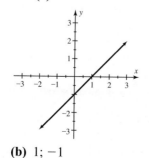

32. x = −1; y = 1   **33.** y = 1   **34.** x = 3
35. y = 3; x = −2   **36. (a)** y-int: 90; x-int: 3   **(b)** The driver is initially 90 miles from home; the driver arrives home after 3 hours.   **37.** 2   **38.** $-\frac{2}{5}$ **39.** 0   **40.** Undefined
41. 3   **42.** $-\frac{1}{2}$
43. (a).

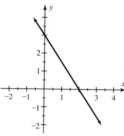

44. (a)

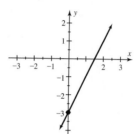

(b) −2; 0

(b) 1; −1

45. (a)

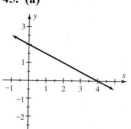

46. (a)

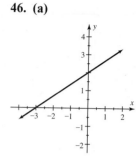

(b) $-\frac{1}{2}$; 2

(b) $\frac{2}{3}$; 2

47.

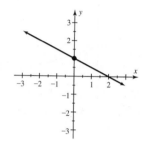

48.

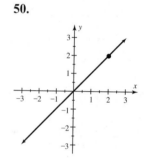

49.

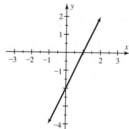

50.

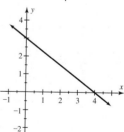

51. slope: 2; x-int: 1; y-int: −2

52.

x	0	1	2	3
y	1	$\frac{3}{2}$	2	$\frac{5}{2}$

53. y = x + 1   **54.** y = −2x + 2
55. y = 2x − 2

56. $y = -\frac{3}{4}x + 3$

57. (a) y = −x + 3   **(b)** −1; 3
58. (a) $y = \frac{3}{2}x - 3$ **(b)** $\frac{3}{2}$; −3
59. (a) y = 2x − 20   **(b)** 2; −20
60. (a) $y = \frac{5}{6}x - 5$ **(b)** $\frac{5}{6}$; −5

61.

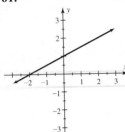

62.

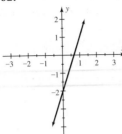

63.

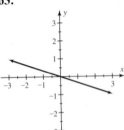

64.

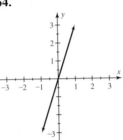

65.

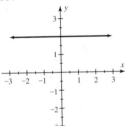

66.

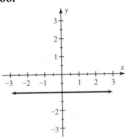

67.

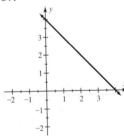

68.

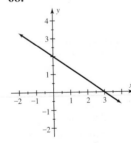

69. $y = 5x - 5$ **70.** $y = -2x$ **71.** $y = -\frac{5}{6}x + 2$

72. $y = -2x - 3$ **73.** $y = \frac{2}{3}x - 2$ **74.** $y = -\frac{1}{5}x - 2$

75. Yes   **76.** No   **77.** $y = 5x - 3$ **78.** $y = 20x - 65$

79. $y = -\frac{2}{3}x - \frac{1}{3}$ **80.** $y = 3x - 90$ **81.** $y = \frac{4}{3}x - 4$

82. $y = 2x - 1$ **83.** $y = 2x - 3$ **84.** $y = -\frac{2}{3}x - \frac{2}{3}$

85. $y = 3x + 5$ **86.** $y = \frac{1}{3}x + 7$

87. $y = 2x + 11$ **88.** $y = -\frac{1}{4}x + 3$ **89.** No

90. Approximate; $y = -x + 5$ **91.** $y = -2x + 40$

92. $y = 20x + 200$ **93.** $y = 50$ **94.** $y = 5x - 20$

95. (a) Yes

(b)

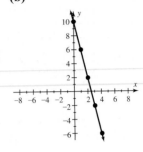

(c) $y = -4x + 10$

96. (a) No

(b)

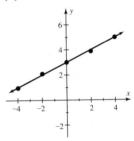

(c) $y = \frac{1}{2}x + 3$

97. (a)

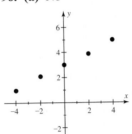

(b) The number of divorces increased significantly between 1960 and 1980, and then remained unchanged from 1980 to 2000.

98. (a) $G = 100t$

(b)

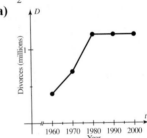

(c) 50 days

99. (a)

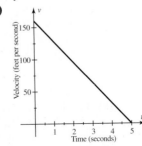

(b) v-int: 160; t-int: 5; the initial velocity was 160 ft/sec, and the velocity after 5 seconds was 0.
100. (a) $m_1 = 0$; $m_2 = -1500$; $m_3 = 500$ **(b)** $m_1 = 0$: The population remained unchanged. $m_2 = -1500$: The population decreased at a rate of 1500 insects per week. $m_3 = 500$: The population increased at a rate of 500 insects per week.   **(c)** For the first week the population did not change from its initial value of 4000. Over the next two weeks the population decreased at a rate of 1500 insects per week until it reached 1000. Finally, the population increased at a rate of 500 per week for two weeks, reaching 2000.
101.

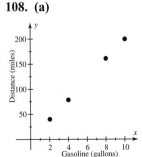

102. (a) -105 **(b)** The number of nursing homes decreased at an average rate of 105 per year.
103. (a) 0.6   **(b)** Enrollment increased at an average rate of 0.6 million students per year.   **(c)** 86 million students
104. (a) $35   **(b)** 20¢   **(c)** 35; the fixed cost of renting the car   **(d)** 0.2; the cost for each mile driven
105. (a) Toward; the slope is negative.   **(b)** After 1 hour the car is 200 miles from home; after 3 hours the car is 100 miles from home.   **(c)** $y = -50x + 250$; the car is moving toward home at 50 mph.   **(d)** 150 miles; 150 miles
106. (a)

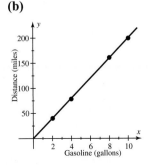

(b) $I = 20t - 39{,}000$; the number of icebergs increased at an average rate of 20 per year.   **(c)** Yes   **(d)** 1100 icebergs
107. (a) 90,100; 65,900   **(b)** -6.05; deaths from pneumonia decreased at an average rate of 6050 per year.
108. (a)   **(b)**

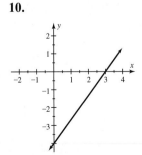

(c) 20; the mileage is 20 miles per gallon.
(d) $y = 20x$; no   **(e)** About 180 miles

CHAPTER 3 TEST (pp. 246–247)

1. $(-2, -2)$: III; $(-2, 1)$: II; $(0, -2)$: none; $(1, 0)$: none; $(2, 3)$: I; $(3, -1)$: IV; $(3, 1)$: I
2.

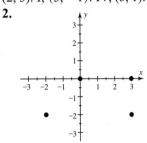

3.

x	-2	-1	0	1	2
y	-8	-6	-4	-2	0

$2; -4$ **4.** Yes

5.

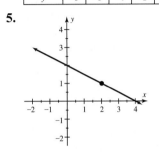

6. x-int: 3; y-int: -5
7.   **8.**

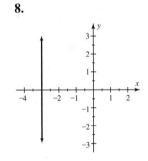

9.   **10.**

11. $y = -x + 2$ **12.** $y = 2x + \frac{1}{2}$; 2; $\frac{1}{2}$
13. $y = -5$; $x = 1$ **14.** $-\frac{2}{9}$ **15.** $y = -\frac{4}{3}x - 5$
16. $y = 3x - 11$ **17.** $y = -3x + 5$ **18.** $y = -\frac{1}{2}x$
19. $y = \frac{1}{2}x + 5$ **20.** Approximate; $y = x - 1$

21.

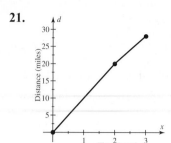

22. (a) $m_1 = 2$; $m_2 = -9$; $m_3 = 2$; $m_4 = 5$
(b) $m_1 = 2$: The population increased at a rate of 2000 fish per year. $m_2 = -9$: The population decreased at a rate of 9000 fish per year. $m_3 = 2$: The population increased at a rate of 2000 fish per year. $m_4 = 5$: The population increased at a rate of 5000 fish per year.   **(c)** For the first year the population increased from an initial value of 8000 to 10,000 at a rate of 2000 fish per year. During the second year the population dropped dramatically to 1000 at a rate of 9000 fish per year. Over the third year the population grew to 3000 at a rate of 2000 fish per year. Finally, over the fourth year, the population grew at a rate of 5000 fish per year to reach 8000.
23. $N = 100x + 2000$

CHAPTERS 1–3 CUMULATIVE REVIEW (pp. 248–250)

1. Composite; $40 = 2 \times 2 \times 2 \times 5$ **2.** Prime
3. $n + 10$ **4.** $n^2 - 2$ **5.** $\frac{7}{6}$ **6.** $\frac{2}{3}$ **7.** 2   **8.** $\frac{17}{30}$ **9.** 14
10. 7   **11.** $\frac{5}{3}$ **12.** -9 **13.** Rational   **14.** Irrational
15.

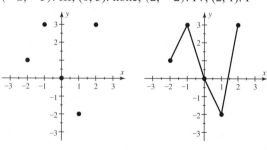

16. 2   **17.** 18   **18.** -4 **19.** $7x + 1$ **20.** $x - 4$
21. -3 **22.** 21   **23.** 2   **24.** 0
25.

x	-2	-1	0	1	2
y	10	8	6	4	2

26. $2n + 2 = n - 5$; -7 **27.** $-26, -25, -24, -23$
28. 8 hours   **29.** 720 ft^2 **30.** 20 ft^2 **31.** $b = \frac{2A}{h}$
32. $L = \frac{P - 2W}{2}$ **33.** $x < 2$
34.

x	0	1	2	3	4
$2x - 3$	-3	-1	1	3	5

$x \le 2$

35. $\{x \mid x > 0\}$ **36.** $\{x \mid x \le \frac{1}{2}\}$
37. $(-2, -3)$: III; $(0, 3)$: none; $(2, -2)$: IV; $(2, 1)$: I
38.

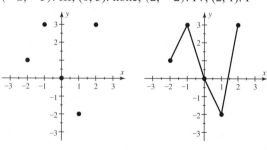

39. Yes
40.

x	-2	-1	0	1	2
y	3	2.5	2	1.5	1

41.

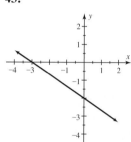

42.

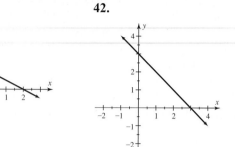

43.

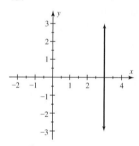

44.

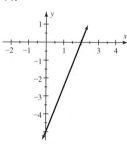

45.

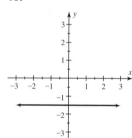

46.

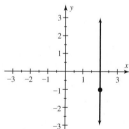

47. $\frac{3}{2}$; -3; $y = 2x - 3$ **48.** 2; 1; $y = -\frac{1}{2}x + 1$
49. x-int: -10; y-int: 8   **50.** $x = \frac{3}{2}$; $y = -2$
51.

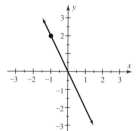

52.

53. $y = 3x - 3$

54. $y = \frac{3}{5}x - 3$;

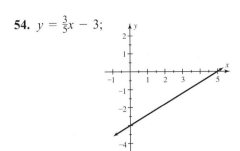

55. $y = -\frac{1}{3}x + 7$ **56.** $y = -\frac{2}{3}x - 3$ **57.** $y = -2x + 1$
58. $y = \frac{1}{4}x + \frac{1}{2}$ **59.** Approximate; $y = \frac{1}{2}x + 1$
60. $y = 5x + 100$ **61.** $I = 5000x + 20{,}000$
62. (a) Yes   **(b)**

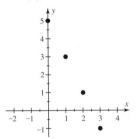   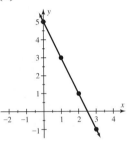

(c) $y = -2x + 5$
63. $C = 8x$ **64.** $\frac{99}{200}$ **65. (a)** \$807.40   **(b)** 1990
66. About 3.1%   **67.** \$1000 at 3%; \$2000 at 4%
68. (a) 0.16   **(b)** Participation increased at an average rate of 0.16 million students per year.   **(c)** 28 million
69. (a) \$85   **(b)** \$25   **(c)** 30¢

CHAPTER 4: SYSTEMS OF LINEAR EQUATIONS IN TWO VARIABLES

SECTION 4.1 (pp. 259–261)

1. ordered   **2.** intersection-of-graphs   **3.** one   **4.** (11, 9)
5. No   **6.** table   **7.** the same   **8.** intercepts   **9.** 1   **11.** 2
13. −2   **15.** 2   **17.** (1, 1)   **19.** (2, −3)   **21.** (2, 0)
23. (2, 1)   **25.** (3, 2)   **27.** (−1, 1)   **29.** (2, 4)   **31.** (3, 1)
33. (1, 3);

x	0	1	2	3
$y = x + 2$	2	3	4	5
$y = 4 - x$	4	3	2	1

35. (a) (−1, 1)   **(b)** (−1, 1)   **37. (a)** (3, 1)   **(b)** (3, 1)
39. (a) (1, 3)   **(b)** (1, 3)   **41.** (−1, −2)   **43.** (1, −1)
45. (1, 2)   **47.** (3, 2)   **49.** (3, 1)   **51.** (1, 2)
53. (a) $x + y = 4, x - y = 0$ **(b)** 2, 2
55. (a) $2x + y = 7, x - y = 2$ **(b)** 3, 1
57. (a) $x = 3y, x - y = 4$ **(b)** 6, 2
59. (a) $C = 0.5x + 50$ **(b)** 60 mi   **(c)** 60 mi
61. (a) $x + y = 36; x = 2y$ **(b)** (24, 12)
63. (a) $x - y = 4; 2x + 2y = 28$ where x is length, y is width   **(b)** (9, 5); the rectangle is 9 in. × 5 in.

SECTION 4.2 (pp. 268–271)

1. substitution   **2.** no solutions; one solution; infinitely many
3. exact   **4.** parentheses   **5.** It has infinitely many solutions.
6. It has no solutions.   **7.** consistent   **8.** inconsistent
9. independent   **10.** dependent   **11.** (3, 6)   **13.** (2, 1)
15. (−1, 0)   **17.** (3, 0)   **19.** $\left(\frac{1}{2}, 0\right)$ **21.** (0, 4)   **23.** (1, 2)
25. (6, −4)   **27.** (4, −3)   **29.** $\left(1, -\frac{1}{2}\right)$ **31.** (1, 3)
33. (0, −2)   **35.** (2, 2)   **37.** (−3, 5)   **39.** One; consistent; independent   **41.** Infinitely many; consistent; dependent   **43.** None; inconsistent   **45.** No solutions; inconsistent   **47.** (2, 1); consistent; independent
49. Infinitely many; consistent; dependent   **51.** No solutions; inconsistent   **53.** No solutions   **55.** Infinitely many
57. (3, 1)   **59.** Infinitely many   **61.** No solutions
63. $\left(-\frac{4}{3}, \frac{11}{9}\right)$ **65.** (0, 0)   **67.** No solutions
69. No solutions   **71.** They are a single line.
73. (a) $L - W = 10, 2L + 2W = 72$ **(b)** (23, 13)
75. (a) $x = \frac{1}{2}y; x + y = 90$ **(b)** (30, 60)   **(c)** (30, 60)
77. (a) $x - y = 1.68; y = 0.98x$ **(b)** (84, 82.32)
79. 94 ft × 50 ft   **81.** 17, 53   **83.** 10 mph; 2 mph
85. $3.\overline{3}$ L of 20% solution, $6.\overline{6}$ L of 50% solution
87. Superior: 32,000 mi²; Michigan: 22,000 mi²

CHECKING BASIC CONCEPTS 4.1 & 4.2 (p. 272)

1. (a) −2   **(b)** 6   **2.** (4, 2)   **3.** (2, 1)   **4. (a)** No solutions
(b) (1, −1); one   **(c)** Infinitely many   **5. (a)** $x + y = 300$, $150x + 200y = 55{,}000$ **(b)** (100, 200)

SECTION 4.3 (pp. 280–283)

1. Substitution; elimination   **2.** addition   **3.** =   **4.** =
5. Add the equations.   **6.** Multiply the first equation by −3.
7. It has infinitely many solutions.   **8.** It has no solutions.
9. (1, 1)   **11.** (−2, 1)   **13.** No solutions   **15.** Infinitely many   **17.** (6, 1)   **19.** (−1, 4)   **21.** (2, 4)   **23.** (2, 1)
25. (−4, 1)   **27.** (5, 8)   **29.** (−2, −5)   **31.** (1, 0)
33. (2, 2)   **35.** (4, −1)   **37.** $\left(\frac{4}{5}, \frac{9}{5}\right)$ **39.** (3, 2)   **41.** (0, 1)
43. (2, 1)   **45.** (4, −3)   **47.** (1, 0)
49. Infinitely many;   **51.** One;

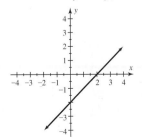   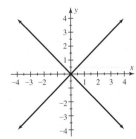

53. No solutions;

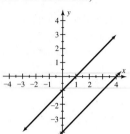

55. One;

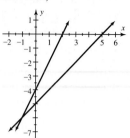

57. Infinitely many;

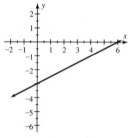

59. Men: 38,000; women: 28,000   **61.** Bicycle:18 min; stair climber: 12 min   **63.** Current: 2 mph; boat: 6 mph
65. $2000 at 3%; $3000 at 5%   **67.** −43, 26
69. (a) 20 in. × 40 in.   **(b)** 20 in. × 40 in.

SECTION 4.4 (pp. 292–295)

1. All points below and including the line $y = k$
2. All points to the right of the line $x = k$.
3. All points above and including the line $y = x$ **4.** test
5. dashed   **6.** solid   **7.** $Ax + By = C$ **8.** both
inequalities   **9.** intersect   **10.** No   **11.** Yes   **13.** No
15. No   **17.** Yes   **19.** Yes   **21.** No   **23.** $x > 1$
25. $y \geq 2$ **27.** $y < x$ **29.** $-x + y \leq 1$
31.

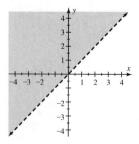

33.

35.

37.

39.

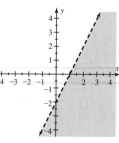

41.

43. Yes   **45.** No   **47.** Yes   **49.** The region containing $(1, 2)$
51. The region containing $(1, 0)$
53.

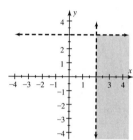

55.

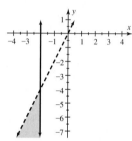

57.

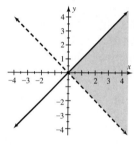

59.

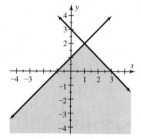

61.

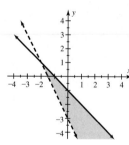

63.

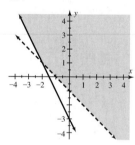

65.

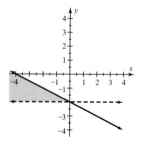

67.

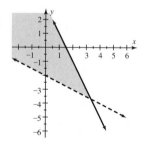

69.

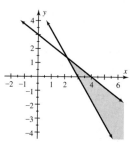

71.

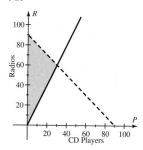

(b) Inconsistent   **24. (a)** One   **(b)** Consistent; independent
25. (a) Infinitely many   **(b)** Consistent; dependent
26. (a) None   **(b)** Inconsistent   **27.** No solutions
28. No solutions   **29.** Infinitely many   **30.** $(1, 1)$
31. $(2, 1)$ **32.** $(-1, 2)$ **33.** $(11, -1)$ **34.** $(1, 0)$
35. $\left(\frac{9}{4}, \frac{7}{4}\right)$ **36.** $\left(\frac{5}{2}, 1\right)$ **37.** $(5, -7)$ **38.** $(8, 2)$
39. $(-2, 3)$ **40.** $(1, 0)$ **41.** $(1, 3)$ **42.** $(2, -1)$
43. Infinitely many   **44.** Infinitely many   **45.** No solutions
46. One   **47.** Yes   **48.** No   **49.** No   **50.** Yes
51. $y > 1$ **52.** $y \leq 2x + 1$

73. (a) 200 bpm; 150 bpm

(b)

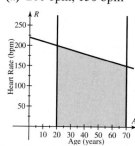

53.

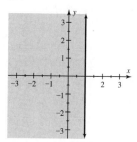

54.

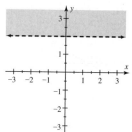

(c) Possible heart rates for ages 20 to 70   **75.** 150 to 200 lb

CHECKING BASIC CONCEPTS 4.3 & 4.4 (p. 295)

1. $(1, 1)$ **2. (a)** $(0, -1)$; one   **(b)** No solutions
(c) Infinitely many   **3.** Infinitely many
4. (a)

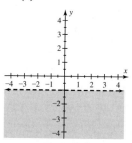

(b)

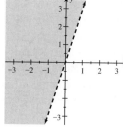

55.

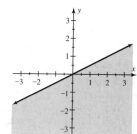

56.

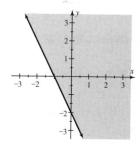

57.

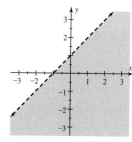

58.

5.

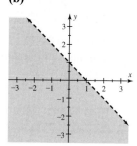

59. Yes   **60.** No   **61.** The region containing $(2, -2)$
62. The region containing $(1, 3)$

63.

64.

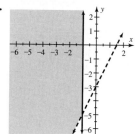

6. (a) $x + y = 11, x - y = 5$ **(b)** $(8, 3)$

CHAPTER 4 REVIEW (pp. 298–301)

1. 3   **2.** 2   **3.** $(1, 2)$ **4.** $(5, 2)$ **5.** $(4, 3)$ **6.** $(2, -4)$
7. $(2, 2)$ **8.** $(1, 2)$ **9.** $(2, 6)$ **10.** $(1, 1)$ **11.** $(4, -3)$
12. $(1, 2)$ **13.** $(1, 1)$ **14.** $(1, 2)$ **15.** $(1, 1)$
16. $(-2, -1)$ **17.** $(2, 6)$ **18.** $(2, -10)$ **19.** $(1, 3)$
20. $(5, 10)$ **21.** $(-2, 1)$ **22.** $(-4, -4)$ **23. (a)** None

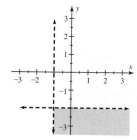
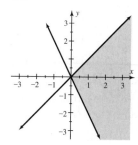

65.

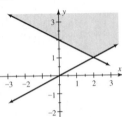

66.

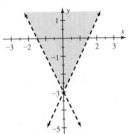

67.

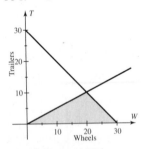

68.

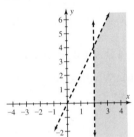

69. 3100 deaths in 1912; 42,625 deaths in 2003
70. Men: 102,500 cases; women: 82,500 cases
71. (a) $C = 0.2x + 40$ **(b)** 250 mi   **(c)** 250 mi
72. 75°, 105°   **73. (a)** $2x + y = 180, 2x - y = 40$
(b) (55, 70)   **(c)** (55, 70)   **74.** 20 ft × 24 ft
75. (a) $x + y = 10, 80x + 120y = 920$; x is $80 rooms,
y is $120 rooms.   **(b)** (7, 3)   **76.** 7 lb of $2 candy; 11 lb of
$3 candy   **77.** Bicycle: 35 min; stair climber: 25 min
78. 2 mph   **79. (a)** 16 ft × 24 ft (answers may vary slightly)
(b) 16 ft × 24 ft
80.

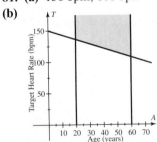

81. (a) 136 bpm; 108 bpm
(b)

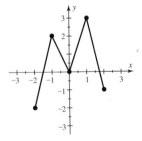

(c) Target heart rates above 70% of the maximum heart rate
for ages 20 to 60

CHAPTER 4 TEST (pp. 302–303)

1. (1, 2)   **2.** (−2, −1)   **3.** (−1, −2)   **4.** (−2, 3)
5. (1, 3)   **6. (a)** (2, 1); one; consistent   **(b)** No solutions;
zero; inconsistent   **7. (a)** Infinitely many   **(b)** Consistent;
dependent   **8. (a)** None   **(b)** Inconsistent   **9.** (−3, 4)
10. No solutions   **11.** Infinitely many solutions
12. (−1, 3)   **13.** $y \le -\frac{1}{2}x$ **14.** Yes
15.

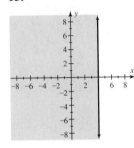

16.

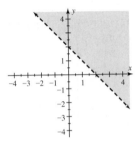

17.

18.

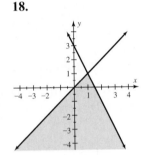

19. $1.9 trillion in 2003; $2.0 trillion in 2004
20. $\frac{2}{3}$ hr at 6 mph; $\frac{1}{3}$ hr at 9 mph

CHAPTERS 1–4 CUMULATIVE REVIEW (pp. 304–305)

1. $2 \times 2 \times 2 \times 3 \times 5$ **2. (a)** 4   **(b)** 8   **3. (a)** Rational
(b) Irrational   **4. (a)** <   **(b)** >   **5.** Associative
(multiplication)   **6.** 3060   **7. (a)** $4x^2$ **(b)** $5x - 2$
8. $\frac{2}{9}$ **9.** $\frac{1}{2}$ **10.** No solutions   **11.** 11, 12, 13, 14
12. 432 in² or 3 ft²   **13.** $x = \frac{W + 7y}{3}$ **14.** $x \ge 1$
15.

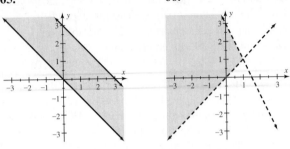

16. (a)   **(b)**

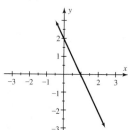

17. $2; -8$

18. (a)   **(b)**

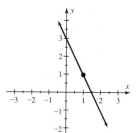 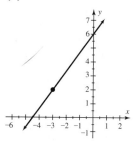

19. (a) $y = -\frac{1}{2}x - 1$ **(b)** $y = 4x - 3$

20. $y = -\frac{1}{4}x + 3$ **21.** $y = -3x + 4$ **22.** $(4, 4)$

23. $(1, -3)$ **24.** No solutions **25.** Infinitely many solutions **26.** $(-1, 2)$

27.   **28.**

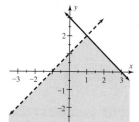 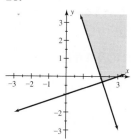

29. $F = 16.5R$ **30.** $94°C$ **31.** $245 **32.** $158.05

33. $G = -3x + 30$ **34.** $1200 at 5%; $1200 at 6%

CHAPTER 5: POLYNOMIALS AND EXPONENTS

SECTION 5.1 (pp. 315–317)

1. base; exponent **2.** 1 **3.** $\left(\frac{1}{2}\right)^3$ **4.** x^4 **5.** -1 **6.** 1

7. a^{m+n} **8.** a^{mn} **9.** $a^n b^n$ **10.** $\frac{a^n}{b^n}$ **11.** 64 **13.** -8

15. -8 **17.** 1 **19.** 32 **21.** 26 **23.** 8 **25.** $\frac{1}{2}$

27. 3^3 or 27 **29.** 4^8 or 65,536 **31.** 2^5 or 32 **33.** x^9

35. z^4 **37.** x^6 **39.** $20x^7$ **41.** $-3x^3 y^4$ **43.** 2^6 or 64

45. n^{12} **47.** x^7 **49.** $49b^2$ **51.** $a^3 b^3$ **53.** $4x^4$

55. $-64b^6$ **57.** $x^{14} y^{21}$ **59.** $x^{12} y^9$ **61.** $a^{10} b^8$ **63.** $\frac{1}{27}$

65. $\frac{a^5}{b^5}$ **67.** $\frac{(x-y)^3}{27}$ **69.** $\frac{25}{(a+b)^2}$ **71.** $\frac{8x^3}{125}$ **73.** $\frac{27x^6}{125y^{12}}$

75. $(x+y)^4$ **77.** $(a+b)^5$ **79.** 6 **81.** $a^3 + 2ab^2$

83. $12a^5 + 6a^3 b$ **85.** $r^2 t + rt^2$

87. $a = 3, b = 1, m = 1, n = 0$ (answers may vary)

89. $10x^4$ **91.** $8x^3$ **93.** $9\pi x^4$ **95.** $1157.63

97.

99.

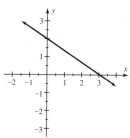

SECTION 5.2 (pp. 325–327)

1. monomial **2.** -3 **3.** polynomial **4.** degree

5. binomial **6.** trinomial **7.** 2; 3 **8.** like **9.** opposite

10. vertically **11.** 2; 3 **13.** 2; -1 **15.** 2; -5

17. 0; -6 **19.** Yes; 1; 1; 1 **21.** Yes; 3; 1; 2 **23.** No

25. Yes; 2; 2; 4 **27.** No **29.** Yes; 1; 3; 6 **31.** Yes; x

33. Yes; $-5x^3$ **35.** No **37.** Yes; $2ab$ **39.** Yes; $4xy^2$

41. $-x + 9$ **43.** $4x^2 + 8x - 5$ **45.** $5a^3 + 1$

47. $4y^3 + 3y^2 + 4y - 9$ **49.** $4xy + 1$ **51.** $2a^2 b^3$

53. $9x^2 + x - 6$ **55.** $x^2 - 7x - 1$ **57.** $-5x^2$

59. $-3a^2 + a - 4$ **61.** $2t^2 + 3t - 4$ **63.** $4x - 2$

65. $-3x^2 + 5x + 10$ **67.** $-3a^2 - 5a$

69. $z^3 - 6z^2 - 6z - 1$ **71.** $3xy + 2x^2 y^2$ **73.** $-a^3 b$

75. $-x^2 - 5x - 4$ **77.** $-2x^3 - 6x - 2$

79. (a) 200 bpm **(b)** 100 bpm **(c)** It decreases quickly at first, then more slowly. **81.** $2z^2$; 200 in^2

83. $2x^2 + x^2$ or $3x^2$; 108 ft^2 **85.** $\pi x^2 + \pi y^2$; 13π ft^2

87. (a) $m_1 \approx 0.077; m_2 \approx 0.083; m_3 \approx 0.077$. A line is reasonable but not exact. **(b)** For the given years, its estimates are reasonable.

CHECKING BASIC CONCEPTS 5.1 & 5.2 (p. 327)

1. (a) -25 **(b)** 1 **2. (a)** 10^8 **(b)** $-12x^7$ **(c)** $a^6 b^2$

(d) $\frac{x^4}{z^{12}}$ **3.** 3; 2; 4 **4. (a)** $2W^2 H$ **(b)** 2880 in^3

5. (a) $3a^2 + 6$ **(b)** $2z^3 + 7z - 8$ **(c)** $4xy$ **6.** 34

SECTION 5.3 (pp. 333–335)

1. The product rule **2.** Distributive **3.** one; two; three

4. term; term **5.** No **6.** vertically **7.** x^7 **9.** $-12a^2$

11. $20x^5$ **13.** $4x^2 y^3$ **15.** $-12x^3 y^3$ **17.** $3x + 12$

19. $-45x - 5$ **21.** $4z - z^2$ **23.** $-5y - 3y^2$

25. $15x^3 - 12x$ **27.** $6x^3 - 6x^2$ **29.** $-32t^2 - 8t - 8$

31. $-5n^4 + n^3 - 2n^2$ **33.** $x^2y + xy^2$ **35.** $x^4y - x^3y^2$
37. $-a^4b + 2ab^4$ **39.** $x^2 + 3x$ **41.** $x^2 + 4x + 4$
43. $x^2 + 9x + 18$ **45.** $x^2 + 8x + 15$
47. $x^2 - 17x + 72$ **49.** $6z^2 - 19z + 10$ **51.** $64b^2 - 1$
53. $10y^2 - 3y - 7$ **55.** $5 - 13a + 6a^2$ **57.** $1 - 9x^2$
59. $x^3 - x^2 + x - 1$ **61.** $4x^3 - 3x^2 + 16x - 12$
63. $2n^3 + n^2 + 6n + 3$ **65.** $m^3 + 4m^2 + 4m + 1$
67. $6x^3 - 7x^2 + 14x - 8$ **69.** $x^3 + 1$
71. $4b^4 + 3b^3 + 19b^2 + 9b + 21$
73. $x^3 - x^2 - 5x + 2$ **75.** $a^3 - 8$
77. $6x^4 - 2x^3 + 5x^2 - x + 1$ **79.** $m \cdot n$
81. (a) $h^3 - 2h^2 - 8h$ (b) $14{,}175$ in^3
83. (a) $50x - x^2$ (b) 625 **85.** $6x^2 + 12x + 6$
87. (a) $64t - 16t^2$ (b) $64; 64$ (c) Yes; yes

SECTION 5.4 (pp. 340–342)

1. $a^2 - b^2$ **2.** Yes; the product is a sum and a difference.
3. $a^2 + 2ab + b^2$ **4.** $a^2 - 2ab + b^2$
5. No; $(x + y)^2 = x^2 + 2xy + y^2$
6. No; $(r - t)^2 = r^2 - 2rt + t^2$
7. No; $(z + 5)^3 = z^3 + 15z^2 + 75z + 125$ **8.** $(a + b)^2$
9. $x^2 - 9$ **11.** $x^2 - 1$ **13.** $16x^2 - 1$ **15.** $1 - 4a^2$
17. $4x^2 - 9y^2$ **19.** $a^2b^2 - 25$ **21.** $a^2b^2 - 16$
23. $a^4 - b^4$ **25.** $x^6 - y^6$ **27.** 9999 **29.** 391
31. 9900 **33.** $x^2 + 2x + 1$ **35.** $a^2 - 4a + 4$
37. $4x^2 + 12x + 9$ **39.** $9b^2 + 30b + 25$
41. $\frac{9}{16}a^2 - 6a + 16$ **43.** $1 - 2b + b^2$
45. $25 + 10y^3 + y^6$ **47.** $a^4 + 2a^2b + b^2$
49. $a^3 + 3a^2 + 3a + 1$ **51.** $x^3 - 6x^2 + 12x - 8$
53. $8x^3 + 12x^2 + 6x + 1$
55. $216u^3 - 108u^2 + 18u - 1$
57. $t^4 + 6t^3 + 12t^2 + 8t$ **59.** $20x + 36$
61. $x^2 + 2x - 35$ **63.** $9x^2 - 30x + 25$
65. $25x^2 + 35x + 12$ **67.** $16b^2 - 25$
69. $-20x^3 + 35x^2 - 10x$ **71.** $64 - 48a + 12a^2 - a^3$
73. $x^3 + 6x^2 + 9x$ **75.** $x^4 - 5x^2 + 4$ **77.** $a^{2n} - b^{2n}$
79. (a) $x^2 + 4x + 4$ (b) $x^2 + 4x + 4$
81. (a) $4x^2 + 12x + 9$ (b) $4x^2 + 12x + 9$
83. (a) $6x^2 + 60x + 150$ (b) $x^3 + 15x^2 + 75x + 125$
85. (a) $1 + 2x + x^2$ (b) 1.21; the money increases by
1.21 times in 2 years if the interest rate is 10%.
87. (a) $1 - 2x + x^2$ (b) 0.25; if the chance of rain on each
day is 50%, then there is a 25% chance that it will not rain on
either day. **89.** (a) $32z + 256$ (b) 2176; the area of an
8-foot-wide sidewalk around a 60×60 foot pool is 2176 ft^2.

CHECKING BASIC CONCEPTS 5.3 & 5.4 (p. 343)

1. (a) $-15x^3y^5$ (b) $-6x + 4x^2$
(c) $3a^3b - 6a^2b^2 + 3ab^3$
2. (a) $4x^2 + 9x - 9$ (b) $2x^4 - 2$ (c) $x^3 + y^3$

3. (a) $25x^2 - 4$ (b) $x^2 + 6x + 9$
(c) $4 - 28x + 49x^2$ (d) $t^3 + 6t^2 + 12t + 8$
4. (a) $4\pi a^2 + 56\pi a + 196\pi$ (b) 900π
5. (a) $m^2 + 10m + 25$ (b) $m^2 + 10m + 25$

SECTION 5.5 (pp. 351–354)

1. $\frac{1}{a^n}$ **2.** $\frac{1}{8}$ **3.** a^n **4.** 8 **5.** a^{m-n} **6.** 8 **7.** $\frac{b^m}{a}$ **8.** $\frac{25}{8}$
9. $\left(\frac{b}{a}\right)^n$ **10.** $\frac{25}{9}$ **11.** $1 \le |b| < 10$ **12.** No **13.** (a) $\frac{1}{4}$
(b) 9 **15.** (a) 2 (b) 100 **17.** (a) $\frac{1}{81}$ (b) $\frac{1}{8}$
19. (a) $\frac{1}{576}$ (b) 64 **21.** (a) 1 (b) 64 **23.** (a) 25
(b) $\frac{49}{4}$ **25.** (a) $\frac{1}{x}$ (b) $\frac{1}{a^4}$ **27.** (a) $\frac{1}{x^2}$ (b) $\frac{1}{a^8}$
29. (a) $\frac{y^3}{x^3}$ (b) $\frac{1}{x^3y^3}$ **31.** (a) $\frac{1}{16t^4}$ (b) $\frac{1}{(x + 1)^7}$
33. (a) a^8 (b) $\frac{1}{r^2t^6}$ **35.** (a) $\frac{b^2}{a^4}$ (b) x^2 **37.** (a) a^{13}
(b) $\frac{2}{z^3}$ **39.** (a) $-\frac{2y^3}{3x^2}$ (b) $\frac{1}{x^3}$ **41.** (a) $2b$ (b) $\frac{a^3}{b^3}$
43. (a) y^5 (b) $2t^3$ **45.** (a) x^2y^2 (b) a^6b^3
47. (a) $\frac{b^2}{a^2}$ (b) $\frac{4v}{u}$ **49.** $\frac{a^n}{a^m} = a^{n-m} = a^{-(m-n)} = \frac{1}{a^{m-n}}$
51. Thousand **53.** Billion **55.** Hundredth **57.** 2000
59. 45,000 **61.** 0.008 **63.** 0.000456 **65.** 39,000,000
67. $-500{,}000$ **69.** 2×10^3 **71.** 5.67×10^2
73. 1.2×10^7 **75.** 4×10^{-3} **77.** 8.95×10^{-4}
79. -5×10^{-2} **81.** $1{,}500{,}000$ **83.** -1.5
85. 2000 **87.** 0.002 **89.** (a) About 5.859×10^{12} mi
(b) About 2.5×10^{13} mi **91.** About 9.2×10^9 mi
93. (a) 1.246×10^{13} (b) About $41{,}812

SECTION 5.6 (pp. 359–360)

1. $\frac{a}{d} + \frac{b}{d}$ **2.** $\frac{a}{d} + \frac{b}{d} - \frac{c}{d}$ **3.** term **4.** No **5.** No
6. $9; 4; 1$ **7.** $x + 1; 2x^2 - 2x + 1; 4$ **8.** $0x^2$ **9.** $2x$
11. $z^3 + z^2$ **13.** $\frac{a^2}{2} - 3$ **15.** $\frac{4}{x} - 7x^2$ **17.** $3x^3 - 1 + \frac{2}{x}$
19. $4y^2 - 1 + \frac{2}{y}$ **21.** $3m^2 - 2m + 4$
23. $2x + 1 + \frac{3}{x - 2}$ **25.** $x + 1$ **27.** $x^2 + 1 + \frac{-1}{x - 1}$
29. $x^2 - x + 2 + \frac{1}{4x + 1}$ **31.** $x^2 + 2x + 3 + \frac{8}{x - 2}$
33. $3x^2 + 3x + 3 + \frac{5}{x - 1}$ **35.** $x + 3 + \frac{-x - 2}{x^2 + 1}$
37. $x + 1$ **39.** $x^2 - 2x + 4$ **41.** They are the same.
43. $4x$
45. $x + 2$;

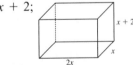

CHECKING BASIC CONCEPTS 5.5 & 5.6 (p. 361)

1. (a) $\frac{1}{81}$ (b) $\frac{1}{2x^7}$ (c) $\frac{b^8}{16a^2}$ **2.** (a) z^5 (b) $\frac{y^6}{x^3}$ (c) $\frac{x^6}{27}$
3. (a) 4.5×10^4 (b) 2.34×10^{-4} (c) 1×10^{-2}

4. (a) 47,100 **(b)** 0.006   **5.** $5a - 3$ **6.** $3x + 2$; R: -2
7. (a) 9.3×10^7 **(b)** 500 sec (8 min 20 sec)

CHAPTER 5 REVIEW (pp. 365–367)

1. 125   **2.** -81 **3.** 4   **4.** 11   **5.** -5 **6.** 1   **7.** 6^5
8. 10^{12} **9.** z^9 **10.** y^6 **11.** $30x^9$ **12.** a^4b^4 **13.** 2^{10}
14. m^{20} **15.** a^3b^3 **16.** x^8y^{12} **17.** x^7y^{11} **18.** 1
19. $(r - t)^9$ **20.** $(a + b)^6$ **21.** $\frac{9}{(x - y)^2}$ **22.** $\frac{(x + y)^3}{8}$
23. 7; 6   **24.** 5; -1 **25.** Yes; 1; 1; 1   **26.** Yes; 4; 1; 3
27. Yes; 3; 2; 2   **28.** No   **29.** $5x^2 - x + 3$
30. $-6x^2 + 3x + 7$ **31.** $3x + 4$ **32.** $-2x^2 - 13$
33. $-2x^2 + 9x + 5$ **34.** $2a^3 - a^2 + 7a$
35. $5y^2 - 3xy$ **36.** $4xy$ **37.** $-x^5$ **38.** $-r^3t^4$
39. $-6t + 15$ **40.** $2y - 12y^2$ **41.** $18x^5 + 30x^4$
42. $-x^3 + 2x^2 - 9x$ **43.** $-a^3b + 2a^2b^2 - ab^3$
44. $a^2 + 3a - 10$ **45.** $8x^2 + 13x - 6$
46. $-2x^2 + 3x - 1$ **47.** $2y^3 + y^2 + 2y + 1$
48. $2y^4 - y^2 - 1$ **49.** $z^3 + 1$
50. $4z^3 - 15z^2 + 13z - 3$ **51.** $z^2 + z$ **52.** $2x^2 + 4x$
53. $z^2 - 4$ **54.** $25z^2 - 81$ **55.** $1 - 9y^2$
56. $25x^2 - 16y^2$ **57.** $r^2t^2 - 1$ **58.** $4m^4 - n^4$
59. $x^2 + 2x + 1$ **60.** $16x^2 + 24x + 9$
61. $y^2 - 6y + 9$ **62.** $4y^2 - 20y + 25$
63. $16 + 8a + a^2$ **64.** $16 - 8a + a^2$
65. $x^4 + 2x^2y^2 + y^4$ **66.** $x^2y^2 - 4xy + 4$
67. $z^3 + 15z^2 + 75z + 125$
68. $8z^3 - 12z^2 + 6z - 1$ **69.** 3599   **70.** 396   **71.** $\frac{1}{9}$
72. $\frac{1}{9}$ **73.** 4   **74.** $\frac{1}{1000}$ **75.** 36   **76.** $\frac{1}{25}$ **77.** $\frac{1}{z^2}$
78. $\frac{1}{y^4}$ **79.** $\frac{1}{a^2}$ **80.** $\frac{1}{x^2}$ **81.** $\frac{1}{4t^2}$ **82.** $\frac{1}{a^3b^6}$ **83.** $\frac{1}{y^3}$
84. x^4 **85.** $\frac{2}{x^3}$ **86.** $\frac{2x^4}{3y^3}$ **87.** $\frac{a^5}{b^5}$ **88.** $4t^4$ **89.** $\frac{27}{x^3}$
90. $2ab$ **91.** $\frac{y^2}{x^2}$ **92.** $\frac{2v}{3u}$ **93.** 600   **94.** 52,400
95. 0.0037   **96.** 0.06234   **97.** 1×10^4
98. 5.61×10^7 **99.** 5.4×10^{-5} **100.** 1×10^{-3}
101. 24,000,000   **102.** 0.2   **103.** $\frac{5}{3}x + 1$
104. $3b^2 - 2 + \frac{1}{b^2}$ **105.** $3x + 2 + \frac{4}{x - 1}$
106. $3x - 4 + \frac{6}{3x + 2}$ **107.** $x^2 - 3x - 1$
108. $x^2 + \frac{-1}{2x - 1}$ **109.** $x - 1 + \frac{-2x + 2}{x^2 + 1}$
110. $x^2 + 2x + 5$ **111. (a)** 60 bpm   **(b)** 160 bpm
(c) It increases.   **112.** $6xy$; 72 ft^2
113. $10z^2 + 25z$; 510 in^2 **114.** x^4y^2 **115.** \$833.71
116. $\frac{4}{3}\pi x^3 + 8\pi x^2 + 16\pi x + \frac{32}{3}\pi$ **117. (a)** $96t - 16t^2$
(b) 128; after 2 sec the ball is 128 ft high.
118. (a) $600L - L^2$ **(b)** 27,500; a rectangular building
with a perimeter of 1200 ft and a side of length 50 ft has an
area of 27,500 ft^2.   **119. (a)** $x^2 + 10x + 25$
(b) $x^2 + 10x + 25$ **120. (a)** $16x$ **(b)** 1600

121. About \$8795 per person   **122.** About 519,590,000 gal
or 5.1959×10^8 gal

CHAPTER 5 TEST (pp. 367–368)

1. Yes; 3; 2; 3   **2.** $x^3 - 4x + 8$ **3.** $4x + 6$
4. $-3y^3 - 2y + 1$ **5.** $x^2 + x - 7$ **6.** $4a^3 + 2ab$
7. (a) -6 **(b)** $\frac{1}{64}$ **(c)** 8   **(d)** -3 **8.** $24y^{11}$
9. a^5b^8 **10.** x^4 **11.** $\frac{a^3}{b^6}$ **12.** $a^3b - ab^3$ **13.** $\frac{4}{9a^4b^6}$
14. $\frac{2y^3}{x}$ **15.** $\frac{16}{(a + b)^4}$ **16.** $12x^5 - 18x^3 + 3x^2$
17. $2z^2 - 2z - 12$ **18.** $49y^4 - 9$ **19.** $9x^2 - 12x + 4$
20. $m^3 + 9m^2 + 27m + 27$ **21.** $y^3 - y + 6$
22. 6396   **23.** 0.0061   **24.** 5.41×10^3 **25.** $3x - 2 + \frac{1}{x}$
26. $x^2 - x + 1 + \frac{-1}{x + 2}$ **27. (a)** $20t$ **(b)** $2t + 2000$
(c) $18t - 2000$; profit from selling t tickets
28. $12x^2$; 1200 ft^2 **29.** $3x^3 + 27x^2 + 54x$
30. (a) $88t - 16t^2$ **(b)** 120; after 3 sec the ball is
120 ft high.

CHAPTERS 1–5 CUMULATIVE REVIEW (pp. 369–370)

1. (a) 8   **(b)** 8   **2. (a)** 29   **(b)** $\frac{3}{2}$ **3. (a)** 7
(b) No solutions   **4. (a)** Infinitely many solutions   **(b)** -2
5. 68 mph   **6.** $-39, -38, -37$ **7. (a)** $\frac{21}{50}$ **(b)** $\frac{19}{250}$
8. $x = \frac{Y}{W - 4}$
9.

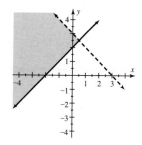

10.

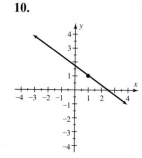

11. $y = -2x + 1$ **12.** 2; -3 **13.** $y = \frac{1}{2}x - 4$
14. $y = 3x + 1$ **15.** No solutions   **16.** $(2, -1)$
17. Infinitely many solutions   **18.** $(0, -6)$
19.

20.

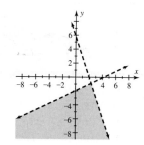

21. (a) $15x^5$ **(b)** $x^{10}y^7$ **22. (a)** $2x^2 - x + 3$
(b) $12a^3 + a - 5$ **23. (a)** $2x^2 - 11x - 21$
(b) $y^3 - 10y - 3$ **(c)** $16x^2 - 49$ **(d)** $25a^2 + 30a + 9$
24. (a) $\frac{1}{x}$ **(b)** $\frac{x^9}{8}$ **(c)** $\frac{x^4}{2y^2}$ **(d)** $\frac{x^7}{y^8}$ **25.** 2.4×10^{10}
26. 0.000000471 **27. (a)** $4x^2 - 1$ **(b)** $2x - 5 + \frac{1}{x+3}$
28. -25% **29.** 200 mL **30. (a)** $2x^2 + 10x$
(b) $x^2 + 7x + 10$ **(c)** $2x^2 + 4x$ **(d)** $10x^2 + 42x + 20$

CHAPTER 6: FACTORING POLYNOMIALS AND SOLVING EQUATIONS

SECTION 6.1 (pp. 378–380)

1. factoring **2.** a **3.** multiplying
4. prime; multiplication **5.** greatest common factor (GCF)
6. grouping **7.** $1, 2, x, 2x$ **8.** $2x$
9. $2(x + 2)$;

2	2x	4
	x	2

11. $z(z + 4)$;

z	z²	4z
	z	4

13. $3x(x + 3)$ **15.** $2y^2(2y - 1)$ **17.** $2z(z^2 + 4z - 2)$
19. $3xy(2x - y)$ **21.** $6x; 6x(1 - 3x)$
23. $4y^2; 4y^2(2y - 3)$ **25.** $3z; 3z(2z^2 + z + 3)$
27. $x^2; x^2(x^2 - 5x - 4)$
29. $5y^2; 5y^2(y^3 + 2y^2 - 3y + 2)$ **31.** $x; x(y + z)$
33. $ab; ab(b - a)$ **35.** $5x^2y^3; 5x^2y^3(y + 2x)$
37. $ab; ab(a + b + 1)$ **39.** $(x - 2)(x + 1)$
41. $(z + 4)(z + 5)$ **43.** $2x(2x^2 - 1)(x - 5)$
45. $(x^2 + 3)(x + 2)$ **47.** $(y^2 + 1)(2y + 1)$
49. $(2z^2 + 5)(z - 3)$ **51.** $(4t^2 + 3)(t - 5)$
53. $(3r^2 - 2)(3r + 2)$ **55.** $(7x^2 - 2)(x + 3)$
57. $(y^2 - 2)(2y - 7)$ **59.** $(z^2 - 7)(z - 4)$
61. $(x^3 + 2)(2x - 3)$ **63.** $(x + y)(a + b)$
65. $3(x^2 + 1)(x + 2)$ **67.** $2y(3y^2 - 1)(y - 4)$
69. $x^2(x^2 - 3)(x + 2)$ **71.** $2x^2(x^2 - 3)(2x + 1)$
73. $xy(x + y)(x - 2)$ **75.** $a(x^2 + \frac{b}{a}x + \frac{c}{a})$
77. (a) $16t$ **(b)** $16t(5 - t)$
79. (a) 168 in³ **(b)** $4x(x^2 - 15x + 50)$

SECTION 6.2 (pp. 385–387)

1. 1 **2.** $x^2 + (m + n)x + mn; m + n; mn$
3. $mn; m + n$ **4.** prime **5.** $1, 12; 2, 6; 3, 4$
6. $-1, -30; -2, -15; -3, -10; -5, -6$ **7.** $4, 7$
9. $3, -10$ **11.** $-5, 10$ **13.** $-7, -4$ **15.** $(x + 1)(x + 2)$
17. $(y + 2)(y + 2)$ **19.** Prime **21.** $(x + 3)(x + 5)$
23. $(m + 4)(m + 9)$ **25.** $(n + 10)(n + 10)$

27. $(x - 1)(x - 5)$ **29.** $(y - 3)(y - 4)$
31. $(z - 5)(z - 8)$ **33.** $(a - 7)(a - 9)$ **35.** Prime
37. $(b - 5)(b - 25)$ **39.** $(x - 5)(x + 18)$
41. $(m - 5)(m + 9)$ **43.** Prime **45.** $(n - 10)(n + 20)$
47. $(x - 1)(x + 23)$ **49.** $(a - 4)(a + 8)$
51. $(b + 4)(b - 5)$ **53.** Prime **55.** $(x + 8)(x - 9)$
57. $(y + 2)(y - 17)$ **59.** $(z + 6)(z - 11)$
61. $5(x - 4)(x + 2)$ **63.** $y(y - 2)(y - 5)$
65. $3a(a + 1)(a + 6)$ **67.** $2x(x^2 - 3x + 4)$
69. $2m^2(m - 7)(m + 2)$ **71.** $(x + 1)(x + 5)$
73. $(x - 1)(x - 3)$ **75.** $(6 - x)(2 + x)$
77. $(8 + x)(4 - x)$ **79.** $(x + 1)(x + k)$
81. $x + 1$;

x	x²	x
1	x	1
	x	1

83. $x + 1, x + 2$;

x	x²	2x
1	x	2
	x	2

85. $x + 1$ **87.** $(x + 2)(x + 6)$

CHECKING BASIC CONCEPTS 6.1 & 6.2 (p. 387)

1. $4x$ **2.** $6z^2(2z - 3)$ **3. (a)** $(6y + 5)(y - 2)$
(b) $(x^2 + 5)(2x + 1)$ **(c)** $4(z^2 + 1)(z - 3)$
4. (a) $(x + 2)(x + 4)$ **(b)** $(x - 7)(x + 6)$ **(c)** Prime
(d) $4a(a + 2)(a + 3)$ **5.** $(x + 5)(x + 5)$

SECTION 6.3 (pp. 393–395)

1. $ac; b$ **2.** $ax^2; c$ **3.** $+; +$ **4.** $+; -$ **5.** $-; -$
6. $-; +$ **7.** $3; x$ **8.** $2; 4x$ **9.** $1; 2x$ **10.** $1; 2x$
11. $(x + 3)(2x + 1)$ **13.** Prime **15.** $(x + 1)(3x + 1)$
17. $(2x + 3)(3x + 1)$ **19.** $(x - 2)(5x - 1)$
21. $(y - 1)(2y - 5)$ **23.** Prime **25.** $(z - 5)(7z - 2)$
27. $(t - 3)(3t + 2)$ **29.** $(3r + 2)(5r - 3)$
31. $(3m - 4)(8m + 3)$ **33.** $(5x - 1)(5x + 2)$
35. $(x + 2)(6x - 1)$ **37.** Prime **39.** $(3n + 1)(7n - 1)$
41. $(2y + 3)(7y + 1)$ **43.** $(4z - 3)(7z - 1)$
45. $(3x - 2)(10x - 3)$ **47.** Prime **49.** $(2t + 3)(9t - 2)$
51. $3(2a - 1)(2a + 3)$ **53.** $y(3y - 2)(4y - 1)$
55. $3x(4x - 3)(2x - 1)$ **57.** $2x^2(4x^2 - 3x + 1)$
59. $7x^2(2x + 1)(2x + 3)$ **61.** $(3x + 1)(x + k)$
63. $(7x + 1)(x + 2)$ **65.** $(2x - 1)(x - 2)$
67. $-(4x + 3)(2x - 1)$ **69.** $-(x + 5)(2x - 3)$
71. $-(x - 3)(5x + 1)$
73. $3x + 2$ by $2x + 1$;

2x	6x²	4x
1	3x	2
	3x	2

75. $(2x + 1)(x + 3)$

SECTION 6.4 (pp. 401–402)

1. $(a - b)(a + b)$ **2.** cannot   **3.** $6x; 7y$ **4.** $(a + b)^2$

5. $(a - b)^2$ **6.** $6x$ **7.** $20rt$ **8.** $(a + b)(a^2 - ab + b^2)$

9. $(a - b)(a^2 + ab + b^2)$ **10.** $2x; 3y$ **11.** $-; +$

12. $+; -$ **13.** $(x - 1)(x + 1)$ **15.** $(z - 10)(z + 10)$

17. $(2y - 1)(2y + 1)$ **19.** $(6z - 5)(6z + 5)$

21. $(3 - x)(3 + x)$ **23.** $(1 - 3y)(1 + 3y)$

25. $(2a - 3b)(2a + 3b)$ **27.** $(6m - 5n)(6m + 5n)$

29. $(9r - 7t)(9r + 7t)$ **31.** $(x + 4)^2$

33. Not possible   **35.** $(x - 3)^2$ **37.** $(3y + 1)^2$

39. $(2z - 1)^2$ **41.** Not possible   **43.** $(3x + 5)^2$

45. $(2a - 9)^2$ **47.** $(x + y)^2$ **49.** $(r - 5t)^2$

51. Not possible   **53.** $(z + 1)(z^2 - z + 1)$

55. $(x + 4)(x^2 - 4x + 16)$ **57.** $(y - 2)(y^2 + 2y + 4)$

59. $(n - 1)(n^2 + n + 1)$ **61.** $(2x + 1)(4x^2 - 2x + 1)$

63. $(m - 4n)(m^2 + 4mn + 16n^2)$

65. $(2x + 5y)(4x^2 - 10xy + 25y^2)$

67. $4(x - 2)(x + 2)$ **69.** $2(y - 7)^2$

71. $5(z + 2)(z^2 - 2z + 4)$ **73.** $xy(x - y)(x + y)$

75. $2m(m^2 - 5m + 9)$ **77.** $7x^2(10x - 3y)(10x + 3y)$

79. $2(2a + b)(4a^2 - 2ab + b^2)$ **81.** $4b^2(b + 3)^2$

83. $4(5r - 2t)(25r^2 + 10rt + 4t^2)$

85. $2x + 3$;

$2x$	$4x^2$	$6x$
3	$6x$	9
	$2x$	3

CHECKING BASIC CONCEPTS 6.3 & 6.4 (p. 402)

1. (a) $(x - 4)(2x + 3)$ **(b)** $(2x + 7)(3x - 2)$

2. (a) Prime   **(b)** $2y(3y + 1)(y - 2)$

3. $(3x + 2)(x + 3)$ **4. (a)** $(z - 8)(z + 8)$

(b) $(3r - 2t)(3r + 2t)$ **5. (a)** $(x + 6)^2$ **(b)** $(3a - 2b)^2$

6. (a) $(m - 3)(m^2 + 3m + 9)$

(b) $(5n + 3)(25n^2 - 15n + 9)$

7. (a) $4(2x - 1)(2x + 1)$ **(b)** $3y(y + 2)(y^2 - 2y + 4)$

SECTION 6.5 (pp. 407–408)

1. Greatest common factor   **2.** GCF   **3.** Grouping

4. No; a sum of squares cannot be factored.

5. Yes; a sum of cubes can be factored.   **6.** completely

7. $2(2x - 1)$ **9.** $2(y^2 - 2y + 2)$ **11.** $(z - 2)(z + 2)$

13. $(a + 2)(a^2 - 2a + 4)$ **15.** $(2b - 3)^2$

17. Not possible   **19.** $(x^2 + 5)(x - 1)$

21. $(y - 4)(y - 1)$ **23.** $(x - 3)(x + 3)(x + 4)$

25. $8(a - 2)(a^2 + 2a + 4)$ **27.** $2x(3x^2 + 1)(2x - 3)$

29. $2t(3t + 2)(9t^2 - 6t + 4)$

31. $2(r - 1)(r + 1)(r + 3)$ **33.** $3z^2(2z + 3)(z - 5)$

35. $2b^2(3b - 1)(2b - 1)$ **37.** $6z(y - 2z)(y + 2z)$

39. $3y(x - 5)^2$ **41.** $(3m - 2n)(9m^2 + 6mn + 4n^2)$

43. $3(x - 2)(x + 2)(x - 1)(x^2 + x + 1)$

45. $(5a + 3)(a - 6)$ **47.** $3r(t + 5)(t + 6)$

49. $(3b^2 + 4)(3b + 2)$ **51.** $2n(3n^2 + n - 5)$

53. $4(x - 3y)(x + 3y)$ **55.** $2a(a - 4)^2$

57. $4x(2y + 1)(4y^2 - 2y + 1)$

59. $2b(2b - 1)(2b + 1)(b + 3)$ **61.** $3x + 1$

SECTION 6.6 (pp. 414–416)

1. $0; 0$ **2.** No; one side of the equation must be zero.

3. $2x = 0; x + 6 = 0$ **4.** Subtract $4x$ from each side.

5. Apply the zero-product property.   **6.** solving

7. $ax^2 + bx + c = 0$ with $a \neq 0$ **8.** $x^2 - 6x + 1 = 0$

9. zero   **10.** 2   **11.** $x = 0$ or $y = 0$ **13.** $-8, 0$

15. $1, 2$ **17.** $\frac{1}{2}, \frac{3}{4}$ **19.** $\frac{1}{3}, \frac{3}{7}$ **21.** $0, 5, 8$ **23.** $0, 1$

25. $0, 5$ **27.** $-\frac{3}{2}, 0$ **29.** $-1, 1$ **31.** $-\frac{1}{2}, \frac{1}{2}$

33. $-2, -1$ **35.** $5, 7$ **37.** $-2, \frac{1}{2}$ **39.** $-\frac{5}{2}, -\frac{2}{3}$

41. $-5, 5$ **43.** $0, 5$ **45.** $-3, 0$ **47.** $-1, 6$ **49.** $-\frac{5}{3}, \frac{3}{4}$

51. $-2, 1$ **53.** $-3, \frac{1}{2}$ **55.** $-\frac{1}{2}$ **57.** $-1, -\frac{2}{3}$ **59.** 12 ft

61. 4   **63.** 4

65. (a) 6 sec

(b)

Time (t)	0	1	2	3	4	5	6
Height (h)	0	80	128	144	128	80	0

; 3 sec

67. (a) 81.8 ft; 327.3 ft; when the speed doubles, the braking distance quadruples.   **(b)** About 19 mph   **(c)** About 19 mph; yes

69. (a) About 11 million; about 66 million   **(b)** About 1981

71. 40 by 50 pixels

CHECKING BASIC CONCEPTS 6.5 & 6.6 (p. 416)

1. (a) $9(a^2 - 2a + 3)$ **(b)** $7x(y^2 + 4)$

2. (a) $2z^2(3z - 2)(z - 4)$ **(b)** $2r^2(t - 3)(t + 3)$

3. (a) $4x(3x - 2)^2$ **(b)** $3(2b - 3)(4b^2 + 6b + 9)$

4. (a) $0, \frac{3}{2}$ **(b)** $-1, \frac{3}{5}$ **5.** $-3, 1$ **6.** After $\frac{11}{2}$ sec

SECTION 6.7 (pp. 421–422)

1. GCF   **2.** zero-product   **3.** factors

4. $(x^2 - y^2)^2$ **5.** $(z^2 + 1)(z^2 + 2)$

6. No; $x^4 - 1 = (x - 1)(x + 1)(x^2 + 1)$

7. Subtract x from each side.   **8.** Factor out x^2.

9. $5(x - 3)(x + 2)$ **11.** $-4(y + 2)(y + 6)$

13. $-10(z + 5)(2z + 1)$ **15.** $-4(t + 3)(7t - 5)$

17. $r(r - 1)(r + 1)$ **19.** $3x(x - 2)(x + 3)$

21. $12z(2z - 1)(3z + 2)$ **23.** $x^2(x - 2)(x + 2)$

25. $t^2(t - 1)(t + 2)$ **27.** $(x^2 - 3)(x^2 - 2)$
29. $(x^2 + 3)(2x^2 + 1)$ **31.** $(y^2 + 3)^2$
33. $(x^2 - 3)(x^2 + 3)$ **35.** $(x - 3)(x + 3)(x^2 + 9)$
37. $z^3(z + 1)^2$ **39.** $(x + y)(2x - y)$
41. $(a + b)^2(a - b)^2$ **43.** $x(x + y)(x - y)$
45. $x(2x + y)^2$ **47. (a)** $x(x - 2)(x + 2)$ **(b)** $-2, 0, 2$
49. (a) $2y(y - 6)(y + 3)$ **(b)** $-3, 0, 6$ **51.** $-8, -3$
53. $-1, 3$ **55.** $-1, 0, 4$ **57.** $-6, 0, 4$ **59.** $-6, 0, 6$
61. $-7, 0, 1$ **63.** $-3, -2, 2, 3$ **65.** $-1, 1$ **67.** $-3, 3$
69. $-1, 1, 2$ **71.** 5 **73. (a)** $x < 7.5$ in. because the width is 15 in.   **(b)** $300 - 4x^2$ **(c)** 2.5 in.
75. 18.6 trillion ft^3 **77.** Factoring is very difficult (answers may vary).

CHECKING BASIC CONCEPTS 6.7 (p. 423)

1. (a) $3(x - 4)(x + 2)$ **(b)** $-5(2y + 1)(y - 1)$
2. (a) $(z^2 - 5)(z^2 + 5)$ **(b)** $7(t - 1)(t + 1)(t^2 + 1)$
3. (a) $(x - 2)^2(x + 2)^2$ **(b)** $y(y + 10)(2y - 3)$
4. $-4, 0, 3$ **5.** 3

CHAPTER 6 REVIEW (pp. 426–428)

1. $4z^2$; $4z^2(2z - 1)$ **2.** $3x^2$; $3x^2(2x^2 + x - 4)$
3. $3y$; $3y(3x + 5z^2)$ **4.** a^2b^2; $a^2b^2(b + a)$
5. $(x - 3)(x + 2)$ **6.** $y(y + 3)(x - 5)$
7. $(z^2 + 5)(z - 2)$ **8.** $(t^2 + 8)(t + 1)$
9. $(x^2 + 6)(x - 3)$ **10.** $(x - y)(a + b)$
11. $x^2(x^2 - 2)(x + 3)$ **12.** $2y(y^2 + 1)(y + 3)$
13. $4, 5$ **14.** $-3, 7$ **15.** $-9, -4$ **16.** $-25, 4$
17. $(x - 4)(x + 3)$ **18.** $(x + 4)(x + 6)$
19. $(x - 2)(x + 8)$ **20.** $(x - 7)(x + 6)$
21. $(x - 1)(x + 3)$ **22.** $(x + 10)(x + 12)$
23. $2x(x - 2)(x + 5)$ **24.** $x^2(x + 4)(x - 7)$
25. $(2 - x)(5 - x)$ **26.** $(6 - x)(4 + x)$
27. $(3x - 1)(3x + 2)$ **28.** $(x - 1)(2x + 5)$
29. $(x + 3)(3x + 5)$ **30.** $(5x - 1)(7x + 1)$
31. $(3x + 1)(8x - 5)$ **32.** $(x + 9)(4x - 3)$
33. $3x(2x + 7)(2x + 1)$ **34.** $2x^2(x + 3)(4x - 5)$
35. $(3 - 2x)(4 + x)$ **36.** $(1 - 2x)(1 + 5x)$
37. $(z - 2)(z + 2)$ **38.** $(3z - 8)(3z + 8)$
39. $(6 - y)(6 + y)$ **40.** $(10a - 9b)(10a + 9b)$
41. $(x + 7)^2$ **42.** $(x - 5)^2$ **43.** $(2x - 3)^2$
44. $(3x + 8)^2$ **45.** $(2t - 1)(4t^2 + 2t + 1)$
46. $(3r + 2t)(9r^2 - 6rt + 4t^2)$
47. $2x(x - 5)(x + 5)$ **48.** $3(2x + 3)(4x^2 - 6x + 9)$
49. $2x(x + 7)^2$ **50.** $2x(x - 4)(x^2 + 4x + 16)$

51. $3(3y^2 - 2y + 2)$ **52.** $y(z - 3)(z + 3)$
53. $x(x - 2)(x + 2)(x + 7)$ **54.** $3x(2x + 3)^2$
55. $3a(b - 2)(b^2 + 2b + 4)$ **56.** $5x(x^2 + 4)$
57. $6x(2x - y)(2x + y)$ **58.** $y(x + 3)(x^2 - 3x + 9)$
59. $m = 0$ or $n = 0$ **60.** 0 **61.** $-9, \frac{3}{4}$ **62.** $-\frac{6}{5}, \frac{1}{4}$
63. $0, 1, 2$ **64.** $0, 7$ **65.** $-8, 8$ **66.** $-7, -2$
67. $-2, 3$ **68.** $-\frac{3}{2}, \frac{2}{5}$ **69.** $-4, 18$ **70.** $-2, \frac{5}{2}$
71. $5(x - 5)(x + 2)$ **72.** $-3(x - 3)(x + 5)$
73. $y(y - 2)(y + 2)$ **74.** $3y(y - 1)(y + 3)$
75. $2z^2(z + 2)(z + 5)$ **76.** $8z^2(z - 2)(z + 2)$
77. $(x^2 - 3)^2$ **78.** $(x - 3)(x + 3)(2x^2 + 3)$
79. $(a + 5b)^2$ **80.** $x(x + y)(x - y)$ **81.** $-\frac{1}{2}, 5$
82. $0, \frac{5}{2}, 3$ **83.** $-5, 0, 5$ **84.** $0, 3, 4$ **85.** $-2, 2$
86. $-4, 4$ **87.** -4 **88.** $-1, 1$
89. $3x + 7$;

$3x$	$9x^2$	$21x$
7	$21x$	49

$3x$ 7

90. $x + 1$ by $x + 5$;

x	x^2	$5x$
1	x	5

x 5

91. $x + 1$ **92.** $(x + 1)(x + 3)$ **93.** $(2x + 3)(x + 6)$
94. 2 **95.** 8 ft by 15 ft   **96.** After 2 sec and 3 sec
97. (a) 390 ft   **(b)** 15 mph   **(c)** 15 mph; yes
98. (a) \$10,000   **(b)** \$50 or \$150   **(c)** \$50 or \$150; yes
99. (a) 372 million   **(b)** 1975   **100.** 50 by 80 pixels
101. (a) $2000 - 4x^2$ **(b)** 5 in.

CHAPTER 6 TEST (p. 429)

1. $4xy$; $4xy(x - 5y + 3)$ **2.** $3a^2b^2$; $3a^2b^2(3a + 1)$
3. $(y + z)(a + b)$ **4.** $(x^2 - 5)(3x + 1)$
5. $(y - 2)(y + 6)$ **6.** $(2x + 5)^2$ **7.** $(z - 4)(4z - 3)$
8. $(t - 7)(2t - 3)$ **9.** $3x(x + 1)(2x - 1)$
10. $2(z - 3)(z + 3)(z^2 + 3)$ **11.** $4y(3y - 5)(3y + 5)$
12. $7x(x + 2)(x^2 - 2x + 4)$ **13.** $a^2(4a + 3)^2$
14. $2(b - 2)(b + 2)(b^2 + 4)$ **15.** $-4, 4$ **16.** $-4, 5$
17. $\frac{4}{3}$ **18.** $-6, 11$ **19.** $-3, 0, 3$ **20.** $-2, -1, 1, 2$
21. $3x + 5$ **22.** $(x + 2)(x + 3)$ **23. (a)** 275 ft
(b) 33 mph   **24.** 1 sec and 2 sec

CHAPTERS 1–6 CUMULATIVE REVIEW (pp. 430–432)

1. $2 \times 2 \times 2 \times 2 \times 3 \times 3$ **2.** 16   **3.** $\frac{3}{7}$ **4.** $\frac{7}{10}$
5. 17   **6.** $-\frac{7}{2}$ **7.** 8

8. 1;

x	-2	-1	0	1	2
$2x + 3$	-1	1	3	5	7

9. $3n - 5 = n - 7$; -1 **10.** $\frac{57}{1000}$; 0.057   **11.** 12.3%
12. $W = \frac{P - 2L}{2}$ **13.** $z > 2$

14.

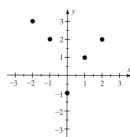

15. x-int: $\frac{2}{3}$; y-int: -2

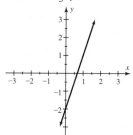

16. y-int: -2

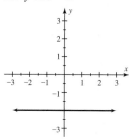

17. -1; -2; $y = -2x - 2$ **18.** $y = -\frac{3}{2}x + \frac{7}{2}$

19. $y = \frac{4}{3}x + \frac{11}{3}$ **20.** $(1, 2)$ **21.** $(1, -1)$ **22.** $(-2, 5)$

23. (a) One   (b) Consistent; independent

24. (a) None   (b) Inconsistent

25.

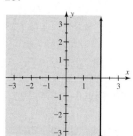

26.

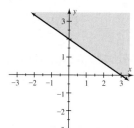

27.

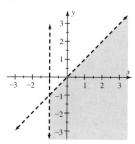

28.

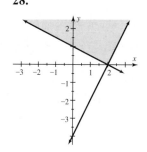

29. -16 **30.** 1   **31.** $\frac{x^{10}}{y^4}$ **32.** $-14x^5 + 21x^4$

33. $21x^2 + 29x - 10$ **34.** $x^2 - 14x + 49$ **35.** $\frac{1}{a^2}$

36. $\frac{1}{4t^6}$ **37.** $\frac{1}{x^2 y^5}$ **38.** $32x^5 y^{10}$ **39.** 0.00623

40. 5.43×10^5 **41.** $2x^2 + 4x$ **42.** $3x + \frac{-4x + 1}{x^2 + 1}$

43. $(2y^2 - 5)(x + 2)$ **44.** $(t^2 + 1)(t + 6)$

45. $(x - 4)(x + 7)$ **46.** $(2y + 3)(3y - 4)$

47. $(5x - 2y)(5x + 2y)$ **48.** $(8x - 1)^2$

49. $(3t - 2)(9t^2 + 6t + 4)$ **50.** $-4(x - 3)(x + 2)$

51. $(x - 3)(x + 3)(x^2 - 3)$ **52.** $x^2 y(x - y)$

53. $-5, 0, 5$ **54.** $-2, 1$ **55.** $-\frac{7}{2}, 0, \frac{7}{2}$ **56.** $-3, 3$

57. (a) $10x + 8x$ or $18x$ (b) 50 min   **58.** $6\frac{11}{12}$ mi

59. \$21,000   **60.** 25 min skiing; 35 min running

61. (a) $C = 0.25x + 20$ (b) 320 mi

62. (a) $2x + y = 180$; $2x - y = 20$ (b) $(50, 80)$, the angles are $50°$, $50°$, and $80°$.

63.

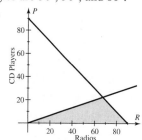

64. $18xy$; 108 yd^2 **65.** $8x^3 y^6$ **66.** (a) $x^2 + 4x + 4$

(b) $x^2 + 2x + 2x + 4 = x^2 + 4x + 4$

67. $x + 6$ **68.** (a) 4 sec   (b) After 1 sec and 3 sec

CHAPTER 7: RATIONAL EXPRESSIONS

SECTION 7.1 (pp. 440–443)

1. $\frac{P}{Q}$; polynomials   **2.** Yes; both x and $2x^2 + 1$ are polynomials.   **3.** the denominator equals zero   **4.** a **5.** $\frac{a}{b}$

6. rational   **7.** $\frac{3}{10}$ **8.** $\frac{3}{2x}$ **9.** -1 **10.** $1 - x$ **11.** $-\frac{3}{7}$

13. $-\frac{4}{9}$ **15.** $-\frac{1}{4}$ **17.** Undefined   **19.** Undefined

21. -1 **23.** 1

25.

x	-2	-1	0	1	2
$\frac{x}{x+1}$	2	—	0	$\frac{1}{2}$	$\frac{2}{3}$

27.

x	-2	-1	0	1	2
$\frac{3x}{2x^2+1}$	$-\frac{2}{3}$	-1	0	1	$\frac{2}{3}$

29. 0   **31.** 3   **33.** $-\frac{4}{5}$ **35.** None   **37.** $-5, 5$

39. $-3, -2$ **41.** $1, \frac{5}{2}$ **43.** $\frac{2}{3}$ **45.** $\frac{1}{2}$ **47.** $-\frac{2}{5}$

49. $-\frac{1}{3}$ **51.** (a) $\frac{1}{2}$ (b) $\frac{1}{2}$ **53.** (a) -1 (b) -1

55. $\frac{1}{2x^2}$ **57.** $\frac{4y}{3x}$ **59.** $\frac{1}{2}$ **61.** $\frac{3}{5}$ **63.** $\frac{x+1}{x+6}$ **65.** $\frac{5y+3}{y+2}$

67. -1 **69.** -1 **71.** $-\frac{1}{3}$ **73.** $-\frac{1}{2}$ **75.** 1

77. $-\frac{3x+5}{3x-5}$ **79.** $\frac{n-1}{n-5}$ **81.** $\frac{x}{6}$ **83.** $\frac{z-2}{z-3}$ **85.** $\frac{x+4}{3x+2}$

87. $\frac{1}{3x-2}$ **89.** 1   **91.** $-\frac{1}{2}$ **93.** (a) Equation   (b) 6

95. (a) Expression   (b) $\frac{1}{x+1}$ **97.** (a) Expression

(b) $x - 2$ **99.** (a) Equation   (b) 8

101. (a) $\frac{1}{2}$; when traffic arrives at a rate of 3 vehicles per minute, the average wait is $\frac{1}{2}$ minute.
(b)

x	2	4	4.5	4.9	4.99
T	$\frac{1}{3}$	1	2	10	100

As x nears 5 vehicles per minute, a small increase in x increases the wait dramatically.
103. $\frac{1}{2}$ **105. (a)** $\frac{3}{n}$ **(b)** $\frac{n-3}{n}$; $\frac{97}{100}$; there is a 97% chance that a winning ball will not be drawn. **107. (a)** 6 hr **(b)** $\frac{M}{60}$
109. (a) $x = 5$ **(b)** As the average rate nears 5 cars per minute, a small increase in x increases the wait dramatically.

SECTION 7.2 (pp. 448–450)

1. $\frac{10}{21}$ **2.** $\frac{14}{15}$ **3.** $\frac{AC}{BD}$ **4.** $\frac{AD}{BC}$ **5.** $\frac{x+7}{x+1}$ **6.** $\frac{4}{B}$ **7.** No; it is equal to $1 + \frac{2}{x}$. **8.** $\frac{1}{x}$; $\frac{z}{y}$ **9.** $\frac{2}{5}$ **11.** $\frac{12}{7}$ **13.** $\frac{2}{3}$
15. $\frac{2}{11}$ **17.** 4 **19.** $\frac{8}{15}$ **21.** 10 **23.** $\frac{12}{5}$ **25.** 1
27. $\frac{z+1}{z+4}$ **29.** $\frac{2}{3}$ **31.** $\frac{x+2}{x-2}$ **33.** $\frac{8(x+1)}{x^2}$ **35.** $\frac{x-3}{x}$
37. $\frac{z+3}{z-7}$ **39.** 1 **41.** $(t+1)(t+2)$ **43.** $\frac{x(x+4)}{x^2+4}$
45. $\frac{z-1}{z+2}$ **47.** $\frac{y}{y-1}$ **49.** $\frac{x+1}{x-3}$ **51.** $\frac{x-3}{x-1}$ **53.** $\frac{2}{2x+3}$
55. -2 **57.** $\frac{z-1}{z+1}$ **59.** $\frac{y+2}{2y+1}$ **61.** $\frac{4(t-1)}{t^2+1}$
63. $\frac{y-3}{y-5}$ **65.** $2x$ **67.** $\frac{2z+1}{z+5}$ **69.** $\frac{1}{t+6}$ **71.** $\frac{2a+3b}{a+b}$
73. $x - y$ **75.** 1 **77. (a)** $D = \frac{30}{x}$ **(b)** 300 ft; 75 ft; one-fourth as far **79. (a)** $\frac{1}{n+1}$ **(b)** $\frac{1}{100}$

CHECKING BASIC CONCEPTS 7.1 & 7.2 (p. 450)

1. Undefined; $\frac{3}{8}$ **2. (a)** $\frac{2x}{5y}$ **(b)** 5 **(c)** $\frac{x+2}{x+4}$
3. (a) $\frac{4}{9}$ **(b)** $\frac{2}{x-1}$ **4. (a)** $\frac{5z}{6}$ **(b)** $x+1$
5. (a)

x	0.5	1.0	1.5	1.9
T	$\frac{2}{3}$	1	2	10

(b) As x nears 2 persons per minute, a small increase in x increases the wait dramatically.

SECTION 7.3 (pp. 457–458)

1. add; numerators; denominators **2.** subtract; numerators; denominators **3.** $\frac{3}{5}$ **4.** $\frac{4}{7}$ **5.** $\frac{A+B}{C}$ **6.** $\frac{A-B}{C}$
7. $4 - x$ **8.** $\frac{4-x}{x-1}$ **9.** $x + 5$ **10.** $\frac{x+5}{7}$ **11.** 1
13. $\frac{6}{5}$ **15.** 1 **17.** $\frac{3}{7}$ **19.** $\frac{1}{2}$ **21.** $\frac{1}{2}$ **23.** $\frac{2}{3}$ **25.** $\frac{3}{x}$
27. $\frac{1}{2}$ **29.** 3 **31.** 1 **33.** $\frac{4z}{4z+3}$ **35.** 2 **37.** 1
39. $\frac{1}{x-1}$ **41.** $z - 1$ **43.** 2 **45.** $y - 1$ **47.** $\frac{x}{2}$
49. $z - 2$ **51.** $x - 3$ **53.** $\frac{7n}{2n^2-n+5}$ **55.** $\frac{6}{x+3}$
57. $\frac{9}{ab}$ **59.** $\frac{1}{x+y}$ **61.** $\frac{16}{a-b}$ **63.** $\frac{b}{2a}$ **65.** 1
67. $a - b$ **69.** $2x - 3y$ **71.** It equals 7.
73. (a) $\frac{14}{n+1}$ **(b)** $\frac{7}{50}$; there are 7 chances out of 50 chances that a defective battery is chosen.

SECTION 7.4 (pp. 465–467)

1. Examples include 36 and 54 (answers may vary). **2.** xy
3. $\frac{3}{3}$ **4.** $\frac{x+1}{x+1}$ **5.** 12 **6.** $6x^2$ **7.** 12 **9.** 6 **11.** 30
13. 72 **15.** $12x$ **17.** $10x^2$ **19.** $x(x+1)$
21. $(2x+1)(x+3)$ **23.** $36x^3$ **25.** $x(x-1)(x+1)$
27. $(x+1)^2$ **29.** $(2x-1)^3(x+3)$
31. $(2x-1)(2x+1)$ **33.** $(x-1)(x+1)$
35. $4x(x^2+4)$ **37.** $(2x+3)(x+2)(x+3)$ **39.** $\frac{3}{9}$
41. $\frac{15}{21}$ **43.** $\frac{2x^2}{8x^3}$ **45.** $\frac{x-2}{x^2-4}$ **47.** $\frac{x}{x^2+x}$
49. $\frac{2x^2+2x}{x^2+2x+1}$ **51.** $\frac{13}{10}$ **53.** $\frac{2}{9}$ **55.** $\frac{14}{25}$ **57.** $\frac{9}{20}$ **59.** $\frac{13}{12x}$
61. $\frac{5z-7}{z^3}$ **63.** $\frac{y-x}{xy}$ **65.** $\frac{a^2+b^2}{ab}$ **67.** $\frac{7}{2(x+2)}$
69. $\frac{t+2}{t(t-2)}$ **71.** $\frac{n^2+4n+5}{(n-1)(n+1)}$ **73.** $-\frac{3}{x-3}$ **75.** 0
77. $\frac{6x-4}{(x-1)^2}$ **79.** $\frac{3}{2y-1}$ **81.** $\frac{x-1}{x(x+2)}$
83. $\frac{3x+5}{(x-2)(x+2)}$ **85.** $\frac{x+9}{x(x-3)(x+3)}$
87. $\frac{x^2+4x-4}{x(x-2)(x+2)}$ **89.** $\frac{2x+2}{(x+2)^2}$
91. $\frac{x^2+2x-1}{(x+1)(x+2)(x+3)}$ **93.** $-\frac{2b}{(a+b)(a-b)}$ **95.** 0
97. $\frac{13}{12a}$ **99.** 0 **101.** $\frac{3}{x}$ **103.** $\frac{1}{75}$; $\frac{1}{75}$; yes **105.** $\frac{D-F}{FD}$

CHECKING BASIC CONCEPTS 7.3 & 7.4 (p. 467)

1. (a) 1 **(b)** $\frac{2-x}{3x}$ **(c)** z **2. (a)** $15x$ **(b)** $4x(x+1)$
(c) $(x+1)(x-1)$ **3. (a)** $\frac{6x+5}{x(x+1)}$ **(b)** $\frac{4}{x-3}$
(c) $-\frac{x+4}{2x+1}$ **4.** $\frac{a^2+b^2}{(a-b)(a+b)}$

SECTION 7.5 (pp. 475–477)

1. $\frac{1}{2} \cdot \frac{4}{3} = \frac{2}{3}$ **2.** $\frac{a}{b} \cdot \frac{d}{c} = \frac{ad}{bc}$ **3.** fractions **4.** $\dfrac{x+\frac{1}{2}}{x-\frac{1}{2}}$
5. Division **6.** $\dfrac{\frac{x}{2}}{\frac{1}{x-1}}$ **7.** $\dfrac{\frac{a}{b}}{\frac{c}{d}}$ **8.** $x(x+2)$ **9.** 30
11. $(x-1)(x+1)$ **13.** $x(2x-1)(2x+1)$
15. $\frac{4}{5}$ **17.** $\frac{10}{7}$ **19.** $\frac{9}{14}$ **21.** $\frac{1}{2}$ **23.** $\frac{3y}{x}$ **25.** 3 **27.** $\frac{p+1}{p+2}$
29. $\frac{5}{z}$ **31.** $\frac{y}{y-3}$ **33.** $\frac{x^2-1}{x^2+1}$ **35.** $\frac{x^2}{3}$ **37.** 1 **39.** $\frac{n+m}{n-m}$
41. $\frac{2x+y}{2x-y}$ **43.** $\frac{1+b}{1-a}$ **45.** $\frac{q+2}{q}$ **47.** $2x+3$
49. $\frac{2x}{(x+1)(2x-1)(2x+1)}$ **51.** $\frac{1}{ab}$ **53.** $\frac{ab}{a+b}$
55. $\dfrac{P\left(1+\frac{r}{26}\right)^{52}-P}{\frac{r}{26}}$ **57.** $R = \frac{ST}{S+T}$

SECTION 7.6 (pp. 487–491)

1. rational **2.** $\frac{2x+5}{3x}$; $\frac{2x+5}{3x} = 9$ (answers may vary)
3. $ad = bc$; b; d **4.** Yes **5.** $12x$ **6.** V; T **7.** $\frac{3}{2}$ **9.** $\frac{5}{2}$
11. $\frac{7}{6}$ **13.** 0 **15.** $\frac{1}{4}$ **17.** $\frac{10}{11}$ **19.** 3 **21.** 25 **23.** 5
25. $\frac{5}{16}$ **27.** 1 **29.** No solutions $\left(\text{extraneous: } -\frac{1}{2}\right)$ **31.** $\frac{1}{2}$

33. -1 **35.** $-\frac{3}{2}, -1$ **37.** 4   **39.** 4   **41.** -6 **43.** 6
45. -3 (extraneous: 1)   **47.** 1 (extraneous: -2)   **49.** $\frac{7}{12}$
51. $\frac{1}{6}, 5$ **53.** 3   **55.** No solutions (extraneous: 1)
57. 3   **59.** No solutions (extraneous: -1)   **61.** No solutions
63. -2 **65.** Expression; 1; 1   **67.** Equation; 2
69. Equation; 4   **71.** Expression; $x + 2$; 4   **73.** $-\frac{1}{2}; \frac{1}{2}$
75. $-1; \frac{1}{2}$ **77. (a)** $-3, 1$ **(b)** $-3, 1$ **79. (a)** -1
(b) -1 **81. (a)** 2   **(b)** 2   **83. (a)** $-2, 2$ **(b)** $-2, 2$
85. 0.300, 1.100   **87.** 1.084   **89.** $a = \frac{F}{m}$
91. $r = \frac{V}{I} - R$ **93.** $b = \frac{2A}{h}$ **95.** $z = \frac{15}{k - 3}$
97. $b = \frac{aT}{a - T}$ **99.** $x = \frac{ky}{3y + 2k}$ **101.** 9 cars per minute
103. $\frac{12}{7} \approx 1.7$ hours   **105.** $\frac{8}{3} \approx 2.7$ days   **107.** 18 mph;
20 mph   **109. (a)** 120; the braking distance is 120 feet when
the slope of the road is -0.05.   **(b)** -0.1 **111.** 15 mph
113. 200 mph   **115.** 10 mph running; 5 mph walking
117. 3   **119.** 32

CHECKING BASIC CONCEPTS 7.5 & 7.6 (p. 491)

1. (a) $\frac{5}{6}$ **(b)** $\frac{1}{9x^2}$ **(c)** $\frac{b - a}{b + a}$ **(d)** $\frac{r + t}{2rt}$ **2. (a)** $\frac{1}{5}$
(b) -4 **3. (a)** 2   **(b)** $-2, \frac{1}{2}$ **4.** -3 (extraneous: 1)
5. (a) $x = \frac{2(b + 3y)}{a}$ **(b)** $m = \frac{k}{2k - 1}$ **6. (a)** 300; when
the slope of the hill is 0.1, the braking distance is 300 feet.
(b) 0.3; the braking distance is 200 feet when the slope of
the road is 0.3.

SECTION 7.7 (pp. 500–504)

1. A statement that two ratios are equal   **2.** $\frac{5}{6} = \frac{x}{7}$
3. It doubles.   **4.** It is halved.   **5.** constant   **6.** constant
7. Directly; if the number being fed doubles, the bill will
double.   **8.** Inversely; doubling the number of painters will
halve the time.   **9.** 15   **11.** 21   **13.** 48   **15.** $\frac{21}{8}$
17. $-4, 4$ **19.** $-\frac{7}{2}, \frac{7}{2}$ **21.** $b = \frac{ad}{c}$ **23. (a)** $\frac{5}{8} = \frac{9}{x}$
(b) $\frac{72}{5}$ **25. (a)** $\frac{4}{8} = \frac{10}{x}$ **(b)** 20   **27. (a)** $\frac{98}{7} = \frac{x}{11}$
(b) $154   **29. (a)** $\frac{3}{180} = \frac{7}{x}$ **(b)** 420 min   **31. (a)** 2
(b) 12   **33. (a)** $\frac{3}{2}$ **(b)** 9   **35. (a)** $-\frac{15}{2}$ **(b)** -45
37. (a) 24   **(b)** 3   **39. (a)** 40   **(b)** 5
41. (a) 400   **(b)** 50
43. (a) Direct   **(b)** $y = \frac{3}{2}x$ **45. (a)** Inverse   **(b)** $y = \frac{36}{x}$
(c)
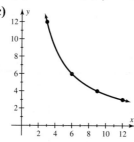
(c)

47. (a) Neither   **(b)** NA   **(c)** NA   **49.** Direct; 2
51. Neither   **53.** Inverse; 8   **55.** 47.6 minutes
57. 1.625 inches   **59.** About 2123 lb   **61.** $9\frac{1}{3}$ c
63. (a) Direct; the ratios   **65. (a)** Direct; the ratios
$\frac{R}{W}$ always equal 0.012. $\frac{G}{A}$ always equal 27.
(b) $R = 0.012W$ **(b)** $G = 27A$

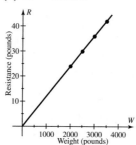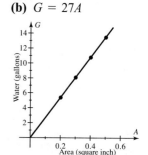

(c) 38.4 pounds   **(c)** For each square-inch
increase in the cross-sectional
area of the hose, the flow
increases by 27 gallons
per minute.
67. (a) $F = \frac{1200}{L}$ **(b)** 80 pounds
69. (a) Direct   **(b)** $y = -19x$ **(c)** Negative; for each
1-mile increase in altitude the temperature decreases by 19°F.
(d) 47.5°F decrease   **71.** 1.8 ohms

CHECKING BASIC CONCEPTS 7.7 (p. 504)

1. (a) $\frac{18}{5}$ **(b)** $\frac{15}{4}$ **2. (a)** $\frac{4}{6} = \frac{8}{x}$; 12   **(b)** $\frac{2}{148} = \frac{5}{x}$;
370 minutes   **3.** 60; 6   **4. (a)** Direct; the ratios $\frac{y}{x}$ always
equal $\frac{3}{2}; \frac{3}{2}$.   **(b)** Inverse; the products xy always equal 24; 24.
5. $160

CHAPTER 7 REVIEW (pp. 508–511)

1. $-\frac{3}{5}$ **2.** -3 **3.** Undefined   **4.** Undefined
5.

x	-2	-1	0	1	2
$\frac{3x}{x - 1}$	2	$\frac{3}{2}$	0	—	6

6. $-2, 2$ **7.** $\frac{5y^3}{3x^2}$ **8.** $x - 6$ **9.** -1 **10.** $\frac{x - 5}{5}$
11. $\frac{x + 3}{x + 1}$ **12.** $\frac{x + 4}{x + 1}$ **13.** 2   **14.** $\frac{1}{x + 5}$ **15.** $\frac{1}{z + 3}$
16. $\frac{x}{x - 2}$ **17.** $\frac{5}{6}$ **18.** $\frac{8}{x(x + 1)}$ **19.** $\frac{1}{2}$ **20.** 1
21. $x + y$ **22.** $2(a^2 + ab + b^2)$ **23.** $\frac{10}{x + 10}$
24. $\frac{1}{x - 1}$ **25.** 1   **26.** 1   **27.** $\frac{1}{x + 1}$ **28.** $\frac{2}{x - 5}$
29. $\frac{2}{xy}$ **30.** $\frac{x}{y}$ **31.** $15x$ **32.** $10x^2$ **33.** $x(x - 5)$
34. $10x^2(x - 1)$ **35.** $(x - 1)(x + 1)^2$
36. $x(x - 4)(x + 4)$ **37.** $\frac{9}{24}$ **38.** $\frac{16}{12x}$ **39.** $\frac{3x^2 + 6x}{x^2 - 4}$
40. $\frac{2x}{x^2 + x}$ **41.** $\frac{3x - 3}{5x^2 - 5x}$ **42.** $\frac{2x^2 + 4x}{2x^2 + x - 6}$ **43.** $\frac{19}{24}$
44. $\frac{7}{4x}$ **45.** $-\frac{1}{9x}$ **46.** $\frac{4x + 3}{x(x - 1)}$ **47.** $\frac{2x}{(x - 1)(x + 1)}$

48. $\frac{8 - 9x}{6x^2}$ **49.** $\frac{2x - 7}{6x}$ **50.** $-\frac{1}{(x - 1)(x + 1)}$

51. $\frac{5y - x}{(x - y)(x + y)}$ **52.** $\frac{13}{6x}$ **53.** $\frac{3x + y}{2xy}$ **54.** -1

55. $\frac{33}{28}$ **56.** $\frac{7}{10}$ **57.** $\frac{n}{2}$ **58.** $\frac{3(p + 1)}{p - 1}$ **59.** $\frac{3(m + 1)}{2(m - 1)^2}$

60. $\frac{2n - 1}{4(2n + 1)}$ **61.** $\frac{1}{3}$ **62.** $\frac{2 - x}{2 + x}$ **63.** $\frac{1}{x^2}$ **64.** $x + 3$

65. $\frac{20}{7}$ **66.** $\frac{8}{3}$ **67.** $\frac{1}{5}$ **68.** -5 **69.** -4 **70.** $-3, -1$

71. 4   **72.** -2 **73.** 5   **74.** 7   **75.** $-2, 5$

76. $-3, 3$ **77.** 8   **78.** No solutions   **79.** No solutions
(extraneous: -1)   **80.** $-\frac{3}{2}$ (extraneous: 0)   **81.** -3

82. -12 **83.** $\frac{4}{5}$ **84.** -3 **85.** $b = \frac{2ac}{3a - c}$

86. $x = \frac{y}{y - 1}$ **87.** 2   **88.** $\frac{15}{7}$ **89.** (a) $\frac{6}{x} = \frac{13}{20}$ (b) $\frac{120}{13}$

90. (a) $\frac{341}{11} = \frac{x}{8}$ (b) \$248   **91.** (a) 4   (b) 20

92. (a) 3   (b) 15   **93.** (a) 10   (b) 2   **94.** (a) 21   (b) $\frac{21}{5}$

95. (a) Inverse   (b) $y = \frac{60}{x}$ **96.** (a) Direct   (b) $y = 3x$

(c) 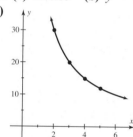   (c)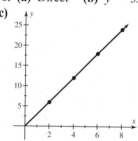

97. Direct; $\frac{1}{2}$ **98.** Inverse; 12   **99.** (a) $\frac{1}{5} = 0.2$; when the
average rate of arrival is 10 cars per minute, the average wait
is 0.2 minute, or 12 seconds.

(b)

x	5	10	13	14	14.9
T	$\frac{1}{10}$	$\frac{1}{5}$	$\frac{1}{2}$	1	10

(c) It increases dramatically.   **100.** 60 mph

101. $\frac{800}{13} \approx 61.5$ hours   **102.** 10 mph and 12 mph

103. 8 mph   **104.** About 33.3 feet   **105.** 200 vehicles

106. 1.6 inches   **107.** 36 pounds   **108.** \$468

109. 750 to 2250 seconds, or 12.5 to 37.5 minutes

110. About 771 lb

CHAPTER 7 TEST (pp. 511–512)

1. $\frac{9}{5}$ **2.** -2 **3.** $x + 5$ **4.** $x - 5$ **5.** 3   **6.** 2

7. $\frac{x - 1}{10x}$ **8.** $\frac{6}{x(x + 3)}$ **9.** $\frac{4x + 1}{x + 4}$ **10.** $\frac{t + 7}{2t - 3}$

11. $-\frac{1}{y + 1}$ **12.** $\frac{x^2y - x + y}{xy^2}$ **13.** $\frac{b}{15}$ **14.** $\frac{p}{p - 2}$

15. $\frac{35}{2}$ **16.** 3   **17.** 1   **18.** $-1, 2$ **19.** $\frac{2}{7}$ **20.** -12

21. No solutions $\left(\text{extraneous: } \frac{1}{2}\right)$ **22.** 0 (extraneous: 5)

23. $x = \frac{2 + 5y}{3y}$ **24.** $b = \frac{a}{a - 1}$ **25.** (a) $\frac{7}{2}$ (b) 21

26. Inversely; 32   **27.** 24 hours   **28.** 67.5 feet

29. $\frac{16}{5} = 3.2$; when the arrival rate is 24 people per hour,
there are about 3 people in line.

CHAPTERS 1–7 CUMULATIVE REVIEW (pp. 513–515)

1. $24\pi \approx 75.4$ **2.** $2x - 2$ **3.** $-\frac{5}{8}$ **4.** $\frac{2}{5}$ **5.** $\frac{3}{4}$

6. $\frac{2}{9}$ **7.** $2x + 2$ **8.** $y - 11$ **9.** -1 **10.** 4.5%

11. $W = \frac{V}{6L}$ **12.** $x \le \frac{1}{4}$

13. 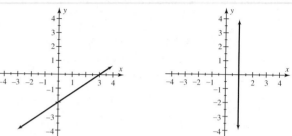   **14.**

15. $y = 3x + 5$

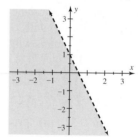

16. $y = 2x - 1$ **17.** $y = -\frac{2}{3}x + \frac{1}{3}$ **18.** $y = \frac{2}{3}x + \frac{8}{3}$

19. $N = 200x + 2000$ **20.** $(1, -2)$ **21.** $(1, 2)$

22. $(2, -8)$ **23.** Infinitely many solutions

24. No solutions   **25.** $(1, 2)$

26. $(0, 0)$ (answers may vary)

27. $4x^2 + 1$ **28.** $15z^8$ **29.** a^3b^3 **30.** $10y^2 - 11y - 6$

31. $x^4 - 2x^2y^2 + y^4$ **32.** $4t^2 + 20t + 25$ **33.** 2

34. 9   **35.** $\frac{1}{27x^6}$ **36.** $\frac{2}{x^2}$ **37.** 1.23×10^{-3}

38. $2x + 1 + \frac{4}{x - 1}$ **39.** $5z^2; 5z^2(4z - 3)$

40. $3xy; 3xy(4x + 5y)$ **41.** $(3 - x)(2 + 5x)$

42. $(3z - 2)(3z + 2)$ **43.** $(t + 8)^2$

44. $x(x - 4)(x + 4)$ **45.** $x(x - 9)(x + 11)$

46. $(a + 3b)^2$ **47.** $-5, \frac{7}{2}$ **48.** $0, 2$ **49.** $-7, 2$

50. $-2, 0, 2$ **51.** $1; 2$ **52.** $\frac{x + 1}{x - 1}$ **53.** 1   **54.** $\frac{x}{x^2 - 1}$

55. $\frac{1}{12x}$ **56.** $\frac{1}{2(x + 2)}$ **57.** $\frac{x + 2}{x - 2}$ **58.** $\frac{40}{7}$ **59.** $\frac{7}{12}$

60. $-1, 2$ **61.** $\frac{z + 2y}{3}$ **62.** 5.5 **63.** 80 **64.** Inverse; $y = \frac{20}{x}$ **65. (a)** $21x$ **(b)** 90 min **66.** 35 min running, 25 min walking **67. (a)** 24.1 **(b)** Head Start participation increased by 24,100 per year, on average. **(c)** 1099 thousand, or 1.1 million
68. (a) $2x + y = 180, 2x - y = 32$ **(b)** $(53, 74)$ or $53°$, $53°$, $74°$ **69. (a)** 21.6; in 1995 natural gas consumption was 21.6 trillion ft^3. **(b)** 2000 **70.** About 500 lb

CHAPTER 8: RADICAL EXPRESSIONS

SECTION 8.1 (pp. 524–527)

1. square root **2.** $3; -3$ **3.** radicand **4.** 2 **5.** 3 **6.** 2
7. Irrational **8.** cube root **9.** 4 **10.** $c^2 = a^2 + b^2$
11. 5 **12.** $\sqrt{(x_2 - x_1)^2 + (y_2 - y_1)^2}$ **13.** ± 2 **15.** ± 7
17. ± 11 **19.** ± 20 **21.** ± 14 **23.** ± 2.828
25. ± 4.899 **27.** 4 **29.** 2 **31.** $\frac{1}{3}$ **33.** $\frac{3}{2}$ **35.** 0.2
37. Real, rational **39.** Real, rational **41.** Real, irrational
43. Real, irrational **45.** None **47.** 12 **49.** -3
51. 3.16 **53.** 2 **55.** -5 **57.** -3 **59.** 1 **61.** 10
63. -7 **65.** 1.710 **67.** -2.520 **69.** -2.080 **71.** a
73. $c = 5$ inches **75.** $c = 13$ meters **77.** $a = \sqrt{5}$ miles
79. $b = 8$ feet **81.** Yes **83.** No **85.** Yes **87.** 5
89. $\sqrt{4100} \approx 64.03$ **91.** 5 **93.** 25
95. $\sqrt{146} \approx 12.083$ **97.** $\sqrt{68} \approx 8.246$
99.   **101.**

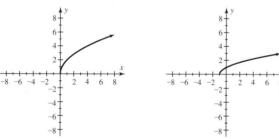

103.

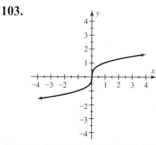

105. 30 inches **107.** 15 feet
109. (a) $\frac{885}{20} \approx 44$ bpm **(b)** 177 bpm **(c)** $\frac{885}{13} \approx 68$ bpm
111. 25 inches **113.** 37.1 feet **115.** 109.2 feet
117. 80 miles **119.** $\frac{a\sqrt{3}}{2}$

SECTION 8.2 (pp. 532–534)

1. 8 **2.** $\sqrt{ab}$ **3.** Yes **4.** 16 **5.** 2 **6.** x **7.** 4 **8.** $\frac{\sqrt{a}}{\sqrt{b}}$
9. b^2 **10.** $x\sqrt{x}$ **11.** 5 **13.** $\sqrt{21}$ **15.** $\frac{1}{2}$
17. $\sqrt{\frac{6}{35}}$ **19.** 8 **21.** 12 **23.** $\sqrt{15x}$ **25.** $\sqrt{5ab}$
27. $\sqrt{21xz}$ **29.** $\sqrt{3} \cdot \sqrt{11}$ **31.** $\sqrt{5} \cdot \sqrt{17}$
33. $\sqrt{7} \cdot \sqrt{z}$ **35.** $\sqrt{5} \cdot \sqrt{6}$ (answers may vary)
37. 4 **39.** 144 **41.** 100 **43.** 64 **45.** 5 **47.** 2
49. 2 **51.** $2\sqrt{5}$ **53.** $3\sqrt{7}$ **55.** $2\sqrt{3}$ **57.** $10\sqrt{3}$
59. $6\sqrt{5}$ **61.** $t\sqrt{t}$ **63.** $2a$ **65.** x^2 **67.** $2x\sqrt{2}$
69. $5t\sqrt{5t}$ **71.** xy^2 **73.** $4xy\sqrt{y}$ **75.** $\frac{2}{7}$ **77.** $\frac{4}{9}$
79. $\frac{\sqrt{17}}{6}$ **81.** $\frac{1}{7}$ **83.** 6 **85.** 5 **87.** $\frac{5}{3}$ **89.** $\frac{2}{3}$ **91.** $\frac{x}{3}$
93. $\frac{y\sqrt{2}}{7}$ **95.** $\frac{2\sqrt{5}}{y^2}$ **97.** $2xy$ **99.** $4xy^2$ **101.** $\frac{\sqrt{2}}{xy}$
103. $5x$ **105.** 5 **107.** $\frac{8}{3}$ **109.** $4x$ **111.** x^n
113. (a) $\frac{9}{2}$ seconds **(b)** $T = \frac{\sqrt{h}}{2}$ **115. (a)** 1.78 in.
(b) $r = \frac{\sqrt{A}}{\sqrt{\pi}}$ **117. (a)** 60 miles per hour **(b)** $V = v\sqrt{\frac{D}{d}}$

CHECKING BASIC CONCEPTS 8.1 & 8.2 (p. 535)

1. (a) 9 **(b)** ± 25 **2.** 2.646 **3. (a)** 2 **(b)** 4 **(c)** -3
4. 10 **5. (a)** $4\sqrt{5}$ **(b)** $4x\sqrt{x}$ **(c)** $2xy\sqrt{5}$ **6. (a)** 4
(b) $\frac{a\sqrt{a}}{10}$ **(c)** $\frac{2}{x}$ **7.** 24 feet **8.** 112 miles

SECTION 8.3 (pp. 540–541)

1. $2\sqrt{2}$ **2.** $\sqrt{b}$ **3.** $\sqrt[3]{b}$ **4.** $5\sqrt[3]{b}$ **5.** Yes. It simplifies to $3\sqrt{3}$. **6.** No. The radicals are unlike. **7.** $\sqrt{2}, 2\sqrt{2}$
9. Not possible **11.** $3\sqrt{7}, 2\sqrt{7}$ **13.** $\sqrt{5}, 5\sqrt{5}$
15. $5\sqrt{5}$ **17.** $9\sqrt{13}$ **19.** $10\sqrt{2}$ **21.** $7\sqrt{3} + 3\sqrt{2}$
23. $58\sqrt{2}$ **25.** $4\sqrt{15}$ **27.** $-10\sqrt{3}$ **29.** $-9\sqrt{7}$
31. $2\sqrt{6}$ **33.** $40\sqrt{2}$ **35.** $7\sqrt[3]{4}$ **37.** $-\sqrt[3]{20}$
39. $15\sqrt{2} + 10\sqrt{3}$ **41.** $(6x - 1)\sqrt{x}$
43. $(2 + x)\sqrt{x} + x\sqrt{2}$ **45.** $5a\sqrt{b}$ **47.** $5\sqrt{t}$
49. $-4\sqrt{x}$ **51.** $5\sqrt{b}$ **53.** $(2t - 10)\sqrt{t}$ **55.** $2\sqrt[3]{x}$
57. $11\sqrt[3]{x}$ **59.** $\sqrt{22} + \sqrt{33} + \sqrt{55}$ inches
61. 850 ft^3/sec

SECTION 8.4 (pp. 545–546)

1. $a^2 - b^2$ **2.** b **3.** -1 **4.** $a - b$ **5.** $4 + \sqrt{11}$
6. $2 - \sqrt{3}$ **7.** $\frac{\sqrt{11}}{\sqrt{11}}$ **8.** $\frac{4 + \sqrt{11}}{4 + \sqrt{11}}$ **9.** 14 **11.** 53
13. -44 **15.** 11 **17.** $x - 25$ **19.** $49 - t$
21. $y - 3\sqrt{y} + 2$ **23.** $3 + 2\sqrt{x} - x$ **25.** $x - y$
27. $r - 9t$ **29.** $\sqrt{17} + 3$ **31.** $20 - \sqrt{5}$
33. $\sqrt{x} + \sqrt{z}$ **35.** $\frac{2\sqrt{5}}{5}$ **37.** $\sqrt{14}$ **39.** $-\sqrt{3}$
41. $\frac{\sqrt{10}}{2}$ **43.** $-\frac{4\sqrt{7}}{21}$ **45.** $\frac{\sqrt{3}}{2}$ **47.** $\frac{2\sqrt{b}}{b}$ **49.** $-\frac{\sqrt{3b}}{b}$

51. $\frac{\sqrt{6x}}{3}$ **53.** $\frac{x}{3}$ **55.** $\frac{\sqrt{3}+1}{2}$ **57.** $5 - 2\sqrt{5}$

59. $2\sqrt{2} + 3$ **61.** $\frac{19 - 5\sqrt{13}}{6}$ **63.** $\frac{\sqrt{x}}{2x}$

65. $\frac{x - 4\sqrt{x} + 4}{x - 4}$ **67.** $\frac{a - 4b\sqrt{a} + 4b^2}{a - 4b^2}$

69. $\frac{a + 2\sqrt{ab} + b}{a - b}$ **71.** $\frac{2x - 2\sqrt{2xy} + y}{2x - y}$

73. $\sqrt{b + 1} - \sqrt{b}$ **75.** $4\sqrt{91}$; 87

77. $\frac{\pi}{2} \approx 1.57$ seconds

CHECKING BASIC CONCEPTS 8.3 & 8.4 (p. 547)

1. (a) $10\sqrt{6}$ **(b)** $9\sqrt{3}$ **(c)** $9\sqrt{k}$ **2.** $16\sqrt{5}$ feet

3. (a) 23 **(b)** $x + 2\sqrt{x} - 15$ **4. (a)** $\frac{\sqrt{5}}{3}$

(b) $3\sqrt{7} + 8$ **(c)** $\frac{\sqrt{x} - \sqrt{y}}{x - y}$

SECTION 8.5 (pp. 553–555)

1. Square each side.   **2.** Check your answers.   **3.** 16
4. Yes   **5.** No   **6.** a^2 **7.** 36 **9.** No solutions   **11.** 16
13. -99 **15.** $\frac{19}{2}$ **17.** $\frac{16}{3}$ **19.** 64 **21.** $\frac{1}{3}$ **23.** No solutions
25. 16 **27.** No solutions   **29.** 18 **31.** 5 **33.** 1
35. 8 **37.** 5 **39.** $\frac{5}{9}, 5$ **41.** 4 **43.** 6 **45.** $-3, 2$
47. $L = T^2$ **49.** $a = \frac{1}{2}b^2$ **51.** $t = \frac{1}{16}D^2 - 1$
53. $b = \sqrt{c^2 - a^2}$ **55.** $d = (k - 4)^2$ **57. (a)** 4 **(b)** 4
59. (a) 3 **(b)** 3 **61. (a)** 4 **(b)** 4 **63. (a)** 1 **(b)** 1
65. 2.21 **67.** $-1.72, 1.32$ **69. (a)** 12.2 miles
(b) About 1680 feet   **(c)** The distance to the horizon doubles.
71. (a) $b = \sqrt{c^2 - a^2}$ **(b)** 24 inches   **73. (a)** $D = \frac{V^2 d}{v^2}$
(b) 312.5 feet   **75.** It increases by a factor of 2, or doubles.
77. It increases to 600 cubic feet per second.

SECTION 8.6 (pp. 561–563)

1. index; radicand   **2.** odd   **3.** even   **4.** No; $\sqrt[3]{3} \approx 1.44$
because $1.44^3 \approx 3$.   **5.** negative number   **6.** real number
7. $\sqrt[n]{a \cdot b}$ **8.** $\frac{\sqrt[n]{a}}{\sqrt[n]{b}}$ **9.** $\sqrt[n]{a}$ **10.** $\sqrt[n]{a^m}$ or $\left(\sqrt[n]{a}\right)^m$
11. -3 **13.** 3 **15.** 2 **17.** 10 **19.** 2 **21.** -1 **23.** 5
25. -5 **27.** $\frac{3}{2}$ **29.** $\frac{\sqrt[4]{20}}{3}$ **31.** Not a real number   **33.** $\frac{2}{3}$
35. $\frac{\sqrt[4]{7}}{2}$ **37.** $\frac{2}{3}$ **39.** 4 **41.** 2 **43.** -2 **45.** $\frac{1}{3}$ **47.** $\frac{1}{2}$
49. -2 **51.** $\sqrt{36} = 6$ **53.** $\sqrt[3]{27} = 3$ **55.** $\sqrt[3]{-1} = -1$
57. $\sqrt[4]{81} = 3$ **59.** $\sqrt[5]{32} = 2$ **61.** $\sqrt{4^3} = \left(\sqrt{4}\right)^3 = 8$
63. $\sqrt[3]{64^2} = \left(\sqrt[3]{64}\right)^2 = 16$ **65.** $\sqrt[4]{16^5} = \left(\sqrt[4]{16}\right)^5 = 32$
67. $\sqrt[5]{(-1)^3} = \left(\sqrt[5]{-1}\right)^3 = -1$ **69.** $\frac{1}{\sqrt{49}} = \frac{1}{7}$
71. $\frac{1}{\sqrt[3]{(-27)^2}} = \frac{1}{\left(\sqrt[3]{-27}\right)^2} = \frac{1}{9}$ **73.** $\frac{1}{\sqrt[4]{625^3}} = \frac{1}{\left(\sqrt[4]{625}\right)^3} = \frac{1}{125}$

75. $\frac{1}{\sqrt[4]{16^5}} = \frac{1}{\left(\sqrt[4]{16}\right)^5} = \frac{1}{32}$ **77.** 900 square inches

79. (a) $r = \left(\frac{4}{\pi}\right)^{1/2}$ **(b)** 3 feet

CHECKING BASIC CONCEPTS 8.5 & 8.6 (p. 563)

1. (a) 4 **(b)** -6 **2.** $t = \frac{(S + 2)^2}{16}$ **3. (a)** -3 **(b)** 2

4. (a) 3 **(b)** $\frac{\sqrt[3]{9}}{4}$ **(c)** 2 **5. (a)** $\sqrt{9} = 3$

(b) $\sqrt[4]{16^3} = \left(\sqrt[4]{16}\right)^3 = 8$ **(c)** $\frac{1}{\sqrt[3]{1000}} = \frac{1}{10}$

CHAPTER 8 REVIEW (pp. 566–569)

1. ± 6 **2.** ± 20 **3.** ± 2.646 **4.** ± 3.464
5. Real, rational   **6.** Real, rational   **7.** Real, irrational
8. None   **9.** 11 **10.** 20 **11.** -7 **12.** -4.12
13. 1.82 **14.** -5 **15.** $\sqrt{41}$ inches   **16.** 24 feet
17. $\sqrt{61} \approx 7.81$ **18.** $5\sqrt{2}$; $4\sqrt{5} + 2\sqrt{30}$ **19.** 13
20. $\sqrt{26} \approx 5.099$ **21.** 11 **22.** 25 **23.** $\sqrt{30}$
24. $\frac{\sqrt{2}}{3}$ **25.** 16 **26.** $2x$ **27.** $9t$ **28.** $\sqrt{14ab}$
29. $3\sqrt{5}$ **30.** $10\sqrt{2}$ **31.** $t^2\sqrt{t}$ **32.** $b\sqrt{5}$ **33.** $3x\sqrt{y}$
34. $7a\sqrt{ab}$ **35.** $\frac{5}{6}$ **36.** $2\sqrt{10}$ **37.** $\frac{9}{x}$ **38.** 4 **39.** 5
40. $r\sqrt{3}$ **41.** $\frac{x\sqrt{x}}{8}$ **42.** $9ab^2$ **43.** 5 **44.** $2y$
45. $3\sqrt{7}$ **46.** $3\sqrt{15}$ **47.** $4\sqrt{2}$ **48.** $5\sqrt{5}$
49. $3\sqrt[3]{9}$ **50.** $10\sqrt[3]{10}$ **51.** $5\sqrt{t}$ **52.** 0 **53.** 75
54. $4 - t$ **55.** $\sqrt{5}$ **56.** $-\frac{3\sqrt{11}}{11}$ **57.** $-\frac{4\sqrt{x}}{3x}$
58. $\frac{\sqrt{5x}}{5}$ **59.** $\sqrt{5} + 2$ **60.** $\frac{16 - 8\sqrt{b} + b}{16 - b}$ **61.** 25
62. No solutions   **63.** -23 **64.** 8 **65.** 32 **66.** 8
67. 6 **68.** 12 **69.** $P = \frac{A^2}{4}$ **70.** $g = \frac{4\pi^2 L}{P^2}$ **71. (a)** 0
(b) 0 **72. (a)** 17 **(b)** 17 **73.** 4 **74.** -10 **75.** -3
76. 2 **77.** 7 **78.** $-\frac{1}{2}$ **79.** -4 **80.** 3 **81.** $\frac{1}{2}$ **82.** $\frac{1}{2}$
83. $\frac{2}{3}$ **84.** $\frac{\sqrt[3]{4}}{3}$ **85.** $\sqrt{25^3} = \left(\sqrt{25}\right)^3 = 125$
86. $\sqrt[3]{1000} = 10$ **87.** $\frac{1}{\sqrt[4]{16^5}} = \frac{1}{\left(\sqrt[4]{16}\right)^5} = \frac{1}{32}$
88. $\frac{1}{\sqrt[3]{(-8)^2}} = \frac{1}{\left(\sqrt[3]{-8}\right)^2} = \frac{1}{4}$ **89.** 85 miles   **90.** 500 feet
91. About 35 pounds   **92.** 64 feet   **93. (a)** $T = \sqrt{D^3}$
(b) About 1.87 years   **94.** $d = \frac{v^2 D}{V^2}$ **95.** 400; 9902
96. About 67 feet   **97. (a)** π seconds   **(b)** About 3.24 feet
98. (a) $L = 27.5W^{1/3}$ **(b)** 55 inches

CHAPTER 8 TEST (pp. 569–570)

1. ± 8 **2.** Real, irrational   **3.** 9 **4.** -4 **5.** 2 **6.** 3.32
7. $\sqrt{25^3} = \left(\sqrt{25}\right)^3 = 125$ **8.** $\frac{1}{8}$ **9.** $\sqrt{74}$ inches
10. $\sqrt{45}$, or $3\sqrt{5}$ **11.** $2\sqrt{10}$ feet; $12\sqrt{2}$ feet

12. $\sqrt{98} = 7\sqrt{2} \approx 9.899$ **13.** $3\sqrt{6}$ **14.** $3t\sqrt{t}$
15. 12   **16.** 3   **17.** $9a$ **18.** $2r$ **19.** $\frac{x}{3}$ **20.** $4ab\sqrt{ab}$
21. $9\sqrt{5}$ **22.** $3\sqrt{2}$ **23.** $-\sqrt[3]{11}$ **24.** $8\sqrt{ab}$
25. $8\sqrt{2} - 3\sqrt{3}$ **26.** $8\sqrt{6}$ **27.** $3\sqrt{b}$
28. $(2x + 4)\sqrt{x}$ **29.** $\frac{2\sqrt{3}}{3}$ **30.** $\frac{4 - 4\sqrt{x} + x}{4 - x}$ **31.** 15
32. 6   **33.** $c = \sqrt{a^2 + b^2}$ **34.** 54 miles
35. **(a)** About 111 beats per minute   **(b)** 25 pounds
(c) $R = 885W^{-1/2}$ **(d)** $W = \left(\frac{885}{R}\right)^2$

CHAPTERS 1–8 CUMULATIVE REVIEW (p. 571)

1. $2 \cdot 2 \cdot 3 \cdot 5$ **2.** 9   **3.** $6xy$ **4.** $\frac{5}{4}$
5. $\frac{9}{20}$; 0.45   **6.** $x \le -\frac{3}{2}$
7.   **8.**

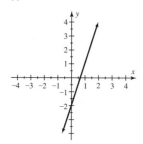   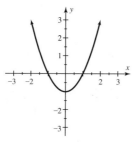

9. $y = -\frac{2}{3}x - \frac{1}{3}$ **10.** $(1, 2)$ **11.** $6x^2 - 3$ **12.** a^5b^6
13. $-24x^3 + 12x^2$ **14.** $x^2 + 3x - 10$ **15.** 4^3 or 64
16. $\frac{1}{27t^6}$ **17.** 65,000   **18.** $3x^2 - 2x$ **19.** $(x + 5)(x - 3)$
20. $(4z - 3)(4z + 3)$ **21.** $(x - y)(x^2 + xy + y^2)$
22. $(x^2 + 3)(x^2 + 1)$ **23.** $-2, 7$ **24.** $-2, 2$ **25.** $\frac{x + 3}{x^2 - 4}$
26. $y(x - y)$ **27.** $\frac{x + y}{x - y}$ **28.** $\frac{1}{2}$ **29.** 11   **30.** $\sqrt{29}$
31. **(a)** $4t$ **(b)** $12\sqrt{2}$ **(c)** $2\sqrt[3]{x}$ **(d)** -1 **(e)** 27   **(f)** $\frac{1}{2}$
32. -2 **33.** $y = \sqrt{4 - x^2}$ **34.** -3 **35.** \$812.50
36. Stair climber: 20 min; rowing: 30 min   **37.** 45º, 45º, 90º
38. 108 mi

CHAPTER 9:   QUADRATIC EQUATIONS

SECTION 9.1 (pp. 580–582)

1. parabola   **2.** The vertex   **3.** axis of symmetry   **4.** $(0, 0)$
5.

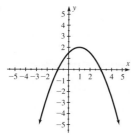

6. $-\frac{b}{2a}$ **7.** narrower   **8.** downward
9. $ax^2 + bx + c$ with $a \ne 0$ **10.** vertex   **11.** Upward
13. Downward   **15.** Downward   **17.** Upward
19. $(1, -2)$; $x = 1$; upward; incr.: $x \ge 1$; decr.: $x \le 1$
21. $(-2, 3)$; $x = -2$; downward; incr.: $x \le -2$;
decr.: $x \ge -2$

23. (a)   **25. (a)**

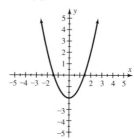   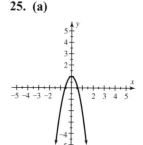

(b) $(0, -2)$; $x = 0$ **(b)** $(0, 1)$; $x = 0$
27. (a)   **29. (a)**

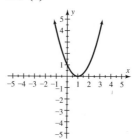   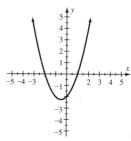 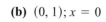

(b) $(1, 0)$; $x = 1$ **(b)** $(-0.5, -2.25)$;
                                         $x = -0.5$
31. (a)   **33. (a)**

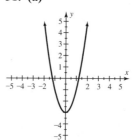   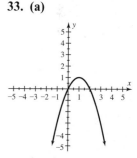 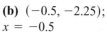

(b) $(0, -3)$; $x = 0$ **(b)** $(1, 1)$; $x = 1$
35. (a)   **37. (a)**

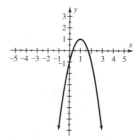   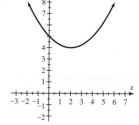

(b) $(1, 1)$; $x = 1$ **(b)** $(2, 4)$; $x = 2$

39. $(2, -6)$ **41.** $(-3, 4)$ **43.** $(0, 3)$ **45.** $(1, 1.4)$
47. (a) $(0, -4); -4$ **49. (a)** $(-1, -2); -1$
(b)   **(b)**

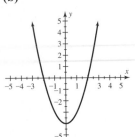   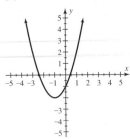

51. (a) $(2, 2); 0$
(b)

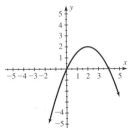

53.   **55.**

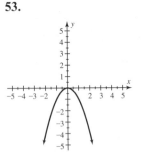

Opens downward   Narrower
57.   **59.**

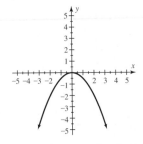

Wider   Opens downward and wider
61. Incr: $x \geq -1$; decr: $x \leq -1$
63. Incr: $x \geq \frac{5}{2}$; decr: $x \leq \frac{5}{2}$
65. Incr: $x \geq -\frac{1}{2}$; decr: $x \leq -\frac{1}{2}$
67. Incr: $x \leq 1$; decr: $x \geq 1$ **69.** Incr: $x \leq 2$; decr: $x \geq 2$
71. Incr: $x \leq \frac{1}{4}$; decr: $x \geq \frac{1}{4}$ **73. (a)** $(2, 66)$ **(b)** The max
imum height is 66 feet after 2 seconds. **75.** 25 ft by 25 ft

SECTION 9.2 (pp. 588–589)

1. $x^2 + 3x - 2 = 0$ (answers may vary); zero, one, or two
solutions **2.** ± 3 **3.** Factoring, square root property
4.   **5.**

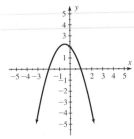   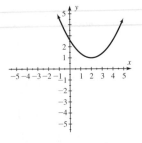

(answers may vary)   (answers may vary)
6. Two; the solutions are the x-intercepts. **7.** Yes **9.** No
11. Yes **13.** No **15.** $-2, 1$ **17.** No real solutions
19. $-2, 3$ **21.** -0.5 **23.** $-1, 5$ **25.** $-3, 1$ **27.** $-3, 3$
29. No real solutions **31.** -1 **33.** $-7, 5$ **35.** $-\frac{1}{3}, \frac{1}{2}$
37. $1, \frac{3}{2}$ **39.** ± 12 **41.** ± 4 **43.** $-6, 4$ **45.** $-7, 9$
47. $-3, 6$ **49.** $3, 5$ **51.** $5, 7$ **53.** $-\frac{1}{2}, \frac{1}{4}$ **55. (a)** 30 miles
per hour **(b)** 40 miles per hour **57.** About 1.9 seconds; no
59. Base: 10 inches; height: 7 inches **61.** 81,200

CHECKING BASIC CONCEPTS 9.1 & 9.2 (p. 590)

1. (a)   **(b)**

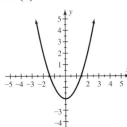   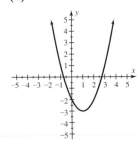

$(0, -2); x = 0$ $(1, -3); x = 1$
2. y_1 opens upward, whereas y_2 opens downward;
y_1 is narrower than y_2. **3.** Incr: $x \leq 2$; decr: $x \geq 2$
4. $\frac{1}{2}, 3$ **5.** $\pm \sqrt{5}$

SECTION 9.3 (p. 594)

1. $x + 3$ **2.** $x - 3$ **3.** $x + \frac{b}{2}$ **4.** $x - \frac{b}{2}$ **5.** x-term
6. completing the square **7.** 9 **8.** $\left(\frac{b}{2}\right)^2$ **9.** No **11.** Yes
13. Yes **15.** No **17.** $16; (x - 4)^2$ **19.** $\frac{81}{4}; \left(x + \frac{9}{2}\right)^2$
21. 9 **23.** $\frac{49}{4}$ **25.** $-3, 5$ **27.** $-3 \pm \sqrt{10}$ **29.** $\frac{3 \pm \sqrt{29}}{2}$
31. $\frac{5 \pm \sqrt{17}}{2}$ **33.** $1 \pm \sqrt{5}$ **35.** $\frac{3 \pm \sqrt{41}}{4}$ **37.** $\frac{2 \pm \sqrt{11}}{2}$
39. $\frac{-3 \pm \sqrt{5}}{12}$ **41.** $-2 + 2\sqrt{5} \approx 2.47$ inches

SECTION 9.4 (pp. 600–602)

1. To solve quadratic equations of the form $ax^2 + bx + c = 0$
2. No solutions, one solution, or two solutions   **3.** $b^2 - 4ac$
4. One solution   **5.** Factoring, square root property, completing the square, and the quadratic formula   **6.** No; not when $b^2 - 4ac < 0$ **7.** $a = 5, b = -4, c = 6$
9. $a = 3, b = -2, c = 5$ **11.** $a = 1, b = -1, c = 0$
13. $a = 1, b = 1, c = -12$ **15.** 1   **17.** $-1, 3$ **19.** $-\frac{1}{2}, 3$
21. $-6, \frac{1}{2}$ **23.** 1   **25.** No real solutions   **27.** $-2, 8$
29. $\frac{1 \pm \sqrt{17}}{8}$ **31.** No real solutions   **33.** $\frac{1}{2}$ **35.** $\frac{3 \pm \sqrt{13}}{2}$
37. $2 \pm \sqrt{3}$ **39.** $\frac{1 \pm \sqrt{37}}{6}$ **41.** $\frac{1 \pm \sqrt{15}}{2}$ **43.** $\frac{5 \pm \sqrt{10}}{3}$
45. (a) $a > 0$ (b) $-1, 2$ (c) Positive   **47.** (a) $a > 0$
(b) No real solutions   (c) Negative   **49.** (a) $a < 0$
(b) 2   (c) Zero   **51.** (a) 25   (b) Two   **53.** (a) 0
(b) One   **55.** (a) $-\frac{7}{4}$ (b) None   **57.** (a) 21   (b) Two
59. $1 \pm \sqrt{2}$ **61.** $-\frac{3}{2}, 1$ **63.** None   **65.** None
67. $\frac{-2 \pm \sqrt{10}}{3}$ **69.** 15 miles per hour   **71.** 45 miles per hour
73. $x \approx 8.04$, or about 1992; this agrees with the graph.
75. 45 ft by 55 ft

CHECKING BASIC CONCEPTS 9.3 & 9.4 (p. 602)

1. (a) No   (b) Yes   **2.** (a) $2 \pm \sqrt{3}$ (b) $3 \pm \sqrt{5}$
3. (a) $\frac{3 \pm \sqrt{17}}{4}$ (b) $\frac{4}{3}$ **4.** (a) 5; two real solutions
(b) -7; no real solutions   (c) 0; one real solution

SECTION 9.5 (p. 609)

1. $2 + 3i$ (answers may vary)   **2.** No; any real
number a can be written as $a + 0i$ **3.** i **4.** -1
5. $i\sqrt{a}$ **6.** $a + bi$ **7.** 4   **8.** -5 **9.** $i\sqrt{3}$
11. $6i$ **13.** $12i$ **15.** $2i\sqrt{3}$ **17.** $3i\sqrt{2}$ **19.** 2
21. $7i$ **23.** $1 - 9i$ **25.** $-10 + 7i$ **27.** $20 - 12i$
29. $-31 - 8i$ **31.** 20   **33.** $1 + 5i$ **35.** $3 + 4i$
37. $-2 - 6i$ **39.** $a^2 + b^2$ **41.** $\pm 3i$ **43.** $\pm 4i\sqrt{5}$
45. $\pm\frac{1}{2}i$ **47.** $\frac{1}{3} \pm i\frac{\sqrt{11}}{3}$ **49.** $\pm i\sqrt{6}$ **51.** $\pm\sqrt{3}$
53. $\pm i\sqrt{2}$ **55.** $\frac{1}{2} \pm i\frac{\sqrt{7}}{2}$ **57.** $-\frac{3}{4} \pm i\frac{\sqrt{23}}{4}$
59. $2 \pm \sqrt{3}$

SECTION 9.6 (pp. 618–620)

1. function   **2.** y equals f of x **3.** 5   **4.** domain   **5.** range
6. one   **7.** To identify graphs of functions   **8.** (3, 4); 3; 6
9. (a, b) **10.** d **11.** $-6; -2$ **13.** $0; \frac{3}{2}$ **15.** $25; \frac{9}{4}$
17. $3; 3$ **19.** $13; -22$ **21.** $-\frac{1}{2}; \frac{2}{5}$
23. (a) $I(x) = 36x$ (b) $I(10) = 360$
25. (a) $A(r) = \pi r^2$ (b) $A(10) = 100\pi \approx 314.2$
27. (a) $A(x) = 43{,}560x$ (b) $A(10) = 435{,}600$

29. $f = \{(1, 3), (2, -4), (3, 0)\}$;
$D = \{1, 2, 3\}; R = \{-4, 0, 3\}$
31. $f = \{(a, b), (c, d), (e, a), (d, b)\}$;
$D = \{a, c, d, e\}; R = \{a, b, d\}$
33.

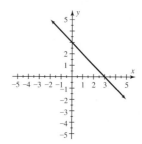

35.

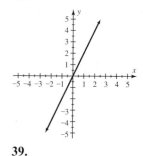

37.

39.

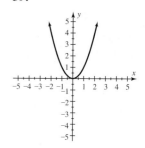

41.

43. $3; -1$ **45.** $0; 2$ **47.** $-4; -3$ **49.** Subtract $\frac{1}{2}$ from the input x to obtain the output y.   **51.** Divide the input x by 3 to obtain the output y.   **53.** 178.5; in 1995 the average price was \$178,500.   **55.** All real numbers   **57.** All real numbers   **59.** $x \neq 5$ **61.** All real numbers   **63.** $x \geq 1$
65. No   **67.** Yes   **69.** Yes   **71.** No   **73.** Yes
75. No   **77.** No

CHECKING BASIC CONCEPTS 9.5 & 9.6 (p. 620)

1. (a) $8i$ (b) $i\sqrt{17}$ **2.** (a) $3 - 4i$ (b) $-2 + 3i$
(c) $5 + i$ **3.** (a) $\pm 5i$ (b) $3 \pm 2i$ (c) $-\frac{1}{2} \pm i\frac{\sqrt{7}}{2}$
4. $f(x) = x^2 - 1$ **5.** 0; 4

CHAPTER 9 REVIEW (pp. 624–626)

1. $(-3, 4); x = -3$; downward; incr: $x \leq -3$; decr: $x \geq -3$ **2.** (1, 0); $x = 1$; upward; incr: $x \geq 1$; decr: $x \leq 1$

3. (a)

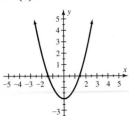

(b) $(0, -2)$; $x = 0$

4. (a)

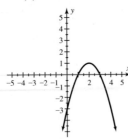

(b) $(2, 1)$; $x = 2$

5. (a)

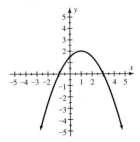

(b) $(1, 2)$; $x = 1$

6. (a)

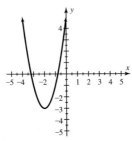

(b) $(-2, -3)$; $x = -2$

7. $(2, -6)$ **8.** $(0, 5)$ **9.** $(2, 1)$ **10.** $(-1, 1)$ **11.** $-2, 3$
12. -1 **13.** No real solutions   **14.** $-4, 6$ **15.** $-10, 5$
16. $-0.5, 0.25$ **17.** $-5, 10$ **18.** $-3, 1$ **19.** $-4, 2$
20. $-1, 3$ **21.** $-5, 4$ **22.** $-8, -3$ **23.** $-\frac{2}{5}, \frac{2}{3}$ **24.** $\frac{4}{7}, 3$
25. ± 10 **26.** $\pm \frac{1}{3}$ **27.** $\pm \frac{\sqrt{6}}{2}$ **28.** No real solutions
29. No   **30.** Yes   **31.** Yes   **32.** No   **33.** $-3 \pm \sqrt{7}$
34. $2 \pm \sqrt{10}$ **35.** $1 \pm \sqrt{6}$ **36.** $\frac{-3 \pm \sqrt{11}}{2}$ **37.** $3, 6$
38. $11, 13$ **39.** $-\frac{1}{2}, \frac{1}{3}$ **40.** $\frac{5 \pm \sqrt{5}}{10}$ **41.** $4 \pm \sqrt{21}$
42. $\frac{3 \pm \sqrt{3}}{2}$ **43. (a)** $a > 0$ **(b)** $-2, 3$ **(c)** Positive
44. (a) $a > 0$ **(b)** 2   **(c)** Zero   **45. (a)** $a < 0$ **(b)** No
real solutions   **(c)** Negative   **46. (a)** $a < 0$ **(b)** $-4, 2$
(c) Positive   **47. (a)** 1   **(b)** Two   **48. (a)** 144   **(b)** Two
49. (a) -23 **(b)** None   **50. (a)** 0   **(b)** One   **51.** 1
52. $-2 + 4i$ **53.** $5 + i$ **54.** $2 - 2i$ **55.** $-\frac{1}{2} \pm i\frac{\sqrt{19}}{2}$
56. $\pm 2i$ **57.** $\frac{1}{4} \pm i\frac{\sqrt{7}}{4}$ **58.** $\frac{1}{7} \pm i\frac{\sqrt{34}}{7}$ **59.** $-7; 0$
60. $-22; 2$ **61. (a)** $P(q) = 2q$ **(b)** $P(5) = 10$
62. (a) $f(x) = 4x - 3$ **(b)** $f(5) = 17$
63.

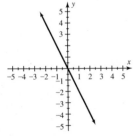

64.

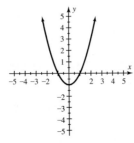

65. $1; 4$ **66.** $1; -2$ **67.** Yes   **68.** No   **69.** All real
numbers   **70.** $x \geq 0$ **71.** $x \neq 0$ **72.** All real numbers

73. (a) $\sqrt{1728} \approx 41.6$ miles per hour   **(b)** 60 miles per
hour   **74. (a)** $x(x + 2) = 143$ **(b)** $x = -13$ or $x = 11$;
the numbers are -13 and -11 or 11 and 13.   **75.** About
11.1 inches by 11.1 inches   **76.** 50 feet

CHAPTER 9 TEST (pp. 626–627)

1. $\left(1, \frac{3}{2}\right)$; $x = 1$ **2.** $-1, 2$ **3.** $-4, \frac{1}{3}$ **4.** $-\frac{1}{2}, \frac{1}{2}$
5. $4 \pm \sqrt{17}$ **6.** $\frac{3 \pm \sqrt{17}}{4}$ **7. (a)** $a < 0$ **(b)** $-3, 1$
(c) Positive   **8. (a)** -44 **(b)** No real solutions
(c) The graph of $y = -3x^2 + 4x - 5$ does not intersect
the x-axis.   **9.** $2 - 19i$ **10.** $-6 + 8i$ **11.** $-6 + 4i$
12. $\frac{5}{4}$ **13.** $\pm i\sqrt{11}$ **14.** $4 \pm 3i$ **15.** $46, (4, 46)$
(answers may vary)   **16.** $C(x) = 4x$; $C(5) = 20$
17. (a)

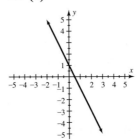

(b)

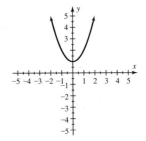

18. $0; -3$ **19.** No, it fails the vertical line test.
20. (a) All real numbers   **(b)** $x \neq 5$
21. $\sqrt{2250} \approx 47.4$ miles per hour

CHAPTERS 1–9 CUMULATIVE REVIEW (pp. 628–630)

1. $2^3 \cdot 3^2 \cdot 5$ **2.** $2n + 7 = n - 2; -9$ **3.** $\frac{23}{21}$ **4.** $-\frac{1}{6}$
5. 18   **6.** $\frac{1}{10}$ **7.** $\frac{6}{5}$ **8.** 2   **9.** $\frac{31}{25}$; 1.24   **10.** 6 miles
11. $b = \frac{2A}{h} - a$ **12.** $\left\{t \mid t < \frac{4}{7}\right\}$
13.

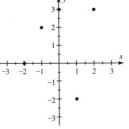

14. x-intercept: 4;
y-intercept: 2

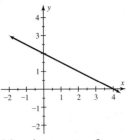

15. x-intercept: -4;
y-intercept: 3

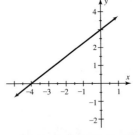

16. x-intercept: -2;
y-intercept: none

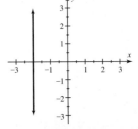

17. x-intercept: 2; y-intercept: 4; $y = -2x + 4$
18. $y = -2x + 1$ **19.** $y = \frac{3}{5}x + \frac{34}{5}$ **20.** $y = 2x + 3$
21. $P = 500x + 4000$ **22.** $(7.5, 10)$
23.   **24.**

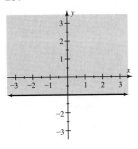   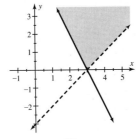

25.   **26.**

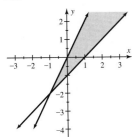

27. $-7ab - 10b^2 - 2$ **28.** -76 **29.** $\frac{27}{4}$ **30.** 1
31. $24t^4$ **32.** $2r^4t^4$ **33.** $\frac{1}{4t^6}$ **34.** $2ab^3$ **35.** $\frac{16r^{12}}{81t^{16}}$
36. $\frac{a^6}{8b^6}$ **37.** $2a^4 - 4a^3 + 6a^2$ **38.** $a^3 + b^3$
39. $5x^2 - 34x - 7$ **40.** $2y^2 - 3y - 9$ **41.** $a^2 - b^2$
42. $4x^2 + 12xy + 9y^2$ **43.** 0.000015 **44.** 2.13×10^6
45. $2x^2 - 4x + 3$ **46.** $x^3 - 4x^2 + 3x - 2 + \frac{1}{x-5}$
47. $5ab^2(2 - 5a^2b^3)$ **48.** $(y - 3)(y^2 + 2)$
49. $(2z + 3)(3z - 1)$ **50.** $(2z - 3)(2z + 3)$
51. $(2y - 5)^2$ **52.** $(a - 3)(a^2 + 3a + 9)$
53. $(z^2 - 3)(4z^2 - 5)$ **54.** $ab(a + b)(2a - b)$
55. $-2, 1$ **56.** $0, 9$ **57.** $-\frac{1}{3}, \frac{3}{2}$ **58.** $-2, 0, 2$ **59.** $x - 4$
60. $\frac{x-6}{3x-4}$ **61.** $-4, 4$ **62.** $\frac{7}{3}$ **63.** $2(x - 1)$ **64.** 1
65. $\frac{2x+3}{(x-1)(x+1)}$ **66.** $\frac{y-x}{y+x}$ **67.** 8   **68.** $\frac{17}{12}$ **69.** 1
70. 10   **71.** $\sqrt{73}$ feet   **72.** 7 miles   **73.** $\sqrt{52} = 2\sqrt{13}$
74. 4   **75.** 4   **76.** $8x^2y$ **77.** $5\sqrt[3]{7}$ **78.** 2   **79.** -45
80. 4   **81.** $(3, 5)$ **82.** $(1, -3)$ **83.** $3, -6$ **84.** $1 \pm \sqrt{3}$
85. $\pm\sqrt{5}$ **86.** $\frac{3 \pm \sqrt{2}}{2}$ **87.** $2 \pm 5i$ **88.** $\frac{1}{3} \pm \frac{4}{3}i$
89. 34   **90. (a)** $\{-3, 0, 2\}$ **(b)** $x \neq -6$ **(c)** $x \geq -4$
91.   **92.**

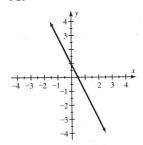   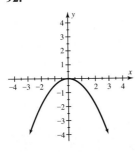

93. $d = 325t$ **94. (a)** $I = \frac{1}{2}x + 4$ **(b)** $\frac{1}{2}$ **(c)** Snow is falling at a rate of $\frac{1}{2}$ inch per hour.   **(d)** 6 inches
95.

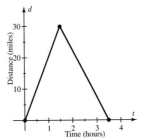

96. 3 children and 5 adults   **97.** $\frac{6}{7}$ hour
98. (a) 112 feet   **(b)** 5.5 seconds   **99.** 61.25 feet
100. (a) $\frac{1}{5}$ minute; the average wait is $\frac{1}{5}$ minute or 12 seconds when vehicles arrive at 15/min.
(b)

x	5	10	15	19	19.9
T	$\frac{1}{15}$	$\frac{1}{10}$	$\frac{1}{5}$	1	10

(c) The waiting time increases dramatically.
101. $73.63 \leq L \leq 74.37$.

APPENDIX: SETS

1. $\{1, 2, 3, 4, 5, 6, 7\}$ **2.** $\{5, 6, 7, 8, 9, 10\}$ **3.** $\{$Sunday, Saturday$\}$ **4.** $\{$Alabama, Alaska, Arizona, Arkansas$\}$
5. $\{$A, B, C, D, E, F, G$\}$ **6.** $\{$D, E, F, G, H, I, J, K, L, M$\}$
7. $\varnothing$ **8.** $\varnothing$ **9.** $\{0, 2, 4, 6, 8, 10\}$ **10.** $\{13, 15, 17, 19\}$
11. $\in$ **13.** $\notin$ **15.** $\in$ **17.** $\notin$ **19.** $\in$
21. $E = \{2, 4, 6, 8, 10\}$ **23.** $E = \{2, 4, 6, 8, 10, \ldots\}$
25. $A = \{$Apple, Apricot$\}$ **27.** $A = \{$Algebra$\}$
29. False   **31.** True   **33.** True   **35.** True   **37.** True
39. True   **41.** False   **43.** True
45.   **47.**

49.   **51.**

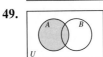

53. $\{1, 2, 3, 4\}$ **55.** $\{a, b\}$ **57.** $\{a, b, c\}$ **59.** $\{a, b, c, d\}$
61. $\{5, 8\}$ **63.** $\{1, 2, 3, 4, 5, 6, 7, 8, 9, 10\}$ **65.** $\{4, 5, 6\}$
67. $\varnothing$ **69.** $\{1, 3, 5, 7, 9\}$ **71.** $\varnothing$ **73.** $\varnothing$
75. $\{1, 3, 5, 7, 9\}$ **77.** $\{1, 2, 3, 5, 7, 8, 9, 10\}$ **79.** $\varnothing$
81. Infinite   **83.** Finite   **85.** Finite

GLOSSARY

absolute value A real number a, written $|a|$, is equal to its distance from the origin on the number line.

addends In an addition problem, the two numbers that are added.

addition property of equality If a, b, and c are real numbers, then $a = b$ is equivalent to $a + c = b + c$.

additive identity The number 0.

additive inverse (opposite) The additive inverse or opposite of a number a is $-a$.

algebraic expression An expression consisting of numbers, variables, operation symbols such as $+, -, \times$, and $\div$, and grouping symbols such as parentheses.

approximately equal The symbol $\approx$ indicates that two quantities are nearly equal.

associative property for addition For any real numbers a, b, and c, $(a + b) + c = a + (b + c)$.

associative property for multiplication For any real numbers a, b, and c, $(a \cdot b) \cdot c = a \cdot (b \cdot c)$.

average The result of adding up the numbers of a set and then dividing the sum by the number of elements in the set.

axis of symmetry The line passing through the vertex of the parabola that divides the parabola into two symmetric parts.

base The value of b in the expression b^n.

basic principle of fractions When simplifying fractions, the principle which states $\dfrac{a \cdot c}{b \cdot c} = \dfrac{a}{b}$.

binary operation An operation that requires two numbers to calculate an answer.

binomial A polynomial with two terms.

braces { }, used to denote a set.

circumference The perimeter of a circle.

coefficient The number that appears in a term.

coefficient of a monomial The number in a monomial.

commutative property for addition For any real numbers a and b, $a + b = b + a$.

commutative property for multiplication For any real numbers a and b, $a \cdot b = b \cdot a$.

complement The set containing all elements in the universal set that are *not* in A, denoted A'.

completing the square method An important technique in mathematics that involves adding a constant to a binomial so that a perfect square trinomial results.

complex fraction A rational expression that contains fractions in its numerator, denominator, or both.

complex number A complex number can be written in standard form as $a + bi$, where a and b are real numbers and i is the imaginary unit.

composite number A natural number greater than 1 that is not a prime number.

conjugate The conjugate of $a + b$ is $a - b$.

consistent system A system of linear equations with at least one solution.

constant of proportionality (constant of variation) In the equation $y = kx$, the number k.

cube root The number b is a cube root of a if $b^3 = a$.

degree A degree (°) is 1/360 of a revolution.

degree of a monomial The sum of the exponents of the variables.

degree of a polynomial The degree of the term (or monomial) with highest degree.

dependent equations Equations in a linear system that have infinitely many solutions.

difference The answer to a subtraction problem.

difference of two cubes Expression in the form $a^3 - b^3$, which can be factored as $(a - b)(a^2 + ab + b^2)$.

difference of two squares Expression in the form $a^2 - b^2$, which can be factored as $(a - b)(a + b)$.

directly proportional A quantity y is directly proportional to x if there is a nonzero number k such that $y = kx$.

discriminant The expression $b^2 - 4ac$ in the quadratic formula.

distance formula The distance d between the points (x_1, y_1) and (x_2, y_2) in the xy-plane is $d = \sqrt{(x_2 - x_1)^2 + (y_2 - y_1)^2}$.

distributive properties For any real numbers a, b, and c, $a(b + c) = ab + ac$ and $a(b - c) = ab - ac$.

dividend In a division problem, the number being divided.

divisor In a division problem, the number being divided *into* another.

elements The members of a set.

elimination method A symbolic method used to solve a system of equations that is based on the property that if "equals are added to equals the results are equal."

empty set (null set) A set that contains no elements.

equal sets If every element in a set A is in set B and every element in set B is in set A, then A and B are equal sets, denoted $A = B$.

equation A mathematical statement that two algebraic expressions are equal.

equivalent equations Equations that have the same solution set.

even root The nth root, $\sqrt[n]{a}$, where n is even.

exponent The value of n in the expression b^n.

exponential expression An expression that has an exponent.

extraneous solution A solution that does not satisfy the given equation.

factors In a multiplication problem, the two numbers multiplied.

factoring The process of writing a polynomial as a *product* of lower degree polynomials.

factoring by grouping A technique that uses the associative and distributive properties by grouping four terms of a polynomial in such a way that the polynomial can be factored even though its greatest common factor is 1.

finite set A set where the elements can be listed explicitly.

FOIL A method for multiplying two binomials $(A + B)$ and $(C + D)$. Multiply First terms AC, Outside terms AD, Inside terms BC, and Last terms BD; then combine like terms.

formula A special type of equation used to calculate one quantity from given values of other quantities.

graphical solution A solution to an equation obtained by graphing.

greater than If a real number b is located to the right of a real number a on the number line, we say that b is greater than a, and write $b > a$.

greater than or equal to If a real number a is greater than or equal to b, denoted $a \geq b$, then either $a > b$ or $a = b$ is true.

greatest common factor (GCF) The term with the highest degree and greatest integer coefficient that is a factor of all terms in the polynomial.

identity property of 1 If any number a is multiplied by 1, the result is a, that is, $a \cdot 1 = 1 \cdot a = a$.

identity property of 0 If 0 is added to any real number, a, the result is a, that is, $a + 0 = 0 + a = a$.

imaginary number A complex number $a + bi$ with $b \neq 0$.

imaginary part The value of b in the complex number $a + bi$.

imaginary unit A number denoted i whose properties are $i = \sqrt{-1}$ and $i^2 = -1$.

improper fraction A fraction whose numerator is greater than its denominator.

inconsistent system A system of linear equations that has no solutions.

independent equations Equations in a linear system that have one solution.

index The value of n in the expression, $\sqrt[n]{a}$.

inequality When the equals sign in an equation is replaced with any one of the symbols $<$, $\leq$, $>$, or $\geq$.

infinite set A set with infinitely many elements.

integers A set of numbers including natural numbers, their opposites, and 0, or $\ldots, -3, -2, -1, 0, 1, 2, 3, \ldots$.

intercept form A linear equation in the form $x/a + y/b = 1$.

intersection Denoted $A \cap B$ and read "A intersect B" is the set containing elements that belong to both A and B.

intersection-of-graphs method A graphical technique for solving two equations.

inversely proportional A quantity y is inversely proportional to x if there is a nonzero number k such that $y = k/x$.

irrational numbers Real numbers that cannot be expressed as fractions, such as π or $\sqrt{2}$.

leading coefficient In a polynomial of one variable, the coefficient of the monomial with highest degree.

least common denominator (LCD) The common denominator with the fewest factors.

least common multiple (LCM) The smallest number that two or more numbers will divide into evenly.

less than If a real number a is located to the left of a real number b on the number line, we say that a is less than b and write $a < b$.

less than or equal to If a real number a is less than or equal to b, denoted $a \leq b$, then either $a < b$ or $a = b$ is true.

like radicals Square roots with the same radicand or cube roots with the same radicand.

like terms Two terms that contain the same variables raised to the same powers.

linear equation An equation that can be written in the form $ax + b = 0$, where $a \neq 0$.

linear equation in two variables An equation that can be written in the form $Ax + By = C$, where A, B, and C are fixed numbers and A and B are not both equal to 0.

linear inequality A linear inequality results whenever the equals sign in a linear equation is replaced with any one of the symbols $<$, $\leq$, $>$, or $\geq$.

linear inequality in two variables When the equals sign in a linear equation of two variables is replaced with $<$, $\leq$, $>$, or $\geq$, a linear inequality in two variables results.

line graph The resulting graph when consecutive data points in a scatterplot are connected with straight line segments.

lowest terms A fraction is in lowest terms if its numerator and denominator have no factors in common.

method of substitution A symbolic method for solving a system of equations in which one equation is solved for one of the variables and then the result is substituted into the other equation.

monomial A number, a variable, or a product of numbers and variables raised to natural number powers.

multiplication property of equality If a, b, and c are real numbers with $c \neq 0$, then $a = b$ is equivalent to $ac = bc$.

multiplicative identity The number 1.

multiplicative inverse (reciprocal) The multiplicative inverse of a nonzero number a is $1/a$.

natural numbers The set of numbers expressed as 1, 2, 3, 4, 5, 6, $\ldots$.

negative slope On a graph, the slope of a line that falls from left to right.

negative square root The negative square root is denoted $-\sqrt{a}$.

nth root The number b is an nth root of a if $b^n = a$, where n is a positive integer.

null set (empty set) A set that contains no elements.

numerical solution A solution often obtained by using a table of values.

odd root The nth root, $\sqrt[n]{a}$, where n is odd.

opposite (additive inverse) The opposite, or additive inverse, of a number a is $-a$.

ordered pair A pair of numbers written in parentheses (x, y), in which the order of the numbers is important.

origin On the number line, the point associated with the real number 0; in the xy-plane, the point where the axes intersect, $(0, 0)$.

parabola A U-shaped graph that opens either upward or downward.

parallel lines Two or more lines in the same plane that never intersect; they have the same slope.

percent change If the quantity changes from x to y, then the percent change is $[(y - x)/x] \times 100$.

perfect square A number with an integer square root.

perfect square trinomial A trinomial that can be factored as the square of a binomial, for example, $a^2 + 2ab + b^2 = (a + b)^2$ and $a^2 - 2ab + b^2 = (a - b)^2$.

perpendicular lines Two lines in a plane that intersect to form a right (90°) angle.

point–slope form The line with slope m passing through the point (x_1, y_1) given by $y - y_1 = m(x - x_1)$, or equivalently, $y = m(x - x_1) + y_1$.

polynomial The sum of one or more monomials.

polynomials in one variable Polynomials that contain one variable.

positive slope On a graph, the slope of a line that rises from left to right.

prime factorization A number written as a product of prime numbers.

prime number A natural number greater than 1 that has *only* itself and 1 as natural number factors.

prime polynomial A polynomial with integer coefficients that cannot be factored by using integer coefficients.

principal nth root of a Denoted $\sqrt[n]{a}$.

principal square root The square root of a that is nonnegative, denoted $\sqrt{a}$.

probability A real number between 0 and 1. A probability of 0 indicates that an event is impossible, whereas a probability of 1 indicates that an event is certain.

product The answer to a multiplication problem.

proportion A statement that two ratios are equal.

Pythagorean theorem If a right triangle has legs a and b with hypotenuse c, then $a^2 + b^2 = c^2$.

quadrants The four regions determined by the xy-plane.

quadratic equation An equation that can be written as $ax^2 + bx + c = 0$, where a, b, and c are real numbers, with $a \neq 0$.

quadratic formula The solutions of the quadratic equation, $ax^2 + bx + c = 0, a \neq 0$, are given by $x = (-b \pm \sqrt{b^2 - 4ac})/2a$.

quadratic polynomial A polynomial of degree 2 that can be written as $ax^2 + bx + c$ with $a \neq 0$.

quotient The answer to a division problem.

radical equation An equation that has a variable as a radicand.

radical expression An expression that contains a radical sign.

radical sign The symbol $\sqrt{\ }$.

radicand The expression under the radical sign.

rate of change Slope can be interpreted as a rate of change. It indicates how fast the graph of a line is changing.

ratio A comparison of two quantities, expressed as a quotient.

rational equation An equation that contains one or more rational expressions.

rational expression A polynomial divided by a nonzero polynomial.

rational number Any number that can be expressed as the ratio of two integers p/q, where $q \neq 0$; a fraction.

rationalizing the denominator The process of removing radicals from a denominator so that the denominator contains only rational numbers.

real numbers All rational and irrational numbers; any number that can be represented by decimal numbers.

real part The value of a in a complex number $a + bi$.

reciprocal (multiplicative inverse) The reciprocal of a nonzero number a is $1/a$.

rectangular coordinate system (xy-plane) The xy-plane used to plot points and graph data.

rise The change in y between two points on a line, that is, $y_2 - y_1$.

run The change in x between two points on a line, that is, $x_2 - x_1$.

scatterplot A graph of distinct points plotted in the xy-plane.

scientific notation A real number a written as $b \times 10^n$, where $1 \leq |b| < 10$ and n is an integer.

set A collection of things.

set-builder notation Notation to describe a set of numbers without having to list all of the elements. For example, $\{x | x > 5\}$ is read as "the set of all real numbers x such that x is greater than 5."

slope The ratio of the change in y (rise) to the change in x (run) along a line. The slope m of a line passing through the points (x_1, y_1) and (x_2, y_2) is $m = (y_2 - y_1)/(x_2 - x_1)$, where $x_1 \neq x_2$.

slope–intercept form The line with slope m and y-intercept b is given by $y = mx + b$.

solution Each value of the variable that makes the equation true.

solution set The set of all solutions to an equation.

solution to a system In a system of two equations in two variables, an ordered pair, (x, y), that makes *both* equations true.

square root The number b is a square root of a number a if $b^2 = a$.

square root property If k is a nonnegative number, then the solutions to the equation $x^2 = k$ are given by $x = \pm \sqrt{k}$. If $k < 0$, then this equation has no real solutions.

squaring property for solving equations The solutions to the equation $a = b$ are among the solutions to the equation $a^2 = b^2$.

standard form of a complex number $a + bi$, where a and b are real numbers.

standard form of an equation for a line The equation given by $Ax + By = C$, where A, B, and C are fixed numbers with A and B not both 0.

standard form of a quadratic equation The equation given by $ax^2 + bx + c = 0$, where $a \neq 0$.

subscript The symbol x_1 has a subscript of 1 and is read "x sub one" or "x one."

subset If every element in a set B is contained in a set A, then we say that B is a subset of A, denoted $B \subseteq A$.

sum The answer to an addition problem.

sum of two cubes Expression in the form $a^3 + b^3$, which can be factored as $(a + b)(a^2 - ab + b^2)$.

symbolic solution A solution to an equation obtained by using properties of equations; the resulting solution set is exact.

system of linear equations in two variables A system of equations in which the equations can be written in the form $ax + by = c$.

term A number, a variable, or a product of numbers and variables raised to powers.

test point When graphing the solution set of an inequality, a point chosen to determine which region of the xy-plane to include in the solution set.

trinomial A polynomial with three terms.

unary operation An operation that requires only one number.

union Denoted $A \cup B$ and read "A union B"; it is the set containing any element that can be found in either A or B.

universal set A set which contains all elements under consideration.

variable A symbol, such as x, y, or z, used to represent any unknown quantity.

varies directly A quantity y varies directly with x if there is a nonzero number k such that $y = kx$.

varies inversely A quantity y varies inversely with x if there is a nonzero number k such that $y = k/x$.

Venn diagrams Diagrams used to depict relationships between sets.

vertex The lowest point on the graph of a parabola that opens upward and the highest point on the graph of a parabola that opens downward.

vertical asymptote A vertical asymptote typically occurs in the graph of a rational function when the denominator of the rational expression is 0, but the numerator is not 0; it can be represented by a vertical line in the graph of a rational function.

whole numbers The set of numbers given 0, 1, 2, 3, 4, 5, ...

***x*-axis** The horizontal axis in the xy-plane.

***x*-intercept** The x-coordinate of a point where a graph intersects the x-axis.

***xy*-plane (rectangular coordinate system)** The system used to plot points and graph data.

***y*-axis** The vertical axis in the xy-plane.

***y*-intercept** The y-coordinate of a point where a graph intersects the y-axis.

zero-product property If the product of two numbers is 0, then at least one of the numbers must be 0, that is, $ab = 0$ implies $a = 0$ or $b = 0$ (or both).

BIBLIOGRAPHY

Brown, D., and P. Rothery. *Models in Biology: Mathematics, Statistics and Computing.* West Sussex, England: John Wiley and Sons Ltd, 1993.

Burden, R., and J. Faires. *Numerical Analaysis.* 5th ed. Boston: PWS-KENT Publishing Company, 1993.

Conquering the Sciences. Sharp Electronics Corporation, 1986.

Eves, H. *An Introduction to the History of Mathematics.* 5th ed. Philadelphia: Saunders College Publishing, 1983.

Garber, N., and L. Hoel. *Traffic and Highway Engineering.* Boston, Mass.: PWS Publishing Co., 1997.

Greenspan, A. *The Economic Importance of Improving Math-Science Education,* Speech before the Committee on Education and the Workforce, U.S. House of Representatives, September 2000.

Haefner, L. *Introduction to Transportation Systems.* New York: Holt, Rinehart and Winston, 1986.

Historical Topics for the Mathematics Classroom, Thirty-first Yearbook. National Council of Teachers of Mathematics, 1969.

Horn, D. *Basic Electronics Theory.* Blue Ridge Summit, Penn.: TAB Books, 1989.

Huffman, R. *Atmospheric Ultraviolet Remote Sensing.* San Diego: Academic Press, 1992.

Kincaid, D., and W. Cheney. *Numerical Analysis.* Pacific Grove, Calif.: Brooks/Cole Publishing Company, 1991.

Mannering, F., and W. Kilareski. *Principles of Highway Engineering and Traffic Analysis.* New York: John Wiley and Sons, 1990.

Mason, C. *Biology of Freshwater Pollution.* New York: Longman Scientific and Technical, John Wiley and Sons, 1991.

Miller, A., and R. Anthes. *Meteorology.* 5th ed. Columbus, Ohio: Charles E. Merrill Publishing Company, 1985.

Miller, A., and J. Thompson. *Elements of Meteorology.* 2nd ed. Columbus, Ohio: Charles E. Merrill Publishing Company, 1975.

Pennycuick, C. *Newton Rules Biology.* New York: Oxford University Press, 1992.

Ronan, C. *The Natural History of the Universe.* New York: MacMillan Publishing Company, 1991.

Taylor, W. *The Geometry of Computer Graphics.* Pacific Grove, Calif.: Wadsworth and Brooks/Cole, 1992.

Thomas, V. *Science and Sport.* London: Faber and Faber, 1970.

Turner, R. K., D. Pierce, and I. Bateman. *Environmental Economics, An Elementary Approach.* Baltimore: The Johns Hopkins University Press, 1993.

Van Sickle, J. *GPS for Land Surveyors.* Chelsey, Mich.: Ann Arbor Press, 1996.

Weidner, R., and R. Sells. *Elementary Classical Physics,* Vol. 2. Boston: Allyn and Bacon, Inc., 1965.

Williams, J. *The Weather Almanac 1995.* New York: Vintage Books, 1994.

Wright, J. *The New York Times Almanac 1999.* New York: Penguin Group, 1998.

PHOTO CREDITS

INDEX OF APPLICATIONS

INDEX